Jean-François Emmenegger, Daniel Chable†, Hassan Ahmed Nour Eldin, and Helmut Knolle

Sraffa and Leontief Revisited

Also of Interest

Stochastic Finance. An Introduction in Discrete Time
Hans Föllmer, Alexander Schied, 2016
ISBN 978-3-11-046344-6, e-ISBN (PDF) 978-3-11-046345-3,
e-ISBN (EPUB) 978-3-11-046346-0

Brownian Motion. An Introduction to Stochastic Processes
René L. Schilling, Lothar Partzsch, 2014
ISBN 978-3-11-030729-0, e-ISBN (PDF) 978-3-11-030730-6,
e-ISBN (EPUB) 978-3-11-037398-1

Combinatorics and Finite Fields
Difference Sets, Polynomials, Pseudorandomness and Applications
Kai-Uwe Schmidt, Arne Winterhof, 2019
ISBN 978-3-11-064179-0, e-ISBN (PDF) 978-3-11-064209-4,
e-ISBN (EPUB) 978-3-11-064196-7

Abelian Groups. Structures and Classifications
Carol Jacoby, Peter Loth, 2019
ISBN 978-3-11-043211-4, e-ISBN (PDF) 978-3-11-042768-4,
e-ISBN (EPUB) 978-3-11-042786-8

Abstract Algebra. Applications to Galois Theory, Algebraic Geometry, Representation Theory and Cryptography
Celine Carstensen-Opitz, Benjamin Fine, Anja Moldenhauer,
Gerhard Rosenberger, 2019
ISBN 978-3-11-060393-4, e-ISBN (PDF) 978-3-11-060399-6,
e-ISBN (EPUB) 978-3-11-060525-9

Jean-François Emmenegger, Daniel Chable†,
Hassan Ahmed Nour Eldin, and Helmut Knolle

Sraffa and Leontief Revisited

Mathematical Methods and Models of a Circular Economy

DE GRUYTER

Mathematics Subject Classification 2010
Primary: 90B30, 91B74, 91B64; Secondary: 91B66, 91B76
Journal of Economic Literature Classification (JEL)
Primary: C02, E23, D57; Secondary: D51, B16

Authors
Dr. Jean-François Emmenegger
Marly
Switzerland
jean-francois.emmenegger@unifr.ch

Daniel Chable†

Prof. Dr. Hassan Ahmed Nour Eldin
Dübendorf
Switzerland
eldin@uni-wuppertal.de

Helmut Knolle
Wohlen b. Bern
Switzerland
helmut.knolle@bluewin.ch

ISBN 978-3-11-063042-8
e-ISBN (PDF) 978-3-11-063509-6
e-ISBN (EPUB) 978-3-11-063199-9

Library of Congress Control Number: 2019947597

Bibliographic information published by the Deutsche Nationalbibliothek
The Deutsche Nationalbibliothek lists this publication in the Deutsche Nationalbibliografie; detailed bibliographic data are available on the Internet at http://dnb.dnb.de.

Cover image: Artwork by Karim Noureldin (*1967), Play, 2016, Coloured pencil on paper, 41.8 × 29.7 cm. Stiftung zur Förderung der Sammlung 'Die Linie', LWL – Museum für Kunst und Kultur – Westfälisches Landesmuseum, Münster. © VG Bild-Kunst, Bonn, 2018
Typesetting: VTeX UAB, Lithuania
Printing and binding: CPI books GmbH, Leck

www.degruyter.com

Notes

Indications for reading

Sections and chapters, marked by an "*" may be skipped on first reading for a primary understanding of the material.

Funding

This publication is granted by the **Swiss National Science Foundation** *(SNSF)* at first as *SCOPES* project Nr. *IZ73Z0*_127962 and further as *Valorization Grant SCOPES* projects Nr. *IZ76Z0*_147586 and Nr. *IZ76Z0*_160605.

https://doi.org/10.1515/9783110635096-201

Table of Symbols

In the Leontief system, economic variables are measured in *physical terms* or in *monetary terms*, as explained throughout the text. In the Sraffa system the economic variables are usually measured in *physical terms*. Some variables are dimensionless. We define the following variables[1]:

a_{ij}: the *input-output coefficient* a_{ij} is the quotient of the *value* z_{ij} of the bundle of commodities i (Input) and the *value* of the total outlays x_j, produced during a considered previous period (year). The amount of commodities i of value a_{ij} is required for the production of one *value unit* of commodities j (Output). It is also called technical coefficient, *coefficient of production*, or direct input-output coefficient, and defined as $a_{ij} = z_{ij}/x_j$, $i,j \in \{1,\ldots,n\}$;

A: the *input-output coefficients* matrix (technical coefficients matrix, direct input-output matrix or *coefficients* matrix) in *monetary terms*, is defined as $\mathbf{A} = (a_{ij}) = \mathbf{Z}\hat{\mathbf{x}}^{-1}$, which is the definition going on from the matrix element: $a_{ij} = z_{ij}/x_j$;

B: the *model generation* matrix (in an interindustrial market), e.g., $\mathbf{Bp} = \mathbf{p}$, here the matrix **B** generates the price vector **p**;

c_{ij}: the *input-output coefficient*; in the context of the Leontief Input-Output economy, c_{ij} indicates the amount of a bundle of commodities i (Input) of sector S_i in *physical terms*, required for the production of one *unit* of a bundle of commodities j of sector S_j (Output) in *physical terms*, and defined as $c_{ij} = s_{ij}/q_j$, $i,j \in \{1,\ldots,n\}$. The coefficients s_{ij}, q_j come from the considered previous period (year) and are measured in {physical unit i/physical unit j}; in the context of the Sraffa price model, each bundle of commodities i contains one and only one commodity i;

C: the *input-output coefficients* matrix in physical terms of single product industries, noted and defined as $\mathbf{C} = (c_{ij}) = \mathbf{S}\hat{\mathbf{q}}^{-1}$;

$\mathbf{C}_T$: the *input-output coefficients* matrix of *joint production*, noted and defined as $\mathbf{C}_T = \mathbf{SF}^{-1}$;

d_i: in the context of the Leontief Input-Output Tables, d_i is the final demand (consumption) of a bundle of commodities i in *physical terms*; in the context of the Sraffa price model, d_i is the produced *surplus* or *net product* of exactly one commodity i, measured in *physical terms*;

d: in the context of the Leontief Input-Output economy, **d** is the vector of *final demand*; in the context of the Sraffa production system, **d** is the vector of *surplus*, also called the vector of *net product*. In both cases the vector coefficients are expressed in *physical terms*;

d_{ij}: the *distribution coefficient* is, in the context of the Leontief Input-Output economy, the *value* of a bundle of commodities i (Input) of sector S_i per unit of *value*

1 The French word *numéraire* is spelled in English as *numeraire*. Both spellings are used.

https://doi.org/10.1515/9783110635096-202

of commodities i, distributed to sector S_j for the production of a *unit* of that bundle of commodities j (Output). It is defined as $d_{ij} = z_{ij}/x_i = s_{ij}/q_i$, $i, j \in \{1, \ldots, n\}$ (monetary of physical units). The coefficients z_{ij}, s_{ij}, x_i, q_i come from the considered previous period (year); in the context of the Sraffa price model, each bundle of commodities i contains one and only one commodity i;

D: the *distribution coefficients* matrix is defined as $\mathbf{D} = (d_{ij}) = \hat{\mathbf{x}}^{-1}\mathbf{Z} = \hat{\mathbf{q}}^{-1}\mathbf{S}$; it is a *stochastic matrix*, if there is no surplus, i.e., $\mathbf{x} = \mathbf{x}_I$ or $\mathbf{q} = \mathbf{q}_I$ (no final demand or no surplus);

D: the total final demand of the economy, given a price vector $\mathbf{p} = [p_1, \ldots, p_n]'$, expressed in a chosen *currency*, $D = \mathbf{d}' \cdot \mathbf{p}$;

e: the summation vector, defined as $\mathbf{e} = [1, 1, \ldots, 1]'$;

f_i: the coefficient of total *final demand (consumption)* of sector i in *monetary terms*, also called *total final use*, e. g., in the Swiss IOT;

f: the vector of the coefficients f_i of total *final demand (use)* for sector i, $\mathbf{f} = [f_1, \ldots, f_n]'$, components in *monetary terms*;

F: the *output* matrix, called also by Schefold ([103], p. 49) the matrix of *outputs* $\mathbf{F} = (f_{ij})$;

F: the total *final demand (consumption)* of the economy, the sum of final demands f_i in *monetary terms*, $F = \mathbf{e}' \cdot \mathbf{f}$;

H: the direct and indirect capital matrix or Pasinetti matrix;

I: the identity matrix of any dimension $n \in \mathbb{N}$, with explicit notation of the dimension: $\mathbf{I}_n$

I – A: the matrix $(\mathbf{I} - \mathbf{A})$ is usually called the *Leontief matrix*, the matrix $(\mathbf{I} - \mathbf{A})^{-1}$ is referred to as the *Leontief Inverse*;

k: usually notes an arbitrary coefficient, i. e., $k \in \mathbb{R}^+$;

K: the total *operating capital*, i. e., the value of the *total means of production*, measured in *numeraire* or in *monetary terms*;

L_j: the annual *quantity of labour* or required working time, employed in industry $S_j, j \in \{1, \ldots, n\}$, to produce the annual quantity q_j of commodity j, usually measured in *man-years*, component of vector **L**;

L: the vector of *labour* $\mathbf{L} = [\mathrm{L}_1, \ldots, \mathrm{L}_n]'$, or vector of the quantity of labour;

L: the *total quantity of labour* (i. e., number of workers) of an economy, $L = \sum_{i=1}^{n} L_i$;

λ_C: the Frobenius number, in connection with the *Perron–Frobenius theorem* for an irreducible and non-negative or positive matrix **C**;

m: the number of sectors in a subsystem of the actual n-sector economy, $m \leq n$, i. e., m may note the number of *non-basic* commodities;

M: Manara's transformation matrix;

n: the number of sectors of the actual economy; this economy may be imbedded in a larger one whose dimension is unknown;

P: the total profits of one year, measured in *numeraire* or in *monetary terms*; profits are usually non-negative, $P = \tilde{r} \cdot Y \geq 0$;

p_i: in the context of the Leontief Input-Output economy, it is the price p_i of the bundle of commodities $i \in \{1,\dots,n\}$ of sector S_i per unit of production; in the context of the Sraffa price model, each bundle of commodities i contains one and only one commodity; the price p_i is measured either in units of a *numeraire*, selected among one of the produced commodities or, alternatively, measured in a unit of a usual *monetary currency*;

p_0: a chosen price unit, acting as an exogenous variable in a price model; if a commodity $i \in \{1,\dots,n\}$ has been selected as a *numeraire*, $p_i = p_0$ sets its price;

$\mathbf{p}$: the vector of prices p_i, $i = 1,\dots,n$, $\mathbf{p} = [p_1, p_2,\dots,p_n]'$;

$\tilde{\mathbf{p}}$: the vector of price indices $\tilde{p}_i$, $i = 1,\dots,n$, $\tilde{\mathbf{p}} = [\tilde{p}_1, \tilde{p}_2,\dots,\tilde{p}_n]'$;

π_i: in the context of the Sraffa price model, π_i is the labour per unit of produced total output q_i, i.e., $\pi_i = L_i/q_i$. Then, $\pi_i^{-1} = (L_i/q_i)^{-1} = q_i/L_i$ is the productivity of labour in industry S_i;

$\boldsymbol{\pi}$: one forms the vector $\boldsymbol{\pi}$ containing the labour components π_i per unit of produced quantity of commodity i, $\boldsymbol{\pi} = [\pi_1,\dots,\pi_n]'$. It is called the vector $\boldsymbol{\pi}$ of *labour per unit of commodities*;

q_i: the total output of commodity i in *physical terms*, produced by sector S_i;

$\mathbf{q}$: the vector of total outputs q_i in *physical terms*, produced by all the sectors S_i together with the final demand d_i;

$\mathbf{q}_I$: the vector of total outputs in *physical terms* of the *interindustrial* part, produced by all the sectors S_i, without final demand d_i;

$\tilde{R}$: share of national income to circulating capital or *surplus ratio*, $\tilde{R} = Y/K$;

$\tilde{r}$: the share of total profits (profit share of total income), $\tilde{r} = P/Y$;

r: the rate of profits, $r = P/K$;

R: the productiveness of an economy $R = (1/\lambda_C) - 1$, in Sraffa's terminology R is called the *maximal rate of profits*, Sraffa [108] Par. 30;

R': in a Standard system $R' := d_j/(q_j - d_j)$, $j = 1,\dots,n$ is the ratio of surplus d_j to the total means of production $(q_j - d_j)$ of commodity j; in a Standard system $R' = \tilde{R} = Y/K$, Sraffa noted R' the *Standard ratio*, [108] Par. 29;

r_0: a fixed value chosen for the *rate of profits* $r = r_0$, $0 \leq r \leq R$, acting as an exogenous variable in a Sraffa price model;

s_{ij}: in the case of Input-Output Tables (IOT) the *commodity flow coefficient* s_{ij} represents the bundle of commodities (Input) of sector S_i in *physical terms*, required for the production of the bundle of commodities (Output) by sector S_j. It presents the *transaction* in *physical terms* from sector S_i to sector S_j (see Miller and Blair [65], p. 11); in the case of the Sraffa price models, the *commodity flow coefficient* s_{ij} represents the quantity (Input) of one commodity i in *physical terms*, required by industry (branch) S_j for the production (Output) of either one commodity in *single product mode* or one or more commodities in *joint production mode*;

$\mathbf{S}$: the *commodity flow* matrix of the transactions s_{ij} (intermediate input) between pairs of sectors (from each sector S_i to each sector S_j), $i, j \in \{1, 2, .., n\}$, expressed in *physical terms*; In the context of Sraffa it is the matrix of the means of production in *physical terms* and represents the *technology*;

S_j: denotes the sector j of industry or of production, $j = 1, \ldots, n$;

T: *total tax revenue*: totality of collected taxes during a year on the various levels of the administration (local, district, state);

$\mathbf{T}$: Schefold's transformation matrix, (basic and non-basic commodities), encountered in Chapter 6;

$\mathbf{T}$: the *price coefficients* matrix (in an interindustrial market): $\mathbf{Te} = \mathbf{p}$; every component t_{ij} of matrix $\mathbf{T}$ is the part of the price p_i, attributed to the sector S_j, $j \in \{1, \ldots, n\}$, for the effort of production of commodity i, encountered in Chapter 9;

U: the (average) national income per unit of quantity of labour, $U = Y/L$;

υ_j: the total value added of sector S_j;

$\boldsymbol{\upsilon}$: the vector of total *value added*, set up by coefficients υ_j, of the buying sectors S_j, $\boldsymbol{\upsilon} = [\upsilon_1, \ldots, \upsilon_n]'$, evaluated in *monetary terms*;

V: the total *value added* of the economy; the sum of the value added υ_j evaluated in *monetary terms*, $V = \mathbf{e}' \cdot \boldsymbol{\upsilon}$;

$\mathbf{V}$: the $n \times n$ adjacency matrix $\mathbf{V} = (\upsilon_{ij})$, $i, j = 1, \ldots, n$, associated with the *commodity flow* matrix $\mathbf{S} = (s_{ij})$;

$\boldsymbol{\upsilon}_c$: the vector of *total value added* of each sector per value unit of produced *commodities* in *monetary terms*, $\boldsymbol{\upsilon}_c = [\upsilon_{c1}, \ldots, \upsilon_{cn}]' = [\upsilon_1/x_1, \upsilon_2/x_2, \ldots, \upsilon_n/x_n]'$;

$\mathbf{v}_c$: the vector of the *labour costs* of each sector per unit of physical output, measured in physical terms, $\mathbf{v}_c = [v_{c1}, \ldots, v_{cn}]' = [w \cdot L_1/q_1, w \cdot L_2/q_2, \ldots, w \cdot L_n/q_n]' = [W_1/q_1, W_2/q_2, \ldots, W_n/q_n]'$;

W: the total wages of one year, Sraffa measures it in *numeraire*, $W = L \cdot w$; sum of all the wages paid for the workers; wages are assumed throughout to be always non-negative $W \geq 0$;

$\mathbf{W}$: the general $n \times n$ adjacency matrix $\mathbf{W} = (w_{ij})$, $i, j = 1, \ldots, n$;

$\mathbf{w}$: in a joint production economy $(\mathbf{S}', \mathbf{L}, \mathbf{F}')$, Chapter 6, the vector $\mathbf{w}$ collects the sectorial wages w_j paid in all the sector S_j, $j = 1, \ldots, n$; in an extension of Sraffa's price model, Chapter 8; the vector $\mathbf{w}$ collects the individual sectorial wage rates w_j;

w: Sraffa ([108], Par. 11) introduces the *wage per unit of labour*, $w = W/L$; it is usually called the *wage rate*. It is the (mean) annual wage of labour for one worker;

$\tilde{w}$: the share of total wages to the national income (wage share of total income), $\tilde{w} = W/Y$. In fact, Sraffa ([108], Par. 30) denotes "*the proportion of net product that goes to wages*";

X: the total *output of the production* of the economy, measured in *numeraire*, when accounting is realised in *physical terms*, but when accounting is performed in *monetary terms*, it measures the value of the total *output of the production* in the actual *currency*, $X = K + Y$;

x_i: the value of the *total output* of commodity i of sector S_i together with the consumption demand f_i, called total output, in *monetary terms*;

y_j: the value of *total outlays* or purchases of sector S_j together with the value added v_i (called the total expenditure or also total input), in *monetary terms*, there is: $x_j = y_j$;

$\mathbf{x}$: the vector of the *total output* of all sectors S_i, $i \in \{1, 2, .., n\}$, in *monetary terms*, also called the vector of *values*;

$\mathbf{y}$: the vector of the *total outlays* of the sectors S_j, $j \in \{1, 2, .., n\}$, in *monetary terms*, there is: $\mathbf{x} = \mathbf{y}$;

Y: the *net income* or the *national income*; Sraffa treats the structure of entire *national economies*. Coming in PCMC, Par. 12, to the question to determine the value of the surplus, Sraffa therefore terms Y the *national income*. In this book we treat many examples of subsystems of *national economies*. Then, it is more appropriated to understand Y simply as the *net income* of such subsystems, which are just a part of a *national income*. We will use either one or the other term, but we remain close to Sraffa's terminology. Sraffa generally uses a *numéraire* as basic standard to express values, like income. Using a numéraire, relative prices appear and Y is measured in this *numéraire*. Sraffa alternatively also normalizes national income, $Y = P + W = 1$. One defines $W = \tilde{w} \cdot Y$;

Y_0: an exogenous value set for the national income within a price model;

z_{ij}: the value of the bundle of commodities in sector S_i (Input), in *monetary terms*, required for the production of the totality of the bundle of commodities in sector S_j (Output). It is also called a transaction (intermediate input) from each sector S_i to each sector S_j;

$\mathbf{Z}$: the *commodity flow* matrix of the transactions z_{ij} (intermediate input) between pairs of sectors (from each sector S_i to each sector S_j), $i, j \in \{1, 2, \ldots, n\}$, evaluated in *monetary terms*, also called the matrix of *intermediate sales* to all sectors S_j. The matrix $\mathbf{Z}$ refers to the *interindustrial market*. We denote the value $z_{ij} \geq 0$ of *a bundle of commodities i* in sector S_i necessary for the production of a bundle of commodities j in sector S_j as *means of production*.

Abbreviations

EBITDA	Earnings Before Interest Taxes Depreciations and Amortization
GDP	Gross Domestic Product
IOT	Input–Output Table
PCMC	Production of Commodities by Means of Commodities
MPK	Marginal Product of Capital
MPL	Marginal Product of Labour
TAL	Total Amount of Labour is an artificial *physical unit* that measures the total amount of labour within a period. We say that one TAL is this total amount of work for one period necessary to realise the production. Other usual *physical units* for measuring labour are *man-years*, *working days*, and *working hours*
WPF	Wage–Profit Frontier

https://doi.org/10.1515/9783110635096-203

Notations

We present the usual conventions for the matrix and vector inequalities, see Bertram Schefold [103], p. 45 and Kurz & Salvadori [100], p. 506.

$\mathbf{o}$ denotes the null vector (zero vector), $\mathbf{o} = [0, \dots, 0]'$;

$\mathbf{x} > \mathbf{o}$ the vector $\mathbf{x} = [x_1, \dots, x_n]'$ is positive, i. e., all its elements are positive, $x_i > 0$, $i \in \{1, \dots, n\}$;

$\mathbf{x} \geq \mathbf{o}$ the vector $\mathbf{x} = [x_1, \dots, x_n]'$ is semi-positive, i. e., there is at least one positive element $x_i > 0$, , $i \in \{1, \dots, n\}$, the other elements are non-negative $\Leftrightarrow$ ($\mathbf{x} \geqq \mathbf{o}$ and $\mathbf{x} \neq \mathbf{o}$);

$\mathbf{x} \geqq \mathbf{o}$ the vector $\mathbf{x} = [x_1, \dots, x_n]'$ is non-negative, i. e. all elements are non-negative, $x_i \geqq 0$, $i \in \{1, \dots, n\}$;

$\mathbf{0}$ denotes the zero matrix (null matrix), i. e., all its elements are 0;

$\mathbf{A} > \mathbf{0}$ the matrix $\mathbf{A} = (a_{ij})$ is positive, i. e., all its elements are positive, $a_{ij} > 0$, $i, j \in \{1, \dots, n\}$;

$\mathbf{A} \geqq \mathbf{0}$ the matrix $\mathbf{A} = (a_{ij})$ is non-negative, i. e., all elements are non-negative, $a_{ij} \geqq 0$, $i, j \in \{1, \dots, n\}$.

$\mathbf{A} \geq \mathbf{0}$ the matrix $\mathbf{A} = (a_{ij})$ is semi-positive, i. e., there is at least one positive element $a_{ij} > 0$, $i, j \in \{1, \dots, n\}$;, the other elements are non-negative $\Leftrightarrow$ ($\mathbf{A} \geqq \mathbf{0}$ and $\mathbf{A} \neq \mathbf{0}$);

$[p]$ denotes the measurement units of variable p. Example: Setting 1 kg = 1 kilogram, consider the price $p = 3$ CHF/kg, then $[p]$ = CHF/kg.

https://doi.org/10.1515/9783110635096-204

Preface

Economics[2] is a decision-based, word-based and number-based science. "Currency and Market Decisions in a Decision-Based Economy" [73]

Piero Sraffa (1898–1983), a classical economist, reformulated in his book *Production of Commodities by Means of Commodities (PCMC) "the theory of value and distribution"*.[3] Wassili Leontief (1906–1999) made early contributions to *input-output analysis* and earned the Nobel Prize in Economics in 1973. Sraffa and Leontief were concerned with the structure of production, considered in its totality as a cyclic process. The exchange between branches of the economy is described quantitatively. On fewer than 100 pages, Sraffa used in *PCMC* mathematical concepts and theorems which remain mainly hidden. He just presented calculus and numerical results. The main purpose of the present book is to reveal, elucidate and illustrate the mathematical background of Sraffa's theory didactically in detail by means of modern *matrix algebra* and the corresponding fundamental theorems. A substantial portion is devoted to computed examples.

Our book is also a contribution to the increasing call for alternative approaches to the understanding of the realities of today's economic activity. To write this book, we have stood on the shoulders of eminent Sraffa connoisseurs: P. Newman [71] (1962); B. Schefold [109] (1976), [103] (1989); L. L. Pasinetti [80] (1977), [83] (1980), [84] (1986), [81], [82]; H. D. Kurz and N. Salvadori [52] (2007); and A. Roncaglia [97] (2009).

Wassili Leontief (1906–1999) modeled economic activity within the context of a circular economy of production and exchange, today expressed in *Input-Output Tables* in *monetary terms*. Leontief proposed to divide the economy into sectors, each one producing a group of products. There is a highly technical process to achieve this partition described by the NACE Rev. 2 report [16], and the CAP nomenclature, leading to *Input-Output Tables* and *Input-Output Analysis*, which are developed today in nearly all countries and state communities of the world (see for example the *European Union* [72]).

Independently of Leontief, Sraffa in PCMC [108], linearly modeled the English production of single commodities, like wheat, iron or pigs, considering the circularity of these production processes expressed in *physical terms*. He solved the distribution problem of David Ricardo (1772–1823), calculating 'costs of production' (PCMC, Par. 7), which he terms as 'values' or 'prices', fulfilling the conditions of production. Sraffa's and Leontief's approach both need *matrix algebra* and a group of theorems belonging to this fundamental mathematical discipline.

---

2 The Webster New Collegiate Dictionary, p. 260, defines the term "Economics" as follows: *"The science that investigates the conditions and laws affecting the production, distribution and consumption of wealth, or material means of satisfying human desires; political economy."*

3 https://static.uni-graz.at/fileadmin/_Persönliche_Webseite/kurz_heinz

https://doi.org/10.1515/9783110635096-205

Sraffa has termed the 'costs of production' as 'prices'. These 'prices' are for him near to the *labour values*. In the case that the total surplus (net product) goes to wages, Sraffa's *prices* are proportional to the *labour values*. When the total surplus is needed for wages and profits then the above proportionality property disappears. The *labour value* of a commodity represents the total amount of labour which has been incorporated into it over all precedent production periods.

With emphasis on Sraffa and Leontief, we address in this book:

- young researchers wishing to explore the foundations of circular economy and possible applications to sustainable economy;
- practitioners wishing to examine the potential of Sraffa's model in connection to Leontief's input-output analysis;
- advanced undergraduate, graduate and PhD students and their instructors in economics, political science, applied mathematics and informatics who seek to understand and master the basic technicalities underlying Sraffian economics and input-output analysis.

Applications requiring numerical calculations are the ultimate aim. The mathematics involved is limited to a fairly basic level of matrix algebra, and occasionally calculus, corresponding to the level attained in Chiang and Wainwright's [19] *Fundamental Methods of Mathematical Economics*.

The required basic graph theory methods are developed as we go along. Our undertaking is set on an operational level. The solution of the numerous examples requires some dexterity to make calculations by hand or using computer facilities,[4] setting a basis for the interpretation of the economic relevance of the obtained numerical results. There are also new results mentioned in the *Introduction* and in the *Conclusions*.

There is an important question to be answered at the beginning of this book: *"Why one uses mathematics for this modelling"*. The answer can be structured as follows: Economics is a decision-based, word-based and number-based science.

Every sector—as a buyer/consumer as well as a seller/producer—has 4 degrees of freedom to decide for: the value x, the price p, the quantity q and the objects (items) e. Every purchase begins with a decision and achieves a result through a number, see [73], *"Currency and Market Decisions in a Decision-Based Economy"*.

There is also the necessity to verify the consistency of the *word-based* approach of economics, see Steenge [110]—we put the question again!

4 For all extended calculations undertaken in this book (basically examples involving equations with more than two unknowns), we have resorted to the software package MATHEMATICA [123]. A part of the computations for Chapter 9 and 10 have been realised with MATLAB (mathworks.com), thanks to the support of the "MathWorks Book Program". Clearly, many other software packages are of course available nowadays to perform these calculations.

"Why are mathematical models being necessary?"

(a) Mathematical models are the most suitable "Knowledge-Base" for number-based decisions.

(b) Decision processes are principally composed of two phases:

(b1) Word-Based Decision Phase:

(1) Motivation Phase.

(2) Goal-Formulation Phase. Decisions of these phases are undertaken in dependence on word-based knowledge. There are principally four sources of deficiency:

- Word-based knowledge should be always updated and corrected.
- Word-based definitions and rules should always be updated and proved to be without contradictions.
- The Inference system for verification should be linguistically powerful in order to decide whether a sentence is true or false (Judicial-system, Judges).
- Word-based pairwise-comparison is limited to 'less than', 'equal to' or 'larger than'.

(b2) Number-Based Decision Phase:

(1) Pairwise-Comparison Phase (objects/quantities/prices/values).

(2) Market-Activity Phase(object/value transactions).

A mathematical model-base is algebraically as well as numerically a powerful tool for modelling the market behaviour. Linear algebra and especially the algebra of non-negative matrices, together with the computations of eigenvalues and eigenvectors, are predestinated to describe and establish the relations between values, quantities and prices. We use in this book the notational conventions applied by the *"International Input-Output Association"* [90] to write the formulas of matrix algebra.

The book is the result of 24 years of work, research and contacts. The authors have the honour to express their gratitude and to acknowledge the following meritorious persons. This book would not exist, if Tamara Bardadym, Kiev, Elena Pervukhina (†), Sevastopol, and Jean-François Emmenegger, Fribourg, had not met in 1995 at the *International Conference on Applied and Industrial Mathematics*, Hamburg. One year later, Heinrich Bortis, Fribourg, met the group. From him came the important impulse to distill this subject in the framework of Keynesian Political Economy. Other impulses came from the group around academician, Prof. Dr. Naum Zuselevich Shor (†), later Petro Stetsyuk, Kiev, V. M. Glushkov Institute of Cybernetics at the National Academy of Sciences of Ukraine, concerning the *Leontief models* and their mathematical properties. Our thanks go to Prof. Dr. Sergio Rossi [93], [94], University of Fribourg, for his guidance during the early stages of this work, and also to Peter Scrivener, Nestlé, Vevey, for his stimulating contribution to cost management in production processes.

The authors thank the organizers of the stimulating *Input-Output Workshops* organized for many years by a group around Prof. Dr. Udo Ludwig, Prof. Dr. Reiner

Stäglin and Dr. Hans-Ulrich Brautzsch [59] at the *Institut für Wirtschaftsforschung Halle*, Germany. In 2014, the organisation of the *Input-Output Workshops* was taken over by the *Gesellschaft für Wirtschaftliche Strukturforschung mbH, GWS*, Osnabrück, Germany, now managed by the dynamic group around Prof. Dr. Tobias Kronenberg, Hochschule Bochum, Prof. Dr. Jutta Günther, University of Bremen, and Anke Mönnig, GWS Osnabrück, held from 2014 to 2017 in Osnabrück, in 2018 in Bremen and in 2019 in Bochum (see Emmenegger [26], [27], [28], [29], [30]). These meetings supplied many impulses for our book. We are indebted to the fruitful contacts and discussions with Prof. Dr. Utz-Peter Reich, Hochschule Mainz, and Prof. Dr. Bert Steenge, University of Groningen. Of course, deep thanks go to the SNSF (Swiss National Science Foundation) for granting this research for many years through the SCOPES programme, project Nr. 127'962 and Nr. 147'586, under the responsibility of Dr. Eveline Glättli. SNSF also granted the meetings hold in Fribourg in 2011 and 2012.

In 2011, a team made out of J.-F. Emmenegger, D. Chable (†) and H. Knolle began to work on the present book. In 2012, H. A. Nour Eldin joined the team concentrating his efforts on Chapters 9 and 10, revealing the strong connection existing between the systems of Leontief and Sraffa. Together, J.-F. Emmenegger and D. Chable contributed to the book as a whole. From 2014 to 2015, H. Knolle wrote Chapter 7 and also elaborated his own book [51] which could be published thanks to an SNSF grant. Daniel Chable died in July 2018.

Without the competent organizational work of Tamara Bardadym, as well as her encouragement and the animation of colleagues of the Kiev-group, our book could not have been written. She also contributed to the work with competent proofreading.

The authors are indebted to their wives for supporting this period of intense work by their husbands. Special thanks go to Françoise, constantly supporting her husband Jean-François Emmenegger, who conducted the book project and also realized the manuscript in LaTeX.

Acknowledgments go to the publishing house de Gruyter-Oldenbourg and Mrs. Kristin Berber-Nerlinger, lecturer, for her encouraging and professional support, to Mrs. Nadja Schedensack and Mrs. Ina Talandienė, technical support, then to the artist Karim Noureldin, who offered us his splendid artwork for the cover of the book, finally to the reviewers of the manuscript for their numerous useful comments.

Fribourg, Switzerland, October 2019

Contents

1 Introduction

1.1 Preliminary comments

This book focuses mainly on Sraffa's theoretical models presented in *Production of Commodities by Means of Commodities* ([108], PCMC, 1960). The questions of cyclicity and production are at the centre of Sraffa's work. Our book interprets and extends PCMC in the footsteps of Newman, Pasinetti and especially Schefold in the German version of PCMC [109]. Matrix algebra is used to apply the mathematical and notational standards, which were established by Miller and Blair [65] and by EUROSTAT [72]. The decisive importance of the Perron–Frobenius theorem, ensuring the existence of a solution to Sraffa's model of production, reformulated as an eigenvalue problem, is stressed, together with the important result contained in Ashmanov's book ([2], the Theorem 1.5, p. 3) concerning the Frobenius number of *Leontief models* and *productive Leontief models*. These results are supplemented by elements of graph theory.

The development of an up-to-date, sophisticated mathematical approach has been absolutely necessary. Newman ([71], p. 58) judged Sraffa's book *"compressed and mathematically incomplete"*. In fact, Sraffa wrote in the preface to PCMC:

> *"My greatest debt is to Prof. A. S. Besicovitch for invaluable mathematical help over many years.*[1] *I am also indebted for similar help at different periods to the late Mr. F. Ramsey and to Mr. A. Watson. It will be only too obvious that I have not always followed the expert advice that was given to me, particularly to the notations adopted, which I have insisted on retaining (although admittedly open to objection in some respects) as being easy to follow for the non-mathematical reader".*

According to Sraffa's comments in that preface, he started to write PCMC in the late 1920s, taking more than 30 years to reach completion. The general economic background in Great Britain during the late 19th and early 20th centuries (Hobson [41]) clearly influenced Sraffa's presentation. Furthermore, he is quite clear: his original basic model is a simplification, a first step in the representation of the real situation. He focuses on short periods, monthly to annual, by today's standards. As the title of his book and the logic of his text indicate, he only considers commodities in the strict sense as goods (in German: *Waren*; in French: *marchandises*), and measurement units, prices, values and even wages are supposed to be expressed in terms of a physical good, the *numéraire*. In this text, we have loosened these restrictions, thus showing how Sraffa's model is flexible enough to be extended to cope with new situations (see in particular Chapter 7, and Chapter 8). A glossary, together with comments, has been added.

As for the numerical examples, the reader should not be upset by their apparent triviality in the various chapters. They have been chosen to highlight the exact mean-

1 For a complete account of the collaboration of Sraffa and Besicovitch, the reader may consult. Kurz and Salvadori [53], Chapter 9.

https://doi.org/10.1515/9783110635096-001

ing of the numerous formulas encountered. Some of them have been taken directly from PCMC without modification and from Pasinetti's original works [80].

Finally, a remark concerning the restriction to n industries (sectors) producing exactly n commodities: In practice, at the microeconomic level, there are usually more commodities produced than existing industries (or in certain circumstances fewer, see the end of Chapter 6, regarding land and agricultural production). This may complicate significantly the mathematical treatment of such situations. Luckily, at the macroeconomic level, dividing the economy into n separate sectors, each producing similar sets of goods, and appropriate bundling of certain goods into composite commodities, enables one to consider $n \times n$ production processes without loss of economic significance. This is the procedure applied and pursued in this text, permitting the systematic use of square matrices and their algebra.

We have deliberately focused our discussion and analysis of Sraffa's PCMC on the *introductory chapters* directly concerned with modeling the production process, i. e., Chapters I–V of Part I: *Single-Product Industries and Circulating Capital*, Chapters VII–IX, of Part II: *Multiple-Product Industries and Fixed Capital* concerning *Joint Production*, and Chapter XI: *Land* with a view to applications.

We have thus not addressed the problems related to *Dated Quantities of Labour* in Chapter VI and *Fixed Capital* in Chapter X (to avoid the complexities involved, which are unsuitable for an introductory text), as well as Chapter XII: *Switch in Methods of Production*, at the centre of the heated *Cambridge capital controversy debates* (see Birner [5], pp. 6–69).

1.2 Summaries of the chapters

Chapter 2 gives a rigorous, detailed and ahead presentation of the set of matrices and vectors used in Input-Output Analysis. The notations and matrix algebra involved permit an advanced presentation of the material. The elements of the Leontief Input-Output Tables (IOT) are then presented. The principles of the system of *Classification of products by activities (CAP)*, respectively the *Nomenclature des activités économiques dans la communauté européenne (NACE)*, are explained. They are at the basis of the determination of the *homogenous branches*,[2] constituting the IOTs. A selection of Leontief Input-Output models is presented. They are commonly labeled as *Leontief quantity models*, *Leontief price models*, or briefly *Leontief models*, either working in monetary or in physical terms, in regards to quantities, to price indices and to prices. The notion of *interindustrial economy* is introduced as the core of a production economy. The relation of Input-Output Tables to the basic framework of *national accounting* is described. Finally, special attention is given to the question of a numéraire as a measurement unit.

2 Also designated as "industries" or "products", see Eurostat Manual of Supply, Use and Input-Output Tables, [72].

Chapter 3 is a complete discussion of the three elementary examples that appear at the beginning of PCMC, now described in terms of matrix algebra, and introduces the **Perron–Frobenius theorem** as the centre piece of the algebraic structure of Sraffa's models.

Novelty: The extension of Sraffa's single commodity examples is developed, replacing numbers by variables with an analysis of loci (curves or planes) generated by the simultaneous variation of the variables involved. For a given technology, the maximal *rate of profits* is presented as a locus plane of the surplus. As such, it is the *productiveness* of this economy. This is exemplified for Sraffa's elementary examples in PCMC, Par. 5.

Chapter 4 develops the complete theory of Sraffa's price model for single-commodity production processes, examining in particular the distinction between basic and non-basic commodities. Examples in which all involved economic variables and economic ratios are determined and are calculated.

Sraffa's price model brings within a cyclic production process of n sectors and n commodities, measured in physical terms, the 'costs of production' of every commodity, termed as 'prices', with positive wages for workers and with positive profits for entrepreneurs into an equilibrium. The Sraffa price model defines a *dynamic system*, because when the *interindustrial market* adopts Sraffa's prices, then the production technology, described by the means of production of the actual period, is recreated for the next period, and this process is going on from period to period.

Essentially, Sraffa's price equations express specific *accounting balances* which can be aggregated to one equation which has to be considered as a complement to the *Balance of National Accounts* contained in the *Input-Output Tables*.

Important novelties: the general relationship between the rate of profits, the surplus ratio and the ratio of total wages to national income (4.36), valid for all Sraffa systems; the introduction of directed graphs and bipartite networks as tools for the analysis of all types of Sraffa production processes; the notion of the calibration of a system; the technology matrix with and without subsistence wages; a formal determination of positive prices in entire single-product industries (single-commodity processes).

Chapter 5 presents the complete theory of the Standard system of production for single-commodity processes, including the famous relationship between the rate of profits, the Standard ratio and the share of total wages to national income (5.54), valid for all Standard systems.

Novelties: an explicit formulation of the fundamental relations at the basis of the Standard system (5.7); the introduction of the notion of the *commodity space* and of the *orthogonal Euler mapping* (Euler affinity), a central piece in the transformation of an actual non-standard System into a *Standard system*.

Chapter 6 is an introduction to *joint production systems*, where the same commodities may be produced by more than one industry.

Novelties: output polyhedrons; a compact algebraic methodology for the distinction between basic and non-basic commodities, which completes the approaches of Manara–Pasinetti–Schefold to this problem; a matrix introduced by Pasinetti [83] shown to be pertinent to determine the number of *basics* in *joint production*; an updated presentation of Manara's conditions for the positivity of prices; an extended model accounting for land and natural resources showing how land rents form a part of national income in addition to profits and wages.

Chapter 7. This chapter has been proposed and developed by H. Knolle. A first part (Section 7.1 and Section 7.2) considers *joint production industries*. This means that the present industries produce in parallel several of the n commodities, a typical situation that has ecological consequences. A second part (Section 7.3–Section 7.8) introduces step by step new approaches and examples treating waste problems and presenting situations involving ecological economics and taxation. The main goal is to show that Sraffa's price model offers an approach to treat ecological problems, considering waste in the whole economic system. The consequences on the prices are then studied.

Chapter 8 is entirely concentrated on novel extensions of Sraffa's price models as indicated in the corresponding item of the Table of Contents.

Chapter 9 is a complete formal, algebraic analysis of the *interindustrial economy*, developed by H. A. Nour Eldin. Tables of matrices presenting synoptically the aspect of *value*, *quantities*, *prices* and *objects* of Leontief's and Sraffa's concepts. It culminates in the statement that the *Interindustrial Market*, together with the *Consumption Market*, are unified within the *Leontief–Sraffa economy*, Figure 9.9. Each one of these three entities is described by a proper set of matrices and vectors. A central novelty is the connection found between the IOTs in *monetary terms* and the Sraffa system in *physical terms*. Indeed, the *productiveness* R of the Sraffa system, described by the *flow commodity matrix* $\mathbf{S}$ in physical terms, is present and calculable from the initial IOT *flow commodity matrix* $\mathbf{Z}$ in monetary terms.

Chapter 10 goes beyond simplified textbook examples and gives a presentation of how Sraffa's approach, together with the IOT apparatus, can be applied to official IOTs. In this case, we apply the developed methodology to the official Swiss IOTs 2008 and 2014 and the German IOT 2013. We also compute the productiveness of these economies. We perform some aggregations of the official IOTs and show the limits of these calculations.

Chapter 11 summarises the results obtained in this book and indicates further avenues of research in an extended Sraffa context.

Appendix A contains all the necessary mathematical tools required for a complete understanding of the present text.

Appendix B is a summary of Schefold's historical contribution to the understanding of Sraffa's PCMC.

Appendix C presents a glossary of terms as they are used in this book.

2 Elements of Input-Output Analysis

The *Eurostat Manual of Supply, Use and Input-Output Tables (IOT)* ([72], p. 479) mentions: *"Input-Output Analysis was founded by Wassily Leontief* [56] *in the thirties of the 20th century, he received the Nobel Prize in Economics in 1973 and is the founding father of a new field for empirical research at the border between microeconomics and macroeconomics."* Generally, the purpose of IOTs is to describe the sale and purchase relationships between producers and consumers within the national economy of an entire country. The OECD tends to harmonise those national IOTs, developed in a first line essentially for national economies.

Input-Output Analysis is, as a first approach, essentially an *equilibrium analysis*, because the time variable does not appear explicitly. Nevertheless, it has to be kept in mind that the entries of IOTs are quantities used or produced during one period (mainly one year as for the Swiss IOT 2008). This means that all these quantities have to be understood as *quantities/period*. The technology does not change within a period. Official IOTs of countries will be explored in Chapter 10.

It is fascinating to understand that *Input-Output Analysis* is based exclusively on quantities that are directly observable and that can be measured using the ordinary instruments for measurements in economics. This objectivist concern is already present in Leontief's PhD thesis: *'Die Wirtschaft als Kreislauf'* [55].

In Section 2.1 the system of classification of products by activities (CAP) based on the European standard classification of productive economic activities (NACE) is sketched. A complete description of this classification process lies outside the scope of this text. The knowledge of these concepts are necessary to understand the construction of IOTs. Classification of products leads to the branches (or sectors) that constitute the IOTs. In Section 2.2 we enter the subject of economic assumptions to be fulfilled by matrices and vectors, issuing from the creation of IOTs. In Section 2.3 we address the subject of the just-mentioned matrices which are often *non-negative*. In matrix algebra, there is the domain of *non-negative* matrices and the mathematical theorems governing this sphere. We will learn of the existence of the group of Lemmas around the Perron–Frobenius theorem. In Section 2.4 we introduce the first models. We present the *Leontief model*, which dates back to Leontief [56], [57], also called *Leontief quantity model* (see Oosterhaven [77], p. 750). The *Leontief model* will be treated in monetary and physical terms. In Section 2.5 we finally come to treat two variants of the *Leontief price model*, issued from Leontief's *Input-Output Analysis*. The *Leontief price model* will as usual be presented in *monetary terms* and in *physical terms*.

2.1 Classification of products, production processes, matrices

We start with a typical description of the economic activity, see NACE Rev. 2 [16], p. 27. *"An economic activity takes place when resources such as capital goods, labour, manu-*

https://doi.org/10.1515/9783110635096-002

facturing techniques or intermediary products are combined to produce specific goods or services. Thus, an economic activity is characterised by an input of resources, a production process and an output of products (goods or services).”

NACE,[a] the European standard classification of productive economic activities, is a classification hierarchy comprising five levels. The first level comprises 21 *sections*, designed by the capital letter form A to U, and the second level comprises 99 *divisions*, numbered from 1 to 99. The construction of input-output tables (IOT), originally developed by Leontief, requires the first two levels.
The headings of the *divisions*, the second level of NACE, lead practically to a one-to-one correspondence of the *branch headings* of the IOTs, describing *palettes of products*, giving as such the CPA = *European Classification of Products by Activities* system.
In other words, the CPA is a classification of products, whose headings are practically identical to those of the *divisions* of NACE, and result in the terms of the *sectors* or the *economic branches* of the IOTs.

a NACE = Nomenclature des activités économiques dans la communauté européenne.

Originally, Leontief developed the IOT in *physical terms*. In this sense, the sectors were generally composed only of one sort of product, for example, wheat or iron, see Miller and Blair ([65], pp. 41–45). When there is more than one product in a sector an appropriated *physical measure* must be found, e. g., *bushels, tons* or *kcal*. In an IOT in *physical terms*, one considers therefore the sector S_j's demand for the input of products from the sector S_i in *physical terms* for one period (generally the period is a year).

At present, following the lines and recommendations of the *Eurostat Manual* [72], input-output data are carried out in *monetary terms*. The countries of the EU, Switzerland and other countries operate in this way, see the Swiss IOT 2008 [68]. The complete palette (bundle) of products of a CPA *economic branch*, and respectively its value, is presented in monetary terms, as sector Nr. 1, e. g., *products of agriculture, forestry and fishing* in the SIOT 2014. See also Chapter 10.

It is necessary to understand the relationships and mathematical transformations between an IOT in *physical terms* (all the products of a sector are measured in the same physical unit) and an IOT in *monetary terms* (all the products are measured in the same currency). There will be new applications, as in Chapter 9 and Chapter 10 in which environmental economics appears (see Miller and Blair [65]). Indeed: “*With the emergence of energy and environmental concerns, mixed-unit models have been developed, where economic transactions are recorded in monetary terms and ecological and/or energy transactions are recorded in physical terms*”, see Miller and Blair [65], p. 41.

For this reason, we clearly distinguish input-output data presented in *physical terms* from their presentation in *monetary terms*.

2.1.1 Notations for transactions in monetary terms

The *Eurostat Manual* ([72], p. 479) says: “*The core of input-output analysis is the input-output table. It describes the flow of goods and services between all sectors of an econ-*

omy over a period of time", describing, briefly speaking, *Leontief economies*. The time period is generally a *year*. Input-output analysis can be applied to observed economic data from regions, nations or countries, Miller and Blair [65], pp. 2–3. For a good survey see also Holub [42].

Leontief economies have to be considered in general as *open economies*, where there are interactions with other economies (import, export). Leontief economies produce various commodities whose prices are expressed in *monetary-value terms* or, briefly speaking, in *value terms*. There are n *homogenous economic branches* (also called sectors) and n groups of similar products (called palette or bundle of products), each one issuing from one *homogenous economic branch*.

Textbooks describe the production process of a *Leontief economy* by resorting to the following variables: the matrix component z_{ij} expresses the *value* of a homogenous *bundle of commodities* of the sector (branch) S_i (Input) required for the production of the homogenous *bundle of commodities* of sector S_j (Output). The result with *value added* v_j[1] leads to *total outlays* x_j of sector S_j, see Figure 2.1.

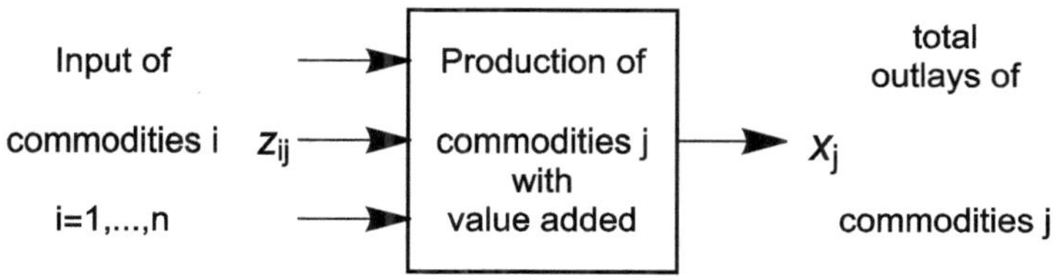

Figure 2.1: Input-output scheme of sector *j* in value terms.

A Leontief economy can be represented in a first attempt by a simplified Input-Output Table (IOT), as illustrated in Table 2.1.

Note that a branch S_i sells its commodities to other branches S_j that use these commodities in the production process. Here we have the *interindustry demand*. The sale of products to final consumers (households, private investors, government), corresponds to what is termed the *final demand* or *final use*. Every branch is therefore a purchaser of input and a seller of output in terms of commodities. Further in the production process, *value added* items (such as wages, gross profits,[2] rents, dividends) are generated. Human resources are required to materialize the production.

More specifically, such a simplified IOT is constructed, comprising three parts: the central *production process* (z_{ij}), $i, j = 1, \ldots, n$, the *final demand (use)* and the *value added*:

1 *Value added "accounts for the other (non-industrial) inputs to production, such as labour, depreciation of capital, indirect business taxes, and imports"*, see Miller and Blair ([65], p. 3).

2 Also termed EBITDA (earnings before interest, taxes, depreciation and amortisation) in business parlance.

Table 2.1: Input-output table of n producing sectors and a sector of final demand.

Selling sectors	Buying sectors						Final demand	Total output
	S_1	S_2	...	S_j	...	S_n		
S_1	z_{11}	z_{12}	...	z_{1j}	...	z_{1n}	f_1	x_1
S_2	z_{21}	z_{22}	...	z_{2j}	...	z_{2n}	f_2	x_2
$\vdots$	$\vdots$	$\vdots$	...	$\vdots$	...	$\vdots$	$\vdots$	$\vdots$
S_i	z_{i1}	z_{i2}	...	z_{ij}	...	z_{in}	f_i	x_i
$\vdots$	$\vdots$	$\vdots$	...	$\vdots$	...	$\vdots$	$\vdots$	$\vdots$
S_n	z_{n1}	z_{n2}	...	z_{nj}	...	z_{nn}	f_n	x_n
value added	v_1	v_2	...	v_j	...	v_n	$V = F$	
total outlays	x_1	x_2	...	x_j	...	x_n		X

(a) A sub-table of *interindustry transactions*, constitutes the core of the system of production: Each sector occupies one row as the selling economic branch of the commodities produced by itself. The same sector occupies a column as the purchasing (demanding) economic branch to acquire commodities.

(b) An additional column contains the *final exogenous demand (use)* for the various commodities, i. e., demand by purchasers outside of *interindustry production.* The single column of *final (exogenous) demand* is replaced in a general IOT by a table of *final use*, mainly composed of *household-consumption expenditures, gross private-domestic investment, government purchases of goods and services and net exports of goods and services*, see Miller and Blair ([65], p. 3).

(c) An additional row contains the *value-added components* per producing sector, indicating further inputs to production.
The single row of *value added* is replaced in a general IOT by a table of *value added*, also called *primary inputs*. The rows of *value added* may "account for the other (non-industrial) inputs of production, such as *labour, depreciation of capital, indirect business taxes*, and *imports*", see Miller and Blair ([65], p. 3). By *labour,* we understand *wages* in money terms. The table of *value added* together with *imports of goods* is "often lumped together as purchases from what is called the *payment sector*" ([65], p. 13). (Some authors understand *primary inputs* as a table to be made up of *value added, services* and *imports of goods.*) We will treat the Swiss input-output tables in detail, see Tables 10.1 and 10.2, in Chapter 10 as an example of a generalized IOT.

(d) An intermediate accounting identity obtained by summation: *total value added = total final demand*, see Swiss IOT 2014, Table 10.2.

(e) A final column comprising *total output* per industry and a right-hand row indicating *total outlays* per industry.
In some vocabularies on IOT's, the term *outlays* correspond to *inputs*, comprising all the accounting items that enter the purchase value of the commodities.

(f) A final accounting identity also obtained by summation: *total output* = *total outlays*.

We are dealing here with double-entry bookkeeping applied at the level of a whole economy. The corresponding national accounting identities will be introduced in further detail in Section 2.7 hereafter.

It is essential to understand the construction and the logic of the IOT, which is the basis of all the following developments of most of the rest of the book.

In this presentation, we will use, according to Miller and Blair ([65], pp. 10–21),[3] the subsequent notations.[4] The epresentation is based on any *previous* period or to the *reporting* period (year):

z_{ij}: the value of a bundle of commodities in sector S_i (Input), in monetary terms, required for the production of the bundle of commodities in sector S_j (Output), i. e., *transactions* (intermediate inputs) from sector S_i to sector S_j;

f_i: total *final demand* of sector S_i for sales (exogenous) in *monetary terms*;

v_j: total *value added* (labour, depreciation of capital, indirect business taxes, imports) of the buying sector S_j, realised for expenditures;

x_i: total *output* of commodity *i* in *monetary terms*, produced by sector S_i, together with the total final consumption f_i;

y_j: total *outlays* of sector S_j in *monetary terms* to produce commodity *j* (total expenditures), together with the total value added v_j, $y_j = x_j$;

X: total *output of the production* in *monetary terms*, $X = \sum_{i=1}^{n} x_i$;

F: total *final demand* of the economy, the sum of final demands f_i, is, in the case of no import, $M = 0$, also the *national income* $Y = F - M = F$;

V: total *value added* of the economy, the sum of the value added v_j, is equal to $F = V$, see Swiss IOT 2014, e. g., Table 10.3.

Again, briefly stated, reading horizontally the rows of Table 2.1 indicate *intermediate output*, sales of each *industry* S_i, $i = 1, \ldots, n$ to the buying sectors S_j, $j = 1, \ldots, n$, and then the *final demand*. Reading vertically, the columns indicate *intermediate inputs*, required by the sectors S_j, $j = 1, \ldots, n$, for the *production* in the sectors S_i, $i = 1, \ldots, n$, and the *value added*, accounting for the other inputs to production, such as labour, depreciation of capital, individual business taxes and imports, Miller and Blair ([65], p. 3).

We are now in a position to write down the two accounting equations referred to above.[5] By definition, we now use the simplified case for the *income approach*, sum-

3 Miller and Blair use the variable X for the total *output of the production of the economy*.

4 The quantities z_{ij} are measured in *monetary values* of the transactions between pairs of sectors or industries (e. g., from each sector S_i to each sector S_j), see Miller and Blair ([65], p. 11) and the *EUROSTAT Manual of Supply, Use and Input-Output Tables*, where the z_{ij} are presented in 'currency units of an economy' ([72], p. 479).

5 The *Eurostat Manual of Supply, Use and Input-Output Tables* ([72], p. 479) speaks of *"input-output tables in the currency of an economy"*. This means that the z_{ij} of Table 2.1 are expressed in *monetary*

ming up the value-added components υ_j, and the *expenditure approach*, summing up the final demands f_i representing the surplus of the production. For the algebraic presentation, consider the $(n \times 1)$ vector of ones, $\mathbf{e} = [1, \ldots, 1]'$, Appendix A, (A.9) the vector of *total final demand (use)* $\mathbf{f} = [f_1, f_2, \ldots, f_n]' \geqq \mathbf{o}$ and the vector of *total value-added* $\boldsymbol{\upsilon} = [\upsilon_1, \upsilon_2, \ldots, \upsilon_n]' \geqq \mathbf{o}$.[6] When there is no final use, $f_i = 0$, then sector S_i produces only interindustrial demand, and when there is $\upsilon_j = 0$, then sector S_j obtains no value-added. These limit cases are not to be excluded. We have therefore the scalar product,

$$V = \mathbf{e}'\boldsymbol{\upsilon} = \sum_{j=1}^{n} \upsilon_j, \quad F = \sum_{i=1}^{n} f_i = \mathbf{e}'\mathbf{f}, \quad F = V.^{7} \tag{2.1}$$

The first accounting condition, which sums up the elements of each row, gives the *total sales* x_i of sector S_i composed of the sum of all the *intermediate outputs* z_{ij} of the commodities j, $j = 1, \ldots, n$, needed to produce the quantity of sector S_i, together with the *final demand* f_i. We obtain, row by row, the *total output* x_i of each commodity sector S_i,

$$x_i = \sum_{j=1}^{n} z_{ij} + f_i, \quad i = 1, \ldots, n. \tag{2.2}$$

The second accounting condition, summing up the columns, over z_{ij}, the value of all the *intermediate inputs*, also called *interindustrial purchases* of the commodities $i = 1, \ldots, n$, and the *value added* υ_j, generated by sector S_j. We obtain, column by column, the *total outlays* y_j to sector S_j,

$$y_j =: x_j = \sum_{i=1}^{n} z_{ij} + \upsilon_j, \quad j = 1, \ldots, n. \tag{2.3}$$

Let us now write the $2n$ equations (2.2) and (2.3) in more compact form, using notations from linear algebra. Consider the *commodity flow* matrix $\mathbf{Z} = (z_{ij})$, composed of the principal coefficients of *interindustrial sales* z_{ij} by sector S_i to all sectors S_j, see Miller and Blair ([65], p. 12) and Eurostat ([72], p. 484). We will later establish that $\mathbf{Z}$ is *semi-positive*, Assumption 2.2.2.,

---

terms or in the *currency of the country* from the very beginning. On the other hand, Leontief had suggested that the coefficients of the input-output table are presented in *physical terms* like *qr. of* wheat, *t. of* iron. These two different ways of presenting input-output tables need to be distinguished. They require a notational differentiation mastered by Miller and Blair [65]. This is one reason why the present text relies on Miller and Blair [65], and we are making this differentiation throughout the text.

6 We exclude negative final use $f_i < 0$ and negative total-value added $\upsilon_j < 0$, $i, j \in \{1, \ldots, n\}$.

7 See German IOT 2013 and Swiss IOT 2014, Chapter 10.

$$\mathbf{Z} = (z_{ij}) = \begin{bmatrix} z_{11} & z_{12} & \dots & z_{1j} & \dots & z_{1n} \\ z_{21} & z_{22} & \dots & z_{2j} & \dots & z_{2n} \\ \dots & \dots & \dots & \dots & \dots & \dots \\ z_{i1} & z_{i2} & \dots & z_{ij} & \dots & z_{in} \\ \dots & \dots & \dots & \dots & \dots & \dots \\ z_{n1} & z_{n2} & \dots & z_{nj} & \dots & z_{nn} \end{bmatrix}. \tag{2.4}$$

From (2.3) with the n total demand components x_j of products j, issued from the sectors $S_j, j = 1, \dots, n$, we set up the *vector* of *total demand of industrial production* or the vector of *total output* $\mathbf{x} = [x_1, x_2, \dots, x_n]'$. We will also need the vector of *total interindustrial production*, $\mathbf{x}_I = [x_{I1}, x_{I2}, \dots, x_{In}]' > \mathbf{o}$, resulting from the *interindustry sales* of each of the n sectors.

With the vector of *interindustrial production* $\mathbf{x}_I$ and the *non-negative* vector of *final demand* $\mathbf{f} \geqq \mathbf{o}$, the set of the n equations (2.2) becomes

$$\mathbf{x}_I = \mathbf{Ze} > \mathbf{o}, \quad \mathbf{x} = \mathbf{Ze} + \mathbf{f} = \mathbf{x}_I + \mathbf{f} > \mathbf{o}. \tag{2.5}$$

The component x_j of the vector $\mathbf{x}$ in (2.5) equals the **total sales** of sector S_j to all sectors S_i, $i \in \{1, \dots, n\}$, comprising also the *final exogenous demand* f_i. It is an accounting identity from which the *total output* X is obtained by summation.

The *commodity-flow* matrix $\mathbf{Z}$ describes the **monetary flow** of *interindustrial production*. Each coefficient z_{ij} represents the value of the *bundle of commodities i* required for the production of a *bundle of commodities j*, evaluated in *monetary terms*, i. e., the transaction (*intermediate input*) from sector S_i to sector S_j. As a whole, matrix $\mathbf{Z}$ represents the **means of production** in monetary terms.

The set of the n equations (2.3) can now be written, using the transposed matrix $\mathbf{Z}'$ and the vector of *total value added* $\boldsymbol{\upsilon} \geqq \mathbf{o}$. At this time we also define the *total production* $\mathbf{y}_I$[8] of sector S_j that will only be used in Chapter 10. At present we obtain,

$$\mathbf{y}_I = \mathbf{Z}'\mathbf{e} > \mathbf{o}, \quad \mathbf{y} = \mathbf{x} = \mathbf{Z}'\mathbf{e} + \boldsymbol{\upsilon} = \mathbf{y}_I + \boldsymbol{\upsilon} > \mathbf{o}, \tag{2.6}$$

and the vector $\mathbf{x}$ contains components x_i, reflecting the *values* of the **purchase** of each *bundle of commodities i*, $i \in \{1, \dots, n\}$, covering also the expenses of the *value added* υ_j required for the production. The components x_i are *accounting identities* for the *total outlays*, $x_i = y_i$.

8 Without loss of generality, we can state that every component of the vectors $\mathbf{x}_I$ and $\mathbf{y}_I$ is greater than zero. If an element of $\mathbf{x}_I$ is zero, then there is no output at all of the corresponding commodities k. The commodities k can be deleted from that economy. If an element of $\mathbf{y}_I$ is zero, then there is no outlay at all of the corresponding sector S_k, and the sector S_k can be deleted from that economy. These conditions will be formulated later as an *economic assumption* used in the sequel.

We thus obtain an accounting identity relating total outlays to total inputs,

$$\mathbf{x} = \mathbf{Ze} + \mathbf{f} = \mathbf{Z}'\mathbf{e} + \mathbf{u} > \mathbf{o}. \tag{2.7}$$

In the column $[z_{1j}, z_{2j}, \ldots, z_{nj}]'$ of the quantities required by sector $S_j, j = 1, \ldots, n$, one is interested in the quantity of the *commodities* i, $i = 1, 2, \ldots, n$, that are necessary to produce one physical unit of the *commodities* j. This gives the definition of the *coefficient of production* or the *technical coefficient*[9] a_{ij}, see Miller and Blair ([65], p. 16)[10]:

a_{ij}: the *input-output coefficient* a_{ij} is the quotient of the *value* z_{ij} of the amount of commodities i (Input) and the value of the total outlays x_j, produced during the present period (year). This amount a_{ij} of commodities i is required for the production of one *value* unit of commodities j (Output). It is also called *direct input-output coefficient*, defined as $a_{ij} = z_{ij}/x_j$, $i, j \in \{1, \ldots, n\}$. It is dimensionless, [currency/currency] = 1.

$$a_{ij} = \frac{z_{ij}}{x_j}, \quad i,j = 1, \ldots, n; \quad \begin{array}{ll} i: & \text{Input index} \\ j: & \text{Output index} \end{array} \quad \text{or}^{11} \quad \mathbf{A} = \mathbf{Z}\hat{\mathbf{x}}^{-1}. \tag{2.8}$$

With the coefficients a_{ij} we then set up the *semi-positive* $n \times n$ *matrix of the technical coefficients*, the *input-output coefficients* matrix or the *input matrix* (for short), see Chiang ([19], p. 117), $\mathbf{A} = (a_{ij})$, Assumption 2.2.2,

$$\mathbf{A} = (a_{ij}) = \mathbf{Z}\hat{\mathbf{x}}^{-1} = \begin{bmatrix} a_{11} & a_{12} & \ldots & a_{1n} \\ a_{21} & a_{22} & \ldots & a_{2n} \\ \ldots & \ldots & \ldots & \ldots \\ a_{n1} & a_{n2} & \ldots & a_{nn} \end{bmatrix}. \tag{2.9}$$

Lemma 2.1.1. *The vector of final use* $\mathbf{f} \geqq \mathbf{o}$ *and the vector of value added* $\mathbf{u} \geqq \mathbf{o}$ *are non-negative. Then, the coefficients* a_{ij} *have dimension* $[a_{ij}] = \frac{\text{currency}}{\text{currency}} = 1$, $i, j = 1, \ldots n$. *The components of the semi-positive input-output coefficients matrix* $\mathbf{A} \geq \mathbf{0}$ *are comprised in the interval* $[0, 1]$, $0 \leq a_{ij} \leq 1$, $i, j \in \{1, \ldots, n\}$, *as well as* $0 \leq \sum_{i=1}^{n} a_{ij} \leq 1$.

Proof. With (2.6) there is, $\mathbf{u} \geqq \mathbf{o}$, $\mathbf{y}_I = \mathbf{Z}'\mathbf{e}$, obtaining $\mathbf{x} = \mathbf{y}_I + \mathbf{u} \geqq \mathbf{y}_I$. We have therefore $x_j \geq y_{Ij}$, $i, j = 1, \ldots, n$, then as $0 \leq a_{ij}$, we get with (A.39),

$$0 \leq \sum_{i=1}^{n} a_{ij} = \sum_{i=1}^{n} \frac{z_{ij}}{x_j} = \frac{1}{x_j} \sum_{i=1}^{n} z_{ij} = \frac{y_{Ij}}{x_j} \leq 1 \Rightarrow 0 \leq a_{ij} \leq 1$$

$$\mathbf{x} = \mathbf{Z}'\mathbf{e} + \mathbf{u} \Rightarrow \hat{\mathbf{x}}^{-1}\mathbf{x} = \hat{\mathbf{x}}^{-1}\mathbf{Z}'\mathbf{e} + \hat{\mathbf{x}}^{-1}\mathbf{u} = \mathbf{A}'\mathbf{e} + \hat{\mathbf{x}}^{-1}\mathbf{u} = \mathbf{e} \geq \mathbf{A}'\mathbf{e}. \tag{2.10}$$

□

9 Some authors use the term *technological coefficient* instead of *technical coefficient*.

10 Because the *flows* z_{ij} *of domestic commodities* of sectors S_i to sectors S_j are expressed in the same monetary value, all the elements a_{ij}, $i \neq j$ are measured in unit 1. That means, all the coefficients a_{ij} are dimensionless. In principle, labour is incorporated in the wage item entering the value added.

11 The *diagonal operator* corresponding to a vector $\mathbf{x}$ leads to a diagonal matrix $\hat{\mathbf{x}}$, see (A.17). The diagonal matrix $\hat{\mathbf{x}}^{-1}$ contains the reciprocals $(1/x_i)$ in its main diagonal.

The *input-output coefficients* matrix **A** describes the **technology** in *monetary terms*. Each coefficient of the column $[a_{1j}, a_{2j}, \ldots, a_{nj}]'$ gives the share in value terms of the *bundle of commodities* i, $i \in \{1, \ldots, n\}$ necessary to produce a *bundle of commodities* j of the value of one unit of the used currency. Thus, expressed in currency CHF, $a_{12} = 0.2$ means that one uses a *bundle of the commodities* 1 of value 0.20 *CHF* to produce a *bundle of the commodities* 2 of value 1 *CHF*.

By analogy, we define a further matrix with the element:

d_{ij}: the *distribution coefficient* is the *value* of a bundle of commodities i (Input) of sector S_i per unit of *value* of commodities i, distributed to sector S_j for the production of a *unit* of commodities j (Output). It is defined as $d_{ij} = z_{ij}/x_i$; $i, j \in \{1, \ldots, n\}$ which is dimensionless, d_{ij} = [currency/currency] = 1:

$$d_{ij} = \frac{z_{ij}}{x_i}; \quad i, j = 1, \ldots, n; \quad \begin{matrix} i: & \text{Input index} \\ j: & \text{Output index} \end{matrix} \quad \text{or} \quad \mathbf{D} = \hat{\mathbf{x}}^{-1}\mathbf{Z}. \tag{2.11}$$

With d_{ij} we then set up the $n \times n$ *distribution coefficients* matrix $\mathbf{D} = (d_{ij})$ in monetary terms:

$$\mathbf{D} = (d_{ij}) = \hat{\mathbf{x}}^{-1}\mathbf{Z} = \begin{bmatrix} d_{11} & d_{12} & \ldots & d_{1n} \\ d_{21} & d_{22} & \ldots & d_{2n} \\ \ldots & \ldots & \ldots & \ldots \\ d_{n1} & d_{n2} & \ldots & d_{nn} \end{bmatrix}. \tag{2.12}$$

The *distribution coefficients* matrix **D** describes the **logistics**. Each coefficient d_{ij} of the row $[d_{i1}, d_{i2}, \ldots, d_{in}]$ presents the share in value terms of commodities i to be distributed to the sectors S_j, $j \in \{1, \ldots, n\}$ per value unit of produced commodities i. Thus, say that the currency *CHF* is used, then the number $d_{32} = 0.2$, e. g., means that sector S_3 distributes for the production of sector S_2, (1/5) = 0.2 of its own production of commodities 3.

The matrix **D** will be used later. For the moment, because we work mainly with the *matrix of the technical coefficients* **A**, we will establish some key equations of input-output analysis after the next subsection.

2.1.2 Notations for transactions in physical terms

As already mentioned, Leontief originally developed the IOT in appropriated *physical terms*, supposed to exist, like *bushels*, *tons* or *joules*. Therefore, it is necessary to treat the notations for *IOTs in physical terms*, meaning that all values are expressed in a chosen *physical numéraire* (explained in Section 2.8 hereafter) to be able to carry out all summations.

Presented in *physical terms*, the production process for the bundle of commodities $j \in \{1, \ldots, n\}$ of sector S_j requires inputs as *intermediate commodities* from the other sectors S_i, $i = 1, \ldots, n$. The *total output* q_i of Sector S_i results from the sum of the partial demands for all the commodities j as intermediate products and from the final demand d_i of end consumers, the government and investors.

In analogy to the earlier Table 2.1 we shall just adopt the notations, where all the quantities are presented in *physical terms*, as will be done in Table 2.2:

s_{ij}: the bundle of commodities (Input) of sector S_i in appropriated *physical terms*, required for the production of the bundle of commodities (Output) by sector S_j. It presents the *transaction* in *physical terms* from sector S_i to sector S_j (see Miller and Blair [65], p. 11);

L_j: *annual quantity of labour* of sector S_j, as a component of value added,

d_i: the *final demand* of sector S_i,

q_i: the *total output* of sector S_i.

Table 2.2: Flows in physical terms, see Table 2.17 in Miller and Blair [65].

Selling sectors	**Buying sectors**					**Final demand**	**Total output**
	S_1	...	S_j	...	S_n		
S_1	s_{11}	...	s_{1j}	...	s_{1n}	d_1	q_1
S_2	s_{21}	...	s_{2j}	...	s_{2n}	d_2	q_2
$\vdots$	$\vdots$	$\vdots$		$\vdots$	$\vdots$	$\vdots$	$\vdots$
S_i	s_{i1}	...	s_{ij}	...	s_{in}	d_i	q_i
$\vdots$	$\vdots$	$\vdots$		$\vdots$	$\vdots$	$\vdots$	$\vdots$
S_n	s_{n1}	...	...	...	s_{nn}	d_n	q_n
value added	...	...	...	...	...		
Labour [working time/period]	L_1	...	L_j	...	L_n	L/D	

The $n \times n$ matrix $\mathbf{S} = (s_{ij})$ is called the *commodity flow* matrix in *physical terms*,

$$\mathbf{S} = \begin{bmatrix} s_{11} & s_{12} & \dots & s_{1j} & \dots & s_{1n} \\ s_{21} & s_{22} & \dots & s_{2j} & \dots & s_{2n} \\ \dots & \dots & \dots & \dots & \dots & \dots \\ s_{i1} & s_{i2} & \dots & s_{ij} & \dots & s_{in} \\ \dots & \dots & \dots & \dots & \dots & \dots \\ s_{n1} & s_{n2} & \dots & s_{nj} & \dots & s_{nn} \end{bmatrix}. \tag{2.13}$$

The basic accounting relationships, expressed with the vector $\mathbf{q} = [q_1, \dots, q_n]'$ and the vector $\mathbf{d} = [d_1, \dots, d_n]'$, see Miller and Blair ([65], pp. 47–48), are:

$$q_i = \sum_{j=1}^{n} s_{ij} + d_i; \quad i = 1, \dots, n. \tag{2.14}$$

In matrix form, in analogy with (2.5) one gets equations in *physical terms* relating the vector of *industrial output* $\mathbf{q}_I$ and the vector of *total output* $\mathbf{q}$ together with the vector of *final demand* $\mathbf{d} \geqq \mathbf{o}$,

$$\boxed{\mathbf{q}_I = \mathbf{Se} > \mathbf{o}, \quad \mathbf{q} = \mathbf{Se} + \mathbf{d} = \mathbf{q}_I + \mathbf{d} > \mathbf{o}; \quad \mathbf{S} = (s_{ij}); \quad i,j = 1,\dots,n.} \tag{2.15}$$

This is a fundamental relationship that we will again use later.[12]

The *commodity flow* matrix **S** describes the **flow of commodities** of *interindustrial production*. Each coefficient s_{ij} represents the bundle of commodities i required for the production of a bundle of commodities j, expressed in *physical terms*, also called the transaction (*intermediate input*) from sector S_i to sector S_j. As a whole, matrix **S** represents the **means of production** in physical terms.

The direct *input-output coefficients* in *physical terms* are now defined:

c_{ij}: the *input-output coefficient* indicates the amount of a bundle of commodities i (Input) of sector S_i in *physical terms*, required for the production of one *unit* of the bundle of commodities j of sector S_j (Output) in *physical terms*, and defined as $c_{ij} = s_{ij}/q_j$, $i,j \in \{1,\dots,n\}$. Their units are: [physical term of Sector S_i/physical term of Sector S_j]:

$$c_{ij} = \frac{s_{ij}}{q_j}; \quad i,j = 1,\dots,n; \quad \begin{matrix} i: & \text{Input index} \\ j: & \text{Output index} \end{matrix} \quad \text{or} \quad \mathbf{C} = \mathbf{S}\hat{\mathbf{q}}^{-1}. \tag{2.16}$$

The $n \times n$ matrix $\mathbf{C} = (c_{ij})$ is called the *input-output coefficients* matrix in *physical terms*,

$$\mathbf{C} = \mathbf{S}\hat{\mathbf{q}}^{-1} = \begin{bmatrix} c_{11} & c_{12} & \dots & c_{1n} \\ c_{21} & c_{22} & \dots & c_{2n} \\ \dots & \dots & \dots & \dots \\ c_{n1} & c_{n2} & \dots & c_{nn} \end{bmatrix}. \tag{2.17}$$

By definition, the units of the *input-output coefficients* c_{ij} are: $[c_{ij}] = \frac{\text{term of commodities i}}{\text{term of commodities j}}$, $i,j = 1,\dots,n$. Thus, the *input-output coefficients* matrix is a *non-negative matrix* of numbers with a dimension in physical units, $c_{ij} \geqq 0$.

The *input-output coefficients* matrix **C** describes the **technology** in physical terms. Each coefficient of the column $[c_{1j}, c_{2j}, \dots, c_{nj}]'$ gives the bundle of the commodities $i \in \{1,\dots,n\}$ necessary to produce one bundle unit of commodities j. Thus, the coefficient $c_{12} = 0.4$ means that one needs (2/5) of a bundle unit of commodities 1 in *physical terms* to produce one bundle unit of commodities 2, measured in an appropriate *physical term*.

---

12 In this context it is useful to diagonalize **Se** and **d** for certain purposes, using the definition (A.17): $\mathbf{Se} \Rightarrow \widehat{\mathbf{Se}}$, the matrix $\widehat{\mathbf{Se}}$ has the row-sums of vector **Se** as diagonal elements. Further, $\mathbf{d} \Rightarrow \hat{\mathbf{d}}$, the matrix $\hat{\mathbf{d}}$ has the components d_i of vector **d** as its diagonal elements.

In an *"... ideal statistical world, integral information on values; quantities, and prices of transactions would be available."*, see Eurostat Manual [72], p. 239.

We suppose that we know the values x_i and, from the statistics, the positive prices p_i of the commodities i, $i = 1, \ldots, n$, put into a price vector $\mathbf{p} = [p_1, \ldots, p_n]' > \mathbf{o}$. Under these assumptions, we can convert the entries from *monetary terms* into *physical terms* and vice versa, see also Miller and Blair ([65], pp. 47–48):

$$\begin{cases} x_i = p_i q_i, \\ z_{ij} = p_i s_{ij}, \\ f_i = p_i d_i, \end{cases} \Leftrightarrow \begin{cases} \mathbf{x} = \hat{\mathbf{p}}\mathbf{q} = \hat{\mathbf{q}}\mathbf{p}; \quad \hat{\mathbf{x}} = \hat{\mathbf{p}}\hat{\mathbf{q}}, \\ \mathbf{Z} = \hat{\mathbf{p}}\mathbf{S} \Leftrightarrow \mathbf{S} = \hat{\mathbf{p}}^{-1}\mathbf{Z} \\ \mathbf{f} = \hat{\mathbf{p}}\mathbf{d} \Leftrightarrow \mathbf{d} = \hat{\mathbf{p}}^{-1}\mathbf{f}. \end{cases} \tag{2.18}$$

Starting with the definition of the *distribution coefficients* matrix $\mathbf{D}$ (2.12), we get:

$$\begin{aligned} \hat{\mathbf{q}}^{-1}\mathbf{S} &= \hat{\mathbf{q}}^{-1}(\hat{\mathbf{p}}^{-1}\hat{\mathbf{p}})\mathbf{S} = \hat{\mathbf{q}}^{-1}\hat{\mathbf{p}}^{-1}(\hat{\mathbf{p}}\mathbf{S}) = (\hat{\mathbf{q}}^{-1}\hat{\mathbf{p}}^{-1})\mathbf{Z} \\ &= \hat{\mathbf{x}}^{-1}\mathbf{Z} =: \mathbf{D} \Rightarrow \mathbf{D} = \hat{\mathbf{q}}^{-1}\mathbf{S} = \hat{\mathbf{x}}^{-1}\mathbf{Z}. \end{aligned} \tag{2.19}$$

In analogy to the distribution coefficients d_{ij} in *monetary terms*, we have then shown that they are equal to the distribution coefficients in *physical terms*.

d_{ij}: the *distribution coefficient* is the *value* of a bundle of commodities i (Input) per unit of *value* of a bundle of commodities i, produced during the reporting period (year), distributed to sector S_j for the production of that bundle of commodities j (Output). It is defined as $d_{ij} = (s_{ij}/q_i)$; $i, j \in \{1, \ldots, n\}$. Having all the prices p_i of commodities i; then we also get $d_{ij} = z_{ij}/x_i = (p_i s_{ij})/(p_i q_i) = s_{ij}/q_i$; $i, j \in \{1, \ldots, n\}$. Its unit is [physical term of i/physical term of i] = 1.

$$d_{ij} = \frac{s_{ij}}{q_i} = \frac{z_{ij}}{x_i}; \quad i, j = 1, \ldots, n; \quad \begin{matrix} i: & \text{Input index} \\ j: & \text{Output index} \end{matrix} \quad \text{or} \quad \mathbf{D} = \hat{\mathbf{q}}^{-1}\mathbf{S}. \tag{2.20}$$

The fully developed $n \times n$ matrix $\mathbf{D} = (d_{ij})$ is:

$$\mathbf{D} = \hat{\mathbf{x}}^{-1}\mathbf{Z} := \hat{\mathbf{q}}^{-1}\mathbf{S} = \begin{bmatrix} d_{11} & d_{12} & \ldots & d_{1n} \\ d_{21} & d_{22} & \ldots & d_{2n} \\ \ldots & \ldots & \ldots & \ldots \\ d_{n1} & d_{n2} & \ldots & d_{nn} \end{bmatrix}. \tag{2.21}$$

By definition, the *distribution coefficients* d_{ij}, $i, j = 1, \ldots n$ are dimensionless. Thus, the *distribution coefficients* matrix is a *non-negative matrix* of pure numbers d_{ij}, $1 \geqq d_{ij} \geqq 0$.

The *distribution coefficients* matrix $\mathbf{D}$ describes the **logistics**. Each coefficient of the row $[d_{i1}, d_{i2}, \ldots, d_{in}]$ represents the bundle of commodities i to be distributed to the sectors $j \in \{1, \ldots, n\}$ per unit of produced bundle of commodities i. Thus, the coefficient $d_{32} = 0.2$ means that sector 3 distributes the part ($0.2 = 1/5$) of the own production precisely for the production of sector 2.

The matrix $\mathbf{D}$ will be extensively used later. For the moment, we use the *matrix of the technical coefficients* $\mathbf{C}$, and we will treat later some key relationships between the variables in *monetary terms* and the variables in *physical terms*.

2.2 Economic assumptions on matrices and vectors, viability

After the presentation of matrices and vectors describing economic processes, we will formulate intrinsic assumptions, corresponding to real-world economic situations. Then we start the discussion on criteria of the *viability of an economy*.

There is the requirement of a positive vector $\mathbf{x} = \mathbf{Ze} + \mathbf{f} = \mathbf{Z}'\mathbf{e} + \boldsymbol{\upsilon} > \mathbf{o}$ (2.5) (2.6) with non-negative vectors $\mathbf{f} \geqq \mathbf{o}$, $\boldsymbol{\upsilon} \geqq \mathbf{o}$ and the requirement of a positive vector $\mathbf{q} = \mathbf{Se} + \mathbf{d} > \mathbf{o}$ (2.15) with a non-negative vector $\mathbf{d} \geqq \mathbf{o}$, in order to be able to calculate the inverse diagonal matrices $\hat{\mathbf{x}}^{-1}$ and $\hat{\mathbf{q}}^{-1}$ and to compute the *input-output coefficients* matrices **A**, **C**, **D**. For this reason, the inequalities $x_j > 0$, $q_j > 0$, $j \in \{1, \dots, n\}$ must hold. We will see that these algebraic facilities relate to properties of countries' official input-output tables, e. g., Germany, Example 2.2.1 or the SWISS IOT 2008, and are therefore elevated by some authors, see Schefold [103], to *conditions* to be fulfilled by the models, fitting therefore to real economies.

Associated to official *input-output tables*, the *commodity flow* matrix **Z** in monetary terms, or if there is a price vector **p**, the *commodity flow* matrix **S** in physical terms, is considered as the primary available matrix, describing the structure of an economy, from which the *input-output coefficients* matrices $\mathbf{A} = \mathbf{Z}\hat{\mathbf{x}}^{-1}$ (2.8), $\mathbf{C} = \mathbf{S}\hat{\mathbf{q}}^{-1}$ (2.16) are then derived.

For example. Kurz and Salvadori [100] used the term *viability of an economy*, and Bertram Schefold ([103], p. 49) formulated *economic assumptions* conceived as *necessary conditions* to be fulfilled by some of the just-defined matrices and vectors, in order to obtain *production economies*, corresponding to reality. These notions are presented here:

Assumption 2.2.1 (Economic assumptions on the positivity of output vectors). Every branch of a production economy in single product industries with or without a surplus has to produce at least a positive amount of a palette of commodities,[a] otherwise this branch does not exist. Consequently, every branch j has its positive value of output $x_j > 0$ or its positive quantity of *output* $q_j > 0$.

Summarizing, the vectors $\mathbf{x} > \mathbf{o}$, $\mathbf{x}_I > \mathbf{o}$ and $\mathbf{q} > \mathbf{o}$, $\mathbf{q}_I > \mathbf{o}$ are *positive*.

Consequently, the inverse diagonal matrices $\hat{\mathbf{x}}^{-1}$, $\hat{\mathbf{x}}_I^{-1}$ and $\hat{\mathbf{q}}^{-1}$, $\hat{\mathbf{q}}_I^{-1}$ exist, so that: $\mathbf{A} = \mathbf{Z}\hat{\mathbf{x}}^{-1}$ (2.8), $\mathbf{C} = \mathbf{S}\hat{\mathbf{q}}^{-1}$ (2.16), $\mathbf{D} = \hat{\mathbf{q}}^{-1}\mathbf{S} \equiv \hat{\mathbf{x}}^{-1}\mathbf{Z}$ (2.21).

---

a Schefold ([103], p. 49) was seemingly the first economist to express this claim clearly and concisely. He said: ...(every process has an input besides labour and an output).

Accordingly, we can formulate the following

Proposition 2.2.1. *According to Assumption* 2.2.1 *the vectors* $\mathbf{x} > \mathbf{o}$ *and* $\mathbf{q} > \mathbf{o}$ *are positive; consequently, the transformation* (2.18) *ensures the existence of a positive price vector* $\mathbf{p} > \mathbf{o}$. *The matrix transformations with the diagonal matrices* $\hat{\mathbf{x}}$, $\hat{\mathbf{q}}$, $\hat{\mathbf{p}}$ *and their inverses also exist* (2.19).

Schefold's further assumptions ([103], p. 49) also imply properties fulfilled by the commodity flow matrices **Z**, **S**.

We apply Definition A.4.3 to the *commodity flow matrix* $\mathbf{Z} = (z_{ij})$, $i, j = 1, \ldots, n$, expressing the quantities in *monetary terms*, as well as to the *commodity flow matrix* $\mathbf{S} = (s_{ij})$, $i, j = 1, \ldots, n$, expressing the quantities in *physical terms*.

Consider the column vectors $\mathbf{z}_{\cdot j} = [z_{1j}, z_{2j}, \ldots, z_{nj}]'$, $\mathbf{s}_{\cdot j} = [s_{1j}, s_{2j}, \ldots, s_{nj}]'$, $j = 1, \ldots, n$, respectively the corresponding row vectors $\mathbf{z}'_{i\cdot} = [z_{i1}, z_{i2}, \ldots, z_{in}]$, $\mathbf{s}'_{i\cdot} = [s_{i1}, s_{i2}, \ldots, s_{in}]$, $i = 1, \ldots, n$. Then the matrices $\mathbf{Z}, \mathbf{S}$ can be written as matrices of column vectors or of row vectors as follows:

$$\mathbf{Z} = \begin{bmatrix} z_{11} & z_{12} & \cdots & \cdots & z_{1n} \\ z_{21} & z_{22} & \cdots & \cdots & z_{2n} \\ \cdots & \cdots & \cdots & \cdots & \cdots \\ \cdots & \cdots & \cdots & \cdots & \cdots \\ z_{n1} & z_{n2} & \cdots & \cdots & z_{nn} \end{bmatrix} = [\mathbf{z}_{\cdot 1}, \mathbf{z}_{\cdot 2}, \ldots, \ldots, \mathbf{z}_{\cdot n}] = \begin{bmatrix} \mathbf{z}'_{1\cdot} \\ \mathbf{z}'_{2\cdot} \\ \cdots \\ \cdots \\ \mathbf{z}'_{n\cdot} \end{bmatrix}, \qquad (2.22)$$

$$\mathbf{S} = \begin{bmatrix} s_{11} & s_{12} & \cdots & \cdots & s_{1n} \\ s_{21} & s_{22} & \cdots & \cdots & s_{2n} \\ \cdots & \cdots & \cdots & \cdots & \cdots \\ \cdots & \cdots & \cdots & \cdots & \cdots \\ s_{n1} & s_{n2} & \cdots & \cdots & s_{nn} \end{bmatrix} = [\mathbf{s}_{\cdot 1}, \mathbf{s}_{\cdot 2}, \ldots, \ldots, \mathbf{s}_{\cdot n}] = \begin{bmatrix} \mathbf{s}'_{1\cdot} \\ \mathbf{s}'_{2\cdot} \\ \cdots \\ \cdots \\ \mathbf{s}'_{n\cdot} \end{bmatrix}. \qquad (2.23)$$

Assumption 2.2.2 (Assumption on the commodity flow matrices). Every branch of the production process has to treat at least one palette of commodities i as a means of production to produce output, otherwise this branch does not exist. Consequently, every branch j must have at least an *input*, so that each column vector is *semi-positive*, $\mathbf{z}_{\cdot j} \geq \mathbf{o}$, $\mathbf{s}_{\cdot j} \geq \mathbf{o}$, $j \in \{1, \ldots, n\}$.[a]

For this reason, the matrices $\mathbf{Z} \geq \mathbf{0}$, $\mathbf{S} \geq \mathbf{0}$, $\mathbf{A} \geq \mathbf{0}$, $\mathbf{C} \geq \mathbf{0}$ are *semi-positive*.

a The condition that each row vector of a flow matrix is semi-positive, $\mathbf{z}_{j\cdot} \geq \mathbf{o}$, $\mathbf{s}_{j\cdot} \geq \mathbf{o}$, $i, j \in \{1, \ldots, n\}$ obviously does not hold when we think of an economy where gold is produced, but where gold does not figure as a *means of production*.

Example 2.2.1 (Exception to Assumption 2.2.2). The Federal Statistical Office of Germany published the input-output table 2013 of the domestic production at basic prices, revised in 2014 and updated to August 2016,[13] website: www.destatis.de. This IOT contains $n = 98$ groups of commodities and the same number of branches of production in terms of 'CAP = Classification of products by Activity' (valid for the 'European Community (EU)', website: http://ec.europa.eu/eurostat/ramon/). There are three tables (use, supply and input-output tables) of the European System of Accounts (ESA 1995), the use table at basic acquisition prices.[14] In this *use table*, corresponding to matrix $\mathbf{Z}$, there is a so-called double-branch,[15] designed by one suf-

13 Statistisches Bundesamt, Fachserie 18, Reihe 2, Volkswirtschaftliche Gesamtrechnung, Input-Output-Rechnung, 2013 (Revision 2014, Stand: August: 2016), date of publication: March 7, 2017.

14 In German: "Verwendungstabelle 2013 zu Anschaffungspreisen".

15 The *divisions (NACE)* 97 and 98 are grouped into one branch, designated by 97–98.

fix $i = 97$–98, called 'Services of private households',[16] with elements $z_{i,97-98} = 0$, $i = 1, \ldots, 98$ [divisions, (NACE)] and $z_{97-98,j} = 0$, $i = 1, \ldots, 98$, but the final consumption is $d_{97-98} = 7{,}247 > 0$, equal to a corresponding value added $v_{97-98} = 7{,}247 > 0$. So there is a null column vector $\mathbf{z}_{\cdot 97-98} = \mathbf{o}$ and a null row vector $\mathbf{z}_{97-98\cdot} = \mathbf{o}$. This is not in accordance with Assumption 2.2.2. In the German IOT 2013, the sector "Services of private households" is a special sector. It does not produce any product nor request any product, but it corresponds to a consumption demand. We will later learn how to treat such exceptions. ▲

Nevertheless, from now on, Assumption 2.2.1 and Assumption 2.2.2 will be adopted for the treated economies. We start a discussion on *sufficient conditions of reproduction.*

At first, we present the notion of a *profitable economy*, which means that the production process produces a surplus. Louis de Mesnard [25] introduces the notion of *profitable economies* and presents it in *monetary terms* ($\mathbf{f} \geqq \mathbf{o}$), and Bertram Schefold introduces the same notion but presents it in *physical terms* ($\mathbf{d} \geqq \mathbf{o}$). Bertram Schefold argues ([109], 6. Anhang, p. 219) that an economy is **profitable**, if the inequality $\mathbf{De} \leqq \mathbf{e}$ holds.

As by Assumption 2.2.1, the vector of total output is positive, $\mathbf{x} > \mathbf{o}$, so with (2.5) and the *input-output coefficients* matrix $\mathbf{A} = \mathbf{Z}\hat{\mathbf{x}}^{-1}$ (2.8) we write $\mathbf{x} = \mathbf{Ze} + \mathbf{f} = (\mathbf{A}\hat{\mathbf{x}})\mathbf{e} + \mathbf{f} = \mathbf{A}(\hat{\mathbf{x}}\mathbf{e}) + \mathbf{f} = \mathbf{Ax} + \mathbf{f}$, and we obtain

$$\begin{aligned}
&\text{monetary terms:} \quad \mathbf{x} \geqq \mathbf{Ax} \Leftrightarrow \mathbf{f} := \mathbf{x} - \mathbf{Ax} = (\mathbf{I} - \mathbf{A})\mathbf{x} \geqq \mathbf{o},\\
&\text{physical terms:} \quad \exists \mathbf{d} \geqq \mathbf{o} : (\mathbf{q} = \mathbf{Se} + \mathbf{d} \wedge \mathbf{D} = \hat{\mathbf{q}}^{-1}\mathbf{S} \Leftrightarrow\\
&\mathbf{q} = \hat{\mathbf{q}}\mathbf{De} + \mathbf{d} \geqq \hat{\mathbf{q}}\mathbf{De} \Leftrightarrow (\hat{\mathbf{q}}^{-1}\mathbf{q}) \geqq (\hat{\mathbf{q}}^{-1}\hat{\mathbf{q}})\mathbf{De} \Leftrightarrow \mathbf{e} \geqq \mathbf{De}.)
\end{aligned} \tag{2.24}$$

Thus Schefold's equivalence is shown: $\mathbf{d} \geqq \mathbf{o} \Leftrightarrow \mathbf{e} \geqq \mathbf{De}$. Second, we present the notion of the *viability of an economy* (see Kurz and Salvadori, ([52], pp. 96–97)).

Definition 2.2.1. An economy in physical or monetary terms is said to be **viable**,
- if the technology at its disposal enables it to reproduce itself (assuming subsistence wages for accomplished labour or labour at no cost [52], p. 96);
- when for vectors $\mathbf{x} > \mathbf{o}$ and $\mathbf{f} \geqq \mathbf{o}$, the equation $(\mathbf{I} - \mathbf{A})\mathbf{x} = \mathbf{f} \Leftrightarrow \mathbf{x} \geqq \mathbf{Ax}$ holds.

An economy is **just viable**, if $\mathbf{f} = \mathbf{o}$. In this case, there are only interindustrial transactions and there is no surplus (no national income). We see that notions are set so that "just viability" is a limiting case of "viability".

From Definition 2.2.1 and equation (2.24), it is immediately obvious that the notions *profitable economy* and *viability of an economy* are equivalent, applied either to economies in *monetary terms* or to economies in *physical terms*. Matrix $(\mathbf{I} - \mathbf{A})$ is named the *Leontief matrix* and its inverse $(\mathbf{I} - \mathbf{A})^{-1}$ the *Leontief Inverse*, see Miller and Blair ([65], pp. 21, 44) and Oosterhaven ([77], p. 751).

16 In German: "Waren und Dienstleistungen privater Haushalte ohne ausgeprägten Schwerpunkt".

For further modelling purposes, we introduce now the mathematical theory of *positive* or *non-negative* matrices, including *semi-positive* matrices, leading us to the **Perron–Frobenius Theorems A.9.3.**

2.3 Positive matrices and non-negative matrices

The mathematical theory of *non-negative*, *positive* or even *irreducible* matrices, see Definition A.8.3, is crucial for Input-Output Analysis, as Schefold, [103], [109], pp. 216–225, Pasinetti [80], Kurz & Salvadori [54], Duchin [23] or Ashmanov [2] and other authors have revealed. *Non-negativity* is less restrictive than the economic Assumption 2.2.1 and Assumption 2.2.2 which hold for the entries of the *commodity flow matrices* **Z** and **S**. But the *non-negativity* and *irreducibility* of matrices are the properties required by the **Perron–Frobenius theorem A.9.3** which guarantees the existence of *positive* eigenvectors, associated with the positive Frobenius number, and needed to model *prices* and *quantities*.

Moreover, there is a weakened version of the seminal Perron–Frobenius theorem. It is Theorem A.10.1 for which only *non-negative* matrices are required, stating the existence of *non-negative* eigenvectors, associated with a unique *non-negative* Frobenius number.

To be safe, we proceed with an example. Consider a *futuristic* border case economy, where a new product such as a *robotic car* does not appear in the means of production of the Leontief *commodity flow* matrix. Even here, Table 2.3, the economic Assumption 2.2.1 and Assumption 2.2.2 being fulfilled, the matrix **Z** is *semi-positive*. In this example, one uses the notions developed in Subsection A.12.

Table 2.3: Transactions, final demand and total output.

Commodities	**Processing sectors (in Mio CHF)**			**Final demand (in Mio. CHF)**	**Total output (in Mio. CHF)**
$i = 1, 2, 3$	S_1	S_2	S_3	f_i	x_i
S_1: hardware	280	180	115	0	575
S_2: software	240	240	120	0	600
S_3: robotic car	0	0	0	250	250
value added	55	180	15	$V = F = 250$	
total outlays	575	600	250		$X = 1425$

Example 2.3.1. Inspired by the **Geneva International Motor Show 2017,** we consider a partial economy with sectors S_1: hardware, S_2: software, S_3: robotic car. There is a final demand only for robotic cars. All the other production is outside this partial economy. We are in presence of a productive economy.

Given Table 2.3 with transaction entries and final demand of the base year and final demand of the current year in monetary terms,

Identify from the Table 2.3 the commodity flow matrix $\mathbf{Z}$ and the vector of final demand $\mathbf{f}$ both in monetary terms.

Compute the vector of total output $\mathbf{x}$, and the input-output coefficients matrix $\mathbf{A} = \mathbf{Z}\hat{\mathbf{x}}^{-1}$, the Frobenius number λ_A, and the Leontief matrix $(\mathbf{I} - \mathbf{A})$. Verify the inequality $\mathbf{A}'\mathbf{e} \leq \mathbf{e}$. Argue on the basis of Theorem A.12.1 and Lemma 2.4.1 (b) about the existence of the *Leontief Inverse* $(\mathbf{I} - \mathbf{A})^{-1}$. Compute the vector $(\mathbf{I} - \mathbf{A})^{-1}\mathbf{f}$ and interpret the result.

Compute the left eigenvectors $\mathbf{p}_A$ and the right eigenvectors $\mathbf{x}_A$ of the matrix $\mathbf{A}$. Formulate your observations (all the calculations can be done by hand).

Solution of Example 2.3.1:
The transaction matrix $\mathbf{Z}$ and the vector of final demand $\mathbf{f}$ are identified from Table 2.3. Matrix $\mathbf{Z}$ is *non-negative*. Then one computes the vector of total demand $\mathbf{x}$, according to

$$\mathbf{Z} = \begin{bmatrix} 280 & 180 & 115 \\ 240 & 240 & 120 \\ 0 & 0 & 0 \end{bmatrix}, \quad \mathbf{f} = \begin{bmatrix} 0 \\ 0 \\ 250 \end{bmatrix}, \quad \mathbf{x} = \mathbf{Z}\mathbf{e} + \mathbf{f} = \begin{bmatrix} 575 \\ 600 \\ 250 \end{bmatrix} > \mathbf{o}. \tag{2.25}$$

Compute the *non-negative input-output coefficients* matrix $\mathbf{A} = \mathbf{Z}\hat{\mathbf{x}}^{-1}$ (2.8),

$$\mathbf{A} = \begin{bmatrix} 280 & 180 & 115 \\ 240 & 240 & 120 \\ 0 & 0 & 0 \end{bmatrix} \begin{bmatrix} \frac{1}{575} & 0 & 0 \\ 0 & \frac{1}{600} & 0 \\ 0 & 0 & \frac{1}{250} \end{bmatrix} = \begin{bmatrix} \frac{56}{115} & \frac{3}{10} & \frac{23}{50} \\ \frac{48}{115} & \frac{2}{5} & \frac{12}{25} \\ 0 & 0 & 0 \end{bmatrix} \geq \mathbf{0}. \tag{2.26}$$

Lemma 2.1.1 holds because the vector of value-added is positive, $\boldsymbol{\upsilon} = [55, 180, 15]'$. There is $\mathbf{A}'\mathbf{e} = [104/115, 7/10, 47/50] \leq \mathbf{e}$. The characteristic polynomial of the *input-output coefficients* matrix is

$$f_3(\lambda) = \det(\mathbf{A} - \lambda\mathbf{I}) = -\lambda\left(\lambda^2 - \frac{102}{115}\lambda + \frac{8}{115}\right) = -\lambda\left(\lambda - \frac{2}{23}\right)\left(\lambda - \frac{4}{5}\right). \tag{2.27}$$

As the vector of *final demand* $\mathbf{f} \geq \mathbf{o}$ is *semi-positive*, consequently the Frobenius number is $\lambda_A = 4/5 < 1$, i. e., less than one. Compute the *Leontief matrix*,

$$\mathbf{I} - \mathbf{A} = \begin{bmatrix} \frac{59}{115} & -\frac{3}{10} & -\frac{23}{50} \\ -\frac{48}{115} & \frac{3}{5} & -\frac{12}{25} \\ 0 & 0 & 1 \end{bmatrix}; \tag{2.28}$$

The economy is productive, applying Theorem A.12.1 the *Leontief Inverse* exists. A computation yields,

$$(\mathbf{I} - \mathbf{A})^{-1} = \begin{bmatrix} \frac{23}{7} & \frac{23}{14} & \frac{23}{10} \\ \frac{16}{7} & \frac{59}{21} & \frac{12}{5} \\ 0 & 0 & 1 \end{bmatrix} \geq \mathbf{0}, \; (\mathbf{I} - \mathbf{A})^{-1}\mathbf{f} = \begin{bmatrix} \frac{23}{7} & \frac{23}{14} & \frac{23}{10} \\ \frac{16}{7} & \frac{59}{21} & \frac{12}{5} \\ 0 & 0 & 1 \end{bmatrix} \begin{bmatrix} 0 \\ 0 \\ 250 \end{bmatrix} = \begin{bmatrix} 575 \\ 600 \\ 250 \end{bmatrix}. \tag{2.29}$$

We have found $\mathbf{x} = (\mathbf{I} - \mathbf{A})^{-1}\mathbf{f}$, see also Lemma 2.4.1, as the Frobenius number is $\lambda_A = 4/5 < 1$. Then we find for the *non-negative* matrix $\mathbf{A} \geq \mathbf{0}$, the *positive* right eigenvector $\mathbf{x}_A = [40/41, 30/41, 1]' > \mathbf{o}$, and finally one obtains the *semi-positive* left eigenvector $\mathbf{p}_A = [23/24, 1, 0]' \geq \mathbf{o}$. ▲

Observation. Theorem A.12.1 applies to Example 2.3.1, because we are in presence of a productive Leontief model, as the vector of *final demand* is *semi-positive*, $\mathbf{f} \geq \mathbf{o}$. Moreover, the *input-output coefficients* matrix is *semi-positive*, $\mathbf{A} \geq \mathbf{O}$, all column vectors of $\mathbf{A}$ being *semi-positive*, the economic *Assumption* 2.2.2 applies. The Frobnius number is smaller than 1, $\lambda_A = 4/5 < 1$, as we have calculated. The Definition A.12.1 is also confirmed: The vector $\mathbf{x} = \mathbf{Ze} + \mathbf{f} = \mathbf{A\hat{x}e} + \mathbf{f} = \mathbf{Ax} + \mathbf{f} = [575, 600, 250]' > \mathbf{o}$ is even positive, because Assumption 2.2.1 is fulfilled by construction of the example. The right eigenvectors are positive, $\mathbf{x}_A > \mathbf{o}$. This is a limit case, where the third row vector $\mathbf{a}_{3.} = \mathbf{o}$ of the matrix $\mathbf{A}$ is a zero vector, the left eigenvectors are *semi-positive*, $\mathbf{p}_A \geq \mathbf{o}$.[17]

The Perron–Frobenius theorem A.9 dominates the field of the Input-Output analysis of Leontief.[18]

2.4 The Leontief quantity model

We assume throughout this section economies with or without *final demand*. Furthermore, we assume that the restrictive Assumption 2.2.1, respectively, Assumption 2.2.2 hold. Thus, we are in the presence of a *semi-positive commodity flow* matrix $\mathbf{Z} \geq \mathbf{O}$, a *semi-positive input-output coefficients* matrix $\mathbf{A} \geq \mathbf{O}$ in *monetary terms*, a *non-negative* vector of *final demand*, $\mathbf{f} \geqq \mathbf{o}$ and a *positive* vector of *total output*, $\mathbf{x} > \mathbf{o}$. For the notations, we rely on Input-Output data representation as in Table 2.1.

In other word, we consider *Leontief models*, also termed *Leontief quantity models* or, *demand-driven input-output quantity model*, see Oosterhaven ([77], p. 751) (in monetary terms), or classically, the *Leontief model* in monetary terms, see Definition A.12.1,

$$\boxed{\mathbf{x} = \mathbf{Ax} + \mathbf{f} \Leftrightarrow \mathbf{f} = (\mathbf{I} - \mathbf{A})\mathbf{x}, \quad \mathbf{x} > \mathbf{o}.} \tag{2.30}$$

Quantities of commodities are here analysed. Prices do not appear, but they will come later. The *Leontief model* (2.30) goes back to Leontief's Input-Output Analysis [56]. The following question has to be solved: What are the conditions that the *Leontief model* (2.30) has as a solution a *positive* vector of *total output* $\mathbf{x} > \mathbf{o}$? The issues will be illustrated by elementary examples.[19]

17 The idea to use *singular vectors* and *singular value decomposition* at the positive rectangular submatrices of a matrix $\mathbf{A}$, as they occur in Example 2.3.1, has been realised (see Stetsyuk and Emmenegger [113]).

18 We will see in Chapter 4 that the **Perron–Frobenius theorem A.9.3** also dominates the field of Sraffa's *single product* industries for the determination of price vectors.

19 The reader should not be put off by the examples. Some are kept in this text in their original simplicity to help the reader to focus on the algebraic and computational aspects. In particular, "wheat", "iron", "turkeys" should be read sectorially as "agriculture", "manufacturing", "food production".

As a key result, we first present the conditions of existence of the *Leontief Inverse*. It has to do with the presence of final demand, expressed by the *semi-positive* vector $\mathbf{f} \geq \mathbf{o}$. We will also obtain the criterion for the unique *positive* solution, $\mathbf{x} > \mathbf{o}$, of the *Leontief quantity model*. We formulate

Lemma 2.4.1. *For the output vectors, the inequalities* $\mathbf{x} > \mathbf{o}$ *and* $\mathbf{x}_I > \mathbf{o}$ *hold:*

(a) *If* $\mathbf{f} = \mathbf{o}$ *and* $\mathbf{Z} \geq \mathbf{0}$, *then the matrix* $\mathbf{A} = \mathbf{Z}\hat{\mathbf{x}}^{-1} \geq \mathbf{0}$ *has a maximal eigenvalue* $\lambda_A = 1$, *the Leontief Inverse* $(\mathbf{I} - \mathbf{A})^{-1}$ *does not exist and the Leontief quantity model* (2.30) *has no solution.*

(b) *If* $\mathbf{f} \geq \mathbf{o}$ *and* $\mathbf{Z} \geq \mathbf{0}$, *then the matrix* $\mathbf{A} = \mathbf{Z}\hat{\mathbf{x}}^{-1} \geq \mathbf{0}$ *has Frobenius number less than 1,* $\lambda_A < 1$. *The Leontief Inverse* $(\mathbf{I} - \mathbf{A})^{-1}$ *exists and the Leontief quantity model* (2.30) *is productive and has a unique positive solution vector,*

$$\mathbf{x} = (\mathbf{I} - \mathbf{A})^{-1}\mathbf{f} > \mathbf{o}. \tag{2.31}$$

Proof. (a) If $\mathbf{f} = \mathbf{o}$ and $\mathbf{Z} \geq \mathbf{0}$, the vector $\mathbf{x}_I = \mathbf{Ze} > \mathbf{o}$ and $\mathbf{A} = \mathbf{Z}\hat{\mathbf{x}}_I^{-1} \geq \mathbf{0}$, $\mathbf{D} = \hat{\mathbf{x}}_I^{-1}\mathbf{Z} \geq \mathbf{0}$. Then $\hat{\mathbf{x}}_I^{-1}(\mathbf{Ze}) = (\hat{\mathbf{x}}_I^{-1}\mathbf{Z})\mathbf{e} = \mathbf{De} = \hat{\mathbf{x}}_I^{-1}\mathbf{x}_I = \mathbf{e}$ (2.12). Then, matrix $\mathbf{D}$ is stochastic with eigenvector $\mathbf{e}$ and maximal eigenvalue $\lambda_D = 1$ (see Lemma A.11.1). Matrices $\mathbf{D}$ and $\mathbf{A}$ are similar, $\lambda_D = \lambda_A = 1$, Lemma A.6.1. Consequently, $\det(\mathbf{A} - \lambda_A\mathbf{I}) = \det(\mathbf{I} - \mathbf{A}) = 0$ and the Leontief Inverse $(\mathbf{I} - \mathbf{A})^{-1}$ does not exist. Then, considering the ranks $n - \operatorname{rank}(\mathbf{A} - \mathbf{I}) = \dim(\mathbb{U}) > 0$, the solution space $\mathbb{U}$ of (2.30) contains an infinity of solution vectors $\mathbf{x}$ (see Kowalsky [48], p. 75), and the *Leontief quantity model* (2.30) has no solution.

(b) If $\mathbf{f} \geq \mathbf{o}$ and $\mathbf{Z} \geq \mathbf{0}$, then $\mathbf{x} = \mathbf{Ze} + \mathbf{f} > \mathbf{0}$, and the matrix is also *semi-positive*, $\mathbf{A} = \mathbf{Z}\hat{\mathbf{x}}^{-1} \geq \mathbf{0}$; then, with $\mathbf{x} = \mathbf{A}\hat{\mathbf{x}}\mathbf{e} + \mathbf{f} = \mathbf{Ax} + \mathbf{f} > \mathbf{o}$, the *Leontief quantity model* becomes a *productive Leontief model*. With Theorem A.12.1 the Frobenius number is less than one, $\lambda_A < 1$. Therefore the Leontief Inverse exists, $(\mathbf{I} - \mathbf{A})^{-1}$, and equation $\mathbf{f} = (\mathbf{I} - \mathbf{A})\mathbf{x}$ (2.30) has the unique solution $\mathbf{x} = (\mathbf{I} - \mathbf{A})^{-1}\mathbf{f} > \mathbf{o}$. □

2.4.1 The Leontief quantity model in monetary terms

We begin with a *Leontief model* in *monetary terms*, where the vector of final demand is *null*, $\mathbf{f} = \mathbf{o}$.[20] Leontief models will later lead to the discussion of *productive economies*, whose performance will be measured by *productiveness*.

Example 2.4.1. Consider the positive 2×2 commodity flow matrix $\mathbf{Z} = \left[\begin{smallmatrix} 20 & 30 \\ 30 & 40 \end{smallmatrix}\right] > \mathbf{0}$. Compute the *input-output coefficients* matrix $\mathbf{A}$ and its eigenvalues. Compute the rank of the matrix $\mathbf{I} - \mathbf{A}$. Compute the dimension of the solution space $\mathbb{U}$ of the corresponding *Leontief quantity model* (2.30). What do you conclude?

---

20 For this case there is Theorem A.12.1 which states that a *Leontief model* is *productive* if and only if the Frobenius number $\lambda_A < 1$. In the present case of $\mathbf{f} = \mathbf{o}$, there is $\lambda_A = 1$ (see also Ashmanov [2], p. 24 and p. 39, and Stetsyuk [111], [112]).

Solution of Example 2.4.1:
We compute the *positive* output vector $\mathbf{x} = \mathbf{Ze} + \mathbf{f} > \mathbf{o}$ (2.7) and diagonalize it. Then, the *input-output coefficients* matrix $\mathbf{A} = \mathbf{Z}\hat{\mathbf{x}}^{-1} > \mathbf{0}$ (2.8) is *positive*,

$$\mathbf{x} = \mathbf{Z}'\mathbf{e} = \mathbf{Ze} = \begin{bmatrix} 20 & 30 \\ 30 & 40 \end{bmatrix} \begin{bmatrix} 1 \\ 1 \end{bmatrix} = \begin{bmatrix} 50 \\ 70 \end{bmatrix}, \quad \hat{\mathbf{x}}^{-1} = \begin{bmatrix} \frac{1}{50} & 0 \\ 0 & \frac{1}{70} \end{bmatrix}, \tag{2.32}$$

concluding that in this case there is $V = F = 0$. Then, we continue with the *input-output coefficients* matrix,

$$\mathbf{A} = \mathbf{Z}\hat{\mathbf{x}}^{-1} = \begin{bmatrix} 20 & 30 \\ 30 & 40 \end{bmatrix} \begin{bmatrix} \frac{1}{50} & 0 \\ 0 & \frac{1}{70} \end{bmatrix} = \begin{bmatrix} \frac{2}{5} & \frac{3}{7} \\ \frac{3}{5} & \frac{4}{7} \end{bmatrix} > \mathbf{0}, \tag{2.33}$$

and the characteristic polynomial,

$$P_2(\lambda) = \det(\mathbf{I} - \lambda\mathbf{A}) = -\frac{1}{35} - \frac{34}{35}\lambda + \lambda^2 = (\lambda - 1)\left(\lambda + \frac{1}{35}\right), \tag{2.34}$$

before determining the rank of matrix $\mathbf{I} - \mathbf{A}$,

$$\operatorname{rank}(\mathbf{I} - \mathbf{A}) = \operatorname{rank}\left(\begin{bmatrix} \frac{3}{5} & -\frac{3}{7} \\ -\frac{3}{5} & \frac{3}{7} \end{bmatrix}\right) = 1. \tag{2.35}$$

As the Frobenius number is $\lambda_A = 1$ because there is no demand, $\mathbf{f} = \mathbf{o}$, the *Leontief Inverse* does not exist, and the *Leontief quantity model* (2.30) has no solution vector. ▲

We will now carry out some **comparative-static analysis** with the *Leontief quantity model* (2.30), implicitly introducing the notion of time in order to compare the states of the system at two different *points in time*. That means, we analyse the evolution of the economy and the variables, during a considered *period*, extending from a *base year* "0" to a *current year* "1".[21] For this purpose, it is necessary to indicate what property of the economic system is carried forward from the base year to the current year.

Assumption 2.4.1 (No change in technology). Unless otherwise stated, we assume that the **technology** of the economy is time-invariant. This assumption has to do with the common observation that the technology of an economy needs generally more time to change than the variability due to *business cycles*, as the "long term" versus the "short and medium terms". This means that the commodity flow matrices $\mathbf{Z}$, respectively $\mathbf{S}$, and the distribution coefficients matrix $\mathbf{D}$ describing the **logistics**, vary from period to period, whereas the *input-output coefficients* matrices $\mathbf{A} = \mathbf{Z}\hat{\mathbf{x}}^{-1}$, respectively $\mathbf{C} = \mathbf{S}\hat{\mathbf{q}}^{-1}$, remain constant.

---

21 The *base year* could be 2008 and the *current year* 2015, thus we would have the period from 2008 to 2015, {2008,2015}.

Now, one sets a positive *final demand* for the base year, $\mathbf{f}_0 > \mathbf{o}$. The final demand changes from $\mathbf{f}_0$ to become the new positive final demand $\mathbf{f}_1 > \mathbf{o}$ of the current year, giving the difference vector $\Delta\mathbf{f} := \mathbf{f}_1 - \mathbf{f}_0$. The question is, how has the *total output* $\mathbf{x}_0$ changed from the base year to the *total output* $\mathbf{x}_1$ of the current year, giving the difference vector $\Delta\mathbf{x} := \mathbf{x}_1 - \mathbf{x}_0$.

Moreover, what happens with the *commodity flow matrix* $\mathbf{Z} > \mathbf{0}$ and the *input-output coefficients* matrix $\mathbf{A} = \mathbf{Z}\hat{\mathbf{x}}^{-1} > \mathbf{0}$? Do both change or are they invariant, independently?

Thus, we assume an invariant technology in what follows. We also know that the positive matrix $\mathbf{A} > \mathbf{0}$ has a Frobenius number $\lambda_A < 1$, Lemma 2.4.1 (b). Then, we get for the *Leontief quantity model* (2.30) a unique solution (2.31) in the difference vector:

$$\Delta\mathbf{x} = (\mathbf{I} - \mathbf{A})^{-1}\Delta\mathbf{f}. \tag{2.36}$$

We are interested in the variations of the entries of the *commodity flow* matrix, $\mathbf{Z} = \mathbf{A}\hat{\mathbf{x}}$ (2.8) and find the following equations to calculate the variation and the new level of matrix $\mathbf{Z}$:

$$\mathbf{Z}_1 = \mathbf{A}\hat{\mathbf{x}}_1 \quad \text{and} \quad \mathbf{Z}_0 = \mathbf{A}\hat{\mathbf{x}}_0 \quad \text{or} \quad \Delta\mathbf{Z} := \mathbf{Z}_1 - \mathbf{Z}_0 = \mathbf{A}(\hat{\mathbf{x}}_1 - \hat{\mathbf{x}}_0) = \mathbf{A}\Delta\hat{\mathbf{x}}. \tag{2.37}$$

Example 2.4.2. Consider an economy with sectors S_1: wheat, S_2: iron, S_3: wood, with corresponding entries and final demands.

Given Table 2.4 with transaction entries and final demand of the base year and final demand of the current year in monetary terms in any currency (all the calculations can be done by hand):

Table 2.4: Transactions and vectors of total demand in monetary terms.

Commodities	**Base year processing sectors**			**Base year final demand**	**Current year final demand**
$i = 1, 2, 3$	S_1	S_2	S_3	f_{0i}	f_{1i}
S_1: wheat	20	30	40	10	20
S_2: iron	20	30	40	60	100
S_3: wood	20	30	40	110	200

Extract the commodity flow matrix $\mathbf{Z}_0$ of the base year from Table 2.4 and both vectors of final demand $\mathbf{f}_0$ and $\mathbf{f}_1$.

Compute the vector of total output $\mathbf{x}_0 > \mathbf{o}$ for the base year, the vector of inter-industrial production $\mathbf{x}_I$ and the total output of the economy X_0. Compute the vector of total value added $\boldsymbol{\upsilon}_0$ of the base year and the total value added V_0. Compute the input-output coefficients matrix $\mathbf{A} = \mathbf{Z}_0\hat{\mathbf{x}}_0^{-1}$, the matrix $\mathbf{I} - \mathbf{A}$ and the *Leontief Inverse* $(\mathbf{I} - \mathbf{A})^{-1}$.

Considering the Leontief quantity model (or the *demand-driven input-output model*) (2.30), compute the unique solution vector of total output, $\mathbf{x}_0 > \mathbf{o}$, for the current year (2.31), using the previously obtained Leontief Inverse. Compute the total production for the current year $\mathbf{x}_1 > \mathbf{o}$.

Compute the difference vector $\Delta\mathbf{x}$ (2.36) showing the change of total output between the base year and the current year.

Compute the commodity flow matrix $\mathbf{Z}_1$ (2.37) of the current year. Compute the vector of total value added $\mathbf{\upsilon}_1$ of the current year and the total value added V_1, the total final demand F_1 and the total output X_1.

Compute the distribution coefficients matrix $\mathbf{D}_0 = \hat{\mathbf{x}}_0^{-1}\mathbf{Z}_0$.

Check the identity $\mathbf{D}_0\mathbf{e} + \hat{\mathbf{x}}_0^{-1}\mathbf{f}_0 = \mathbf{e}$.

Check also the identity $\mathbf{D}_1\mathbf{e} + \hat{\mathbf{x}}_1^{-1}\mathbf{f}_1 = \mathbf{e}$.

Explain the variability of the matrices $\mathbf{Z}_i$ and $\mathbf{D}_i$, $i = 0, 1$, during this period from the base year to the current year.

Solution of Example 2.4.2:

The entries of the transaction matrix $\mathbf{Z}_0$, and the vectors of final demand $\mathbf{f}_0$, $\mathbf{f}_1$ and $\Delta\mathbf{f} = \mathbf{f}_1 - \mathbf{f}_0$ are taken from Table 2.4,

$$\mathbf{Z}_0 = \begin{bmatrix} 20 & 30 & 40 \\ 20 & 30 & 40 \\ 20 & 30 & 40 \end{bmatrix}, \quad \mathbf{f}_0 = \begin{bmatrix} 10 \\ 60 \\ 110 \end{bmatrix}, \quad \mathbf{f}_1 = \begin{bmatrix} 20 \\ 100 \\ 200 \end{bmatrix}, \quad \Delta\mathbf{f} = \begin{bmatrix} 10 \\ 40 \\ 90 \end{bmatrix}. \tag{2.38}$$

We calculate the vector of *interindustrial production* $\mathbf{x}_I$ and the vector of *total demand of industrial production*, i. e., the vector of *total output* $\mathbf{x}_0$,

$$\mathbf{x}_I := \mathbf{Z}_0\mathbf{e} = \begin{bmatrix} 20 & 30 & 40 \\ 20 & 30 & 40 \\ 20 & 30 & 40 \end{bmatrix}\begin{bmatrix} 1 \\ 1 \\ 1 \end{bmatrix} = \begin{bmatrix} 90 \\ 90 \\ 90 \end{bmatrix},$$

$$\mathbf{x}_0 := \mathbf{x}_I + \mathbf{f}_0 = \begin{bmatrix} x_1 \\ x_2 \\ x_3 \end{bmatrix} = \begin{bmatrix} 90 \\ 90 \\ 90 \end{bmatrix} + \begin{bmatrix} 10 \\ 60 \\ 110 \end{bmatrix} = \begin{bmatrix} 100 \\ 150 \\ 200 \end{bmatrix}. \tag{2.39}$$

We now compute the vector of total *value added* for the base year,

$$\mathbf{\upsilon}_0 = \mathbf{x}_0 - \mathbf{Z}_0'.\mathbf{e} = \begin{bmatrix} 100 \\ 150 \\ 200 \end{bmatrix} - \begin{bmatrix} 20 & 20 & 20 \\ 30 & 30 & 30 \\ 40 & 40 & 40 \end{bmatrix}\begin{bmatrix} 1 \\ 1 \\ 1 \end{bmatrix} = \begin{bmatrix} 40 \\ 60 \\ 80 \end{bmatrix}, \tag{2.40}$$

and compute the *total final demand* and the *total value added* of the base year,

$$F_0 = \mathbf{e}'\mathbf{f}_0 = [1,1,1]\begin{bmatrix} 10 \\ 60 \\ 110 \end{bmatrix} = [1,1,1]\begin{bmatrix} 40 \\ 60 \\ 80 \end{bmatrix} = \mathbf{e}'\mathbf{\upsilon}_0 = V_0 = 180. \tag{2.41}$$

The total production output of the economy during the base year is $X_0 = \mathbf{e}'\mathbf{x}_0 = 100 + 150 + 200 = 450$, presented in Table 2.5.

Table 2.5: Final demand and value added, total output and outlays in the base year.

	Processing sectors			Final demand	Total
	S_1	S_2	S_3	f_{0i}	output
S_1: wheat	20	30	40	10	100
S_2: iron	20	30	40	60	150
S_3: wood	20	30	40	110	200
value added v_j	40	60	80	$V_0 = F_0 = 180$	
total outlays	100	150	200		$X_0 = 450$

Then the *input-output coefficients* matrix $\mathbf{A} = (a_{ij})$ is computed with equation (2.8),

$$\mathbf{A} = \mathbf{Z}_0\hat{\mathbf{x}}_0^{-1} = \begin{bmatrix} 20 & 30 & 40 \\ 20 & 30 & 40 \\ 20 & 30 & 40 \end{bmatrix} \begin{bmatrix} \frac{1}{100} & 0 & 0 \\ 0 & \frac{1}{150} & 0 \\ 0 & 0 & \frac{1}{200} \end{bmatrix} = \begin{bmatrix} 0.2 & 0.2 & 0.2 \\ 0.2 & 0.2 & 0.2 \\ 0.2 & 0.2 & 0.2 \end{bmatrix}, \quad (2.42)$$

indicating that in the base year the constant *input-output coefficients* $a_{ij} = 0.2$ are the *values* of the amounts of each commodity i, required for the production of one *value unit* of each commodity j. The Frobenius number is $\lambda_A = 0.6$.

Now the Leontief matrix $\mathbf{I} - \mathbf{A}$ and its determinant are computed,

$$\mathbf{I} - \mathbf{A} = \begin{bmatrix} 0.8 & -0.2 & -0.2 \\ -0.2 & 0.8 & -0.2 \\ -0.2 & -0.2 & 0.8 \end{bmatrix}, \quad \det(\mathbf{I} - \mathbf{A}) = 0.4, \quad (2.43)$$

giving the positive *Leontief Inverse* $(\mathbf{I} - \mathbf{A})^{-1} = \text{adj}(\mathbf{I} - \mathbf{A})/\det(\mathbf{I} - \mathbf{A})$, (A.29) because the matrix $\mathbf{A}$ is *irreducible*,

$$(\mathbf{I} - \mathbf{A})^{-1} = \begin{bmatrix} 1.5 & 0.5 & 0.5 \\ 0.5 & 1.5 & 0.5 \\ 0.5 & 0.5 & 1.5 \end{bmatrix} > \mathbf{0}, \quad (2.44)$$

and we calculate the positive vectors of *total output* using the above *Leontief Inverse*,

$$\mathbf{x}_0 = (\mathbf{I} - \mathbf{A})^{-1}\mathbf{f}_0 = \begin{bmatrix} 1.5 & 0.5 & 0.5 \\ 0.5 & 1.5 & 0.5 \\ 0.5 & 0.5 & 1.5 \end{bmatrix} \begin{bmatrix} 10 \\ 60 \\ 110 \end{bmatrix} = \begin{bmatrix} 100 \\ 150 \\ 200 \end{bmatrix},$$

$$\mathbf{x}_1 = (\mathbf{I} - \mathbf{A})^{-1}\mathbf{f}_1 = \begin{bmatrix} 1.5 & 0.5 & 0.5 \\ 0.5 & 1.5 & 0.5 \\ 0.5 & 0.5 & 1.5 \end{bmatrix} \begin{bmatrix} 20 \\ 100 \\ 200 \end{bmatrix} = \begin{bmatrix} 180 \\ 260 \\ 360 \end{bmatrix}. \quad (2.45)$$

Then we compute the difference vector

$$\Delta\mathbf{x} = (\mathbf{I} - \mathbf{A})^{-1}\Delta\mathbf{f} = \begin{bmatrix} 1.5 & 0.5 & 0.5 \\ 0.5 & 1.5 & 0.5 \\ 0.5 & 0.5 & 1.5 \end{bmatrix} \begin{bmatrix} 10 \\ 40 \\ 90 \end{bmatrix} = \begin{bmatrix} 80 \\ 110 \\ 160 \end{bmatrix}, \tag{2.46}$$

and the *commodity flow* matrix $\mathbf{Z}_1$ of the current year is given through the invariant *input-output* coefficients matrix **A**, Assumption 2.4.1,

$$\mathbf{Z}_1 = \mathbf{A}\hat{\mathbf{x}}_1 = \begin{bmatrix} 0.2 & 0.2 & 0.2 \\ 0.2 & 0.2 & 0.2 \\ 0.2 & 0.2 & 0.2 \end{bmatrix} \begin{bmatrix} 180 & 0 & 0 \\ 0 & 260 & 0 \\ 0 & 0 & 360 \end{bmatrix} = \begin{bmatrix} 36 & 52 & 72 \\ 36 & 52 & 72 \\ 36 & 52 & 72 \end{bmatrix}, \tag{2.47}$$

the equations (2.46) and (2.47) justify the term *demand-driven input-output model.* Indeed, the vectors of *final demand* for the *base year* and *current year* determine the difference vector $\Delta\mathbf{f}$ and the difference vector $\Delta\mathbf{x}$ upgrading the *total output* of the economy and the *commodity flow* matrix $\mathbf{Z}_1$. This is a typical issue in a *comparative-static analysis.*

We compute the vector of total *value added* for the current year,

$$\boldsymbol{\upsilon}_1 = \mathbf{x}_1 - \mathbf{Z}_1'.\mathbf{e} = \begin{bmatrix} 180 \\ 260 \\ 360 \end{bmatrix} - \begin{bmatrix} 36 & 36 & 36 \\ 52 & 52 & 52 \\ 72 & 72 & 72 \end{bmatrix} \begin{bmatrix} 1 \\ 1 \\ 1 \end{bmatrix} = \begin{bmatrix} 72 \\ 104 \\ 144 \end{bmatrix}, \tag{2.48}$$

and the *total final demand* and the *total value added* of the current year,

$$F_1 = \mathbf{e}'\mathbf{f}_1 = [1,1,1] \begin{bmatrix} 20 \\ 100 \\ 200 \end{bmatrix} = [1,1,1] \begin{bmatrix} 72 \\ 104 \\ 144 \end{bmatrix} = \mathbf{e}'\boldsymbol{\upsilon}_1 = V_1 = 320. \tag{2.49}$$

The total production output of the economy during the base year is $X_1 = \mathbf{e}'\mathbf{x}_0 = 180 + 260 + 360 = 800$, presented in Table 2.6.

Table 2.6: Current year—final demand, value added, total output and outlays.

	Processing sectors			Final demand	Total
	S_1	S_2	S_3	f_{1i}	output
S_1: wheat	36	52	72	20	180
S_2: iron	36	52	72	100	260
S_3: wood	36	52	72	200	360
value added υ_j	72	104	144	$V_1 = F_1 = 320$	
total outlays	180	260	360		$X_1 = 800$

Finally, we calculate the *distribution coefficients* matrix (2.12) in the *base year*,

$$\mathbf{D}_0 = \hat{\mathbf{x}}_0^{-1}\mathbf{Z}_0 = \begin{bmatrix} \frac{1}{100} & 0 & 0 \\ 0 & \frac{1}{150} & 0 \\ 0 & 0 & \frac{1}{200} \end{bmatrix} \begin{bmatrix} 20 & 30 & 40 \\ 20 & 30 & 40 \\ 20 & 30 & 40 \end{bmatrix} = \begin{bmatrix} \frac{1}{5} & \frac{3}{10} & \frac{2}{5} \\ \frac{2}{15} & \frac{1}{5} & \frac{4}{15} \\ \frac{1}{10} & \frac{3}{20} & \frac{1}{5} \end{bmatrix}. \tag{2.50}$$

The *distribution coefficients* d_{ij} indicate in the *base year* the amounts of the palette of commodities i, from either the wheat, iron or wood sectors, distributed per *value unit* of the production of the palette of commodities i to sector j.[22] For instance, $d_{13} = (2/5)$ means that the quantity of wheat commodities 1 of value 0.4 CHF is required to produce a unit of wood commodities 3 of value 1 CHF.

At the moment, we also mention the important identity,

$$\mathbf{D}_0\mathbf{e} + \hat{\mathbf{x}}_0^{-1}\mathbf{f}_0 = \begin{bmatrix} 0.9 \\ 0.6 \\ 0.45 \end{bmatrix} + \begin{bmatrix} 0.1 \\ 0.4 \\ 0.55 \end{bmatrix} = \begin{bmatrix} 1 \\ 1 \\ 1 \end{bmatrix} = \mathbf{e}. \tag{2.51}$$

Finally, we calculate the *distribution coefficients* matrix (2.12) in the *current year*,

$$\mathbf{D}_1 = \hat{\mathbf{x}}_1^{-1}\mathbf{Z}_1 = \begin{bmatrix} \frac{1}{180} & 0 & 0 \\ 0 & \frac{1}{260} & 0 \\ 0 & 0 & \frac{1}{360} \end{bmatrix} \begin{bmatrix} 36 & 52 & 72 \\ 36 & 52 & 72 \\ 36 & 52 & 72 \end{bmatrix} = \begin{bmatrix} \frac{1}{5} & \frac{13}{45} & \frac{2}{5} \\ \frac{9}{65} & \frac{1}{5} & \frac{18}{65} \\ \frac{1}{10} & \frac{13}{90} & \frac{1}{5} \end{bmatrix} \tag{2.52}$$

and recognize that $\mathbf{D}_0 \neq \mathbf{D}_1$. This means, even if the *technology* matrix $\mathbf{A}$ remains invariant for this period, that the increase of final demand, climbing from $\mathbf{f}_0$ to $\mathbf{f}_1$, generates a change of the *distribution* matrix from $\mathbf{D}_0$ to $\mathbf{D}_1$.

For the *current year*, we also verify the requested identity,

$$\mathbf{D}_1\mathbf{e} + \hat{\mathbf{x}}_1^{-1}\mathbf{f}_1 = \begin{bmatrix} \frac{8}{9} \\ \frac{8}{13} \\ \frac{4}{9} \end{bmatrix} + \begin{bmatrix} \frac{1}{9} \\ \frac{5}{13} \\ \frac{5}{9} \end{bmatrix} = \begin{bmatrix} 1 \\ 1 \\ 1 \end{bmatrix} = \mathbf{e}. \tag{2.53}$$

Summary. According to Assumption 2.4.1 the **technology** remains invariant. For this reason, the *input-output coefficients* matrix $\mathbf{A}$ is constant. The two other matrices change from *base year* to *current year*: $\mathbf{Z}_0 \neq \mathbf{Z}_1$, $\mathbf{D}_0 \neq \mathbf{D}_1$. The increase in final demand $\Delta\mathbf{f} > 0$ corresponds to an upswing of a business cycle. The processing sectors have to increase their production in order to cover this additional demand, the total output climbs from $X_0 = 450$ to $X_1 = 800$, as seen through the increase of the entries of the *commodity flow* matrix $\mathbf{Z}_1 > \mathbf{Z}_0$. Even if the **technology** remains constant, the **logistic**

22 The matrix $\mathbf{D}_0$ is also used to describe *supply-driven input-output models* (see Oosterhaven, [77]).

changes, $\mathbf{D}_0 \neq \mathbf{D}_1$, here visible through the variability of the *distribution coefficients* matrix.[23] Such variabilities are observed in real economies: when there is an up-swing or down-swing in a business cycle, transport activity on the roads changes. ▲

We now continue with the *Leontief quantity model* based on physical terms.

2.4.2 The Leontief quantity model in physical terms

Here we again rely on the *Leontief model*, Definition A.12.1. From Duchin and Steenge [23], we know that this model is also applied to data in physical terms. The Assumption 2.2.1 and Assumption 2.2.2 continue to hold. The presentation follows Table 2.2. Consider a *semi-positive commodity flow* matrix $\mathbf{S} \geq \mathbf{O}$ in *physical terms* and a *non-negative* vector $\mathbf{d} \geqq \mathbf{o}$ of *final demand* as given, describing a production process. From equation (2.16) and (2.15), one obtains

$$\mathbf{q} = \mathbf{S}\mathbf{e} + \mathbf{d} > \mathbf{o} \Rightarrow \mathbf{q} = (\mathbf{C}\hat{\mathbf{q}})\mathbf{e} + \mathbf{d} = \mathbf{C}(\hat{\mathbf{q}}\mathbf{e}) + \mathbf{d} = \mathbf{C}\mathbf{q} + \mathbf{d} > \mathbf{o}. \tag{2.54}$$

Consider now the semi-positive *input-output coefficients* matrix $\mathbf{C} = \mathbf{S}\hat{\mathbf{q}}^{-1} \geq \mathbf{O}$ and the vector of *demand* $\mathbf{d} \geqq \mathbf{o}$ describing the underlying production process. We seek a unique *positive* vector of *total output* $\mathbf{q} > \mathbf{o}$, as a solution of the following equation:

$$\boxed{\mathbf{q} = \mathbf{C}\mathbf{q} + \mathbf{d} \Leftrightarrow \mathbf{d} = (\mathbf{I} - \mathbf{C})\mathbf{q}, \quad \mathbf{q} > \mathbf{o}.} \tag{2.55}$$

Here, the model (2.55) is referred to as the *demand-driven input-output quantity model* in physical terms, see Oosterhaven ([77], p. 751), or in short the *Leontief model* in *physical terms*.

We now discuss some further **comparative-static** applications of the demand-driven *Leontief quantity model in physical terms*. We again assume an **invariant technology**, Assumption 2.4.1 the *input-output coefficients* matrix **C** remaining unchanged, whereas the *demand* is subject to change during the period from the *base year* to the *current year*.

We illustrate the *Leontief quantity model* (2.55) with data, presented by Pasinetti ([80], pp. 36–37) and start with a *static analysis* of one equilibrium. Analogue data will be used subsequently to develop further examples.

Example 2.4.3. Consider an economy with $n = 3$ sectors, each sector distinct from the others, producing wheat (w), iron (i), turkeys (t), respectively. During one year, the production is as follows: The wheat sector needs 240 tons of wheat, 12 tons of iron

23 Remark: If the final demand F is composed of *consumption* C, *investment* I and *government expenditure* G, $F = C + I + G$, and if there is no import and no export, $E = M = 0$, then the *national income* is $Y = F = V$, see Section 2.7.

and 18 dozens of turkeys to attain the total production of 450 tons of wheat. The iron sector needs 90 tons of wheat, 6 tons of iron *and* 12 dozens of turkeys to attain the total production of 21 tons of iron. The turkey sector needs 120 tons of wheat, 3 tons of iron and 30 dozens of turkeys to attain the production of 60 dozens of turkeys.

Determine the commodity flow matrix **S**, the vector of total output **q**, and the input-output coefficients matrix **C** and apply Lemma 2.4.1. Calculate the distribution coefficients matrix **D** and verify the identity **De** = **e**.

Solution to Example 2.4.3:
We set up the *commodity flow matrix* and the vector of *total output* and then present the issues in the following Table 2.7:

$$\mathbf{S} = \begin{bmatrix} 240 & 90 & 120 \\ 12 & 6 & 3 \\ 18 & 12 & 30 \end{bmatrix}, \quad \mathbf{q} = \mathbf{Se} = \begin{bmatrix} 240 & 90 & 120 \\ 12 & 6 & 3 \\ 18 & 12 & 30 \end{bmatrix}\begin{bmatrix} 1 \\ 1 \\ 1 \end{bmatrix} = \begin{bmatrix} 450 \\ 21 \\ 60 \end{bmatrix} > \mathbf{o}. \tag{2.56}$$

Table 2.7: Flow of commodities in physical terms, total output, no final demand.

Selling sectors	**Buying sectors**			**Total production**
	wheat	**iron**	**turkeys**	
S_1: wheat	240	90	120	450 tons of wheat
S_2: iron	12	6	3	21 tons of iron
S_3: turkeys	18	12	30	60 dozens turkeys

We can now calculate the *input-output coefficients* matrix

$$\mathbf{C} = \mathbf{S}\hat{\mathbf{q}}^{-1} = \begin{bmatrix} 240 & 90 & 120 \\ 12 & 6 & 3 \\ 18 & 12 & 30 \end{bmatrix}\begin{bmatrix} \frac{1}{450} & 0 & 0 \\ 0 & \frac{1}{21} & 0 \\ 0 & 0 & \frac{1}{60} \end{bmatrix} = \begin{bmatrix} \frac{8}{15} & \frac{30}{7} & 2 \\ \frac{2}{75} & \frac{2}{7} & \frac{1}{20} \\ \frac{1}{25} & \frac{4}{7} & \frac{1}{2} \end{bmatrix}. \tag{2.57}$$

Its characteristic polynomial is

$$f_3(\lambda) = \det(\mathbf{C} - \lambda\mathbf{I}) = -\lambda^3 + \frac{277}{210}\lambda^2 - \frac{178}{525}\lambda + \frac{1}{50} = -(\lambda - 1)\left(\lambda - \frac{3}{35}\right)\left(\lambda - \frac{7}{30}\right). \tag{2.58}$$

The *input-output coefficients* matrix **C** is positive and there is exclusively *interindustrial production* and no *final demand*, $\mathbf{d} = \mathbf{o}$, therefore the Frobenius number is $\lambda_C = 1$. Consequently, the *Leontief Inverse* does not exist, see Lemma 2.4.1.

Finally, we calculate the *distribution coefficients matrix* **D** and verify the identity **De** = **e**. We obtain

$$\mathbf{D} = \hat{\mathbf{q}}^{-1}\mathbf{S} = \begin{bmatrix} \frac{1}{450} & 0 & 0 \\ 0 & \frac{1}{21} & 0 \\ 0 & 0 & \frac{1}{60} \end{bmatrix} \begin{bmatrix} 240 & 90 & 120 \\ 12 & 6 & 3 \\ 18 & 12 & 30 \end{bmatrix} = \begin{bmatrix} \frac{8}{15} & \frac{1}{5} & \frac{4}{15} \\ \frac{4}{7} & \frac{2}{7} & \frac{1}{7} \\ \frac{3}{10} & \frac{1}{5} & \frac{1}{2} \end{bmatrix},$$

$$\mathbf{De} = \begin{bmatrix} \frac{8}{15} & \frac{1}{5} & \frac{4}{15} \\ \frac{4}{7} & \frac{2}{7} & \frac{1}{7} \\ \frac{3}{10} & \frac{1}{5} & \frac{1}{2} \end{bmatrix} \begin{bmatrix} 1 \\ 1 \\ 1 \end{bmatrix} = \begin{bmatrix} 1 \\ 1 \\ 1 \end{bmatrix} = \mathbf{e}. \quad \blacktriangle \tag{2.59}$$

In the next example, the final demand is introduced, defining a *productive Leontief model*. Thus, the vector of *final demand* changes from $\mathbf{d}_0 \geq \mathbf{o}$ in the base year to $\mathbf{d}_1 \geq \mathbf{o}$ in the current year, giving the difference vector $\Delta\mathbf{d} := \mathbf{d}_1 - \mathbf{d}_0$. The *Leontief matrix* $\mathbf{I} - \mathbf{C}$ and the *Leontief Inverse* $(\mathbf{I} - \mathbf{C})^{-1}$ exist and are termed in analogy to the matrices used in Subsection 2.4.1. The matrices $\mathbf{Z}$, $\mathbf{A}$ and the vectors $\mathbf{x}$, $\mathbf{f}$ just have to be replaced by the matrices $\mathbf{S}$, $\mathbf{C}$ and the vectors $\mathbf{q}$, $\mathbf{d}$ in that order.

The question is: How does the vector of *total output* $\mathbf{q}_0$ of the base year change to become the vector of *total output* $\mathbf{q}_1$ of the current year, giving the difference vector $\Delta\mathbf{q} := \mathbf{q}_1 - \mathbf{q}_0$. We get from model (2.55), since the *Leontief Inverse* exists, because evidently Lemma 2.4.1 (b) applies to the model (2.55) in physical terms, $\mathbf{d}_0 > \mathbf{o}$, obtaining the Frobenius number $\lambda_C < 1$, the equation,

$$\Delta\mathbf{q} = (\mathbf{I} - \mathbf{C})^{-1}\Delta\mathbf{d}. \tag{2.60}$$

We are further interested in the variations of the entries of the *commodity flow* matrix $\mathbf{S} = \mathbf{C}\hat{\mathbf{q}}$ (2.16). We find the following equation to calculate the new levels of matrix $\mathbf{S}_1$ from matrix $\mathbf{S}_0$, given $\mathbf{C}$ = constant,

$$\mathbf{S}_1 = \mathbf{C}\hat{\mathbf{q}}_1 \quad \text{and} \quad \mathbf{S}_0 = \mathbf{C}\hat{\mathbf{q}}_0 \quad \text{or} \quad \Delta\mathbf{S} := \mathbf{S}_1 - \mathbf{S}_0 = \mathbf{C}(\hat{\mathbf{q}}_1 - \hat{\mathbf{q}}_0) = \mathbf{C}\Delta\hat{\mathbf{q}}. \tag{2.61}$$

We consider a first *base year* vector of *final demand* $\mathbf{d}_0$ and then a second *current year* vector of *final demand* $\mathbf{d}_1$. We suppose an invariant technology, Assumption 2.4.1 the *input-output coefficients* matrix $\mathbf{C}$ remaining time-invariant.

The entries of the illustrating example are taken from a problem treated by Pasinetti ([80], pp. 38–40, Table II.2).

Example 2.4.4. The means of production and the vectors of *final demand* of the base year and of the current year are presented in Table 2.8. Determine the commodity flow matrix $\mathbf{S}_0$ and the vector of final demand $\mathbf{d}_0$, the vector of total output $\mathbf{q}_0$, the input-output coefficients matrix $\mathbf{C} = \mathbf{S}_0\hat{\mathbf{q}}_0^{-1}$ of the base year. Determine the vector of final demand $\mathbf{d}_1$ of the current year. Compute the Leontief Inverse (for data based in physical terms), and apply (2.55) to calculate the total output vectors $\mathbf{q}_0$ and $\mathbf{q}_1$. Apply Lemma 2.4.1.

Compute the commodity flow matrix $\mathbf{S}_1$ (2.61) of the current year.

Give an interpretation of the results obtained with the Leontief quantity model.

Table 2.8: Flow of commodities in physical terms with final demands.

Commodity	Processing sectors			Base year final demand	Base year total output	Current year, final demand
	wheat	iron	turkeys	d_{1i}		d_{2i}
wheat	186	54	30	180	450 tons of wheat	228
iron	12	6	3	–	21 tons of iron	25
turkeys	9	6	15	30	60 doz. turkeys	45

Calculate the distribution coefficients matrices $\mathbf{D}_0 = \hat{\mathbf{q}}_1^{-1}\mathbf{S}_0$ and $\mathbf{D}_1 = \hat{\mathbf{q}}_1^{-1}\mathbf{S}_1$.

Compute the difference vector $\Delta\mathbf{q}$ (2.60), which shows the change of total output between the base year and the current year.

Verify the identities $\mathbf{D}_0\mathbf{e}+\hat{\mathbf{q}}_0^{-1}\mathbf{d}_0 = \mathbf{e}$ and $\mathbf{D}_1\mathbf{e}+\hat{\mathbf{q}}_1^{-1}\mathbf{d}_1 = \mathbf{e}$ and explain the variability of the matrices $\mathbf{S}$ and $\mathbf{D}$.

Solution to Example 2.4.4:
We can again set up the *commodity flow matrix*, the vector of final demand and the vector of total output,

$$\mathbf{S}_0 = \begin{bmatrix} 186 & 54 & 30 \\ 12 & 6 & 3 \\ 9 & 6 & 15 \end{bmatrix}, \quad \mathbf{d}_0 = \begin{bmatrix} 180 \\ 0 \\ 30 \end{bmatrix}, \quad \mathbf{d}_1 = \begin{bmatrix} 228 \\ 25 \\ 45 \end{bmatrix}, \quad \Delta\mathbf{d} = \begin{bmatrix} 48 \\ 25 \\ 15 \end{bmatrix},$$

$$\mathbf{q}_0 = \mathbf{S}_0\mathbf{e} + \mathbf{d}_0 = \begin{bmatrix} 186 & 54 & 30 \\ 12 & 6 & 3 \\ 9 & 6 & 15 \end{bmatrix}\begin{bmatrix} 1 \\ 1 \\ 1 \end{bmatrix} + \begin{bmatrix} 180 \\ 0 \\ 30 \end{bmatrix} = \begin{bmatrix} 450 \\ 21 \\ 60 \end{bmatrix} > \mathbf{o}, \tag{2.62}$$

from which we can calculate the *input-output coefficients* matrix, assumed to be time invariant,

$$\mathbf{C} = \mathbf{S}_0\hat{\mathbf{q}}_0^{-1} = \begin{bmatrix} 186 & 54 & 30 \\ 12 & 6 & 3 \\ 9 & 6 & 15 \end{bmatrix}\begin{bmatrix} \frac{1}{450} & 0 & 0 \\ 0 & \frac{1}{21} & 0 \\ 0 & 0 & \frac{1}{60} \end{bmatrix} = \begin{bmatrix} \frac{31}{75} & \frac{18}{7} & \frac{1}{2} \\ \frac{2}{75} & \frac{2}{7} & \frac{1}{20} \\ \frac{1}{50} & \frac{2}{7} & \frac{1}{4} \end{bmatrix}, \tag{2.63}$$

with the characteristic polynomial

$$\begin{aligned} f_3(\lambda) = \det(\mathbf{C} - \lambda\mathbf{I}) &= -\lambda^3 + \frac{1{,}993}{2{,}100}\lambda^2 - \frac{1}{5}\lambda + \frac{1}{100} \\ &= -(\lambda - 0.675)(\lambda - 0.201)(\lambda - 0.079). \end{aligned} \tag{2.64}$$

The *input-output coefficients* matrix $\mathbf{C} > 0$ is positive, the **Perron–Frobenius theorem** A.9.2 therefore applies, the Frobenius number is $\lambda_C = 0.675 < 1$ and the *Leontief Inverse* $(\mathbf{I} - \mathbf{C})^{-1}$ exists, according to Lemma 2.4.1 (b).

Now, the matrix **I** – **C** is computed

$$\mathbf{I}-\mathbf{C}=\begin{bmatrix} \frac{44}{75} & -\frac{18}{7} & -\frac{1}{2} \\ -\frac{2}{75} & \frac{5}{7} & -\frac{1}{20} \\ -\frac{1}{50} & -\frac{2}{7} & \frac{3}{3} \end{bmatrix}, \tag{2.65}$$

giving the positive *Leontief Inverse* in physical terms $(\mathbf{I}-\mathbf{C})^{-1}$, as stated by Theorem A.10.2,

$$(\mathbf{I}-\mathbf{C})^{-1}=\begin{bmatrix} \frac{1{,}095}{506} & \frac{2{,}175}{253} & \frac{510}{253} \\ \frac{441}{5{,}060} & \frac{903}{506} & \frac{224}{1{,}265} \\ \frac{1}{11} & \frac{10}{11} & \frac{16}{11} \end{bmatrix}=\begin{bmatrix} 2.16 & 8.60 & 2.02 \\ 0.0872 & 1.78 & 0.177 \\ 0.0909 & 0.909 & 1.45 \end{bmatrix}>\mathbf{0}. \tag{2.66}$$

We then calculate the positive vectors of *total output* using the *Leontief Inverse* (2.66) with $\mathbf{d}_k$, $k=0,1$ (the figures are rounded off).

$$\mathbf{q}_0=(\mathbf{I}-\mathbf{C})^{-1}\mathbf{d}_0=\begin{bmatrix} 2.16 & 8.60 & 2.02 \\ 0.0872 & 1.78 & 0.177 \\ 0.0909 & 0.909 & 1.45 \end{bmatrix}\begin{bmatrix} 180 \\ 0 \\ 30 \end{bmatrix}=\begin{bmatrix} 450 \\ 21 \\ 60 \end{bmatrix},$$

$$\mathbf{q}_1=(\mathbf{I}-\mathbf{C})^{-1}\mathbf{d}_1=\begin{bmatrix} 2.16 & 8.60 & 2.02 \\ 0.0872 & 1.78 & 0.177 \\ 0.0909 & 0.909 & 1.45 \end{bmatrix}\begin{bmatrix} 228 \\ 25 \\ 45 \end{bmatrix}=\begin{bmatrix} 799 \\ 72.5 \\ 109 \end{bmatrix}. \tag{2.67}$$

Then, we compute the diagonal matrix $\hat{\mathbf{q}}_1$ and the difference vector

$$\hat{\mathbf{q}}_1=\begin{bmatrix} 799 & 0 & 0 \\ 0 & 72.5 & 0 \\ 0 & 0 & 109 \end{bmatrix}, \quad \Delta\mathbf{q}=\mathbf{q}_1-\mathbf{q}_0=\begin{bmatrix} 799 \\ 72.5 \\ 109 \end{bmatrix}-\begin{bmatrix} 450 \\ 21 \\ 60 \end{bmatrix}=\begin{bmatrix} 349 \\ 51.5 \\ 49 \end{bmatrix} \tag{2.68}$$

and the commodity flow matrix of the current year, see Table 2.9,

$$\mathbf{S}_1=\mathbf{C}\hat{\mathbf{q}}_1=\begin{bmatrix} \frac{31}{75} & \frac{18}{7} & \frac{1}{2} \\ \frac{2}{75} & \frac{2}{7} & \frac{1}{20} \\ \frac{1}{50} & \frac{2}{7} & \frac{1}{4} \end{bmatrix}\begin{bmatrix} 799 & 0 & 0 \\ 0 & 72.5 & 0 \\ 0 & 0 & 109 \end{bmatrix}$$

$$=\begin{bmatrix} 330.266 & 186.311 & 54.455 \\ 21.308 & 20.701 & 5.445 \\ 15.981 & 20.701 & 27.227 \end{bmatrix}>\begin{bmatrix} 186 & 54 & 30 \\ 12 & 6 & 3 \\ 9 & 6 & 15 \end{bmatrix}=\mathbf{S}_0. \tag{2.69}$$

The calculations (2.68), (2.69) justify the designation *demand-driven input-output model*. Indeed, the two given vectors of *final demand* for the *base year* and *current year*

Table 2.9: Sector outputs, total final demand and output in the current year.

Current year	Processing sectors S_1	S_2	S_3	Final demand d_{1i}	Total output
S_1: wheat	330.266	186.311	54.455	228	799.032
S_2: iron	21.308	20.701	5.445	25	72.526
S_3: wood	15.981	20.701	27.227	45	108.909

determine the difference vector $\Delta\mathbf{d}$, and then the vector $\Delta\mathbf{q}$, the growing *total output* of the economy and the *commodity flow* matrix $\mathbf{S}_1$. This is a typical *comparative-statics analysis*.

Then, we compute the *distribution coefficients* matrix for the *base year*,

$$\mathbf{D}_0 = \hat{\mathbf{q}}_0^{-1}\mathbf{S}_0 = \begin{bmatrix} \frac{1}{450} & 0 & 0 \\ 0 & \frac{1}{21} & 0 \\ 0 & 0 & \frac{1}{60} \end{bmatrix} \begin{bmatrix} 186 & 54 & 30 \\ 12 & 6 & 3 \\ 9 & 6 & 15 \end{bmatrix} = \begin{bmatrix} 0.41 & 0.12 & 0.07 \\ 0.57 & 0.29 & 0.14 \\ 0.15 & 0.10 & 0.25 \end{bmatrix} = (d_{ij0}), \tag{2.70}$$

each *distribution coefficient* d_{ij0} indicates in the *base year* the *part* of the total amount of commodity i (Input) required for the production of one *unit* of commodity j (Output). Finally, we calculate for the *current year*, the components of matrix $\mathbf{S}_1$ presented as *common fractions*, like $417{,}787/1{,}265 = 330.266$,

$$\mathbf{D}_1 = \hat{\mathbf{q}}_1^{-1}\mathbf{S}_1 = \begin{bmatrix} \frac{253}{202{,}155} & 0 & 0 \\ 0 & \frac{2{,}530}{183{,}309} & 0 \\ 0 & 0 & \frac{11}{1{,}198} \end{bmatrix} \begin{bmatrix} \frac{417{,}787}{1{,}265} & \frac{235{,}683}{1{,}265} & \frac{599}{11} \\ \frac{26{,}954}{1{,}265} & \frac{26{,}187}{1{,}265} & \frac{599}{110} \\ \frac{40{,}431}{2{,}530} & \frac{26{,}187}{1{,}265} & \frac{599}{22} \end{bmatrix}$$
$$= \begin{bmatrix} \frac{31}{75} & \frac{78{,}561}{336{,}925} & \frac{13{,}777}{202{,}155} \\ \frac{53{,}908}{183{,}309} & \frac{2}{7} & \frac{13{,}777}{183{,}309} \\ \frac{40{,}431}{275{,}540} & \frac{26{,}187}{137{,}770} & \frac{1}{4} \end{bmatrix} = \begin{bmatrix} 0.41 & 0.23 & 0.07 \\ 0.29 & 0.29 & 0.08 \\ 0.15 & 0.19 & 0.25 \end{bmatrix}. \tag{2.71}$$

Because the matrices $\mathbf{D}_0$ and $\mathbf{D}_1$ are different, the **logistic** has changed between the *current year* and the *base year*, even if the **technology** remained unchanged during this period.

We finish the example noting the property:

$$\mathbf{D}_0\mathbf{e} + \hat{\mathbf{q}}_0^{-1}\mathbf{d}_0 = \begin{bmatrix} 0.6 \\ 1 \\ 0.5 \end{bmatrix} + \begin{bmatrix} 0.4 \\ 0 \\ 0.5 \end{bmatrix} = \begin{bmatrix} 1 \\ 1 \\ 1 \end{bmatrix} = \mathbf{e}, \tag{2.72}$$

$$\mathbf{D}_1\mathbf{e} + \hat{\mathbf{q}}_1^{-1}\mathbf{d}_1 = \begin{bmatrix} 0.715 \\ 0.655 \\ 0.587 \end{bmatrix} + \begin{bmatrix} 0.285 \\ 0.345 \\ 0.413 \end{bmatrix} = \begin{bmatrix} 1 \\ 1 \\ 1 \end{bmatrix} = \mathbf{e}. \quad ▲ \tag{2.73}$$

The *distribution coefficients* matrix is used to describe *supply-driven input-output models*, the so-called Ghosh models (see Oosterhaven [77], pp. 753–755).

Final observation for Section 2.4:

1 The coefficients of every $n \times n$ *input coefficients matrix* in physical terms $\mathbf{C}$ (independently of *base year* or *current year*) are non-negative: $0 \le c_{ij}$.

2 The coefficients of every $n \times n$ *input coefficients matrix* $\mathbf{A}$ in monetary terms lie between 0 and 1: $0 \le a_{ij} \le 1$, according to Lemma 2.1.1.

2.5 The Leontief cost-push input-output price models*

This section focuses on a variant of *Leontief price models*, leading to the calculation of *price indices* and *prices*. *Comparative-static analysis* over a reporting period between a *base year* to a *current year* is performed for the obtained price indices and prices.

There are two types of specific models. On the one hand, the Leontief *cost-push input-output price models* in *monetary terms*, which are in fact specific *price-index* models, comparing a *value added* of a *current year* to a *value added* of a *base year*. The obtained price indices are presented in vector form. On the other hand, the Leontief *cost-push input-output price models* in *physical terms* are *price* models for all the commodities of the investigated production economy (see for the price index and price models Oosterhaven [77]). Miller and Blair [65], pp. 41–51, explicitly use the notion *index price* in relation with the results of the *price models* based on *monetary data*!

The *cost-push input-output price models* have been developed and applied by Schumann [104],[24] Oosterhaven [77] and Dietzenbacher [22].

2.5.1 The Leontief cost-push input-output price model in monetary terms

Throughout this subsection, we posit a *productive Leontief model*, Definition A.12.1, meaning $\mathbf{f} \ge \mathbf{o}$. Therefore the Frobenius number is smaller than one, $\lambda_A < 1$, Lemma 2.4.1 (b). Both economic prerequisites hold, Assumption 2.2.1 for the vector of total output, $\mathbf{x} > \mathbf{o}$, and Assumption 2.2.2 for the matrix $\mathbf{Z} \ge \mathbf{0}$. Therefore, we can define $\mathbf{A} = \mathbf{Z}\hat{\mathbf{x}}^{-1}$. Consequently and necessarily, the *value-added* vector is *semi-positive*, $\boldsymbol{\upsilon} = [\upsilon_1, \ldots, \upsilon_n]' \ge \mathbf{o}$, see Miller and Blair [65], pp. 43–47,[25] indicating the

24 Schumann uses the term "Schattenpreis" (shadow price) for the prices having the characteristic of *price indices* and writes that these *shadow prices* have clearly to be distinguished from *market prices*, obtained from the intersection of the supply- and demand-curves.

25 For didactic reasons, Miller and Blair [65] restrict value added to labour. We shall first follow this procedure where no ambiguities arise, in particular, in the introductory steps of this text.

monetary value of *total value added.* For the notations, we rely on the IOT representation as in Table 2.1. The *semi-positive* matrix $\mathbf{Z} \geq \mathbf{0}$ represents interindustrial sales and purchases in value (monetary) terms.

We rely on the equality between *total output* and *total outlays* and start from equation (2.7) which relates to the accounting identity, bringing together the vector of total *final demand* and the vector of *value added*,

$$\mathbf{x} = \mathbf{Z}\mathbf{e} + \mathbf{f} = \mathbf{Z}'\mathbf{e} + \boldsymbol{\upsilon} > \mathbf{o}. \tag{2.74}$$

Then, as we shall see, the identity $\mathbf{x} = \mathbf{Z}'\mathbf{e} + \boldsymbol{\upsilon}$ will lead to specific *price indices.* Taking $\mathbf{A} = \mathbf{Z}\hat{\mathbf{x}}^{-1}$ gives the identity $\mathbf{x} = \mathbf{Z}'\mathbf{e} + \boldsymbol{\upsilon} = (\mathbf{A}\hat{\mathbf{x}})'\mathbf{e} + \boldsymbol{\upsilon} = \hat{\mathbf{x}}\mathbf{A}'\mathbf{e} + \boldsymbol{\upsilon}$; after multiplying the obtained equation by $\hat{\mathbf{x}}^{-1}$ from the left side, with $\boldsymbol{\upsilon}_c := \hat{\mathbf{x}}^{-1}\boldsymbol{\upsilon} \geq \mathbf{o}$, we get

$$\hat{\mathbf{x}}^{-1}\mathbf{x} = \hat{\mathbf{x}}^{-1}(\hat{\mathbf{x}}\mathbf{A}'\mathbf{e}) + \hat{\mathbf{x}}^{-1}\boldsymbol{\upsilon} = (\hat{\mathbf{x}}^{-1}\hat{\mathbf{x}})\mathbf{A}'\mathbf{e} + \boldsymbol{\upsilon}_c, \tag{2.75}$$

giving, with $\hat{\mathbf{x}}^{-1}\mathbf{x} = \mathbf{e}$ from (A.39) observing under the stated conditions the inequality equivalence $\mathbf{x} > \mathbf{o} \Leftrightarrow \mathbf{e} > \mathbf{o}$,

$$\mathbf{e} = \mathbf{A}'\mathbf{e} + \boldsymbol{\upsilon}_c. \tag{2.76}$$

Analysing the peculiar equation (2.76), we see that it leads to the concept of the *Laspeyres price index*, and we make this interlude to explain the concept.

The *Laspeyres price index* is used to measure the price evolution of a basket of consumption goods. Consider a basket $\mathcal{B}$ of k goods and services in a given region, during a given period spanning the *base year* to the *current year*. Let the prices and quantities of the goods $i \in \{1, \dots, k\}$ be $p_{i|0}$ and $q_{i|0}$ in the *base year*, respectively $p_{i|1}$ and $q_{i|1}$ in the *current year*. The *Laspeyres price indexes* from the *base year* to the *current year* and for the *base year* are defined as follows:

$$P_{L,0|1} = \frac{\sum_{i=1}^{k} p_{i|1} \cdot q_{i|0}}{\sum_{i=1}^{k} p_{i|0} \cdot q_{i|0}}; \quad P_{L,0|0} = \frac{\sum_{i=1}^{k} p_{i|0} \cdot q_{i|0}}{\sum_{i=1}^{k} p_{i|0} \cdot q_{i|0}} = 1. \tag{2.77}$$

We see that the *Laspeyres price index* is a *weighted average of prices for the given basket $\mathcal{B}$ of goods or services.* Indeed, the *Laspeyres price index* of the *base year* is always $P_{L,0|0} = 1$. Today, the concept of the *Laspeyres price index* is applied by most national economies to develop their consumer price index (CPI).

After this interlude, we return to equation (2.76) and start with the case where *the current year* and *base year* are identical. Consequently, the cost of production of each commodity is referred to itself. With this idea in mind, we go on and interpret the three terms of identity (2.76).

The *j-th* component of the first vector $\mathbf{A}'\mathbf{e}$, $j \in \{1, \dots, n\}$, is the sum $\sum_{i=1}^{n} a_{ij}$, $0 \leq \sum_{i=1}^{n} a_{ij} \leq 0$, Lemma 2.1.1 each term a_{ij} of which represents the monetary value of the amount of the palette of commodities $i \in \{1, \dots, n\}$ required for the production of one *value unit* of the palette of commodities $j \in \{1, \dots, n\}$ of sector S_j.

The components υ_j/x_j of the vector $\mathbf{\upsilon}_c := \hat{\mathbf{x}}^{-1}\mathbf{\upsilon} = [\upsilon_1/x_1, \dots, \upsilon_n/x_n]'$ represent the *total value added*, respectively, the *total labour*, necessary to produce one *value unit* of the palette of commodities j, $j \in \{1, \dots, n\}$, see Miller and Blair ([65], pp. 43–44). The vector $\mathbf{\upsilon}_c$ is called the vector of the *total value added of every sector S_j, per value unit of produced commodities* in monetary terms.

The right-hand side of (2.76) gives the sum of the requirements for a produced unit of every sector S_j, composed of the sum of the costs of production, represented by vector $\mathbf{A}'\mathbf{e}$, together with the costs of value added (labour) υ_j/x_j for each produced unit of every sector S_j, together placed within the vector $\mathbf{\upsilon}_c$.

The left hand side of (2.76) is the dimensionless summation vector $\mathbf{e}$. The unit coefficients are the unit *costs of production* of output x_j per unit *costs of production* of output x_j in sector S_j, meaning $x_j/x_j = 1$. They are also designated as "(index) prices standardized to one" (see Oosterhaven [77], p. 751).

The expression between the brackets, such as $[\upsilon_j/x_j]$, containing *only one algebraic term*, means *a dimension* of fraction υ_j/x_j. Clearly, the vector $\mathbf{\upsilon}_c = [\upsilon_1/x_1, \dots, \upsilon_n/x_n]'$ has dimensionless components, $[\upsilon_j/x_j] = 1$. The vector $\mathbf{A}'\mathbf{e}$ is also dimensionless. Then, the sum of $\mathbf{A}'\mathbf{e}$ and $\mathbf{\upsilon}_c$ gives the summation vector $\mathbf{e}$ of unit weights (2.76).

Summarizing, the identity (2.76) contains on the left hand-side the *costs of production* of sector S_j for the total output x_j of the palette of commodities j per unit of total output, meaning the unit weights $x_j/x_j = 1$. On the right-hand side, the term $\mathbf{A}'\mathbf{e}$ gives the costs for *interindustrial production per value unit of produced commodities*; the vector $\mathbf{\upsilon}_c$ represents the total *value added of each sector per value unit of produced commodities* (called the vector of *total labour per value unit of produced commodities*), summed up and designated as "*weights* for each industry equal one" (see Oosterhaven [77], p. 752).

We now present the value-based *Leontief cost-push input-output price model* which is an *index-price model*, see Oosterhaven [77]. Schumann [104], Dietzenbacher [22]. Miller and Blair ([65], p. 44, (2.33)) term it the *input-output price model*.

We start transforming (2.76), using the identity $\mathbf{\upsilon}_c = \hat{\mathbf{\upsilon}}_c\mathbf{e} \geq \mathbf{o}$. We obtain

$$\mathbf{e} = \mathbf{A}'\mathbf{e} + \mathbf{\upsilon}_c \Rightarrow (\mathbf{I} - \mathbf{A}')\mathbf{e} = (\mathbf{I} - \mathbf{A})'\mathbf{e} = \mathbf{\upsilon}_c = \hat{\mathbf{\upsilon}}_c\mathbf{e}. \tag{2.78}$$

Because, by the previously stated model assumptions, the Frobenius number is $\lambda_A < 1$, by Lemma 2.4.1 the *transposed Leontief Inverse* exists and we get

$$\mathbf{e} = (\mathbf{I} - \mathbf{A}')^{-1}(\hat{\mathbf{\upsilon}}_c\mathbf{e}) = ((\mathbf{I} - \mathbf{A})^{-1})'(\hat{\mathbf{\upsilon}}_c\mathbf{e}). \tag{2.79}$$

We mention here that Miller and Blair ([65], pp. 43–44) comment on the identity (2.76) as follows:

> *"This [equation (2.76)] illustrates the unique measurement units in the base year table – amounts that can be purchased for 1.00 USD".*

In the second equation of (2.78) the summation, vector $\mathbf{e} = [1, \dots, 1]'$ is interpreted as a vector of *index prices* relative to the *base year*; this means the *current year* is identical with the *base year*, see Miller and Blair [65] or Oosterhaven [77].[26]

We now take up the specific *price index* model (2.79) and continue explaining the ideas of Oosterhaven ([77], p. 752) and Miller and Blair ([65], p. 44).

Consider the *value added* (labour costs) v_{j1} of sector S_j in the *current year*, respectively, the *value added* (labour costs) v_{j0} in the *base year* "0". Then $\tilde{p}_{vj,0|1} := v_{1j}/v_{0j}$ is the dimensionless *index price* of value added (labour costs) of sector S_j. Set up the vector $\tilde{\mathbf{p}}_{v,0|1} = [\tilde{p}_{v1,0|1}, \dots, \tilde{p}_{vn,0|1}]'$ of *index prices* of value added.

We replace the vector $\mathbf{e}$ on the **right-hand side** of (2.79) by the vector $\tilde{\mathbf{p}}_{v,0|1}$.

When the *current year* is identical to the *base year*, $\tilde{\mathbf{p}}_{v,0|0} = \mathbf{e}$, we fall back on the identity (2.76). Now we continue, assuming that the *current year* is different from the *base year*.

We continue applying the notations defined for the *Laspeyres price index* (2.77). Consequently, p_{0i}, respectively p_{1i}, determine the price of commodity i in the *base year* "0", respectively in the *current year* "1". We define for the palette of commodities i the *index price* of the *base year* "0" relative to the *base year* "0" by $\tilde{p}_{i,0|0} = p_{0i}/p_{0i} = 1$ and set the vector $\tilde{\mathbf{p}}_{0|0} = [\tilde{p}_{1,0|0}, \dots, \tilde{p}_{n,0|0}]' := \mathbf{e} = [1, 1, \dots, 1]'$. By analogy, we then define for the palette of commodities i the *index price* of the *current year* "1" relative to the *base year* "0" by $\tilde{p}_{i,0|1} = p_{1i}/p_{0i}$ and set the vector of the *index prices* of all the commodities of the *current year* relative to the *base year* as $\tilde{\mathbf{p}}_{0|1} = [\tilde{p}_{1,0|1}, \dots, \tilde{p}_{n,0|1}]'$.

On the **left-hand side** of (2.79), we replace $\mathbf{e}$ by the unknown vector of the *index prices* of the n commodities $\tilde{\mathbf{p}}_{0|1} = [\tilde{p}_{1,0|1}, \dots, \tilde{p}_{n,0|1}]'$. This means, considering the *period* from the *base year* to the *current year*, both vectors $\mathbf{e}$ on both **right-hand sides** of (2.79) are replaced by the vector $\tilde{\mathbf{p}}_{v,0|1}$. The vector $\hat{\mathbf{v}}_c$ from now on is denoted as vector $\hat{\mathbf{v}}_{c,0}$, emphasizing that we are in presence of a *base year*, giving

$$\mathbf{v}_{c,1} := \hat{\mathbf{v}}_{c,0}\tilde{\mathbf{p}}_{v,0|1}, \tag{2.80}$$

the vector of *value added* in units of the *current year*. We thus obtain the vector of the *index prices* of all the commodities of the *current year*[27] with respect to the *base year*

---

26 Remark: Both authors, Oosterhaven ([77], p. 752) and Miller and Blair ([65], p. 44) use the designation "index price" or "(index) price". Oosterhaven uses also the term "unit price-indices of the base-year. Hence, for this base-year: $\tilde{\mathbf{p}}'_{0|0} = \mathbf{e}' = \tilde{\mathbf{p}}'_{v,0|0}$" ([77], p. 752), for the signification of the variables, see (2.81) hereafter. The common understanding of these terms is that both authors consider dimensionless *price indices* of a *base year* relative to itself with value 1. Following up on this concept, there are *price indices* of a *current year* relative to a *base year* that are also dimensionless but can have values different than 1. Empirically, the term "(index) price" may be understood as an estimation of a *price index*.

27 We thus observe the appearance of a formalism in the multiplication of a matrix of "quantities" or "relative quantities", like $\hat{\mathbf{v}}_{c,0}$, by a vector of *price indices*, resulting in another vector of *price indices*. Two subsequent "0", "0" give one "0", like in (2.81) {"0"}, {"0|1"} → {"0|1"}. This is the transformation of a *base year to current-year price indices* within the same period.

by the Leontief *cost-push input-output price model* in *monetary terms*:

$$\boxed{\tilde{\mathbf{p}}_{0|1} = ((\mathbf{I} - \mathbf{A})^{-1})' \cdot \hat{\boldsymbol{\upsilon}}_{c,0} \cdot \tilde{\mathbf{p}}_{\upsilon,0|1}.} \tag{2.81}$$

The economic interpretation of the *index price* model (2.81) is a topic of its own. Oosterhaven [77] and Miller and Blair ([65], p. 44, (2.33)) present it as the value-based *cost-push input-output price model* because it can be interpreted for a given period as the transformation from the *base year* to the *current year* of the *index prices* of *value-added* into the *index prices* of the *sectors* of the economy.

Miller and Blair comment:

> *"This model is generally used to measure the impact on prices through the economy of new-primary input costs (or a change in those costs) in one or more sectors" (cited from Miller & Blair [65], p. 45).*

Oosterhaven ([77], p. 752) commented: It is used to "simulate the cost-push inflatory processes".[28]

This model construction gives rise to **comparative-static analysis**, summarized hereafter before tackling an example. We show how to calculate the levels of *value added*, whose variations in the period (base year = 0, current year = 1) is described by a *price index* vector $\tilde{\mathbf{p}}_{\upsilon,0|1}$. Then, the *price index* vector $\tilde{\mathbf{p}}_{0|1}$ of the whole economy can be calculated for the *current year* (2.81). Finally, the levels of all entries can be upgraded from the *base year* to the *current year* with corresponding diagonalized vectors, in such a way that all the levels are multiplied by the corresponding sector *price index*. We compute the *commodity flow* matrix for the *current year*, where the *index price* $\tilde{p}_{j,0|1}$ acts on industry S_j, as $z_{ij,1} = \tilde{p}_{i,0|1} \cdot z_{ij,0}$, $i,j = 1,\dots,n$. By analogy, we have for the upgraded *final consumption* $f_{i,1} = \tilde{p}_{i,0|1} \cdot f_{i,0}$ and for the upgraded *total value added* $\upsilon_{j,1} = \tilde{p}_{\upsilon j,0|1} \cdot \upsilon_{j,0}$. Setting for the levels of *total value added* $\boldsymbol{\upsilon}_0 = \hat{\mathbf{x}}_0 \cdot \boldsymbol{\upsilon}_{c,0}$, we get

$$\boxed{\mathbf{Z}_1 = \hat{\tilde{\mathbf{p}}}_{0|1}\mathbf{Z}_0; \quad \mathbf{f}_1 = \hat{\tilde{\mathbf{p}}}_{0|1}\mathbf{f}_0; \quad \boldsymbol{\upsilon}_1 = \hat{\boldsymbol{\upsilon}}_0\tilde{\mathbf{p}}_{\upsilon,0|1} = (\hat{\mathbf{x}}_0 \cdot \hat{\boldsymbol{\upsilon}}_{c,0})\tilde{\mathbf{p}}_{\upsilon,0|1}.} \tag{2.82}$$

Here is a further justification of the current economic interpretation of $\tilde{\mathbf{p}}_{0|1}$ and $\tilde{\mathbf{p}}_{\upsilon,0|1}$ as vectors of *index prices*. We observe that the *units of measurement*, i. e., the *currency* used in matrix $\mathbf{Z}_0$ and matrix $\mathbf{Z}_1$ are the same, as, e. g., $[\mathbf{Z}_0] = [\mathbf{Z}_1] = [\mathbf{f}_0] = [\mathbf{f}_1] = [\boldsymbol{\upsilon}_0] = [\boldsymbol{\upsilon}_1] = \text{CHF}$; consequently $[\tilde{\mathbf{p}}_{0|1}] = [\tilde{\mathbf{p}}_{\upsilon,0|1}] = 1$. This fact confirms that with $\tilde{\mathbf{p}}_{0|1}, \tilde{\mathbf{p}}_{\upsilon,0|1}$ we do not have *prices*, but ratios of prices in the same *units of measurement*, resulting in *price indices*, or specifically *index prices*.

---

28 As an example, consider in a three-sector economy the vector of index prices $\tilde{\mathbf{p}}_{\upsilon,0|1} = [1.2, 1.1, 1.05]'$, stating that from the *base year* period to the *current year* period the *growth rate* of the *labour costs* are 20 % for commodity 1, 10 % for commodity 2 and 5 % for commodity 3. The notations with the *tilde-sign* is introduced by Miller and Blair ([65], p. 44), and the *index* υ of the index prices is proposed by Oosterhaven ([77], p. 752). Moreover, Oosterhaven ([77], p. 752) considers only index-price vectors of constant *growth rate*, like $\tilde{\mathbf{p}}_{\upsilon,0|1} = [1.2, 1.2, 1.2]'$.

Starting with the identity $\mathbf{p}_{\upsilon,0|0} := \mathbf{e}$, we now present the following introductory problem, applying the *cost-push input-output model* (2.81) to calculate the vector of *index prices* of commodities $\tilde{\mathbf{p}}_{0|1}$, given the vector $\tilde{\mathbf{p}}_{\upsilon,0|1}$ of *index prices* of *value added* (labour costs). Finally, we will calculate the levels of all the entries in the *current year* (2.82).

Example 2.5.1. Given the following table of entries relative to a base year "0": The Input-Output Table 2.10 represents an economy with sectors S_1: wheat, S_2: iron, S_3: wood, and a *commodity flow* matrix $\mathbf{Z}_0$ with entries z_{ij}, $i,j = 1,\dots,3$ and final demand f_{0i}.

Table 2.10: Transactions and *final demand* for the *base year*.

Commodities	**Processing sectors**			**Final demand**
***base year* entries**	S_1	S_2	S_3	f_{0i}
S_1: wheat	40	30	20	10
S_2: iron	30	30	30	60
S_3: wood	25	30	35	110

Consider separately the positive commodity flow matrix $\mathbf{Z}_0 > \mathbf{O}$ of Table 2.10 and the positive vector of final demand $\mathbf{f}_0 > \mathbf{o}$.

Compute the total production vector $\mathbf{x}_0$, the vector of *monetary value* of *total value added* (labour costs) $\boldsymbol{\upsilon}_0$, the input-output coefficients matrix $\mathbf{A}_0 = \mathbf{Z}_0\hat{\mathbf{x}}_0^{-1}$, the Frobenius number λ_A, the transposed Leontief matrix $(\mathbf{I} - \mathbf{A}_0)'$ and the transposed *Leontief Inverse* $((\mathbf{I} - \mathbf{A}_0)^{-1})'$.

Compute the vector of *value added per value unit of produced commodities*, $\boldsymbol{\upsilon}_c = \hat{\mathbf{x}}^{-1}\boldsymbol{\upsilon} = [\upsilon_1/x_1,\dots,\upsilon_3/x_3]'$, expressing the value added per unit of a produced commodity, and confirm the identity (2.76). Examine two cases of index prices:

(a) (Osterhaven) Consider in the period from the base year to the current year a total increase of value added (labour costs) of 20 % and calculate the *index prices* $\tilde{\mathbf{p}}_{0|1}$ of the whole economy, and for the *current year* matrix and vector entries.

(b) (Miller and Blair) Consider in the period from the base year to the current year the sectorial growth of the value added (labour costs), in sector S_1 of 30 %, in sector S_2 of 20 % and in sector S_3 of 10 % and calculate the *index prices* $\tilde{\mathbf{p}}_{0|1}$ of the whole economy, and for the *current year* the matrix and vector entries.

We also make a comparative-static analysis, and we will discuss the question of the invariance of the technology.

Solution to Example 2.5.1:

We rely on the *base-year* data. The entries of the positive transaction matrix $\mathbf{Z}_0$ and the positive vector of final demand $\mathbf{f}_0$ are taken from Table 2.10,

$$\mathbf{Z}_0 = \begin{bmatrix} 40 & 30 & 20 \\ 30 & 30 & 30 \\ 25 & 30 & 35 \end{bmatrix} > \mathbf{0}, \quad \mathbf{f}_0 = \begin{bmatrix} 10 \\ 60 \\ 110 \end{bmatrix} > \mathbf{o}. \tag{2.83}$$

We calculate the positive vector of *total output*,

$$\mathbf{x}_0 := \mathbf{Z}_0\mathbf{e} + \mathbf{f}_0 = \begin{bmatrix} 40 & 30 & 20 \\ 30 & 30 & 30 \\ 25 & 30 & 35 \end{bmatrix}\begin{bmatrix} 1 \\ 1 \\ 1 \end{bmatrix} + \begin{bmatrix} 10 \\ 60 \\ 110 \end{bmatrix} = \begin{bmatrix} 100 \\ 150 \\ 200 \end{bmatrix} > \mathbf{o}. \tag{2.84}$$

The vector of *total value added* in *monetary terms* (2.6) is

$$\boldsymbol{\upsilon}_0 = \mathbf{x}_0 - \mathbf{Z}_0'\mathbf{e} = \begin{bmatrix} 100 \\ 150 \\ 200 \end{bmatrix} - \begin{bmatrix} 40 & 30 & 25 \\ 30 & 30 & 30 \\ 20 & 30 & 35 \end{bmatrix}\begin{bmatrix} 1 \\ 1 \\ 1 \end{bmatrix} = \begin{bmatrix} 5 \\ 60 \\ 115 \end{bmatrix} > \mathbf{o}. \tag{2.85}$$

Then we compute the diagonal matrix $\hat{\mathbf{x}}_0^{-1}$ and then the vector of *value added of each sector per value unit of produced commodities*,

$$\boldsymbol{\upsilon}_{c,0} = \hat{\mathbf{x}}_0^{-1}\boldsymbol{\upsilon}_0 = \begin{bmatrix} \frac{1}{100} & 0 & 0 \\ 0 & \frac{1}{150} & 0 \\ 0 & 0 & \frac{1}{200} \end{bmatrix}\begin{bmatrix} 5 \\ 60 \\ 115 \end{bmatrix} = \begin{bmatrix} 0.05 \\ 0.4 \\ 0.575 \end{bmatrix}. \tag{2.86}$$

The positive *input-output coefficients* matrix $\mathbf{A}_0 = (a_{ij})$ is computed along (2.8), specified here for the *base year* "0":

$$\mathbf{A}_0 = \mathbf{Z}_0\hat{\mathbf{x}}_0^{-1}$$
$$= \begin{bmatrix} 40 & 30 & 20 \\ 30 & 30 & 30 \\ 25 & 30 & 35 \end{bmatrix}\begin{bmatrix} \frac{1}{100} & 0 & 0 \\ 0 & \frac{1}{150} & 0 \\ 0 & 0 & \frac{1}{200} \end{bmatrix} = \begin{bmatrix} 0.4 & 0.2 & 0.1 \\ 0.3 & 0.2 & 0.15 \\ 0.25 & 0.2 & 0.175 \end{bmatrix}, \tag{2.87}$$

which gives the quantity of the commodities *i*, required for the production of the quantity of one *value unit* each of the commodities *j*.

Then we confirm the identity (2.76):

$$\mathbf{A}_0'\mathbf{e} + \boldsymbol{\upsilon}_{c,0} = \begin{bmatrix} 0.4 & 0.3 & 0.25 \\ 0.2 & 0.2 & 0.2 \\ 0.1 & 0.15 & 0.175 \end{bmatrix}\begin{bmatrix} 1 \\ 1 \\ 1 \end{bmatrix} + \begin{bmatrix} 0.05 \\ 0.4 \\ 0.575 \end{bmatrix} = \begin{bmatrix} 1 \\ 1 \\ 1 \end{bmatrix} = \mathbf{e}. \tag{2.88}$$

The transposed Leontief matrix $(\mathbf{I} - \mathbf{A}_0') = (\mathbf{I} - \mathbf{A}_0)'$ is

$$(\mathbf{I} - \mathbf{A}_0)' = \begin{bmatrix} 0.6 & -0.3 & -0.25 \\ -0.2 & 0.8 & -0.2 \\ -0.1 & -0.15 & 0.825 \end{bmatrix}. \tag{2.89}$$

We calculate the characteristic polynomial

$$f_3(\lambda) = \det(\mathbf{A_0} - \lambda\mathbf{I}) = \det(\mathbf{A_0}' - \lambda\mathbf{I})$$
$$= -\frac{1}{200}\lambda(200\lambda^2 - 155\lambda + 14) = -\lambda(\lambda - 0.6706)(\lambda - 0.1044). \tag{2.90}$$

The *input-output coefficients* matrix $\mathbf{A}_0 > 0$ is positive, the Perron–Frobenius theorem A.9.2 applies, the Frobenius number is $\lambda_A = 0.6706 < 1$ and the *transposed Leontief Inverse* $((\mathbf{I} - \mathbf{A_0})^{-1})'$ exists, Lemma 2.4.1 and is positive, see Theorem A.10.2,

$$((\mathbf{I} - \mathbf{A}_0)^{-1})' = (\mathbf{I} - \mathbf{A}_0')^{-1} = \frac{1}{59} \cdot \begin{bmatrix} 126 & 57 & 52 \\ 37 & 94 & 34 \\ 22 & 24 & 84 \end{bmatrix} > \mathbf{0}. \tag{2.91}$$

(a) (Oosterhaven) Consider now the uniform 20 % increase of *value added* (labour costs) from the base year to the current year, resulting in the vector of *index prices* of *value added* $\mathbf{p}_{\upsilon,0|1} = [1.2, 1.2, 1.2]'$. Compute the vector of value added per value unit of produced commodities for the current year (2.80),

$$\boldsymbol{\upsilon}_{c,1} := \hat{\boldsymbol{\upsilon}}_{c,0}\tilde{\mathbf{p}}_{\upsilon,0|1} = \begin{bmatrix} 0.05 & 0 & 0 \\ 0 & 0.4 & 0 \\ 0 & 0 & 0.575 \end{bmatrix} \begin{bmatrix} 1.2 \\ 1.2 \\ 1.2 \end{bmatrix} = \begin{bmatrix} 0.06 \\ 0.48 \\ 0.69 \end{bmatrix} > \mathbf{o}. \tag{2.92}$$

Then, we compute with (2.92) and (2.91) the vector of *index prices* for all the commodities (2.81),

$$\tilde{\mathbf{p}}_{0|1} = ((\mathbf{I} - \mathbf{A}_0)^{-1})'\boldsymbol{\upsilon}_{c,1} = \frac{1}{59} \cdot \begin{bmatrix} 126 & 57 & 52 \\ 37 & 94 & 34 \\ 22 & 24 & 84 \end{bmatrix} \begin{bmatrix} 0.06 \\ 0.48 \\ 0.69 \end{bmatrix} = \begin{bmatrix} 1.2 \\ 1.2 \\ 1.2 \end{bmatrix} > \mathbf{o}. \tag{2.93}$$

This result is not a fluke. It can be shown that if the *growth rate* $\rho > 0$ of the value added (labour cost) is constant in all the sectors, then the *growth rate* of the prices of all the commodities i, $i = 1, \ldots, n$ is the same and vice versa:

$$\tilde{\mathbf{p}}_{\upsilon,0|1} = [1+\rho, \ldots, 1+\rho]' \Leftrightarrow \tilde{\mathbf{p}}_{0|1} = [1+\rho, \ldots, 1+\rho]'. \tag{2.94}$$

Finally, we compute the *commodity flow* matrix for the *current year* with (2.82) as

$$\mathbf{Z}_1 = \hat{\tilde{\mathbf{p}}}_{0|1}\mathbf{Z}_0 = \begin{bmatrix} 1.2 & 0 & 0 \\ 0 & 1.2 & 0 \\ 0 & 0 & 1.2 \end{bmatrix} \begin{bmatrix} 40 & 30 & 20 \\ 30 & 30 & 30 \\ 25 & 30 & 35 \end{bmatrix} = \begin{bmatrix} 48 & 36 & 24 \\ 36 & 36 & 36 \\ 30 & 36 & 42 \end{bmatrix} > \mathbf{0}, \tag{2.95}$$

as well as *final consumer demand* and the vector of *total value added* in *monetary terms* in the *current year* with (2.82),

$$\mathbf{f}_1 = \hat{\tilde{\mathbf{p}}}_{0|1}\mathbf{f}_0 = \begin{bmatrix} 1.2 & 0 & 0 \\ 0 & 1.2 & 0 \\ 0 & 0 & 1.2 \end{bmatrix} \begin{bmatrix} 10 \\ 60 \\ 110 \end{bmatrix} = \begin{bmatrix} 12 \\ 72 \\ 132 \end{bmatrix} > \mathbf{o},$$

$$\boldsymbol{\upsilon}_1 = \hat{\boldsymbol{\upsilon}}_0\tilde{\mathbf{p}}_{\upsilon,0|1} = \begin{bmatrix} 5 & 0 & 0 \\ 0 & 60 & 0 \\ 0 & 0 & 115 \end{bmatrix} \begin{bmatrix} 1.2 \\ 1.2 \\ 1.2 \end{bmatrix} = \begin{bmatrix} 6 \\ 72 \\ 138 \end{bmatrix} > \mathbf{o}. \tag{2.96}$$

This gives Table 2.11. As expected, with $V = \boldsymbol{\upsilon}_1 \cdot \mathbf{e} = 216$ and $F = \mathbf{f}_1 \cdot \mathbf{e} = 216$ we obtain. We also find

$$\mathbf{x}_1 = \hat{\tilde{\mathbf{p}}}_{0|1}\mathbf{x}_0 = \begin{bmatrix} 1.2 & 0 & 0 \\ 0 & 1.2 & 0 \\ 0 & 0 & 1.2 \end{bmatrix} \begin{bmatrix} 100 \\ 150 \\ 200 \end{bmatrix} = \begin{bmatrix} 120 \\ 180 \\ 240 \end{bmatrix} \Rightarrow \hat{\mathbf{x}}_1 = \hat{\tilde{\mathbf{p}}}_{0|1}\hat{\mathbf{x}}_0, \tag{2.97}$$

Table 2.11: Example 2.5.1—entries for the current year (Oosterhaven).

Commodities	**Processing sectors**			**Final demand**	**Total output**
current year	S_1	S_2	S_3	f_{1i}	
S_1: wheat	48	36	24	12	120
S_2: iron	36	36	36	72	180
S_3: wood	30	36	42	132	240
total value added per sector, υ_{1j}	6	72	138	$V = 216/F = 216$	
total Outlays	120	180	240		$X = 540$

We observe that in this example the vector of *index prices* $\tilde{\mathbf{p}}_{0|1} = [1.2, 1.2, 1.2]'$ of this economy is identical to the given vector of the *index prices* of the *value added* $\tilde{\mathbf{p}}_{\upsilon,0|1} = [1.2, 1.2, 1.2]'$ because we are in presence of a constant homogeneous increase of all the *value added* labour costs.

We invite the reader to carry out the following control operations:

1. $\Delta\tilde{\mathbf{p}}_{\upsilon,0|1} = \tilde{\mathbf{p}}_{\upsilon,0|1} - \tilde{\mathbf{p}}_{\upsilon,0|0} = [0.2, 0.2, 0.2]'$.
2. $\Delta\tilde{\mathbf{p}}_{0|1} = \tilde{\mathbf{p}}_{0|1} - \tilde{\mathbf{p}}_{0|0} = [0.2, 0.2, 0.2]'$.
3. $\Delta\boldsymbol{\upsilon}_{c,0|1} := \hat{\boldsymbol{\upsilon}}_{c,0} \cdot \Delta\tilde{\mathbf{p}}_{\upsilon,0|1} = \boldsymbol{\upsilon}_{c,1} - \boldsymbol{\upsilon}_{c,0} = [0.01, 0.08, 0.115]'$.
4. $\Delta\tilde{\mathbf{p}}_{0|1} = ((\mathbf{I} - \mathbf{A}_0)^{-1})' \cdot \Delta\boldsymbol{\upsilon}_{c,0|1} = [0.2, 0.2, 0.2]'$.
5. $\mathbf{x}_0 = (\mathbf{I} - \mathbf{A}_0)^{-1}\mathbf{f}_0 = [100, 150, 200]'$.
6. $\mathbf{x}_0 = \mathbf{Z}_0\mathbf{e} + \mathbf{f}_0 = [100, 150, 200]'$ and $\mathbf{x}_1 = \mathbf{Z}_1\mathbf{e} + \mathbf{f}_1 = [120, 180, 240]'$.
7. $\mathbf{e} := \tilde{\mathbf{p}}_{0|0} = ((\mathbf{I} - \mathbf{A}_0)^{-1})' \cdot \hat{\boldsymbol{\upsilon}}_{c,0} \cdot \tilde{\mathbf{p}}_{\upsilon,0|0} = [1, 1, 1]'$.
8. Check the invariance of technology, respectively, of the *input-output coefficients matrix* $\mathbf{A}_1$. We have here $\hat{\tilde{\mathbf{p}}}_{0|1}\mathbf{Z}_0 = \mathbf{Z}_0\hat{\tilde{\mathbf{p}}}_{0|1}$ and with (2.97), (2.94) the equality

$$\mathbf{A}_1 = \mathbf{Z}_1\hat{\mathbf{x}}_1^{-1} := (\hat{\tilde{\mathbf{p}}}_{0|1}\mathbf{Z}_0)(\hat{\tilde{\mathbf{p}}}_{0|1}\hat{\mathbf{x}}_0)^{-1} = (\mathbf{Z}_0\hat{\tilde{\mathbf{p}}}_{0|1})(\hat{\mathbf{x}}_0\hat{\tilde{\mathbf{p}}}_{0|1})^{-1}$$
$$= (\mathbf{Z}_0\hat{\tilde{\mathbf{p}}}_{0|1})(\hat{\tilde{\mathbf{p}}}_{0|1}^{-1}\hat{\mathbf{x}}_0^{-1}) = \mathbf{Z}_0(\hat{\tilde{\mathbf{p}}}_{0|1}\hat{\tilde{\mathbf{p}}}_{0|1}^{-1})\hat{\mathbf{x}}_0^{-1} = \mathbf{Z}_0\hat{\mathbf{x}}_0^{-1} = \mathbf{A}_0. \qquad (2.98)$$

We now continue with the generalized model.

(b) (Miller and Blair) Consider now the *index prices* of *value added* (labour costs) from the base year to the current year, given by the vector $\tilde{\mathbf{p}}_{\upsilon,0|1} = [1.1, 1.2, 1.3]'$. We compute the vector of the *value added of each sector per value unit of produced commodities* for the *current year* (2.80),

$$\boldsymbol{\upsilon}_{c,1} := \hat{\boldsymbol{\upsilon}}_{c,0}\tilde{\mathbf{p}}_{\upsilon,0|1} = \begin{bmatrix} 0.05 & 0 & 0 \\ 0 & 0.4 & 0 \\ 0 & 0 & 0.575 \end{bmatrix}\begin{bmatrix} 1.1 \\ 1.2 \\ 1.3 \end{bmatrix} = \begin{bmatrix} 0.055 \\ 0.48 \\ 0.7475 \end{bmatrix} > \mathbf{o}. \qquad (2.99)$$

The vector of *index prices* of the whole economy, i. e., (2.81) with (2.99), (2.91), including the consumer market (final demand), then is

$$\tilde{\mathbf{p}}_{0|1} = (\mathbf{I} - \mathbf{A}_0)^{-1}\boldsymbol{\upsilon}_{c,1}$$
$$= \frac{1}{59}\begin{bmatrix} 126 & 57 & 52 \\ 37 & 94 & 34 \\ 22 & 24 & 84 \end{bmatrix}\begin{bmatrix} 0.055 \\ 0.48 \\ 0.7475 \end{bmatrix} = \begin{bmatrix} 1.24 \\ 1.23 \\ 1.28 \end{bmatrix} > \mathbf{o}. \qquad (2.100)$$

Finally, we compute the diagonal matrix $\hat{\tilde{\mathbf{p}}}_{0|1}$, taking matrix $\mathbf{Z}_0$ (2.83), we compute the *commodity flow* matrix for the *current year*, do

$$\mathbf{Z}_1 = \hat{\tilde{\mathbf{p}}}_{0|1}\mathbf{Z}_0$$
$$= \begin{bmatrix} 1.24 & 0 & 0 \\ 0 & 1.23 & 0 \\ 0 & 0 & 1.28 \end{bmatrix}\begin{bmatrix} 40 & 30 & 20 \\ 30 & 30 & 30 \\ 25 & 30 & 35 \end{bmatrix} = \begin{bmatrix} 49.6 & 37.2 & 24.8 \\ 36.9 & 36.9 & 36.9 \\ 32 & 38.4 & 44.8 \end{bmatrix} > \mathbf{O}, \qquad (2.101)$$

as well as the *final consumer demand* and the vector of *total value added* in monetary terms (here total labour costs) for the *current year*:

$$\mathbf{f}_1 = \hat{\tilde{\mathbf{p}}}_{0|1}\mathbf{f}_0 = \begin{bmatrix} 1.24 & 0 & 0 \\ 0 & 1.23 & 0 \\ 0 & 0 & 1.28 \end{bmatrix}\begin{bmatrix} 10 \\ 60 \\ 110 \end{bmatrix} = \begin{bmatrix} 12.4 \\ 73.8 \\ 140.8 \end{bmatrix} > \mathbf{o},$$
$$\boldsymbol{\upsilon}_1 = \hat{\boldsymbol{\upsilon}}_0\tilde{\mathbf{p}}_{\upsilon,0|1} = \begin{bmatrix} 5 & 0 & 0 \\ 0 & 60 & 0 \\ 0 & 0 & 115 \end{bmatrix}\begin{bmatrix} 1.1 \\ 1.2 \\ 1.3 \end{bmatrix} = \begin{bmatrix} 5.5 \\ 72 \\ 149.5 \end{bmatrix} > \mathbf{o}. \qquad (2.102)$$

We also find, as given at the bottom of Table 2.12,

$$\mathbf{x}_1 = \hat{\tilde{\mathbf{p}}}_{0|1}\mathbf{x}_0 = \begin{bmatrix} 1.24 & 0 & 0 \\ 0 & 1.23 & 0 \\ 0 & 0 & 1.28 \end{bmatrix}\begin{bmatrix} 100 \\ 150 \\ 200 \end{bmatrix} = \begin{bmatrix} 124 \\ 184.5 \\ 256 \end{bmatrix} > \mathbf{o}. \qquad (2.103)$$

Table 2.12: Example 2.5.1—entries of the current year (Miller and Blair) ▲.

	Processing sectors			Final demand	Total
	S_1	S_2	S_3	f_{1i}	output
S_1: wheat	49.6	37.2	24.8	12.4	124
S_2: iron	36.9	36.9	36.9	73.8	184.5
S_3: wood	32	38.4	44.8	140.8	256
total value added per sector, v_{1j}	5.5	72	149.5	$V = 227/F = 227$	
total Outlays	124	184.5	256		$X = 564.5$

Therefore we have $\hat{\mathbf{x}}_1 = \hat{\mathbf{x}}_0 \hat{\tilde{\mathbf{p}}}_{0|1}$. We observe that in the case (b) (Miller and Blair) the vector of *index prices* of the whole economy $\tilde{\mathbf{p}}_{0|1} = [1.24, 1.23, 1.28]'$ is different from the given vector of *index price* of *value added* $\tilde{\mathbf{p}}_{v,0|1} = [1.1, 1.2, 1.3]'$ because we are in the presence of different sectorial increases of the labour costs. Therefore, the *input-output coefficients matrix* $\mathbf{A}_1 = \mathbf{Z}_1 \hat{\mathbf{x}}_1^{-1} \neq \mathbf{A}_0$ is no longer invariant:

$$\mathbf{A}_0 \neq \mathbf{A}_1 := \mathbf{Z}_1 \hat{\mathbf{x}}_1^{-1}$$
$$= \begin{bmatrix} 49.6 & 37.2 & 24.8 \\ 36.9 & 36.9 & 36.9 \\ 32 & 38.4 & 44.8 \end{bmatrix} \begin{bmatrix} \frac{1}{124} & 0 & 0 \\ 0 & \frac{2}{369} & 0 \\ 0 & 0 & \frac{1}{256} \end{bmatrix} = \begin{bmatrix} 0.4 & 0.202 & 0.097 \\ 0.298 & 0.2 & 0.144 \\ 0.258 & 0.208 & 0.175 \end{bmatrix}. \tag{2.104}$$

We enter here the subject of *dynamic Input-Output Analysis* that is not treated any further this book. We again invite the reader to carry out the following eight control operations:

1. $\Delta\tilde{\mathbf{p}}_{v,0|1} = \tilde{\mathbf{p}}_{v,0|1} - \tilde{\mathbf{p}}_{v,0|0} = [0.1, 0.2, 0.3]'$.
2. $\Delta\tilde{\mathbf{p}}_{0|1} = \tilde{\mathbf{p}}_{0|1} - \tilde{\mathbf{p}}_{0|0} = [0.24, 0.23, 0.28]'$.
3. $\Delta\mathbf{v}_{c,1} := \hat{\mathbf{v}}_{c,0} \cdot \Delta\tilde{\mathbf{p}}_{v,0|1} = [0.005, 0.08, 0.1725]'$.
4. $\Delta\tilde{\mathbf{p}}_{0|1} = ((\mathbf{I} - \mathbf{A}_0)^{-1})' \cdot \Delta\mathbf{v}_{c,1} = [0.24, 0.23, 0.28]'$.
5. $\mathbf{e} := \tilde{\mathbf{p}}_{0|0} = ((\mathbf{I} - \mathbf{A}_0)^{-1})' \cdot \hat{\mathbf{v}}_{c,0} \cdot \tilde{\mathbf{p}}_{v,0|0} = [1, 1, 1]'$.
6. $\mathbf{x}_1 = \mathbf{Z}_1 \mathbf{e} + \mathbf{f}_1 = [124, 184.5, 256]'$.
7. Check the variability of the *input-output coefficients matrix* $\mathbf{A}_0 \neq \mathbf{A}_1 := \mathbf{Z}_1 \hat{\mathbf{x}}_1^{-1}$.
8. $\mathbf{x}_1 = (\mathbf{I} - \mathbf{A}_1)^{-1} \mathbf{f}_1 = [124, 184.5, 256]'$. ▲

2.5.2 The Leontief cost-push input-output price model in physical terms

In this subsection, we present a further *Leontief price model*, the *Leontief cost-push input-output price model* in *physical terms*. We propose the evident economic Assumption 2.5.1 of positive amounts of *working time* in every sector. The limiting case of purely robotic production sectors are here excluded.

Assumption 2.5.1 (Assumption on labour forces). Every sector employs a certain number of workers, $L_j > 0, j = 1, \dots, n$, comprising labour for the sector. No sector employs only industrial robots for manufacturing. The unit of L_j is man-years.

For the notations, we rely on the Input-Output data representation as on Tables 2.1 and 2.2. The *semi-positive* matrices $\mathbf{Z} \geq \mathbf{0}$, respectively $\mathbf{S} \geq \mathbf{0}$ represent interindustrial sales and purchases in monetary, respectively, in physical terms. The economic Assumption 2.2.1 and 2.2.2 hold. We need a measurement unit, a *currency* or a *numéraire*, to define prices for all commodities. Here we choose as the currency the *Swiss franc* (CHF) and present the entries in *thousands of Swiss francs* (kCHF). Then we need for each industry S_j the quantity of annual labour, expressed here in *man-years*. So we have the number of workers $L_j > 0$, Assumption 2.5.1 holds, leading to a positive vector of *labour* $\mathbf{L} = [L_1, \dots, L_n]' > \mathbf{o}$. All sectors have workers. We also need the row of *total labour costs* (*total wages*), $W_j := \upsilon_j > 0$, identical to *value added* in *monetary terms*, and finally we can express the *wage rates* $w_j = W_j/L_j > 0, j = 1, \dots, n$.

A consequence of the condition: $\mathbf{x} > \mathbf{o}$, $\mathbf{q} > \mathbf{o}$, is that Proposition 2.2.1 on price vectors holds, so there is a positive price vector $\mathbf{p} > \mathbf{o}$. The conversion of the entries from *monetary terms* into *physical terms* and vice versa, see also Miller and Blair ([65], p. 48), exists. We come back to (2.18),

$$\begin{cases} \mathbf{x} = \hat{\mathbf{p}}\mathbf{q} = \hat{\mathbf{q}}\mathbf{p}; \quad \hat{\mathbf{x}} = \hat{\mathbf{p}}\hat{\mathbf{q}}, \\ \mathbf{Z} = \hat{\mathbf{p}}\mathbf{S}, \quad \Leftrightarrow \mathbf{S} = \hat{\mathbf{p}}^{-1}\mathbf{Z}, \\ \mathbf{f} = \hat{\mathbf{p}}\mathbf{d}, \quad \Leftrightarrow \mathbf{d} = \hat{\mathbf{p}}^{-1}\mathbf{f}, \end{cases} \tag{2.105}$$

and write separately, setting up the vector of *wage rates* $\mathbf{w} = [w_1, \dots, w_n]' > \mathbf{o}$,

$$\upsilon_j := W_j = w_j \cdot L_j \Leftrightarrow \boldsymbol{\upsilon} = \hat{\mathbf{w}}\mathbf{L} > \mathbf{o}. \tag{2.106}$$

We again rely on the identity between *total output* and *total outlays* and start from equation (2.7),

$$\mathbf{x} = \mathbf{Z}'\mathbf{e} + \boldsymbol{\upsilon} = \mathbf{Z}\mathbf{e} + \mathbf{f} = \mathbf{A}\mathbf{x} + \mathbf{f} > \mathbf{o}. \tag{2.107}$$

We use the definitions $\mathbf{S} = \hat{\mathbf{p}}^{-1}\mathbf{Z}$ and $\hat{\mathbf{q}} = \hat{\mathbf{p}}^{-1}\hat{\mathbf{x}}$, which leads to $\mathbf{C} = \mathbf{S}\hat{\mathbf{q}}^{-1}$ (2.16), $\mathbf{A} = \mathbf{Z}\hat{\mathbf{x}}^{-1}$ (2.8). Throughout this subsection, we again posit that model (2.107) is *productive*, Definition A.12.1 as $\mathbf{f} \geq \mathbf{o}$. Therefore, the Frobenius number $\lambda_A < 1$ (see Lemma 2.4.1). We then have the Frobenius number of the matrix $\mathbf{C}$, $\lambda_C = \lambda_A < 1$, Lemma A.6.1 and obtain with (2.107) and (2.105),

$$\mathbf{x} = \hat{\mathbf{q}}\mathbf{p} = \mathbf{Z}'\mathbf{e} + \boldsymbol{\upsilon} = (\hat{\mathbf{p}}\mathbf{S})'\mathbf{e} + \hat{\mathbf{w}}\mathbf{L} = \mathbf{S}'\hat{\mathbf{p}}\mathbf{e} + \hat{\mathbf{w}}\mathbf{L} = \hat{\mathbf{q}}\mathbf{C}'\hat{\mathbf{p}}\mathbf{e} + \hat{\mathbf{w}}\mathbf{L} > \mathbf{o}. \tag{2.108}$$

We pre-multiply (2.108) from the left with the vector $\hat{\mathbf{q}}^{-1}$ and get

$$\mathbf{p} = \mathbf{C}'\mathbf{p} + \hat{\mathbf{q}}^{-1}(\hat{\mathbf{w}}\mathbf{L}). \tag{2.109}$$

For each industry S_j, the parts of *total labour costs* $W_j = w_j \cdot L_j$ *per unit of physical output* q_j, identified as $v_{cj} := (w_j \cdot L_j)/q_j$, is now written as a vector, called the vector of *labour costs of each sector per unit of physical output*, $\mathbf{v}_c = [v_{c1}, \dots, v_{cn}]' := \hat{\mathbf{q}}^{-1}(\hat{\mathbf{w}}\mathbf{L})$.[29] Taking (2.109), we get the Leontief *cost-push input-output price model* in *physical terms*:

$$\mathbf{p} := \mathbf{C}'\mathbf{p} + \mathbf{v}_c \Rightarrow (\mathbf{I} - \mathbf{C}')\mathbf{p} = \mathbf{v}_c. \tag{2.110}$$

As the Frobenius number of $\mathbf{C}$ is less than 1, $\lambda_C < 1$, its transposed *Leontief Inverse* exists, and we have a unique solution for the *price vector* $\mathbf{p} = [p_1, p_2, \dots, p_n]'$,

$$\boxed{\mathbf{p} = ((\mathbf{I} - \mathbf{C})^{-1})'\mathbf{v}_c = (\mathbf{I} - \mathbf{C}')^{-1}\mathbf{v}_c,} \tag{2.111}$$

by analogy with the *price indices* of the Leontief *cost-push input-output price model* in *monetary terms* (2.81) see also Miller and Blair ([65], p. 49, (2.52)).

Observing that the transformation between variables in *monetary terms* and variables in *physical terms* exists, we now concentrate on the relationships between $\mathbf{A}$ from (2.9) and $\mathbf{C}$ from (2.16). From (2.8) and (2.105) we obtain

$$a_{ij} = \frac{z_{ij}}{x_j} = \frac{p_i s_{ij}}{p_j q_j} = c_{ij}\left(\frac{p_i}{p_j}\right), \quad i, j = 1, \dots, n. \tag{2.112}$$

Expressed in matrix terms, this gives:

$$\begin{aligned} \mathbf{A} := \mathbf{Z}\hat{\mathbf{x}}^{-1} &= (\hat{\mathbf{p}}\mathbf{S})(\hat{\mathbf{p}}\hat{\mathbf{q}})^{-1} \Rightarrow \\ \mathbf{A} = \hat{\mathbf{p}}\mathbf{S}(\hat{\mathbf{q}}^{-1}\hat{\mathbf{p}}^{-1}) = \hat{\mathbf{p}}(\mathbf{C}\hat{\mathbf{q}})(\hat{\mathbf{q}}^{-1}\hat{\mathbf{p}}^{-1}) &= \hat{\mathbf{p}}\mathbf{C}(\hat{\mathbf{q}}\hat{\mathbf{q}}^{-1})\hat{\mathbf{p}}^{-1} = \hat{\mathbf{p}}\mathbf{C}\hat{\mathbf{p}}^{-1} \Rightarrow \\ \mathbf{A} &= \hat{\mathbf{p}}\mathbf{C}\hat{\mathbf{p}}^{-1}. \end{aligned} \tag{2.113}$$

We continue with the transformation equations (2.105) and find with $\hat{\mathbf{x}} = \hat{\mathbf{p}}\hat{\mathbf{q}}$,

$$\begin{aligned} \mathbf{A} = \hat{\mathbf{p}}\mathbf{C}\hat{\mathbf{p}}^{-1} &= (\hat{\mathbf{x}}\hat{\mathbf{q}}^{-1})\mathbf{C}(\hat{\mathbf{q}}\hat{\mathbf{x}}^{-1}) = \hat{\mathbf{x}}\hat{\mathbf{q}}^{-1}(\mathbf{C}\hat{\mathbf{q}})\hat{\mathbf{x}}^{-1} = \hat{\mathbf{x}}\hat{\mathbf{q}}^{-1}\mathbf{S}\hat{\mathbf{x}}^{-1} \\ &= \hat{\mathbf{x}}(\hat{\mathbf{q}}^{-1}\mathbf{S})\hat{\mathbf{x}}^{-1} = \hat{\mathbf{x}}\mathbf{D}\hat{\mathbf{x}}^{-1} \Rightarrow \mathbf{A} = \hat{\mathbf{x}}\mathbf{D}\hat{\mathbf{x}}^{-1}. \end{aligned} \tag{2.114}$$

This means that the three matrices of *input-output coefficients* **A, C** and **D** are *similar*. The crossing between the *transaction matrices* **Z** and **S** is obtained through the use of the diagonal price matrix $\hat{\mathbf{p}} > \mathbf{o}$ and its inverse $\hat{\mathbf{p}}^{-1} > \mathbf{o}$.

We continue with the elementary input-output example presented in L. L. Pasinetti ([80], pp. 35–47), also used in Example 2.4.4.

Example 2.5.2. Consider the economy described by Table 2.8 where there is a sector of *final demand*. The activity of the workers is explicitly described. The economy employs

29 Here *value added* includes only *labour costs*.

60 workers, 18 to produce wheat, 12 to produce iron and 30 to produce turkeys. Each worker consumes three tons of wheat a year and half a dozen turkeys.

Identify the commodity flow matrix **S**, the vector of total output **q** and the input-output coefficients matrix **C** and discuss the applicability of the **Perron theorem A.9.1.** Why is it not possible to calculate prices at this stage?

Table 2.8 is extended to Table 2.13.

Table 2.13: Closed flow of commodities in physical terms with labour included.

Commodity sectors	Processing sectors wheat	iron	turkeys	Final demand of consumers	Total output
wheat	186	54	30	180	450 tons of wheat
iron	12	6	3	–	21 tons of iron
turkeys	9	6	15	30	60 dozen turkeys
workers (man-years)	18	12	30	$L = 60/D = 210$	

Solution to Example 2.5.2:

The positive *commodity flow matrix* **S**, the vector of *final demand* **d** and the positive vector of *total output* **q** are identified:

$$\mathbf{S} = \begin{bmatrix} 186 & 54 & 30 \\ 12 & 6 & 3 \\ 9 & 6 & 15 \end{bmatrix}, \quad \mathbf{d} = \begin{bmatrix} 180 \\ 0 \\ 30 \end{bmatrix},$$

$$\mathbf{q} = \mathbf{Se} + \mathbf{d} = \begin{bmatrix} 186 & 54 & 30 \\ 12 & 6 & 3 \\ 9 & 6 & 15 \end{bmatrix} \begin{bmatrix} 1 \\ 1 \\ 1 \end{bmatrix} + \begin{bmatrix} 180 \\ 0 \\ 30 \end{bmatrix} = \begin{bmatrix} 450 \\ 21 \\ 60 \end{bmatrix} > \mathbf{o}. \tag{2.115}$$

Consequently, we compute the diagonal matrix $\hat{\mathbf{q}}^{-1}$ and then calculate the positive *input-output coefficients* matrix

$$\begin{aligned} \mathbf{C} = \mathbf{S}\hat{\mathbf{q}}^{-1} &= \begin{bmatrix} 186 & 54 & 30 \\ 12 & 6 & 3 \\ 9 & 6 & 15 \end{bmatrix} \begin{bmatrix} \frac{1}{450} & 0 & 0 \\ 0 & \frac{1}{21} & 0 \\ 0 & 0 & \frac{1}{60} \end{bmatrix} \\ &= \begin{bmatrix} \frac{31}{75} & \frac{18}{7} & \frac{1}{2} \\ \frac{2}{75} & \frac{2}{7} & \frac{1}{20} \\ \frac{1}{50} & \frac{2}{7} & \frac{1}{4} \end{bmatrix} = \begin{bmatrix} 0.413 & 2.571 & 0.500 \\ 0.027 & 0.286 & 0.050 \\ 0.020 & 0.286 & 0.250 \end{bmatrix} > \mathbf{0}, \end{aligned} \tag{2.116}$$

and the characteristic polynomial of the *input-output coefficients* matrix

$$\begin{aligned} f_3(\lambda) = \det(\mathbf{C} - \lambda\mathbf{I}) &= -\lambda^3 + \frac{1{,}993}{2{,}100}\lambda^2 - \frac{1}{5}\lambda - \frac{1}{100} \\ &= (\lambda - 0.675)(\lambda - 0.201)(\lambda - 0.079). \end{aligned} \tag{2.117}$$

Because the *input-output coefficients* matrix $\mathbf{C}$ is positive, the **Perron theorem A.9.1** applies, the Frobenius number is $\lambda_C = 0.675 < 1$ and the *transposed Leontief Inverse* $((\mathbf{I}-\mathbf{C})^{-1})' = (\mathbf{I}-\mathbf{C}')^{-1}$ exists, see Lemma 2.4.1.

As long as there is no defined *numéraire* or *currency*, the vector of *wage rates* $\mathbf{w}$ and consequently the vector *labour costs of each sector per unit of physical output* $\mathbf{v}_c = \hat{\mathbf{q}}^{-1}(\hat{\mathbf{w}}\mathbf{L})$ cannot be defined. ▲

We go on with a **comparative-static analysis**, for a period from a *base year* "0" to a *current year* "1". We analyse the evolution of the matrices and vectors, from the *base year* "0" to the *current year* "1".

For this purpose, we again adapt L. L. Pasinetti's ([80], pp. 35–47) Example 2.5.2 introducing the currency CHF in units of 1,000 CHF = 1 kCHF, in order to be able to calculate prices and wages.[30]

Example 2.5.3. For the economy presented by Example 2.5.2, Table 2.13, adopt the currency CHF.

The following indications are given for the prices: 20/11 man-years of labour have the value of 1 kCHF. This means that there is a constant wage rate, expressed as $w_0 = (11/20)\,\frac{\text{kCHF}}{\text{man-years}}$.

Establish then for the base year "0" the vector of wage rates $\mathbf{w}_0$. Compute the transposed Leontief Inverse, using the matrix $\mathbf{C}_0 := \mathbf{C}$ (2.116).

Compute the vector $\boldsymbol{\upsilon}_0$ of labour costs (total wages). Compute the vector $\mathbf{v}_{c,0}$ of labour costs per unit of produced quantity and indicate the units of measurements of all its components.

Compute the vector of the absolute prices[31] $\mathbf{p}_0 = ((\mathbf{I}-\mathbf{C}_0)^{-1})'\mathbf{v}_{c,0}$, set up in kCHF, the commodity flow matrix $\mathbf{Z}_0$, the vector of *final consumer demand* $\mathbf{f}_0$ and the vector of *total output* $\mathbf{x}_0$.

Compute the input-output coefficients matrix $\mathbf{A}_0 = \mathbf{Z}_0\hat{\mathbf{x}}_0^{-1}$ and confirm the identity $\mathbf{A}_0 = \hat{\mathbf{p}}_0\mathbf{C}_0\hat{\mathbf{p}}_0^{-1}$.

Consider then an increase of the uniform wage rate from $w_0 = 0.55\,\frac{\text{kCHF}}{\text{man-years}}$ (base year "0") to $w_1 = 0.65\,\frac{\text{kCHF}}{\text{man-years}}$ (current year "1").

Repeat all the above calculations, compute especially the vectors $\boldsymbol{\upsilon}_1, \mathbf{v}_{c,1}, \mathbf{p}_1, \mathbf{f}_1, \mathbf{x}_1$, the matrices $\mathbf{Z}_1$, $\mathbf{A}_1$, $\mathbf{S}_1$ and $\mathbf{C}_1$ and confirm finally the equalities between the matrices $\mathbf{S}_0 = \mathbf{S}_1$, $\mathbf{C}_0 = \mathbf{C}_1$, $\mathbf{A}_0 = \mathbf{A}_1$ and the vectors $\mathbf{d}_0 = \mathbf{d}_1$ and $\mathbf{q}_0 = \mathbf{q}_1$.

Show why the labour vector $\mathbf{L}$, as well as the input coefficients matrices $\mathbf{S}_0$, $\mathbf{d}_0$, $\mathbf{q}_0$, $\mathbf{A}_0$ and $\mathbf{C}_0$, remain invariant in this problem within this period.

30 In L. L. Pasinetti's presentation, the prices are indicated on the basis of the *numéraire*, i. e., *tons of* iron. In our presentation, the prices are calculated with *Leontief's cost-push input-output price model* in physical terms (2.111), using the currency CHF with the unit kCHF. The *quantity of labour* (working time) and a currency are needed.

31 An absolute price gives the price in *currency* per *unit of commodity*.

Solution to Example 2.5.3:
We use, wherever necessary to enhance understanding, the index "0" for the *base year* and "1" for the *current year* and start with the vectors of *wage rates*, respectively, the constant wage rate

$$\mathbf{w}_0 = [0.55, 0.55, 0.55]'; \quad w_0 = 0.55. \tag{2.118}$$

We also have to transform the *labour* vector $\mathbf{L} = [18, 12, 30]'$, $[\mathbf{L}] = \textit{man-years}$ into the vector of *labour costs*. We calculate with the diagonalised vector of wage rates $\mathbf{w}_0$ the vector of *total labour costs* for the *basic year*, with the measurement unit $[\boldsymbol{\upsilon}_0] = \text{kCHF}$. We calculate

$$\begin{aligned}\boldsymbol{\upsilon}_0 &= \hat{\mathbf{w}}_0\mathbf{L} := w_0\mathbf{L}\\ &= \begin{bmatrix} 0.55 & 0 & 0 \\ 0 & 0.55 & 0 \\ 0 & 0 & 0.55 \end{bmatrix}\begin{bmatrix} 18 \\ 12 \\ 30 \end{bmatrix} = 0.55 \cdot \begin{bmatrix} 18 \\ 12 \\ 30 \end{bmatrix} = \begin{bmatrix} 9.9 \\ 6.6 \\ 16.5 \end{bmatrix}\end{aligned} \tag{2.119}$$

The total *labour costs* (*value added*) are $W_0 = \boldsymbol{\upsilon}_0'\mathbf{e} = 33$ kCHF (closed economy), and equal to the total *final demand*, i. e., $D_0 = \mathbf{f}_0'\mathbf{e} = 33$ kCHF.

We then calculate with the diagonalised vector of total output $\mathbf{q}_0$ and the vector of *labour costs* $\boldsymbol{\upsilon}_0 = w_0\mathbf{L} = [9.9, 6.6, 16.5]'$ the *labour costs of each sector per unit of produced quantities*, defined as:

$$\mathbf{v}_{c,0} = \hat{\mathbf{q}}_0^{-1}\boldsymbol{\upsilon}_0 = \begin{bmatrix} \frac{1}{450} & 0 & 0 \\ 0 & \frac{1}{21} & 0 \\ 0 & 0 & \frac{1}{60} \end{bmatrix}\begin{bmatrix} 9.9 \\ 6.6 \\ 16.5 \end{bmatrix} = \begin{bmatrix} 0.022 \\ 0.314 \\ 0.275 \end{bmatrix}, \tag{2.120}$$

with units of measurements:

$$v_{c,01} = 0.22\,\frac{\text{kCHF}}{\text{tons of wheat}}, \quad v_{c,02} = 0.314\,\frac{\text{kCHF}}{\text{tons of iron}}, \quad v_{c,03} = 0.275\,\frac{\text{kCHF}}{\text{dozen of turkeys}}.$$

Then, we can compute the *transposed Leontief Inverse* with matrix $\mathbf{C}_0 := \mathbf{C}$ from (2.116) because the Frobenius number is $\lambda_C = 0.675 < 1$,

$$((\mathbf{I} - \mathbf{C}_0)^{-1})' = \begin{bmatrix} \frac{1{,}095}{506} & \frac{441}{5{,}060} & \frac{1}{11} \\ \frac{2{,}175}{253} & \frac{903}{506} & \frac{10}{11} \\ \frac{510}{253} & \frac{224}{1{,}265} & \frac{16}{11} \end{bmatrix} = \begin{bmatrix} 2.16 & 0.0872 & 0.0909 \\ 8.60 & 1.78 & 0.909 \\ 2.02 & 0.177 & 1.45 \end{bmatrix} > \mathbf{0}, \tag{2.121}$$

and we calculate with the *Leontief cost-push input-output price model* in physical terms (2.111) the vector of prices

$$\mathbf{p}_0 = ((\mathbf{I} - \mathbf{C}_0)^{-1})' \cdot \mathbf{v}_{c,0} = \begin{bmatrix} 2.16 & 0.0872 & 0.0909 \\ 8.60 & 1.78 & 0.909 \\ 2.02 & 1.77 & 1.45 \end{bmatrix}\begin{bmatrix} 0.022 \\ 0.314 \\ 0.275 \end{bmatrix} = \begin{bmatrix} 0.1 \\ 1 \\ 0.5 \end{bmatrix}. \tag{2.122}$$

So the absolute prices are:

$$p_{10} = 0.1 \frac{\text{kCHF}}{\text{tons of wheat}}, \quad p_{20} = 1 \frac{\text{kCHF}}{\text{tons of iron}}, \quad p_{30} = 0.5 \frac{\text{kCHF}}{\text{dozen turkeys}}.$$

We apply the *diagonal operator* to $\mathbf{p}_0$, getting $\operatorname{diag}(\mathbf{p}_0) = \hat{\mathbf{p}}_0$, and calculate the *commodity flow matrix* $\mathbf{Z}_0$ in *numéraire* by the matrix transformation (2.113):

$$\mathbf{Z}_0 = \hat{\mathbf{p}}_0 \mathbf{S}_0 = \begin{bmatrix} 0.1 & 0 & 0 \\ 0 & 1 & 0 \\ 0 & 0 & 0.5 \end{bmatrix} \begin{bmatrix} 186 & 54 & 30 \\ 12 & 6 & 3 \\ 9 & 6 & 15 \end{bmatrix} = \begin{bmatrix} 18.6 & 5.4 & 3 \\ 12 & 6 & 3 \\ 4.5 & 3 & 7.5 \end{bmatrix}. \tag{2.123}$$

Then the vectors of *final demand* and *total output* in *monetary terms* (2.105) are computed,

$$\mathbf{f}_0 = \hat{\mathbf{p}}_0 \mathbf{d}_0 = \begin{bmatrix} 0.1 & 0 & 0 \\ 0 & 1 & 0 \\ 0 & 0 & 0.5 \end{bmatrix} \begin{bmatrix} 180 \\ 0 \\ 30 \end{bmatrix} = \begin{bmatrix} 18 \\ 0 \\ 15 \end{bmatrix} \geq \mathbf{o}, \tag{2.124}$$

$$\mathbf{x}_0 = \hat{\mathbf{p}}_0 \mathbf{q}_0 = \begin{bmatrix} 0.1 & 0 & 0 \\ 0 & 1 & 0 \\ 0 & 0 & 0.5 \end{bmatrix} \begin{bmatrix} 450 \\ 21 \\ 60 \end{bmatrix} = \begin{bmatrix} 45 \\ 21 \\ 30 \end{bmatrix} > \mathbf{o}. \tag{2.125}$$

This gives for the *base year* “0” Table 2.14, where the three sectors S_1, S_2, S_3 replace “wheat”, “iron”, “turkeys”. It is a sommaire *national accounting system.*

Table 2.14: Flow of commodities in physical terms in the *base year*.

Commodity sectors	Processing sectors			Final demand	Total
	S_1	S_2	S_3	f_{0i}	(*kCHF*)
S_1	18.6	5.4	3	18	$x_{01} = 45$
S_2	12	6	3	–	$x_{02} = 21$
S_3	4.5	3	7.5	15	$x_{03} = 30$
total (partial)	35.1	14.4	13.5	–	
labour costs W_{0j} (total wages)	9.9	6.6	16.5	$W_0 = D_0 = 33$	
total outlays (*kCHF*)	$x_{01} = 45$	$x_{02} = 21$	$x_{03} = 30$		$X_0 = 129$

Table 2.14 contains the entries of the interindustrial sectors, the sector of final consumer demand, the value added generated by labour and the *total output* of the economy. The table comprises the essence of the *supply-and-use tables* in accordance with Eurostat ([72], p. 21).

Then we calculate the *input-output coefficients matrix* $\mathbf{A}_0$ in monetary terms for the *base year* “0”:

$$\mathbf{A}_0 = \mathbf{Z}_0\hat{\mathbf{x}}_0^{-1}$$
$$= \begin{bmatrix} 18.6 & 5.4 & 3 \\ 12 & 6 & 3 \\ 4.5 & 3 & 7.5 \end{bmatrix} \begin{bmatrix} \frac{1}{45} & 0 & 0 \\ 0 & \frac{1}{21} & 0 \\ 0 & 0 & \frac{1}{30} \end{bmatrix} = \begin{bmatrix} 0.413 & 0.257 & 0.100 \\ 0.270 & 0.286 & 0.1 \\ 0.1 & 0.143 & 0.25 \end{bmatrix}, \quad (2.126)$$

and use the identity (2.113) to recalculate

$$\mathbf{A}_0 = \hat{\mathbf{p}}_0\mathbf{C}_0\hat{\mathbf{p}}_0^{-1}$$
$$= \begin{bmatrix} 0.1 & 0 & 0 \\ 0 & 1 & 0 \\ 0 & 0 & 0.5 \end{bmatrix} \begin{bmatrix} 0.413 & 2.571 & 0.500 \\ 0.027 & 0.286 & 0.050 \\ 0.020 & 0.286 & 0.250 \end{bmatrix} \begin{bmatrix} 10 & 0 & 0 \\ 0 & 1 & 0 \\ 0 & 0 & 2 \end{bmatrix}$$
$$= \begin{bmatrix} 0.413 & 0.257 & 0.100 \\ 0.270 & 0.286 & 0.1 \\ 0.1 & 0.143 & 0.25 \end{bmatrix}, \quad (2.127)$$

recognizing that $\mathbf{A}_0$ and $\mathbf{C}_0$ have the same diagonal elements. Now we perform the **comparative-static analysis.**

Within the considered period, the constant wage rate is increased to attain in the *current year* $w_1 = 0.65 \frac{\text{kCHF}}{\text{man-years}}$, giving naturally $\mathbf{w}_1 = [0.65, 0.65, 0.65]'$, the vector of *wage rates* which we diagonalise. We come back to the labour vector of Example 2.5.2, $\mathbf{L} = [18, 12, 30]'$, with units $[\mathbf{L}]$ = man-year.

From here on, we transform data from the *base year* to the *current year.* We signal this operation by a "0" index on the left-hand side and a "1" on the right-hand side of the equations. So, we restart, calculating the *labour costs* with the unchanged labour vector **L**,

$$\boldsymbol{\upsilon}_1 = \hat{\mathbf{w}}_1\mathbf{L} := w_1\mathbf{L} = \begin{bmatrix} 0.65 & 0 & 0 \\ 0 & 0.65 & 0 \\ 0 & 0 & 0.65 \end{bmatrix} \begin{bmatrix} 18 \\ 12 \\ 30 \end{bmatrix} = \begin{bmatrix} 11.7 \\ 7.8 \\ 19.5 \end{bmatrix}, \quad (2.128)$$

with measurement unit $[\boldsymbol{\upsilon}_1]$ = kCHF. Then, we calculate the *labour costs of each sector per unit of produced quantity* for the *current year* on the basis of the output vector (2.115) of the *base year.*

$$\boldsymbol{\upsilon}_{c,1} = \hat{\mathbf{q}}_0^{-1}.\boldsymbol{\upsilon}_1 = \begin{bmatrix} \frac{1}{450} & 0 & 0 \\ 0 & \frac{1}{21} & 0 \\ 0 & 0 & \frac{1}{60} \end{bmatrix} \begin{bmatrix} 11.7 \\ 7.8 \\ 19.5 \end{bmatrix} = \begin{bmatrix} 0.026 \\ 0.371 \\ 0.325 \end{bmatrix}, \quad (2.129)$$

and again calculate the vector of prices (2.111),

$$\mathbf{p}_1 = ((\mathbf{I} - \mathbf{C}_0)^{-1})' \cdot \boldsymbol{\upsilon}_{c,1} = \begin{bmatrix} 2.16 & 0.0872 & 0.0909 \\ 8.60 & 1.78 & 0.909 \\ 2.02 & 1.77 & 1.45 \end{bmatrix} \begin{bmatrix} 0.026 \\ 0.371 \\ 0.325 \end{bmatrix} = \begin{bmatrix} 0.118 \\ 1.182 \\ 0.591 \end{bmatrix}. \quad (2.130)$$

We determine the *physical units* of $[\mathbf{p}_1] = [\boldsymbol{\upsilon}_{c,1}] = \frac{\text{kCHF}}{\text{unit of commodity}}$; the *physical units* of the elements of the *transposed Leontief Inverse* are neutral, $[((\mathbf{I} - \mathbf{C})^{-1})'] = 1$.

Now we calculate the updated matrix with the diagonalised price vector $\mathbf{p}_1$,

$$\mathbf{Z}_1 = \hat{\mathbf{p}}_1 \mathbf{S}_0 = \begin{bmatrix} 0.118 & 0 & 0 \\ 0 & 1.182 & 0 \\ 0 & 0 & 0.591 \end{bmatrix} \begin{bmatrix} 186 & 54 & 30 \\ 12 & 6 & 3 \\ 9 & 6 & 15 \end{bmatrix} = \begin{bmatrix} 21.98 & 6.38 & 3.55 \\ 14.18 & 7.09 & 3.55 \\ 5.32 & 3.55 & 8.86 \end{bmatrix}, \tag{2.131}$$

and proceed to calculate the vectors of *total output* and of *final demand* (2.115),

$$\mathbf{x}_1 = \hat{\mathbf{p}}_1 \mathbf{q}_0 = \begin{bmatrix} 0.118 & 0 & 0 \\ 0 & 1.182 & 0 \\ 0 & 0 & 0.591 \end{bmatrix} \begin{bmatrix} 450 \\ 21 \\ 60 \end{bmatrix} = \begin{bmatrix} 53.18 \\ 24.81 \\ 35.45 \end{bmatrix} > \mathbf{o},$$
$$\mathbf{f}_1 = \hat{\mathbf{p}}_1 \mathbf{d}_0 = \begin{bmatrix} 0.118 & 0 & 0 \\ 0 & 1.182 & 0 \\ 0 & 0 & 0.591 \end{bmatrix} \begin{bmatrix} 180 \\ 0 \\ 30 \end{bmatrix} = \begin{bmatrix} 21.27 \\ 0 \\ 17.73 \end{bmatrix} \geq \mathbf{o}. \tag{2.132}$$

This gives in a first step the following Table 2.15. Then, the total labour costs are $W_1 = \boldsymbol{\upsilon}_1' \mathbf{e} = 39$ kCHF (a closed economy).

Table 2.15: Flow of commodities in physical terms in the *current year*.

Commodity sectors	Processing sectors S_1	S_2	S_3	Final demand f_{1i}	Total (*kCHF*)
S_1	21.98	6.38	3.55	21.27	$x_{11} = 53.18$
S_2	14.8	7.09	3.55	–	$x_{12} = 24.82$
S_3	5.32	3.55	8.86	17.73	$x_{13} = 35.45$
total (partial)	42.48	17.02	15.95	–	
labour costs W_{1j} (value added)	11.7	7.8	19.5	$W_1 = D_1 = 39$	
total outlays (*kCHF*)	$x_{11} = 53.18$	$x_{12} = 24.82$	$x_{13} = 35.45$		$X_1 = 152.46$

The total consumer demand is $D_1 = \mathbf{f}_1' \mathbf{e} = 39$ kCHF. The *total output* of the economy is $X_1 = x_{11} + x_{12} + x_{13} + W_1 = x_{11} + x_{12} + x_{13} + D_1 = 152.46$ kCHF.

Then we continue to update the matrices and vectors in *physical terms* to the level of the *current year*, using the initial definitions, starting with (2.131); it shows clearly that there is no change in *interindustrial production*,

$$\mathbf{S}_1 := \hat{\mathbf{p}}_1^{-1} \mathbf{Z}_1 = \hat{\mathbf{p}}_1^{-1} (\hat{\mathbf{p}}_1 \mathbf{S}_0) = (\hat{\mathbf{p}}_1^{-1} \hat{\mathbf{p}}_1) \mathbf{S}_0 = \mathbf{S}_0. \tag{2.133}$$

We continue with the *surplus* and *total output*,

$$\mathbf{d}_1 = \hat{\mathbf{p}}_1^{-1}\mathbf{f}_1 = \begin{bmatrix} \frac{1}{0.118} & 0 & 0 \\ 0 & \frac{1}{1.182} & 0 \\ 0 & 0 & \frac{1}{0.591} \end{bmatrix} \begin{bmatrix} 21.275 \\ 0 \\ 17.73 \end{bmatrix} = \begin{bmatrix} 180 \\ 0 \\ 30 \end{bmatrix} = \mathbf{d}_0,$$

$$\mathbf{q}_1 = \mathbf{S}_1\mathbf{e} + \mathbf{d}_1 = \begin{bmatrix} 186 & 54 & 30 \\ 12 & 6 & 3 \\ 9 & 6 & 15 \end{bmatrix} \begin{bmatrix} 1 \\ 1 \\ 1 \end{bmatrix} + \begin{bmatrix} 180 \\ 0 \\ 30 \end{bmatrix} = \begin{bmatrix} 450 \\ 21 \\ 60 \end{bmatrix} = \mathbf{q}_0. \tag{2.134}$$

Now we can calculate the *input-output coefficients matrix* in *physical terms* for the *current year*,

$$\mathbf{C}_1 = \mathbf{S}_1\hat{\mathbf{q}}_1^{-1} = \mathbf{S}_0\hat{\mathbf{q}}_0^{-1} = \begin{bmatrix} 186 & 54 & 30 \\ 12 & 6 & 3 \\ 9 & 6 & 15 \end{bmatrix} \begin{bmatrix} \frac{1}{450} & 0 & 0 \\ 0 & \frac{1}{21} & 0 \\ 0 & 0 & \frac{1}{60} \end{bmatrix}$$

$$= \begin{bmatrix} \frac{31}{75} & \frac{18}{7} & \frac{1}{2} \\ \frac{2}{75} & \frac{2}{7} & \frac{1}{20} \\ \frac{1}{50} & \frac{2}{7} & \frac{1}{4} \end{bmatrix} = \begin{bmatrix} 0.413 & 2.571 & 0.500 \\ 0.027 & 0.286 & 0.050 \\ 0.020 & 0.286 & 0.250 \end{bmatrix} = \mathbf{C}_0. \tag{2.135}$$

Finally, we discover that the *input-output coefficients matrix* $\mathbf{A}_1$, where quantities are expressed in *monetary units*, is invariant in the period from the *base year* to the *current year* (2.127),

$$\mathbf{A}_1 = \mathbf{Z}_1\hat{\mathbf{x}}_1^{-1} = \begin{bmatrix} 21.98 & 6.38 & 3.55 \\ 14.8 & 7.09 & 3.55 \\ 45.32 & 3.55 & 8.86 \end{bmatrix} \begin{bmatrix} \frac{1}{53.18} & 0 & 0 \\ 0 & \frac{1}{24.82} & 0 \\ 0 & 0 & \frac{1}{35.45} \end{bmatrix}$$

$$= \begin{bmatrix} 0.413 & 0.257 & 0.100 \\ 0.267 & 0.286 & 0.1 \\ 0.1 & 0.143 & 0.25 \end{bmatrix} = \mathbf{A}_0. \tag{2.136}$$

The identity (2.113) and the equations (2.116), (2.122) are used to recalculate the equality

$$\mathbf{A}_1 = \hat{\mathbf{p}}_1\mathbf{C}_0\hat{\mathbf{p}}_1^{-1} = \begin{bmatrix} 0.118 & 0 & 0 \\ 0 & 1.182 & 0 \\ 0 & 0 & 0.591 \end{bmatrix} \begin{bmatrix} 0.413 & 2.571 & 0.500 \\ 0.027 & 0.286 & 0.050 \\ 0.020 & 0.286 & 0.250 \end{bmatrix}$$

$$\cdot \begin{bmatrix} 8.46 & 0 & 0 \\ 0 & 0.846 & 0 \\ 0 & 0 & 1.692 \end{bmatrix} = \begin{bmatrix} 0.413 & 0.257 & 0.100 \\ 0.267 & 0.286 & 0.1 \\ 0.1 & 0.143 & 0.25 \end{bmatrix} = \mathbf{A}_0, \tag{2.137}$$

recognizing that $\mathbf{A}_0$ and $\mathbf{A}_1$ are identical as well in this case.

We recognize that the increase of the wage rate from $\mathbf{w}_0$ to $\mathbf{w}_1$ has only an effect on the prices. The quantities produced, the input-output coefficients and the number of workers necessary to realize the production remain unchanged during this period from “0” to “1”.

This example shows how to navigate with matrix algebra and how to realize all the necessary calculations.

Note that some coefficients of matrix $\mathbf{C}_0$ can be larger than one: for instance we get with (2.112) $c_{21} = (p_2/p_1)a_{21} = (1/0.1)0.257 = 2.571 > 1$. This happens because the ratio of the prices $p_i > 0, p_j > 0$ in different *physical units*, p_i/p_j, determines the magnitude of the coefficients $c_{ij} \geq 0$, which may exceed 1. This is not the case for the elements of $\mathbf{A}_0 = \mathbf{Z}_0\hat{\mathbf{x}}_0^{-1}$, Lemma 2.1.1. ▲

Example 2.5.4 (Left to the reader). Additional questions to Example 2.5.3
- What happens, when the increase of the wage rate starts from $\mathbf{w}_0 = [0.55, 0.55, 0.55]'$ with specific sectorial increases according to $\mathbf{w}_1 = [0.65, 0.75, 0.60]'$?
- What happens if in the whole Example 2.5.3 the currency kCHF is replaced by a *numéraire*, like iron?

Table 2.16 compares the results obtained with comparative-static analyses of the Leontief *index price models* based on *monetary terms* and of the Leontief *price models* based on physical terms.

2.6 The interindustrial and the final consumption markets

It is time to provide an overview of the various price models encountered in Section 2.5. For this purpose, we present a synoptic table. As usual, the index "0" is for the *base year* and the index "1" is for the *current year*, see Table 2.15.

In this section, we introduce some notions that we will develop in Chapter 9. In the considered production economies, there is on one hand the market of *interindustrial production*, consisting of the interchange of the *intermediate products* among the n production sectors, and on the other hand there is the *final demand* of all the consumers on the *final consumption market*.[32]

Consider a *Leontief quantity model* in *monetary terms* or in *physical terms*. Then, the market of *interindustrial production* is characterized by the *industrial output*, which is measured in *value terms*, $\mathbf{x}_I = \mathbf{Ze}$, or in *physical terms*, $\mathbf{q}_I = \mathbf{Se}$. The *final consumer demand* on the *final consumption market* is measured by the vector of final demand $\mathbf{f}$ in *value terms* (2.5), or by the vector of final demand $\mathbf{d}$ in *physical terms* (2.15).

In Chapter 9 we will investigate in particular the market of *interindustrial production*. That means, we will set the quantities of the *final consumption market* equal to zero. There will be no produced *surplus* nor *final demand*. We have to set $\mathbf{f} = \mathbf{d} = \mathbf{o}$. The *production cycle* of the economy can be represented twofold, by the matrix $\mathbf{Z}$, giving the presentation in *monetary value terms* of the *means of production*, or by the matrix $\mathbf{S}$, giving the presentation in *physical terms* of the *means of production*.

32 In IOTs, the final consumption market comprises: the consumption of households, government expenditures, investment and export.

Table 2.16: Overview on the Leontief index prices and prices.

Comparative-static analysis with the *cost-push input output price models*		
	Leontief *index prices* based on monetary terms	**Leontief *prices* based on physical terms**
references	Miller and Blair ([65], pp. 45–46)	Oosterhaven ([77], p. 752)
initial	Table 2.1 (general)	Table 2.2 (general)
base year	$\tilde{\mathbf{p}}_{\upsilon,0\vert 0} = [1,1,1]'$ (Subsection 2.5.1)	$\mathbf{w}_0 = [0.55, 0.55, 0.55]'$ (2.118)
current year	$\tilde{\mathbf{p}}_{\upsilon,0\vert 1} = [1.1, 1.2, 1.3]'$ (Subsection 2.5.1)	$\mathbf{w}_1 = [0.65, 0.65, 0.65]'$ (Subsection 2.5.2)
value added	$\boldsymbol{\upsilon}_c = \mathbf{e} - \mathbf{A}'\mathbf{e}$ (2.76)	$\boldsymbol{\upsilon}_0 = \hat{\mathbf{w}}_0\mathbf{L}$ (2.119)
base year	$\boldsymbol{\upsilon}_{c,1} := \hat{\boldsymbol{\upsilon}}_{c,0}\tilde{\mathbf{p}}_{\upsilon,0\vert 1}$ (2.80)	$\mathbf{v}_{c,0} = \hat{\mathbf{q}}_0^{-1}\boldsymbol{\upsilon}_0$ (2.120)
price models	(2.81)	
index prices	$\tilde{\mathbf{p}}_{0\vert 1} = ((\mathbf{I} - \mathbf{A})^{-1})' \cdot \boldsymbol{\upsilon}_{c,1}$	(2.111)
prices	–	$\mathbf{p}_0 = ((\mathbf{I} - \mathbf{C})^{-1})' \cdot \mathbf{v}_{c,0}$
effect on index prices		
base year	$\tilde{\mathbf{p}}_{0\vert 0} = [1,1,1]'$ (Subsection 2.5.1)	
current year	$\tilde{\mathbf{p}}_{0\vert 1} = [1.24, 1.23, 1.28]'$ (2.100)	
effect on prices		
base year		$\mathbf{p}_0 = [0.1, 1, 0.5]'$ (2.122)
current year		$\mathbf{p}_1 = [0.118, 1.182, 0.591]'$ (2.130)
measurement units	$[\tilde{\mathbf{p}}_{0\vert 0}] = [\tilde{\mathbf{p}}_{0\vert 1}] = 1$	$[\mathbf{p}_0] = [\mathbf{p}_0] = \frac{\text{kCHF}}{\text{units of commodity}}$

This gives the complete *production cycle* as illustrated by Figure 2.2 in a twofold manner: as a cycle in *physical terms* and as a cycle in (monetary) *value terms* with the *interindustrial market* and the *final consumption market*. The *interindustrial market* has to generate the quantity $\mathbf{q}_I = \mathbf{Se}$, necessary to replace the *means of production* used by the technology in the present period, described by the *commodity flow* matrix $\mathbf{S}$. The same *means of production* are presented by the cycle in *value terms* and have the monetary value $\mathbf{x}_I = \mathbf{Ze}$. In the case of self-replacement, the *interindustrial economy* produces the quantity $\mathbf{q} = \mathbf{q}_I + \mathbf{d} \geqq \mathbf{q}_I$, whose value is $\mathbf{x} = \mathbf{x}_I + \mathbf{f} \geqq \mathbf{x}_I$. So, there can be described at a second stage the *final consumption market*, where the *non-negative* surplus quantity vector $\mathbf{d} = \mathbf{q} - \mathbf{Se} \geqq \mathbf{o}$ appears. Presented in *value terms*, this vector becomes the *non-negative* vector of *final demand* $\mathbf{f} = \mathbf{x} - \mathbf{Ze} \geqq \mathbf{o}$.

The commodities are produced in different sectors. At the end of the production process a part of them is exchanged in the interindustrial market at the costs of production, clearing this market. The surplus becomes consumption goods for the consumers and is treated at the costs of production in the final consumption market.

All the material necessary for production comes from the environment. After having been used, it goes back into the environment. Thus, the production cycle is closed,

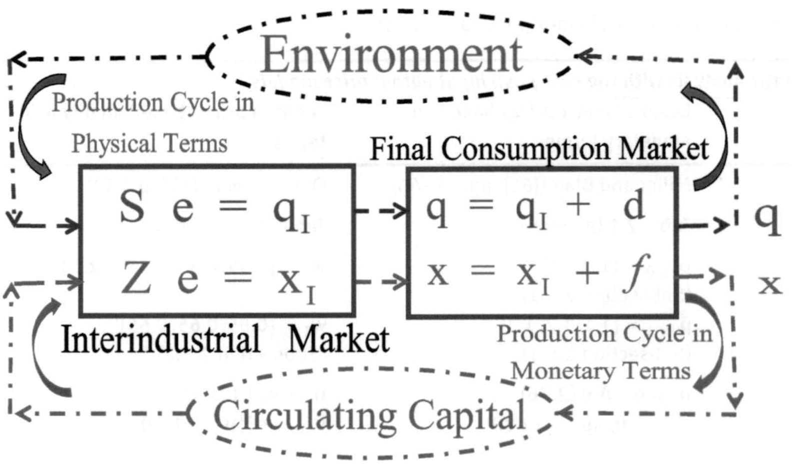

Figure 2.2: Production cycle in monetary and physical terms.

because there are no other sources of material for the production. The same cycle has the second presentation in *value terms*, generating the *circulating capital* K.[33]

Note in passing that using money is not mandatory. The Input-Output framework can be usefully extended by employing a *physical numéraire* for models based exclusively on physical units. In such areas as in industrial ecology or in ecological economy, and as Sraffa did it extensively in PCMC (see the next Chapter 4), a *physical numéraire* is used.

Summarizing, Input-Output Analysis presents the sectorial structure of an economy and is primarily interested in calculating the *total output* of an economy based on the *interindustrial demand* of a sectorial industrial production. It also presents the *final demand* of the consumers.

Finally, if one aims at operating within the framework of national accounts, or more generally social accounting, one must resort to values expressed in numerical monetary units calibrated to the corresponding currency. In short, Input-Output Analysis in monetary values is a model of national accounting. This is the subject of the next section.

2.7 The Input-Output framework and National Accounting

The *national accounting identities* in macroeconomics are fundamental mathematical relationships expressing the symmetry of double-entry bookkeeping which registers incomes and expenditures at given conventional calender dates (usually quarterly or annually) in an ongoing economic process. National accounts are thus a by-product

33 Remark: In every modern economy, the *money-flow cycle* goes in the inverse direction of the *production cycle* in *value terms*. The *money-flow cycle* is not treated in this text.

of the economic process. To make sense, macroeconomic models of production and exchange must be consistent with these identities.

Input-Output Tables (IOT) provide the data required to set up the national accounts in the first place. In fact, the IOT model represents the most detailed analysis of an ongoing economic process of production and exchange in value terms, i. e., the production and use of commodities and the income generated in that process. The compilation of IOT tables is a complex statistical exercise, and one must bear in mind that the numerical results finally published in the official national accounts are approximations of what has happened in the economic process during the reporting period. For expository reasons, we limit ourselves here to presenting the essentials for a conceptual understanding. In practice, IOT and national accounts are far more elaborated, as we shall see later in Chapter 10, with the Swiss IOT 2008.

So by construction, the IOT model is consistent with the *national accounting identities* and in the subsequent chapters we shall see that, for a closed economy, this is also the case for Sraffa's model presented in PCMC. It is therefore essential to present these identities that are implicit in all the following developments.

Three vector entities, expressed here in value terms, i. e., quantities multiplied by a price, and their numerical aggregates which we have already encountered, intervene here (see Table 2.1): Consider the vector $\mathbf{\upsilon}$ of *value added*, defined by the adopted model, the vector $\mathbf{f}$ of *final demand* and the vector $\mathbf{x}$ of *total output*.[34]

These vectors can be aggregated, giving V = *total value added*, F = *total final demand* and X = *total output*;[35]

$$\begin{aligned} &\mathbf{\upsilon}; && \text{aggregate to} && V = \mathbf{e}'\mathbf{\upsilon},\\ &\mathbf{f} = \hat{\mathbf{d}}\mathbf{p}; && \text{aggregate to} && F = \mathbf{e}'\mathbf{f},\\ &\mathbf{x} = \hat{\mathbf{q}}\mathbf{p}; && \text{aggregate to} && X = \mathbf{e}'\mathbf{x}. \end{aligned} \tag{2.138}$$

As already mentioned, Subsection 2.1.1, the value added components υ_j comprise such essential items as[36]:

W: wages and pensions;
P: gross entrepreneurial profits or, more precisely, EBITDA;
R_L: rental incomes, in particular land rents;
M: imports (in open economies).

---

34 Depending on what circumstances in which these entities are considered, we deal *ex ante* for example with either expected, projected or planned values for forecasting, *ex post* with statistical values collected for reporting purposes.

35 We are dealing here with single product industries. In joint production, see Chapter 6, $\mathbf{f}$ is replaced by a matrix, unless final demand per industry is defined as the combined output (commodity mix) of that industry.

36 One distinguishes between competitive imports (imports of commodities in competition with the same commodities produced domestically) and non-competitive imports (for which there is no domestic source). In this book, we shall consider exclusively the latter in Section 8.1.

By aggregation, passing by sub-summation over these various items, one obtains *total value added*

$$V = W + P + R_L + M. \tag{2.139}$$

From macroeconomics, we know that final demand may be split into essential sales components external to interindustry sales:
C: private consumption;
I: private investments;
G: government expenditures;
E: exports.

Again by aggregation and sub-summation, one obtains *total final demand*

$$F = C + I + G + E. \tag{2.140}$$

Now as may be seen from Table 2.1 we have for *total output*,

$$X = \sum_{i=1}^{n} x_i + F, \tag{2.141}$$

and for *total outlays*,

$$X = \sum_{j=1}^{n} x_j + V. \tag{2.142}$$

Finally from equations (2.2) and (2.3) the identity $V = F$, written out, gives

$$P + W + R_L + M = C + I + G + E. \tag{2.143}$$

Introducing by definition a new important entity Y, and setting $R_L = 0$ for the time being, we have,

$$Y \equiv P + W = C + I + G + (E - M) = \text{GDP}. \tag{2.144}$$

The *Gross Domestic Product* (GDP) measures the total income produced domestically, see Mankiw ([63], p. 27). The equation GDP $= C + I + G + (E - M)$ is the "expenditure method" to obtain GDP, whereas the equation $Y = P + W$ comes from the "National Income Account" and is equal to *Gross National Income* Y.[37] In the present text, Y is a proxy for GDP.[38]

---

37 A further identity (not used in the sequel) is $Y = C + S + T_R$, where S designates savings and T_R tax revenues. Equated to the right-hand side of (2.143), this gives $S + T_R = I + G$ or $S = I + (G - T_R)$, i. e., total investments. In a strictly balanced government budget, $G = T_R$ and therefore $I = S$, an identity of historical fame.

38 The national accounting identity can be refined. For our purposes we shall however retain this presentation.

For a closed economy (as expressed in Table 2.1), no activity is conducted with outside economies. A closed economy is self-sufficient: No imports are brought in, no exports are sent out. We have: $E = M = 0$, so

$$\boxed{Y = W + P = C + I + G.} \tag{2.145}$$

In other words, Y is the *social surplus* which will be referred to in Chapter 10 in more elaborate IOTs, especially the SWISS IOT 2008.

In conclusion, by examining IOTs in connection with national accounting, four fundamental entities appear in the periodic macroeconomic description of a monetary economic production and exchange process, which until further notice we consider as closed:

1. National Income;
2. Gross Domestic Product (GDP);
3. Final Demand;
4. Total Value Added.

The points 1. and 2. are national accounting items and constitute the basic national accounting identity established *ex post* for each accounting period:

$$\text{national income} = \text{GDP}. \tag{2.146}$$

The points 3 and 4 represent *national accounting identities* which govern the production process during the period under consideration and in fine determine prices.[39] For a given period, Table 2.1 can accordingly be interpreted in two ways:

A. As a dynamic table describing the ongoing economic process, in which prices adjust to demand and value added items, in particular wages (labour).
B. As an accounting table, established ex post, covering values registered over a complete period.

2.8 Measuring value: the numéraire

The reader has undoubtedly realised that measurement issues are central to the construction of an IOT and reporting for national accounting. In fact, we live in a world of measurements, a translation of our observations (here about the economic world) into numerical form for calculation. We are mapping the observed world to an artificial one in which we can apply mathematical tools such as those presented in this text.

---

39 A word of caution: In a closed economy with $E = M = 0$, aggregate final demand F is equal to *national income* Y. This is usually not the case in an open economy: aggregate exports will of course be included in F and Y. But as aggregate imports M are regarded, they will figure in the aggregate value added V. Then, simplifying $V = P + W + M = F$, one obtains $Y = P + W = F - M$ $(M > 0)$ in an open economy. Note that M will usually be split into a part M_I entering *inter industry production* and a part M_F going to *final consumer demand*, both catered to by import activities.

Measurements are also conventions, as we have seen in this section. For deeper understanding, we invite the reader to become thoroughly familiar with measurement issues, such as those superbly presented by D. J. Hand [40].

This being said, we have seen that the quantities entering an IOT are measured in various units depending on the commodity involved. For example: quarters of wheat, tons of iron, litres of water, barrels of petrol, kWh of electricity, heads of cattle, man-hours of labour, etc. Note that in the context under consideration commodities include products (or goods) and services.

Obviously, it is possible to construct vectors $\mathbf{q} = [q_1, q_2, \dots, q_n]'$, the components q_i of which represent quantities of commodities, each with its specific unit of measurement. But it is as obvious that it is logically impossible to define an aggregate "bundle" of such commodities by simply adding the quantities expressed in their specific units. This type of aggregation or summation, involving linear equations, is only meaningfully possible by expressing each quantity in a common unit of measurement providing a uniform measure of value to each commodity. This is a basic tenant of dimensional analysis, see Chable [13], or de Jong [24]. All linear equations used in economic analysis must be tested according to this criterion.

We refer to a particular good as the *numéraire*, typically wheat. One then says that all other prices are normalized by the price of that good. For example, set as a unit of wheat: "*1 quarter of* wheat". This unit is then the *monetary unit*, a monetary numéraire or an appropriate physical unit (physical numéraire). The latter expression is not in the commodity as such, but in the measurement unit of that commodity (see Schmitt in Gnos and Rossi [95], pp. 36–37, [38]), e. g., not in wheat, but in *quarters of* wheat.

A monetary accounting unit is usually expressed in a specific currency: USD, EURO, CHF, etc. or, even more exotically, in *bitcoins*. Such accounting units do not constitute money, rather they are just a numerical counter. If money is taken as the numéraire, one implicitly assumes a sophisticated banking system underlying the monetary economy of production and exchange. Money becomes then a twofold inseparable entity formed of money as such (the numerical form) and bank deposits (the economic substance).[40]

The price of a commodity is then defined as an equivalence relation of the type: "1 quantity unit of commodity x is equivalent to n_y units of numéraire y".

Example 2.8.1. The quantity of 1 ton of iron equals 10 quarters of wheat. The relative price of iron is $p_{\text{iron}} = 10\frac{\text{qr. of wheat}}{\text{tons of iron}}$ and for wheat $p_{\text{wheat}} = 1\frac{\text{qr. of wheat}}{\text{qr. of wheat}} = 1$, in this latter case a dimensionless entity. ▲

40 This highlights the increasing rift between the "real" *production economy* reflected in the national accounts expressed in money as numéraire and the surging "virtual" world of *financial markets* and commodity trading, where transactions and accounts are expressed in numerical accounting units, an accounting numéraire accepted at face value, see Chesney [18].

Formally, the *value* of a given quantity of a commodity is then the quantity of that commodity multiplied by its price, expressed in terms of the chosen numéraire, and is usually designated as the cost of that specific quantity.

Example 2.8.2. The price of a barrel of petrol is, say $p_{\text{barrel}} = 40\ \frac{\text{USD}}{\text{barrel}}$, the quantity $M = 10$ barrels of petrol costs $C = p_{\text{barrel}} \times M = 10 \times 40 = 400$ USD, the currency USD taken as accounting numéraire. ▲

In input-output analysis, the link with the national accounting identities is obtained by expressing all transactions in value terms expressed in the currency related to the corresponding national economy, i. e., in terms of money as the numéraire.

As for Sraffa, and as mentioned already, his prices are expressed in a physical numéraire, but he goes on to adopt in PCMC, Par. 12 and Par. 34, as the numéraire a composite commodity: *national income* expressed in *physical terms* normalized to 1, taken as unit of measurement, i. e., the *National Income Unit* (NIU), to express wages and prices. The reason of such a choice will appear in Chapter 5 below.

3 Sraffa's first examples of single- product industries

We first concentrate our attention on Sraffa's subject of PCMC and summarize the main results. Sraffa's price model brings, within a cyclic production process of n sectors and n commodities, the 'costs of production' of every commodity, termed as 'prices', with positive wages for workers and with positive profits of entrepreneurs into an equilibrium. If the interindustrial market accepts the 'costs of production', then the means of production are exchanged at the end of the period and the production cycle continues. If there is no surplus, then there is no consumption market. If there is a surplus, then Sraffa's prices apply to the commodities, which are used as consumption goods in the adjacent consumption market, according to the existing demand. In this sense Sraffa brings together the interindustrial market and the consumption market. This theme will now be developed from the very beginning, presenting at first Sraffa's elementary examples. We will also learn that Sraffa's models are accountable balanced. This is a reason why the treatment of market prices, resulting from an intersection of a demand and a supply curve is out of scope of Sraffa's model. Indeed, Sraffa does not treat market prices.

3.1 Production for subsistence and production with surplus

Sraffa starts in the first and second chapters of PCMC (Par. 1, Par. 2, Par. 5) with elementary *production economies* of *single-product industries*[1] for partial production systems of extremely simple societies, contrary to IOTs which are developed for whole countries; the aim is to determine *production prices*. Oosterhaven [77] calls such a system a *price model*; we specify *it as a Sraffa price model*. As every one of the n industries (or branches) produces exactly one of the n commodities, Bertram Schefold ([103], p. 13) specifies a *single-product Sraffa system*. Sraffa ultimately solved with his price model the economically important question of the *transformation problem*, i. e., transformation of *labour values* (4.184), a notion going back to Adam Smith [106], into prices (see Bortis [6], pp. 67–68, [7] and Pasinetti [80], pp. 122–150).

We treat now the elementary examples of PCMC (Par. 1, Par. 2, Par. 5) the first industry producing only wheat, the second only iron and the third only pigs. Only wheat, iron and pigs are used as means of production. In addition, sustenance for workers is provided as a certain amount of wheat, and, at the end of each production period (say one year), some amounts of goods are interchanged between the producers. Production appears as a social and circular process. Hence the concept of *social surplus* or, more concisely, a *surplus* emerges, which was important in classical theory (from

1 We repeat that the terms "product" and "commodity" are used synonymously in this text.

https://doi.org/10.1515/9783110635096-003

William Petty to David Ricardo). There is a *social surplus* if some of the quantities produced are greater than the quantities used up in the process of production. The surplus is conceived by Sraffa as a *semi-positive* vector of *surplus* $\mathbf{d} \geq \mathbf{o}$. Sraffa however begins with a simplified situation, where the quantities produced and the quantities used are exactly equal, $\mathbf{d} = \mathbf{o}$. He calls such processes *production for subsistence*. There is accordingly no surplus.[2]

Formally, Sraffa's production technology is described by the *semi-positive commodity flow* matrix $\mathbf{S} \geq \mathbf{O}$ (2.13) and the vector of *total output* $\mathbf{q} = \mathbf{Se} + \mathbf{d} > \mathbf{o}$. Throughout this chapter, Assumption 2.2.1, Assumption 2.2.2 hold.

Bertram Schefold ([109], pp. 216–225) published, in the German translation of PCMC, a short description of the mathematical background of the single-product Sraffa system, revealing that its solution leads to eigenvalue–eigenvector problems in connection with the theorem of Perron–Frobenius (1907, 1912), as is the case for various Leontief models, treated in Chapter 2. In this text, we will often refer to the theorem of Perron–Frobenius and use matrix algebra to solve the numerical problems arising in the framework of single-product Sraffa systems.

In fact, PCMC evokes four *distribution scenarios*, which will be examined in detail in this and the following chapters:

(1) *Garden of Eden economy*. There is no economic surplus. Entrepreneurs and workers form one and only one group of households living on a bundle of subsistence commodities produced by the economy. The quantities of commodities are fully absorbed by households who work to produce the amounts required in the next production period, thus attaining the goal of sustainability of the economy, which is thus self-replacing, see Schefold [103], p. 49.

(2) *Exploitation of a labour economy*. There is an economic surplus. Entrepreneurs hoard the surplus generated by the economy at the expense of workers, leaving them just with the bundle of commodities required for survival (subsistence wages paid in kind), which is, in fact, a slave economy.

(3) *Domination of a labour economy*. There is an economic surplus. All the surplus goes to workers who act in regard to producers in the same way as producers in scenario (2). The means of production are fully in the hands of workers. Entrepreneurs can contribute to the economy through the proceeds of their private estates.

(4) *Uniform distributive economy*. There is an economic surplus. The surplus is distributed between producers and workers, but there is no specification of the rules applicable for such distribution. The *wage rates* w of the workers are the same in all the sectors of production, as well as the *rates of profits* r are the same for all entrepreneurs. The means of production remain in the hands of producers, which grants them leverage in bargaining situations with workers.

2 Sraffa's first examples contain no *surplus* and no *labour in man-years*.

3.1.1 Sraffa's first example: conditions of production

Let's present Sraffa's first example of a production economy limited to wheat and iron. We will go through Sraffa's first example, giving detailed comments.

Example 3.1.1 ((PCMC, Par. 1)). Suppose at first that only two commodities are produced, wheat and iron. Both are used, in part as sustenance for those who work, and for the rest as means of production—wheat as seed, and iron in the form of tools. Suppose that, all in all, and including the necessaries for the workers, 280 quarters of wheat and 12 tons of iron are used to produce 400 quarters of wheat, while 120 quarters of wheat and 8 tons of iron are used to produce 20 tons of iron. The production process[3] (indicated by arrows) is symbolised as follows: the first line representing the wheat production, the second line representing the iron production:

$$\begin{aligned}&(280\text{ qr. wheat}, 12\text{ t. iron}) \rightarrow (400\text{ qr. wheat}, 0),\\ &(120\text{ qr. wheat}, 8\text{ t. iron}) \rightarrow (0, 20\text{ t. iron}).\end{aligned} \tag{3.1}$$

Nothing is added by the production[4] to the wealth of society as a whole. Four hundred qr. of wheat and 20 t. of iron have been used (during an annual period), and the same quantities are produced (during this same period). In one year, the entire input into each industry for that year is used for production requirements, and the initial amount for each industry is reproduced as output of that year. There is *no surplus*.

Solution to Example 3.1.1:
Sraffa (PCMC, Par. 1) notes, without presenting the calculations, that

> *"there is a unique set of exchange-values which if adopted by the market*[5] *restores the original distribution of the products and makes it possible for the process to be repeated. In the particular example the required exchange value is 10 qr. of* wheat *for 1 t. of* iron."

This text is absolutely essential for the understanding of Sraffa's price model. In addition a comment: In Chapter 9 we will describe the concept of the *interindustrial market*, especially the algebraic properties in the background. Indeed, if the *interindustrial market* adopts these prices, then the production technology, described by the means

3 To be more precise, this represents an Input-Output process of production, inputs on the left-hand side, outputs on the right-hand side. Schefold's notation ([102], p. 7) has been adopted to describe the production process.

4 Sraffa's representation (3.1) of the production process must be read from left to right. It is a representation of a *scheme of production*, a production process. The numbers that appear in this scheme will be subject to calculations and mathematics, essentially linear algebra. In the present text, the mathematics for the calculation of all the amounts that enter Sraffa's production processes will be fully developed.

5 Sraffa evokes here the *interindustrial market* which has to adopt 'exchange-values' or 'prices'. In PCMC, Par. 7, he will give different possible terms for these prices. He says: *"Such classical terms as 'necessary prices', 'natural price' or 'price of production' would meet the case, but value and price have been preferred as being shorter and in the present context (which contains no reference to market prices) no more ambiguous."* Throughout this text we will apply and use the term 'price', adopted by Sraffa.

of production of the actual or present period, is recreated for the next period, ensuring the same technology, and this from period to period. Obviously, these "exchange values", "prices" or "production costs" do not depend on subjective preferences, but only on the technology of production.

It becomes evident that we may have to distinguish between different equilibrium processes: here, the *technological adaption process* for the determination of the Sraffa 'prices', also called 'costs of production', and other equilibrium processes, as the *behavioural demand-supply adaption process* for the determination of the Walras 'prices', often assimilated to 'market prices', see also Section 8.8. In this text we exclusively treat Sraffa 'prices'.

Now, we return to our example. Considering a quarter of wheat as the measure of exchange (numéraire), i. e., a measure expressed in *physical terms*,[6] the exchange value corresponds to the relative price for wheat as $p_1 = 1(\frac{\text{qr. wheat}}{\text{qr. wheat}}) = 1$ (dimensionless) and the relative price for iron as $p_2 = 10(\frac{\text{qr. wheat}}{\text{t. iron}})$, which we will use to write the production equations.

It is now easily verified that these prices set both processes in equilibrium, satisfying the following first two equations. In the third and the fourth equations the prices have been introduced and the equilibrium values calculated in "qr. wheat":

$$\begin{aligned}
&280\text{ (qr. wheat)}p_1\left(\frac{\text{qr. wheat}}{\text{qr. wheat}}\right) + 12\text{ (t. iron)}p_2\left(\frac{\text{qr. wheat}}{\text{t. iron}}\right)\\
&\quad = 400\text{ (qr. wheat)}p_1\left(\frac{\text{qr. wheat}}{\text{qr. wheat}}\right),\\
&120\text{ (qr. wheat)}p_1\left(\frac{\text{qr. wheat}}{\text{qr. wheat}}\right) + 8\text{ (t. iron)}p_2\left(\frac{\text{qr. wheat}}{\text{t. iron}}\right)\\
&\quad = 20\text{ (t. iron)}p_2\left(\frac{\text{qr. wheat}}{\text{t. iron}}\right) \quad \Rightarrow\\
&280 \cdot 1\text{ qr. wheat} + 12 \cdot 10\text{ qr. wheat} = 400 \cdot 1\text{ qr. wheat},\\
&120 \cdot 1\text{ qr. wheat} + 8 \cdot 10\text{ qr. wheat} = 20 \cdot 10\text{ qr. wheat.}
\end{aligned} \tag{3.2}$$

We now analyse the steps to the next period. Sector 1 acquires 12 *t.* iron and pays for it with 120 *qr.* wheat to Sector 2. On the other hand, Sector 2 sells 12 *t.* iron retaining the needed 120 qr. wheat, and 8 t. iron remain in Sector 2. These transactions are possible because we have equilibrium prices with relation $p_2/p_1 = 10$.

Here Sraffa ends the explanation of his first numerical example. We intend to unravel the *matrix algebra* behind this example. Matrix algebra is much more powerful than only numerics to explain and set the structure of a problem. We will then be able to generalize and define a class of problems to be solved with the presented mathematical methods.

6 The *numéraire* is here the unit "*qr.* wheat", which is then multiplied by a number. Only after introduction of this common measure of exchange, i. e., the *numéraire,* can the process be expressed in terms of equations.

At first, we complete Sraffa's indications in PCMC, Par. 1, and use elementary matrix algebra to calculate the relative prices p_1, p_2.

Let us identify the matrices. From the given data, one finds the *commodity flow* matrix as in (2.13),

$$\mathbf{S} = \begin{bmatrix} s_{11} & s_{12} \\ s_{21} & s_{22} \end{bmatrix} = \begin{bmatrix} 280 & 120 \\ 12 & 8 \end{bmatrix}, \tag{3.3}$$

in physical terms. Then we calculate the *total output* vector,

$$\mathbf{q} = \begin{bmatrix} q_1 \\ q_2 \end{bmatrix} = \mathbf{Se} = \begin{bmatrix} 280 + 120 \\ 12 + 8 \end{bmatrix} = \begin{bmatrix} 400 \\ 20 \end{bmatrix} > \mathbf{o}, \tag{3.4}$$

in physical terms. The price vector $\mathbf{p} = [p_1, p_2]'$ has to be determined.

Sraffa's price model (3.2) can then be written as follows: first, the *left side*, then the *right side*, giving useful matrix equations,

$$\begin{aligned} \begin{bmatrix} 280p_1 + 12p_2 \\ 120p_1 + 8p_2 \end{bmatrix} &= \begin{bmatrix} s_{11} & s_{21} \\ s_{12} & s_{22} \end{bmatrix} \begin{bmatrix} p_1 \\ p_2 \end{bmatrix} = \begin{bmatrix} 280 & 12 \\ 120 & 8 \end{bmatrix} \begin{bmatrix} p_1 \\ p_2 \end{bmatrix} = \mathbf{S}'\mathbf{p} \\ &= \begin{bmatrix} 400p_1 \\ 20p_2 \end{bmatrix} = \begin{bmatrix} p_1 & 0 \\ 0 & p_2 \end{bmatrix} \begin{bmatrix} 400 \\ 20 \end{bmatrix} \\ &= \begin{bmatrix} p_1 & 0 \\ 0 & p_2 \end{bmatrix} \begin{bmatrix} 280 & 120 \\ 12 & 8 \end{bmatrix} \begin{bmatrix} 1 \\ 1 \end{bmatrix} = \hat{\mathbf{p}}\mathbf{Se}, \\ &\Rightarrow \mathbf{S}'\mathbf{p} = \hat{\mathbf{p}}\mathbf{Se} = \hat{\mathbf{p}}\mathbf{q}. \end{aligned} \tag{3.5}$$

We now have Sraffa's *price model* (3.2) in condensed matrix form, where every industry produces exactly one commodity. Later, this *Sraffa price model* will be called a *single-product Sraffa system*. Remember that Sraffa's first example (PCMC, Par. 3) contains no *surplus*. The vector of *final demand* vanishes, $\mathbf{d} = \mathbf{o}$. We then refer to equation (2.15) $\mathbf{q} = \mathbf{Se} + \mathbf{d}$. Therefore, for the vector of *total output* of a production economy with no surplus,[7] we have,

$$\mathbf{q} = \mathbf{Se} > \mathbf{o}.^{8} \quad \blacktriangle \tag{3.6}$$

Here we make a formal link to the presentation of the values in *monetary terms*. We can establish the following equality:

$$\hat{\mathbf{p}}\mathbf{Se} = \hat{\mathbf{p}}(\mathbf{Se}) = \hat{\mathbf{p}}\mathbf{q} = \begin{bmatrix} q_1 p_1 \\ q_2 p_2 \end{bmatrix} = \begin{bmatrix} x_1 \\ x_2 \end{bmatrix} = \mathbf{x}, \tag{3.7}$$

which define the total output $\mathbf{x}$ in *value terms*.

7 We are dealing here with a closed economic process in a self-replacing state, embedded in a natural environment and operating over the short and medium terms. Such processes are viable, contrary to closed self-replacing processes in thermodynamics that contradict the *Second Law* and cannot subsist. Note that we are not dealing here with the very long-term perspective adopted by Georgescu-Roegen [35].

8 If we want to emphasize that there is no surplus, we write $\mathbf{q}_I = \mathbf{Se}$ (the Index I means *an interindustrial economy*).

As there is no *final demand* in physical terms, i. e., $\mathbf{d} = \mathbf{o}$, there is with $f_i = p_i d_i$ (2.105) also no *final demand* in value terms, i. e., $\mathbf{f} = \mathbf{o}$. One then derives, with equations (3.5) and (2.18),

$$\mathbf{Z} = \hat{\mathbf{p}}\mathbf{S} \Rightarrow \hat{\mathbf{p}}\mathbf{S}\mathbf{e} = \hat{\mathbf{p}}\mathbf{q} = \mathbf{Z}\mathbf{e} = \mathbf{x}, \tag{3.8}$$

giving here a link between presentations of the data in *physical terms* and in *value terms*. This gives also the link between the monetary circle and the production circle of the economic process, visualized in Figure 2.2. This subject will be treated algebraically in Chapter 9. Note that at this stage Sraffa's construction operates only within the *price model* based on physical terms.

With elementary rules of matrix algebra using (3.5), one finds

$$\mathbf{S}'\mathbf{p} = \hat{\mathbf{p}}\mathbf{q} = \hat{\mathbf{q}}\mathbf{p} = \mathbf{x}. \tag{3.9}$$

Alternatively, taking the transposed equation on the left of (3.9):

$$\boxed{\mathbf{S}'\mathbf{p} = \hat{\mathbf{q}}\mathbf{p} = \mathbf{x} \quad \text{or} \quad \mathbf{p}'\mathbf{S} = \mathbf{p}'\hat{\mathbf{q}} = \mathbf{x}'.} \tag{3.10}$$

Sraffa (PCMC, Par. 3) termed the left equation (3.10) the *conditions of production*, which we formulate here in two steps:

Definition 3.1.1. In order to guarantee sustainability of a production process (a self-replacing process), the following equation must hold in terms of quantities: $\mathbf{S}\mathbf{e} = \mathbf{q}$,[9] more explicitly written as

$$(\hat{\mathbf{q}} - \mathbf{S})\mathbf{e} = \mathbf{o} = \mathbf{d}, \quad \text{no surplus.} \quad \blacktriangle \tag{3.11}$$

We will analyse some implications of Definition 3.1.1.

Equation (3.11) represents a *homogeneous system of linear equations* in a n-dimensional vector space. The $n \times 1$ summation vector $\mathbf{e} = [1, \ldots, 1]'$ spans a vector subspace with $\dim(\mathbf{e}) = 1$. Therefore, the matrix $\hat{\mathbf{q}} - \mathbf{S}$ has rank $n - 1$ because of the property $n - \text{rank}(\hat{\mathbf{q}} - \mathbf{S}) = \dim(\mathbf{e}) = 1$ (see Nef [69], Theorem 1, p. 121). In other words, we have,

$$(\hat{\mathbf{q}} - \mathbf{S})\mathbf{e} = \mathbf{o} \Rightarrow \text{rank}(\hat{\mathbf{q}} - \mathbf{S}) = n - 1 \Rightarrow \det(\hat{\mathbf{q}} - \mathbf{S}) = 0. \tag{3.12}$$

This obviously means that the n equations (3.11) are not linearly independent.

Now we come to the conditions of production expressed in value terms:

Definition 3.1.2. The conditions of production in value terms mean that, given the vector of total output $\mathbf{q} = \mathbf{S}\mathbf{e} > \mathbf{o}$, the price vector $\mathbf{p}$ is determined by the equation $\mathbf{p}'\mathbf{S} = \mathbf{p}'\hat{\mathbf{q}}$ (or transposed $\mathbf{S}'\mathbf{p} = \hat{\mathbf{q}}\mathbf{p}$). ▲

9 This is nothing other than "Say's Law" for this type of productive economic activity. Jean-Baptiste Say (1767–1832) formulated the law of markets found in *classical economics*, which states that aggregate production necessarily creates an equal quantity of aggregate demand.

We see that the expressions "An economy fulfills *Sraffa's conditions of production*" and an "*economy is just viable*" (Definition 2.2.1) are equivalent statements.

Starting from (3.11), we distinguish three subsequent cases of *self-replacement*, in analogy to Schefold ([103], p. 49). We calculate the vector of *surplus* $\mathbf{d} = (\hat{\mathbf{q}} - \mathbf{S})\mathbf{e}$, also termed vector of *net product*, and we have

$$\begin{aligned} \mathbf{d} = \mathbf{o}: &\quad \text{no surplus,} \\ \mathbf{d} \geq \mathbf{o}: &\quad \text{self-replacement,} \\ \mathbf{d} > \mathbf{o}: &\quad \text{positive self-replacement.} \end{aligned} \tag{3.13}$$

Clearly, the case of no surplus is equivalent to a *just-viable economy*, Definition 2.2.1, the cases of self-replacement and positive self-replacement are equivalent to a *viable economy*.

We shall now proceed with the formal development of further implications of Sraffa's *conditions of production*, expressed by equation (3.10), leading to a first version of the seminal *Sraffa price model*. Matrix algebra is applied.[10]

We have to multiply the equation on the right side of (3.10) by the diagonal matrix $\hat{\mathbf{q}}^{-1}$. Using elementary rules of *matrix algebra*, especially about diagonal matrices, and using the *input coefficients matrix in physical terms* given in (2.16), i. e., $\mathbf{C} = \mathbf{S}\hat{\mathbf{q}}^{-1} \geq \mathbf{0}$, one finds,

$$\begin{aligned} (\mathbf{p}'\mathbf{S})\hat{\mathbf{q}}^{-1} = \mathbf{p}'(\mathbf{S}\hat{\mathbf{q}}^{-1}) = \mathbf{p}'\mathbf{C} = (\mathbf{p}'\hat{\mathbf{q}})\hat{\mathbf{q}}^{-1} = \mathbf{p}'(\hat{\mathbf{q}}\hat{\mathbf{q}}^{-1}) = \mathbf{p}' \\ \mathbf{p}'\mathbf{C} = \mathbf{p}' \Leftrightarrow \mathbf{C}'\mathbf{p} = \mathbf{p}, \end{aligned} \tag{3.14}$$

which is therefore equivalent to the *left eigenvector* equation,

$$\mathbf{p}'\mathbf{C} = \mathbf{p}', \tag{3.15}$$

with *left eigenvector* $\mathbf{p}$ for the matrix $\mathbf{C}$ associated to the Frobenius number $\lambda_C = 1$.[11]

In the first numerical example (PCMC, Par. 1), Sraffa has hidden the calculation of the prices as a solution to a specific *left eigenvector equation*.

The question arises under what conditions does this eigenvector problem (3.15) have the desired solution of a positive *price vector* $\mathbf{p} > \mathbf{o}$. The answer is of fundamental importance, constituting the core of the various representations of the economic process of production examined in this text: the *Leontief models*, the *Sraffa price model* and the Weintraub model, see Section 8.4.

To our knowledge, in the year 1976, Bertram Schefold seemingly was the first economist to justify the occurrence of positive prices in *Sraffa price models* by the

---

10 Nowhere does Sraffa treat in PCMC the mathematics operating on the background of his *price model* $\mathbf{p}'\mathbf{S} = \mathbf{p}'\hat{\mathbf{q}}$ which guarantees the existence of strictly positive price vectors $\mathbf{p} > \mathbf{o}$.

11 For calculation purposes, we will however in general take the equivalent equation for right eigenvectors $\mathbf{C}'\mathbf{p} = \mathbf{p}$ obtained in (3.14).

Perron–Frobenius theorem A.9.3, a fundamental theorem that governs this field, (see his comments in the German edition of PCMC ([109], pp. 216–225)).[12]

After these explanations, we concentrate again on the calculation of the prices of Sraffa's first example, using the *positive* input-output coefficients matrix **S** (3.3) and the *positive* vector of total output **q** (3.4).

(1) Prices. Taking as *numéraire* a quarter of wheat, following Sraffa, the price p_1 of wheat and the price p_2 of iron are expressed in this numéraire, the units being $[p_1] =$ (qr. wheat/qr. wheat) = 1 and $[p_2] =$ (qr. wheat/t. iron).

Then, let us calculate the positive *input coefficients matrix* in physical terms,

$$\mathbf{C} = \mathbf{S}\hat{\mathbf{q}}^{-1} = \begin{bmatrix} 280 & 120 \\ 12 & 8 \end{bmatrix} \begin{bmatrix} \frac{1}{400} & 0 \\ 0 & \frac{1}{20} \end{bmatrix} = \begin{bmatrix} 0.7 & 6 \\ 0.03 & 0.4 \end{bmatrix} > \mathbf{0}. \tag{3.16}$$

As the conditions of the **Perron–Frobenius theorem A.9.3** are fulfilled, we calculate the eigenvalues λ in order to get the left positive price eigenvector $\mathbf{p} > \mathbf{o}$. We set the usual eigenvector equation for any eigenvalue λ,

$$\mathbf{p}'\mathbf{C} = \lambda\mathbf{p}'. \tag{3.17}$$

Now, transpose equation (3.17) and introduce the numerical entries of matrix **C** (3.16) and the price variables p_1 and p_2,

$$\mathbf{C}'\mathbf{p} = \lambda\mathbf{p} \Rightarrow \begin{bmatrix} 0.7 & 0.03 \\ 6 & 0.4 \end{bmatrix} \begin{bmatrix} p_1 \\ p_2 \end{bmatrix} = \lambda \begin{bmatrix} p_1 \\ p_2 \end{bmatrix}. \tag{3.18}$$

Taking the vector of prices $\mathbf{p} = [p_1, p_2]'$ and the *identity matrix* $\mathbf{I}_2 = \left[\begin{smallmatrix} 1 & 0 \\ 0 & 1 \end{smallmatrix}\right]$, we compute the eigenvalues of matrix **C** (the same as those of the transposed matrix $\mathbf{C}'$), as the roots of the characteristic polynomial,

$$\begin{aligned} P_2(\lambda) = \det(\mathbf{C} - \lambda\mathbf{I}_2) &= \begin{vmatrix} 0.7 - \lambda & 6 \\ 0.03 & 0.4 - \lambda \end{vmatrix} = (0.7 - \lambda)(0.4 - \lambda) - 0.03 \cdot 6 \\ &= \lambda^2 - 1.1\lambda + 0.1 = (\lambda - 1)(\lambda - 0.1). \end{aligned} \tag{3.19}$$

The quadratic equation $P_2(\lambda) = 0$ confirms that one of the eigenvalues is the Frobenius number, $\lambda_1 := \lambda_C = 1$, the other one is $\lambda_2 = 0.1$.

12 It is interesting to note that Sraffa in PCMC never mentions the theorem of Perron (1907) nor the theorem of Frobenius (1912), later known in a modern formulation as the Perron–Frobenius theorem. Historians of economics, like Kurz and Salvadori [52] mention that Sraffa did not know the work of Perron and Frobenius. As already pointed out, Bertram Schefold [109] published his seminal work initially as a short Appendix in the German translation of PCMC; he was seemingly the first economist to discover and treat profoundly the significance and importance of the Perron–Frobenius theorem for Sraffa's work. It is assumed that Ladislaus von Bortkewicz (1868–1931) [9], a Russian economist, was the first scientist to discover the role of eigenvalues and eigenvectors in the context of classical production economics and models, see Knolle [50]. See also Parys [79]..

With these two eigenvalues λ_1, λ_2, one starts with the *left eigenvectors* of (3.17), setting specifically

$$\mathbf{C}'\mathbf{p} = \lambda_i \mathbf{p}, \quad i = 1, 2, \tag{3.20}$$

and transposing, we obtain the system of equations to determine the eigenvectors:

$$\mathbf{C}'\mathbf{p} = \lambda_i \mathbf{p} \Rightarrow \mathbf{C}'\mathbf{p} - \lambda_i \mathbf{p} = (\mathbf{C}' - \lambda_i \mathbf{I}_2)\mathbf{p} = \mathbf{o}. \tag{3.21}$$

As the Frobenius number is $\lambda_C = 1$, we plug it into the eigenvalue equations (3.20), and can set $p_{12} = 1$ and compute p_{11},[13] which is necessarily positive,

$$\begin{aligned} 0.7p_{11} + 0.03p_{12} &= p_{11} \\ 6p_{11} + 0.4p_{12} &= p_{12} \end{aligned} \Rightarrow \begin{aligned} -0.3p_{11} + 0.03p_{12} &= 0 \\ 6p_{11} - 0.6p_{12} &= 0 \end{aligned} \to p_{11} = 0.1. \tag{3.22}$$

Both prices are positive, according to the Perron–Frobenius theorem.

Basis vectors of the eigenspace[14] of $\lambda_C = 1$ for any $k \in \mathbb{R}$ are thus described by $\mathbf{p}_1 = \left[\begin{smallmatrix} p_{11} \\ p_{12} \end{smallmatrix}\right] = k\left[\begin{smallmatrix} 0.1 \\ 1 \end{smallmatrix}\right]$. With $k = 10$ the prices are $\mathbf{p}_1 = \left[\begin{smallmatrix} p_{11} \\ p_{12} \end{smallmatrix}\right] = \left[\begin{smallmatrix} 1 \\ 10 \end{smallmatrix}\right]$.

The relationship between the prices is $p_{11} = 0.1 \cdot p_{12}$.

To satisfy curiosity, let's now continue with the second eigenvalue λ_2. Plugging $\lambda_2 = 0.1$ into (3.20) and taking the numerical values of the matrix coefficients, setting $p_{21} = 1$, one gets here a negative value of the solution,

$$\begin{aligned} 0.7p_{21} + 0.03p_{22} &= 0.1p_{21} \\ 6p_{21} + 0.4p_{22} &= 0.1p_{22} \end{aligned} \Rightarrow \begin{aligned} 0.6p_{21} + 0.03p_{22} &= 0 \\ 6p_{21} + 0.3p_{22} &= 0 \end{aligned} \to p_{22} = -20. \tag{3.23}$$

Only the first *eigenvector*, associated with the Frobenius number $\lambda_C = 1$, has an economic meaning, i. e., necessarily positive prices. This is not the case for the second eigenvalue $\lambda_2 = 0.1$: No theorem guarantees both prices to be positive.

Concluding, we summarize that for the Frobenius number $\lambda_C = 1$, we obtain the price relationship 1:10, $p_{11} = 0.1 \cdot p_{12}$, between wheat and iron (PCMC, Par. 1).

With $p_{12} = 10$, one gets Sraffa's price vector $\mathbf{p}_1 = [p_{11} = 1, p_{12} = 10]'$; quoting Sraffa (PCMC, Par. 1): "*In the particular example we have taken, the exchange-value required is 10 qr. of* wheat *for 1 t. of* iron."

Written out in physical terms, starting from 1 t. iron and its price p_{12} is 10 qr. wheat, the equality mentioned by Sraffa reads:

$$1\text{ t. iron} \cdot p_{12}\left(\frac{\text{qr. wheat}}{\text{t. iron}}\right) = 1\text{ t. iron} \cdot 10\left(\frac{\text{qr. wheat}}{\text{t. iron}}\right) = 10\text{ qr. wheat} \tag{3.24}$$

Let's now formally analyse the quantities.

---

13 We can choose one vector component arbitrarily because $\mathbf{p}$ is an eigenvector of the eigenvalue equation (3.21). The vector $k \cdot \mathbf{p}$, $k \in \mathbb{R}$, is also a solution of the system (3.21). This means we are dealing here with relative prices.

14 The $n \times n$ square matrix $\mathbf{C}$ has the Frobenius number $\lambda_C = 1$, and the associated eigenvector $\mathbf{p}_1$ then generates a subspace in $\mathbb{R}^n$ known as the eigenspace of $\lambda_C = 1$.

(2) Quantities. To calculate the quantities, we will rely by analogy on Sraffa's conditions of production (Definition 3.1.2) that indicate that the production process (3.1) reproduces the same quantities of wheat and iron every year.

Consider the *positive input-output coefficients* matrix **C** (3.16) of Sraffa's example as given. The eigenvalues of **C** and **C**′ are the same and have been calculated as $\lambda_C = 1$, $\lambda_2 = 0.1$. Then, remembering that here *final demand* vanishes, $\mathbf{d} = \mathbf{o}$, one immediately gets from equation (2.54),

$$\mathbf{Cq} = \mathbf{q}. \tag{3.25}$$

Equation (3.25) is the *right eigenvector* equation for the Frobenius number $\lambda_C = 1$. To describe the general *right eigenvector* problem for both eigenvalues $\lambda_C = 1$ and $\lambda_2 = 0.1$ with matrix **C** (3.16), we again write a general eigenvalue equation,

$$\mathbf{Cq} = \lambda\mathbf{q}, \tag{3.26}$$

and proceed as follows:

For the Frobenius number $\lambda_C = 1$, calculate the *right eigenvector* $\mathbf{q}_1 = [q_{11}, q_{12}]'$

$$\begin{matrix} 0.7q_{11} + 6q_{12} = q_{11} \\ 0.03q_{11} + 0.4q_{12} = q_{12} \end{matrix} \Rightarrow \begin{matrix} 6q_{12} = 0.3q_{11} \\ 0.03q_{11} = 0.6q_{12} \end{matrix}$$

$$\Rightarrow q_{11} = \frac{6}{0.3}q_{12} = \frac{0.6}{0.03}q_{12} = 20q_{12}. \tag{3.27}$$

The *right eigenvector* associated to the Frobenius number $\lambda_C = 1$ is a basis vector of the eigenspace for any $k \in \mathbb{R}$, $\mathbf{q}_1 = \left[\begin{smallmatrix} q_{11} \\ q_{12} \end{smallmatrix}\right] = k\left[\begin{smallmatrix} 20 \\ 1 \end{smallmatrix}\right] \Rightarrow q_{11} = 20 \cdot q_{12}$. For the second eigenvalue $\lambda_2 = 0.1$, we also calculate the *right eigenvector* $\mathbf{q}_2 = [q_{21}, q_{22}]'$

$$\begin{matrix} 0.7q_{21} + 6q_{22} = 0.1q_{21} \\ 0.03q_{21} + 0.4q_{22} = 0.1q_{22} \end{matrix} \Rightarrow \begin{matrix} 0.6q_{21} = -6q_{22}, \\ 0.03q_{21} = -0.3q_{22}, \end{matrix}$$

$$\Rightarrow q_{21} = \frac{-6}{0.6}q_{22} = \frac{-0.3}{0.03}q_{22} = -10q_{22}. \tag{3.28}$$

The *eigenvector*, belonging to the second eigenvalue $\lambda_2 = 0.1$, is also a basis vector of the eigenspace for any $k \in \mathbb{R}$, $\mathbf{q}_2 = \left[\begin{smallmatrix} q_{21} \\ q_{22} \end{smallmatrix}\right] = k\left[\begin{smallmatrix} -10 \\ 1 \end{smallmatrix}\right] \Rightarrow q_{21} = -10 \cdot q_{22}$.

Only the *right eigenvector* associated to the Frobenius number $\lambda_C = 1$ has an economic meaning because both components represent *positive* produced quantities, in accordance with the Perron–Frobenius theorem. This is not the case for the *eigenvector* associated with the second eigenvalue $\lambda_2 = 0.1$, where no mathematical theorem assures all components to be positive.

Concluding, we summarize that, for the Frobenius number $\lambda_C = 1$, we obtain the quantity relation 1:20, $q_{11} = 20 \cdot q_{12}$, between wheat and iron, where q_{11} is the quantity of wheat and q_{12} is the quantity of iron that have to be produced.

With these numerical values, (3.26) becomes

$$\mathbf{C}\cdot\mathbf{q}_1 = \begin{bmatrix} 0.7 & 6 \\ 0.03 & 0.4 \end{bmatrix}\begin{bmatrix} 20q_{12} \\ q_{12} \end{bmatrix} = \lambda_1\mathbf{q}_1 = 1\cdot\begin{bmatrix} 20q_{12} \\ q_{12} \end{bmatrix}, \tag{3.29}$$

giving with $q_{12} = 20$ (*t. iron*) the values implicitly postulated by Sraffa (PCMC, Par. 1). We now obtain:

$$\begin{aligned} 280\ (\text{qr. wheat}) + 120\ (\text{qr. wheat}) &= 400\ (\text{qr. wheat}), \\ 12\ (\text{t. iron}) + 8\ (\text{t. iron}) &= 20\ (\text{t. iron}) \end{aligned} \tag{3.30}$$

These equations can be presented as Input-Output Table 3.1. ▲

Table 3.1: Sraffa's first numerical example as a flow of commodities in physical terms.

	Processing sectors (3.2)		**Total**
	wheat	**iron**	**output**
wheat (in qr. wheat)	$s_{11} = 280$	$s_{12} = 120$	$q_1 = 400$
iron (in t. iron)	$s_{21} = 12$	$s_{21} = 8$	$q_2 = 20$

We have at this stage achieved the *eigenvalue-eigenvector analysis* of Sraffa's first numerical example, based on the Perron–Frobenius theorem A.9.3.

We have found the following result:

Proposition 3.1.1. *In a sustainable system of production, with no surplus, which fulfills the formal conditions of production (Definition 3.1.2), we observe that for a positive commodity flow matrix* **S**, *one obtains the positive vector of total output* $\mathbf{q} = \mathbf{Se} > \mathbf{o}$, *and one calculates the input-output coefficients matrix* $\mathbf{C} = \mathbf{S}\hat{\mathbf{q}}^{-1}$. *The Frobenius number* $\lambda_C = 1$,

(a) *associated to the left positive price eigenvector,* $\mathbf{p} > \mathbf{o}$, *solves equation (3.15),* $\mathbf{p}'\mathbf{C} = \mathbf{p}'$, *which is an eigenvector of constant sign;*

(b) *associated to the right positive quantity eigenvector,* $\mathbf{q} > \mathbf{o}$, *solves equation (3.25),* $\mathbf{Cq} = \mathbf{q}$, *which is an eigenvector of constant sign.*

The uniqueness of the price or quantity vectors is obtained as usual by initially imposing additional information on prices and quantities.

3.1.2 Sraffa's second example

In the present subsection, we show how the foregoing methodology, using *matrix algebra*, and summarized in Proposition 3.1.1, is further applicable to Sraffa's second example with three commodities. There is *no surplus*.

Example 3.1.2 (PCMC, Par. 2). Now three commodities (wheat, iron, pigs) are considered, one industry for each commodity. In fact, any number n of commodities and the same number of single industry processes can be set up. The production scheme is described as follows, each line corresponding to an industry:

$$\begin{aligned}(240 \text{ qr. wheat}, 12 \text{ t. iron}, 18 \text{ pigs}) &\rightarrow (450 \text{ qr. wheat}, 0, 0),\\ (90 \text{ qr. wheat}, 6 \text{ t. iron}, 12 \text{ pigs}) &\rightarrow (0, 21 \text{ t. iron}, 0),\\ (120 \text{ qr. wheat}, 3 \text{ t. iron}, 30 \text{ pigs}) &\rightarrow (0, 0, 60 \text{ pigs}).\end{aligned} \tag{3.31}$$

Compute the price vector $\mathbf{p}$ and the quantity vector $\mathbf{q}$, in analogy to the method of solution used for Example 3.1.1. Compute $\det(\hat{\mathbf{q}} - \mathbf{S}')$ and $\operatorname{rank}([\mathbf{S}', \hat{\mathbf{q}}])$.

Solution to Example 3.1.2:

One first identifies the *positive commodity flow matrix* $\mathbf{S} > \mathbf{O}$ (2.13). Here, the *surplus* in physical terms vanishes, $\mathbf{d} = \mathbf{o}$. Then the *positive total output* per sector is calculated, giving the *positive* vector of *total output* $\mathbf{q} = \mathbf{Se} > \mathbf{o}$ in (2.15).

$$\mathbf{S} = \begin{bmatrix} s_{11} & s_{12} & s_{13} \\ s_{21} & s_{22} & s_{23} \\ s_{31} & s_{32} & s_{33} \end{bmatrix} = \begin{bmatrix} 240 & 90 & 120 \\ 12 & 6 & 3 \\ 18 & 12 & 30 \end{bmatrix}, \quad \mathbf{q} = \mathbf{Se} = \begin{bmatrix} 450 \\ 21 \\ 60 \end{bmatrix} > \mathbf{o}. \tag{3.32}$$

Then, we compute the *positive input-output coefficients* matrix $\mathbf{C} = \mathbf{S}\hat{\mathbf{q}}^{-1}$ (2.16):

$$\mathbf{C} = \begin{bmatrix} 240 & 90 & 120 \\ 12 & 6 & 3 \\ 18 & 12 & 30 \end{bmatrix} \begin{bmatrix} \frac{1}{450} & 0 & 0 \\ 0 & \frac{1}{21} & 0 \\ 0 & 0 & \frac{1}{60} \end{bmatrix} = \begin{bmatrix} \frac{8}{15} & \frac{30}{7} & 2 \\ \frac{4}{150} & \frac{2}{7} & \frac{1}{20} \\ \frac{1}{25} & \frac{4}{7} & \frac{1}{2} \end{bmatrix} > \mathbf{0}. \tag{3.33}$$

As matrix $\mathbf{C}$ is *positive* and there is no surplus, Lemma 4.1.1 (a) applies.

(1) Prices. One again writes the *left eigenvector* equation (3.17) to calculate the eigenvalues of matrix $\mathbf{C}$:

$$\mathbf{p}'\mathbf{C} = \lambda \mathbf{p}'. \tag{3.34}$$

For this purpose, we set the characteristic function

$$P_3(\lambda) = \det(\mathbf{C} - \lambda \mathbf{I}) = -\lambda^3 + \frac{277}{210}\lambda^2 - \frac{178}{525}\lambda + \frac{1}{50}. \tag{3.35}$$

The characteristic polynomial $P_3(\lambda)$ is factorized, as $P_3(\lambda) = -(\lambda - 1)(\lambda - \frac{3}{35})(\lambda - \frac{7}{30})$. We identify the Frobenius number $\lambda_C = 1$ and have now associated positive price vectors $\mathbf{p} > \mathbf{o}$. We calculate $\mathbf{p} = k \cdot [\frac{1}{5}, 2, 1]$. With $k = 5$, the price vector is $\mathbf{p} = [p_1 = 1, p_2 = 10, p_3 = 5]'$, corresponding to Sraffa's exchange-values (production costs), formulated in PCMC, Par. 2, as: "*The exchange-values which ensure replacement all around are 10 qr.* wheat = *1 t.* iron = 2 pigs."

(2) Quantities. Then, we write again the *right eigenvector* equation (3.26).

$$\mathbf{C} \cdot \mathbf{q} = \lambda \mathbf{q}. \tag{3.36}$$

The eigenvalues are known. The *positive* matrix $\mathbf{C} > \mathbf{0}$ has the Frobenius number $\lambda_C = 1$, associated to the positive right quantity *eigenvector*, $\mathbf{q} > 0$.

We calculate $\mathbf{q} = m \cdot [\frac{15}{2}, \frac{7}{20}, 1]$. Thus, these positive eigenvectors have an economic meaning.[15] With $m = 60$, one gets then exactly Sraffa's quantity vector $\mathbf{q} = [450, 21, 60]'$, see PCMC, Par. 2.

Summarizing, Sraffa states the equality expressed in physical terms, written out, as:

$$\begin{aligned} 10 \text{ qr. wheat} \cdot p_1\left(\frac{\text{qr. wheat}}{\text{qr. wheat}}\right) &= 1 \text{ t. iron} \cdot p_2\left(\frac{\text{qr. wheat}}{\text{t. iron}}\right) = 2 \text{ pigs} \cdot p_3\left(\frac{\text{qr. wheat}}{\text{pigs}}\right), \\ 10 \text{ qr. wheat} \cdot 1\left(\frac{\text{qr. wheat}}{\text{qr. wheat}}\right) &= 1 \text{ t. iron} \cdot 10\left(\frac{\text{qr. wheat}}{\text{t. iron}}\right) = 2 \text{ pigs} \cdot 5\left(\frac{\text{qr. wheat}}{\text{pigs}}\right) \\ &= 10 \text{ qr. wheat}. \end{aligned} \tag{3.37}$$

The quantities may again be presented in an Input-Output Table 3.2:

Table 3.2: Sraffa's example (PCMC, Par. 2), a flow of commodities.

	Processing sectors (3.32)			**Total**
	wheat	**iron**	**pigs**	**output**
wheat (in qr. wheat)	$s_{11} = 240$	$s_{12} = 90$	$s_{13} = 120$	$q_1 = 450$
iron (in t. iron)	$s_{21} = 12$	$s_{21} = 6$	$s_{23} = 3$	$q_2 = 21$
pigs (in numbers)	$s_{31} = 18$	$s_{32} = 12$	$s_{33} = 30$	$q_3 = 60$

Finally, we compute

$$\det(\hat{\mathbf{q}} - \mathbf{S}') = \det\left(\begin{bmatrix} 210 & -12 & -18 \\ -90 & 15 & -12 \\ -120 & -3 & 30 \end{bmatrix}\right) = 0, \tag{3.38}$$

reflecting a singular matrix, and the rank of the composed matrix $[\mathbf{S}', \hat{\mathbf{q}}]$,

$$\operatorname{rank}([\mathbf{S}', \hat{\mathbf{q}}]) = \operatorname{rank}\left(\begin{bmatrix} 240 & 12 & 8 & 450 & 0 & 0 \\ 90 & 6 & 12 & 0 & 21 & 0 \\ 120 & 3 & 30 & 0 & 0 & 60 \end{bmatrix}\right) = 3, \tag{3.39}$$

showing *linear independence* of the $n = 3$ production processes. ▲

15 The other eigenvectors are indeed not positive. The two other price vectors are $\mathbf{p_2} = [-\frac{10}{9}, -\frac{7}{18}, 1]'$ and $\mathbf{p_3} = [\frac{15}{2}, -\frac{5}{4}, 1]'$. The two other quantity vectors are $\mathbf{q_2} = [-\frac{2}{15}, 0, 1]'$ and $\mathbf{q_3} = [-\frac{19}{65}, \frac{310}{91}, 1]'$.

We have now to explain the two last results, starting with the singularity of the matrix $\hat{\mathbf{q}} - \mathbf{S}'$ in (3.38). By Assumption 2.2.1 the vector of total output is positive, $\mathbf{q} > \mathbf{o}$, consequently, the determinant of the diagonal matrix is different from 0, $\det(\hat{\mathbf{q}}) > 0$. Then, as there is *no surplus* in the economy presented in Example 3.1.1, Sraffa's conditions of production hold, Definition 3.1.1 and with (3.11) and (3.12) we obtain:

$$\begin{aligned}(\hat{\mathbf{q}} - \mathbf{S})\mathbf{e} = \mathbf{o} \Rightarrow \ & \operatorname{rank}(\hat{\mathbf{q}} - \mathbf{S}) \le n - 1 \Rightarrow \det(\hat{\mathbf{q}} - \mathbf{S}) = 0 \\ &= \det(\hat{\mathbf{q}} - \mathbf{S}') = \det(\hat{\mathbf{q}} - \hat{\mathbf{q}}\mathbf{C}') = \det(\hat{\mathbf{q}}(\mathbf{I} - \mathbf{C}')) \\ &= \det(\hat{\mathbf{q}}) \cdot \det(\mathbf{I} - \mathbf{C}') = 0 \Leftrightarrow \det(\mathbf{I} - \mathbf{C}') = 0 \Leftrightarrow \exists \text{ eigenvalue } \lambda = 1. \qquad (3.40)\end{aligned}$$

This calculation shows that an economy *without surplus* has an *input-output coefficients* matrix $\mathbf{C}$ with an eigenvalue $\lambda = 1$. We generalize this result in Lemma 4.1.1, Chapter 4.

Now we explain the rank of matrix $[\mathbf{S}', \hat{\mathbf{q}}]$ in (3.39). For this purpose we need the notion of column and row vectors of matrices, Definition A.4.3.

The row vectors $\mathbf{s}'_{i\cdot} = [s_{i1}, s_{i2}, \ldots, s_{in}]$, $i = 1, \ldots, n$, of $\mathbf{S}$ (the column vectors of $\mathbf{S}'$) are set up of the quantities s_{ij} of commodity i, necessary as inputs for the production in each sector S_j, $j \in \{1, \ldots, n\}$. For this reason these row vectors $\mathbf{s}'_{i\cdot}$ are called i-th *commodity vectors*. The column vectors $\mathbf{s}_{\cdot j} = [s_{1j}, s_{2j}, \ldots, s_{nj}]'$ of $\mathbf{S}$ (the row vectors of $\mathbf{S}'$) indicate the quantities s_{ij} of each commodity $i \in \{1, \ldots, n\}$ entering the production of sector S_j. For this reason, these column vectors $\mathbf{s}_{\cdot j}$ are called input vectors to the j-th *production process*, corresponding to the output q_j of that *production process*.

Consider then the column vectors $\mathbf{s}_{\cdot j} = [s_{1j}, s_{2j}, \ldots, s_{nj}]'$, $\hat{\mathbf{q}}_{\cdot j} = [0, \ldots, 0, q_j, 0, \ldots, 0]'$ of matrices $\mathbf{S}$ and $\hat{\mathbf{q}}$ to describe production processes.

Notation 3.1.1. The j-th production process is represented by the $(1\times 2n)$ vector $[\mathbf{s}_{\cdot j}, \hat{\mathbf{q}}_{\cdot j}]$, $j \in \{1, \ldots, n\}$, where the vector $\mathbf{s}_{\cdot j}$ represents the quantities of input necessary to produce the quantity of output q_j of commodity j, represented in the vector $\hat{\mathbf{q}}_{\cdot j}$.

Schefold ([103], p. 50) has postulated the *linear independence* of the production processes for joint production, see Chapter 6. We apply this notion for *single-product industries*. For the notion of rank, see Definition A.2.4.

Lemma 3.1.1. *If the n production processes* $[\mathbf{s}_{\cdot j}, \hat{\mathbf{q}}_{\cdot j}]$, *collected in the matrix* $[\mathbf{S}', \hat{\mathbf{q}}]$, *are linearly independent, then* $\operatorname{rank}([\mathbf{S}', \hat{\mathbf{q}}]) = n$.

Proof. As the output vector is positive, $\mathbf{q} > \mathbf{o}$, the $n \times n$ diagonal matrix $\hat{\mathbf{q}}$ is regular and is a submatrix of $[\mathbf{S}', \hat{\mathbf{q}}]$. Thus, $\operatorname{rank}([\mathbf{S}', \hat{\mathbf{q}}]) = n$ and the n production processes $[\mathbf{s}_{\cdot j}, \hat{\mathbf{q}}_{\cdot j}]$ are linearly independent. □

We can now summarize the solution procedure to obtain prices and quantities in the case of *no surplus*, applying the Perron–Frobenius theorem A.9.3.

Proposition 3.1.2. *Solution procedure for the right eigenvector equation* $\mathbf{Cq} = \lambda\mathbf{q}$ *and the left eigenvector equation* $\mathbf{p}'\mathbf{C} = \lambda\mathbf{p}'$*:*

(a) *Check that the matrix* **C** *is positive;*
(b) *check that the Frobenius number is* $\lambda_C = 1$*;*
(c) *calculate the associated, necessarily positive, left price eigenvector* $\mathbf{p} > \mathbf{o}$ *and the positive right quantity eigenvector* $\mathbf{q} > \mathbf{o}$ *of matrix* **C***;*
(d) *determine the eigenvectors from the initially given price and quantity information.* ▲

3.1.3 Sraffa's third example and productiveness

In this subsection, we extend the construction to economies with *positive* surplus. In the present production economies, we assume subsistence wages and equal rate of profits in all branches.[16]

Example 3.1.3. Sraffa (PCMC, Par. 5) considers the same model as presented in Example 3.1.1, but now the output of farmers is 575 *qr.* instead of 400 *qr. of* wheat, with the same inputs as before. Hence, there is a *surplus* and the possibility of *profit*. The process of production is symbolized by the well-known scheme:

$$\begin{aligned} &(280 \text{ qr. wheat}, 12 \text{ t. iron}) \to (575 \text{ qr. wheat}, 0), \\ &(120 \text{ qr. wheat}, 8 \text{ t. iron}) \to (0, 20 \text{ t. iron}). \end{aligned} \tag{3.41}$$

Compute the price vector **p** and the quantity vector **q** in applying matrix algebra, as it has been realised for the Example 3.1.2. Compute $\det(\hat{\mathbf{q}} - \mathbf{S}')$ and $\operatorname{rank}([\hat{\mathbf{q}}, \mathbf{S}'])$. Compute the vector of values **x**.

Solution to Example 3.1.3:
The uniform *rate of profits r* (see PCMC, Par. 4) is maximal because there is a *positive* surplus and only *subsistence wages*. The *rate of profits* is written as $r = R > 0$ (PCMC, Par. 22), and the prices satisfy the following equations,

$$\begin{aligned} (280p_1 + 12p_2)(1 + R) &= 575p_1 = x_1 \text{ (qr. wheat)}, \\ (120p_1 + 8p_2)(1 + R) &= 20p_2 = x_2 \text{ (qr. wheat)}. \end{aligned} \tag{3.42}$$

Again, in (3.42) the physical term "qr. wheat" is considered as the measure of exchange, i. e., is the numéraire.[17]

---

16 In fact, we mean that the total output exceeds the means of production, that is, $\mathbf{d} = \mathbf{q} - \mathbf{Se} \geq \mathbf{o}$, producing a surplus or a net product for the present partial economies of Sraffa's elementary examples. In PCMC, Par. 12, Sraffa early introduced the term "national income" in relation with the term *surplus* or *net product*. Indeed, at this moment Sraffa has in mind the *complete production scheme* of an economy, as presented by an *input-output* table (IOT) in physical terms. Then the net product is the "gross national income" and can be associated to the *accounting balance* (2.145) $Y = W + P = C + I + G$, see also Chapter 10. The *national income* is then a proxy of *Gross Domestic product* (GDP).

17 One may ask how we now may obtain a surplus without changing the inputs. Economically, this would correspond to the following types of situation: Either rationalization measures have been un-

(1) Prices. The system (3.42), Sraffa's price model, is now transcribed to matrix form as

$$\boxed{\mathbf{S}'\mathbf{p}(1+R) = \hat{\mathbf{q}}\mathbf{p} = \mathbf{x}.} \tag{3.43}$$

Then equation (3.43) is multiplied by $1/(1+R)$ and the diagonal matrix $\hat{\mathbf{q}}^{-1}$, giving the eigenvector equation with matrix $\mathbf{C}'$ for the *right eigenvectors*,[18]

$$\hat{\mathbf{q}}^{-1}(\mathbf{S}'\mathbf{p}) = (\hat{\mathbf{q}}^{-1}\mathbf{S}')\mathbf{p} = \mathbf{C}'\mathbf{p} = \frac{1}{1+R}(\hat{\mathbf{q}}^{-1}\hat{\mathbf{q}})\mathbf{p} = \lambda_C\mathbf{p} \Rightarrow \mathbf{C}'\mathbf{p} = \lambda_C\mathbf{p}. \tag{3.44}$$

As in Example 3.1.1 we get an eigenvector equation to calculate the price vector $\mathbf{p}$, but here the eigenvector equation (3.44) exhibits the Frobenius number $\lambda_C = 1/(1+R) < 1$ of economies with *positive* surplus, different from eigenvector equation (3.14) $\mathbf{C}'\mathbf{p} = \mathbf{p}$, revealing the Frobenius number $\lambda_C = 1$ of economies *without* surplus.

Now, we treat the numerical solutions:

We identify again the *positive commodity flow matrix* $\mathbf{S} > \mathbf{0}$ (3.42). We know that the *semi-positive* vector of *surplus* in physical terms is $\mathbf{d} = \mathbf{q} - \mathbf{S}\mathbf{e} \geq \mathbf{o}$. We use the same physical units as in *Example* 3.1.1. Then, the detailed calculations are as follows,

$$\mathbf{S} = \begin{bmatrix} s_{11} & s_{12} \\ s_{21} & s_{22} \end{bmatrix} = \begin{bmatrix} 280 & 120 \\ 12 & 8 \end{bmatrix}, \quad \mathbf{S}\mathbf{e} = \begin{bmatrix} 280 & 120 \\ 12 & 8 \end{bmatrix}\begin{bmatrix} 1 \\ 1 \end{bmatrix} = \begin{bmatrix} 400 \\ 20 \end{bmatrix},$$
$$\mathbf{q} = \begin{bmatrix} q_1 \\ q_2 \end{bmatrix} = \begin{bmatrix} 575 \\ 20 \end{bmatrix}, \quad \mathbf{d} = \mathbf{q} - \mathbf{S}\mathbf{e} = \begin{bmatrix} 575 \\ 20 \end{bmatrix} - \begin{bmatrix} 400 \\ 20 \end{bmatrix} = \begin{bmatrix} 175 \\ 0 \end{bmatrix}. \tag{3.45}$$

Then, we compute the positive matrix $\mathbf{C} = \mathbf{S}\hat{\mathbf{q}}^{-1} > \mathbf{0}$ (2.16):

$$\mathbf{C} = \begin{bmatrix} 280 & 120 \\ 12 & 8 \end{bmatrix}\begin{bmatrix} \frac{1}{575} & 0 \\ 0 & \frac{1}{20} \end{bmatrix} = \begin{bmatrix} \frac{56}{115} & 6 \\ \frac{12}{575} & \frac{2}{5} \end{bmatrix} > \mathbf{0}. \tag{3.46}$$

The matrix $\mathbf{C}$ and the vector of surplus are *positive*. The Perron–Frobenius theorem A.9.3 applies. The eigenvalues are computed, along with

$$P_2(\lambda) = \det(\mathbf{C} - \lambda\mathbf{I}) = \lambda^2 - \frac{102}{115}\lambda + \frac{8}{115} = \left(\lambda - \frac{4}{5}\right)\left(\lambda - \frac{2}{23}\right). \tag{3.47}$$

The Frobenius number $\lambda_C = 1/(1+R) = 4/5 < 1$, associated with the positive *price* eigenvector $\mathbf{p} = k[1, 15]'$, is calculated. The *productiveness* $R = (1/\lambda_C) - 1$, the notion has been proposed by Knolle [49], is calculated as $R = 0.25$.

Indeed, the value R characterizes the overall productivity of an economy in the sense of Krugman [47]. Sraffa called R the maximal rate of profit.

---

dertaken to make the existing process more efficient, or there is additional production capacity left after fulfillment of the conditions of production. The benefits revert solely to the producers in the present case.

18 This equation is equivalent to equation $\mathbf{p}'\mathbf{C} = \lambda\mathbf{p}'$ for a *left eigenvector*.

Finally, with $k = 1$, one gets the price vector $\mathbf{p} = [p_1, p_2]' = [1, 15]'$, corresponding to Sraffa's exchange-values[19]:

> "*The exchange-ratio which enables the advances to be replaced and the profits to be distributed to both industries in proportion to their advances is 15 qr. of* wheat *for 1 t. of* iron; *and the corresponding rates of profits in each industry is 25 %.*" (PCMC, Par. 5).[20]

Sraffa has shown that with a surplus the price model generates a profit. As there are no explicit wages, the profits are consequently maximal. The productiveness R, called by Sraffa the *maximal rate of profits*, is in the present case $R = (5/4) - 1 = 0.25$. The vector of *values* is $\mathbf{x} = \hat{\mathbf{q}}\mathbf{p} = [575, 300]'$ in qr. wheat.

(2) Quantities. In analogy with the realised calculations, we set the *right eigenvector* equation (3.26) $\mathbf{Cq} = \lambda\mathbf{q}$. Taking the matrix $\mathbf{C}$ from (3.46), we compute the *right eigenvector*, to obtain the quantities

$$\mathbf{Cq} = \begin{bmatrix} \frac{56}{115} & 6 \\ \frac{12}{575} & \frac{2}{5} \end{bmatrix} \mathbf{q} = \lambda_1 \mathbf{q} = \frac{4}{5}\mathbf{q}. \tag{3.48}$$

We get $\mathbf{q} = k \cdot [115, 6]'$. The proportion 115 : 6 is neither the initial quantity proportion, 400 : 20 = 20 : 1 accordingly to (3.25), nor the proportion 575 : 20 after the surplus has been introduced!.[21]

Finally, we compute

$$\det(\hat{\mathbf{q}} - \mathbf{S}') = \det\left(\begin{bmatrix} 295 & -12 \\ -120 & 12 \end{bmatrix}\right) = 2{,}100, \tag{3.49}$$

which shows that the matrix is regular. With Lemma 3.1.1 we confirm that the rank of matrix $[\mathbf{S}', \hat{\mathbf{q}}]$ is equal to $n = 2$,

$$\operatorname{rank}([\mathbf{S}', \hat{\mathbf{q}}]) = \operatorname{rank}\left(\begin{bmatrix} 280 & 12 & 575 & 0 \\ 120 & 8 & 0 & 20 \end{bmatrix}\right) = 2, \tag{3.50}$$

certifying the linear independence of both production processes ▲.

The task remains to give an explanation of the regularity of matrix $\hat{\mathbf{q}} - \mathbf{S}'$ in (3.49). Considering $\det(\hat{\mathbf{q}}) > 0$, we have the following equivalence:

$$\begin{aligned} 0 \neq \det(\hat{\mathbf{q}} - \mathbf{S}') &= \det(\hat{\mathbf{q}} - \hat{\mathbf{q}}\mathbf{C}') = \det(\hat{\mathbf{q}}(\mathbf{I} - \mathbf{C}')) = \det(\hat{\mathbf{q}}) \cdot \det(\mathbf{I} - \mathbf{C}') \\ &\Leftrightarrow \det(\mathbf{I} - \mathbf{C}') \neq 0 \Leftrightarrow \lambda = 1 \text{ is no eigenvalue of } \mathbf{C}. \end{aligned} \tag{3.51}$$

19 The other vectors $\mathbf{p}_1 = k[-6, 115]'$ actually have no economic meaning.

20 "Advances" means in Sraffa's quotation that each industry sets up and finances the means of production at the beginning of each reference period. It will be shown that $0 < \lambda_C < 1$ occurs when there is a *semi-positive* surplus $\mathbf{d} \geq \mathbf{o}$ and that then, consequently for that *positive commodity flow* matrix $\mathbf{S} = \mathbf{C}\hat{\mathbf{q}} > \mathbf{0}$, the price vector is *positive*, $\mathbf{p} > \mathbf{o}$, see Lemma 4.1.1 (b).

21 We will later see that the proportion 115 : 6 represents the proportion between the means of production of a *Standard system* based on the matrix $\mathbf{C}$, see equation (5.96).

This calculation shows that an economy *with surplus*, described by the *input-output coefficients* matrix **C**, has no eigenvalue $\lambda = 1$. This preliminary result is fully worked out in Lemma 4.1.1, Chapter 4.

Proposition 3.1.2 will now be extended to *production economies* with a surplus. We get the following preliminary result, applying the Perron–Frobenius theorem A.9.3.

Proposition 3.1.3. *Solution procedure for the Sraffa price model (3.43):*
(a) *Check that the matrix* **C** *is positive;*
(b) *compute the Frobenius number* $\lambda_C := 1/(1+R) < 1$ *of the matrix* **C**, *which is necessarily smaller than 1;*
(c) *determine the positive left eigenvector (price vector* $\mathbf{p} > \mathbf{0}$*) of the matrix* **C** *with initially given price information;*
(d) *compute the necessarily positive productiveness* $R = (1/\lambda_C) - 1 > 0$. ▲

We have thus largely described the way to solve Sraffa's first three numerical examples (PCMC, Par. 1, Par. 2, Par.5). We made systematic use of the Perron–Frobenius theorem A.9.3.

By analogy with a *productive Leontief model* (see Definition A.12.1) defined by Ashmanov [2] and treated later, we define a *productive Sraffa model* for *semi-positive* vectors of surplus $\mathbf{d} \geq \mathbf{o}$ and *positive* vectors of output $\mathbf{q} > \mathbf{o}$ with a *semi-positive input-output coefficients* matrix $\mathbf{C} \geq \mathbf{0}$, Assumptions 2.2.1 and 2.2.2 hold,

$$\mathbf{q} = \mathbf{Se} + \mathbf{d} = \mathbf{C\hat{q}e} + \mathbf{d} = \mathbf{Cq} + \mathbf{d} > \mathbf{o}. \tag{3.52}$$

The notion of *productive Sraffa model* is applied in the proof of Lemma 4.1.1.

We conclude this section with a further *proposition that* summarizes the obtained formal mathematical properties of *production economies* with surplus. For this purpose, we consider at the moment a positive $n \times n$ *commodity flow matrix*[22] $\mathbf{S} > \mathbf{0}$ and the diagonal matrix $\hat{\mathbf{q}}$, with $\mathbf{q} > \mathbf{o}$, which are both split into columns and row vectors, applying Definition A.4.3.

Proposition 3.1.4. *The n production processes* $[\mathbf{s}_{.j}, \hat{\mathbf{q}}_{.j}]$, $j \in \{1,\ldots,n\}$, *represented by the* $n \times 2n$ *matrix* $[\mathbf{S}', \hat{\mathbf{q}}]$ *are linearly independent (see Schefold [103], p. 50). The matrix* $(\hat{\mathbf{q}} - \mathbf{S}')$ *represents in the columns the difference between produced and industrially used commodities. The condition for a surplus is*

$$\det(\hat{\mathbf{q}} - \mathbf{S}') = \det(\hat{\mathbf{q}}(\mathbf{I} - \mathbf{C}')) \neq 0 \Leftrightarrow \det(\mathbf{I} - \mathbf{C}') \neq 0 \Leftrightarrow \text{has no eigenvalue } \lambda = 1. \tag{3.53}$$

All the output being concentrated in the diagonal matrix $\hat{\mathbf{q}}$, *we get* $\text{rank}([\mathbf{S}', \hat{\mathbf{q}}]) = n$.

---

22 Subsequently, matrix **S** will be assumed to be *semi-positive* and *irreducible* or only *semi-positive*.

In this section, we have gained a milestone, because we have identified the *productiveness* $R = (1/\lambda_C) - 1$, derived from the Frobenius number $0 < \lambda_C < 1$ of the *positive* matrix $\mathbf{C} = \mathbf{S}\hat{\mathbf{q}}^{-1} > \mathbf{O}$ as a measure of productivity of a production economy of type (3.41). *Productiveness* will accompany us throughout this text with the purpose of deeply questioning the measurement of the productivity of a *circular economic production process*.

In the next section, we develop the concept of the production schemes.

3.2 Symbolic representation of the production scheme

We continue to treat *single-product industries* and consider the previously discussed way to write the *commodity flow matrix* **S**

$$\mathbf{S} = \begin{bmatrix} s_{11} & s_{12} & \dots & \dots & s_{1n} \\ s_{21} & s_{22} & \dots & \dots & s_{2n} \\ \dots & \dots & \dots & \dots & \dots \\ \dots & \dots & \dots & \dots & \dots \\ s_{n1} & s_{n2} & \dots & \dots & s_{nn} \end{bmatrix} = [\mathbf{s}_{.1}, \mathbf{s}_{.2}, \dots, \dots, \mathbf{s}_{.n}] = \begin{bmatrix} \mathbf{s}'_{1.} \\ \mathbf{s}'_{2.} \\ \dots \\ \dots \\ \mathbf{s}'_{n.} \end{bmatrix}. \tag{3.54}$$

As Assumption 2.2.1 holds, $\mathbf{q} = \mathbf{S}\mathbf{e} + \mathbf{d} > \mathbf{o}$, and according to Schefold ([103], p. 7), the representation of the production scheme takes the following form:

$$\begin{aligned} (s_{11}, s_{21}, s_{31}, \dots, s_{n1}) &\rightarrow (q_1, 0, 0, \dots, 0), \\ (s_{12}, s_{22}, s_{32}, \dots, s_{n2}) &\rightarrow (0, q_2, 0, \dots, 0), \\ (s_{13}, s_{23}, s_{33}, \dots, s_{n3}) &\rightarrow (0, 0, q_3, \dots, 0), \\ (\dots, \dots, \dots, \dots) &\rightarrow (0, 0, 0, \dots, 0), \\ (s_{1n}, s_{2n}, s_{3n}, \dots, s_{nn}) &\rightarrow (0, 0, 0, \dots, q_n), \\ (\mathbf{S}') &\rightarrow (\hat{\mathbf{q}}). \end{aligned} \tag{3.55}$$

As the Assumption 2.2.2 holds, the column vectors $\mathbf{s}_{.j} \geq \mathbf{o}$, $j \in \{1, \dots, n\}$, and the *commodity flow* matrix $\mathbf{S} \geq \mathbf{O}$ are *semi-positive*.

3.3 A generalization of Sraffa's first example*

We shall now proceed to a generalization of two of the treated elementary Sraffa models (PCMC, Par. 1, Par. 2, Par. 5). This generalization is done in several steps. We take first a *positive commodity flow matrix* **S** with $n = 2$ sectors, where the coefficients, represented by *numerical values* are now replaced by any *literal values*, permitting to advance the *numerical analysis* towards an *algebraic analysis*. Then, we will be prepared to increase the number $n \in \{3, 4, \dots, \}$ of the sectors.

3.3.1 A global economy

But before proceeding, we will have another look at Sraffa's first example (PCMC, Par. 1), see Example 3.1.1. This example is not that trivial; let's apply it by analogy to an elementary model of a fully globalized world economy, which by definition is closed.

Replace "wheat" by a composite commodity of goods and services required for interindustry demands (and ultimately consumer consumption), let's call it here "Industrial Necessities". We restrict ourselves here to interindustry transactions, in other words, to the conditions of production. "Iron" will then be replaced by "Crude Oil", assumed to be the sole source covering energy requirements.

The components of the flow matrix **S** for "Industrial Necessities" will be assumed to remain unchanged, but the composition of the commodity "Industrial Necessities" can change (not addressed here). The price of that commodity will however also be assumed to remain constant.

Industrial necessities and "Crude Oil" will be expressed in appropriate measurement units (which we will not go into here): say, "Necessity Units" and "Crude Oil Units". As for the components of the flow matrix for Crude Oil, they will now be made to increase twofold, indicating a vast increase in the supply of that commodity (pollution issues set aside), which can have multiple causes not analysed here. The economic process then reads

No: 1, before changes in oil production

$$\begin{gathered}(280 \text{ Necessity Units}, 12 \text{ Crude Oil Units}) \rightarrow (400 \text{ Necessity Units}, 0),\\ (120 \text{ Necessity Units}, 8 \text{ Crude Oil Units}) \rightarrow (0, 20 \text{ Crude Oil Units}).\end{gathered} \tag{3.56}$$

No: 2, after changes in oil production

$$\begin{gathered}(280 \text{ Necessity Units}, 24 \text{ Crude Oil Units}) \rightarrow (400 \text{ Necessity Units}, 0),\\ (120 \text{ Necessity Units}, 16 \text{ Crude Oil Units}) \rightarrow (0, 40 \text{ Crude Oil Units}).\end{gathered} \tag{3.57}$$

Setting the price for "Industrial Necessities" $p_{11} = 1$, we already know that for **No: 1** process the price of "Crude Oil" will be: $p_{12} = 10$. For **No: 2** process, Sraffa's price model will now read:

$$\begin{aligned}280p_{11} + 24p_{12} &= 400p_{11},\\ 120p_{11} + 16p_{12} &= 40p_{12},\end{aligned} \tag{3.58}$$

and here, with $p_{11} = 1$, we obtain $p_{12} = 5$. The increase in "Crude Oil" supply thus reflects a change in technology, represented by the new **S** matrix, and a significant drop in oil price (ceteris paribus).

This elementary example shows how Sraffa's model is complementary to the neoclassical view. Marshall's scissors for supply-and-demand analysis would only give tendencies at the macroeconomic level; here we obtain precise numerical values, and the following Subsection 3.3.2 shows how these values change with variable components of the *commodity flow* matrix.

3.3.2 Variables replace numbers in the commodity flow matrix

Let's start with the following exercise. Its aim is to find the rules of construction for a *general* economy where there is no surplus, given the *positive commodity flow matrix* $\mathbf{S} > \mathbf{O}$ with $n = 2$ sectors. Let us simplify the notations of the elements of the *commodity flow matrix* throughout Sections 3.3, 3.4 and 3.5.

We set up matrix $\mathbf{S} > \mathbf{O}$ in physical terms and compute the vector of *total output*

$$\mathbf{S} = \begin{bmatrix} a & b \\ c & d \end{bmatrix} := \begin{bmatrix} s_{11} & s_{12} \\ s_{21} & s_{22} \end{bmatrix}, \quad \mathbf{q} = \begin{bmatrix} q_1 \\ q_2 \end{bmatrix} = \mathbf{Se} = \begin{bmatrix} a+b \\ c+d \end{bmatrix}. \tag{3.59}$$

Example 3.3.1 (First generalization of PCMC, Par. 1). Suppose at first that only two commodities are produced, wheat and iron. Both are used, in part, as sustenance for those who work, and for the rest as means of production—wheat as seed and iron in the form of tools. Suppose that, all in all, and including the necessaries for the workers, there is a positive commodity flow matrix $\mathbf{S} > \mathbf{O}$. Here we have set $a > 0$ quarters of wheat and $c > 0$ tons of iron are used to produce $a + b$ quarters of wheat, while $b > 0$ quarters of wheat and $d > 0$ tons of iron are used to produce $c + d$ tons of iron. The process of production (indicated by arrows) is symbolised as follows, the first line represents the wheat production, the second line the iron production (note the transpose $\mathbf{S}'$ on the left side):

$$\begin{aligned} &(a \text{ qr. wheat}, c \text{ t. iron}) \to ((a+b) \text{ qr. wheat}, 0), \\ &(b \text{ qr. wheat}, d \text{ t. iron}) \to (0, (c+d) \text{ t. iron}). \end{aligned} \tag{3.60}$$

Determine the *eigenvalues* and the *left* and *right eigenvectors* of the *input coefficient matrix.*

Solution to Example 3.3.1:

Using again elementary rules of matrix algebra, especially those concerning diagonal matrices and the definition (2.16) of the *input-output coefficients matrix* in physical terms $\mathbf{C} = \mathbf{S}\hat{\mathbf{q}}^{-1}$, one finds the matrix,

$$\mathbf{C} = \mathbf{S}\hat{\mathbf{q}}^{-1} = \begin{bmatrix} a & b \\ c & d \end{bmatrix} \begin{bmatrix} \frac{1}{a+b} & 0 \\ 0 & \frac{1}{c+d} \end{bmatrix} = \begin{bmatrix} \frac{a}{a+b} & \frac{b}{c+d} \\ \frac{c}{a+b} & \frac{d}{c+d} \end{bmatrix} > \mathbf{O}. \tag{3.61}$$

The eigenvalues of matrix $\mathbf{C}$ (3.61) (the same as those of the transposed matrix $\mathbf{C}'$) are computed. One finds

$$\det(\mathbf{C}-\lambda\mathbf{I}_2) = \begin{vmatrix} \frac{a}{a+b}-\lambda & \frac{b}{c+d} \\ \frac{c}{a+b} & \frac{d}{c+d}-\lambda \end{vmatrix}$$
$$= \left(\frac{a}{a+b}-\lambda\right)\left(\frac{d}{c+d}-\lambda\right) - \frac{b}{c+d}\cdot\frac{c}{a+b}$$
$$= \lambda^2 - \left(\frac{a}{a+b}+\frac{d}{c+d}\right)\lambda + \frac{ad-bc}{(a+b)(c+d)}$$
$$= (\lambda-1)\left(\lambda - \frac{ad-bc}{(a+b)(c+d)}\right) = 0, \tag{3.62}$$

giving the eigenvalues, see Proposition 3.1.2,

$$\lambda_1 = 1; \quad \lambda_2 = \frac{ad-bc}{(a+b)(c+d)}. \tag{3.63}$$

Now we have the general price vectors $\mathbf{p} = [p_1, p_2]'$ and quantity vectors $\mathbf{q} = [q_1, q_2]'$. We then compute the right eigenvectors of the transposed *input coefficients matrix* in physical terms $\mathbf{C}$ (3.61)

$$\mathbf{C}'\mathbf{p} = \lambda_i\mathbf{p} \Rightarrow \mathbf{C}'\mathbf{p} - \lambda_i\mathbf{p} = \mathbf{o} \Rightarrow (\mathbf{C}' - \lambda_i\mathbf{I}_2)\mathbf{p} = \mathbf{o}. \tag{3.64}$$

The matrix $\mathbf{C}$ (3.61) is *positive* for positive coefficients $a > 0$, $b > 0$, $c > 0$, $d > 0$. Later we will see that these conditions guarantee the existence of a positive price eigenvector $\mathbf{p} > \mathbf{o}$, associated for the existing Frobenius number $\lambda_C = 1$ (see Lemma 4.1.1 (a)). One finds:

$$\mathbf{p} = [p_1, p_2]' = \left[p_1, p_1\left(\frac{b}{c}\right)\right]' = p_1 \cdot \left[1, \frac{b}{c}\right]', \quad \frac{b}{c} \in \mathbb{R}^+. \tag{3.65}$$

Then we compute the right eigenvectors of the *input coefficients matrix* $\mathbf{C}$ (3.61) in physical terms,

$$\mathbf{C}\mathbf{q} = \lambda_i\mathbf{q} \Rightarrow \mathbf{C}\mathbf{q} - \lambda_i\mathbf{q} = \mathbf{o} \Rightarrow (\mathbf{C} - \lambda_i\mathbf{I}_2)\mathbf{q} = \mathbf{o}. \tag{3.66}$$

By analogy, there exists a positive quantity vector for $\lambda_C = 1$. One finds

$$\mathbf{q} = [q_1, q_2]' = \left[q_2\left(\frac{a+b}{c+d}\right), q_2\right]' = q_2 \cdot \left[\frac{a+b}{c+d}, 1\right]', \quad \frac{a+b}{c+d} \in \mathbb{R}^+. \tag{3.67}$$

Summarizing:

(1) As there is no surplus, the positive *input coefficients matrix* $\mathbf{C}$ (3.61) has Frobenius number $\lambda_C = 1$, guaranteed by Lemma 4.1.1 (a). The second eigenvalue λ_2 is smaller than λ_C,[23] Perron theorem A.9.1.

23 The second eigenvalue λ_2 can even be a negative real number.

(2) As the matrix **C** (3.61) is *positive*, its right and left eigenvectors, associated to the Frobenius number $\lambda_C = 1$, are positive. They represent the quantity vectors and the price vectors. They are entirely determined by the coefficients of matrix **C** (3.61), equations (3.65) and (3.67).
The proportions of the prices and quantities are:

$$p_2 = p_1\left(\frac{b}{c}\right), \qquad q_1 = q_2\left(\frac{a+b}{c+d}\right). \tag{3.68}$$

Formally, this means that the proportions of prices and quantities are constants,

$$\frac{p_1}{p_2} = \frac{c}{a}, \qquad \frac{q_1}{q_2} = \frac{a+b}{c+b}. \tag{3.69}$$

Applying this to *Example* 3.1.1 we can formulate the following result:
Given the coefficients of the *commodity flow matrix*,

$$\mathbf{S} = \begin{bmatrix} a & b \\ c & d \end{bmatrix} = \begin{bmatrix} 280 & 120 \\ 12 & 8 \end{bmatrix}: \tag{3.70}$$

(1) The price proportions are, see Sraffa ([108], Par. 1):

$$\frac{p_1}{p_2} = \frac{c}{b} = \frac{12}{120} = \frac{1}{10}. \tag{3.71}$$

Setting for the numéraire wheat $p_1 = 1\,\frac{\text{qr. wheat}}{\text{qr. wheat}} = 1$, $p_2 = 10\,\frac{\text{qr. wheat}}{\text{t. iron}}$.

(2) The quantity of $b = 120$ qr. wheat is used by the iron sector to produce the quantity of $(c+d) = 20$ t. iron. The quantity of $c = 12$ t. iron is used by the wheat sector to produce $(a+b) = 400$ qr. wheat. With the price proportion (3.71) one has: $b \cdot p_1 = 120 \cdot 1$ qr. wheat $= c \cdot p_2 = 12 \cdot 10$ qr. wheat $= 120$ qr. wheat.

(3) In this production economy with no surplus, the quantity relationship between the total amount of wheat and the total amount of iron produced every period is given by the proportion of the components of the quantity vector **q** (3.67), namely

$$\frac{q_1}{q_2} = \frac{a+b}{c+d} = \frac{280+120}{12+8} = \frac{400}{20} = 20. \quad \blacktriangle \tag{3.72}$$

3.3.3 Fixing rules to construct the commodity flow matrix

A further exercise consists in fixing sufficient conditions, like the price and quantity proportions, for an $n = 2$ sector economy, where there is no surplus as in PCMC, Par. 1. We assume a *positive commodity flow matrix* $\mathbf{S} > \mathbf{0}$.

Example 3.3.2 (Second generalization of PCMC, Par. 1). Two industries produce two commodities, wheat and iron. Both are used, in part as sustenance for those who work and for the rest as means of production—wheat as seed and iron in the form of tools. Suppose that, all in all, and including the necessaries for the workers, there are four positive quantities, represented by the real numbers $a > 0$, $b > 0$, $c > 0$, $d > 0$, namely a quarters of wheat and c tons of iron are used to produce $a + b$ quarters of wheat; while b quarters of wheat and d tons of iron are used to produce $c + d$ tons of iron. The process of production (indicated by arrows) is again symbolised as follows by a production scheme:

$$\begin{aligned}(a \text{ qr. wheat}, c \text{ t. iron}) &\to ((a+b) \text{ qr. wheat}, 0),\\ (b \text{ qr. wheat}, d \text{ t. iron}) &\to (0, (c+d) \text{ t. iron}).\end{aligned} \tag{3.73}$$

As there is no surplus, verify that the Frobenius number of the input-output coefficients matrix $\mathbf{C} = \mathbf{S}\hat{\mathbf{q}}^{-1}$ (3.61) is equal to one, $\lambda_C = 1$.

The *numéraire* is again wheat. This means: $p_1 = 1 \frac{\text{qr. wheat}}{\text{qr. wheat}} = 1$.

Consider as exogenously given:

- one of the matrix coefficients, let's work here with given $a = A > 0$;
- the price relationship $m = \frac{p_1}{p_2}$;
- the quantity relationship between wheat and iron $k = \frac{q_1}{q_2}$;
- and the smaller eigenvalue $0 < \lambda_2 < 1$ of the *input coefficient matrix* $\mathbf{C}$;

Determine then the other positive coefficients b, c, d of the *commodity flow matrix* $\mathbf{S}$ and discuss the smaller eigenvalue λ_2 of matrix $\mathbf{C}$.

Solution to Example 3.3.2:

Consider the *positive commodity flow matrix* $\mathbf{S} = \left[\begin{smallmatrix} a & b \\ c & d \end{smallmatrix}\right] > \mathbf{O}$ and the *positive* vector of *total output* with no surplus, $\mathbf{q} = \left[\begin{smallmatrix} q_1 \\ q_2 \end{smallmatrix}\right] = \mathbf{Se} = \left[\begin{smallmatrix} a+b \\ c+d \end{smallmatrix}\right] > \mathbf{o}$, both in physical terms.

Then the *positive input-output coefficients matrix* (3.61) in physical terms is computed, $\mathbf{C} = \mathbf{S}\hat{\mathbf{q}}^{-1} > \mathbf{o}$, followed by the eigenvalues of that matrix. One finds:

$$\begin{aligned}\det(\mathbf{C} - \lambda \mathbf{I}_2) &= \begin{vmatrix} \frac{a}{a+b} - \lambda & \frac{b}{c+d} \\ \frac{c}{a+b} & \frac{d}{c+d} - \lambda \end{vmatrix}\\ &= \left(\frac{a}{a+b} - \lambda\right)\left(\frac{d}{c+d} - \lambda\right) - \frac{b}{c+d} \cdot \frac{c}{a+b}\\ &= \lambda^2 - \left(\frac{a}{a+b} + \frac{d}{c+d}\right)\lambda + \frac{ad - bc}{(a+b)(c+d)}\\ &= (\lambda - 1)\left(\lambda - \frac{ad - bc}{(a+b)(c+d)}\right) = 0,\end{aligned} \tag{3.74}$$

giving the eigenvalues, see Proposition 3.1.2,

$$\lambda_1 := \lambda_C = 1, \quad \lambda_2 = \frac{ad - bc}{(a+b)(c+d)}. \tag{3.75}$$

Now we set the general vector of prices $\mathbf{p} = [p_1, p_2]'$ and the general vector of *total output* $\mathbf{q} = [q_1, q_2]'$ in physical terms. We compute the right eigenvectors of the positive transposed *input coefficient matrix* $\mathbf{C}' > \mathbf{O}$ (3.61),

$$\mathbf{C}'\mathbf{p} = \lambda_i \mathbf{p} \Rightarrow \mathbf{C}'\mathbf{p} - \lambda_i \mathbf{p} = \mathbf{o} \Rightarrow (\mathbf{C}' - \lambda_i \mathbf{I}_2)\mathbf{p} = \mathbf{o}. \tag{3.76}$$

As the coefficients $a > 0$, $b > 0$, $c > 0$, $d > 0$ are positive, the matrix $\mathbf{C}$ (3.61) is *positive*.[24] Setting $m = b/c$, one finds[25]

$$\mathbf{p} = [p_1, p_2]' = [p_1, p_2 = m \cdot p_1]' = p_1 \cdot [1, m]', \quad m \in \mathbb{R}^+. \tag{3.77}$$

Then we compute the right eigenvectors of the *input-output coefficients matrix* $\mathbf{C}$ (3.61) in physical terms,

$$\mathbf{C}\mathbf{q} = \lambda_i \mathbf{q} \Rightarrow \mathbf{C}\mathbf{q} - \lambda_i \mathbf{q} = \mathbf{o} \Rightarrow (\mathbf{C} - \lambda_i \mathbf{I}_2)\mathbf{q} = \mathbf{o}. \tag{3.78}$$

For positive coefficients $a > 0$, $b > 0$, $c > 0$, $d > 0$, the matrix $\mathbf{C}$ (3.61) is *positive*. By analogy, there is a positive quantity vector associated to $\lambda_C = 1$. Setting by analogy $k = (a + b)/(c + d)$, one finds,

$$\mathbf{q} = [q_1, q_2]' = [q_1 = k \cdot q_2, q_2]' = q_2 \cdot [k, 1]', \quad k \in \mathbb{R}^+. \tag{3.79}$$

Knowing the coefficients $A > 0$, $m > 0$, $k > 0$ and $\lambda_2 < 1$ (3.75), four equations and inequalities can be set with equations (3.72), (3.77) and (3.79). We get:

$$\begin{cases} k = \frac{q_1}{q_2} = \frac{a+b}{c+d} > 0, \\ m = \frac{p_2}{p_1} = \frac{b}{c} > 0, \\ \lambda_2 = \frac{ad-bc}{(a+b)(c+d)} < 1, \\ A = a > 0. \end{cases} \tag{3.80}$$

The system (3.80) then gives the following positive solutions for a, b, c, d, as can easily be verified:

$$a = A > 0, \quad b = \frac{Am(1-\lambda_2)}{k + m\lambda_2} > 0,$$
$$c = \frac{A(1-\lambda_2)}{k + m\lambda_2} > 0, \quad d = \frac{A(m + k\lambda_2)}{k(k + m\lambda_2)} > 0. \tag{3.81}$$

Summarizing:
The *numéraire* is again wheat. This means: $p_1 = 1 \, \frac{\text{qr. wheat}}{\text{qr. wheat}} = 1$.

(1) Any *positive commodity flow matrix* $\mathbf{S} = \left[\begin{smallmatrix} a & b \\ c & d \end{smallmatrix}\right] > \mathbf{0}$ can be taken to calculate an *input-output coefficients matrix* $\mathbf{C} = \mathbf{S}\hat{\mathbf{q}}^{-1}$ (3.61) in physical terms.

---

24 Lemma 4.1.1 (a) guarantees positive price vectors $\mathbf{p}$ for $\lambda_C = 1$, as will be shown later.

25 The price p_1 calibrates the measurement of the numéraire, here wheat. Therefore Sraffa sets as unit $p_1 = 1 \, \frac{\text{qr. wheat}}{\text{qr. wheat}} = 1$. There would be other possibilities, like $p_1 = 100 \, \frac{\text{kg. wheat}}{\text{qr. wheat}}$.

(2) Given the coefficient $a = A > 0$, the price relationship $m = p_2/p_1$, the quantity relationship between wheat and iron $k = q_1/q_2$, the smaller eigenvalue $\lambda_2 < 1$ of the *input-output coefficients matrix* **C**, the coefficients of the *commodity flow matrix* **S** are

$$a = A > 0, \quad b = \frac{Am(1-\lambda_2)}{k+m\lambda_2}, \quad c = \frac{A(1-\lambda_2)}{k+m\lambda_2}, \quad d = \frac{A(m+k\lambda_2)}{k(k+m\lambda_2)}. \quad \blacktriangle \tag{3.82}$$

3.3.4 Numerical examples

Let's illustrate these results with two numerical examples. The *numéraire* is again wheat. This means: $p_1 = 1\ \frac{\text{qr. wheat}}{\text{qr. wheat}} = 1$. By analogy to price calibration, we calibrate the quantities: $q_2 = c + d$.

Example 3.3.3 (Once again PCMC, Par. 1). Based on *Example* 3.3.2, we set values for the variables A, k, m and λ_2:

- the coefficient $A = 280$;
- the price proportion fixing the slope of the *price eigenvector*, $m = p_2/p_1 = 10$;
- the quantity proportion between wheat and iron, giving the slope of the *quantity eigenvector*, $q_2/q_1 = 1/k = 0.05$;
- the smaller eigenvalue $\lambda_2 = 0.1$ of the *input-output coefficients matrix* **C**.

Compute the coefficients a, b, c, d of the *commodity flow matrix* **S**. For an illustration of the fixed values m and k, see Figure 3.1.

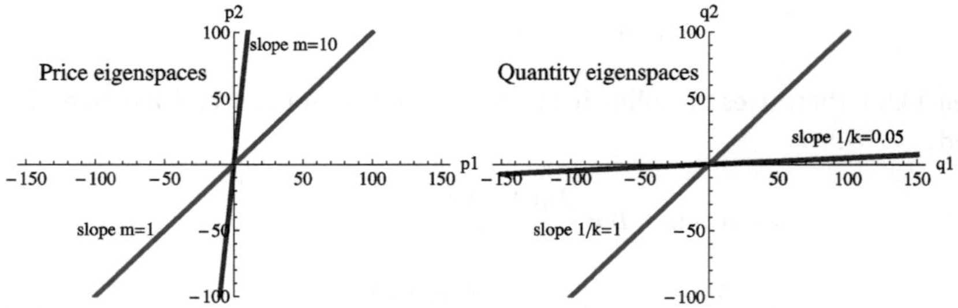

Figure 3.1: Price and quantity eigenspaces of Example 3.3.3 and Example 3.3.4.

Solution to Example 3.3.3:
We obtain, as expected, exactly Sraffa's *commodity flow matrix* **S** of his first numerical example ([108], Par. 1),

$$b = \frac{280 \cdot 10(1-0.1)}{20 + 10 \cdot 0.1} = 120, \quad c = \frac{280(1-0.1)}{20 + 10 \cdot 0.1} = 12,$$

$$d = \frac{280(10 + 20 \cdot 0.1)}{20(20 + 10 \cdot 0.1)} = 8, \quad a = 280. \quad \blacktriangle \tag{3.83}$$

Example 3.3.4 (Further generalization of PCMC, Par. 1). Based on *Example* 3.3.2, we set values for the variables m and k:

- the price relationship fixing the slope of the *price eigenspace* $m = p_2/p_1 = 1$;
- the quantity relationship between wheat and iron, i. e., the slope of the *quantity eigenspace* $1/k = q_2/q_1 = 1$.

Compute the coefficients a, b, c, d of the *commodity flow matrix* **S**. Express them by the unknown variables A and λ_2 only. For an illustration of the fixed values m and k, see Figure 3.1.

Solution to Example 3.3.4:
The *commodity flow matrix* **S** is symmetric because the price and quantity relationships are $m = k = 1$.

$$a = A, \quad b = \frac{A(1-\lambda_2)}{1+\lambda_2}, \quad c = \frac{A(1-\lambda_2)}{1+\lambda_2}, \quad d = A. \quad \blacktriangle \tag{3.84}$$

3.4 A generalization of Sraffa's second example*

This generalization shows how we can replace in Example 3.1.2, see PCMC, Par. 2, the numerical values of the *positive* 3×3 *commodity flow matrix* **S** by any *literal values* and then compute the *characteristic polynomial* of the *input-output coefficients matrix* **C**. We also verify that for an economy without surplus the maximal eigenvalue is $\lambda_C = 1$, as will be stated later by Lemma 4.1.1 (a).

Example 3.4.1 (Generalization of PCMC, Par. 2). Consider three commodities: wheat, iron, pigs, and any positive numbers $a > 0$, $b > 0$, $c > 0$, $d > 0$, $e > 0$, $f > 0$, $g > 0$, $h > 0$, $i > 0$ indicating quantities, describing the subsequent production process:

$$\begin{aligned}(a\text{ qr. wheat}, d\text{ t. iron}, g\text{ pigs}) &\to ((a+b+c)\text{ qr. wheat}, 0, 0),\\ (b\text{ qr. wheat}, e\text{ t. iron}, h\text{ pigs}) &\to (0, (d+e+f)\text{ t. iron}, 0),\\ (c\text{ qr. wheat}, f\text{ t. iron}, i\text{ pigs}) &\to (0, 0, (g+h+i)\text{ pigs}).\end{aligned} \tag{3.85}$$

Set up the *positive commodity flow matrix* **S**, the *positive* vector of *total output* $\mathbf{q} = \mathbf{Se}$ (2.15), the *positive input-output coefficients matrix* $\mathbf{C} = \mathbf{S}\hat{\mathbf{q}}^{-1}$ (2.16) and verify that matrix **C** has Frobenius number $\lambda_C = 1$.

Solution of Example 3.4.1:
One first sets up the *positive commodity flow matrix* $\mathbf{S} > \mathbf{0}$. Here, also the *surplus* vanishes, $\mathbf{d} = \mathbf{0}$. Then the *total output* per sector is calculated, giving the vector of *total*

output $\mathbf{q} = \mathbf{Se} > \mathbf{o}$ (2.15).

$$\mathbf{S} = \begin{bmatrix} a & b & c \\ d & e & f \\ g & h & i \end{bmatrix} := \begin{bmatrix} s_{11} & s_{12} & s_{13} \\ s_{21} & s_{22} & s_{23} \\ s_{31} & s_{32} & s_{33} \end{bmatrix} > \mathbf{0},$$

$$\mathbf{q} = \mathbf{Se} = \begin{bmatrix} q_1 \\ q_2 \\ q_3 \end{bmatrix} = \begin{bmatrix} a+b+c \\ d+e+f \\ g+h+i \end{bmatrix} > \mathbf{o}. \tag{3.86}$$

Then, compute the *positive* matrix $\mathbf{C} = \mathbf{S}\hat{\mathbf{q}}^{-1}$ (2.16):

$$\mathbf{C} = \begin{bmatrix} a & b & c \\ d & e & f \\ g & h & i \end{bmatrix} \begin{bmatrix} \frac{1}{a+b+c} & 0 & 0 \\ 0 & \frac{1}{d+e+f} & 0 \\ 0 & 0 & \frac{1}{g+h+i} \end{bmatrix} = \begin{bmatrix} \frac{a}{a+b+c} & \frac{b}{d+e+f} & \frac{c}{g+h+i} \\ \frac{d}{a+b+c} & \frac{e}{d+e+f} & \frac{f}{g+h+i} \\ \frac{g}{a+b+c} & \frac{h}{d+e+f} & \frac{i}{g+h+i} \end{bmatrix} > \mathbf{0}. \tag{3.87}$$

Matrix **C** is *positive*, because Matrix **S** is *positive*.[26]

It is not cumbersome to verify that $\lambda_C = 1$ is a root of the characteristic polynomial $P_3(\lambda) = 0$. You just need to attain the same numerator everywhere in the fractions of (3.88), namely $(a+b+c)(d+e+f)(g+h+i)$.[27] The *left eigenvector* is a positive price vector $\mathbf{p} > \mathbf{o}$, and respectively the *right eigenvector* is a positive quantity vector $\mathbf{q} > \mathbf{o}$. The eigenvectors associated with the Frobenius number have an economic meaning: they are price and quantity vectors. ▲

3.5 A generalization of Sraffa's third example*

Let's generalise *Example* 3.1.3. The idea is to take for the surplus of wheat not just a number, as in PCMC, Par. 5, but to vary the surplus of wheat, and to replace the

26 The eigenvalues of matrix **C** are computed by setting up the characteristic polynomial

$$\begin{aligned} P_3(\lambda) = \det(\mathbf{C} - \lambda\mathbf{I}_3) = -\lambda^3 &+ \frac{a}{a+b+c}\lambda^2 + \frac{e}{d+e+f}\lambda^2 + \frac{i}{g+h+i}\lambda^2 \\ &+ \frac{bd}{(a+b+c)(d+e+f)}\lambda - \frac{ae}{(a+b+c)(d+e+f)}\lambda + \frac{cg}{(a+b+c)(g+h+i)}\lambda \\ &+ \frac{fh}{(d+e+f)(g+h+i)}\lambda - \frac{ai}{(a+b+c)(g+h+i)}\lambda - \frac{ei}{(d+e+f)(g+h+i)}\lambda \\ &- \frac{ceg}{(a+b+c)(d+e+f)(g+h+i)} + \frac{bfg}{(a+b+c)(d+e+f)(g+h+i)} \\ &+ \frac{cdh}{(a+b+c)(d+e+f)(g+h+i)} - \frac{afh}{(a+b+c)(d+e+f)(g+h+i)} \\ &- \frac{bdi}{(a+b+c)(d+e+f)(g+h+i)} + \frac{aei}{(a+b+c)(d+e+f)(g+h+i)}. \end{aligned} \tag{3.88}$$

27 As the surplus vector vanishes, the Frobenius number λ_C is equal to 1, Lemma 4.1.1 (a).

quantity of 175 qr. of wheat by the variable $x \in \mathbb{R}^+$. We consider also the surplus of iron no longer only as a number, but replace it by the variable $y \in \mathbb{R}^+$. The profit rate $r \in \mathbb{R}$ is the third variable. Thus, we see that we are in presence of a new type of question. We are confronted namely with the analysis of geometric *locus curves and planes*, of the three variables x, y, r. These curves can be implicitly or explicitly presented as a function of one or more variables, for the notations see Chiang ([19], p. 174). Usually, the following cases occur, using explicit functions:

$$r = F(x, y), \quad r = f(x), \quad r = g(y), \quad y = h(x). \tag{3.89}$$

The interest of a geometric *locus* is that it tells how a variable r varies when other variables $x, y, \ldots$ vary.[28] Actually, there are only the three variables x, y, r in our problem. But, if we consider an economy with $n \in \mathbb{N}$ sectors, the question is generalised to a problem of more variables. At the moment, we analyse the geometric *locus* for at most three variables x, y, r. We present some examples with only two of the three variables x, y, r to represent functions in the plane.

3.5.1 Variation of the wheat surplus

In a first example, we vary the surplus of wheat production. Where Sraffa fixed in *Example* 3.1.3, this surplus at 175 qr. wheat, we introduce a *positive* surplus variable for wheat, $x \in \mathbb{R}_0^+$, keeping a zero surplus of iron production, setting $y = 0$. We will study the influence of the wheat surplus x on the rate of profit, which will depend on x as the first variable and on $y = 0$ as the second variable; we then set $r = f(x) > 0$, necessarily positive, because $x = 0$ is excluded. The function $r = f(x)$ is now a geometric *locus curve*.

Example 3.5.1. Consider a similar model to that presented in Example 3.1.3, but now the output of farmers is $(400 + x)$ qr. wheat, $x > 0$, instead of 400 qr. wheat, with the same entries as before. Hence there is a *surplus* and the possibility of *profit*. The iron production is kept at 20 t. iron without surplus, $y = 0$. In analogy to equation (3.1), the production process is symbolized as:

$$\begin{aligned} &(280 \text{ qr. wheat}, 12 \text{ t. iron}) \to ((400 + x) \text{ qr. wheat}, 0), \\ &(120 \text{ qr. wheat}, 8 \text{ t. iron}) \to (0, 20 \text{ t. iron}). \end{aligned} \tag{3.90}$$

In this economy, in which the wage at this stage enters the means of production, calculate the rate of profit $r = f(x) > 0$, depending of the surplus of wheat x, then analyse the prices.

28 The most famous geometric *locus curves* are surely the orbits of the planets around the Sun.

Solution to Example 3.5.1:
We keep the production of iron at 20 t. iron and set for the wheat production the amount of $(400 + x)$ qr. wheat. This gives the vector of *surplus* $\mathbf{d} = [x, 0]'$. Then, we compute the vector of *total output* (2.15), $\mathbf{q} = [400 + x, 20]'$, and the *input-output coefficients matrix* $\mathbf{C} = \mathbf{S}\hat{\mathbf{q}}^{-1}$ (2.16), becoming a function of the wheat surplus x:

$$\mathbf{C}(x) = \begin{bmatrix} 280 & 120 \\ 12 & 8 \end{bmatrix} \begin{bmatrix} \frac{1}{400+x} & 0 \\ 0 & \frac{1}{20} \end{bmatrix} = \begin{bmatrix} \frac{280}{400+x} & 6 \\ \frac{12}{400+x} & \frac{2}{5} \end{bmatrix} > \mathbf{0}. \tag{3.91}$$

Again, because there is a positive surplus, we can apply Lemma 4.1.1 (b) after verifying that the matrix $\mathbf{C}(x)$ is *positive*. This is the case, because for $x \geq 0$ all its elements are strictly positive. Then, the eigenvalues of matrix $\mathbf{C}(x)$ (3.91) are computed.

The characteristic polynomial, now dependent of the variable x, reads:

$$P_2(\lambda, x) = \det(\mathbf{C}(x) - \lambda\mathbf{I}) = \lambda^2 - \frac{280\lambda}{400 + x} - \frac{2\lambda}{5} + \frac{40}{400 + x}. \tag{3.92}$$

The polynomial $P_2(\lambda, x)$ is then factorised, obtaining,

$$\begin{aligned} P_2(\lambda, x) &= (\lambda - \lambda_1(x))(\lambda - \lambda_2(x)) \\ &= \left(\lambda - \frac{1{,}100 + x - \sqrt{810{,}000 + 1{,}200x + x^2}}{5(400 + x)}\right) \\ &\quad \times \left(\lambda - \frac{1{,}100 + x + \sqrt{810{,}000 + 1{,}200x + x^2}}{5(400 + x)}\right). \end{aligned} \tag{3.93}$$

Remember that the eigenvalue is related to the profit rate $r = f(x) > 0$ and is now a function of the surplus x, giving the function $\lambda(x) = 1/(1+f(x)) < 1$ for the existing positive Frobenius number of the positive matrix $\mathbf{C}(x)$ (3.91), corresponding to a *positive left price eigenvector*. Referring to (3.93), one sets for the Frobenius number,

$$\lambda_C(x) = \frac{1}{1 + f(x)} = \frac{1{,}100 + x + \sqrt{810{,}000 + 1{,}200x + x^2}}{5(400 + x)} < 1, \tag{3.94}$$

obtaining the locus for the rate of profit with respect to the wheat surplus $x > 0$,

$$r = f(x) = \frac{900 + 4x - \sqrt{810{,}000 + 1{,}200x + x^2}}{1{,}100 + x + \sqrt{810{,}000 + 1{,}200x + x^2}}. \tag{3.95}$$

We easily verify the values $r_1 = f(0) = 0$, $r_2 = f(175) = 0.25$, as it has to be, due to *Example* 3.1.3.

One also calculates the limits,

$$\begin{aligned} \lim_{x\to\infty} f(x) &= \lim_{x\to\infty} \frac{900 + 4x - \sqrt{810{,}000 + 1{,}200x + x^2}}{1{,}100 + x + \sqrt{810{,}000 + 1{,}200x + x^2}} = \frac{3}{2}, \\ \lim_{x\to 0} f(x) &= \lim_{x\to 0} \frac{900 + 4x - \sqrt{810{,}000 + 1{,}200x + x^2}}{1{,}100 + x + \sqrt{810{,}000 + 1{,}200x + x^2}} = 0. \end{aligned} \tag{3.96}$$

The limits (3.96) are positive rates of profit, $r = f(x) > 0$. The function of the wheat surplus $x > 0$ with no iron surplus, $y = 0$, has an upper bound of 150 % and a lower bound of 0 %. Sraffa's $r = 25\,\%$ is just a special case for $x = 175$ and $y = 0$.

The limits (3.96) of the rate of profits $r = f(x) > 0$ also determine the limits of the Frobenius numbers of the *input coefficients matrix* in physical terms $\mathbf{C}(x)$ (3.91). We get indeed:

$$\lim_{x\to\infty} \lambda_1(x) = \lim_{x\to\infty} \frac{1{,}100 + x + \sqrt{810{,}000 + 1{,}200x + x^2}}{5(400 + x)} = \frac{2}{5},$$

$$\lim_{x\to 0} \lambda_1(x) = \lim_{x\to 0} \frac{1{,}100 + x + \sqrt{810{,}000 + 1{,}200x + x^2}}{5(400 + x)} = 1. \tag{3.97}$$

Sraffa's *rate of profits* $r = 1/4$ is just a special case for the wheat surplus $x = 175$ and the corresponding Frobenius number $\lambda = 1/(1 + \frac{1}{4}) = 4/5$ that is of course within the calculated open interval $]0.4, 1[$.

We now proceed to the *left eigenvector* equation for matrix $\mathbf{C}(x)$, (3.91), and calculate for the Frobenius number $\lambda_C(x)$ the positive price vector $\mathbf{p} > \mathbf{o}$, at present also a function of the wheat surplus x, therefore noted as $\mathbf{p}(x) > \mathbf{o}$,

$$\mathbf{p}(x)'\mathbf{C}(x) = \mathbf{p}(x)' \begin{bmatrix} \frac{280}{400+x} & 6 \\ \frac{12}{400+x} & \frac{2}{5} \end{bmatrix} = \mathbf{p}(x)'\lambda_1(x) = \frac{1}{1 + f(x)}\mathbf{p}(x)'. \tag{3.98}$$

With the calibration $p_1(x) = 1$, we compute the price component $p_2(x)$ as a function of the wheat surplus x, thus making it a *relative price*. We obtain for the positive price vector $\mathbf{p}(x) = [1, p_2(x)]'$ explicitly,

$$\mathbf{p}(x) = \begin{bmatrix} 1 \\ \frac{30(400+x)}{72{,}000+1{,}800x}(-300 + x + \sqrt{810{,}000 + 1{,}200x + x^2}) \end{bmatrix}. \tag{3.99}$$

Only the relative price $p_2(x)$ varies in function of the wheat surplus x,

$$p_2(x) = \frac{30(400 + x)}{72{,}000 + 1{,}800x}(-300 + x + \sqrt{810{,}000 + 1{,}200x + x^2}), \tag{3.100}$$

which has the asymptote:

$$a(x) = \frac{1}{30}x + 5. \tag{3.101}$$

Calculating the relative price $p_2(x)$ (3.100) and the rate of profits $r = f(x)$ (3.95) for two specific values of x, namely $x = 0$ and $x = 175$ and the zero surplus of iron $y = 0$, gives the triplets: ($p_1(0) = 1$, $p_2(0) = 10$, $r_1 = f(0) = 0$), ($p_1(175) = 1$, $p_2(175) = 15$, $r_2 = f(175) = 0.25$), as it has to be, according to *Example* 3.1.1 *and Example* 3.1.3. Then we can go on and find round numbers for the wheat surplus $x = 825$: ($p_1(825) = 1$, $p_2(825) = 35$, $r_3 = f(825) = 0.75$).

In *Example* 3.1.3 (PCMC, Par. 5), the wheat surplus is 175 qr. of wheat and there is no iron surplus, $y = 0$. Sraffa presents numerical values for the exchange-values of that example as follows:

"The exchange-ratio which enables the advances to be replaced and the profits to be distributed to both industries in proportion to their advances, namely 15 qr. *of* wheat *for 1* t. *of* iron; *and the corresponding rates of profits in each industry is 25 %".*

We have attained a generalization, where the numbers are replaced by variables. It works as follows: the surplus 175 qr. of wheat is replaced by a general surplus of x qr. of wheat. There is still no iron surplus. We have been able to calculate the exchange-ratio as (3.100) $p_2(x)$ qr. of wheat to 1 t. of iron, depending on the variable surplus of wheat x, instead Sraffa's constant proportion: 15 qr. of wheat to 1 t. of iron for his constant surplus of 175 qr. of wheat.

Consequently, Sraffa's fixed rate of profits of 25 % in each industry is replaced by a calculable rate of profits, namely by the *geometric locus curve* $r = f(x)$ (3.95), a function of the wheat surplus x, see Figure 3.2 (left). ▲

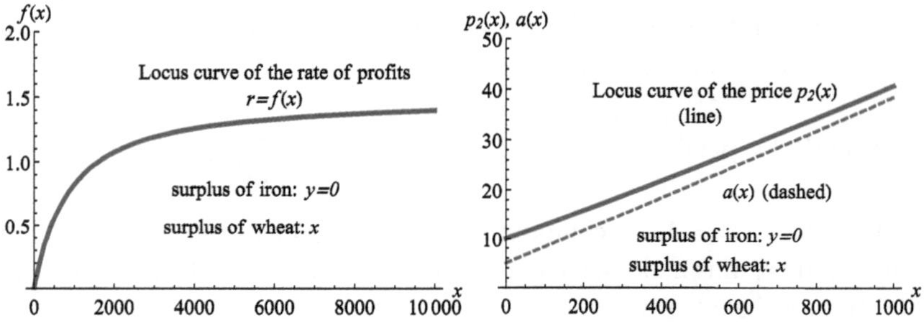

Figure 3.2: Locus curves—the rate of profits $r = f(x)$ (3.95) (left), the price $p_2(x)$ (3.100) and its asymptote $a(x)$ (3.101) (right).

3.5.2 Variation of the iron surplus

In a second attempt, we vary the surplus of iron production and keep the wheat production at 400 qr. wheat without surplus, $x = 0$. We introduce a surplus y of iron, measured in tons, $y > 0$. We will study the influence of the iron surplus on the rate of profit, and we set for this purpose $r = g(y)$, giving a geometric locus.

Example 3.5.2. Consider a similar model to that presented in Example 3.5.1, but now the output of farmers is 400 qr. wheat. It is kept constant, whereas the surplus of iron varies. It is $(20 + y)$ t. iron, with the same entries as in *Example* 3.1.1. Hence there is a *surplus* and the possibility of *profit*.

In analogy to equation (3.41), the production process is symbolised as follows:

$$\begin{aligned}(280 \text{ qr. wheat}, 12 \text{ t. iron}) &\rightarrow (400 \text{ qr. wheat}, 0),\\ (120 \text{ qr. wheat}, 8 \text{ t. iron}) &\rightarrow (0, (20 + y) \text{ t. iron}).\end{aligned} \tag{3.102}$$

In this economy, exhibiting again subsistence wages, calculate the rate of profit $r = g(y)$, depending of the surplus of iron y, and analyse the prices.

Solution to Example 3.5.2:
We keep the production of wheat at 400 qr. wheat and set for the iron production the amount $(20+y)$ t. iron. This gives the vector of *surplus* $\mathbf{d} = [0, y]'$. We then compute the vector of *total output* (2.15), $\mathbf{q} = [400, 20+y]'$, and the *input-output coefficients matrix* $\mathbf{C} = \mathbf{S}\hat{\mathbf{q}}^{-1}$ (2.16), making it a function of the iron surplus y:

$$\mathbf{C}(y) = \begin{bmatrix} 280 & 120 \\ 12 & 8 \end{bmatrix} \begin{bmatrix} \frac{1}{400} & 0 \\ 0 & \frac{1}{20+y} \end{bmatrix} = \begin{bmatrix} \frac{280}{400} & \frac{120}{20+y} \\ \frac{12}{400} & \frac{8}{20+y} \end{bmatrix} > \mathbf{0}. \tag{3.103}$$

As there is a positive surplus, Lemma 4.1.1 (b) applies, and as the elements of matrix $\mathbf{C}(y)$ (3.103) are with $y > 0$ strictly *positive*, the matrix itself is *positive*. Then, the eigenvalues of matrix $\mathbf{C}(y)$ (3.103) are computed, using the characteristic polynomial, where y is now the second variable,

$$P_2(\lambda, y) = \det(\mathbf{C}(y) - \lambda\mathbf{I}) = \lambda^2 - \frac{8\lambda}{20+y} - \frac{7\lambda}{10} + \frac{2}{20+y}. \tag{3.104}$$

The polynomial $P_2(\lambda, y)$ is factorised, obtaining,

$$\begin{aligned} P_2(\lambda, y) &= (\lambda - \lambda_1(y))(\lambda - \lambda_2(y)) \\ &= \left(\lambda - \frac{220 + 7y - \sqrt{32{,}400 + 2{,}280y + 49y^2}}{20(20+y)}\right) \\ &\quad \times \left(\lambda - \frac{220 + 7y + \sqrt{32{,}400 + 2{,}280y + 49y^2}}{20(20+y)}\right). \end{aligned} \tag{3.105}$$

We now relate the eigenvalues to the profit rate $r = g(y) > 0$ as a function of the surplus y, giving the locus equation $0 < \lambda(y) = 1/(1 + g(y)) < 1$. We choose the maximal eigenvalue of matrix $\mathbf{C}(y)$ (3.103), easily recognizable in the factorised polynomial $P_2(\lambda, y)$ (3.105), setting thus the Frobenius number,

$$\lambda_C(y) = \frac{1}{1 + g(y)} = \frac{220 + 7y + \sqrt{32{,}400 + 2{,}280y + 49y^2}}{20(20+y)} < 1. \tag{3.106}$$

The locus curve for the rate of profits depending on the iron surplus y is,

$$r = g(y) = \frac{180 + 13y - \sqrt{32{,}400 + 2{,}280y + 49y^2}}{220 + 7y + \sqrt{32{,}400 + 2{,}280y + 49y^2}} > 0. \tag{3.107}$$

Then, one also calculates the limits,

$$\lim_{y\to\infty} g(y) = \lim_{y\to\infty} \frac{180 + 13y - \sqrt{32{,}400 + 2{,}280y + 49y^2}}{220 + 7y + \sqrt{32{,}400 + 2{,}280y + 49y^2}} = \frac{3}{7},$$

$$\lim_{y\to 0} g(y) = \lim_{y\to 0} \frac{180 + 13y - \sqrt{32{,}400 + 2{,}280y + 49y^2}}{220 + 7y + \sqrt{32{,}400 + 2{,}280y + 49y^2}} = 0. \tag{3.108}$$

The limits (3.108) mean that the rate of profits $r = g(y)$, a function of the iron surplus y without wheat surplus $x = 0$, has the lower bound $r_{\min} = 0$ and the upper bound $r_{\max} = 3/7 = 42.857\,\%$, corresponding to the Frobenius numbers $\lambda_C = 1$ and the second eigenvalue $\lambda_2 = 1/(1+\frac{3}{7}) = 7/10 = 0.7$. Remember that if we have no wheat surplus $x = 0$ and an absence of iron surplus $y = 0$, we easily verify the *rate of profits* $r = g(0) = 0$, giving the Frobenius number (3.104) $\lambda_C = 1/(1 + g(0)) = 1$, as expected. We also verify that, for an iron surplus of $y = 35$, we get $r = g(35) = 0.25$ with Frobenius number $\lambda_C = 1/(1 + \frac{1}{4}) = (4/5) \in\,](3/7), 1[$.

Now we consider the *left eigenvector* equation for matrix $\mathbf{C}(y)$ (3.103), associated to the Frobenius number $\lambda_C(y)$, and the positive price vector, expressed as a function of the iron surplus y. We set up for this price vector:

$$\mathbf{p}(y)'\mathbf{C}(y) = \begin{bmatrix} \frac{280}{400} & \frac{120}{20+y} \\ \frac{12}{400} & \frac{8}{20+y} \end{bmatrix} = \lambda_C(y)\mathbf{p}(y)'$$
$$= \mathbf{p}(y)'\lambda_C(y) = \frac{1}{1 + r(y)}\mathbf{p}(y)'. \tag{3.109}$$

We have explicitly the positive price vectors,

$$\mathbf{p}(y) = [p_1(y), p_2(y)]' = \left[\frac{60 + 7y + \sqrt{32{,}400 + 2{,}280y + 49y^2}}{2{,}400}, 1\right]', \tag{3.110}$$

calibrating with $p_2(y) = 1$ (iron is the numéraire), we get the relative price of wheat $p_1(y)$ as a function of the iron surplus y,

$$p = p_1(y) = \frac{60 + 7y + \sqrt{32{,}400 + 2{,}280y + 49y^2}}{2{,}400}. \tag{3.111}$$

The asymptote of the function $p_1(y)$ calculated, giving:

$$a(y) = \frac{7}{1{,}200}y + \frac{13}{140}. \tag{3.112}$$

Computing the relative price $p_1(y)$ (3.111) and the rate of profits $r = g(y)$ (3.107) for one specific value of y, namely $y = 0$, and also no surplus of wheat, $x = 0$, we obtain the triplet: $(p_1(0) = 0.1, p_2(0) = 1, r = g(0) = 0)$, as it has to be according to *Example* 3.1.1. Then we can go on and find for the arbitrary chosen iron surpluses $y = 35$ and $y = 58$

two other triplets: ($p_1(35) = 1$, $p_2(35) = 1$, $r = g(35) = 0.25$), respectively ($p_1(58) = 39/90$, $p_2(58) = 1$, $r = g(58) = 0.3$), giving in both cases relative prices $p_1(35) = 1$, respectively, $p_1(58) = 39/90$ and rates of profits $r = g(y = 35) = 0.25$, respectively, $r = g(y = 58) = 0.3$.

The rate of profits is the *geometric locus curve* $r = g(y)$ (3.107), Figure 3.3 (left). The relative price of iron is the *geometric locus curve* $p = p_1(y)$ (3.111), Figure 3.3 (right). ▲

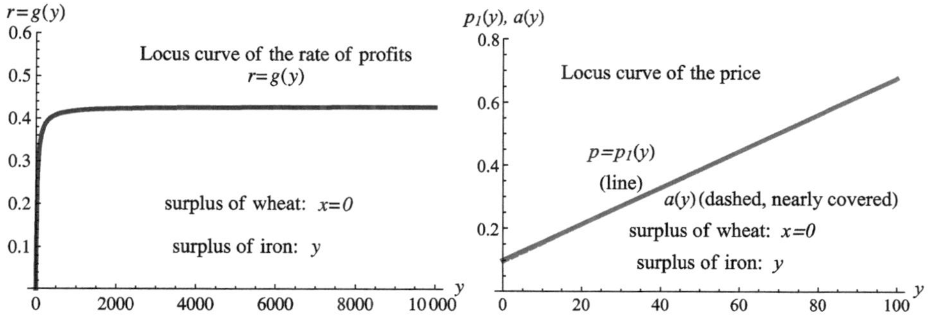

Figure 3.3: Locus curves—the rate of profits $r = g(y)$ (3.107) (left), the price $p = p_1(y)$ (3.111) and its asymptote $a(y)$ (3.112) (right).

3.5.3 Simultaneous variation of the wheat and iron surplus

Here, we examine a further generalization of PCMC, Par. 5, in connection with the surplus. We vary the surplus production of wheat and iron, considering both a surplus x of wheat and a surplus y of iron, $x > 0, y > 0$. We study the influence of the simultaneous wheat and iron surplus on the rate of profit r. We will specially treat the geometric *locus planes* for a constant profit rate, $r = F(x, y) =$ constant, being now a geometric locus of both surplus variables x and y.

We will then analyse the prices in dependence of the surpluses x of wheat and y iron.

Example 3.5.3. Let's start with the model as presented in Example 3.1.1, but now the output of farmers is $(400 + x)$ qr. wheat instead of 400 qr. wheat, the output of iron production is $(20 + y)$ t. iron instead of 20 t. iron, for $x > 0$, $y > 0$. The left entries are unchanged.

By analogy to equation (3.1) the production process presents as follows:

$$\begin{aligned} &(280 \text{ qr. wheat}, 12 \text{ t. iron}) \to ((400 + x) \text{ qr. wheat}, 0), \\ &(120 \text{ qr. wheat}, 8 \text{ t. iron}) \to (0, (20 + y) \text{ t. iron}). \end{aligned} \tag{3.113}$$

We assume again that the rate of profits r is equal in all branches and the prices satisfy the equations and compute the vector of values $\mathbf{x}$:

$$
\begin{aligned}
(280p_1 + 12p_2)(1 + r) &= (400 + x)p_1 = x_1 \text{ (qr. wheat)},\\
(120p_1 + 8p_2)(1 + r) &= (20 + y)p_2 = x_2 \text{ (qr. wheat)}.
\end{aligned} \tag{3.114}
$$

Solution to Example 3.5.3:
In (3.114), "qr. wheat" is again the (numéraire). The prices p_1, p_2 are accordingly measured in the same physical units as in equation (3.2).

By analogy, the *positive commodity flow matrix* **S** is unchanged relative to *Example* 3.1.3. Here, again the vector of *surplus* in physical terms is set up, $\mathbf{d} = [x, y]'$, x meaning qr. wheat surplus and y meaning t. iron surplus. Then the vector of *total output* per sector is calculated,

$$
\mathbf{S} = \begin{bmatrix} s_{11} & s_{12} \\ s_{21} & s_{22} \end{bmatrix} = \begin{bmatrix} 280 & 120 \\ 12 & 8 \end{bmatrix} > \mathbf{0},
$$

$$
\mathbf{q} = \mathbf{q}(x,y) = \mathbf{Se} + \mathbf{d} = \begin{bmatrix} 280 & 120 \\ 12 & 8 \end{bmatrix}\begin{bmatrix} 1 \\ 1 \end{bmatrix} + \begin{bmatrix} x \\ y \end{bmatrix} = \begin{bmatrix} 400 + x \\ 20 + y \end{bmatrix} > \mathbf{o}, \tag{3.115}
$$

the vector of *total output* becoming dependent of the variables x, y, and consequently also the *input-output coefficients* matrix $\mathbf{C}(x,y) = \mathbf{S}\hat{\mathbf{q}}(x,y)^{-1}$ (2.16) in physical terms:

$$
\mathbf{C}(x,y) = \begin{bmatrix} 280 & 120 \\ 12 & 8 \end{bmatrix}\begin{bmatrix} \frac{1}{400+x} & 0 \\ 0 & \frac{1}{20+y} \end{bmatrix} = \begin{bmatrix} \frac{280}{400+x} & \frac{120}{20+y} \\ \frac{12}{400+x} & \frac{8}{20+y} \end{bmatrix} > \mathbf{0}. \tag{3.116}
$$

As there is a positive surplus, we resort to Lemma 4.1.1 (b) and know that the Frobenius number is smaller than one, $\lambda_C < 1$. We have just to verify that matrix $\mathbf{C}(x,y)$ is *positive*. This is the case with $x > 0$, $y > 0$.

The eigenvalues of matrix $\mathbf{C}(x,y)$ are then computed. Setting up the characteristic polynomial, $P_2(\lambda) = \det(\mathbf{C}(x,y) - \lambda\mathbf{I})$ gives the eigenvalues, depending on the variables x, y,

$$
\lambda_{1,2}(x,y) = \frac{4(1{,}100 + x + 35y \pm \sqrt{810{,}000 + 1{,}200x + x^2 + 57{,}000y + 20xy + 1{,}225y^2})}{8{,}000 + 20x + 400y + xy}. \tag{3.117}
$$

The polynomial $P_2(\lambda)$ is factorised, obtaining $P_2(\lambda) = (\lambda - \lambda_1)(\lambda - \lambda_2)$. We choose the Frobenius number $0 < \lambda_C = 1/(1 + r) < 1$ and relate it immediately to the rate of profits $r > 0$.

(1) Setting $r = 1/4$, following Sraffa (PCMC, Par. 5), gives for the Frobenius number

$$
\begin{aligned}
\lambda_C(x,y) &= \frac{1}{1 + 0.25} = 0.8 = H(x,y)\\
&= \frac{4(1{,}100 + x + 35y + \sqrt{810{,}000 + 1{,}200x + x^2 + 57{,}000y + 20xy + 1{,}225y^2})}{8{,}000 + 20x + 400y + xy},
\end{aligned} \tag{3.118}
$$

which is the *geometric locus* in implicit form that we are looking for. Solving the equation (3.118) for y, one gets a hyperbolic function as *locus curve* for the constant rate of profits $r = 1/4$, see Figure 3.4 (left).

$$y = h(x) = \frac{1{,}750 - 10x}{50 + x}. \tag{3.119}$$

(2) We repeat the operation with the constant rate of profits $r = 3/7$, obtaining

$$\begin{aligned}\lambda_C(x,y) &= \frac{1}{1+r} = \frac{1}{1+\frac{3}{7}} = 0.7 = J(x,y)\\ &= \frac{4(1{,}100 + x + 35y + \sqrt{810{,}000 + 1{,}200x + x^2 + 57{,}000y + 20xy + 1{,}225y^2})}{8{,}000 + 20x + 400y + xy},\end{aligned} \tag{3.120}$$

solving again (3.120) for the variable y, one gets another hyperbolic function as *locus curve* for the constant rate of profits $r = 3/7$, see Figure 3.4 (right),

$$y = h(x) = \frac{2{,}938.7755 - 8.57143x}{x}. \tag{3.121}$$

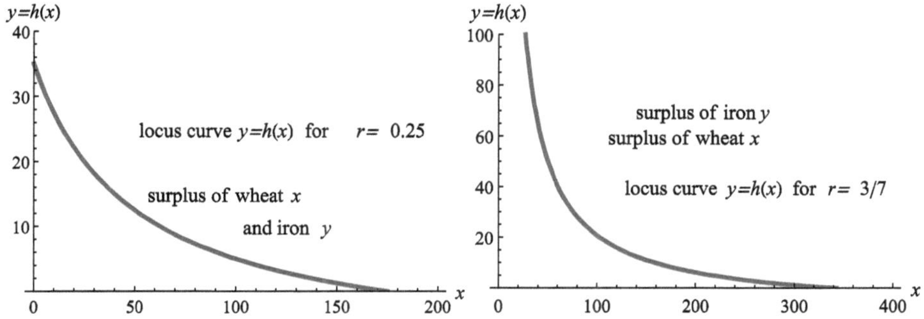

Figure 3.4: Geometric *locus curves* (PCMC, Par. 5) with constant rates of profit (3.119) and (3.121).

Obviously, every constant profit rate r generates its own geometric *locus curve*, $y = h(x)$ (3.119) for $r = 1/4$, and (3.121) for $r = 3/7$.

Examples of points, measured in the units [qr. of wheat], [t. of iron] are $P_1(175, 0)$, $P_2(0, 35)$, $P_3(25, 20)$ situated on the *locus curve* (3.119) of Figure 3.4 (left).

Now consider the constant rate of profits $r = 0.25$ for which we will investigate the price vector $\mathbf{p}$. The *left eigenvector* equation with the *input-output coefficients matrix* $\mathbf{C}(x,y)$ (3.116), now depending on the variables x, y, is set up to calculate this price vector $\mathbf{p} = \mathbf{p}(x,y)$. As we are on the locus curve (3.119), we immediately set $\lambda(x,y)\mathbf{p}(x,y)' = 1/(1+r) = 1/(1+0.25) = 0.8$, following (3.118), getting

$$\mathbf{p}(x,y)'\mathbf{C}(x,y) = \mathbf{p}(x,y)'\begin{bmatrix} \frac{280}{400+x} & \frac{120}{20+y} \\ \frac{12}{400+x} & \frac{8}{20+y} \end{bmatrix} = \lambda(x,y)\mathbf{p}(x,y)' = 0.8\mathbf{p}(x,y)'. \tag{3.122}$$

We now calculate the positive price vector $\mathbf{p}(x,y) = [p_1(x,y), p_2(x,y)]'$ on the *locus curve* (3.119). We calibrate $p_1(x,y) = 1$ (wheat is the numéraire) and calculate the proportion:

$$p_2(x,y) : p_1(x,y) = p_2(x,y) : 1 = p_2(x,y),$$

being the price, specific to any chosen point $P(x,y)$ on that locus (3.119).

We get for the reciprocal of the relative price $p_2(x,y)$

$$\frac{1}{p_2(x,y)} = -\frac{1}{30}\left(2 - \frac{1{,}100 + x + 35x + \sqrt{810{,}000 + 1{,}200x + x^2 + 57{,}000y + 20xy + 1{,}225y^2}}{400 + x}\right). \tag{3.123}$$

We have then solved the following problem:

For a given rate of profits r, say $r = 1/4$, and any arbitrary surplus x of wheat production, compute the *locus curve* $y = h(x)$ (3.119), giving the surplus of iron production. Then, setting 1 qr. of wheat again as the *numéraire*, we calculate the relative price $p_1 = 1$ (qr. of wheat)/(qr. of wheat) $= 1$ with respect to its physical unit (dimensionless) and compute the relative price of iron $p_2(x,y)$ (3.123). This relative price is also the value of the price proportion $p_2(x,y) : p_1(x,y) = p_2(x,y)$. The economic meaning is: For a fixed rate of profits r and one of the surpluses, say the wheat surplus x, the locus curve $y = h(x)$ is determined and therefore also the price proportion $p_2(x,y) : p_1(x,y)$ between iron and wheat. With (3.117), you may also control that you have obtained the correct Frobenius number, which is $\lambda_C = 0.8 = 1/(1+r)$, giving again the maximal profit rate $R = r = 1/4$. The vector of *values* is $\mathbf{x} = \hat{\mathbf{q}}\mathbf{p} = [(400 + x)p_1(x,y), (20 + y)p_2(x,y)]'$. ▲

We now see that, in the context of PCMC, Par. 5, there are four variables, operating to determine profit rate r: the wheat surplus x, the iron surplus y and the prices p_1 and p_2. Considering x, y, p_1, p_2 as independent, we have a *four-dimensional locus* problem which is difficult to treat. As we have decided to present *locus curves* in the plane, we accordingly select two of these four variables in order to define a *two-dimensional locus curve* problem.

Example 3.5.4. We investigate the dependence of the rate of profits r on the wheat surplus x, given a constant iron surplus $y = y_0$ and consider the surface $r = I(x,y)$. We carry out three cuts through the surface $r = I(x,y)$. We realize three cuts for the specific values $y_1 = 0$, $y_2 = 10$, $y_3 = 20$ parallel to the r-axis and x-axis through the curved surface $r = I(x,y)$ and obtain *locus curves*.

- Compute then the limits $\lim_{x\to\infty} I(x, y_i)$ for $i = 1, 2, 3$.
- Represent the curves $r = f(x) := I(x, y = y_1)$, $r = g(x) := I(x, y = y_2)$, $r = h(x) := I(x, y = y_3)$ in the (r,x)-plane. Compute $R = I(x = 175, y = 0)$.

Solution to Example 3.5.4:
Setting the relation between the Frobenius number $\lambda_C = \lambda_C(x,y)$ and the rate of profits r, $\lambda_C = 1/(1+r) = 1/(1+I(x,y))$, we get with equation (3.117),

$$\lambda(x,y) = \frac{1}{1+I(x,y)} = \frac{4(1{,}100 + x + 35y \pm \sqrt{810{,}000 + 1{,}200x + x^2 + 57{,}000y + 20xy + 1{,}225y^2})}{8{,}000 + 20x + 400y + xy}. \tag{3.124}$$

Then, we consider the different iron surplus productions:
(a) $y_1 = 0$ and find:

$$r = f(x) = I(x, y = y_1 = 0) = \frac{(900 + x - \sqrt{810{,}000 + 1{,}200x + x^2})}{200}; \tag{3.125}$$

(b) $y_2 = 10$ and find:

$$r = g(x) = I(x, y = y_2 = 10) = \frac{(1{,}250 + x - \sqrt{1{,}502{,}500 + 1{,}400x + x^2})}{200}; \tag{3.126}$$

(c) $y_3 = 20$ and find:

$$r = h(x) = I(x, y = y_3 = 20) = \frac{(1{,}600 + x - \sqrt{2{,}440{,}000 + 1{,}600x + x^2})}{200}. \tag{3.127}$$

For the limits, we obtain

$$\lim_{x\to\infty} f(x) = \frac{3}{2} = 1.5, \quad \lim_{x\to\infty} g(x) = \frac{11}{4} = 2.75, \quad \lim_{x\to\infty} h(x) = 4. \tag{3.128}$$

The three *locus curves* $r = f(x)$, $r = g(x)$ and $r = h(x)$ are horizontal cuts along the vertical y-axis and parallel to the (r,x)-plane through the curved surface $r = I(x,y)$. They are visualized in the (r,x)-plane, see Figure 3.5. The value $R = I(175, 0) = 0.25$ is exactly the profit rate of PCMC, Par. 5, indicated by Sraffa. ▲

Concluding, the locus plane $I(x,y)$ represents the *rate of profits* of the Sraffa example PCMC, Par. 5, and as such it is the *productiveness* of this economy with given technology. We recognize that the productiveness is a function of the surplus, given a technology of a production system.

3.6 Subsistence wages and wages as a part of the surplus

Up to now, no separate vector of *labour* has been used, although labour is the essential ingredient required for the generation of value added. Resorting to labour means paying wages and the explicit inclusion of a wage element in Sraffa's model is the

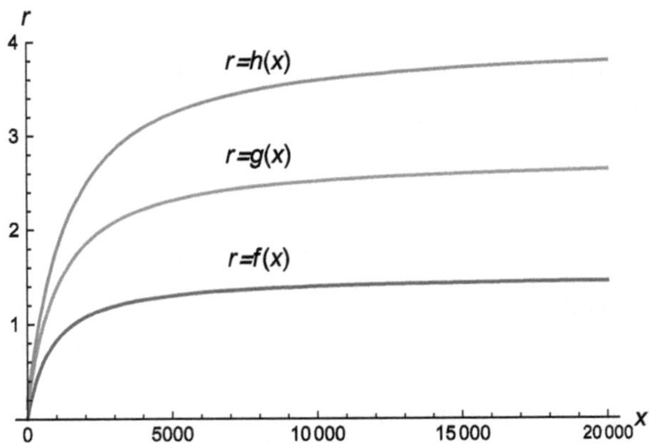

Figure 3.5: Geometric *locus curves* (3.140) $f(x)$, (3.126) $g(x)$ and (3.127) $h(x)$.

next important step to be undertaken. Up to now labour, paid at "subsistence wages", has been assumed as being a constituent of the *commodity flow* matrix $\mathbf{S}$ (respectively $\mathbf{S}'$), included in parts of some of the means of production. Sraffa (PCMC, Par. 8) writes:

> *"...In view of the double character of wages it would be appropriate, when we come to consider the division of the surplus between producers and workers, to separate the two component part of wages and regard only the surplus part as variable; whereas the goods necessary for the subsistence of the workers would continue to appear with the fuel, etc,.... among the means of production. We shall nevertheless, ..., follow the usual practice of treating the whole of wages as variable".*

By the double character of wages, Sraffa refers to the concept of wages as consisting of a basic constant income in the form of goods, initially required for survival, and of a variable part as income participation in the surplus. "Wage as a variable" means here that the wage participates in the creation of the surplus and is in principle considered as a variable in the same sense as the *prices* or the *rate of profits* entering his model[29] Initially, the *subsistence* part of wages does not contribute to any added value.[30]

A straightforward way of operating the separation of wages from the means of production is to proceed as follows. Assume that in *single-product industry* S_j, "subsistence wages" are paid out in kind with commodity $i = j$ produced by that industry S_j, say as quantity σ_j, which then has value $\sigma_j p_j$. Historical examples are payments in "salt" or "corn". This approach implies that there exists a market where commodities are exchanged or sold, a market that by the way must exist if the surplus introduced in the *complete single-product Sraffa system* is to be absorbed in order for producers

29 Note that in Leontief's framework labour is incorporated in the input matrix $\mathbf{A}$ (see (2.8) and (2.9)).

30 The classic example is agriculture: labour and wages paid to workers just to produce the harvest required for sustainability of production in the next season do not contribute to the production of a surplus and thus do not create *value added*.

to see their profits effectively materialize. Sraffa alludes to this in PCMC, Par. 1. The payment of a subsistence wage, say σ_j, $j = 1, \ldots, n$, is therefore included as a part of the diagonal components of the matrix $\mathbf{S}'$ which then, with the price vector $\mathbf{p}$, takes the following form:

$$\begin{bmatrix} \dot{s}_{11} + \sigma_1 & s_{21} & \cdots & s_{(n-1)1} & s_{n1} \\ s_{12} & \dot{s}_{22} + \sigma_2 & \cdots & s_{(n-1)2} & s_{n2} \\ \cdots & \cdots & \cdots & \cdots & \cdots \\ \cdots & \cdots & \cdots & \cdots & \cdots \\ s_{1(n-1)} & s_{2(n-1)} & \cdots & \dot{s}_{(n-1)(n-1)} + \sigma_{n-1} & s_{n(n-1)} \\ s_{n1} & s_{n2} & \cdots & s_{(n-1)n} & \dot{s}_{nn} + \sigma_n \end{bmatrix}, \begin{bmatrix} p_1 \\ p_2 \\ \cdots \\ \cdots \\ p_{n-1} \\ p_n \end{bmatrix}. \tag{3.129}$$

The component s_{jj} has now been split into two parts: $s_{jj} = \dot{s}_{jj} + \sigma_j$, $j \in \{1, \ldots, n\}$. We can now split $\mathbf{S}'$ into a square residual matrix $\dot{\mathbf{S}}'$ and a diagonal matrix $\hat{\boldsymbol{\sigma}}$:

$$\mathbf{S}' = \dot{\mathbf{S}}' + \hat{\boldsymbol{\sigma}}, \tag{3.130}$$

and multiplication of the diagonal matrix $\hat{\boldsymbol{\sigma}}$ by the price vector $\mathbf{p}$ gives a column vector $\hat{\boldsymbol{\sigma}}\mathbf{p} = [\sigma_1 p_1, \ldots, \sigma_j p_j, \ldots, \sigma_n p_n]'$ on the left-hand side of the system of equations defining the conditions of production.

Now, $\sigma_j p_j$ has as a counter-value the expression $w_s \dot{L}_j$,[31] representing the *labour employed* in industry S_j, at subsistence wage rate w_s, required to realise the conditions of production for each industry (see PCMC, Par. 3). These expressions are gathered to form the vector $w_s \dot{\mathbf{L}} = [w_s \dot{L}_1, \ldots, w_s \dot{L}_n]'$.

The situation changes radically with the production of a surplus $\mathbf{d}' = [d_1, \ldots, d_n]$ bringing added value, which will then entail for each industry: either described for each industry,

- an increase in wages $w_a \dot{L}_j$ paid to the labour force, realising labour $\dot{L}_j$, which now is recognised as contributing to producing a surplus, i. e., value added, at *wage rate* w_a, and possibly;
- an additional increase in the labour force, measured by $\Delta\dot{L}_j$, at the total wage rate $w = w_a + w_s$ incurring additional wages $w\Delta\dot{L}_j = (w_a + w_s)\Delta\dot{L}_j$ required by the production process.

We will accordingly have then as total wages W_j payable in industry j:

$$\begin{aligned} W_j &= w_s \dot{L}_j + w_a \dot{L}_j + (w_a + w_s)\Delta\dot{L}_j \\ &= w(\dot{L}_j + \Delta\dot{L}_j) = wL_j. \end{aligned} \tag{3.131}$$

If $\Delta\dot{L}_j = 0$ then $\dot{\mathbf{L}} = \mathbf{L}$.

31 We follow Sraffa by assuming uniform wage rates throughout the industries. Remember in this connection that the measurement unit in Sraffa is in principle a *physical numéraire*, so this also applies to wages. We shall introduce a *monetary numéraire* later on.

In vector form:

$$w\mathbf{L} = \begin{bmatrix} (w_s + w_a)(\dot{\mathbf{L}} + \Delta\dot{\mathbf{L}}), & \Delta\dot{\mathbf{L}} > \mathbf{o}, \\ (w_s + w_a)\mathbf{L}, & \Delta\dot{\mathbf{L}} = \mathbf{o}. \end{bmatrix} \quad (3.132)$$

But in Subsection 3.1.3 we have seen that a surplus is assumed, without participation of labour in value added.

The "subsistence wage" approach raises therefore a number of problems and in addition complicates the task of addressing the question of the split of the surplus between producers and workers when a surplus has been generated. So, as Sraffa proposes (PCMC, Par. 8), and subsequently materialises by introducing *profits* and *wages* in his model construction, a qualitative change in outlook is undertaken and labour and wages are regrouped de facto in a separate vector $w\mathbf{L}$ (also expressible as a diagonal matrix $w\hat{\mathbf{L}}$), both w and $\mathbf{L}$ assumed known.

The uniform rate of profits r allotted to producers is then obtained from Sraffa's price model, which we shall analyse in the following Chapter 4,

$$\dot{\mathbf{S}}'\mathbf{p}(1 + r) = \hat{\mathbf{q}}\mathbf{p} - w\hat{\mathbf{L}}. \quad (3.133)$$

These equations restrict the matrix $\mathbf{S}'$ to the technical means of production excluding labour (in fact now $\mathbf{S}'$ should become $\dot{\mathbf{S}}'$).[32] This means that the profit rate, $r = R$, is the *productiveness* obtained here by setting $w = 0$ and is actually not the same as the profit rate R obtained above on the matrix $\mathbf{S}'$ including subsistence labour. Sraffa did not point out explicitly this fact. To determine $\dot{\mathbf{S}}$, we must calculate the diagonal elements σ_j of $\hat{\boldsymbol{\sigma}}$. Referring to the general form (3.42), setting $r = R$,

$$\mathbf{S}'\mathbf{p}(1 + R) = \hat{\mathbf{q}}\mathbf{p} = \mathbf{x}, \quad (3.134)$$

we resort to the following sequence of equations, having set $\dot{\mathbf{L}} = \mathbf{L}$,

$$\begin{aligned} \hat{\mathbf{q}}\mathbf{p} = \mathbf{S}'(1 + R)\mathbf{p} &= [\dot{\mathbf{S}}' + \hat{\boldsymbol{\sigma}}](1 + R)\mathbf{p} \\ &= \dot{\mathbf{S}}'(1 + R)\mathbf{p} + (1 + R)\hat{\boldsymbol{\sigma}}\mathbf{p} \\ &= \dot{\mathbf{S}}'(1 + R)\mathbf{p} + w_s\mathbf{L}, \end{aligned} \quad (3.135)$$

because it is assumed that $w_s\mathbf{L}$ has now produced $(1 + R)\hat{\boldsymbol{\sigma}}\mathbf{p}$ instead of $\hat{\boldsymbol{\sigma}}\mathbf{p}$, i. e., participates here in creating a surplus without benefitting from value added at that stage.

32 By convention however, the notation $\mathbf{S}'$ is usually retained for (3.133) as is the case in PCMC, Chapter II, although this $\mathbf{S}'$ no longer contains labour elements for the period under consideration. One must keep this in mind that, for numerical calculations, the structure of $\mathbf{S}'$ remains and only the values of the diagonal coefficients change. Of course, the non-labour commodities include indirectly the labour from proceeding periods. One can go back by induction over many periods to account for "dated quantities of labour". This is the cycle presented in PCMC, Chap. VI. We shall not pursue this refinement in order to avoid overloading our text.

Now R and $\mathbf{p}$ are known from (3.134) above, and w_s, $\mathbf{L}$ are exogenous parameters, so we can then calculate σ_j from

$$\sigma_j(1+R)p_j = w_s L_j \quad \Rightarrow \quad \sigma_j = \frac{w_s L_j}{(1+R)p_j}. \tag{3.136}$$

Example 3.6.1. Example 3.1.3 presents commodity flow matrix $\mathbf{S} = \left[\begin{smallmatrix} 280 & 120 \\ 12 & 8 \end{smallmatrix}\right]$ and the vector of total output $\mathbf{q} = \left[\begin{smallmatrix} 575 \\ 20 \end{smallmatrix}\right]$. The productiveness is equal to the maximal rate of profits $r = R = 0.25$ and the price vector $\mathbf{p} = \left[\begin{smallmatrix} 1 \\ 15 \end{smallmatrix}\right]$ have been calculated. We assume $w_s = 0.2$, $\mathbf{L}' = [L_1, L_2]' = [100, 20]'$. Calculate $\dot{\mathbf{S}}$ with equations (3.130) and σ_j, $j = 1, 2$ with (3.136).

Solution to Example 3.6.1:
We start with the single-product Sraffa system (3.134), establishing numerically the following identity

$$\mathbf{S}'\mathbf{p}(1+r) = \hat{\mathbf{q}}\mathbf{p} = \begin{bmatrix} 280 & 12 \\ 120 & 8 \end{bmatrix}(1+0.25)\begin{bmatrix} 1 \\ 15 \end{bmatrix} = \begin{bmatrix} 575 & 0 \\ 0 & 20 \end{bmatrix}\begin{bmatrix} 1 \\ 15 \end{bmatrix}, \tag{3.137}$$

then we obtain with (3.136)

$$\begin{aligned} \sigma_1 &= \frac{w_s L_1}{(1+R)p_1} = \frac{0.2 \cdot 100}{(1+0.25)\cdot 1} = 16, \\ \sigma_2 &= \frac{w_s L_2}{(1+R)p_2} = \frac{0.2 \cdot 20}{(1+0.25)\cdot 15} = 0.2133, \end{aligned} \tag{3.138}$$

giving with equation (3.130),

$$\begin{aligned} \mathbf{S}' &= \begin{bmatrix} 280 & 12 \\ 120 & 8 \end{bmatrix} = \dot{\mathbf{S}}' + \hat{\boldsymbol{\sigma}} = \dot{\mathbf{S}}' + \begin{bmatrix} 16 & \\ & 0.2133 \end{bmatrix} \\ &= \begin{bmatrix} \dot{s}_{11} + 16 & 12 \\ 120 & \dot{s}_{22} + 0.2133 \end{bmatrix} \Rightarrow \dot{\mathbf{S}}' = \begin{bmatrix} 264 & 12 \\ 120 & 7.7867 \end{bmatrix}. \quad \blacktriangle \end{aligned} \tag{3.139}$$

Sraffa's formulation (3.133) is a novel extension of Leontief's Input-Output Tables. It opens up the discussion of the allocation of surplus and is the object of Chapter 4 hereafter.

Now, keeping the *total quantity of labour L* fixed and regrouping *total wages W* of one year into an initial part W_s corresponding to *subsistence wages* contributing to the surplus, but not benefitting initially from that surplus, and a second part, corresponding to total wages W_a of one year also contributing to the surplus, recognized as benefitting from value added, the natural accounting relations can be written, simplifying (2.144). The term "national income" is used in relation with *complete production schemes*, like *input-output tables* (IOT) in physical terms, associated to national economies, see also footnote 16 of this chapter,

$$\begin{aligned} \text{national income:} \quad & Y = P + W, \\ \text{total output:} \quad & X = Y + K, \end{aligned} \tag{3.140}$$

with K as the total value of the *means of production*, designated in this text as *circulating capital*, obtained from Sraffa's conditions of production. W regroups W_s and W_a,

$$W = W_s + W_a = (w_s + w_a)L = w \cdot L. \tag{3.141}$$

So, Sraffa correctly incorporates subsistence wages into "national income" because, based on (3.134) and (3.135), they contribute in fact to value added in the case, where all the surplus goes into profits without paid wages ($w_a = 0$) and therefore should participate in the final allocation of surplus to workers. We address this matter in detail in Chapter 4.

We have examined this issue for the sake of completeness. Bearing in mind this caveat, we shall in the sequel assume (3.141) and use by convention, to avoid complications, the notation $\mathbf{S}'$ instead of $\dot{\mathbf{S}}'$. Wage rates will accordingly be designated by w, and both w and L are assumed to fully contribute to the surplus.

Subsistence wages not contributing to *value added*, if the term is ever used at present, play a secondary role in modern economies of production and exchange.[33] One must be aware that this is however not the case in many emerging economies around the world.

33 With the rampant pauperization of Western societies in the wake of ultra-liberal economic policies, discussions are however emerging in certain countries (France, Finland and Switzerland, for example) concerning the introduction of a generalized "minimal income", which would replace unemployment benefits and the like, guaranteeing a subsistence income to all citizens.

4 Sraffa's single-product industries with wages and profits

In this chapter, we analyse in more formal details the *Sraffa price model* (3.5), also called the *single-product Sraffa system*, where every industry or branch produces exactly one commodity.

The Sraffa price model results in the calculation of 'values' or 'prices' (PCMC, Par. 7), *"rather than, as might be thought 'costs of production'"*. Later in (PCMC, Par. 7) Sraffa gives his reasons to avoid the term 'cost of production' and justifies his choice: *"Such classical terms as 'necessary price', 'natural price' or 'price of production' would meet the case, but value and price have been preferred as being shorter and in the context (which contains no reference to market prices) no more ambiguous"*.

The economic Assumptions 2.2.1,, 2.2.2, and 2.4.1 hold for the whole chapter, even if this is not explicitly mentioned in every specific model. The economic Assumption 2.5.1 on existing labour forces in each industrial sector also holds, except if the contrary is formulated. These *economic assumptions* have to be clearly distinguished from the *mathematical conditions*, like *positivity, semi-positivity, non-negativity, irreducibility* and *reducibility*. Some of them have to be fulfilled in situations, where one of the Theorems around the group of the Perron–Frobenius theorem is applied.

4.1 The surplus as the sum of profits and wages

Sraffa followed the idea of David Ricardo considering the surplus of an economy as the sum of *profits* for entrepreneurs and of *wages* for workers. We have first to treat the situation, where all the human beings in an economy live only on *subsistence wages*, then comes the case of *wages* paid by means of a *numéraire* or a *currency*.

4.1.1 Economies delivering only subsistence wages

Calibration of quantities of produced commodities to annual production units

In economics, one generally measures quantities of products in physical units, such as wheat in *quarters of* wheat or iron in *tons of* iron and *working time* may be measured in *hours*, in *days* or *man-years*. Then currencies are used as the means of payment, like USD or CHF. But as we want to elaborate the essence of the means of payment, the simplest possible unit is chosen, calibrating the measurements as *entities* normalized to 1. Sraffa used this technique, e. g., in the case of *total labour* (PCMC, Par. 10), for the *national income* (PCMC, Par. 12) and for *Standard systems* (PCMC, Par. 26) exhibiting *Standard commodities*, and especially a *Standard net product*, where the total *labour* is equal to that used by the initial *system of production*, Section 5.3. Later, Schefold ([109], p. 217) introduced this technique to simplify the argumentation of mathemati-

https://doi.org/10.1515/9783110635096-004

cal statements, linked to the **Perron–Frobenius theorem.** He introduced *units of annual production* for products, e. g., the total annual production of wheat, measured in *annual* wheat *crops* (AWC) or the total annual production of iron, measured in *annual* iron *production* (AIP). We will follow this technique to simplify proofs of Lemmas.

Sraffa's elementary examples

(a) Garden of Eden economy

Sraffa begins PCMC with an example of an elementary production process of only two branches. There are no wages and no profits. We call this economy a *Garden of Eden economy*. It was presented in *Example* 3.1.1 (PCMC, Par. 1). We further develop this example here together with appropriate comments. We begin with the production scheme,

$$\begin{aligned} &(280 \text{ qr. wheat}, 12 \text{ t. iron}) \to (400 \text{ qr. wheat}, 0), \\ &(120 \text{ qr. wheat}, 8 \text{ t. iron}) \to (0, 20 \text{ t. iron}). \end{aligned} \tag{4.1}$$

We identify the positive *commodity flow* matrix. The positive vector of *total output* $\mathbf{q}_I$ is equal to the total *means of production* $\mathbf{Se}$,

$$\mathbf{S} = \begin{bmatrix} 280 & 120 \\ 12 & 8 \end{bmatrix}, \quad \mathbf{q}_I = \mathbf{Se} = \begin{bmatrix} 280 & 120 \\ 12 & 8 \end{bmatrix} \begin{bmatrix} 1 \\ 1 \end{bmatrix} = \begin{bmatrix} 400 \\ 20 \end{bmatrix}. \tag{4.2}$$

Now we introduce a different calibration which measures both commodities in units of one annual output, that means, $q_1 = 400$ qr. of wheat = 1 *annual* wheat *crop* (AWC) and $q_2 = 20$ *tons of* iron = 1 *annual* iron *production* (AIP). This operation is realized by calculating the *distribution coefficients* matrix $\mathbf{D} = \hat{\mathbf{q}}_I^{-1}\mathbf{S}$ (2.20). We multiply the right equation of (4.2) by $\hat{\mathbf{q}}_I^{-1}$ and get

$$\hat{\mathbf{q}}_I^{-1}(\mathbf{Se}) = (\hat{\mathbf{q}}_I^{-1}\mathbf{S})\mathbf{e} = \mathbf{De} = \hat{\mathbf{q}}_I^{-1}\mathbf{q}_I = \mathbf{e} \Rightarrow \mathbf{De} = \mathbf{e} \Rightarrow \mathbf{e}'\mathbf{De} = \mathbf{e}'\mathbf{e} = n. \tag{4.3}$$

Summarising, we get an eigenvalue equation and a quadratic form

$$\mathbf{De} = \mathbf{e} \Rightarrow \mathbf{e}'\mathbf{De} = n, \tag{4.4}$$

meaning that therefore matrix **D** is a stochastic matrix (see Section A.11).

Let us now introduce as a measurement unit, the *annual output of a commodity* $i \in \{1, \ldots, n\}$. As a matter of fact, this is another *calibration*. In doing so, every commodity has as output i per year an *entity* 1 which appears as *one object* of that commodity i (see Section 9.2.2). For this reason, the summation vector **e** represents the vector of *total output* in this mode of calibration. Consequently, vector **e** is also called the vector of **objects.** Further, the vector **e** is also the positive *right eigenvector* of matrix **D**, associated to the Frobenius number $\lambda_D = 1$. Equation $\mathbf{De} = \mathbf{e}$ (4.3) is the *object representation* of the *Interindustrial Economy*. Algebraic properties of stochastic matrices in *Sraffa–Leontief-Interindustrial economies* are treated in Chapter 9.

Besides Sraffa's initial *production scheme* in initial physical units (4.5) (1), we obtain a second production scheme in annual production units (4.5) (2), presented in parallel. Both schemes are written without wages (3.55) in matrix form, $\hat{\mathbf{e}} = \mathbf{I}$.

$$\begin{aligned} &(1) \quad (\mathbf{S}', \mathbf{o}) \to (\hat{\mathbf{q}}_I), \\ &(2) \quad (\mathbf{D}', \mathbf{o}) \to (\mathbf{I}). \end{aligned} \tag{4.5}$$

Remember that the balances of both sectors, wheat and iron are in equilibrium.

We start from Sraffa's *conditions of production*, Definition 3.1.2, a *single-product Sraffa system* without surplus, the *Sraffa price model*, $\mathbf{S}'\mathbf{p} = \hat{\mathbf{q}}_I\mathbf{p}$. We need the matrix $\mathbf{C} = \mathbf{S}\hat{\mathbf{q}}_I^{-1}$.

We set then the left eigenvector equation $\mathbf{x}'\mathbf{D} = \mathbf{x}'$ in analogy to the right eigenvector equation $\mathbf{De} = \mathbf{e}$, associated with the Frobenius number $\lambda_C = \lambda_D = 1$ and put both eigenvector equations together,

$$\begin{aligned} &(1) \quad \mathbf{S}'\mathbf{p} = \hat{\mathbf{q}}_I\mathbf{p} = \mathbf{x} \Leftrightarrow \mathbf{C}'\mathbf{p} = \mathbf{p}, \\ &(2) \quad \mathbf{D}'\mathbf{x} = \mathbf{x}. \end{aligned} \tag{4.6}$$

The question arises: What is the meaning of vector $\mathbf{x}$? The answer is given in the following

---

Proposition 4.1.1. *We apply the single-product Sraffa system to determine the prices of commodities (4.6) (1), setting the equation* $\mathbf{x} := \hat{\mathbf{q}}_I\mathbf{p}$ *(2.18) between the corresponding price vectors* $\mathbf{p}$ *and vector* $\mathbf{x}$ *with a positive vector of outputs* $\mathbf{q}_I > \mathbf{o}$. *The circle is closed, passing from matrix* $\mathbf{C}$ *over matrix* $\mathbf{S}$ *to matrix* $\mathbf{D}$,

$$\begin{aligned} \mathbf{x} &:= \hat{\mathbf{q}}_I\mathbf{p} = \hat{\mathbf{q}}_I(\mathbf{C}'\mathbf{p}) = \hat{\mathbf{q}}_I\hat{\mathbf{q}}_I^{-1}\mathbf{S}'\mathbf{p} = \mathbf{S}'\mathbf{p} = \mathbf{S}'(\hat{\mathbf{q}}_I^{-1}\hat{\mathbf{q}}_I)\mathbf{p} = (\mathbf{S}'\hat{\mathbf{q}}_I^{-1})\hat{\mathbf{q}}_I\mathbf{p} \\ &= \mathbf{D}'(\hat{\mathbf{q}}_I\mathbf{p}) = \mathbf{D}'\mathbf{x} = \mathbf{x} \Leftrightarrow \mathbf{p} = \hat{\mathbf{q}}_I^{-1}\mathbf{x} \Leftrightarrow \mathbf{x} = \hat{\mathbf{q}}_I\mathbf{p}. \end{aligned} \tag{4.7}$$

The vector $\mathbf{x} := \hat{\mathbf{q}}_I\mathbf{p}$ *is the vector of values!*

The **Perron theorem A.9.1** guarantees for matrix $\mathbf{C}$ the existence of a real, positive, unique, maximal eigenvalue λ_C, the Frobenius number, associated with a positive eigenvector.[1] We discover that an economic meaning can be attributed to the eigenvectors of $\mathbf{C}$, associated with the Frobenius number, namely the *right eigenvector* of matrix $\mathbf{C}$ gives the *quantities*, the *left eigenvector* the *prices* and the *left eigenvector* of matrix $\mathbf{D}$ gives the *values*. We shall illustrate this finding on the basis of Sraffa's elementary Example 3.1.1 (PCMC, Par. 1).

Example 4.1.1. Consider the production scheme (4.5) (1). The numéraire is wheat. Identify the positive matrix $\mathbf{S} > \mathbf{O}$, and the positive vector of output $\mathbf{q}_I = \mathbf{Se} > \mathbf{o}$ (4.2). Compute the positive matrices $\mathbf{C} = \mathbf{S}\hat{\mathbf{q}}_I^{-1}$ and $\mathbf{D} = \hat{\mathbf{q}}_I^{-1}\mathbf{S}$, the Frobenius numbers λ_C and λ_D, the positive left eigenvector $\mathbf{p}$ of matrix $\mathbf{C}$ and the positive left eigenvector $\mathbf{x}$

---

1 Eigenvectors are determined up to a scalar factor.

of matrix **D**, associated with the Frobenius numbers λ_C and λ_D. Verify that matrix **D** is stochastic, namely, $\mathbf{e} = \mathbf{De}$ (4.3).

Confirm here the transformation equation, $\mathbf{x} = \hat{\mathbf{q}}_I\mathbf{p}$ (2.18).

Solution to Example 4.1.1:
We first present and compute the matrices. We identify matrix **S** and vector $\mathbf{q}_I$ (4.2),

$$\mathbf{C} = \mathbf{S}\hat{\mathbf{q}}_I^{-1} = \begin{bmatrix} 280 & 120 \\ 12 & 8 \end{bmatrix}\begin{bmatrix} \frac{1}{400} & 0 \\ 0 & \frac{1}{20} \end{bmatrix} = \begin{bmatrix} \frac{7}{10} & 6 \\ \frac{3}{100} & \frac{2}{5} \end{bmatrix} > \mathbf{0},$$

$$\mathbf{D} = \hat{\mathbf{q}}_I^{-1}\mathbf{S} = \begin{bmatrix} \frac{1}{400} & 0 \\ 0 & \frac{1}{20} \end{bmatrix}\begin{bmatrix} 280 & 120 \\ 12 & 8 \end{bmatrix} = \begin{bmatrix} \frac{7}{10} & \frac{3}{10} \\ \frac{3}{5} & \frac{2}{5} \end{bmatrix} > \mathbf{0}. \tag{4.8}$$

Then we compute the characteristic polynomials based on the equality between the determinants:

$$P_2(\lambda) = \det[\mathbf{C} - \lambda\mathbf{I}] := \det[\mathbf{D} - \lambda\mathbf{I}] = \frac{1}{10} - \frac{11}{10}\cdot\lambda + \lambda^2 = (\lambda - 1)\left(\lambda - \frac{1}{10}\right), \tag{4.9}$$

confirming that matrices **C** and **D** have the same eigenvalues, see Lemma A.6.1. As matrices **C** and **D** are positive, in application of the **Perron theorem A.9.1** we obtain, setting $P_2(\lambda) = 0$, the Frobenius number $\lambda_C = \lambda_D = 1$. The corresponding positive *left* eigenvectors up to the scalar factors $a, b \in \mathbb{R}$ are vectors of prices and of values,

$$\mathbf{C}'\mathbf{p} = \begin{bmatrix} 0.7 & 0.03 \\ 6 & 0.4 \end{bmatrix}\mathbf{p} = \mathbf{p} \Rightarrow \mathbf{p} = a\cdot\begin{bmatrix} 1 \\ 10 \end{bmatrix},$$

$$\mathbf{D}'\mathbf{x} = \begin{bmatrix} 0.7 & 0.6 \\ 0.3 & 0.4 \end{bmatrix}\mathbf{x} = \mathbf{x} \Rightarrow \mathbf{x} = b\cdot\begin{bmatrix} 2 \\ 1 \end{bmatrix}. \tag{4.10}$$

We identify
- the prices, in the case of *physical units*, where the *numéraire* is wheat, $a = 1$:
  - $p_1 = 1$ qr. wheat/qr. wheat = 1;
  - $p_2 = 10$ qr. wheat/t. iron;
- the values, in the case of *annual production units* (AWC) and (AIP), $b = 200$:
  - $x_1 = q_1 \cdot p_1 = 400 \cdot 1 = 400$ qr. wheat,
  - $x_2 = q_2 \cdot p_2 = 20 \cdot 10 = 200$ qr. wheat.

Then, we confirm the stochastic matrix property (4.3)

$$\mathbf{De} = \begin{bmatrix} 0.7 & 0.3 \\ 0.6 & 0.4 \end{bmatrix}\begin{bmatrix} 1 \\ 1 \end{bmatrix} = \begin{bmatrix} 1 \\ 1 \end{bmatrix} = \mathbf{e}. \tag{4.11}$$

Finally, we analyse the transformation of the price vector by the diagonal matrix $\hat{\mathbf{q}}_I$ into the vector of values **x**, we get

$$\mathbf{x} = \hat{\mathbf{q}}_I\mathbf{p} = \begin{bmatrix} 400 & 0 \\ 0 & 20 \end{bmatrix}\begin{bmatrix} 1 \\ 10 \end{bmatrix} = \begin{bmatrix} 400 \\ 200 \end{bmatrix}. \quad \blacktriangle \tag{4.12}$$

Remark: We get the presentation of the prices in *physical units*, $\mathbf{p} = [1, 10]'$ and the vector $\mathbf{x} = [400, 200]'$ of values, expressed by the numéraire wheat.

(b) Exploitation of labour economy

Sraffa continues (PCMC, Par. 5) to develop the previous example, presented in this text as *Example* 3.1.3. We remember that the output of farmers is increased from 400 qr. wheat to 575 qr. wheat, while the output of iron remains unchanged. The technology is again described by the matrix **S** of Example 4.1.1. Consequently, there is a surplus. We take this example again and reproduce in a first step the production scheme (3.41) expressed in the initial *physical units*. We apply the analytical tools with the aim to become familiar with it and to render it transparent,

$$\begin{aligned}(280 \text{ qr. wheat}, 12 \text{ t. iron}) &\to (575 \text{ qr. wheat}, 0),\\ (120 \text{ qr. wheat}, 8 \text{ t. iron}) &\to (0, 20 \text{ t. iron}).\end{aligned} \tag{4.13}$$

We then identify the positive vector of *total output* and the positive *commodity flow matrix*

$$\mathbf{q} = \begin{bmatrix} 575 \\ 20 \end{bmatrix} > \mathbf{o}, \quad \mathbf{S} = \begin{bmatrix} 280 & 120 \\ 12 & 8 \end{bmatrix} > \mathbf{0}, \quad \mathbf{d} = \mathbf{q} - \mathbf{Se} = \begin{bmatrix} 175 \\ 0 \end{bmatrix} \geq \mathbf{o}. \tag{4.14}$$

We remember at present the important identity to compute the vector of *total output* $\mathbf{q} > \mathbf{o}$, using either the *commodity flow* matrix **S** or the *input-output coefficients* matrix $\mathbf{C} = \mathbf{S}\hat{\mathbf{q}}^{-1}$,

$$\boxed{\mathbf{q} = \mathbf{Se} + \mathbf{d} = \mathbf{Cq} + \mathbf{d}.} \tag{4.15}$$

We then return to the calibration where both commodities are measured in units of one annual output, that means, $q_1 = 575$ qr. wheat = 1 annual wheat crop (AWC) and $q_2 = 20$ tons iron = 1 annual iron production (AIP). As the vector of *total output* is positive, $\mathbf{q} > \mathbf{o}$, we compute the *distribution coefficients* matrix $\mathbf{D} = \hat{\mathbf{q}}^{-1}\mathbf{S}$ (2.20). We have the case of *self-replacement* with the semi-positive vector of surplus $\mathbf{d} \geq \mathbf{o}$ containing at least one component greater than 0. We multiply the equation $\mathbf{q} = \mathbf{Se} + \mathbf{d}$ from the right by $\hat{\mathbf{q}}^{-1} > \mathbf{o}$ and get with (A.11) a further inequality, in analogy to (4.3),

$$\begin{aligned}\mathbf{q} = \mathbf{Se} + \mathbf{d} \Rightarrow \hat{\mathbf{q}}^{-1}\mathbf{q} = \mathbf{e} = \hat{\mathbf{q}}^{-1}(\mathbf{Se}) + \hat{\mathbf{q}}^{-1}\mathbf{d} = (\hat{\mathbf{q}}^{-1}\mathbf{S})\mathbf{e} + \hat{\mathbf{q}}^{-1}\mathbf{d}\\ = \mathbf{De} + \hat{\mathbf{q}}^{-1}\mathbf{d} \Rightarrow \mathbf{De} \leq \mathbf{e} \Rightarrow \mathbf{e}'\mathbf{De} < \mathbf{e}'\mathbf{e} = n.\end{aligned} \tag{4.16}$$

In the *exploitation of labour economy*, all the surplus goes into profits of entrepreneurs. The wages are exclusively *subsistence wages* and are included in the means of production. We continue in this line, assuming furthermore that the rate of profits r is equal in both sectors, so we have consequently a "maximal rate of profits" $R > 0$. The production scheme in initial physical units (4.17) (1) is presented by analogy to (4.5) in parallel to the production scheme in annual production units (4.17) (2) of produced commodities, both in matrix form, giving

$$\begin{aligned}&(1)\quad (\mathbf{S}',\mathbf{o}) \to (\hat{\mathbf{q}}),\\ &(2)\quad (\mathbf{D}',\mathbf{o}) \to (\mathbf{I}).\end{aligned} \tag{4.17}$$

We now start from the *single-product Sraffa system* with surplus (3.43) in initial physical units (1), and state also a model to determine the vector of value $\mathbf{x}$ in analogy to (4.5) together with (2.18),

$$\begin{aligned}&(1)\quad \mathbf{S}'\mathbf{p}(1+R) = \hat{\mathbf{q}}\mathbf{p} = \mathbf{x} \Leftrightarrow \mathbf{C}'\mathbf{p}(1+R) = \mathbf{p}.\\ &(2)\quad \mathbf{D}'\mathbf{x}(1+R) = \mathbf{x}.\end{aligned} \tag{4.18}$$

In analogy to Proposition 4.1.1 one obtains

Proposition 4.1.2. *We start from the single-product Sraffa system to determine the prices of commodities (4.18) (1), setting the equation* $\mathbf{x} := \hat{\mathbf{q}}\mathbf{p}$ *(2.18) between the corresponding price vector* $\mathbf{p}$ *and the vector of value* $\mathbf{x}$ *with a positive vector of output* $\mathbf{q} > \mathbf{o}$*. The circle is closed, passing from matrix* $\mathbf{S}$ *to matrix* $\mathbf{D}$,

$$\begin{aligned}\mathbf{x} &:= \hat{\mathbf{q}}\mathbf{p} = \mathbf{S}'\mathbf{p}(1+R) = \mathbf{S}'(\hat{\mathbf{q}}^{-1}\hat{\mathbf{q}})\mathbf{p}(1+R) = (\mathbf{S}'\hat{\mathbf{q}}^{-1})\hat{\mathbf{q}}\mathbf{p}(1+R)\\ &= \mathbf{D}'(\hat{\mathbf{q}}\mathbf{p})(1+R) = \mathbf{D}'\mathbf{x}(1+R) = \mathbf{x} \Leftrightarrow \mathbf{D}'\mathbf{x}(1+R) = \mathbf{x}.\end{aligned} \tag{4.19}$$

The vector $\mathbf{x} := \hat{\mathbf{q}}\mathbf{p}$ *is the vector of values!*

$$\mathbf{p} = \hat{\mathbf{q}}^{-1}\mathbf{x} \Leftrightarrow \mathbf{x} = \hat{\mathbf{q}}\mathbf{p}. \tag{4.20}$$

Setting $\lambda_C = \lambda_D = 1/(1+R)$, we summarize (4.18) and (4.19), the *single-product Sraffa system* with a surplus rewritten under both calibration aspects as

$$\boxed{\begin{aligned}&(1)\quad \mathbf{C}'\mathbf{p} = \lambda_C\mathbf{p},\\ &(2)\quad \mathbf{D}'\mathbf{x} = \lambda_D\mathbf{x}.\end{aligned}} \tag{4.21}$$

We can now present a further version of Sraffa's example PCMC, Par. 5, in our text it is Example 3.1.3, to illustrate the foregoing presentations.

Example 4.1.2. The numéraire is 1 qr. wheat. Consider the production scheme (1) (4.17). Identify the positive matrix $\mathbf{S} > \mathbf{O}$ and the positive vector of total output $\mathbf{q} = \mathbf{Se} + \mathbf{d} > \mathbf{o}$ (4.15).

Compute the positive matrices $\mathbf{C} = \mathbf{S}\hat{\mathbf{q}}^{-1} > \mathbf{O}$ and $\mathbf{D} = \hat{\mathbf{q}}^{-1}\mathbf{S} > \mathbf{0}$, the Frobenius numbers λ_C and λ_D, and the positive left eigenvector $\mathbf{p}$ of matrix $\mathbf{C}$, associated with λ_C, and the positive left eigenvector $\mathbf{x}$ of matrix $\mathbf{D}$ associated with λ_D.

Confirm here the transformation equation, $\mathbf{x} = \hat{\mathbf{q}}\mathbf{p}$ (2.18). Confirm the inequality $\mathbf{De} \leq \mathbf{e}$. Compute the productiveness R.

Solution to Example 4.1.2:

Consider the production scheme (4.13) and identify and compute the matrices. We already know matrix $\mathbf{S}$ and vector $\mathbf{q}$ (4.14),

$$\mathbf{C} = \mathbf{S}\hat{\mathbf{q}}^{-1} = \begin{bmatrix} 280 & 120 \\ 12 & 8 \end{bmatrix} \begin{bmatrix} \frac{1}{575} & 0 \\ 0 & \frac{1}{20} \end{bmatrix} = \begin{bmatrix} \frac{56}{115} & 6 \\ \frac{12}{575} & \frac{2}{5} \end{bmatrix} > \mathbf{0},$$

$$\mathbf{D} = \hat{\mathbf{q}}^{-1}\mathbf{S} = \begin{bmatrix} \frac{1}{575} & 0 \\ 0 & \frac{1}{20} \end{bmatrix} \begin{bmatrix} 280 & 120 \\ 12 & 8 \end{bmatrix} = \begin{bmatrix} \frac{56}{115} & \frac{24}{115} \\ \frac{3}{5} & \frac{2}{5} \end{bmatrix} > \mathbf{0}. \quad (4.22)$$

Then, we compute the characteristic polynomial, of matrices **C**, respectively **D**,

$$P_2(\lambda) = \det(\mathbf{C} - \lambda\mathbf{I}) = \det(\mathbf{D} - \lambda\mathbf{I}) = \frac{8}{115} - \frac{102}{115} \cdot \lambda + \lambda^2 = \left(\lambda - \frac{4}{5}\right)\left(\lambda - \frac{2}{23}\right). \quad (4.23)$$

Setting $P_2(\lambda) = 0$, we get the common Frobenius number $\lambda_C = \lambda_D = 4/5$. Because matrices **C** and **D** are positive, this gives in application of the **Perron theorem A.9.1**, the corresponding positive eigenvector **p** of the *single-product Sraffa system* and the value vector **x** (4.21) up to factors $a, b \in \mathbb{R}$.

$$(1) \quad \mathbf{C}'\mathbf{p} = \begin{bmatrix} \frac{56}{115} & \frac{12}{575} \\ 6 & \frac{2}{5} \end{bmatrix} \mathbf{p} = \frac{4}{5}\mathbf{p} \Rightarrow \mathbf{p} = a \cdot \begin{bmatrix} 1 \\ 15 \end{bmatrix},$$

$$(2) \quad \mathbf{D}'\mathbf{x} = \begin{bmatrix} \frac{56}{115} & \frac{3}{5} \\ \frac{24}{115} & \frac{2}{5} \end{bmatrix} \mathbf{x} = \frac{4}{5}\mathbf{x} \Rightarrow \mathbf{x} = b \cdot \begin{bmatrix} 575 \\ 300 \end{bmatrix}. \quad (4.24)$$

One gets with $a = 1$ the vector of relative prices $\mathbf{p} = [1, 15]'$, expressed in units of the numéraire per commodity, corresponding to Sraffa's exchange-values:

Sraffa writes: "*The exchange-ratio which enables the advances to be replaced and the profits to be distributed to both industries in proportion to their advances is 15 qr. wheat for 1 t. iron; and the corresponding rate of profits in each industry is R = 25 %.*" (PCMC, Par. 5, p. 7).

On the other hand, setting $b = 1$ and keeping the *numéraire* wheat, we compute now the *vector of values*, expressed by the numéraire wheat.

$$\mathbf{x} = \hat{\mathbf{q}}\mathbf{p} = \begin{bmatrix} 575 & 0 \\ 0 & 20 \end{bmatrix} \begin{bmatrix} 1 \\ 15 \end{bmatrix} = \begin{bmatrix} 575 \\ 300 \end{bmatrix}. \quad (4.25)$$

Equation (4.25) performs the transformation from the price vector $\mathbf{p} = [1, 15]'$ of the commodities, into the vector of values $\mathbf{x} = [575, 300]'$.

We now come to the inequality $\mathbf{De} \le \mathbf{e}$ (4.16) and we get indeed,

$$\mathbf{De} = \begin{bmatrix} \frac{56}{115} & \frac{24}{115} \\ \frac{3}{5} & \frac{2}{5} \end{bmatrix} \begin{bmatrix} 1 \\ 1 \end{bmatrix} = \begin{bmatrix} \frac{16}{23} \\ 1 \end{bmatrix} \le \begin{bmatrix} 1 \\ 1 \end{bmatrix} = \mathbf{e}, \quad \mathbf{De} \le \mathbf{e}. \quad (4.26)$$

Then we compute the right eigenvector **r** of matrix **C**, with $k \in \mathbb{R}$,

$$\mathbf{Cr} = \lambda\mathbf{r} = \mathbf{Cr} = \begin{bmatrix} \frac{56}{115} & 6 \\ \frac{12}{575} & \frac{2}{5} \end{bmatrix} \mathbf{r} = \frac{4}{5}\mathbf{r} \Rightarrow \mathbf{r} = k \cdot \begin{bmatrix} 575 \\ 30 \end{bmatrix}. \quad (4.27)$$

This eigenvector equation gives as solution the positive right eigenvector $\mathbf{r} = [575, 30]'$, associated with the Frobenius number $\lambda_C = 0.8$.

Finally, we compute the *maximal rate of profits* $R = (1/\lambda_D) - 1 = (5/4) - 1 = (1/4)$. ▲

After the calculation of these two examples, we compare both obtained right eigenvectors of matrix **C** with the corresponding vector of total output **q** and answer the question:

What is the meaning of this *right eigenvector* of **C**?

When there is no surplus, as in the numerical *Example* 4.1.1, the *right eigenvector* of **C**, $\mathbf{Cq} = \mathbf{q}$ (3.25), gives the *vector of total output* $\mathbf{q} = [400, 20]' > \mathbf{o}$. We observe that the *initial quantity proportion* of production 400:20 = 20:1 between the quantity of *wheat* and the quantity of iron is equal to the proportion between the components of the computed right eigenvector.

When there is a surplus, as in the numerical *Example* 4.1.2, the *right eigenvector* **r** of **C**, $\mathbf{Cr} = \lambda\mathbf{r}$ (4.27), is no longer parallel to the *vector of total output*, $\mathbf{r} = [575, 30]' \nparallel \mathbf{q} = [575, 20]' > \mathbf{o}$. The proportion of production $575 : 30 = 19.166$ is no longer respected by the components of **q** (575:20). Why? We will elucidate this question in Chapter 5, dealing with Sraffa's *Standard systems*.

For the moment, we just note that the vectors **q**, **Se**, **d** and **p** of Example 4.1.2 are pairwise not parallel, see Figure 4.1.

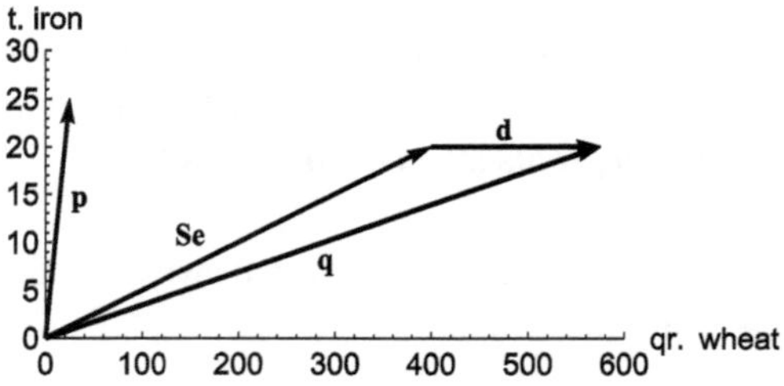

Figure 4.1: Representation of the vectors **q**, **d**, **Se** and **p**.

Generalizations to *n* dimensions[2]

The calculations carried out in the previous subsection can be generalized to an economy with $n \in \mathbb{N}$ *single product industries*, each sector producing only one commodity.[3] Later, we will treat *multi-product industries* or *joint production*, where one sector produces more than one of the n commodities, see Chapter 6.

2 This subsection is inspired by the "Nachwort" of Schefold in Sraffa's German edition of PCMC ([109], pp. 216–225).

3 Algorithms to calculate eigenvalues and eigenvectors of $n \times n$ matrices are available in many software packages, like *Mathematica*, from Wolfram Research, Inc., Champaign, IL (2016).

So, in the sequel we consider either a *semi-positive* and *irreducible* (Definition A.8.3) or a *positive* $n \times n$ *commodity flow matrix* **S**, see Table 2.2 and equation (2.13).

Having computed the *input-output coefficients* matrix $\mathbf{C} = \mathbf{S}\hat{\mathbf{q}}^{-1}$ and the *distribution coefficients* matrix $\mathbf{D} = \hat{\mathbf{q}}^{-1}\mathbf{S}$, both either *non-negative* and *irreducible* or *positive*, consider the corresponding production scheme, where there are only *subsistence wages* and no explicit wages paid via a *numéraire* or a *currency*:

$$\begin{aligned} &(1) \quad (\mathbf{S}', \mathbf{o}) \rightarrow (\hat{\mathbf{q}}), \\ &(2) \quad (\mathbf{D}', \mathbf{o}) \rightarrow (\mathbf{I}). \end{aligned} \tag{4.28}$$

If the produced commodities are measured in usual *physical units*, like the qr. wheat or t. iron, then the technology is described by (4.28) (1).

If the produced commodities are measured in units of annual production, like the annual wheat crop (AWC) or the annual iron production (AIP), then the technology is described by (4.28) (2), choosing the numéraire as (AWC).

The economy may produce a *surplus* or *no surplus* at all. Remember that according to (4.3), matrix **D** in the case of *no surplus* has the summation vector **e** as positive right eigenvector, associated with the Frobenius number $\lambda_D = 1$. In the case of production of a *surplus*, Example 4.1.2, illustrated that matrix **D** has Frobenius number $\lambda_D < 1$. We will now show here that this result is general.

The Frobenius numbers of matrices **C** and **D** coincide, $\lambda_C = \lambda_D$, see Lemma A.6.1. *Single-product Sraffa systems*, only with subsistence wages and no paid *wages*, with *surplus* or without *surplus* are presented. The transformation equations (4.20), (4.7) between the price vector **p** and the vector of values **x**, were justified previously.

$$\boxed{\begin{aligned} &(1) \quad \mathbf{S}'\mathbf{p} = \lambda_C\hat{\mathbf{q}}\mathbf{p} = \lambda_C\mathbf{x} \Leftrightarrow \mathbf{C}'\mathbf{p} = \lambda_C\mathbf{p}, \\ &(2) \quad \mathbf{D}'\mathbf{x} = \lambda_D\mathbf{x}. \qquad\qquad\qquad \mathbf{p} = \hat{\mathbf{q}}^{-1}\mathbf{x}, \quad \lambda_C = \lambda_D. \end{aligned}} \tag{4.29}$$

We formulate the obtained results concerning the Frobenius numbers.

Lemma 4.1.1. *Assumption 2.2.1 and Assumption 2.2.2 hold. Consider the single-product Sraffa price model, (4.29) (1), only with subsistence wages and a given numéraire.*

(a) *A just-viable economy, an economy without surplus, Definition 2.2.1.*
Consider the production scheme (4.5), described by the semi-positive matrices $\mathbf{S} \geq \mathbf{0}$ *and* $\mathbf{C} = \mathbf{S}\hat{\mathbf{q}}_I^{-1} \geq \mathbf{0}$ *with vector of output* $\mathbf{q}_I = \mathbf{S}\mathbf{e} > \mathbf{o}$ *and the stochastic matrix* $\mathbf{D} = \hat{\mathbf{q}}_I^{-1}\mathbf{S} \geq \mathbf{o}$. *Then the maximal eigenvalues of* **C** *and* **D** *coincide and are equal to 1,* $\lambda_C = \lambda_D = 1$.

(b) *A viable economy with surplus, Definition 2.2.1.*
Consider the production scheme (4.28), described by the semi-positive matrices $\mathbf{S} \geq \mathbf{0}$ *and* $\mathbf{C} = \mathbf{S}\hat{\mathbf{q}}^{-1} \geq \mathbf{0}$, *a semi-positive vector of surplus* $\mathbf{d} \geq \mathbf{o}$, *the vector of output* $\mathbf{q} = \mathbf{S}\mathbf{e} + \mathbf{d} > \mathbf{o}$ *and the matrix* $\mathbf{D} = \hat{\mathbf{q}}^{-1}\mathbf{S} \geq \mathbf{o}$. *Then the real, positive, maximal eigenvalues of* **C** *and* **D** *coincide and are less than 1,* $0 < \lambda_C = \lambda_D < 1$.

Proof. (a) Starting with $\mathbf{q}_I = \mathbf{Se} > \mathbf{o}$, one gets the stochastic matrix $\mathbf{D} = \hat{\mathbf{q}}_I^{-1}\mathbf{S}$, with eigenvector $\mathbf{e}$, $\mathbf{De} = \mathbf{e}$ (4.3) and associated maximal eigenvalue $\lambda_D = 1$, Lemma A.11.1. Matrices $\mathbf{C}$ and $\mathbf{D}$ are similar with identical eigenvalues, Lemma A.6.1. Thus, the maximal eigenvalue of $\mathbf{C}$, coincides with the maximal eigenvalue $\lambda_D = 1$ of $\mathbf{D}$. There is $\lambda_C = \lambda_D = 1$.

(b) Starting with the *input-output coefficients* matrix $\mathbf{C} = \mathbf{S}\hat{\mathbf{q}}^{-1} \geq \mathbf{0}$, we are in the presence of a productive Sraffa model, $\mathbf{d} \geq \mathbf{o}$, according to Definition A.12.1, $\mathbf{q} = \mathbf{Se} + \mathbf{d} = \mathbf{Cq} + \mathbf{d} > \mathbf{o}$ (3.52). The Theorem A.12.1 states that the Frobenius number is less than 1, $\lambda_C < 1$. The matrices $\mathbf{C}$ and $\mathbf{D}$ are similar with identical eigenvalues, Lemma A.6.1. Thus, the real, maximal, eigenvalues coincide, $\lambda_C = \lambda_D < 1$. With Lemma A.10.3, the real, maximal eigenvalue of the semi-positive *input-output coefficients* matrix $\mathbf{C}$ is positive, $\lambda_C > 0$. □

The next subsection presents Sraffa price models with wages paid in numéraire.

4.1.2 Economies with wages paid in numéraire

Up to now, the wages for the labour of the workers were covered by the *means of production* as "subsistence wages", described by the *input-output* matrix $\mathbf{S}$. If on the other hand wages are covered by the surplus, then the initial "subsistence wages" become a part of the surplus. So with Sraffa (PCMC, Par. 8), we *"follow the practice of treating the whole of the wage as variable"* and do not enter the discussion of wage formation.

Human labour in Sraffa's model

Economically speaking, by Assumption 2.5.1, the quantity of human labour necessary for the production of a commodity j is positive in all sectors S_j, $L_j > 0$, $j \in \{1, \ldots, n\}$. But here, we only need a weaker condition, namely, the semi-positivity of the vector of *labour*, $\mathbf{L} \geq \mathbf{o}$,[4] to obtain the goal of a positive quantity of labour $L = \mathbf{e}'\mathbf{L} > 0$. If $L_j = 0$, for a specific sector S_j, then this means economically that no *wages* are paid in relation with an *annual working time*. This may occur in the case when all the production of that sector S_j is generated by robots controlled by another sector.

How is labour measured? We have among various measures the possibility of choosing as *unit of labour*, the work one man realizes in one hour, a so-called *man-hour*, or the work one man realizes in one year, a so-called *man-year*.

We may also consider the annual output of economic variables, like *total labour*, as a *unit*. We have already encountered this technique (Subsection 4.1.1), in relation with matrix $\mathbf{D}$ (4.3). Such *normalisation* is a process of *calibration*. Sraffa speaks of

4 There exists at least one $j' \in \{1, \ldots, n\}$, such that $L_{j'} > 0$.

"annual quantities of labour" (PCMC, Par. 10) and calls the resulting total entity "*the standard in terms of which the wage and the ...prices are expressed*" (PCMC, Par. 12). He thus considers the *total quantity of labour* L as the "*total annual labour of society*". We denote this *physical unit* as 1 TAL (= Total Amount of Labour) which becomes the *standard*. Accordingly, the *annual quantities of labour* L_j, employed respectively in the various industries j, are fractions of 1 TAL. Therefore, taking the vector of *labour* $\mathbf{L} = [L_1, \dots, L_n]' \geq 0$ (required working-time per industrial sector), its *non-negative* components $L_j \geq 0, j \in \{1, \dots, n\}$, usually measured in a unit like *man-hours*, are now fractions of the TAL.[5] This normalization gives:

$$L = \mathbf{e}'\mathbf{L} = L_1 + L_2 + \cdots + L_n = 1\,\text{TAL}. \tag{4.30}$$

We continue with the presentation of *economic variables*, which will appear in the *complete single-product Sraffa system*, Section 4.1.3.

National accounting and other economic variables, ratios

We consider as solutions *positive* price vectors $\mathbf{p} = [p_1, \dots, p_n]' > \mathbf{o}$, containing the prices p_i conceived as the quantity of the *numéraire* per unit of produced quantity of commodity i, measuring the *value* of a unit of commodity i.

Then, we consider the *single-product Sraffa system* (4.18) (1), consisting of three parts:

(a) the vector $\mathbf{S}'\mathbf{p}$ of components giving the value of the means of production of each sector;
(b) the vector $R\mathbf{S}'\mathbf{p}$ of components, giving the surplus produced by each sector;
(c) the vector $\hat{\mathbf{q}}\mathbf{p}$ of components, giving the total production of each sector.

All the vector components have as *a physical unit* the chosen *numéraire*. In order to get the key economic variables, we multiply the right equation (4.18) (1) by the vector $\mathbf{e}'$, summing up all the sectorial parts, getting the *circulating capital* K, e. g., the value of the total means of the production or the total value of the interindustrial economy, the amount of *total output* X, the *national income* Y, i. e., the total value of the surplus. All three variables are measured in *numéraire* quantities. We have obtained:

$$\begin{aligned} K &= \mathbf{e}'(\mathbf{S}'\mathbf{p}) = (\mathbf{S}\mathbf{e})'\mathbf{p}, \\ X &= \mathbf{e}'(\hat{\mathbf{q}}\mathbf{p}) = \mathbf{e}'\mathbf{x} = (\mathbf{e}'\hat{\mathbf{q}})\mathbf{p} = \mathbf{q}'\mathbf{p} = \mathbf{e}'(\mathbf{S}'\mathbf{p})(1+R) = K(1+R). \end{aligned} \tag{4.31}$$

5 The measurement unit *man-hour* is replaced here by the new measurement unit Total Amount of Labour = TAL, giving, e. g., the equivalence: 1 TAL $\simeq$ 96,000 man-hours for 50 workers, working 48 weeks a year, 40 hours a week ($96{,}000 = 50 \cdot 48 \cdot 40$).

Then, there is the national income which is the difference between *total output* X and *circulating capital* K,

$$\begin{aligned} Y &= X - K = R \cdot K = R\mathbf{e}'(\mathbf{S}'\mathbf{p}) = \mathbf{e}'(\hat{\mathbf{q}}\mathbf{p}) - \mathbf{e}'(\mathbf{S}')\mathbf{p} \\ &= \mathbf{q}'\mathbf{p} - (\mathbf{Se})'\mathbf{p} = (\mathbf{q}' - (\mathbf{Se})')\mathbf{p} = \mathbf{d}'\mathbf{p}. \end{aligned} \tag{4.32}$$

One of the aims of Ricardo's and Sraffa's work is to solve the *problem of distribution*. Kaldor [45] describes this as follows:

> *"According to the Preface of Ricardo's Principles, the discovery of the laws which regulate distributive shares is the 'principal problem in Political Economy' ".*

Thus, Sraffa's idea is to divide the whole surplus Y into two parts: the profits P for entrepreneurs and the wages W for workers;[6] Y is here Sraffa's *national income*. In Sraffa (PCMC, Chapters III–IV), all salary components are part of the surplus, giving $Y = P + W$. Here, again according to Ricardo and Sraffa, a uniform *rate of profits* $r = P/K$, $0 \le r \le R$, for all the sectors is introduced, meaning that every sector has the same *profit rate* for the various amounts of capital engaged in the production process. We shall relax this constraint in Chapter 8.

When $r < R$ is chosen, the *total profit* is defined as $P = r \cdot \mathbf{e}'(\mathbf{S}'\mathbf{p}) = r \cdot K$. It follows that the *total wages* are equal to $W = Y - P > 0$. According to what has been developed just above, the components of the vector $r \cdot (\mathbf{S}'\mathbf{p})$ represent the parts of profit generated by each sector. Sraffa also considers a uniform *rate of wages* $w = W/L$ applicable to all sectors, meaning that every worker has the same wage per unit of working time.[7] Introducing also the *share of total wages to national income* $\tilde{w} = W/Y$, considering further that $L_j \ge 0$ is the *annual quantity of labour* required in sector S_j, $j \in \{1, \ldots, n\}$, then $w \cdot L_j$ is the *total amount of wages* necessary to pay the workers of sector S_j. Thus, we can formulate both vectors, describing the *distribution* of *total profits* and *total wages*, as follows

$$\text{total profits} : r \cdot (\mathbf{S}'\mathbf{p}), \quad \text{total wages} : w \cdot \mathbf{L} = \frac{W}{L} \cdot \mathbf{L} = \frac{\tilde{w} \cdot Y}{L} \cdot \mathbf{L}. \tag{4.33}$$

The vectors $r \cdot (\mathbf{S}'\mathbf{p})$ and $[(\tilde{w} \cdot Y)/L] \cdot \mathbf{L}$ have components, describing the surplus distributed to the branches, of the *complete single-product Sraffa system* which will be treated below, Section 4.1.3. We conclude this subsection, presenting some important

---

6 Remember that Sraffa's total wages W is the sum of subsistence wages W_s and the wages part of the surplus W_a, $W = W_s + W_a$, with W_s extracted from the production coefficients. Thus, initially (PCMC, Par. 1–8) W_s is part of K and $P = Y$ if there is a surplus. Later (PCMC, Par. 9), W_s becomes part of Y, together with W_a, so we get $W = W_s + W_a$ and P is also a part of the surplus, giving $Y = W + P$. This yields finally $X = Y + K = W + P + K = W_s + W_a + P + K$.

7 In this case, the *physical unit* of total wages W is again easily verified as $[W] = [L] \cdot [w] = \frac{\text{numeraire}}{\text{TAL}} \cdot \frac{\text{TAL}}{1}$ = numéraire. The unit of wages W is the *numéraire*. Note that the constraint of uniform wages may also be relaxed.

relations, existing between the treated economic variables and ratios (see the *Table of Symbols*). Consider especially the pair constituted by the *share of total wages to national income* $\tilde{w}$ and the *share of total profits*[8] $\tilde{r}$:

$$W = \tilde{w} \cdot Y \Leftrightarrow \tilde{w} = \frac{W}{Y}, \quad P = \tilde{r} \cdot Y \Leftrightarrow \tilde{r} = \frac{P}{Y},$$
$$Y = P + W \Rightarrow \frac{P}{Y} + \frac{W}{Y} = \tilde{r} + \tilde{w} = 1. \tag{4.34}$$

From the definitions, we can also immediately write down the following relations:

$$w =: \frac{W}{L}, \quad \tilde{w} := \frac{W}{Y}, \quad \tilde{R} := \frac{Y}{K}, \quad r := \frac{P}{K}, \quad U := \frac{Y}{L},$$
$$w = \frac{\tilde{w} \cdot L}{Y} = \frac{W}{Y}\frac{Y}{L} = \tilde{w}U, \quad r = \frac{P}{Y}\frac{Y}{K} = \tilde{r}\tilde{R} \quad \text{and trivially}$$
$$P = rK = \tilde{r}Y, \quad W = wL = \tilde{w}Y. \tag{4.35}$$

We now state a key result, based on the foregoing definitions and ratios, which will be used soon:

$$\boxed{\tilde{R}(1 - \tilde{w}) = \frac{Y}{K}\left(1 - \frac{W}{Y}\right) = \frac{Y}{K} \cdot \frac{Y - W}{Y} = \frac{Y}{K} \cdot \frac{P}{Y} = \frac{P}{K} = r.} \tag{4.36}$$

The *national income* Y is occasionally measured by a *physical unit* chosen as *the numéraire*. Sraffa normalized the *national income* Y, also called “composite commodity” (PCMC, Par. 12). The national income Y appears then as the *numéraire*. The term “GPD = Gross National Product” is chosen. Therefore one sets,

$$Y = 1\,\text{GPD}. \tag{4.37}$$

The relationships between the variables $\mathbf{L}$, w, W, L, Y, $\tilde{w}$ and the *average national income* $U = Y/L$ appear in equation (4.35). The *labour* vector $\mathbf{L}' = [L_1, \dots, L_n] \geq \mathbf{o}$ gives the *distribution* of the wages $w \cdot L_j \geq 0$ paid to all the workers of the sectors j of this production economy, and we have[9]:

$$\mathbf{L} \cdot w = \mathbf{L} \cdot \frac{W}{L} = \mathbf{L} \cdot \frac{\tilde{w} \cdot Y}{L} = \mathbf{L} \cdot \tilde{w} \cdot U \Rightarrow w = \frac{\tilde{w} \cdot Y}{L} = \tilde{w} \cdot U. \tag{4.38}$$

Then, the total wages W are calculated by summing up the components (4.33) of the vector of *labour* times the *wage per unit of labour* w, $\mathbf{L} \cdot w$, giving

$$W = \mathbf{e}'(\mathbf{L} \cdot w) = \mathbf{e}'\frac{\mathbf{L}}{L} \cdot (L \cdot w) = \mathbf{e}'\mathbf{L} \cdot \frac{W}{L}. \tag{4.39}$$

In the next subsection, we present the complete *single-product Sraffa system*, including labour as a part of surplus.

8 The identity for $GDP \sim Y = P + W$ coincides with Sraffa’s concept of dividing the value of wages W for workers and profit P for producers by the total surplus (or national income).

9 The *physical unit* of vector components in (4.38) is $[\mathbf{L} \cdot w] = [\text{TAL} \cdot \frac{\text{numeraire}}{\text{TAL}}]$ = numéraire, as is easily verified.

4.1.3 The complete single-product Sraffa system

Having discussed labour, and in order to guarantee a *positive* surplus, we consider a *semi-positive* vector of *surplus* $\mathbf{d} \geq \mathbf{o}$, together with either a *semi-positive* and *irreducible* or a *positive commodity flow matrix* $\mathbf{S}$, and the vector of *total output* $\mathbf{q} = \mathbf{Se} + \mathbf{d}$ (4.14). We are now able to describe the corresponding Sraffa production scheme, completed by the *semi-positive* vector of labour $\mathbf{L} \geq \mathbf{o}$, as an extension of the initial production scheme (3.55). We compute the *input-output coefficients* matrix $\mathbf{C} = \mathbf{S}\hat{\mathbf{q}}^{-1}$ and the *distribution coefficients* matrix $\mathbf{D} = \hat{\mathbf{q}}^{-1}\mathbf{S}$, both consequently either *semi-positive* and *irreducible* or *positive*, the economy now producing a positive surplus.

If the produced commodities are measured in initial *physical units*, then the technology is described by (4.40) (1). If the produced commodities are measured in units of *annual production*, then the technology is described by (4.40) (2). We get

$$\begin{aligned} &(1) \quad (\mathbf{S}', \mathbf{L}) \rightarrow (\hat{\mathbf{q}}), \\ &(2) \quad (\mathbf{D}', \mathbf{L}) \rightarrow (\mathbf{I}). \end{aligned} \tag{4.40}$$

We now present the *complete single-product Sraffa system.*[10] Starting from *the single-product Sraffa system* $\mathbf{S}'\mathbf{p}(1 + R) = \hat{\mathbf{q}}\mathbf{p}$ (3.43), the *productiveness* $R > 0$ is replaced by the smaller *profit rate* r, $0 < r < R$, and the vector $\mathbf{S}'\mathbf{p}R$ is replaced by the vectors $\mathbf{S}'\mathbf{p}r$ (4.33) and $w \cdot \mathbf{L}$ (4.38) respectively, realising the distribution of the *surplus* between *profits* and *wages*.

We need the *wage rate* w in the developed form (4.33), $w = (\tilde{w} \cdot Y)/L$ together with an equation to calculate the *national income* Y, as mentioned by Kurz and Salvadori ([52], p. 98).[11] The commodities are measured in initial *physical units*, respectively in units of *annual production*, if we want to point out the *object* character of the produced goods (see for this purpose Chapter 9).

We recognize that the *complete Sraffa price model* (4.41) expresses the *produced value* with $\mathbf{x} = \hat{\mathbf{q}}\mathbf{p}$ (2.18). One undertakes the following transformation,

---

10 Pasinetti ([80], p. 72) called Sraffa's equations (PCMC, Par. 11) *the Sraffa system* or *the price system*.
11 Sraffa (PCMC, Par. 11) presents this equation with normalised national income $Y = 1$ and normalised labour $L = 1$, consequently, with $W = w \cdot L = w = |\tilde{w}|$ (We have to pay attention to the units of measurements) (4.46), it becomes

$$\mathbf{S}'\mathbf{p}(1 + r) + \mathbf{L} \cdot w = \hat{\mathbf{q}}\mathbf{p}. \tag{4.41}$$

We have introduced the unit of *physical measure* GDP, in order to obtain the normalised national income $Y = 1$ GDP. The *physical measure* GDP acts as a *numéraire*, therefore the prices are presented in the units $[p_i] = $ (GDP/quantity of commodity i), $i = 1, \ldots, n$. It is important to note that independently of the measurement unit of *labour*, either [TAL] or [man-years] or another unit, the vector $(\mathbf{L} \cdot W)/L$ represents the distribution of the total wages W among all the branches. Sraffa discussed the required number of equations and variables constituting a solvable linear system of equations for the calculation of the prices.

$$\mathbf{S}'\mathbf{p}(1+r)+\mathbf{L}\cdot\frac{\tilde{w}\cdot Y}{L}=\hat{\mathbf{q}}\mathbf{p}=\mathbf{x}=\mathbf{S}'(\hat{\mathbf{q}}^{-1}\hat{\mathbf{q}})\mathbf{p}(1+r)+\mathbf{L}\cdot\frac{\tilde{w}\cdot Y}{L}$$
$$=(\mathbf{S}'\hat{\mathbf{q}}^{-1})(\hat{\mathbf{q}}\mathbf{p})(1+r)+\mathbf{L}\cdot\frac{\tilde{w}\cdot Y}{L}=\mathbf{D}'\mathbf{x}(1+r)+\mathbf{L}\cdot\frac{\tilde{w}\cdot Y}{L}. \qquad (4.42)$$

We obtain the *Sraffa price model* to compute the price vector $\mathbf{p}$ and an equation to compute the vector of *values* $\mathbf{x}$:

$$\boxed{\begin{array}{ll}(1) & \mathbf{S}'\mathbf{p}(1+r)+\dfrac{\tilde{w}\cdot Y}{L}\cdot\mathbf{L}=\hat{\mathbf{q}}\mathbf{p}=\mathbf{x}, \quad Y=\mathbf{d}'\mathbf{p}, \quad L=\mathbf{e}'\mathbf{L},\\ (2) & \mathbf{D}'\mathbf{x}(1+r)+\dfrac{\tilde{w}\cdot Y}{L}\cdot\mathbf{L}=\mathbf{x}.\end{array}} \qquad (4.43)$$

We can further easily compute the values of the five economic variables K, X, (4.31) Y (4.32), P (4.35), W (4.34),

$$X=\mathbf{e}'(\hat{\mathbf{q}}\mathbf{p})=\mathbf{e}'\mathbf{x}, \quad K=\mathbf{e}'\mathbf{S}'\mathbf{p}, \quad Y=X-K, \quad P=r\cdot K, \quad W=Y-P. \qquad (4.44)$$

It is now easy to show that Sraffa's complete price model (4.43) hides an *accountable identity*. Reminding $\tilde{w}=\frac{W}{L}$, we just have to multiply it from the left by vector $\mathbf{e}'$,

$$\mathbf{e}'\mathbf{S}'\mathbf{p}(1+r)+\mathbf{e}'\frac{\tilde{w}\cdot Y}{L}\cdot\mathbf{L}=\mathbf{e}'(\hat{\mathbf{q}}\mathbf{p})=(\mathbf{S}\mathbf{e})'\mathbf{p}(1+r)+\frac{\tilde{w}\cdot Y}{L}\cdot\mathbf{e}'\mathbf{L}=\mathbf{e}'(\hat{\mathbf{q}}\mathbf{p})$$
$$=K+P+L(\frac{\tilde{w}\cdot Y}{L})=K+P+\tilde{w}Y=K+P+W=K+Y=X. \qquad (4.45)$$

We now develop Sraffa's Example 4.1.2 (PCMC, Par. 5). As the Frobenius number is $\lambda_C=4/5$, the *productiveness* is $R=0.25$. The profit rate is set to $r=0.05$. We introduce further a vector of *labour* $\mathbf{L}$.[12]

We illustrate equations (4.43) by the following example.

Example 4.1.3. Consider the production scheme (4.13) with matrix $\mathbf{S}>\mathbf{O}$, the output vector $\mathbf{q}=\mathbf{S}\mathbf{e}+\mathbf{d}>\mathbf{o}$ and $\mathbf{D}=\hat{\mathbf{q}}^{-1}\mathbf{S}>\mathbf{O}$ (4.22). Set up the complete single-product

12 The *physical unit* of *wage per unit of labour* w is $[w]$ = (numeraire/man-years), if labour L is measured in *man-years*. The *physical unit* of total wages W is: $[W]=[L]\cdot[w]=$ man-years $\cdot\,\frac{\text{numeraire}}{\text{man-years}}=$ numeraire. In that case, the total wages are expressed by equation

$$W=L\cdot w. \qquad (4.46)$$

If labour L is measured in TAL, then the *total quantity of labour* is $L=1$ TAL and the *physical unit* of *wage per unit of labour* w is $[w]=\frac{\textit{numeraire}}{\text{TAL}}=$ *numeraire*. The *physical unit* of total wages W is: $[W]=[L]\cdot[w]=\text{TAL}\cdot\frac{\textit{numeraire}}{\text{TAL}}=$ *numeraire*, the total wages are then

$$W=L\cdot w=1\,\text{TAL}\cdot w \quad \text{or} \quad |W|=|w|. \qquad (4.47)$$

This latter numerical equality (4.47) is used by Sraffa (PCMC, Par. 11).

Sraffa price model (4.43) with the rate of profits $r = 0.05$ and the positive vector of labour $\mathbf{L} = [50, 100]' > \mathbf{o}$ in *man-years*. The *numéraire* is wheat. Calculate the price vector $\mathbf{p}$ and the vector of values $\mathbf{x}$. Calculate the national income Y, the total wages W and the total profit P.

Solution to Example 4.1.3:
We identify matrix $\mathbf{S}$ (4.14), and vector $\mathbf{q}$ (4.16), as well as matrix $\mathbf{D}$ (4.22) and the *rate of profits* $r = 0.05$.

$$\mathbf{S} = \begin{bmatrix} 280 & 120 \\ 12 & 8 \end{bmatrix}, \quad \mathbf{q} = \begin{bmatrix} 575 \\ 20 \end{bmatrix}, \quad \mathbf{D} = \begin{bmatrix} \frac{56}{115} & \frac{24}{115} \\ \frac{3}{5} & \frac{2}{5} \end{bmatrix}. \quad \blacktriangle \tag{4.48}$$

Thus, we set up the *complete single-product Sraffa price model* (4.43) (1) with the unknown price vector $\mathbf{p} = [p_1, p_2]'$,

$$(1) \quad \mathbf{S}'\mathbf{p}(1 + 0.05) + \frac{\tilde{w} \cdot Y}{L} \cdot \mathbf{L} = \hat{\mathbf{q}}\mathbf{p} = \mathbf{x}, \quad Y = \mathbf{d}'\mathbf{p}, \quad L = \mathbf{e}'\mathbf{L}. \tag{4.49}$$

We have in (4.49) four equations and five unknown variables p_1, p_2, $\tilde{w}$, Y and L. There are two commodities, wheat and iron; wheat is measured in qr. wheat and iron is measured in t. iron. As is the case in this economy, the prices have to be expressed in the chosen *numéraire* wheat. For this reason, the price of the chosen unit of wheat is determined, setting $p_1 = 1$ qr. wheat/qr. wheat = 1 (dimensionless). The physical units of the price of iron is $[p_2]$ = qr. wheat/t. iron. This gives the *complete single-product Sraffa price model*. The corresponding equations are with the vector of surplus $\mathbf{d} = [175, 0]'$ and $Y = \mathbf{d}'\mathbf{p}$:

$$\begin{aligned} (280 \cdot 1 + 12p_2)(1 + 0.05) + 50 \cdot \frac{\tilde{w} \cdot Y}{L} &= 575 \cdot 1, \\ (120 \cdot 1 + 8p_2)(1 + 0.05) + 100 \cdot \frac{\tilde{w} \cdot Y}{L} &= 20p_2, \\ Y &= 175 \cdot 1 + 0 \cdot p_2, \\ L &= 50 + 100. \end{aligned} \tag{4.50}$$

The solutions are: the price of iron is $p_2 = 18.6957$ qr. wheat/t. iron, the quantity of labour is $L = 150$ *man-years* and the *national income* is $Y_1 = 175$ qr. wheat. The dimensionless *share of total wages to national income* $\tilde{w} = W/Y = 0.778882$ is an invariant, as wages and national income are both measured in the same *physical units*, and the ratio W/Y is by construction a constant in a *single-product Sraffa system*. Then the *total wages* are $W = \tilde{w} \cdot Y = 136.304$ qr. wheat and the *total profits* are $P = Y - W = 38.694$ qr. wheat.

We now move on to the value model (4.43) (2) to determine the vector of *values* $\mathbf{x} = [x_1, x_2]'$,

$$(2) \quad \mathbf{D}'\mathbf{x}(1 + 0.05) + \frac{\tilde{w} \cdot Y}{L} \cdot \mathbf{L} = \mathbf{x}, \quad Y = 175, \quad L = \mathbf{e}'\mathbf{L}. \tag{4.51}$$

Again, (4.51) has four equations and five unknown variables x_1, x_2, $\tilde{w}$, Y and L. In this case, we set $x_1 = 575$ qr. of wheat as the unit measure of the numéraire. This gives the explicitly written system, determining the vector of *values*,

$$\begin{aligned}\left(\frac{56}{115}\cdot 575+\frac{3}{5}x_2\right)(1+0.05)+\frac{\tilde{w}\cdot Y}{L}\cdot 50 &= 575,\\ \left(\frac{24}{115}\cdot 575+\frac{2}{5}x_2\right)(1+0.05)+\frac{\tilde{w}\cdot Y_2}{L}\cdot 100 &= x_2,\\ Y &= 175,\\ L &= 50+100.\end{aligned}\tag{4.52}$$

We obtain the following solutions: the value $x_2 = 373.913$ qr. of wheat, the share of total wages $\tilde{w} = 0.778882$ and the quantity of labour $L = 150$ *man-years*. We further calculate the total output $X = \mathbf{e}'\mathbf{x} = [1,1].[575,373.913]' = 948.913$ qr. wheat, the circulating capital $K = X - Y = 773.913$ qr. of wheat, the total profit $P = r \cdot K = 0.05 \cdot 773.913 = 38.696$ qr. wheat and the total wages $W = Y - P = 136.304$ qr. of wheat.

We can state now an important result:

Proposition 4.1.3. *We start from the single-product Sraffa price model to determine the prices of commodities (4.41),* $w = \frac{\tilde{w}\cdot Y}{L}$*, setting the equation* $\mathbf{x} := \hat{\mathbf{q}}\mathbf{p}$ *(2.18) between the corresponding price vector* $\mathbf{p}$ *and the vector of values* $\mathbf{x}$ *with a positive vector of output* $\mathbf{q} > \mathbf{o}$*. The circle is closed, passing from matrix* $\mathbf{S}$ *to matrix* $\mathbf{D}$*,*

$$\begin{aligned}\mathbf{S}'\mathbf{p}(1+r)+\mathbf{L}\cdot\frac{\tilde{w}\cdot Y}{L} &= \hat{\mathbf{q}}\mathbf{p} = \mathbf{x} = \mathbf{S}'(\hat{\mathbf{q}}^{-1}\hat{\mathbf{q}})\mathbf{p}(1+r)+\mathbf{L}\cdot\frac{\tilde{w}\cdot Y}{L}\\ &= (\mathbf{S}'\hat{\mathbf{q}}^{-1})(\hat{\mathbf{q}}\mathbf{p})(1+r)+\mathbf{L}\cdot\frac{\tilde{w}\cdot Y}{L} = \mathbf{D}'\mathbf{x}(1+r)+\mathbf{L}\cdot\frac{\tilde{w}\cdot Y}{L}.\end{aligned}\tag{4.53}$$

The vector $\mathbf{x} := \hat{\mathbf{q}}\mathbf{p}$ *is the vector of values!*

$$\mathbf{p} = \hat{\mathbf{q}}^{-1}\mathbf{x} \Leftrightarrow \mathbf{x} = \hat{\mathbf{q}}\mathbf{p}.\tag{4.54}$$

We will now show that the *complete single Sraffa price model* (4.43) (1), formulated on the basis of the *commodity flow* matrix $\mathbf{S}$, can also be expressed on the basis of the *input-output coefficients* matrix $\mathbf{C}$. We start writing down the *Sraffa price model* with the wage rate per unit of labour $w = (\tilde{w}\cdot Y)/L$,

$$\boxed{\mathbf{S}'\mathbf{p}(1+r)+\mathbf{L}\cdot w = \hat{\mathbf{q}}\mathbf{p} = \mathbf{x}.}\tag{4.55}$$

The demonstration of this result is obtained as follows:

We multiply equation (4.55) from the left with the diagonal matrix $\hat{\mathbf{q}}^{-1}$. Using again elementary rules of matrix algebra, especially those concerning diagonal matrices and the product rule (A.39). We take the *input coefficients matrix* $\mathbf{C} = \mathbf{S}\hat{\mathbf{q}}^{-1}$ (2.16), respectively the transpose $\mathbf{C}' = \hat{\mathbf{q}}^{-1}\mathbf{S}'$, and find

$$\begin{aligned}\hat{\mathbf{q}}^{-1}(\mathbf{S}'\mathbf{p})(1+r)+\hat{\mathbf{q}}^{-1}(\mathbf{L}\cdot w) &= (\hat{\mathbf{q}}^{-1}\mathbf{S}')\mathbf{p}(1+r)+(\hat{\mathbf{q}}^{-1}\mathbf{L})w\\ &= \mathbf{C}'\mathbf{p}(1+r)+(\hat{\mathbf{q}}^{-1}\mathbf{L})w = \hat{\mathbf{q}}^{-1}(\hat{\mathbf{q}}\mathbf{p}) = (\hat{\mathbf{q}}^{-1}\hat{\mathbf{q}})\mathbf{p} = \mathbf{p},\end{aligned}\tag{4.56}$$

resulting in,

$$\boxed{\mathbf{C}'\mathbf{p}(1+r) + (\hat{\mathbf{q}}^{-1}\mathbf{L})w = \mathbf{p}.} \tag{4.57}$$

In the presentation (4.55) of the *complete single-product Sraffa system*, the entities are *commodity flows*, whereas in the presentation (4.57) the entities are *technical coefficients*. In future developments, we will refer to one or the other of the equivalent representations (4.55) and (4.57) of the *complete single-product Sraffa system*.

We continue defining the vector $\boldsymbol{\pi}$ of *labour per unit of commodity*,

$$\boldsymbol{\pi} = \hat{\mathbf{q}}^{-1}\mathbf{L} = \begin{bmatrix} \pi_1 \\ \pi_2 \\ \dots \\ \pi_n \end{bmatrix} = \begin{bmatrix} \frac{1}{q_1} & \dots & \dots & 0 \\ 0 & \frac{1}{q_2} & \dots & 0 \\ \dots & \dots & \dots & \dots \\ 0 & \dots & \dots & \frac{1}{q_n} \end{bmatrix} \begin{bmatrix} L_1 \\ L_2 \\ \dots \\ L_n \end{bmatrix} = \begin{bmatrix} \frac{L_1}{q_1} \\ \frac{L_2}{q_2} \\ \dots \\ \frac{L_n}{q_n} \end{bmatrix}. \tag{4.58}$$

The ratio $\pi_i = (L_i/q_i)$, $i = 1, \dots, n$, is the labour required for the production of one unit of each commodity i, including replacements to satisfy the conditions of production. After reverting to vector representations, $\boldsymbol{\pi}$ for the *labour per unit of commodity* and $\mathbf{p}$ for the *prices*, we obtain with equations (4.56), (4.58) the *complete single-product Sraffa system* expressed in quantities per *commodity units* $c_{ij} = s_{ij}/q_j$,

$$\mathbf{C}'\mathbf{p}(1+r) + \boldsymbol{\pi} \cdot w = \mathbf{p} \quad \text{or} \quad (\mathbf{I} - (1+r)\mathbf{C}')\mathbf{p} = \boldsymbol{\pi} \cdot w. \tag{4.59}$$

We arrive at an extension of Lemma 4.1.1.

We treat *stability conditions* to guarantee the existence of positive price vectors $\mathbf{p} > \mathbf{o}$ in the case of *complete single-product Sraffa system*. Such *stability conditions* are sufficient but not necessary conditions. We come to

Lemma 4.1.2. *Assumption 2.2.1 and Assumption 2.2.2 hold.*

Consider a commodity flow matrix $\mathbf{S} \geq \mathbf{0}$*, a vector of surplus* $\mathbf{d} \geq \mathbf{o}$ *both semi-positive and a vector of labour* $\mathbf{L} \geq \mathbf{o}$*, at least semi-positive. There is therefore a positive quantity of labour* $L = \mathbf{e}'\mathbf{L}$*. Compute the vector of output* $\mathbf{q} = \mathbf{Se} + \mathbf{d} > \mathbf{o}$*, leading to the production scheme* $(\mathbf{S}', \mathbf{L}) \rightarrow (\hat{\mathbf{q}})$*, the input-output coefficients matrix* $\mathbf{C} = \mathbf{S}\hat{\mathbf{q}}^{-1}$ *(2.16) and the complete single-product Sraffa system (4.55) or (4.57). The positive surplus is split into profits with an appropriate positive rate of profits* $r > 0$ *and into wages with a positive wage rate* $w > 0$.

(a) *When the matrices* $\mathbf{S}$ *(4.55) and, equivalently,* $\mathbf{C}$ *are either semi-positive and irreducible or positive and the vector of labour is semi-positive,* $\mathbf{L} \geq \mathbf{o}$*, then the price vector of the model is positive,* $\mathbf{p} > \mathbf{o}$.

(b) *Additionally, Assumption 2.5.1 holds, here* $\mathbf{L} > \mathbf{o}$ *and so positive. When the matrix* $\mathbf{S} \geq \mathbf{0}$ *and therefore* $\mathbf{C} \geq \mathbf{0}$ *are only semi-positive, then the price vector of the model is positive,* $\mathbf{p} > \mathbf{o}$.

Proof. In the cases (a) and (b), the matrices **S** (4.55) and **C** are both *semi-positive*, so that Lemma 4.1.1 (a) applies and ensures for matrix **C** the existence of a Frobenius number greater than 0, $\lambda_C > 0$. Moreover, Lemma 4.1.1 (b) applies, and we also have an upper bounded Frobenius number, $\lambda_C < 1$.

One sets $\lambda_C = \frac{1}{1+R} < 1$ getting the *productiveness* $R > 0$. A *rate of profits* r, $0 < r < R$ is chosen. Define $\lambda := \frac{1}{1+r} \leq 1$ and get $1 \geq \lambda > \lambda_C > 0$,

(a) The Theorem A.10.2 applies because matrix **S** is *irreducible*. Take the just defined number λ, $1 \geq \lambda > \lambda_C$, and get with (A.103),

$$(\lambda\mathbf{I} - \mathbf{C}')^{-1} := \left(\frac{1}{1+r}\mathbf{I} - \mathbf{C}'\right)^{-1} > \mathbf{0} \Rightarrow (\mathbf{I} - (1+r)\mathbf{C}')^{-1} > \mathbf{0}. \tag{4.60}$$

As there is the inequality $0 < r < R$, a part of the surplus consists of the wages to pay the realized quantity of labour $L > 0$, applying the positive wage rate $w > 0$. As the vector of labour is by assumption *semi-positive*, $\mathbf{L} \geq \mathbf{o}$, the vector $\boldsymbol{\pi} \geq \mathbf{o}$ (4.58), is also *semi-positive*. We obtain with the positive matrices (4.60) a positive price vector[13]:

$$\boxed{\mathbf{p} = (\mathbf{I} - (1+r)\mathbf{C}')^{-1}\boldsymbol{\pi} \cdot w > \mathbf{o}.} \tag{4.61}$$

(b) The Theorem A.10.2 applies, as the matrix $\mathbf{C} \geq \mathbf{0}$ is *semi-positive* and the Frobenius number is positive with Lemma 4.1.1, $\lambda_C > 0$. Choose λ, $1 \geq \lambda > \lambda_C > 0$, and get with Theorem A.10.2, (A.100) the matrices of full rank n, where every row of matric $\mathbf{C}'$, see Assumption 2.2.2, has at least one positive entry, see footnote 13, one gets,

$$(\lambda\mathbf{I} - \mathbf{C}')^{-1} := \sum_{j=0}^{\infty} \frac{\mathbf{C}'^j}{\lambda^{j+1}} = \left(\frac{1}{1+r}\mathbf{I} - \mathbf{C}'\right)^{-1} \geq \mathbf{0} \Rightarrow (\mathbf{I} - (1+r)\mathbf{C}')^{-1} \geq \mathbf{0}. \tag{4.62}$$

Given the inequality $0 < r < R$, a part of the surplus consists of the wages to pay the positive quantity of labour $L > 0$ with positive wage rate $w > 0$. As the vector of labour is *positive* by assumption, $\mathbf{L} > \mathbf{o}$, this property prevails also for the vector $\boldsymbol{\pi} > \mathbf{o}$ (4.58) and we obtain the *positive* price vector (4.61). □

Let's comment on these results, at first by illustrating Theorem A.10.2.

Example 4.1.4. Given the 4×4 *semi-positive* and *irreducible commodity flow* matrix **S** and a positive vector of *total output* **q** (4.63),

$$\mathbf{S} = \begin{bmatrix} 280 & 120 & 150 & 250 \\ 50 & 0 & 0 & 0 \\ 0 & 30 & 20 & 50 \\ 20 & 0 & 0 & 30 \end{bmatrix} \geq \mathbf{0}, \quad \mathbf{q} = \begin{bmatrix} 1000 \\ 100 \\ 200 \\ 100 \end{bmatrix} > \mathbf{o}. \tag{4.63}$$

13 A positive $n \times n$ matrix [a semi-positive $n \times n$ matrix, where every row has at least one positive entry] times a semi-positive [positive] $n \times 1$ vector yields a positive $n \times 1$ vector.

Calculate the *input-output coefficients* matrix $\mathbf{C} = \mathbf{S}\hat{\mathbf{q}}^{-1}$, the Frobenius number λ_C and the productiveness $R = (1/\lambda_C) - 1$, and choose any rate of profits r, $0 \leq r < R$ and confirm that the inverse matrix $[\mathbf{I} - (1+r)\mathbf{C}']^{-1}$ is positive.

Solution of Example 4.1.4:
Compute

$$(\mathbf{I}+\mathbf{S})^3 = \begin{bmatrix} 28{,}906{,}041 & 12{,}192{,}660 & 14{,}445{,}450 & 27{,}405{,}750 \\ 4{,}512{,}150 & 1{,}923{,}001 & 2{,}272{,}500 & 4{,}287{,}500 \\ 787{,}500 & 313{,}890 & 384{,}262 & 727{,}650 \\ 1{,}992{,}660 & 841{,}200 & 999{,}000 & 1{,}894{,}791 \end{bmatrix} > \mathbf{0}, \tag{4.64}$$

verifying that $(\mathbf{S}+\mathbf{I})^3$ is positive. Then $\mathbf{S}$ is *irreducible* (Lemma A.8.2). We continue computing

$$\mathbf{C} = \mathbf{S}\hat{\mathbf{q}}^{-1} = \begin{bmatrix} \frac{7}{25} & \frac{6}{5} & \frac{3}{4} & \frac{5}{2} \\ \frac{1}{20} & 0 & 0 & 0 \\ 0 & \frac{3}{10} & \frac{1}{10} & \frac{1}{2} \\ \frac{1}{50} & 0 & 0 & \frac{3}{10} \end{bmatrix} \geq \mathbf{0}. \tag{4.65}$$

The Frobenius number of $\mathbf{C}$ is smaller than one, $\lambda_C = 0.616506 < 1$, and the *productiveness* is positive, $R = (1/0.61650) - 1 = 0.622045 > 0$. We choose the *rate of profits* $r = 0.5 < R$. We conclude with a positive inverse matrix,

$$(\mathbf{I} - (1+0.5)\mathbf{C}')^{-1} = \begin{bmatrix} 7.0601 & 0.5295 & 0.6201 & 0.3851 \\ 16.9131 & 2.2685 & 2.0150 & 0.9225 \\ 9.3446 & 0.7008 & 1.9972 & 0.5097 \\ 60.8792 & 4.5660 & 6.9516 & 5.1389 \end{bmatrix} > \mathbf{0}, \tag{4.66}$$

in accordance with Theorem A.10.2, equation (A.103). ▲

We now present an illustration of Lemma 4.1.2 (a). For an economy composed of two sectors we investigate – ceteris paribus – the relationship between the *profit rate* r and the *wage rate* w.

Example 4.1.5. An economy is composed of a sector of wheat (physical unit: "qr. of wheat") where labour is measured in $[\mathbf{L}]$ = man-years, and a sector of wood (physical unit: "cubic meter of wood"), based on subsistence wages. There is the positive matrix $\mathbf{S}$ and the positive vector of surplus, $\mathbf{d} > \mathbf{o}$, together with the vectors $\mathbf{L}$ and $\mathbf{q} = \mathbf{S}\mathbf{e} + \mathbf{d} > \mathbf{o}$, leading to the production scheme $(\mathbf{S}', \mathbf{L}) \rightarrow (\hat{\mathbf{q}})$.

$$\mathbf{S} = \begin{bmatrix} 500 & 500 \\ 500 & 500 \end{bmatrix} > \mathbf{0}, \quad \mathbf{L} = \begin{bmatrix} 200 \\ 0 \end{bmatrix} \geq \mathbf{o}, \quad \mathbf{q} = \begin{bmatrix} 1{,}500 \\ 1{,}500 \end{bmatrix}, \quad \mathbf{d} = \begin{bmatrix} 500 \\ 500 \end{bmatrix} > \mathbf{o}. \tag{4.67}$$

Compute the input-output coefficients matrix $\mathbf{C} = \mathbf{S}\hat{\mathbf{q}}^{-1} > \mathbf{0}$, the Frobenius number λ_C of matrix $\mathbf{C}$, *and* the productiveness $R = (1/\lambda_C) - 1$.

The price vector is $\mathbf{p} = [p_1, p_2]'$. Choose as the numéraire the commodity wheat, so that $p_1 = 1$ (qr. wheat/qr. wheat) = 1. Compute the matrix $(\mathbf{I} - (1+r)\mathbf{C}')^{-1}$ and the vector $\boldsymbol{\pi} = \hat{\mathbf{q}}^{-1}\mathbf{L}$. Set up (4.61)

$$\mathbf{p} = \begin{bmatrix} 1 \\ p_2 \end{bmatrix} = (\mathbf{I} - (1+r)\mathbf{C}')^{-1}\boldsymbol{\pi} \cdot w. \tag{4.68}$$

(a) Choose $r = 0.1$ and determine p_2 and w from the system (4.68).
(b) Determine the relationship between r and w from the first equation of (4.68).

Compute then the national income Y, the total output X, the circulating capital K, the total profit P, the total wages W and the total quantity of labour L, as well as the ratios: share of total wages to national income $\tilde{w}$, share of total profits to national income $\tilde{r}$, average national income per unit of total quantity of labour U, and surplus ratio $\tilde{R}$.

Solution to Example 4.1.5:
Let us calculate the matrix

$$\mathbf{C} = \mathbf{S}\hat{\mathbf{q}}^{-1} = \begin{bmatrix} 500 & 500 \\ 500 & 500 \end{bmatrix} \begin{bmatrix} \frac{1}{1{,}500} & 0 \\ 0 & \frac{1}{1{,}500} \end{bmatrix} = \begin{bmatrix} \frac{1}{3} & \frac{1}{3} \\ \frac{1}{3} & \frac{1}{3} \end{bmatrix} > \mathbf{0}. \tag{4.69}$$

The Frobenius number of matrix $\mathbf{C}$ is $\lambda_C = 2/3$ and the *maximal rate of profits* is $R = (3/2) - 1 = 1/2$. If the rate of profits $r \in [0, R]$, then all the coefficients of the inverse matrix $(\mathbf{I} - (1+r)\mathbf{C}')^{-1}$ are positive (see Theorem A.10.2, equation (A.103)):

$$\boldsymbol{\pi} = \hat{\mathbf{q}}^{-1}\mathbf{L} = \begin{bmatrix} \frac{2}{15} \\ 0 \end{bmatrix} \geq \mathbf{o}; \quad (\mathbf{I} - (1+r)\mathbf{C}')^{-1} = \begin{bmatrix} \frac{-2+r}{-1+2r} & \frac{1+r}{1-2r} \\ \frac{1+r}{1-2r} & \frac{-2+r}{-1+2r} \end{bmatrix} > \mathbf{0} \Rightarrow$$

$$\mathbf{p} = (\mathbf{I} - (1+r)\mathbf{C}')^{-1}\boldsymbol{\pi} \cdot w = \begin{bmatrix} p_1 = 1 \\ p_2 \end{bmatrix} = \begin{bmatrix} \frac{-2+r}{-1+2r} & \frac{1+r}{1-2r} \\ \frac{1+r}{1-2r} & \frac{-2+r}{-1+2r} \end{bmatrix} \begin{bmatrix} \frac{2}{15} \\ 0 \end{bmatrix} w > \mathbf{o}. \tag{4.70}$$

Setting $r = 0.1$, we solve the second matrix equation of the system (4.70) and get then $p_2 = 0.578947$ or $\mathbf{p} = [1, 0.575947]' > \mathbf{o}$ and $w = 3.15789$. Then we consider its first equation for $p_1 = 1$ and get a relationship between r and w

$$p_1 = 1 = \frac{2(2-r)w}{15(1-2r)} \Rightarrow 3.75 \geq w = \frac{15(1-2r)}{2(2-r)} \geq 0, \quad 0 \leq r < 0.5. \tag{4.71}$$

Now we list the economic variables dependent on $r = 0.1$ and get
- the *national income* $Y = \mathbf{d}'\mathbf{p} = [500, 500][1, 0.5789]' = 789.47$ qr. wheat;
- the *circulating capital* $K = (\mathbf{Se})'\mathbf{p} = 1{,}578.95$ qr. wheat;
- the *total quantity of labour* $L = \mathbf{e}'\mathbf{L} = 200$ man-year;

- the *total output* $X = \mathbf{q}'\mathbf{p} = [1{,}500, 1{,}500][1, 0.5789]' = 2{,}368.42$ qr. wheat;
- the *national income per unit of quantity of labour* $U = Y/L = 3.9474$ qr. wheat/man-years;
- the *surplus ratio* $\tilde{R} = Y/K = 789.47/1{,}578.95 = 0.5$ is equal to R in this case.[14] We can compute: $Y = R \cdot K = 0.5 \cdot 1{,}578.95 = 789.47$.

Then we list the economic variables dependent on r in general and obtain
- the *total profit* $P = K \cdot r = 1{,}578.95 \cdot r = 157.90$ qr. wheat;
- the *total wages* $W = Y - P = (789.47 - 1{,}578.95 \cdot 0.1)$ qr. wheat $= Y \cdot (1 - 2r) = 631.58$ qr. wheat;
- the *share of total profits to national income* $\tilde{r} = P/Y = (K{\cdot}r)/Y = 1{,}578.85{\cdot}r/789.47 = 2r = 0.2$;
- the *share of total wages to national income* $\tilde{w} = W/Y = Y \cdot (1 - 2r)/Y = 1 - 2r = 0.8$, so we have correctly $\tilde{r} + \tilde{w} = 1$.

Finally, we confirm $\tilde{R}(1 - \tilde{w}) = 0.5(1 - 0.8) = 0.1 = r$ and get for this example the relationship:

$$\boxed{\begin{array}{l} w = f(r) = \dfrac{15(1-2r)}{2(2-r)} > 0, \quad 0 \le r < R = \dfrac{1}{2}, \quad 0 < w \le \dfrac{15}{4}, \\ Y = P + W, \quad 0 < P, \quad 0 < W. \end{array}} \tag{4.72}$$

The function $w = f(r)$ to calculate the *wage rate*, dependent on the *rate of profit* r, is specific to each production scheme. In the present case, the wage rate w is limited by the interval, $w \in \,]0, 3.75]$, when $r \in [0, 0.5[$. The *national income* $Y = \mathbf{d}'\mathbf{p}$ is determined by the vector of prices and the vector of surplus and is always the sum of *total wages* W plus *total profits* P. ▲

We continue illustrating Lemma 4.1.2 (b). The next example shows that we can get positive prices without needing the **Perron–Frobenius theorem A.9.3** just requiring additional economic conditions to be fulfilled by some vectors and matrices of the Sraffa price model.

Example 4.1.6. Consider a partial economy producing iron and wheat both measured in tons (t.) and gold measured in kilograms (kg). There is a total number of $L = 25$ workers in the three sectors, working the whole year. Labour is measured in man-years. The vector of *labour* $\mathbf{L} = [5, 10, 10]' > \mathbf{o}$ in man-years shows the repartition of the workers among the sectors. The value is measured by the currency CHF. The production

14 The result $\tilde{R} = R$ is a special case. We are in the presence of a *Standard system* that will be treated in Chapter 5, namely, $\mathbf{d} \parallel \mathbf{q}$.

scheme is as follows:

$$\begin{aligned}(30\text{ t. wheat}, 30\text{ t. iron}, 5\text{ MJ}) &\to (400\text{ qr. wheat}, 0, 0),\\ (15\text{ t. wheat}, 90\text{ t. iron}, 10\text{ MJ}) &\to (0, 300\text{ t. iron}, 0),\\ (5\text{ t. wheat}, 30\text{ t. iron}, 10\text{ MJ}) &\to (0, 0, 6\text{ kg gold}).\end{aligned} \tag{4.73}$$

Identify the matrix $\mathbf{S}$ and the vector $\mathbf{q} = \mathbf{Se} + \mathbf{d}$. Calculate the vector of surplus $\mathbf{d}$ and the input-output matrix $\mathbf{C} = \mathbf{S}\hat{\mathbf{q}}^{-1}$. Verify Proposition 3.1.4 (3.53). Set up the Sraffa price model in the form (4.61). Compute the productiveness R, choose the rate of profits $r = R/2$. Finally, calculate the wage rate w, using the exogenously given price of wheat: $p_1 = 500$ (CHF/tons of wheat), and then the vectors of absolute prices $\mathbf{p}$.

Solution to Example 4.1.6:
Identify now the *semi-positive* matrix $\mathbf{S}$, and compute the vector of output $\mathbf{q}$ and the vector of surplus $\mathbf{d}$,

$$\mathbf{S} = \begin{bmatrix} 30 & 15 & 5 \\ 30 & 90 & 30 \\ 0 & 0 & 0 \end{bmatrix}, \quad \mathbf{q} = \mathbf{Se} + \mathbf{d} = \begin{bmatrix} 30 & 15 & 5 \\ 30 & 90 & 30 \\ 0 & 0 & 0 \end{bmatrix}\begin{bmatrix} 1 \\ 1 \\ 1 \end{bmatrix} + \mathbf{d}$$

$$= \begin{bmatrix} 50 \\ 150 \\ 0 \end{bmatrix} + \mathbf{d} = \begin{bmatrix} 400 \\ 300 \\ 6 \end{bmatrix} \Rightarrow \mathbf{d} = \begin{bmatrix} 350 \\ 150 \\ 6 \end{bmatrix} > \mathbf{o}. \tag{4.74}$$

Compute the *semi-positive input-output* matrix, $\mathbf{C} = \mathbf{S}\hat{\mathbf{q}}^{-1}$,

$$\mathbf{C} = \begin{bmatrix} 30 & 15 & 5 \\ 30 & 90 & 30 \\ 0 & 0 & 0 \end{bmatrix}\begin{bmatrix} \frac{1}{400} & 0 & 0 \\ 0 & \frac{1}{300} & 0 \\ 0 & 0 & \frac{1}{6} \end{bmatrix} = \begin{bmatrix} \frac{3}{40} & \frac{1}{20} & \frac{5}{6} \\ \frac{3}{40} & \frac{3}{10} & 5 \\ 0 & 0 & 0 \end{bmatrix} \geq \mathbf{0}. \tag{4.75}$$

With the regularity, $\det(\mathbf{I} - \mathbf{C}') = 103/160$, Proposition 3.1.4 is verified. There is no eigenvalue $\lambda = 1$ of matric $\mathbf{C}'$. Computing the *semi-positivity* of matrix,

$$(\mathbf{I} + \mathbf{C})^2 = \begin{bmatrix} \frac{371}{320} & \frac{19}{160} & \frac{95}{48} \\ \frac{57}{320} & \frac{271}{160} & \frac{185}{16} \\ 0 & 0 & 1 \end{bmatrix} \geq \mathbf{0}, \tag{4.76}$$

confirms that matrix $\mathbf{C}$ is *reducible* (see Lemma A.8.2). Continuing with Lemma 4.1.1 (b) we now know that matrix $\mathbf{C}$ has a Frobenius number λ_C smaller than 1, which is positive (see Lemma A.10.3). One gets the Frobenius number λ_C, $0 < \lambda_C < 1$, calculating the characteristic polynomial of matrix $\mathbf{C}$,

$$P_2(\lambda) = \det(\mathbf{C} - \lambda\mathbf{I}) = \frac{-1}{160}\lambda(160\lambda^2 - 60\lambda + 3) = \frac{-1}{160}\lambda(\lambda - 0.0594)(\lambda - 0.3156). \tag{4.77}$$

One obtains $\lambda_C = 0.3156$, the *maximal profit rate* $R = (1/\lambda_C) - 1 = 2.1687$ and the profit rate $r = R/2 = 1.0844$, $\lambda := \frac{1}{1+r} > \lambda_C$. With Theorem A.10.2 one gets the *semi-positive* matrix:

$$(\mathbf{I}-(1+r)\mathbf{C}')^{-1}=\sum_{j=0}^{\infty}\frac{\mathbf{C}'^{j}}{\lambda^{j}}=\begin{bmatrix}1.250 & 0.521 & 0\\ 0.348 & 2.814 & 0\\ 5.793 & 30.231 & 1\end{bmatrix}\geq \mathbf{o}. \tag{4.78}$$

One computes the vector $\boldsymbol{\pi}$ of labour per unit of commodity expressed in the units $[\boldsymbol{\pi}]$ = (MJ/unit of commodity),

$$\boldsymbol{\pi}:=\hat{\mathbf{q}}^{-1}\mathbf{L}=\begin{bmatrix}\frac{1}{400} & 0 & 0\\ 0 & \frac{1}{300} & 0\\ 0 & 0 & \frac{1}{6}\end{bmatrix}\begin{bmatrix}5\\ 10\\ 10\end{bmatrix}=\begin{bmatrix}\frac{1}{80}\\ \frac{1}{30}\\ \frac{5}{3}\end{bmatrix}>\mathbf{o}, \tag{4.79}$$

and finally, considering the price p_1 = 500 CHF/t. of wheat, one gets with Lemma 4.1.2 (b) the *positive* price vector $\mathbf{p}=[500,p_2,p_3]'$ and computes with the system of equation

$$\mathbf{p}=(\mathbf{I}-(1+r)\mathbf{C}')^{-1}\boldsymbol{\pi}w=\begin{bmatrix}1.250 & 0.521 & 0\\ 0.348 & 2.814 & 0\\ 5.793 & 30.231 & 1\end{bmatrix}\begin{bmatrix}\frac{1}{80}\\ \frac{1}{30}\\ \frac{5}{3}\end{bmatrix}w=\begin{bmatrix}500\\ p_2\\ p_3\end{bmatrix}>\mathbf{o}. \tag{4.80}$$

the variables p_2, p_3, w, obtaining w = 15,151 CHF/MJ, p_2 = 1,487 CHF/t. of iron and p_3 = 41,617 CHF/ kg of gold. ▲

Example 4.1.6 is of central importance in this Chapter 4 because it shows that a complete Sraffa price model with positive surplus, distributed between the workers as wages and the sectors as profits, generate positive prices for all commodities, if suitable additional economic assumptions are fulfilled by the matrices and vectors of the *Sraffa price model* under discussion. The Lemmas belonging to the group of the Perron-Frobenius theorem can be applied. We see that there is an inter connecetion between economic conditions and matrix theorems to obtain such positive prices.

4.1.4 Variables of national accounting in Sraffa's price model

Let us now go back to the Subsection 4.1.2, where the various variables in connection with *national income* are defined and to the *complete single-product Sraffa system* (PCMC, Par. 11, 12), i. e., (4.55) with the vector of *labour* $\mathbf{L}\geq\mathbf{o}$. The quantity of labour is measured in *man-years*.[15] The *wage rate* $w=\tilde{w}\cdot Y/L>0$ is usually positive, as well as the *national income* $Y>0$ (4.32), measuring the surplus, and the *total quantity of labour* $L=\mathbf{e}'\mathbf{L}>0$. Again, matrix $\mathbf{S}$ is either *non-negative* and *irreducible* or *positive*,

15 The *total quantity of labour* $L>0$ may also be presented as *the total amount of labour* (TAL).

the *input-output coefficients* matrix $\mathbf{C} = \mathbf{S}\hat{\mathbf{q}}^{-1}$ is computed, the vector of surplus is *semi-positive*, $\mathbf{d} \geq \mathbf{o}$, the vector of total output is *positive*, $\mathbf{q} = \mathbf{Se}+\mathbf{d} > \mathbf{o}$ and the production scheme is described with matrix **S** or matrix **C** by

$$(\mathbf{S}', \mathbf{L}) \rightarrow (\hat{\mathbf{q}}) \Leftrightarrow (\mathbf{C}', \hat{\mathbf{q}}^{-1}\mathbf{L}) \rightarrow (\mathbf{I}). \tag{4.81}$$

The Frobenius number of matrix **C** is in the known limiting interval, $0 < \lambda_C < 1$ (see Lemma 4.1.1 (b)).[16] The positive *productiveness* $R = (1/\lambda_C) - 1 > 0$ is computed, and we get the *complete single-product Sraffa system*, where the *rate of profits r* can freely be chosen within the range, $0 \leq r < R$, here again w is replaced by $w = (\tilde{w} \cdot Y)/L > 0$,

$$\boxed{\begin{aligned} \mathbf{S}'\mathbf{p}(1+r) + \mathbf{L} \cdot \frac{\tilde{w} \cdot Y}{L} &= \hat{\mathbf{q}}\mathbf{p} = \mathbf{x}, \\ Y &= (\mathbf{q} - \mathbf{Se})'\mathbf{p}, \\ L &= \mathbf{e}'\mathbf{L}. \end{aligned}} \tag{4.82}$$

Finally, we write down the explicit equations of system (4.82):

$$\boxed{\begin{aligned} s_{11}(1+r)p_1 + s_{21}(1+r)p_2 + \cdots + s_{n1}(1+r)p_n + wL_1 &= q_1p_1, \\ s_{12}(1+r)p_1 + s_{22}(1+r)p_2 + \cdots + s_{n2}(1+r)p_n + wL_2 &= q_2p_2, \\ \ldots &= \cdots, \\ s_{1n}(1+r)p_1 + s_{2n}(1+r)p_2 + \cdots + s_{nn}(1+r)p_n + wL_n &= q_np_n, \\ d_1p_1 + d_2p_2 + \cdots + d_np_n &= Y, \\ L_1 + L_2 + \cdots . + L_n &= L, \\ w &= \frac{\tilde{w} \cdot Y}{L}. \end{aligned}} \tag{4.83}$$

In the sequel, we shall concentrate on computational aspects. The Sraffa price model (4.82) has $n + 2$ equations and $n + 4$ unknown variables, the n absolute *prices* p_i, $i \in \{1, \ldots, n\}$, the *share of total wages to national income* $\tilde{w}$, the *total quantity of labour* L, the *rate of profits* r and the *national income* Y.[17] Lemma 4.1.2 gives the necessary conditions to obtain *positive* price vectors for the *complete single-product Sraffa system* (4.82), $\mathbf{p} = [p_1, \ldots, p_n]' > \mathbf{o}$. These conditions are fulfilled for the following Example 4.1.7 because matrix **S** and vector **L** are both positive.

---

16 Remark: The required *semi-positiveness* of vector $\mathbf{d} \geq \mathbf{o}$ and *positivity* of vector $\mathbf{q} = \mathbf{Se} + \mathbf{d}$ are sufficient for the Frobenius number of matrix $\mathbf{C} = \mathbf{S}\hat{\mathbf{q}}^{-1}$ to be smaller that 1, $\lambda_C < 1$. One verifies that the splitting of Y in *profits* P and wages W, $Y = P + W$ has no influence on this property, even if $W = 0$ or $P = 0$.

17 The solution of (4.82) is usually performed by the following method: The rate of profits r is exogenously given, $0 \leq r \leq R$. Then, either a measurement unit of commodity i is taken as the *numéraire* and $p_i = p_0$ is exogenous and given, the $n + 2$ remaining variables p_j, $(j \neq i)$, Y, $\tilde{w}$, L are endogenous, or *national income* $Y = Y_0$ is taken as exogenous and given and the remaining $n + 2$ variables $p_1, \ldots, p_n$, $\tilde{w}$, L are endogenous. The remaining system (4.82) of $n + 2$ equations and the same number

Example 4.1.7. Consider Sraffa's two sector micro-economy (PCMC, Par. 5), an extension of the Example 3.1.3. In accordance with the production scheme (4.81), a vector of *labour* $\mathbf{L} = [152, 152]$, measured in working hours, is introduced, see Table 4.1. The workers get their salary at the end of the year.

Table 4.1: Input-Output Table in physical terms, Sraffa's example PCMC, Par. 5.

Commodities	**Buying sectors**		**Surplus**	**Total**
	wheat	**iron**		**output**
wheat (qr. wheat)	$s_{11} = 280$	$s_{12} = 120$	$d_1 = 175$	$q_1 = 575$
iron (t. iron)	$s_{21} = 12$	$s_{21} = 8$	$d_2 = 0$	$q_2 = 20$
labour (*hours*)	$L_1 = 152$	$L_2 = 152$		$L = 304$
	$\downarrow$	$\downarrow$		
production	$q_1 = 575$	$q_2 = 20$		

The *wage per unit of labour* is $w = (\tilde{w} \cdot Y)/L$ (4.38). Then we obtain the single product Sraffa system (4.82) together with the national income Y and the total quantity of labour L.[18] The working time is adapted in a way to get contemporary wages.

The *productiveness* $R = 0.25$ of this system is known from the calculation of the Frobenuis number (3.47). We choose as profit rate $r = r_0 = 0.05$, $0 \leq r \leq R$.

Calibrate the system according to the following possibilities: Either we set the national income $Y = Y_0$, where Y_0 acts as the freely chosen exogenous variable (see PCMC, Par. 11), or, we choose a *numéraire* and set the corresponding price $p_i = p_0$, $i \in \{1, 2\}$, where p_0 acts as a freely chosen exogenous variable (PCMC, Par. 3).

We summarise the entries (4.14) of the *complete single-product Sraffa system* (4.82) for the present case.

$$\mathbf{S} = \begin{bmatrix} 280 & 120 \\ 12 & 8 \end{bmatrix}, \quad \mathbf{q} = \begin{bmatrix} 575 \\ 20 \end{bmatrix}, \quad \mathbf{d} = \begin{bmatrix} 175 \\ 0 \end{bmatrix}, \quad \mathbf{L} = \begin{bmatrix} 152 \\ 152 \end{bmatrix}. \tag{4.84}$$

Present a table with the economic variables X, K, Y, P, W, $U = Y/L$ and the ratios r, w, $\tilde{r}$, $\tilde{w}$, $\tilde{R}$ (4.32), (4.33), (4.34), (4.35).

---

of endogenous variables is comfortably solved as a system of simultaneous equations with software packages, like *MATHEMATICA* or *MATLAB*. Note that fixing again a measurement unit of commodity i as the numéraire, the price $p_i = p_0$ and the *rate of profits* r are exogenous and given; further setting the *wage rate* $w = (\tilde{w} \cdot Y)/L$ as a variable, the system becomes linear and we can determine the $n - 1$ variables p_j, $(j \neq i)$, together with the n-th variable w in (4.55).

18 For calculation purposes we have taken here the same commodity flow matrix **S** as previously, although to be quite consistent this matrix should be modified to account for the fact that all labour is now regrouped in the second left-hands terms (see PCMC Par. 8).

Solution to Example 4.1.7:
We begin, setting up the *complete single-product Sraffa system* (*Sraffa price model*) with the matrix and vectors (4.84). We are now ready for computational illustrations. We treat *three cases* of parametrisation, keeping in mind that these examples are not exhaustive.

$$\begin{bmatrix} 280 & 12 \\ 120 & 8 \end{bmatrix}\begin{bmatrix} p_1 \\ p_2 \end{bmatrix}(1+r) + \frac{\tilde{w}\cdot Y}{L}\begin{bmatrix} 152 \\ 152 \end{bmatrix} = \begin{bmatrix} 575\cdot p_1 \\ 20\cdot p_2 \end{bmatrix}, \quad Y = [175, 0]\begin{bmatrix} p_1 \\ p_2 \end{bmatrix}, \quad L = [152, 152]'\begin{bmatrix} 1 \\ 1 \end{bmatrix}. \tag{4.85}$$

Case 1. We set Sraffa's normalisation of the *national income* (4.37), $Y = 1$ GDP, and Sraffa's normalisation of the *total quantity of labour* (4.30), $L = 1$ TAL. Compute p_1, p_2, $\tilde{w}$. We will now start to solve the *complete single-product Sraffa system* (4.85).[19] Having set $Y = 1$ GDP and the "*artificial numéraire*" is GDP = currency/year, we calculate the prices p_1, p_2 and the *share of total wages to national income* $\tilde{w}$.[20] See Table 4.2.

$$\begin{cases} (280p_1 + 12p_2)(1.05) + \frac{1}{2}\cdot\tilde{w} = 575p_1, \\ (120p_1 + 8p_2)(1.05) + \frac{1}{2}\cdot\tilde{w} = 20p_2, \\ (575 - 280 - 120)p_1 + (20 - 12 - 8)p_2 = 1. \end{cases} \tag{4.89}$$

19 It is also possible to compute the *complete single-product Sraffa system* based on the entities *per unit of commodity* (4.58). Using here *labour*, measured in working *hours*, we have to determine the vector $\boldsymbol{\pi}$ of *labour per unit of commodity* (4.58) giving,

$$\boldsymbol{\pi} = \hat{\mathbf{q}}^{-1}\mathbf{L} = \begin{bmatrix} w_1 \\ w_2 \end{bmatrix} = \begin{bmatrix} \frac{1}{575} & 0 \\ 0 & \frac{1}{20} \end{bmatrix}\begin{bmatrix} 152 \\ 152 \end{bmatrix} = \begin{bmatrix} \frac{152}{575} \\ \frac{152}{20} \end{bmatrix}, \tag{4.86}$$

the units in physical terms being: $[w_1]$ = hour/qr. of wheat, $[w_2]$ = hour/t. of iron. Continuing with the computation of the matrix of *input coefficients* $\mathbf{C}$ (2.16) in *physical terms*

$$\mathbf{C} = \mathbf{S}\hat{\mathbf{q}}^{-1} = \begin{bmatrix} 280 & 120 \\ 12 & 8 \end{bmatrix}\begin{bmatrix} \frac{1}{575} & 0 \\ 0 & \frac{1}{20} \end{bmatrix} = \begin{bmatrix} \frac{56}{115} & 6 \\ \frac{12}{575} & \frac{2}{5} \end{bmatrix}, \quad \mathbf{C}' = \begin{bmatrix} \frac{56}{115} & \frac{12}{575} \\ 6 & \frac{2}{5} \end{bmatrix} \tag{4.87}$$

Sraffa's *price model* (4.82) is then expressed in commodity units as:

$$\begin{bmatrix} \frac{56}{115} & \frac{12}{575} \\ 6 & 0.4 \end{bmatrix}\begin{bmatrix} p_1 \\ p_2 \end{bmatrix}(1+r) + \begin{bmatrix} \frac{152}{575} \\ \frac{152}{20} \end{bmatrix}\cdot\frac{\tilde{w}\cdot Y}{304} = \begin{bmatrix} p_1 \\ p_2 \end{bmatrix}, \quad Y = [175, 0]\begin{bmatrix} p_1 \\ p_2 \end{bmatrix}. \tag{4.88}$$

20 This is exactly *Sraffa's price model* with normalised labour $L = 1$ TAL and normalised national income $Y = 1$ GDP giving $\tilde{w} = |w| = |W|$.

Solving the system (4.89), we obtain the relative prices $p_1 = (1/175)$ GDP/qr. wheat, $p_2 = (37/385)$ GDP/t. iron and directly the *share of total wages to national income* $\tilde{w} = (304/385) = 0.7896$.

We use the obtained values of p_1, p_2 and $\tilde{w}$ to calculate the economic variables (4.31), (4.32) and ratios (4.34), (4.35),

$$X = \mathbf{q}'\mathbf{p} = [575, 20] \begin{bmatrix} \frac{1}{175} \\ \frac{37}{385} \end{bmatrix} = \frac{401}{77} \text{ GDP},$$

$$K = (\mathbf{Se})'\mathbf{p} = [400, 20] \begin{bmatrix} \frac{1}{175} \\ \frac{37}{385} \end{bmatrix} = \frac{324}{77} \text{ GDP},$$

$$P = (\mathbf{Se})'\mathbf{p} \cdot r = K \cdot r = [400, 20] \begin{bmatrix} \frac{1}{175} \\ \frac{37}{385} \end{bmatrix} \cdot 0.05 = \frac{81}{385} \text{ GDP},$$

$$Y = \mathbf{d}'\mathbf{p} = X - K = \frac{401}{77} - \frac{324}{77} = 1 \text{ GDP}, \tag{4.90}$$

$$W = Y - P = 1 - \frac{81}{385} = \frac{304}{385} \text{ GDP},$$

$$w = \frac{\tilde{w} \cdot Y}{L} = \frac{304}{385} \cdot \frac{1}{1} = \frac{304}{385} \frac{\text{GDP}}{\text{TAL}},$$

$$U = \frac{Y}{L} = \frac{1}{1} = 1 \frac{\text{GDP}}{\text{TAL}}. \tag{4.91}$$

Then we compute the rates and shares

$$\tilde{r} = \frac{P}{Y} = \frac{81}{385 \cdot 1} = \frac{81}{385} = 0.2104,$$

$$\tilde{w} = \frac{W}{Y} = \frac{304}{385 \cdot 1} = \frac{304}{385} = 0.7896,$$

$$\tilde{R} = \frac{Y}{K} = \frac{1 \cdot 77}{324} = \frac{77}{324} = 0.2377,$$

$$r = \tilde{R}(1 - \tilde{w}) = \left(\frac{Y}{K}\right)(1 - \tilde{w}) = 1 \cdot \left(\frac{77}{324}\right)\left(1 - \frac{304}{385}\right) = 0.05. \tag{4.92}$$

The *rate of profits* $r = P/K = \frac{81}{385} / \frac{324}{77} = 0.05$, the economic identity for the *national income* $Y = W + P = L \cdot w + P = 1 \cdot (304/385) + (81/385) = 1$ GDP and the economic identity for the *total output* $X = Y + K = 1 + (324/77) = (401/77)$, as well as $Y = \tilde{R} \cdot K = (77/324)(324/77) = 1$ GDP are confirmed.

We confirm the identity $|w| = |\tilde{w}| = |W| = 0.7896$. Sraffa uses variable w in his *price model* ([108], Par. 11), and we also see that $R = 0.25 \neq \tilde{R} = 0.2377$. This will be discussed later. The numerical equality of wage per unit of labour and the share of total wages is here confirmed, $\tilde{w} = W/Y = W/L = w = (304/385)$, but the units are different, as mentioned.

Note that the relative prices are given in the physical units:

$$[p_1] = (\text{GDP/qr. wheat}) \quad \text{and} \quad [p_2] = (\text{GDP/t. iron}).$$

Table 4.2: Calculated variants and invariants of **Case 1**.

Variable terms		**Invariant terms**	
notion	**value**	**notion**	**value**
total output	$X = 5.2078$ GDP	productiveness	$R = 0.25$
circulating capital	$K = 4.2078$ GDP	rate of profits	$r = 0.05$
total profits	$P = 0.2104$ GDP	share of t. profits	$\tilde{r} = 0.2104$
total wages	$W = 0.7896$ GDP	surplus ratio	$\tilde{R} = 0.2377$
national income (NI)	$Y = 1$ GDP	share of t. wages	$\tilde{w} = 0.7896$
average NI per labour	$U = 1\ \frac{\text{GDP}}{\text{TAL}}$		
wage per unit of labour	$w = 0.7896\ \frac{\text{GDP}}{\text{TAL}}$		

Case 2. The "*numéraire*" is 1 qr. wheat and the price of wheat is set up. and we obtain $p_1 = 1$ (qr. wheat/qr. wheat) = 1 and the *rate of profit* is $r = 0.05$. Sraffa's normalisation of the *total quantity of labour* (4.30), $L = 1$ TAL, is applied. This being done, we calculate the *complete single-product Sraffa system* (4.85) with normalised labour $L = 1$ TAL, getting p_2, Y, $\tilde{w}$. See Table 4.3.

$$\begin{cases} (280 \cdot 1 + 12p_2)(1.05) + \frac{1}{2} \cdot \tilde{w} \cdot Y = 575 \cdot 1, \\ (120 \cdot 1 + 8p_2)(1.05) + \frac{1}{2} \cdot \tilde{w} \cdot Y = 20p_2, \\ (575 - 280 - 120) \cdot 1 + (20 - 12 - 8)p_2 = Y. \end{cases} \tag{4.93}$$

We thus obtain by solving (4.93) the national income $Y = 175$ qr. wheat, the relative price $p_2 = (185/11) = 16.8182$ qr. wheat/t. iron and the *share of total wages to national income* $\tilde{w} = (304/385) = 0.7896$.

We introduce the obtained variables p_1, p_2 and $\tilde{w}$ to calculate the economic variables (4.31), (4.32) and ratios (4.34), (4.35),

$$\begin{aligned} X &= \mathbf{q}'\mathbf{p} = [575, 20] \begin{bmatrix} 1 \\ \frac{185}{11} \end{bmatrix} = \frac{10{,}025}{11} \text{ qr. wheat}, \\ K &= (\mathbf{Se})'\mathbf{p} = [400, 20] \begin{bmatrix} 1 \\ \frac{185}{11} \end{bmatrix} = \frac{8{,}100}{11} \text{ qr. wheat}, \\ P &= (\mathbf{Se})'\mathbf{p} \cdot r = K \cdot r = [400, 20] \begin{bmatrix} 1 \\ \frac{185}{11} \end{bmatrix} \cdot 0.05 = \frac{405}{11} \text{ qr. wheat}, \\ Y &= \mathbf{d}'\mathbf{p} = X - K = \frac{10{,}025}{11} - \frac{8{,}100}{11} = 175 \text{ qr. wheat}, \\ W &= Y - P = \tilde{w} \cdot Y = 175 - \frac{405}{11} = \frac{304}{385} \cdot 175 = \frac{1{,}520}{11} \text{ qr. wheat}, \end{aligned} \tag{4.94}$$

$$\begin{aligned} w &= \frac{\tilde{w} \cdot Y}{L} = \frac{304}{385} \cdot \frac{175}{1} = \frac{1{,}520}{11} \frac{\text{qr. wheat}}{\text{TAL}}, \\ U &= \frac{Y}{L} = \frac{175}{1} = 175 \frac{\text{qr. wheat}}{\text{TAL}}, \end{aligned} \tag{4.95}$$

Table 4.3: Calculated variants and invariants of **Case 2**.

Variable terms		Invariant terms	
notion	**value**	**notion**	**value**
total output	$X = 911.36$ qr. wheat	productiveness	$R = 0.25$
circulating capital	$K = 736.36$ qr. wheat	rate of profits	$r = 0.05$
total profits	$P = 36.82$ qr. wheat	share of t. profits	$\tilde{r} = 0.2104$
total wages	$W = 138.18$ qr. wheat	surplus ratio	$\tilde{R} = 0.2377$
national income (NI)	$Y = 175$ qr. wheat	share of t. wages	$\tilde{w} = 0.7896$
average NI per labour	$U = 175 \frac{\text{qr. wheat}}{\text{TAL}}$		
wage p. unit of labour	$w = 138.18 \frac{\text{qr. wheat}}{\text{TAL}}$		

and the rates and shares

$$\tilde{r} = \frac{P}{Y} = \frac{405}{11 \cdot 175} = \frac{81}{385} = 0.2104,$$
$$\tilde{w} = \frac{W}{Y} = \frac{1{,}520}{11 \cdot 175} = \frac{304}{385} = 0.7896,$$
$$\tilde{R} = \frac{Y}{K} = \frac{175 \cdot 11}{8{,}100} = \frac{77}{324} = 0.2377,$$
$$r = \tilde{R}(1 - \tilde{w}) = \left(\frac{Y}{K}\right)(1 - \tilde{w}) = \left(\frac{77}{324}\right)\left(1 - \frac{304}{385}\right) = 0.05. \tag{4.96}$$

The *rate of profit* $r = P/K = (\frac{405}{11})/(\frac{8{,}100}{11}) = 0.05$, the economic identities for the *national income* $Y = W + P = L \cdot w + P = 1 \cdot (1{,}520/11) + (405/11) = 175$ qr. wheat and for the *total output* $X = Y + K = 175 + 736.36 = 911.36$ qr. wheat are confirmed. Here, we have $|w| = 138.18 \neq |\tilde{w}| = 0.7896$ and also $R = 0.25 \neq \tilde{R} = 0.2377$ and finally $Y = \tilde{R} \cdot K = (77/324)(8{,}100/11) = 7 \cdot 25 = 175$ qr. wheat.

Case 3. The *monetary numéraire* is 1 CHF. Consequently, we estimate the price of *wheat* to $p_1 = 55$ (CHF/qr. wheat). We abandon the idea of the normalised *labour*, and we estimate that in this micro-economy (PCMC, Par. 5) the needed *total quantity of labour* is $L = 304$ *working hours* to produce with modern technology the relatively small amounts of wheat and iron. This being said, taking the *complete single-product Sraffa system* (4.85), we then calculate $p_2, Y, \tilde{w}$. See Table 4.4.

$$\begin{cases} (280 \cdot 55 + 12p_2)(1.05) + 152 \cdot \frac{\tilde{w} \cdot Y}{304} = 575 \cdot 55, \\ (120 \cdot 55 + 8p_2)(1.05) + 152 \cdot \frac{\tilde{w} \cdot Y}{304} = 20p_2, \\ (575 - 280 - 120) \cdot 55 + (20 - 12 - 8)p_2 = Y. \end{cases} \tag{4.97}$$

We thus obtain the relative price $p_2 = 925$ CHF, the national income $Y = 9{,}625$ CHF and the *share of total wages to national income* $\tilde{w} = (304/385) = 0.7896$.

Table 4.4: Calculated variants and invariants of **Case 3**.

Variable terms		**Invariant terms**	
notion	**value**	**notion**	**value**
total output	$X = 50{,}125$ CHF	productiveness	$R = 0.25$
circulating capital	$K = 40{,}500$ CHF	rate of profits	$r = 0.05$
total profits	$P = 2{,}025$ CHF	share of t. profits	$\tilde{r} = 0.2104$
total wages	$W = 7{,}600$ CHF	surplus ratio	$\tilde{R} = 0.2377$
national income (NI)	$Y = 9{,}625$ CHF	share of t. wages	$\tilde{w} = 0.7896$
average NI per labour	$U = 31.66\ \frac{\text{CHF}}{\text{hour}}$		
wage/unit of labour	$w = 25\ \frac{\text{CHF}}{\text{hour}}$		

We use all the available variables p_1, p_2 and $\tilde{w}$ to calculate the remainder of the economic variables (4.31), (4.32) and ratios (4.34), (4.35),

$$
\begin{aligned}
X &= \mathbf{q}'\mathbf{p} = [575, 20]\begin{bmatrix} 55 \\ 925 \end{bmatrix} = 50{,}125 \text{ CHF},\\
K &= (\mathbf{Se})'\mathbf{p} = [400, 20]\begin{bmatrix} 55 \\ 925 \end{bmatrix} = 40{,}500 \text{ CHF},\\
P &= (\mathbf{Se})'\mathbf{p}\cdot r = K\cdot r = [400, 20]\begin{bmatrix} 55 \\ 925 \end{bmatrix} = 2{,}025 \text{ CHF},\\
Y &= \mathbf{d}'\mathbf{p} = X - K = 50{,}125 - 40{,}500 = 9{,}625 \text{ CHF},\\
W &= Y - P = 9{,}625 - 2{,}025 = 7{,}600 \text{ CHF},\\
w &= \frac{\tilde{w}\cdot Y}{L} = \tilde{w}\frac{Y}{L} = \frac{304}{385}\cdot\frac{9{,}625}{304} = 25\,\frac{\text{CHF}}{\text{hour}},\\
U &= \frac{Y}{L} = \frac{9{,}625}{304} = 31.66\,\frac{\text{CHF}}{\text{hour}},
\end{aligned}
\tag{4.98}
$$

then the rates and shares,

$$
\begin{aligned}
\tilde{r} &= \frac{P}{Y} = \frac{2{,}025}{9{,}625} = \frac{81}{385} = 0.2104,\\
\tilde{w} &= \frac{W}{Y} = \frac{7{,}600}{9{,}625} = \frac{304}{385} = 0.7896,\\
\tilde{R} &= \frac{Y}{K} = \frac{9{,}625}{40{,}500} = \frac{77}{324} = 0.2377,\\
r &= \tilde{R}(1-\tilde{w}) = \left(\frac{Y}{K}\right)(1-\tilde{w}) = \left(\frac{77}{324}\right)\left(1-\frac{304}{385}\right) = 0.05.
\end{aligned}
\tag{4.99}
$$

The *rate of profit* $r = (P/K) = (2{,}025/40{,}500) = 0.05$, the economic identity for the *national income* $Y = W + P = L\cdot w + P = 304\cdot(7{,}600/304) + 2{,}025 = 9{,}625$ CHF and the economic identity for the *total output* $X = Y + K = 7{,}600 + 2{,}025 = 9{,}625$ CHF are confirmed.

In this example, we found: $|w| = 25 \neq |\tilde{w}| = 0.7896$ and $R = 0.25 \neq \tilde{R} = 0.2377$ and $Y = \tilde{R} \cdot K = (77/324)(40{,}500) = 77 \cdot 125 = 9{,}625$ qr. wheat.

Conclusion: We observe that in the three cases the *productiveness* $R = 0.25$ depends on the Frobenius number $\lambda_C = 0.8 = 1/(1+R)$ of the positive *input-output coefficients matrix* $\mathbf{C}$ (4.87). The *rate of profits* being set equal to $r = 0.05$, the *share of total profits* $\tilde{r} = P/Y = 0.2104$, the *share of total wages* $\tilde{w} = W/Y = 0.7896$ also remain unchanged, confirming the rule $\tilde{r} + \tilde{w} = 1$ (4.34).

Finally, the share of national income to circulating capital $\tilde{R} = Y/K = 0.2377$ (dimensionless) also remains unchanged, whereas the average national income per unit of quantity of labour $U = Y/L$, and the *wage per unit of labour* $w = W/L$ depend on the measurement unit of labour L.

Example 4.1.7 illustrates the important general relation (4.36),

$$\tilde{R}(1 - \tilde{w}) = r = (77/324)(1 - (304/385)) = 0.05.$$

Moreover, if there are no wages, then $r = R = 0.25 \Rightarrow \tilde{w} = 0 \Rightarrow R = \tilde{R}$.

4.2 From a productive Sraffa model to a productive Leontief model

In this subsection, we investigate the *productive Sraffa model*, as defined in (3.52), and transform it in a *productive Leontief model*, Definition A.12.1, (A.110).

For this purpose, we start from the *Sraffa price model* (4.82), i. e., the *complete single-product Sraffa system*. With the extension of Sraffa's example (PCMC, Par. 5), Example 4.1.7, **Case 3**, including now the calculation of the economic variables X, K, W, P, W, having absolute positive prices in the CHF currency, we have prepared the foundations for what follows.

We transform all the entries of Sraffa's example (PCMC, Par. 5), which are in physical terms, into value terms, using appropriated prices. Then, we pour the obtained values into a *Leontief Input-Output Table*. Moreover, we are in presence of the comfortable *positivity* of the *commodity flow* matrix $\mathbf{S} > \mathbf{0}$ and *semi-positivity* of the vector of *surplus* $\mathbf{d} \geq \mathbf{o}$, giving the vector of *total output* $\mathbf{q} = \mathbf{Se} + \mathbf{d} > \mathbf{o}$, Table 4.1. With the vector of labour $\mathbf{L} > \mathbf{o}$, we formulate the production scheme $(\mathbf{S}', \mathbf{L}) \rightarrow (\hat{\mathbf{q}})$.

Example 4.2.1. The currency is CHF. Take the entries of Example 4.1.7, (PCMC, Par. 5), namely matrix $\mathbf{S} > \mathbf{0}$, vectors $\mathbf{q} > \mathbf{o}$, $\mathbf{d} \geq \mathbf{o}$, $\mathbf{L} > \mathbf{o}$ (4.84), and from the third case (4.97) the price vector $\mathbf{p} = [55, 925]'$, obtained with the rate of profits $r = 0.05$, $w = 25$.

Compute the positive commodity flow matrix $\mathbf{Z} = \hat{\mathbf{p}}\mathbf{S}$ in monetary terms, the positive vector of total output $\mathbf{x} = \hat{\mathbf{q}}\mathbf{p}$ and the semi-positive vector of final demand $\mathbf{f} = \hat{\mathbf{d}}\mathbf{p}$. Establish the input-output coefficients matrix $\mathbf{A} = \mathbf{Z}\hat{\mathbf{x}}^{-1}$. Show that the Frobenius number λ_A of matrix $\mathbf{A}$ is in the range, $0 < \lambda_A < 1$. Compute the productiveness $R = (1/\lambda_A)-1$. Calculate the Leontief Inverse $(\mathbf{I} - \mathbf{A})^{-1}$ and the solution $\mathbf{x} = \mathbf{Ax} + \mathbf{f}$ (2.30). Compute the values of the economic variables X, K, Y, P and W in CHF.

Solution of Example 4.2.1:
We begin by computing the *commodity flow* matrix $\mathbf{Z} = \hat{\mathbf{p}}\mathbf{S}$ (2.113), the vector of labour $\mathbf{L}$, the vectors of *total output* $\mathbf{x}$ and *final demand* $\mathbf{f}$, all in monetary terms (CHF):

$$\mathbf{Z} = \begin{bmatrix} z_{11} & z_{12} \\ z_{21} & z_{22} \end{bmatrix} = \hat{\mathbf{p}}\mathbf{S} = \begin{bmatrix} p_1 & 0 \\ 0 & p_2 \end{bmatrix}\begin{bmatrix} s_{11} & s_{12} \\ s_{21} & s_{22} \end{bmatrix}$$
$$= \begin{bmatrix} 55 & 0 \\ 0 & 925 \end{bmatrix}\begin{bmatrix} 280 & 120 \\ 12 & 8 \end{bmatrix} = \begin{bmatrix} 15{,}400 & 6{,}600 \\ 11{,}100 & 7{,}400 \end{bmatrix} > \mathbf{0}, \quad \mathbf{L} = \begin{bmatrix} 152 \\ 152 \end{bmatrix}. \tag{4.100}$$

We determine with equations (2.105) the elements $x_i = p_i q_i$, $z_{ij} = p_i s_{ij}$, $f_i = p_i d_i$. The calculation can be presented in condensed matrix equations as follows,

$$\mathbf{x} = \begin{bmatrix} x_1 \\ x_2 \end{bmatrix} = \hat{\mathbf{q}}\mathbf{p} = \begin{bmatrix} q_1 p_1 \\ q_2 p_2 \end{bmatrix} = \begin{bmatrix} 575 \cdot 55 \\ 20 \cdot 925 \end{bmatrix} = \begin{bmatrix} 31{,}625 \\ 18{,}500 \end{bmatrix} > \mathbf{o}, \tag{4.101}$$

$$\mathbf{f} = \begin{bmatrix} f_1 \\ f_2 \end{bmatrix} = \hat{\mathbf{d}}\mathbf{p} = \begin{bmatrix} d_1 & 0 \\ 0 & d_2 \end{bmatrix}\begin{bmatrix} p_1 \\ p_2 \end{bmatrix} = \begin{bmatrix} 175 & 0 \\ 0 & 0 \end{bmatrix}\begin{bmatrix} 55 \\ 925 \end{bmatrix} = \begin{bmatrix} 9{,}625 \\ 0 \end{bmatrix} \geq \mathbf{o}. \tag{4.102}$$

Then we compute the *input-output coefficients* matrix, which is consequently *positive*:

$$\mathbf{A} = \mathbf{Z}\hat{\mathbf{x}}^{-1} = \begin{bmatrix} 15{,}400 & 6{,}600 \\ 11{,}100 & 7{,}400 \end{bmatrix}\begin{bmatrix} \frac{1}{31{,}625} & 0 \\ 0 & \frac{1}{18{,}500} \end{bmatrix} = \begin{bmatrix} \frac{56}{115} & \frac{66}{185} \\ \frac{444}{1{,}265} & \frac{2}{5} \end{bmatrix} > \mathbf{0}. \tag{4.103}$$

We calculate the characteristic polynomial

$$P_2(\lambda) = \det(\mathbf{A} - \lambda\mathbf{I}) = \frac{8}{115} - \frac{102}{115}\lambda + \lambda^2 = \left(\lambda - \frac{2}{23}\right)\left(\lambda - \frac{4}{5}\right). \tag{4.104}$$

As there is a surplus, due to Lemma 4.1.1 (b), the Frobenius number of matrix $\mathbf{C} = \mathbf{S}\hat{\mathbf{q}}^{-1}$ is smaller than 1, $\lambda_C < 1$. Moreover with Lemma A.6.1, there is equality between the Frobenius numbers, $\lambda_A = \lambda_C = 4/5 < 1$. The productiveness is $R = (5/4) - 1 = 0.25$, identical to the value obtained from the *input-output coefficients* matrix $\mathbf{C}$ (3.47) (PCMC, Par.5). The Leontief Inverse exists because $\lambda_A < 1$. Due to Theorem A.10.2, the Leontief Inverse is *positive*,

$$(\mathbf{I} - \mathbf{A})^{-1} = \begin{bmatrix} \frac{23}{7} & \frac{506}{259} \\ \frac{148}{77} & \frac{59}{21} \end{bmatrix} > \mathbf{0}. \tag{4.105}$$

Consequently, we obtain a *positive* productive solution:

$$\mathbf{x} = (\mathbf{I} - \mathbf{A})^{-1}\mathbf{f} = \begin{bmatrix} \frac{23}{7} & \frac{506}{259} \\ \frac{148}{77} & \frac{59}{21} \end{bmatrix}\begin{bmatrix} 9{,}625 \\ 0 \end{bmatrix} = \begin{bmatrix} 31{,}625 \\ 18{,}500 \end{bmatrix} > \mathbf{o},$$
$$\mathbf{x} = \mathbf{A}\mathbf{x} + \mathbf{f} = \begin{bmatrix} \frac{56}{115} & \frac{66}{185} \\ \frac{444}{1{,}265} & \frac{2}{5} \end{bmatrix}\begin{bmatrix} 31{,}625 \\ 18{,}500 \end{bmatrix} + \begin{bmatrix} 9{,}625 \\ 0 \end{bmatrix} = \begin{bmatrix} 31{,}625 \\ 18{,}500 \end{bmatrix} > \mathbf{o}. \tag{4.106}$$

Now, we compute the values of the economic variables X, K, Y, P and W, based on matrix $\mathbf{Z}$ and vectors $\mathbf{x}$ and $\mathbf{f}$ in monetary terms.

$$X = \mathbf{e}'\mathbf{x} = [1,1]\begin{bmatrix} 31{,}625 \\ 18{,}500 \end{bmatrix} = 50{,}125 \text{ CHF},$$
$$K = \mathbf{e}'\mathbf{Z}\mathbf{e} = [1,1]\begin{bmatrix} 15{,}400 & 6{,}600 \\ 11{,}100 & 7{,}400 \end{bmatrix}\begin{bmatrix} 1 \\ 1 \end{bmatrix} = 40{,}500 \text{ CHF},$$
$$Y = \mathbf{e}'\mathbf{f} = [1,1]\begin{bmatrix} 9{,}625 \\ 0 \end{bmatrix} = 9{,}625 \text{ CHF},$$
$$P = K \cdot r = 40{,}500 \cdot 0.05 = 2{,}025 \text{ CHF},$$
$$W = Y - P = 9{,}625 - 2{,}025 = 7{,}600 \text{ CHF}. \tag{4.107}$$

Taking the single-product Sraffa system (4.97) with positive *commodity flow*, an *input-output* vector and *labour*, we obtain *positive prices*, obtain the Input-Output Table 4.5, set up on the basis of Table 2.2 in Miller and Blair ([65], p. 14), then the value-added $\mathbf{\upsilon} = \mathbf{x} - \mathbf{Z}\mathbf{e} = [5{,}125, 4{,}500]'$, the wages $l_i = w \cdot L_i = 25 \cdot 152 = 3{,}800$, $i = 1, 2$, resulting in the vector of wages $\mathbf{W} = [3{,}800, 3{,}800]'$ and further the vector of profits $\mathbf{P} = \mathbf{\upsilon} - \mathbf{W} = [n_1, n_2]' = [1{,}325, 700]'$. There are the identities $l_i + n_i = \upsilon_i$, $i \in \{1, 2\}$, $W = l_1 + l_2 = 7{,}600$, $P = n_1 + n_2 = 2{,}025$, $Y = W + P = 7{,}600 + 2{,}025 = 9{,}625$, $L = 304$. Remember the wage rate $w = \tilde{w}(Y/L) = (W/Y)(Y/L) = (W/L) = (7{,}600/304) = 25$ CHF/hour. ▲

Table 4.5: Input-Output Table in monetary terms for Sraffa's model with $r = 0.05$.

Currency = CHF	Buying sectors		Final demand	Total output
	wheat	iron		
wheat	$z_{11} = 15{,}400$	$z_{12} = 6{,}600$	$f_1 = 9{,}625$	$x_1 = 31{,}625$
iron	$z_{21} = 11{,}100$	$z_{21} = 7{,}400$	$f_2 = 0$	$x_2 = 18{,}500$
wages	$l_1 = L_1 w = 3{,}800$	$l_2 = L_2 w = 3{,}800$	$W = 7{,}600$	
profits	$n_1 = 1{,}325$	$n_2 = 700$	$P = 2{,}025$	
value-added	$\upsilon_1 = 5{,}125$	$\upsilon_2 = 4{,}500$	$Y = 9{,}625$	
production	$x_1 = 31{,}625$	$x_2 = 18{,}500$		$X = 50{,}125$

This result is generalised.

(A) We exclude the cases of *no surplus* or *no final demand*, $\mathbf{d} = \mathbf{f} = \mathbf{o}$ because we are in the presence of *productive* models.

(B) We consider a semi-positive $n \times n$ commodity flow matrix $\mathbf{S} \geq \mathbf{0}$, a $n \times 1$ *semi-positive* vector of surplus $\mathbf{d} \geq \mathbf{o}$ in physical units.

We compute the vector of *total output* $\mathbf{q} = \mathbf{S}\mathbf{e} + \mathbf{d}$. The matrix $\mathbf{C} = \mathbf{S}\hat{\mathbf{q}}^{-1}$ and the *productive Sraffa model*, $\mathbf{q} = (\mathbf{C}\hat{\mathbf{q}})\mathbf{e} + \mathbf{d} = \mathbf{C}\mathbf{q} + \mathbf{d} > \mathbf{o}$, correspond to the *production scheme* $(\mathbf{S}, \mathbf{L}) \Rightarrow (\hat{\mathbf{q}})$. We have chosen an appropriate currency or numéraire.

In the case of requested additional *irreducibility* of matrices $\mathbf{S}$ and $\mathbf{C}$ and of a *semi-positive* vector of labour, $\mathbf{L} \geq \mathbf{o}$, the $n \times 1$ price vector of the *complete single-product Sraffa system* (4.82), $\mathbf{p} > \mathbf{o}$, is positive due to Lemma 4.1.2 (a). In the case of *semi-positivity* only of matrices $\mathbf{S}$ and $\mathbf{C}$ and the presence of a positive vector of labour $\mathbf{L} > \mathbf{o}$, the Lemma 4.1.2 (b) guarantees the existence of a *positive* price vector $\mathbf{p} > \mathbf{o}$.

With the established *productive Sraffa model* $\mathbf{q} = \mathbf{Cq} + \mathbf{d} > \mathbf{o}$ and the obtained *positive* price vector $\mathbf{p} > \mathbf{o}$, we then consequently get a semi-positive $n \times n$ commodity flow matrix $\mathbf{Z} = \hat{\mathbf{p}}\mathbf{S} \geq \mathbf{0}$, a semi-positive $n \times 1$ vector of final demand $\mathbf{f} = \hat{\mathbf{p}}\mathbf{d} \geq \mathbf{o}$ and a positive $n \times 1$ vector of *total output* $\mathbf{x} = \hat{\mathbf{p}}\mathbf{q} > \mathbf{o}$ in monetary terms. Finally, we calculate the semi-positive *input-output coefficients* matrix $\mathbf{A} = \mathbf{Z}\hat{\mathbf{x}}^{-1} \geq \mathbf{0}$ and set up the *productive Leontief model* $\mathbf{x} = \mathbf{Ax} + \mathbf{f}$.

Lemma 4.2.1. *The productive Sraffa model* $\mathbf{q} = \mathbf{Cq} + \mathbf{d}$, *for matrices in physical terms and the productive Leontief model* (A.110)*, Definition* A.12.1, $\mathbf{x} = \mathbf{Ax} + \mathbf{f}$, *for matrices in monetary terms are equivalent with the transformation equations* (2.18), $\mathbf{x} = \hat{\mathbf{p}}\mathbf{q}$, $\mathbf{Z} = \hat{\mathbf{p}}\mathbf{S}$, $\mathbf{f} = \hat{\mathbf{p}}\mathbf{d}$ *for a positive price vector* $\mathbf{p} > \mathbf{o}$ *and the equalities* $\mathbf{S} = \mathbf{C}\hat{\mathbf{q}}$, $\mathbf{Z} = \mathbf{A}\hat{\mathbf{x}}$.

Proof. *Equivalence* has to be proved, for this reason we show both, the *sufficient* and the *necessary* conditions.

$\Rightarrow$ (sufficiency): We start with $\mathbf{q} = \mathbf{Cq} + \mathbf{d}$ and multiply this identity from the left with the diagonal matrix $\hat{\mathbf{p}}$, in order to get,

$$\begin{aligned} \mathbf{q} &= \mathbf{Cq} + \mathbf{d} = (\mathbf{C}\hat{\mathbf{q}})\mathbf{e} + \mathbf{d} = \mathbf{Se} + \mathbf{d} \Rightarrow \hat{\mathbf{p}}\mathbf{q} = \mathbf{x} = (\hat{\mathbf{p}}\mathbf{S})\mathbf{e} + \hat{\mathbf{p}}\mathbf{d} \\ &= \mathbf{Ze} + \mathbf{f} = (\mathbf{A}\hat{\mathbf{x}})\mathbf{e} + \mathbf{f} = \mathbf{Ax} + \mathbf{f} > \mathbf{o} \Leftrightarrow \mathbf{f} = (\mathbf{I} - \mathbf{A})\mathbf{x}. \end{aligned} \tag{4.108}$$

$\Leftarrow$ (necessity): We start now from the other side, taking $\mathbf{x} = \mathbf{Ax} + \mathbf{f}$ and multiply this identity from the left with the inverse diagonal matrix $\hat{\mathbf{p}}^{-1}$, and we get,

$$\begin{aligned} \mathbf{x} &= \mathbf{Ax} + \mathbf{f} = (\mathbf{A}\hat{\mathbf{x}})\mathbf{e} + \mathbf{f} = \mathbf{Ze} + \mathbf{f} = \hat{\mathbf{p}}\mathbf{q} = (\hat{\mathbf{p}}\mathbf{S})\mathbf{e} + \hat{\mathbf{p}}\mathbf{d} = \hat{\mathbf{p}}(\mathbf{Se} + \mathbf{d}) \\ &\Rightarrow \hat{\mathbf{p}}^{-1}(\hat{\mathbf{p}}\mathbf{q}) = \hat{\mathbf{p}}^{-1}(\hat{\mathbf{p}}(\mathbf{Se} + \mathbf{d})) = ((\mathbf{C}\hat{\mathbf{q}})\mathbf{e} + \mathbf{d}) = \mathbf{q} = \mathbf{Cq} + \mathbf{d} \\ &\Leftrightarrow \mathbf{d} = (\mathbf{I} - \mathbf{C})\mathbf{q}. \end{aligned} \tag{4.109}$$

□

Complement: (Assumption 2.2.1 and Assumption 2.2.2 hold.) We assume that matrices **A** and **C** are *irreducible*. Furthermore, they are similar and have therefore the same Frobenius number $\lambda_C = \lambda_A < 1$, according to Lemma A.6.1. The Leontief Inverses $(\mathbf{I}-\mathbf{A})^{-1}$, $(\mathbf{I}-\mathbf{C})^{-1}$ exist, because $\lambda_A < 1$, according to Lemma 4.1.1 (b). Furthermore, they are positive $(\mathbf{I} - \mathbf{A})^{-1} > \mathbf{0}$, $(\mathbf{I} - \mathbf{C})^{-1} > \mathbf{0}$, according to Theorem A.10.2 equation (A.103). Thus, the *productive Leontief model* $\mathbf{x} = \mathbf{Ax} + \mathbf{f}$ has with $\mathbf{f} \geq \mathbf{o}$ the unique *positive* solution vector of *total output* in *monetary terms*, $\mathbf{x} = (\mathbf{I} - \mathbf{A})^{-1}\mathbf{f} > \mathbf{o}$. The *productive Sraffa model* $\mathbf{q} = \mathbf{Cq} + \mathbf{d}$ has with $\mathbf{d} \geq \mathbf{o}$ the unique *positive* solution vector of *total output in physical terms*, $\mathbf{q} = (\mathbf{I} - \mathbf{C})^{-1}\mathbf{d} > \mathbf{o}$.

4.3 The complete single-product Sraffa system and national accounting

The *single-product Sraffa system* (PCMC, Par. 11) (4.41) is again reproduced here:

$$\mathbf{S}'\mathbf{p}(1 + r) + \mathbf{L}w = \hat{\mathbf{q}}\mathbf{p}, \tag{4.110}$$

and is now correlated with the *national accounting identities*. *National income* Y, here approximately assimilated to GDP, is either expressed in its components of factor payments, having been simplified to $Y = W + P$, or in components of expenditure, i. e., total consumption C plus total investment I plus total government purchases G and net exports $(E - M)$, in short: $Y = C + I + G + (E - M)$, see also Miller and Blair [65], p. 15, and Mankiw [63], pp. 24–26,

$$Y = P + W = C + I + G + (E - M). \tag{4.111}$$

We summarise the variables we will use in this context:
Y: national income (~ GDP) = surplus in value terms;
P: total gross profits;
W: total wages;
$\tilde{r}$: share of total profits (profit share of total (factor) income), $\tilde{r} = P/Y$;
$\tilde{w}$: share of total wages (wage share of total (factor) income), $\tilde{w} = W/Y$;
$\tilde{R}$: share of national income to circulating capital, $\tilde{R} = Y/K$.

In normalised form, we obtain with (4.111)

$$\frac{P}{Y} + \frac{W}{Y} = \tilde{r} + \tilde{w} = 1, \tag{4.112}$$

$\tilde{r}$ again representing the share of total profits P and $\tilde{w}$ the share of wages W to national income Y.

One should note here in passing that Sraffa (PCMC, Par. 12) makes no reference to effective demand, or to any demand function for that matter: the total surplus $\mathbf{d}$ is implicitly supposed to be consumed and then split into *wages* and *profits*. We now proceed with a left-multiplication of the *single-product Sraffa system* (4.110) by the summation vector $\mathbf{e}'$ to finally attain the *circulating capital* K, plus the value of the *surplus*, the *national income* Y, the sum of the total *wages* W and the total *profits* P, forming all three constituents of *total output* X. We obtain:

$$\begin{aligned} &\mathbf{e}'\mathbf{S}'\mathbf{p}(1+r) + \mathbf{e}'\mathbf{L}w = \mathbf{e}'\mathbf{S}'\mathbf{p} + \mathbf{e}'\mathbf{S}'\mathbf{p}r + \mathbf{e}'\mathbf{L}w = \mathbf{e}'\hat{\mathbf{q}}\mathbf{p}, \\ &X = K + (P + W) = K + Y. \end{aligned} \tag{4.113}$$

As an illustration, we present the national accounting entities of the foregoing Example 4.2.1, **Case 3**.

Example 4.3.1. From Table 4.5 we obtain the following national accounting entities, economic variables and ratios. We have *total wages* $W = 7{,}600$, *total profits* $P = 2{,}025$, the *national income* $Y = 9{,}625$ and *circulating capital* $K = 40{,}500$, so there is $\tilde{r} = P/Y = 2{,}025/9{,}625 = 0.21104$, $\tilde{w} = W/Y = 7{,}600/9{,}625 = 0.7896$, $\tilde{R} = Y/K = 9{,}625/40{,}500 = 0.2377$ and $r = \tilde{R} \cdot (1 - \tilde{w}) = 0.2377 \cdot (1 - 0.7896) = 0.05$. ▲

Then, the *national income* Y (or *gross national income*, see Miller and Blair [65], p. 15), is aggregated and normalised (PCMC, Par. 12, p. 11), see also Kurz and Salvadory [52], p. 98, (4.5b). Together with relationship (4.37) we get, normalising,

$$Y := \mathbf{d}'\mathbf{p} = (\mathbf{q} - \mathbf{Se})'\mathbf{p} = 9{,}625 \text{ CHF} := 1 \text{ GDP}. \tag{4.114}$$

Applying (4.111) and (4.114), considering the total circulating capital K, the part of profits P and the part of wages W, as well as the national income Y, we find the identities

$$\begin{aligned} K = (\mathbf{Se})'\mathbf{p}, \quad P = ((\mathbf{Se})'\mathbf{p})r = K \cdot r, \quad W = (\mathbf{e}'\mathbf{L})w = L \cdot w \\ Y = P + W = r(\mathbf{Se})'\mathbf{p} + \mathbf{e}'\mathbf{L}w = \mathbf{d}'\mathbf{p} = 1 \text{ GDP}. \end{aligned} \tag{4.115}$$

Sraffa proceeded with the normalisation of *national income* and of the *total amount of labour*. Following these lines, we study now the resulting algebraic simplifications. Let us start with the normalisation of labour by Sraffa (PCMC, Par. 10),

$$L := \mathbf{e}'\mathbf{L} = 1 \text{ TAL}. \tag{4.116}$$

Now with the normalisation (4.116), $L = 1$, and (4.114), $Y = 1$, skipping the new *physical units*, some expressions are greatly simplified. We consider equation (4.46) $W = w \cdot L = w$, the *rate of profits* $r = P/K$, the share of total wages, $\tilde{w} = W/Y = W$, the share of total profits, $\tilde{r} = P/Y = P$ and the ratio of national income to total capital, $\tilde{R} = Y/K = 1/K$. We further have without measurement units:

$$\tilde{w} = \frac{W}{Y} = W = \frac{wL}{Y} = w \cdot \frac{1}{U} = w, \quad U = \frac{Y}{L} = 1, \tag{4.117}$$

the *average national income per unit of quantity of labour*, and we keep

$$r = \frac{P}{K} = \frac{P}{Y} \cdot \frac{Y}{K} = \tilde{r} \cdot \tilde{R}. \tag{4.118}$$

We also keep from (4.36),

$$r = \tilde{R} \cdot \frac{P}{Y} = \tilde{R} \cdot \frac{Y - W}{Y} = \tilde{R} \cdot \left(1 - \frac{W}{Y}\right) = \tilde{R} \cdot (1 - \tilde{w}), \tag{4.119}$$

and find again

$$w = \frac{W}{L} = \frac{W}{Y} \cdot \frac{Y}{L} = \tilde{w} \cdot U = \tilde{w}. \tag{4.120}$$

The foregoing enables us quite trivially to write the relationship between r and $\tilde{r}$, and w and $\tilde{w}$,

$$rK = \tilde{r}Y = \tilde{r}, \quad W = wL = \tilde{w}Y. \tag{4.121}$$

Closing this section, we remark that normalisation of Y and L induces identities such as $|w| = |W|$ and $|w| = |(\tilde{w} \cdot Y)/L| = |\tilde{w}|$ in the *single-product Sraffa system* (4.110).

4.4 Basic commodities and non-basic commodities

In this section, we introduce the notions of *basic* commodities and *non-basic* commodities (PCMC, Par. 6, 7,) for *single-product industries* along the lines of Sraffa's definitions, e. g., every industry or branch producing exactly one commodity.

> (PCMC, Par. 6) "*The criterion is whether a commodity enters (no matter whether directly or indirectly) into the production of all commodities. Those that do so we shall call basic, and those that do not, non-basic commodity.*
> *We shall assume throughout that any system contains at least one basic commodity.*"

Earlier on, Sraffa characterises rather quaintly *non-basic* commodities also as *luxury* products. He provides examples of racehorses, ostriches or ostrich eggs.[21] As usual, we will use the terms *basic products* and *basic commodities*, respectively *non-basic products* and *non-basic commodities* synonymously, as do many authors.

Sraffa then introduces the notion of *basic industries* (PCMC, Par. 25) and of *non-basic industries* (PCMC, Par. 35). In the context of *single-product industries*, each *basic commodity* is produced by exactly one *basic industry* and each *non-basic commodity* is produced by exactly one *non-basic industry*. Consequently, as the correspondence is one to one, commodities may be formally identified with the industry producing them. We shall use this possibility in the developments presented next.

Roncaglia ([97], p. 60) has reformulated more precisely the definition of *basic products* and *non-basic products* and defines:

> *Basic products are "commodities that enter directly or indirectly as means of production in every and each sector of production", and on the contrary, non-basic commodity are "commodities which do not serve as means of production or which are used, directly or indirectly, only in a limited number of processes".*[22]

Until further notice, we deal with *single-product industries* and their representation by the *adjacency matrix* and the *associated digraph*. We will see that it is necessary to have strict criteria to determine if products in a given *process of production* are *basic*, entering *directly* or *indirectly* into the various industries constituting the process or if, on the contrary, they are *non-basic*. Fortunately, all these criteria exist. They have been precisely established since the publication of Sraffa's book PCMC ([109], 1960).

At present, we introduce the mathematical notions, Lemmas, Theorems and criteria that serve as the basis to determine whether the products entering *directly* or

---

21 The example is of course exotic: even in Sraffa's original model based on the production of goods only in a closed economy, beer and other alcoholic beverages are non-basic, but they are certainly not luxury products.

22 This definition will be later illustrated by *Sraffa Networks* (Section 4.6, respectively Definition A.14.11), e. g., Example 6.4.2 and Figure 6.5.

indirectly into a process are *basic* or are *non-basic*. They are systematically presented in Appendix A.

The ongoing presentation of *basics* and *non-basics* in relation with the notion of *reducible* and *irreducible* matrices has been developed by Pasinetti [80], Steedman [114] and brought to maturity by Schefold [103]. We start by presenting this concept and corresponding ideas.

Consider a process of production, presented in terms of *n single-product industries*, and described by the entries s_{ij} of the $n \times n$ *commodity flow* matrix $\mathbf{S} = (s_{ij})$, $i, j = 1, \ldots, n$. This means, economically speaking, commodity i enters into the production of commodity j.

The matrix $\mathbf{S}$ is *semi-positive*, containing eventually some zero elements. From Sraffa's description of *basic* commodities and *non-basic* commodities, it is immediately evident that an appropriate *commodity flow* matrix $\mathbf{S}$ of an economy has to be brought in the following form, called the "canonical form", after eventually reordering the industries and products,

$$\mathbf{S}' = \begin{bmatrix} \mathbf{S}'_{11} & \mathbf{0} \\ \mathbf{S}'_{12} & \mathbf{S}'_{22} \end{bmatrix}, \tag{4.122}$$

where $\mathbf{S}_{11}$ is a square matrix that refers to *basic* commodities entering the production of other basics only. $\mathbf{S}_{22}$ is also a square matrix that refers to *non-basic* commodities only, in the same sense. $\mathbf{S}_{12}$ refers to basic commodities entering the production of *non-basics*. If $\mathbf{S}_{12} = \mathbf{0}$, then the economy is split into two independent economies that have no connection with one another.

But generally, the *basic* commodities and the *non-basic* commodities of an economy are mixed in the process of production. Then it cannot be expected for the *commodity flow* matrix $\mathbf{S}$ to have the aspect shown in equation (4.122). The mixture of the *basic* and *non-basic* commodities will entail the zero elements of the matrix $\mathbf{S}$ to be disseminated over the whole matrix. See as an example the matrix, called $\mathbf{Z}_1$, in Exercise A.8.1, and its transformation into "canonical form".

This suggests the idea to formally separate in a given economy the *basic* commodities from the *non-basic* commodities, thus forming two categories, in order to get the *basic* commodities in some set, separated from the *non-basic* commodities in another set. But various *non-basics* are however dependent for their production on *basics*. See hereafter for the characterisation of the two categories and their representative sets.

Fortunately, there exists a procedure, described by Definition A.8.3 which details how to transform the given *commodity flow* matrix $\mathbf{S}$, by multiplication with a permutation matrix $\mathbf{P}$, Definition A.8.2 into a "canonical form" (4.122), realising the separation of *basics* from *non-basics*. If this is possible, the initial matrix $\mathbf{S}$ is called *reducible*.

A permutation matrix **P** exists, so that the submatrix $\tilde{\mathbf{S}}_{11}$ is *irreducible.*[23] We obtain,

$$\mathbf{S} = \begin{bmatrix} \mathbf{S}_{11} & \mathbf{S}_{12} \\ \mathbf{S}_{21} & \mathbf{S}_{22} \end{bmatrix} \Rightarrow \mathbf{P}^{-1}\mathbf{S}'\mathbf{P} := \tilde{\mathbf{S}}' = \begin{bmatrix} \tilde{\mathbf{S}}'_{11} & \mathbf{0} \\ \tilde{\mathbf{S}}'_{12} & \tilde{\mathbf{S}}'_{22} \end{bmatrix}. \tag{4.123}$$

The distinction between *basic* and *non-basic* commodities can be described by the properties of *irreducible matrices* and *reducible matrices* of the corresponding production process. Indeed, if matrix **S** is an *irreducible* matrix, then *"all the commodities in the economic system are basic commodities"*, see Pasinetti ([80], p. 104), and he adds: *"...on the other hand if the matrix is reducible, some of the commodities are basic commodities, while others are non-basic commodities"*.

The relationship between the property *basics* and *irreducibility*, respectively *non-basics*, and *reducibility* is treated in Lemma A.15.1 and Lemma A.15.2.

Remark. If matrix $\mathbf{S} > \mathbf{0}$ is positive, then this matrix is *irreducible* and all the commodities represented by **S** are *basic* and enter directly as means of production into the production process.[24]

We continue to present and develop some essential notions that are needed for that purpose, such as the *adjacency matrix* and the *digraph.*

We represent the $n \times n$ *commodity flow* matrix $\mathbf{S} = (s_{ij})$, by an $n \times n$ *adjacency matrix* $\mathbf{V} = (v_{ij})$, $i, j = 1, \ldots, n$, $v_{ij} = 1 \Leftrightarrow s_{ij} > 0$ and entry $v_{ij} = 0 \Leftrightarrow s_{ij} = 0$, Definition A.14.8. The *directed digraph*, associated with the *adjacency matrix* **V**, is called the *commodity digraph* $G(\mathbf{V})$.

Let's illustrate the newly presented notions by the following example.

Example 4.4.1. Consider here the four *commodity flow* matrices $\mathbf{S}_l = (s_{lij})$, $i, j = 1, \ldots, n$, $l = 1, \ldots, 4$ of four different processes of production.

(1) Establish the *adjacency matrices* and the corresponding *commodity digraphs*.
(2) Determine the processes of production containing *non-basic* commodities.

$$\mathbf{S}_1 = \begin{bmatrix} 90 & 50 & 40 \\ 120 & 125 & 40 \\ 60 & 150 & 200 \end{bmatrix}, \quad \mathbf{S}_2 = \begin{bmatrix} 280 & 180 & 115 \\ 240 & 240 & 120 \\ 0 & 0 & 0 \end{bmatrix},$$
$$\mathbf{S}_3 = \begin{bmatrix} 90 & 0 & 40 \\ 50 & 125 & 0 \\ 0 & 60 & 200 \end{bmatrix}, \quad \mathbf{S}_4 = \begin{bmatrix} 90 & 50 & 0 \\ 120 & 125 & 0 \\ 0 & 0 & 200 \end{bmatrix}. \tag{4.124}$$

Solution to Example 4.4.1:
First of all, we determine the *adjacency matrices* $\mathbf{V}_i$, $i = 1, \ldots, 4$ of these four processes of production.

23 A matrix is *irreducible* if it is not *reducible*.
24 The question of the construction of the permutation matrix **P** is not treated in this text.

$$\mathbf{V}_1 = \begin{bmatrix} 1 & 1 & 1 \\ 1 & 1 & 1 \\ 1 & 1 & 1 \end{bmatrix}, \quad \mathbf{V}_2 = \begin{bmatrix} 1 & 1 & 1 \\ 1 & 1 & 1 \\ 0 & 0 & 0 \end{bmatrix},$$
$$\mathbf{V}_3 = \begin{bmatrix} 1 & 0 & 1 \\ 1 & 1 & 0 \\ 0 & 1 & 1 \end{bmatrix}, \quad \mathbf{V}_4 = \begin{bmatrix} 1 & 1 & 0 \\ 1 & 1 & 0 \\ 0 & 0 & 1 \end{bmatrix}. \tag{4.125}$$

Now we present the *commodity digraphs* of the *adjacency matrices* $\mathbf{V}_1$, $\mathbf{V}_2$, $\mathbf{V}_3$, $\mathbf{V}_4$ of these production economies.

Theorem A.14.1 (see Varga [118], p. 20), says that a matrix $\mathbf{V}_i$ is *irreducible*, if and only if its corresponding commodity digraphs $G(\mathbf{V}_i)$ is strongly connected, according to Definition A.14.9.

The *commodity digraph* $G(\mathbf{V}_1)$, see Figure 4.2 (left), is a *node-complete digraph*, according to Definition A.14.3. It is also a strongly connected digraph, according to Definition A.14.9. Therefore $\mathbf{V}_1$, respectively $\mathbf{S}_1$, are irreducible (see Lemma A.15.3). This economy contains accordingly only *basic* commodities.

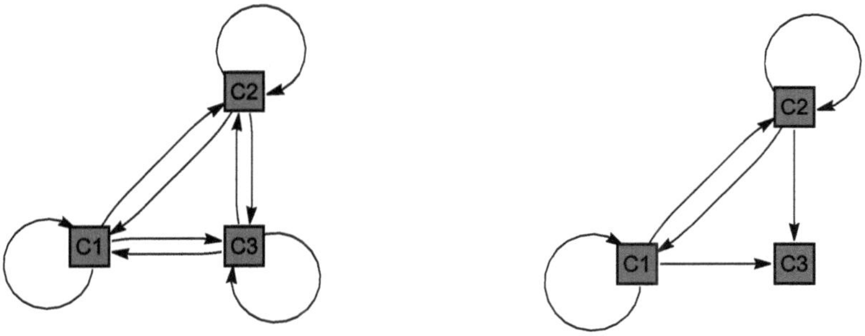

Figure 4.2: (Left) *Commodity digraph* $G(\mathbf{V}_1)$; (right) *commodity digraph* $G(\mathbf{V}_2)$.

The commodity digraph $G(\mathbf{V}_2)$, see Figure 4.2 (right), is only a weakly connected digraph (see Definition A.14.10). Therefore $\mathbf{V}_2$, respectively $\mathbf{S}_2$, are reducible (see Lemma A.15.2) and this economy thus contains therefore some *non-basic* commodities. In fact in this case, commodity 3 is *non-basic.*

The *commodity digraph* $G(\mathbf{V}_3)$, Figure 4.3 (left), is also a strongly connected digraph, because it contains a *directed circle*,[25] covering the whole set of the digraph, see also Skiena [105]. The matrix $\mathbf{V}_3$ is *irreducible* and the corresponding production economy contains only *basic* commodities (see Lemma A.15.3).

25 A directed cycle in a digraph may be defined either as a closed walk with no repetitions of vertices and edges allowed, other than the repetition of the starting and ending vertices, and possibly some intermediate vertices, or as the set of edges in such a walk.

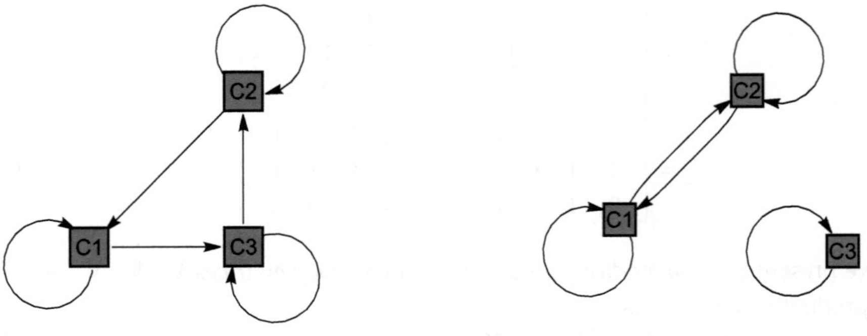

Figure 4.3: (Left) *commodity digraph* $G(\mathbf{V}_3)$; (right) *commodity digraph* $G(\mathbf{V}_4)$.

The commodity digraph $G(\mathbf{V}_4)$, Figure 4.3 (right), is *disconnected*, according to Definition A.14.10 and has two *components*.[26] It is only a weakly connected digraph. Therefore $\mathbf{V}_4$, respectively $\mathbf{S}_4$, are reducible (see Lemma A.15.2) fact, even completely reducible, and this economy contains therefore some *non-basic* commodities. As expected, we can even say more.

The processes of production, represented by the positive matrix $\mathbf{S}_1$ has exclusively *basic* commodities because all matrix elements are positive and matrix $\mathbf{S}_1$ is irreducible, according to Definition A.8.3.

In the process of production, represented by the non-negative matrices $\mathbf{S}_2$ and $\mathbf{S}_4$, which are in "canonical form" (4.122), commodity 3 is *non-basic.*

The *commodity digraph* $G(\mathbf{S}_4)$, according to *Definition* A.14.7 is separated in two disjoint components. ▲

Now, we will deal with some algebraic properties of matrices, enabling us to decide whether we are in presence of *basic* commodities, entering directly or indirectly into the process of production, or enabling us to decide whether we are in presence of *non-basic* commodities entering the process of production.

Appendix A, Section A.15 presents Lemma A.15.1 and Lemma A.15.3, containing algebraic criteria, applicable to matrices, enabling determination of the nature of the commodities (*basic* or *non-basic*), entering *directly* or *indirectly* into an industry.

Applying Lemma A.15.1 and Lemma A.15.3 to the extended Example 4.4.1 just treated, we want to show that we can answer precise questions as to whether a given commodity is a *basic* commodity, entering either directly or indirectly into the producing industries. But before proceeding, let us solve an example.

26 In graph theory, a connected component (or just component) of an undirected graph is a subgraph in which any two vertices are connected to each other by paths, and which is not connected to any additional vertices in the supergraph. A graph that is connected to itself has exactly one connected component, consisting of the whole graph.

Example 4.4.2 (Refer to Example 4.4.1).

(1) Show that all commodities of the production economy, described by the commodity flow matrix $\mathbf{S}_1$, enter directly into the production of all commodities and all the commodities, are thus *basic*.
(2) Decide if commodity 3 of the economy, described by the commodity flow matrix $\mathbf{S}_2$, enters indirectly into the production of commodities 1 and 2. Argue that commodity 3 is non-basic. Show that matrix $\mathbf{S}_2$ is reducible.
(3) Show that all commodities of the economy, described by the commodity flow matrix $\mathbf{S}_3$, are basic.
(4) Analyse the nature of the economy described by matrix $\mathbf{S}_4$.

Solution to Example 4.4.2:

(1) All the elements $s_{ij} > 0$ of matrix $\mathbf{S}_1$ are positive. With Lemma A.15.1 (1), all the commodities of this economy enter directly into the production and are therefore all *basic* commodities. Moreover, the *commodity digraph* $G(\mathbf{V}_1)$ is a node-complete digraph and therefore also *strongly connected*, Definition A.14.9. With Lemma A.15.3, matrix $\mathbf{S}_1$ is *irreducible*, confirming that all commodities are *basic*.
(2) Note the matrix power notation, e. g., $\mathbf{V}_2^2 = \mathbf{V}_2 \cdot \mathbf{V}_2$, $\mathbf{V}_2^3 = \mathbf{V}_2 \cdot \mathbf{V}_2 \cdot \mathbf{V}_2$, etc. One applies Lemma A.15.1 (2) to the quadratic forms,

$$\mathbf{e}_1'\mathbf{V}_2\mathbf{e}_3 = [1,0,0]\begin{bmatrix} 1 & 1 & 1 \\ 1 & 1 & 1 \\ 0 & 0 & 0 \end{bmatrix}\begin{bmatrix} 0 \\ 0 \\ 1 \end{bmatrix} = 1 > 0,$$

$$\mathbf{e}_1'(\mathbf{V_2}^2 + \mathbf{V}_2^3)\mathbf{e}_3 = [1,0,0]\begin{bmatrix} 6 & 6 & 6 \\ 6 & 6 & 6 \\ 0 & 0 & 0 \end{bmatrix}\begin{bmatrix} 0 \\ 0 \\ 1 \end{bmatrix} = 6 > 0.. \tag{4.126}$$

Both quadratic forms are positive. This means that the commodity 3 of the economy, described by the commodity flow matrix $\mathbf{S}_2$, does not enter *indirectly* into the production of commodity 1 (see Lemma A.15.1 (2)). Thus, commodity 3 is *non-basic*.

Then one applies Lemma A.15.1 (5),

$$(\mathbf{V}_2 + \mathbf{V}_2^2 + \mathbf{V}_2^3) = \begin{bmatrix} 7 & 7 & 7 \\ 7 & 7 & 7 \\ 0 & 0 & 0 \end{bmatrix} \geq \mathbf{0}. \tag{4.127}$$

This matrix is *not* positive, therefore matrix $\mathbf{S}_2$ is *reducible*.

(3) One applies Lemma A.15.1 (5),

$$(\mathbf{V}_3 + \mathbf{V}_3^2 + \mathbf{V}_3^3) = \begin{bmatrix} 4 & 4 & 6 \\ 6 & 4 & 4 \\ 4 & 6 & 4 \end{bmatrix} > \mathbf{0}. \tag{4.128}$$

This matrix is positive, therefore all commodities of the economy, described by the commodity flow matrix $\mathbf{S}_3$, are *basic* commodities.

(4) Matrix $\mathbf{S}_4$ represents two separated economies, described by the two positive sub-matrices $\mathbf{S}_{11}$ and $\mathbf{S}_{22}$, because the *digraph* $G(\mathbf{S}_4)$ is separated into two components.

$$\mathbf{S}_{11} = \begin{bmatrix} 90 & 50 \\ 120 & 125 \end{bmatrix} > \mathbf{0}, \quad \mathbf{S}_{22} = [\ 200\ ] > \mathbf{0}. \tag{4.129}$$

By applying Lemma A.15.1 (5), separately to both economies, we find that only *basic* commodities are represented by both sub-matrices. ▲

The solution to Example 4.4.2, presented here, shows that we have the tools to determine the nature of the commodities of an economy, *basics* or *non-basics*, which enter *directly* or *indirectly* into the process of production. We take one of the transaction matrices, the *commodity flow matrix* **S** or the *adjacency matrix* **V**, consider its *irreducible* or *reducible* properties, associated with the *digraphs* $G(\mathbf{S})$ or $G(\mathbf{V})$, and then argue with the algebraic criteria stated in the Lemmas A.15.1–A.15.3.

To further illustrate the type of relation that exists between the category of *basic* commodities and the category of *non-basic* commodities, let us consider the following specific production process composed of *n single-product industries*.

For any $n \in \mathbb{N}$, there are $i = 1, \ldots n-1$ *basic* commodities, the n-th commodity is a *non-basic* commodity. Moreover the commodity n is not used in any of the industries $j \in \{1, 2, \ldots, n-1\}$. The matrix $\mathbf{S} \geq \mathbf{0}$ is *semi-positive*. The means of production of the n-th *non-basic* industry are composed of basic commodities provided by the entire initial surplus of each of these commodities. The net surplus of each *basic* industry is accordingly equal to 0. Only the n-th industry produces an effective output $d_n > 0$ that is the unique surplus generated by the entire process. The surplus vector is therefore, $\mathbf{d} = [0, \ldots, 0, d_n]'$, $d_n = q_n$, and the economy is self-reproducing, $\mathbf{q} = \mathbf{Se} + \mathbf{d} \geq \mathbf{Se} > \mathbf{0}$. The vector of labour is noted as $\mathbf{L} \geqq \mathbf{o}$ and can even be the null vector. The production scheme is, in detail, as follows:

$$(\mathbf{S}', \mathbf{L}) \rightarrow (\hat{\mathbf{q}}), \tag{4.130}$$

where $\mathbf{S}'$ is the transpose of the *commodity flow* matrix **S** (4.123) (right, without *tilde-sign*):

$$\mathbf{S} = \begin{bmatrix} s_{11} & s_{12} & \cdots & s_{1(n-1)} & s_{1n} \\ \cdots & \cdots & \cdots & \cdots & \cdots \\ \cdots & \cdots & \cdots & \cdots & \cdots \\ s_{(n-1)1} & s_{(n-1)2} & \cdots & s_{(n-1)n-1} & s_{(n-1)n} \\ 0 & \cdots & \cdots & 0 & 0 \end{bmatrix}. \tag{4.131}$$

Written out, we have the production scheme:

$$
\begin{aligned}
&(s_{11}, \dots \quad \dots, s_{(n-1)1}, s_{n1} = 0, \quad L_1) \rightarrow \left(q_1 = \sum_{j=1}^{n-1} s_{1j}, \quad 0, \dots \quad \dots, 0\right),\\
&(s_{12}, \dots \quad \dots, s_{(n-1)2}, s_{n2} = 0, \quad L_2) \rightarrow \left(0, q_2 = \sum_{j=1}^{n-1} s_{2j}, 0, \dots \quad \dots, 0\right),\\
&(\dots, \quad \dots, \quad \dots, \quad \dots) \rightarrow (0, \dots \quad \dots, 0),\\
&(s_{1(n-1)}, \dots \quad \dots, s_{n(n-1)} = 0, L_{n-1}) \rightarrow \left(0, \dots \quad \dots, 0, q_{n-1} = \sum_{j=1}^{n-1} s_{(n-1)j}, 0\right),\\
&(s_{1n}, \dots \quad \dots, s_{(n-1)n}, s_{nn} = 0, \quad L_n) \rightarrow (0, \dots \quad \dots, 0, \quad q_n).
\end{aligned}
\tag{4.132}
$$

Now we continue to consider the special case of an *exploitation of labour economy*, only with subsistence wages, $\mathbf{L} = \mathbf{o}$, giving the production scheme $(\mathbf{S}', \mathbf{o}) \rightarrow (\hat{\mathbf{q}})$. In consequence, the resulting specific production scheme (4.132) gives in fact four different calculation methods to determine the *national income*. Starting from the definition, giving $Y = \mathbf{d}'\mathbf{p}$ (a), we get the term $Y = q_n \cdot p_n$ (b) and then the total sum $\mathbf{e}'(R\mathbf{S}'\mathbf{p})$ of the *vectorial* surplus part $R\mathbf{S}'\mathbf{p}$ of the Sraffa price model (3.43), or below (4.175) (c), that is therefore equal to $R \cdot K$ (d). We have:

$$
Y = \mathbf{d}'\mathbf{p} = R(\mathbf{Se})'\mathbf{p} = R\sum_{i=1}^{n}\left(\sum_{j=1}^{n} s_{ij}\right)p_i = R \cdot K = q_n \cdot p_n. \tag{4.133}
$$

We adapt Sraffa's third example (PCMC, Par. 5), Example 3.1.3, modifying the quantities of commodities to produce a *non-basic* commodity, namely gold.

Example 4.4.3. The three single-product industries produce corn (agricultural production), iron (manufacturing) as basic commodities and gold (gold extraction instead of Sraffa's racehorses, PCMC, Par. 6) as a non-basic commodity. This process produces an amount of q_3 kg of gold. First, gold is chosen as means of payment, setting it as numéraire "1 kg of gold". We get relative prices. Second, we choose then the currency CHF, setting the price of gold equal to $p_3 = 33{,}000$ CHF/kg and generate absolute prices.

There are no wages, $w = 0$. The production scheme is established as follows:

$$
\begin{aligned}
&(2{,}800 \text{ qr. wheat}, 240 \text{ t. iron}, 0, 0) \rightarrow (5'750 \text{ qr. wheat}, 0, 0),\\
&(1{,}800 \text{ qr. wheat}, 240 \text{ t. iron}, 0, 0) \rightarrow (0, 600 \text{ t. iron}, 0),\\
&(1{,}150 \text{ qr. wheat}, 120 \text{ t. iron}, 0, 0) \rightarrow \left(0, 0, q_3 \text{ kg gold}\right).
\end{aligned}
\tag{4.134}
$$

Identify matrix $\mathbf{S}$ and vector $\mathbf{q}$. Calculate the input-output matrix $\mathbf{C}$, set up the Sraffa price model and calculate the vectors of absolute and relative prices $\mathbf{p}$ and the *national income* Y using the four ways of calculation (4.133). Set later $q_3 = 3/2$.

Calculate the total output X and the circulating capital K, expressing Y, X, K in monetary values.

Solution to Example 4.4.3:
We can now identify with $q_3 = \frac{3}{2}$, $p_3 = 33{,}000$,

$$\mathbf{S} = \begin{bmatrix} 2{,}800 & 1{,}800 & 1{,}150 \\ 240 & 240 & 120 \\ 0 & 0 & 0 \end{bmatrix},$$

$$\mathbf{q}_I = \mathbf{Se} = \begin{bmatrix} 2{,}800 & 1{,}800 & 1{,}150 \\ 240 & 240 & 120 \\ 0 & 0 & 0 \end{bmatrix} \begin{bmatrix} 1 \\ 1 \\ 1 \end{bmatrix} = \begin{bmatrix} 5{,}750 \\ 600 \\ 0 \end{bmatrix},$$

$$\mathbf{q} = \begin{bmatrix} 5{,}750 \\ 600 \\ q_3 \end{bmatrix} > \mathbf{o}, \quad \mathbf{d} = \mathbf{q} - \mathbf{q}_I = \begin{bmatrix} 0 \\ 0 \\ q_3 \end{bmatrix} \geq \mathbf{o}. \tag{4.135}$$

Calculate the *input-output* matrix $\mathbf{C} = \mathbf{S}\hat{\mathbf{q}}^{-1}$,

$$\mathbf{C} = \begin{bmatrix} 2{,}800 & 1{,}800 & 1{,}150 \\ 240 & 240 & 120 \\ 0 & 0 & 0 \end{bmatrix} \begin{bmatrix} \frac{1}{5{,}750} & 0 & 0 \\ 0 & \frac{1}{600} & 0 \\ 0 & 0 & \frac{1}{q_3} \end{bmatrix} = \begin{bmatrix} \frac{56}{115} & 3 & \frac{1{,}150}{q_3} \\ \frac{24}{575} & \frac{2}{5} & \frac{120}{q_3} \\ 0 & 0 & 0 \end{bmatrix} \geq \mathbf{0}, \tag{4.136}$$

and its characteristic polynomial, which is independent of q_3,

$$P_3(\lambda) = \det(\mathbf{C} - \lambda\mathbf{I}) = \lambda^3 - \frac{102}{115}\lambda^2 + \frac{8}{115}\lambda = \lambda\left(\lambda - \frac{2}{23}\right)\left(\lambda - \frac{4}{5}\right). \tag{4.137}$$

Thus, the Frobenius number is $\lambda_C = 0.8$. We establish the Sraffa price model, leading to the eigenvalue equation,

$$\mathbf{S}'\mathbf{p}(1 + R) = \hat{\mathbf{q}}\mathbf{p} \Rightarrow \mathbf{C}'\mathbf{p} = \frac{1}{1 + R}\mathbf{p} = \frac{4}{5}\mathbf{p}, \tag{4.138}$$

and to the price eigenvector of the transposed matrix $\mathbf{C}'$, associated to the Frobenius number $\lambda_C = 0.8$,

$$\mathbf{p} = [\ p_1, \ \ p_2, \ \ p_3\ ] = p_3 \left[\ \frac{2q_3}{5{,}125}, \ \ \frac{3q_3}{1{,}025}, \ \ 1\ \right], \qquad \frac{p_2}{p_1} = \frac{15}{2}. \tag{4.139}$$

The ratio $p_2/p_2 = 7.5$ is independent of q_3. The *productiveness* is here equal to the maximal *rate of profit* $r = R = (1/\lambda_C) - 1] = (1/0.8) - 1 = 0.25$, as there is no wage rate $w = 0$ and a null vector of *labour*, $\mathbf{L} = \mathbf{o}$.

Note that the eigenvalue equation (4.138) leads to *relative prices*, when "1 kg of gold" is set as nuḿméraire, giving $p_3 = 1$. On the other hand, when we introduce a currency, like CHF, setting $p_3 = 33{,}000$ CHF/kg, we obtain absolute prices,

$$\begin{aligned}&\begin{bmatrix}2{,}800 & 240 & 0\\ 1{,}800 & 240 & 0\\ 1{,}150 & 120 & 0\end{bmatrix}\begin{bmatrix}p_1\\ p_2\\ p_3\end{bmatrix}\frac{5}{4}\\ &=\begin{bmatrix}5{,}750 & 0 & 0\\ 0 & 600 & 0\\ 0 & 0 & \frac{3}{2}\end{bmatrix}\begin{bmatrix}p_1\\ p_2\\ p_3\end{bmatrix}=\begin{bmatrix}111{,}073\\ 86{,}927\\ 49{,}500\end{bmatrix},\quad p_3 = 33{,}000\ \text{CHF/kg}.\end{aligned} \tag{4.140}$$

Indeed, one obtains the vector of *absolute prices*, the positive (left) eigenvector of matrix **C**,

$$[\mathbf{p}] = [19.32\ \text{CHF/qr. wheat}, 144.88\ \text{CHF/t. iron}, 33{,}000\ \text{CHF/kg gold}]'.$$ [27]

Having got the vector **p** of *absolute prices*, we are able to compute the national income Y (4.133), using definition (a), then, calculating the product of q_3 and p_3 (b), one obtains:

$$Y = \mathbf{d}'\mathbf{p} = \left[0, 0, \frac{3}{2}\right]\begin{bmatrix}19.32\\ 144.88\\ 33{,}000.00\end{bmatrix} = q_3 \cdot p_3 = 49{,}500\ \text{CHF}. \tag{4.141}$$

We then confirm the calculations by the total sum (c) and the product (d):

$$\begin{aligned}Y = R(\mathbf{Se})'\mathbf{p} &= R\sum_{i=1}^{3}\left(\sum_{j=1}^{3} s_{ij}\right)p_i = 0.25\left(575\frac{7{,}920}{41} + 600\frac{5{,}940}{41}\right)\\ &= R \cdot K = 0.25 \cdot 198{,}000 = 49{,}500\ \text{CHF}.\end{aligned} \tag{4.142}$$

Finally, we compute the *total output* $X = \mathbf{q}'\mathbf{p} = 247{,}500$ CHF and the *circulating capital* $K = (\mathbf{Se})'\mathbf{p} = 198{,}000$ CHF. Again, the *national income* can be calculated. It is the difference $Y = X - K = 49{,}500$ CHF.

The commodity digraph of the process is given by $G(\mathbf{V}_2)$ of Figure 4.2 (right) with the adjacency matrix $\mathbf{V}_2$ (4.125). The node C_1 depicts wheat, the node depicts C_2 iron and the node C_3 represents gold. The digraph clearly shows that gold is a *non-basic* commodity. ▲

This instructive example calls for a number of important comments.

(1) The *non-basic* commodity gold has been chosen here as means of payment and "1 kg of gold" as the *numéraire*, $p_3 = 1$. The ratio p_2/p_1 (4.139) of the prices of wheat an iron is independent of q_3. Relative and absolute prices can be determined.
(2) If one considers an open economy (see Section 8.3 hereafter), assuming in this example that gold is exported, the export market fixing the price of gold, the export industry, $j = 3$, may also vary the quantity produced q_3. It is easily shown that the

27 The vector of relative prices is $\mathbf{p}_0 = [6/1{,}025, 9/2{,}050, 1]'$ is based on gold as the means of payment.

characteristic polynomial $P_3(\lambda)$ (4.137) remains independent of q_3. For this reason, the Frobenius number and the *productiveness* remain unchanged, $R = 0.25$. In this case, the quantity of the *non-basic* gold will determine the absolute prices of wheat and iron. Changes in the gold parameters p_3, q_3 will determine the rise and fall of the absolute prices of wheat and iron, the ratio p_1/p_2 remaining constant (4.139).

(3) We see that *non-basic* commodities participate in national income ($Y = p_3 \cdot q_3$) and can influence the prices of other commodities (4.139), including basic commodities.[28] Then, the national income Y is directly proportional to the quantity of gold produced! But gold does here not determine the technology. This is not a one-way street as Sraffa implies in PCMC (Par. 6) when commenting on *non-basics*:

"These products have no part in the determination of the system. Their role is purely passive".

In fact, to the extent that basics enter the means of production of non-basics, the latter can have an impact on the absolute prices of basics, but not on the relative prices of basics (4.139).

(4) The foregoing examples illustrate the peculiar status of *non-basic* commodities and, as mentioned, the characteristic: *not entering into the production of other commodities, directly or indirectly,* must be analysed in each specific situation. Indeed, in the foregoing situations if gold is produced, there must be a demand: for jewelery, for certain industries, for exports, for payments, etc, unless gold is considered as waste. Commodities which are considered *non-basic* may turn out to become *basic* on further examination. We revert to these matters in the next Chapter 6 in relation with *joint production* analyses.

The cases must be examined from the economic point of view. Consider Example 4.4.3.

(a) Closed economy. The surplus of commodity n accumulates period after period if the basic commodities do not generate an additional surplus (beyond that required for the production of commodity n) to absorb the gold production.

(b) Open economy (with exports only). The gold surplus generated by the n-th *non-basic* industry can be exported, export revenue constituting the entire national income, if no additional surpluses are generated by the *basic* industries and provides foreign exogenous demand for gold, sufficient to absorb the gold production.

(5) A fundamental result, easily proved, follows from Example 4.4.3: In a closed economy of single-product processes, the economic existence of *non-basic* commodities depends on the surplus generated by the basic commodities. If the latter generate no surplus, no *non-basics* can be produced.[29]

---

28 This was, for example, confirmed by Steedman [114].

29 This characteristic of non-basic commodities seems to have been overlooked by Sraffa.

(6) Let us state once again that *non-basic* commodities which contains no *basics* in their means of production are then completely separated from the *basic* commodities and form an autonomous system. The corresponding economy considered as a whole is then composed of separate sub-systems of production for which in general no uniform *rate of profit* can be defined.[30]

In the next section, we introduce directed digraphs to represent Sraffa production processes composed of *basic* and *non-basic* commodities.

4.5 Representation of production processes by bipartite directed graphs

There is a great tradition in economics to represent production processes by *directed graphs*. The most famous historical example indirectly embodying a directed graph, are the "Tableaux Economiques" of François Quesnay (1694–1774) [88]. We shall present this model in detail in Section 4.11.

We now introduce the useful tool of *directed graphs* or *digraphs* already encountered in Section 4.4 to represent *commodity flow* matrices of *single-product industries*.[31] Formal definitions and properties of directed graphs are presented in Section A.14. Simply stated, a digraph (Definition A.14.1) is a set of nodes, linked by arrows, the direction of the arrows indicating causal links between the nodes.

The digraphs envisaged here to represent Sraffa's production schemes are *bipartite digraph* $G = (\mathcal{N}, \mathcal{A})$. Its node set $\mathcal{N}$ are partitioned into two disjoint subsets, $\mathcal{F}$ for the n industries and $\mathcal{C}$ for the n commodities, $|\mathcal{F}| = n$ and $|\mathcal{C}| = n$, $\mathcal{F} \cup \mathcal{C} = \mathcal{N}$, $\mathcal{F} \cap \mathcal{C} = \emptyset$, and $|\mathcal{N}| = 2n$, such that each arrow of $G = (\mathcal{N}, \mathcal{A})$ is exclusively directed either from a node of $\mathcal{F}$ to a node of $\mathcal{C}$ or vice versa. The arrows of the bipartite digraph G are in the set $\mathcal{A} \subseteq \mathcal{F} \times \mathcal{C}$ (Definition A.14.5). A *digraph* G' may be imbedded in a larger digraph G, $G' \prec G$, representing a larger economy.

At the moment, we need two types of nodes and two types of arrows (also called arcs).

Definition 4.5.1. Description of nodes and arcs of bipartite digraphs to represent Sraffa's economies.

- Each **productive entity** or industry S_j is represented by one **round yellow node.** These nodes are numbered $j = 1, \ldots, n$, and all of them are elements of the node set $\mathcal{F}$.

---

30 Sraffa in PCMC, Par. 35, seems to have had this scenario in mind.

31 Subsequent *bipartite digraphs* will also be used to represent Sraffa's production scheme of *joint production*, also called *joint production economies* or *multi-product industries*.

- Each **commodity** i is represented by one **square green node.**[32] The nodes are numbered bold $i = \mathbf{1}, \ldots, \mathbf{n}$ or identified by a letter to designate the corresponding commodity i. All nodes designing a commodity are elements of the node set $\mathcal{C}$.
- An **arrow** $(j \leftarrow i)$ of digraph G', pointing from a commodity i to an industry S_j, shows that commodity i is required by industry S_j. The arrow $(j \leftarrow i)$ can be labeled by the quantity s_{ij}.
- An **arrow** $(j \rightarrow j)$ of digraph G', pointing from an industry S_j to a commodity j, shows in the case of single product industries that industry S_j produces commodity j. The arrow $(j \rightarrow j)$ can be labeled by the quantity q_j.
- An **arrow** $(j \rightarrow i)$, pointing from an industry S_j to a commodity i, shows in the case of joint production that industry S_j produces commodity i. The arrow $(j \rightarrow i)$ can be labeled by the quantity f_{ji}.
- The **dashed arrow** $(j \leftarrow i)$, pointing from a node $i \in G$ to a node $j \in G'$, $G' \preceq G$, or vice versa, shows that only a part of the economy is represented by G'. Clearly, G' is imbedded in a greater economy, represented by digraph G.

For a good understanding of this visualization, consider the following mnemotechnical presentation: (a) the *round nodes* correspond to the wheels of machines in a productive entity of an industry, (b) the *square nodes* correspond to a storage depot, where the commodities are temporarily stored,[33] (c) the *arrows* $(j \leftarrow i)$ correspond to the required commodity i driven to industry S_j, (d) the *arrows* $(j \rightarrow i)$ correspond to a productive entity of industry S_j having produced commodity i that is driven to the warehouse storing commodity i (see Figure 4.4).

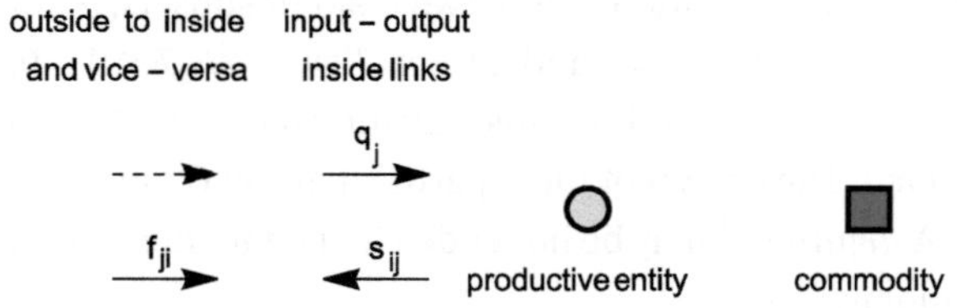

Figure 4.4: Elements of construction of the *bipartite digraph.*

Later we will also need the notions of degrees of nodes on a digraph, respectively indegree and outdegree of nodes of a digraph (see Definition A.14.6 and also Wagner [119], p. 68), in order to describe some economic properties.

For a node of a digraph G, the number of head endpoints of arrows pointing to that node is called the *indegree* of the node. The number of tail endpoints of arrows pointing out of that node is its *outdegree*.

32 Only the numbers of commodities are written as bold signs. The variable i will not be bolded.

33 The only exceptions are places for permanent disposal of certain products such as waste that cannot be recycled, or, more subtly, dissipated energy.

For a node $\upsilon \in \mathcal{N}$, $G = (\mathcal{N}, \mathcal{A})$, the *indegree* is denoted $\gamma^-(\upsilon)$ and its *outdegree* as $\gamma^+(\upsilon)$. A node with $\gamma^-(\upsilon) = 0$ is called a *source* because it is the origin of each of its incident arcs. Similarly, a node with $\gamma^+(\upsilon) = 0$ is called a *sink*. The sum $\gamma^+(\upsilon) + \gamma^-(\upsilon) = \gamma(\upsilon)$ is called the *degree* of the node.

Before going on, let's look at the following illustration for a *single-product industry*, where both commodities are *basic*:

Example 4.5.1. We represent Sraffa's third numerical example in PCMC, Par. 5 (see *Example* 3.1.3, equation (3.41)) by a bipartite digraph characterising this process of production.

The bipartite digraph (Fig. 4.5) reveals that:
(1) The *productive entities* 1 and 2 are single-product industries, 1 producing the commodity wheat W (generic for agricultural products), represented by arrow $(1 \rightarrow \mathrm{W})$, and 2 producing the productive entities commodity iron I (generic for manufactural goods) as outputs, represented by arrow $(2 \rightarrow \mathrm{I})$;
(2) The *productive entities* 1 and 2 both require the commodity wheat and commodity iron as inputs for production (interindustrial market), represented by arrows $(\mathrm{W} \rightarrow 1)$ and $(\mathrm{I} \rightarrow 1)$, respectively $(\mathrm{W} \rightarrow 2)$ and $(\mathrm{I} \rightarrow 2)$.
(3) The surplus is clearly visualised. The output quantities minus the input quantities of both commodities yield the surplus per commodity. Indeed, for wheat, we have $q_1 - (s_{11} + s_{12}) = 575 - (280 + 120) = 175$ and for iron $q_2 - (s_{21} + s_{22}) = 20 - (8 + 12) = 0$.

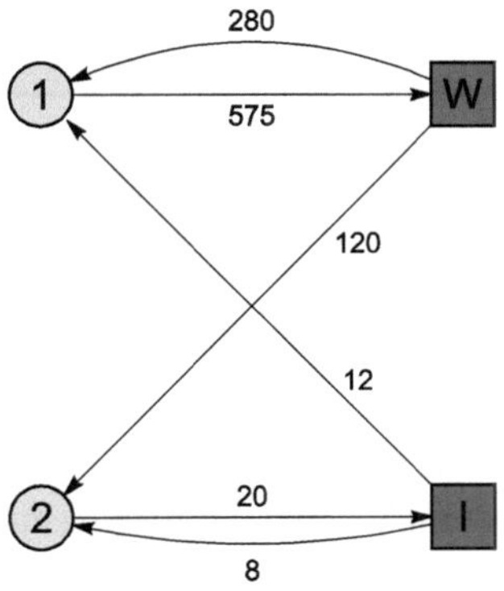

Figure 4.5: Bipartite digraph of Sraffa's example (PCMC, Par. 5), (3.41).

In principle, the *bipartite digraph* $G = (\mathcal{N}, \mathcal{A})$ can be constructed with node sets $\mathcal{F}$ and $\mathcal{C}$ of different cardinality, $|\mathcal{F}| = m \neq |\mathcal{C}| = n$, $\mathcal{F} \cup \mathcal{C} = \mathcal{N}$, $\mathcal{F} \cap \mathcal{C} = \emptyset$. In the present context, we only consider the case $n = m$[34]:

34 This is a vital prerequisite to be in accordance with PCMC: In single-product industries, each industry produces exactly one commodity (PCMC, Par. 1–3), and in joint production (Chapter 6), some of

---

The basic assumption throughout the single-product Sraffa systems and joint production Sraffa systems (see Chapter 6) is that the total number of commodities equals the total number of industries.

Sraffa Network: Figure 4.5 shows the *bipartite digraph* $G = (\mathcal{N}, \mathcal{A})$ of the *production economy* figuring in PCMC, Par. 5, and Example 3.1.3. Such a *bipartite digraph* is called a **Sraffa Network** for short (see Definition A.14.11).

There is the *same number n of industries as there are commodities*, but, of course, a given industry will usually not require all commodities for production; accordingly for some commodities, there may be fewer arrows outgoing from the corresponding commodity node than there are industries.

In the next section, we further illustrate the properties of Sraffa Networks of *single-product industries* for single-commodity industries, which are quite simple, as an introduction to the use of this tool that will also be used in the more complicated situations of joint production treated in Chapter 6.

4.6 The Sraffa Network: a bipartite directed graph

Sraffa Networks are furthermore constructed on the *principle of production*, formulated by Assumption 2.2.1 and Assumption 2.2.2. A formal presentation of the *Sraffa Network* is given in Section A.14:

Every industry produces a positive output quantity of at least one commodity (exactly one concerns only single-commodity processes) and has to use at least one positive quantity of the n commodities as means of production, otherwise this industry does not exist and is eliminated from the production economy.

With a view to subsequent developments the term "industry"[35] should be considered in a large sense as an organised entity encompassing notably a *productive entity*, i. e., a center of production regrouping the "means of production", to use Sraffa's terminology, such as raw materials, tools, machinery, factories etc., productive land and

---

the industries $j = 1, \ldots, n$ may produce more than one of the commodities $i = 1, \ldots, n$, so a commodity may be produced by more than one industry, but the total number of industries is always equal the total number n of commodities (PCMC, Par. 51).

35 Even simple commodities such as wild fruit for consumption require a primitive industry for output: gathering or harvesting. Land providing raw materials is productive industrial land and requires intensive human and technological input. To our knowledge, there exists only one commodity which does not depend in any way on an industry for production, transformation and distribution until further notice: the air we all breath, under the condition that it is not polluted, so that it does no require treatment.

human resources.[36] Commodities will be understood to include goods, including for example composite goods such as machinery (whether depreciated or not) and services and, depending on the economic process, equipped land (e. g., providing indispensable infrastructures for all types of buildings).

In *single-product industries*, some commodities will enter, directly or indirectly, the production of all commodities, including themselves. They will be linked to all other commodities by one or more uninterrupted sequences of arrows passing one way or another through all the productive entities. As we know, such commodities are called *basic* commodities. Commodities for which this does not apply are called *non-basic* commodities, which we have already encountered in *Section* 4.4.

Example 4.6.1. This example shows the mode of construction of a Sraffa Network. We have chosen here $n = 4$ productive entities 1, 2, 3, 4. The number of commodities, tagged a, b, c, d is also $n = 4$.

The dependencies show up as follows:

$$\begin{aligned} &\mathrm{a} \to 2 \to \mathrm{b} \to 1 \to \mathrm{a} \to 3 \to \mathrm{c} \to 4 && \text{.Basic, indirect} \\ &\mathrm{b} \to 1, 2, 3, 4 && \text{.Basic, direct (4x)} \\ &\mathrm{c} \to 1 \to \mathrm{a} \to 2 \to \mathrm{b} \to 3 \to \mathrm{c} \to 4 && \text{.Basic, indirect} \\ &\mathrm{d} \to 4 \to \mathrm{d} && \text{.Non-basic} \end{aligned} \tag{4.143}$$

So, following the traced paths in the bipartite digraph of Figure 4.6, the Sraffa Network, commodities a, b, c are *basic*, and d is *non-basic*. The *indegrees* of the nodes $\upsilon \in C = \{\mathrm{a}, \mathrm{b}, \mathrm{c}, \mathrm{d}\}$ are all $\gamma^{-}(\upsilon) = 1$. We are dealing with *single-product industries*, (Lemma A.14.1). Indeed, from round nodes to square nodes, there is exactly one *unique* arrow because each industry, represented by a round yellow node, produces exactly one commodity, represented by a square green node. ▲

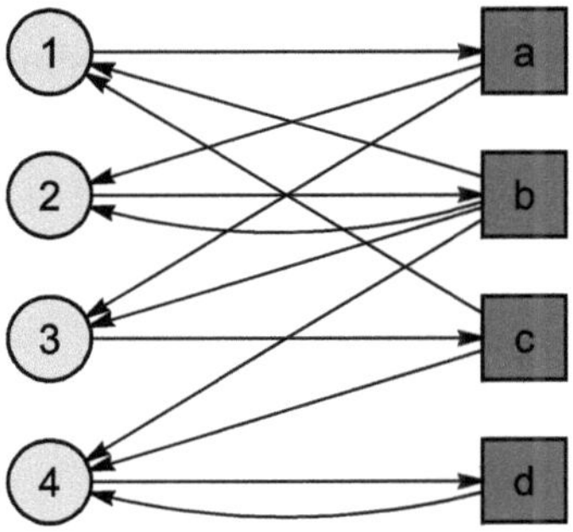

Figure 4.6: *Sraffa Network* of Example 4.6.1.

36 In the sequel, where no confusion should arise, we shall, abusing language, use indifferently the terms industry, industrial sector or productive entity, to avoid pedantry.

We continue by giving examples of *basic* commodities in today's economies, before starting to look more closely at the nature of *non-basic* commodities.

Example 4.6.2. Examples of basic commodities: steel, water prepared by humans as drinking water or to produce electricity, electric power, food, land, computer facilities, education, health care and credit services for production firms. Such commodities enter one way or another, directly or indirectly, into the production of all commodities in the real economies of any country.

The nature of *basic* and *non-basic* commodities will be illustrated soon by some elementary examples and partial digraphs as preparation for the understanding of a complete example with the Sraffa Network G', see Figure 4.14 which is a subgraph embedded in a wider digraph G of a larger economy. Note the clear cut separation between both categories of *basic* and *non-basic products*.

Sraffa mentions three types of *non-basic* commodities[37] We exemplify these three types in today's economies (PCMC, Par. 35 and Par. 58):

(i) *Non-basics* completely excluded from the means of production (sinks):

Example 4.6.3. Unrecycled waste W from industries 1 and 2, left over as toxic products and radioactive nuclear deposits, Figure 4.7. ▲

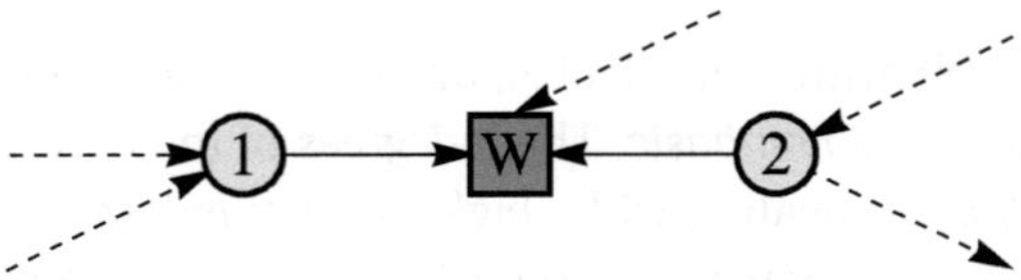

Figure 4.7: Non-basics completely excluded.

(ii) *Non-basics* which only enter their own production:

Example 4.6.4. Race horses which only involve their own production for breeding and racing (Sraffa's example, PCMC, Par. 6). ▲

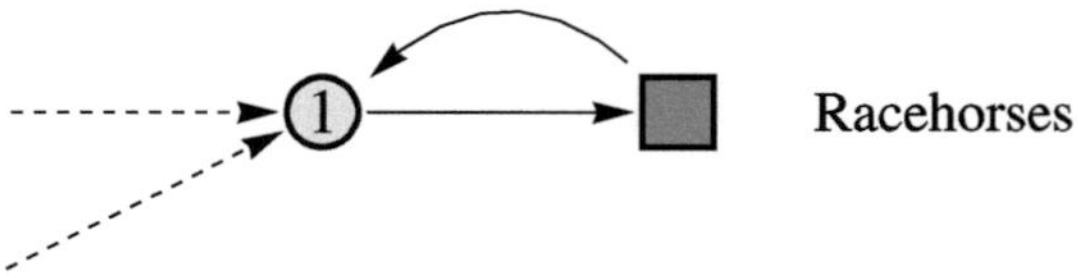

Figure 4.8: Non-basics entering only their own production.

37 Sraffa's comments in PCMC, Par. 35 and 58, must be reconsidered bearing in mind the following presentation that also illustrates Roncaglia's definition of *basic* and *non-basic products* ([97], p. 60) given in Section 4.4.

(iii) *Non-basics* used in the production of one or more other non-basics:

Example 4.6.5. Consider in a modern economy the commodity of stored data S: The commodity of stored data S used by data-processing productive entity ② releasing thermal energy E and used by specialised services L required by another productive entity ① which creates waste W. See Figure 4.9. ▲

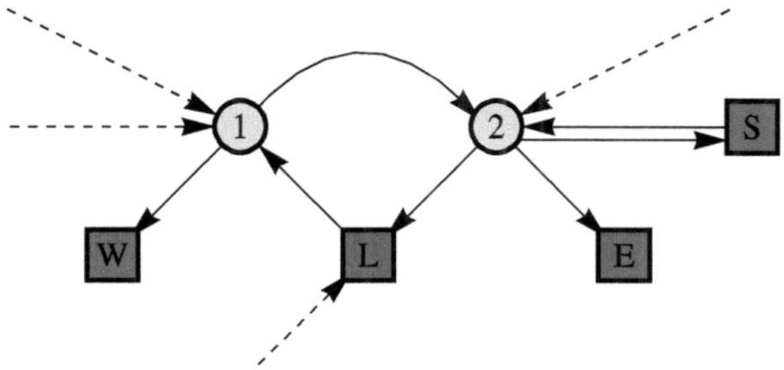

Figure 4.9: Non-basics L, S in the production of other non-basics E, W and S itself.

Incoming, respectively outgoing *dashed arrows* in Figure 4.7, Figure 4.8 and Figure 4.9 indicate input and output of commodities *from* or *to* other sources.

Basic commodities are important insofar as no production economy can operate without them. Remember that they are the only commodities on which sustainable closed economies may operate. In fact the whole present day discussions on global sustainable growth, circular economics and overcoming excessive consumerism, implicitly rely on giving priority to *basic* commodities. This is the main reason behind the important distinction between *basic* and *non-basic* commodities.

Example 4.6.6. Set up a list of *basic* commodities entering the economy of your country, containing (a) 10 items, (b) 20 items, (c) 30 items, (d) more? Is there a limit? Then draw up a list of important non-basics in your economy. Do some of them have cultural significance?

We will now introduce a known tool to represent Sraffa production schemes. It is the *adjacency matrix*. We will later learn how to visualize with *Sraffa Networks* the presence of *basic* and *non-basic* commodities in an economy.

4.7 The adjacency matrix in the case of single-product industries

Consider an economy of *n single-product industries*. Set up Sraffa's corresponding production scheme $(\mathbf{S}', \mathbf{L}) \rightarrow (\hat{\mathbf{q}})$ and the single product Sraffa system (4.82),

$$\boxed{\begin{aligned} \mathbf{S}'\mathbf{p}(1+r) + \mathbf{L}\frac{\tilde{w}Y}{L} &= \hat{\mathbf{q}}\mathbf{p} = \mathbf{x}, \\ Y &= (\mathbf{q} - \mathbf{S}\mathbf{e})'\mathbf{p}, \\ L &= \mathbf{e}'\mathbf{L}. \end{aligned}} \tag{4.144}$$

We take the *commodity flow* matrix **S**, describing the technology, and the vector of *total output* **q** of this *single-product industry*.

We set up the following square $(2n \times 2n)$ matrix, comprising matrices **S** and $\hat{\mathbf{q}}$ in one matrix,

$$\mathbf{\Sigma} = (\sigma_{kl}) = \begin{bmatrix} \mathbf{0} & \hat{\mathbf{q}} \\ \mathbf{S} & \mathbf{0} \end{bmatrix}, \quad k, l = 1, \ldots, 2n. \tag{4.145}$$

Then for matrix $\mathbf{\Sigma}$, we set up the $(2n \times 2n)$ *adjacency matrix*, where the diagonal matrix $\hat{\mathbf{q}}$ is replaced by the identity matrix **I**, and matrix **S** is replaced by its Boolean representation, matrix $\mathbf{P} = (p_{ij})$, consisting only of '0' and '1', in accordance to Definition A.8.5, getting

$$w_{n+i,j} := p_{ij} = \begin{cases} 1 & \text{if } s_{ij} > 0, \\ 0 & \text{if } s_{ij} = 0, \end{cases} \qquad i, j \in \{1, \ldots, n\}. \tag{4.146}$$

We have now defined the $(2n \times 2n)$ *adjacency matrix*

$$\mathbf{W} = (w_{kl}) = \begin{bmatrix} \mathbf{0} & \mathbf{I} \\ \mathbf{P} & \mathbf{0} \end{bmatrix}; \quad k, l = 1, \ldots, 2n, \tag{4.147}$$

corresponding to matrix $\mathbf{\Sigma}$, directly composed of matrices **S** and $\hat{\mathbf{q}}$, issued from the production scheme $(\mathbf{S}', \mathbf{L}) \to (\hat{\mathbf{q}})$, describing the Sraffa production system. Now we show the corresponding *Sraffa Network*.

Commodity flows

Clearly, the directions of the arrows in the Sraffa Networks show the *commodity flow* of the production process of the represented economy. We illustrate this with the following Sraffa Network:

Example 4.7.1. Consider $n = 4$ virtual *single-product* industries, represented by the Sraffa Network of Figure 4.10. Set up the corresponding adjacency matrix, applying Definition A.8.5.

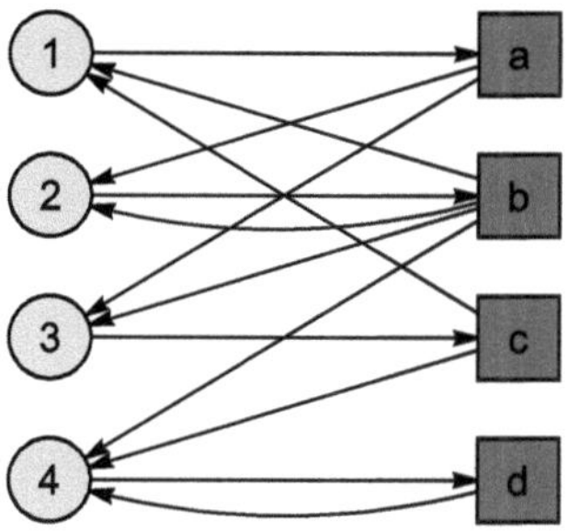

Figure 4.10: *Sraffa Network* of Example 4.6.1, p. 161: commodity flows.

Solution to Example 4.7.1:
The specific form of the $(2n \times 2n)$ adjacency matrix **W** of this Sraffa Network is reproduced here.[38]

$$
\mathbf{W} = \begin{array}{ll} & \\ \text{Industry} & \begin{array}{c} 1 \\ 2 \\ 3 \\ 4 \end{array} \\ \text{Commodity} & \begin{array}{c} a \\ b \\ c \\ d \end{array} \end{array}
\begin{array}{c} \begin{array}{cccc|cccc} 1 & 2 & 3 & 4 & a & b & c & d \end{array} \\
\left[\begin{array}{cccc|cccc}
0 & 0 & 0 & 0 & 1 & 0 & 0 & 0 \\
0 & 0 & 0 & 0 & 0 & 1 & 0 & 0 \\
0 & 0 & 0 & 0 & 0 & 0 & 1 & 0 \\
0 & 0 & 0 & 0 & 0 & 0 & 0 & 1 \\ \hline
0 & 1 & 1 & 0 & 0 & 0 & 0 & 0 \\
1 & 1 & 1 & 1 & 0 & 0 & 0 & 0 \\
1 & 0 & 0 & 1 & 0 & 0 & 0 & 0 \\
0 & 0 & 0 & 1 & 0 & 0 & 0 & 0
\end{array}\right] \\ \mathbf{P} \end{array} \; \mathbf{I}
\tag{4.148}
$$

Note that the arrows in the *Sraffa Network*, pointing from green square nodes, representing commodities, to yellow circle nodes, representing industries, are pictured by '1' in the lower left submatrix **P** of the adjacency matrix **W** (4.148).

For industries, there is a one-to-one correspondence between the coefficients of **P** and the *incoming* arrows of the *Sraffa Network* of Figure 4.10, e. g., arrows entering the *yellow circle nodes*. We can define: $w_{ij} = 1 \leftrightarrow \exists(i \to j)$, respectively $w_{ij} = 0 \leftrightarrow \nexists(i \to j)$, $i \in \{a, \ldots, d\}, j = 1, \ldots, 4$.[39]

For *single-product industries*, the upper right square submatrix is the identity matrix **I**. Again for yellow nodes, representing industries, there is a one-to-one correspondence between the coefficients of **I** and the *outgoing* arrows of the *Sraffa Network* of

38 The adjacency matrix **W** can be of help in understanding Pasinetti's concept of "vertical integration" (Pasinetti [80], Chapter 5, p. 123), an algebraic artefact.

39 The sign ∃ means "there exists at least one", this operator indicates *existential quantification*, the sign ∄ means "there is no", the negation of ∃. The Boolean matrix reflects this.

Figure 4.10, e. g., arrows entering the *green square nodes*. We can therefore define: $w_{ij} = 1 \leftrightarrow \exists(i \to j)$, respectively $w_{ij} = 0 \leftrightarrow \nexists(i \to j)$, $i = 1, \dots, 4$, and $j \in \{a, \dots, d\}$. ▲

We have presented the *commodity flow* in the Sraffa Network and its adjacency matrix **W**. We again specify in practical terms the meaning of the directed arrows in such a network:
- outgoing arrows from production entity S_j, $j \in \{1, \dots, 4\}$ (yellow circle node) mean: entity S_j produces commodity $i \in \{a, \dots, d\}$ (represented by a green square node) (*outdegree* of j: $\gamma^+(j) = 1$) (single-product industries!);
- incoming arrows to a production entity S_j, $j \in \{1, \dots, 4\}$, (yellow circle node) mean: entity S_j requires (purchases) commodities $i \in \{a, \dots, d\}$ for production (from a green square node) (*indegree* of j: $\gamma^-(j) \geq 1$).

Each directed arrow is uniquely represented in the adjacency matrix by a coefficient equal to one.

Payment flows

Now, to be complete in economic terms, one considers the *payment flow* which is *inverse* to the *commodity flow*. This means: that to each *commodity flow*, represented by a *Sraffa Network*, there corresponds the inverse *Sraffa Network* where all arrows show in the opposite direction. See Figure 4.11.

Referring to Figure 4.10, the *payment flow* is represented by the corresponding *inverse Sraffa Network*:

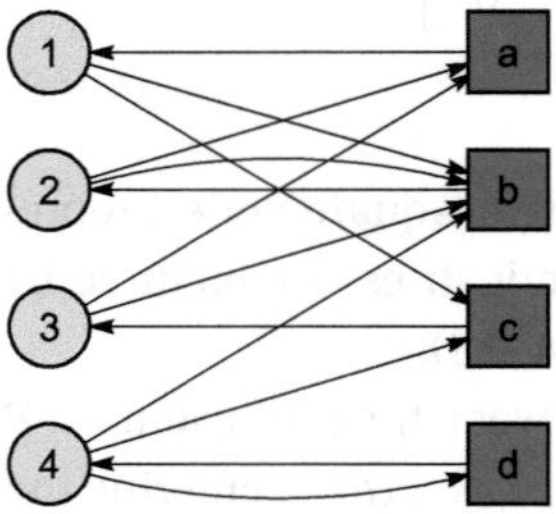

Figure 4.11: *Sraffa Network* of Example 4.6.1: payment (outlays) flows.

In practical terms, the meaning of the *inverse* directed arrows in the *inverse Sraffa Network* is:
- outgoing arrows mean that the industry S_j, $j \in \{1, \dots, 4\}$, has outlays (payments) for purchases of commodities $i \in \{a, \dots, d\}$, entering into its production (*outdegree* of j: $\gamma^-(j) \geq 1$);[40]

40 Internal industry payments also exist, and industrial accounting includes internal transfer prices.

- incoming arrows mean that the industry S_j, $j \in \{1,\ldots,4\}$ receives payments for sales from the commodities $i \in \{a,\ldots,d\}$ that it has produced. For this reason, there is an *indegree* of j: $\gamma^+(j) = 1$, in the case of single-product industries!

The adjacency matrix of the *payment flow* Sraffa Network is the *transpose* of the adjacency matrix $\mathbf{W}$ (4.148) of the corresponding *Sraffa Network* of the initial *commodity flow*.

$$\mathbf{W}' = \begin{array}{cc} & \\ \text{Industry} & 1 \\ & 2 \\ \text{(outlays)} & 3 \\ & 4 \\ & a \\ \text{Commodity} & b \\ \text{(income)} & c \\ & d \end{array} \overset{\begin{array}{cccc|cccc} 1 & 2 & 3 & 4 & a & b & c & d \end{array}}{\left[\begin{array}{cccc|cccc} 0 & 0 & 0 & 0 & 0 & 1 & 1 & 0 \\ 0 & 0 & 0 & 0 & 1 & 1 & 0 & 0 \\ 0 & 0 & 0 & 0 & 1 & 1 & 0 & 0 \\ 0 & 0 & 0 & 0 & 0 & 1 & 1 & 1 \\ \hline 1 & 0 & 0 & 0 & 0 & 0 & 0 & 0 \\ 0 & 1 & 0 & 0 & 0 & 0 & 0 & 0 \\ 0 & 0 & 1 & 0 & 0 & 0 & 0 & 0 \\ 0 & 0 & 0 & 1 & 0 & 0 & 0 & 0 \end{array}\right]} \begin{array}{c} \mathbf{P}' \\ \\ \\ \\ \\ \\ \\ \end{array} \qquad \underset{\mathbf{I}}{} \tag{4.149}$$

After this intermezzo, we continue with our discussion of *commodity flows* represented by *Sraffa Networks* and also *commodity digraphs*, explained in Section A.14, which are associated to adjacency matrices (4.146).

Example 4.7.2. Consider Sraffa's third example (PCMC, Par. 5), *Example* 3.1.3, equation (3.41), and its *Sraffa Network*, Figure 4.5. Set up the adjacency matrix $\mathbf{W}$ (4.147) and establish the associated commodity digraph $G(\mathbf{P})$, see Figure 4.12.

Solution to Example 4.7.2:
In this example, there are $n = 2$ economic sectors, one producing wheat (W) and the other iron (I). There is consequently the production scheme $(\mathbf{S}', \mathbf{o}) \to (\hat{\mathbf{q}})$. Consider matrix $\boldsymbol{\Sigma}$ (4.145), composed of $\mathbf{S}$ and $\hat{\mathbf{q}}$ and set up the Boolean matrix $\mathbf{P}$,

$$\mathbf{q} = \begin{bmatrix} 575 \\ 20 \end{bmatrix}, \quad \mathbf{S} = \begin{bmatrix} 280 & 120 \\ 12 & 8 \end{bmatrix} \Rightarrow \mathbf{P} = \begin{bmatrix} 1 & 1 \\ 1 & 1 \end{bmatrix}. \tag{4.150}$$

Then we establish the adjacency matrix (4.147),

$$\mathbf{W} = (w_{kl}) = \begin{array}{c} \\ 1 \\ 2 \\ W \\ I \end{array} \overset{\begin{array}{cc|cc} 1 & 2 & W & I \end{array}}{\left[\begin{array}{cc|cc} 0 & 0 & 1 & 0 \\ 0 & 0 & 0 & 1 \\ \hline 1 & 1 & 0 & 0 \\ 1 & 1 & 0 & 0 \end{array}\right]}, \quad k, l = 1, \ldots, 4. \tag{4.151}$$

We set up the associated *commodity digraph* $G(\mathbf{P})$ corresponding to the *Sraffa Network*, Figure 4.5. ▲

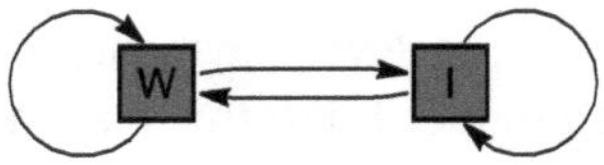

Figure 4.12: Associated digraph $G(\mathbf{P})$ of Example 4.7.2, (4.150).

The associated *commodity digraph* $G(\mathbf{P})$ (explanations in Section A.14) is a complement to the *Sraffa Network*. Notice that the upper right hand matrix, $\mathbf{I}$ (4.147), indicates the one to one relationship between an industry and the unique commodity it produces. In fact, $G(\mathbf{P})$ is a compression of the *Sraffa Network*, indicating admissible interconnections commodities.

In *joint production processes*, we shall see that the situation is not as simple, and the *commodity digraph* associated to a *Sraffa Network* will be generated by the product of two adjacency matrices, see Example A.14.6.

We continue now to illustrate the foregoing explanations by the following Sraffa Network of *single-product industries* with labelled arrows.

Example 4.7.3. Consider the Sraffa Network, Figure 4.13, of $n = 5$ commodities and sectors. Numbers attached to arrows indicate commodity flows. The flow in brackets of the commodity No. **8** produced by industry No. **3** is the flow restricted to the system constituted by the first three commodities, after removal of the *non-basic* industries No. **4** and No. **5** together with their commodities No. **9** and No. **10**. Set up the adjacency matrix $\mathbf{W}$ and analyse the subgraphs G', G''.

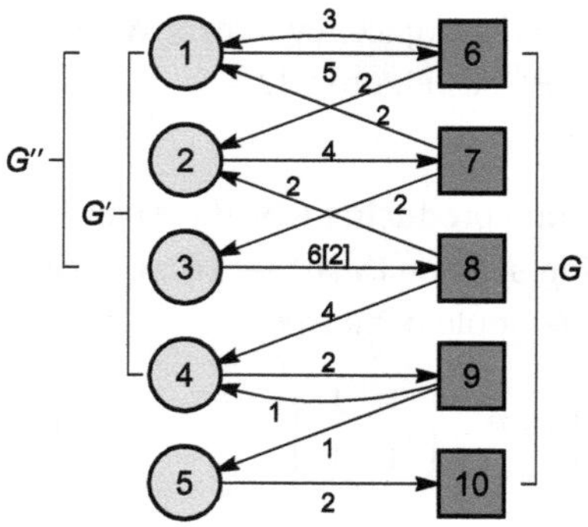

Adjacency matrix $\mathbf{W}$

$$
\begin{array}{c|ccccc|ccccc}
 & 1 & 2 & 3 & 4 & 5 & 6 & 7 & 8 & 9 & 10 \\
1 & 0 & 0 & 0 & 0 & 0 & 1 & 0 & 0 & 0 & 0 \\
2 & 0 & 0 & 0 & 0 & 0 & 0 & 1 & 0 & 0 & 0 \\
3 & 0 & 0 & 0 & 0 & 0 & 0 & 0 & 1 & 0 & 0 \\
4 & 0 & 0 & 0 & 0 & 0 & 0 & 0 & 0 & 1 & 0 \\
5 & 0 & 0 & 0 & 0 & 0 & 0 & 0 & 0 & 0 & 1 \\
\hline
6 & 1 & 1 & 0 & 0 & 0 & 0 & 0 & 0 & 0 & 0 \\
7 & 1 & 0 & 1 & 0 & 0 & 0 & 0 & 0 & 0 & 0 \\
8 & 0 & 1 & 0 & 1 & 0 & 0 & 0 & 0 & 0 & 0 \\
9 & 0 & 0 & 0 & 1 & 1 & 0 & 0 & 0 & 0 & 0 \\
10 & 0 & 0 & 0 & 0 & 0 & 0 & 0 & 0 & 0 & 0
\end{array}
\tag{4.152}
$$

Figure 4.13: Sraffa Network of single-product industries.

Solution to Example 4.7.3:

We set up the adjacency matrix $\mathbf{W}$ of the Sraffa Network G, which thus has to be considered as generated by matrix $\mathbf{W}$, $G = G(\mathbf{W})$. Now, one analyses its subgraphs G', G'',

Figure 4.13.

G: is the complete not fully connected network,

G': is the network restricted to the commodities Nos: **6, 7, 8**, including the link of commodity No. **8** to the *non-basic* industry No. **4**;

G'': is the fully connected network comprising the basic structure; but note that in G'' the flow changes for the industry No. **3** producing commodity **8** in order to fulfill, for this "miniature system", the conditions of production in quantity terms presented in the following scheme.

We present the commodity flows of the "miniature system", represented by G''.

Entries of the flow matrix for G''

	1	2	3	Output
6	3	2	0	5
7	2	0	2	4
8	0	2	0	2
Input	5	4	2	11

(4.153)

Figure 4.14: Conditions of production for the "miniature system" G''.

The adjacency matrix $\mathbf{W}''$, corresponding to the "miniature system", is imbedded as a submatrix in the adjacency matrix $\mathbf{W}$,

$$\mathbf{W}'' = \begin{array}{c} \\ 1 \\ 2 \\ 3 \\ 6 \\ 7 \\ 8 \end{array} \begin{array}{c} \begin{array}{ccc|ccc} 1 & 2 & 3 & 6 & 7 & 8 \end{array} \\ \left[\begin{array}{ccc|ccc} 0 & 0 & 0 & 1 & 0 & 0 \\ 0 & 0 & 0 & 0 & 1 & 0 \\ 0 & 0 & 0 & 0 & 0 & 1 \\ \hline 1 & 1 & 0 & 0 & 0 & 0 \\ 1 & 0 & 1 & 0 & 0 & 0 \\ 0 & 1 & 0 & 0 & 0 & 0 \end{array}\right] \end{array} . \tag{4.154}$$

The "miniature system" (4.153) represented here has no surplus. Inclusion of a surplus modifies the output vector (assuming means of production unchanged), see Section 3.1.3. ▲

Consider the production scheme $(\mathbf{S}', \mathbf{L}) \to (\hat{\mathbf{q}})$, describing the Sraffa production system, as presented in the foregoing chapters. The entries of the matrices $\mathbf{S}$ and $\hat{\mathbf{q}}$ are represented on the *Sraffa Network*, acting as a graphic *representation of matrices*[41]

41 In fact, we will see later that algebraically the k basic commodities form the basis of a k-dimensional vector space of commodities ($k \le n$).

for the system and fixes the constraints to which Sraffa's price model will comply. The submatrix **P** of the adjacency matrix **W** (4.147) will define the matrix structure of the left-hand side of the corresponding *Sraffa price model* (4.144) and the submatrix **I** the right-hand side.

A commodity flow *Sraffa Network* is associated with a *production economy*, represented by the logical structure of the production scheme $(\mathbf{S}', \mathbf{L}) \rightarrow (\hat{\mathbf{q}})$. It is built on the basis of the *adjacency matrix* **W**. Its dual, the *payment flow Sraffa Network*, is built on the basis of the *transposed adjacency matrix* $\mathbf{W}'$.

Finally, let us mention that the *adjacency matrix* will be suited for the determination of *basic* and *non-basic* commodities by algebraic means, whereas the corresponding *Sraffa Network* acts as a graphic representation suited to visualize *basic* and *non-basic* commodities.

4.8 Distinction between basic and non-basic commodities

In this section, we will learn how to distinguish *basic* from *non-basic* commodities in economies, given by a production scheme $(\mathbf{S}', \mathbf{L}) \rightarrow (\hat{\mathbf{q}})$ and composed strictly of *single-product industries*. We will use two pairs of tools:

(a) the $(2n \times 2n)$ *adjacency matrix* **W** (4.147), corresponding to the $(2n \times 2n)$ matrix $\mathbf{\Sigma}$ (4.145), associated to the Sraffa Network, presented in Section 4.7;
(b) the $n \times n$ *adjacency matrix* **P** (4.150), corresponding to the $n \times n$ matrix **S**, associated with the *digraph* $G(\mathbf{P})$ (Definition A.14.7).

Assumptions 2.2.1, and Assumptions 2.2.2 apply to matrix **S**. Consequently, the $n \times n$ *transaction matrix* **S** is *semi-positive*.

If, furthermore, matrix **S** is *irreducible*, then $(\mathbf{I} + \mathbf{S})^{(n-1)} > 0$ (see Lemma A.8.2). Lemma A.8.2 also applies to the corresponding *adjacency matrix* (Definition A.8.5).

Here we have two possibilities: We consider either the $n \times n$ *adjacency matrix* **P** (4.150), and the associated digraph $G(\mathbf{P})$ (Definition A.14.7), or the $(2n \times 2n)$ *adjacency matrix* **W** and the corresponding *Sraffa Network* (Definition A.14.11).

Varga's Theorem A.14.1 states that the adjacency matrix **W** is irreducible if and only if the associated digraph $G(\mathbf{W})$ is *strongly connected*.

When the digraph is *strongly connected*, then every node can be linked to all the other nodes, including itself, by at least one connected sequence of arrows all of the same direction. Such nodes may be termed *basic* nodes because they correspond to *basic* commodities and industries.

When the *adjacency matrix* is *irreducible*, then all the production centres and all the commodities are *basic*, Lemma A.15.3 (v).

We will now illustrate these properties by a Sraffa example (PCMP, Par. 5).

Example 4.8.1. We consider Sraffa's third numerical example (PCMC, Par.5); equation (3.41) and the adjacency matrix **P** (4.150) of the *semi-positive* commodity flow matrix **S**. Show that this production economy consists only of *basics*. Show that the associated commodity digraph $G(\mathbf{P})$ and the corresponding *Sraffa Network* are strongly connected.

Solution to Example 4.8.1:
We consider the commodity flow matrix and its adjacency matrix

$$\mathbf{S} = \begin{bmatrix} 280 & 120 \\ 12 & 8 \end{bmatrix} \Rightarrow \mathbf{P} = \begin{bmatrix} 1 & 1 \\ 1 & 1 \end{bmatrix}, \quad \mathbf{I} = \begin{bmatrix} 1 & 0 \\ 0 & 1 \end{bmatrix}, \tag{4.155}$$

then we set up the adjacency matrix **W** (4.147), according to the *Sraffa Network*,

$$\mathbf{W} = (w_{kl}) = \begin{bmatrix} \mathbf{0} & \mathbf{I} \\ \mathbf{P} & \mathbf{0} \end{bmatrix} = \left[\begin{array}{cc|cc} 0 & 0 & 1 & 0 \\ 0 & 0 & 0 & 1 \\ \hline 1 & 1 & 0 & 0 \\ 1 & 1 & 0 & 0 \end{array}\right], \quad k,l = 1,\ldots,4. \tag{4.156}$$

We apply Lemma A.8.2 on the *non-negative* matrix **W** and find,

$$(\mathbf{W}+\mathbf{I})^3 = \begin{bmatrix} 4 & 3 & 4 & 1 \\ 3 & 4 & 1 & 4 \\ 5 & 5 & 4 & 3 \\ 5 & 5 & 3 & 4 \end{bmatrix} > 0, \tag{4.157}$$

matrix **W** is *irreducible*, as is matrix **P**, because it is *positive*. With Theorem A.14.1 we conclude that the *Sraffa Network* and the associated digraph $G(\mathbf{P})$ are *strongly connected* as indicates Figure 4.15. We conclude with Lemma A.15.3 (i) and (v) that all the products are basic. ▲

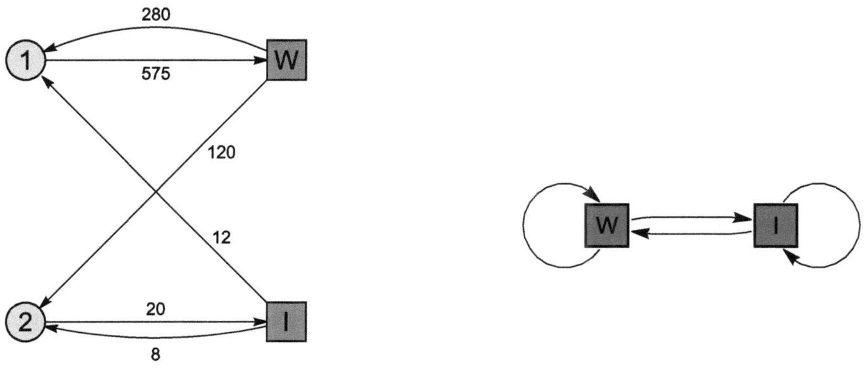

Figure 4.15: Sraffa Network of PCMC, Par. 5, (3.41) and associated digraph $G(\mathbf{P})$.

When the matrix **S** is *reducible*, the production economy is depicted, e. g., by one of the *adjacency matrices* including *basic* and *non-basic* commodities. In this case, we apply Lemma A.15.2 and accordingly realise the permutation to bring the *commodity flow* matrix **S** into "canonical form", to separate the *basics* from the *non-basics*.

This is the subject of the next illustration with the *Sraffa Network* and the corresponding digraph $G(\mathbf{S})$.

Example 4.8.2. We consider Example 4.7.3, presented by the Sraffa Network, Figure 4.13. Identify the commodity flow matrix **S**, transform it into "canonical form" and calculate the corresponding adjacency matrix **P**.

Determine the basics and non-basics and determine the connectivity of the *Sraffa Network* and the associated commodity digraph $G(\mathbf{P})$.

Solution to Example 4.8.2:
We identify the commodity flow matrix, its adjacency matrix and the vector of total output,

$$\mathbf{S} = \begin{bmatrix} 3 & 2 & 0 & 0 & 0 \\ 2 & 0 & 2 & 0 & 0 \\ 0 & 2 & 0 & 4 & 0 \\ 0 & 0 & 0 & 1 & 1 \\ 0 & 0 & 0 & 0 & 0 \end{bmatrix} \Rightarrow \mathbf{P} = \begin{bmatrix} 1 & 1 & 0 & 0 & 0 \\ 1 & 0 & 1 & 0 & 0 \\ 0 & 1 & 0 & 1 & 0 \\ 0 & 0 & 0 & 1 & 1 \\ 0 & 0 & 0 & 0 & 0 \end{bmatrix}, \quad \mathbf{q} = \begin{bmatrix} 5 \\ 4 \\ 6 \\ 2 \\ 2 \end{bmatrix} \Rightarrow \mathbf{I}. \quad (4.158)$$

Assumption 2.2.1 and Assumption 2.2.2 are fulfilled, but commodity 5 does not belong to the means of production. We set up the associated digraph $G(\mathbf{P})$, Figure 4.16.[42]

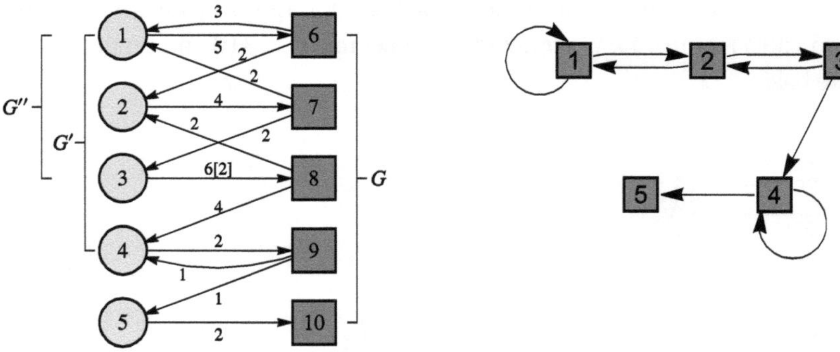

Figure 4.16: Sraffa Network (left) and associated commodity digraph (right).

42 A check confirms that the number of arcs on the associated digraph corresponds to the number of outgoing connections from the square commodity nodes in the Sraffa Network, namely 8. Note further that node 5 in the digraph contains no outgoing arc, and so it is a sink (e. g., consumed commodities, waste, etc).

Matrix **S** is in "canonical form", so we identify the (3×3) the sub-matrices $\mathbf{S}_{11}$, respectively the (3×3) matrix $\mathbf{P}_{11}$ and apply Lemma A.8.2,

$$\mathbf{S}_{11} = \begin{bmatrix} 3 & 2 & 0 \\ 2 & 0 & 2 \\ 0 & 2 & 0 \end{bmatrix} \Rightarrow \mathbf{P}_{11} = \begin{bmatrix} 1 & 1 & 0 \\ 1 & 0 & 1 \\ 0 & 1 & 0 \end{bmatrix},$$

$$(\mathbf{P}_{11} + \mathbf{I})^2 = \begin{bmatrix} 5 & 3 & 1 \\ 3 & 3 & 2 \\ 1 & 2 & 2 \end{bmatrix} > 0. \tag{4.159}$$

We continue with matrix **P** and find,

$$(\mathbf{P} + \mathbf{I})^4 = \begin{bmatrix} 35 & 26 & 13 & 6 & 1 \\ 26 & 22 & 13 & 13 & 5 \\ 13 & 13 & 9 & 20 & 12 \\ 0 & 0 & 0 & 16 & 15 \\ 0 & 0 & 0 & 0 & 1 \end{bmatrix} \geq 0. \tag{4.160}$$

Thus, matrix **P** is *reducible*, whereas submatrix $\mathbf{P}_{11}$ is *irreducible*. With Theorem A.14.1 we conclude that the sub-network G'', respectively $G(\mathbf{P}_{11})$ in Figure 4.16 (right) is strongly connected, whereas the whole digraph G, respectively $G(\mathbf{P})$ are only *weakly connected*.

(a) See Figure 4.16 (left): In the Sraffa Network there is exactly one path from an industry j to the corresponding commodity $i = j + 5$, $i \in \{1, \ldots, 5\}$, thus we are in presence of *single-product industries*.

(b) See Figure 4.16 (left): In the Sraffa Network, there is no directed path from the industries **4** and **5** to the industries **1, 2** and **3**, consequently industries **4** and **5** are *non-basic*, whereas there are directed pathes form any of the three industries **1, 2** and **3** to industries **1, 2, 3, 4** and **5.** For this reason industries **1, 2** and **3** are *basics*.

(c) See Figure 4.16 (right): In the digraph $G(\mathbf{P})$, there is no directed path from the commodities **4** and **5** to the commodities **1, 2** and **3**, consequently commodities **4** and **5** are *non-basic*, whereas there are directed paths from any of the three commodities **1, 2** and **3** to commodities **1, 2, 3, 4** and **5**, so **1, 2** and **3** are *basics*. There is no distinction between industries and commodities.

Then, we conclude with Lemma A.15.2 (i)–(iii) that the commodities **6, 7, 8** are *basic*, whereas **9, 10** are *non-basic*. Indeed, matrix **S** is reducible with the upper left (3×3) submatrix $\mathbf{S}_{11}$, Lemma A.15.2 (i). We have just shown with (4.159) that the (3×3) upper left submatrix $\mathbf{S}_{11}$ is *irreducible* and therefore refers to basics, Lemma A.15.2 (ii). Finally the remaining commodities **9, 10** are *non-basic*, Lemma A.15.2 (iii). Indeed, matrix **S** is *reducible* by (i); $\mathbf{S}_{11}$ is *irreducible* by (ii) and thus refers to *basics* (upper left (3×3) submatrix) and (iii) confirms that **9** and **10** are *non-basics*. ▲

Proposition 4.8.1. *Consider a production scheme* $(\mathbf{S}', \mathbf{L}) \to (\hat{\mathbf{q}})$ *with a semi-positive completely reducible* $n \times n$ *commodity flow matrix* $\mathbf{S}$ *in canonical form, Definition A.8.4,* $\mathbf{q} \geq \mathbf{Se} > \mathbf{o}$, $\mathbf{L} \geq \mathbf{o}$. *The upper-left submatrix* $\mathbf{S}_{11}$ *(noted as* $\tilde{\mathbf{S}}_{11}$ *(A.74)) is irreducible by construction. Compute for* $m = n - 1$ *the matrix power* $\mathbf{S}^m$, *which then is moreover completely reducible. The economy is thus decomposed into separate economies.*

We illustrate Proposition 4.8.1 with an example:

Example 4.8.3. Consider the (5×5) semi-positive and not completely reducible matrix $\mathbf{S}$ of the production scheme $(\mathbf{S}', \mathbf{L}) \to (\hat{\mathbf{q}})$ in Example 4.8.2 and compute,

$$\mathbf{S}^4 = \begin{bmatrix} 3 & 2 & 0 & 0 & 0 \\ 2 & 0 & 2 & 0 & 0 \\ 0 & 2 & 0 & 4 & 0 \\ 0 & 0 & 0 & 1 & 1 \\ 0 & 0 & 0 & 0 & 0 \end{bmatrix}^4 = \begin{bmatrix} 221 & 126 & 68 & 64 & 16 \\ 126 & 100 & 24 & 72 & 8 \\ 68 & 24 & 32 & 20 & 20 \\ 0 & 0 & 0 & 1 & 1 \\ 0 & 0 & 0 & 0 & 0 \end{bmatrix}. \tag{4.161}$$

The present economy of $n = 5$ commodities has $k = 3$ *basics* and $(n-k) = 2$ *non-basics*.

Consider now a slightly modified matrix $\mathbf{S}$ in which one element is changed, $s_{34} = 4$ becoming $s_{34} = 0$. This means that the technology of process **4** is modified, indeed commodity **8** is no longer used by process **4**. The effect is that the present economy is separated into two distinct independent economies, which is immediately visible in the corresponding *Sraffa Network* Figure 4.16 (left), where arc (**8** → **4**) has to be skipped to describe the new production process. We compute

$$\mathbf{S}^4 = \begin{bmatrix} 3 & 2 & 0 & 0 & 0 \\ 2 & 0 & 2 & 0 & 0 \\ 0 & 2 & 0 & 0 & 0 \\ 0 & 0 & 0 & 1 & 1 \\ 0 & 0 & 0 & 0 & 0 \end{bmatrix}^4 = \begin{bmatrix} 221 & 126 & 68 & 0 & 0 \\ 126 & 100 & 24 & 0 & 0 \\ 68 & 24 & 32 & 0 & 0 \\ 0 & 0 & 0 & 1 & 1 \\ 0 & 0 & 0 & 0 & 0 \end{bmatrix}. \tag{4.162}$$

The present economy of $n = 5$ commodities is decomposed into an economy of $k = 3$ *basic* economies and a second economy containing $(n - k) = 2$ commodities. In this second economy, the commodity No. **9** is *basic*, and the commodity No. **10** is *non-basic because* it does not belong to the means of production, see Figure 4.16 (left). ▲

We accordingly have devised the following *method* to distinguish the set of *basic* commodities from the set of *non-basic* commodities:

(i) When the *semi-positive* $n \times n$ commodity flow matrix $\mathbf{S}$ is in "canonical form", the order k of the upper left matrix $\mathbf{S}_{11}$ gives the number of *basics*, and $(n-k)$ is the number of *non-basics*, according to Lemma A.15.2.

(ii) Consider the *non-negative* and not completely reducible $n\times n$ commodity flow matrix $\mathbf{S}$ in "canonical form". Setting $m = n - 1$, compute $\mathbf{S}^m$ to check the rows of the iterated matrix. A strictly positive row of $\mathbf{S}^m$ corresponds to a *basic* commodity. A row of matrix $\mathbf{S}^m$ corresponds to a *non-basic* commodity if it is not *strictly positive*, as illustrated by Example 4.8.3.

4.9 Recapitulation of Sraffa's price calculations, the labour value

It is now time to sum up the foregoing developments. PCMC, Par. 10, treats single product industries and uses a particular notation for the *semi-positive* and *irreducible* or *positive* $n \times n$ *commodity flow* matrix **S** in *physical units*, which we transcribe in this text to the notations of Miller and Blair [65]:

$$\mathbf{S} = \begin{bmatrix} A_a & A_b & \dots & A_k \\ B_a & B_b & \dots & B_k \\ \dots & \dots & \dots & \dots \\ K_a & K_b & \dots & K_k \end{bmatrix} := \begin{bmatrix} s_{11} & s_{12} & \dots & s_{1n} \\ s_{21} & s_{22} & \dots & s_{2n} \\ \dots & \dots & \dots & \dots \\ s_{n1} & s_{n2} & \dots & s_{nn} \end{bmatrix} = (s_{ij}). \tag{4.163}$$

Remember Sraffa's notation for the annual amounts of labour is $\{L_a, L_b, \dots, L_k\}$; our notation is: L_i, $i = 1, \dots, n$, presented as a vector $\mathbf{L} = [L_1, L_2, \dots, L_n]' \geqq \mathbf{o}$. We have the vector of surplus $\mathbf{d} \geqq \mathbf{o}$. The total outputs q_j of each industry S_j are collected in the *output* vector $\mathbf{q} = \mathbf{Se} + \mathbf{d} > \mathbf{o}$. So we write the *production scheme*, also described for each industry S_j with the row vectors $\mathbf{s}_{.j}$ (Definition A.4.3), in the form:

$$\begin{aligned} &(\mathbf{S}', \mathbf{L}) \to (\hat{\mathbf{q}}), \\ &(\mathbf{s}_{.j}, L_j) \to (0, \dots, q_j, \dots, 0), \quad j = 1, \dots, n. \end{aligned} \tag{4.164}$$

The *Sraffa price model*, also called the *single-product Sraffa system* (4.82), primarily aimed at fixing prices, expressed in some given *numéraire*, is set up in four steps. The matrix **S** represents the level of technology attained by the economic system. The economy is driven by:

- the conditions of production fixing the required *means of production* designated here as circulating capital, and
- final demand, which initiates the production of a surplus [national income following the national accounting identities].

Step I (Sraffa, PCMC, Par. 3). No surplus means no profit and no wages paid by surplus, $w = 0$, only subsistence wages, see Section 3.6.

Setting up the conditions of production in a *self-replacement* process (production for subsistence), there is no surplus produced, i. e., $\mathbf{d} := \mathbf{q} - \mathbf{Se} = \mathbf{o}$. This phase represents the constitution of the *means of production*, which are used up and have to be replaced after each reference period. Start with the *commodity flow matrix* **S** in physical terms, assuming at this stage that **S** indirectly incorporates labour in the means of production s_{ij} (see PCMC, Par. 8) with the prices p_i, $i = 1, \dots, n$, constituting the vector $\mathbf{p} = [p_1.p_2, \dots, p_n]'$,

We have the identity, a reduced *Sraffa price model*, which figures in PCMC, Par. 3; presented in matrix form

$$\boxed{\mathbf{S}'\mathbf{p} = \hat{\mathbf{q}}\mathbf{p} = \mathbf{x},} \tag{4.165}$$

representing the *required circulating capital* in each sector. Also, we may aggregate *circulating capital* sector by sector and constitute its *total amount as*,

$$K := \mathbf{e}'\mathbf{S}'\mathbf{p} = (\mathbf{Se})'\mathbf{p} = \mathbf{q}'\mathbf{p} = \mathbf{e}'\mathbf{x} =: X, \tag{4.166}$$

which is here equal to the *total output* because there is no surplus.

Due to *Lemma* 4.1.1 *(a)* the Frobenius number of the *semi-positive* and *irreducible* or *positive* matrix $\mathbf{C} = \mathbf{S}\hat{\mathbf{q}}^{-1}$ is $\lambda_C = 1$, and there is no *productiveness*, $R = (1/\lambda_C) - 1 = 0$. The *rate of profits* r and the share of national income to circulating capital $\tilde{R}$ also vanishes. With (4.36) we have the expected triple equalities $\tilde{R} = r = R = 0$ and consequently $Y = P = W = 0$.

We return now to Sraffa's Example 3.1.1 completing the calculations.

Example 4.9.1. Consider equation (3.5),

$$\mathbf{S}'\mathbf{p} = \begin{bmatrix} 280p_1 + 12p_2 \\ 120p_1 + 8p_2 \end{bmatrix} = \begin{bmatrix} 400p_1 \\ 20p_2 \end{bmatrix} = \hat{\mathbf{q}}\mathbf{p}, \tag{4.167}$$

with solutions $p_1 = 1$, $p_2 = 10$. The value of the circulating capital K is obtained by taking these relative prices, giving $p_1 = 1$ and $p_2 = 10$, producing then the capital $K = \mathbf{e}'\mathbf{S}'\mathbf{p} = 400p_1 + 20p_2 = 400 \cdot 1 + 20 \cdot 10 = 600 = X$, and the national income (the term is introduced in PCMC, Par. 12, but it is more convenient to call Y *net income*), giving $Y = (\mathbf{q} - \mathbf{Se})'\mathbf{p} = \mathbf{o}'\mathbf{p} = 0$. ▲

Step II (PCMC, Par. 4). All the surplus goes into profits, no wages are paid out of the surplus, i. e., $w = 0$, only subsistence wages.

Sraffa introduced production with a surplus in addition to the required means of production. Either $\mathbf{d} := \mathbf{q} - \mathbf{Se} > \mathbf{o}$, called *positive self-replacement*, where there is a positive surplus produced by each industry, or $\mathbf{d} := \mathbf{q} - \mathbf{Se} \geq \mathbf{o}$, called *self-replacement*, where it is admitted that some industries produce no surplus.

Here, the totality of the surplus goes to entrepreneurs, with *maximal rate of profits R*, assumed uniform for all industries.[43] The corresponding new equations to determine $\mathbf{p}$ are extended with respect to (4.165),

$$\mathbf{S}'\mathbf{p}(1 + R) = \hat{\mathbf{q}}\mathbf{p} = \mathbf{x}. \tag{4.168}$$

Aggregating (4.168), we obtain the equation, relating *total output* X to the *circulating capital* $K = (\mathbf{Se})'\mathbf{p}$ (4.31) and to *national income* Y (4.32):

---

43 Sraffa indeed assumes here, and in **Step III** below that the profit rate r is equal in all industries $j = 1, 2, \ldots, n$ so that $r = \frac{P_1}{K_1} = \cdots = \frac{P_j}{K_j} = \cdots = \frac{P_n}{K_n} = \frac{P}{K}$ and he designates r as the "uniform rate of profits". In other words, the surplus is distributed in proportion to the capital value of the means of production (PCMC, Par. 4). This uniformity assumption renders the Sraffa model mathematically more tractable for large n. In fact mathematically, the subsequent(4.171) can be handled even assuming profit rates r_j and wage rates w_j, differing for each industry $j = 1, \ldots, n$ (see Section 8.3 hereafter).

$$X := \mathbf{e}'\hat{\mathbf{q}}\mathbf{p} = \mathbf{q}'\mathbf{p} = \mathbf{e}'\mathbf{S}'\mathbf{p}(1+R) = (\mathbf{Se})'\mathbf{p} + (\mathbf{Se})'\mathbf{p}\cdot R$$
$$= K(1+R) = (\mathbf{Se})'\mathbf{p} + \mathbf{d}'\mathbf{p} = K + Y \Rightarrow Y = R\cdot K \Rightarrow R = \frac{Y}{K}. \qquad (4.169)$$

With $\tilde{w} = 0$ and (4.36), $\tilde{R}(1-\tilde{w}) = \tilde{R} = r$, having in this case $r = R$, we get $R = \tilde{R} = Y/K$.

The determination of the prices of the *Sraffa production scheme* $(\mathbf{S}', \mathbf{L}) \to (\hat{\mathbf{L}})$ does not require an explicit presentation of the surplus $\mathbf{d} \geq \mathbf{o}$ (or $\mathbf{d} > \mathbf{o}$).[44] However, in the logic of *single-product industries*, presented in Sraffa (PCMC, Part I), an identification of $\mathbf{d}$ is useful to calculate the *national income*:

(a) The scalar product of the surplus vector $\mathbf{d}$ with the price vector $\mathbf{p}$ gives national income $Y = \mathbf{d}'\mathbf{p}$.
(b) Sraffa's price model: There are *relative prices*, where the value of each commodity is measured by a *physical numéraire* (i. e., a quantity of a commodity, like a *quarter of* wheat), expressed in a *physical unit* of that commodity. But there are *absolute prices*, where the value of each commodity is expressed by a freely chosen *currency* (like CHF) per *physical unit* of that commodity.
(c) The reference to a separate surplus vector $\mathbf{d}$ is systematically used in the construction of the *Standard system* (PCMC, Chapters IV–V) presented in Chapter 5.

And finally, the surplus vector $\mathbf{d}$ in Sraffa's price model provides the link with the vector of *final demand* $\mathbf{f}$ in the Leontief models.

According to *Lemma* 4.1.1 *(b)*, the Frobenius number of the irreducible non-negative matrix $\mathbf{C}$ is $\lambda_C < 1$, and the productiveness is $R = (1/\lambda_C) - 1 > 0$ for $\lambda_C > 0$. With (4.36) we have the expected triple equalities $\tilde{R} = r = R > 0$, $Y = P > 0$, for this system without wages, i. e., $W = 0$.

Example 4.9.2. We continue with Sraffa's Example 3.1.3 (PCMC, Par. 5) (3.42) and (3.45), a system of production (3.43) with surplus and no wages, $W = 0$. Taking Sraffa's solutions $p_1 = 1$, $p_2 = 15$, we obtain with equations (4.33) and (4.31) and equation (4.34) and calculate the following values for the economic variables X, K, Y and P.

Solution to Example 4.9.2:

$$K = 400p_1 + 20p_2 = 400\cdot 1 + 20\cdot 15 = 700,$$
$$X = 575p_1 + 20p_2 = 575\cdot 1 + 20\cdot 15 = 875,$$
$$Y = (575-400)p_1 + (20-20)p_2 = 175,$$
$$W = 0 \Rightarrow Y = P + W = 175 \Rightarrow P = 175. \qquad (4.170)$$

44 This will appear quite clear later in connection with *joint production*, Chapter 6, where $\hat{\mathbf{q}}$ is replaced by an non-diagonal matrix $\mathbf{F}$, the surplus $\mathbf{d}$ being replaced by $(\mathbf{F} - \mathbf{S})$.

Then we calculate $\tilde{R} = Y/K = 175/700 = 0.25$, $r = P/K = 175/700 = 0.25$ and from the Frobenius number of the *input-output coefficients* matrix **C** (3.46), $\lambda_C = 0.8$, we also get $R = (1/\lambda_C) - 1 = (1/0.8) - 1 = 0.25$, leading to the triple equality $r = R = \tilde{R} = 0.25$. Finally we confirm: $Y = R \cdot K = 0.25 \cdot 700 = 175$. ▲

Step III (PCMC, Par. 10–12) (profits and wages). Inclusion of wages and labour directly in the production process leads to sharing, the surplus (national income) $Y > 0$ between profits and wages.

Step I and **Step II** include labour as *means of production* in the *semi-positive* and *irreducible* or *positive commodity flow matrix* **S** (at the so-called subsistence level). In **Step III,** one now introduces a *semi-positive* vector of labour $\mathbf{L} \geq \mathbf{o}$,[45] representing the annual amount of labour, e. g., *man-hours per annum per industry*, which then enables us to regroup separately total wages, i. e., subsistence wages plus wages contributing to *value added* (see PCMC, Par. 8 and 9). That means from here on, that the *commodity flow matrix* **S** no longer contains any labour component, as was the case earlier, specially in equations (4.165), (4.168).

We set up the *complete single-product Sraffa system* (4.55) which then reads:

$$\mathbf{S}'\mathbf{p}(1+r) + \mathbf{L}\frac{\tilde{w} \cdot Y}{L} = \hat{\mathbf{q}}\mathbf{p} = \mathbf{x}. \tag{4.171}$$

In (4.171) r, $0 < r < R$, is still the rate of profits, and $\tilde{w}$ is designated as the ratio of *total wages to national income*. We then consider the *semi-positive* vector $\boldsymbol{\pi} = \hat{\mathbf{q}}^{-1}\mathbf{L} \geq \mathbf{o}$ (4.58) and apply Lemma 4.1.2 (a) obtaining a positive price vector (4.61) reproduced here for the *complete single-product Sraffa system*,

$$\mathbf{p} = [\mathbf{I} - (1+r)\mathbf{C}']^{-1}\boldsymbol{\pi} \cdot \frac{\tilde{w} \cdot Y}{L} > \mathbf{o}. \tag{4.172}$$

This leaves us with a vector of prices to be determined.

Step IV (PCMC, Par. 12) (*Sraffa's price model*). In a final step, *national income* Y, or equivalently GDP,[46] is explicitly as

$$Y = (\mathbf{q}' - (\mathbf{Se})')\mathbf{p} =: \mathbf{d}'\mathbf{p} > 0. \tag{4.173}$$

We keep Assumption 2.2.1 and Assumption 2.2.2 and consider now the *commodity flow matrix* **S** to be only *semi-positive* and eventually *reducible*. We consequently retain Assumption 2.5.1 for the *vector of labour*, which must be positive, $\mathbf{L} = [L_1, L_2, \dots, L_n]' > \mathbf{o}$; all workers get their salary at the end of the period. Lemma 4.1.2 (b), applies, the Frobenius number $\lambda_C = 1/(1+R) < 1$ giving a positive *productiveness* $R > 0$. We consider two cases.

---

45 Assumption 2.5.1 is again here not necessary.

46 Recall that *national income* = total sum of added values = profits + wages.

(a) At first, we recall that Sraffa *normalized* the quantity of labour, to be measured as $L = 1\,\text{TAL}$. This is done, calculating from the given vector of labour $\mathbf{L} = [L_1, L_2, \ldots, L_n]' > \mathbf{o}$ the sum $L = \mathbf{e}'\mathbf{L} > 0$ and then normalising to a new labour vector $[L_1/L, L_2/L, \ldots, L_n/L]'$, whose sum is $1 = \mathbf{e}'(\mathbf{L}/L)$.
(b) Then, if we calculate labour in the physical unit *man-years*, we also start by setting the vector of labour $\mathbf{L} = [L_1, L_2, \ldots, L_n]'$. Then we just calculate the *total quantity of labour* $L = \mathbf{e}'\mathbf{L}$ (4.34), measured in the unit *man-years* and introduce these results.

We describe the *complete single-product Sraffa system* (4.171) and replace Sraffa's *wage per unit of labour* w, referring to equation (4.38), by $w = (\tilde{w} \cdot Y)/L$, entering explicitly the *national income* Y (4.173), the total quantity of labour L (4.34), the *share of total wages to national income* $\tilde{w}$.

There are $n + 2$ equations and $n + 4$ variables $r, \tilde{w}, L, p_1, p_2, \ldots, p_n$ and Y.[47]

$$\boxed{\begin{aligned} \mathbf{S}'\mathbf{p}(1+r) + \mathbf{L}\frac{\tilde{w} \cdot Y}{L} &= \hat{\mathbf{q}}\mathbf{p} = \mathbf{x}, \\ Y &= (\mathbf{q}' - (\mathbf{Se})')\mathbf{p} =: \mathbf{d}'\mathbf{p}, \\ L &= \mathbf{e}'\mathbf{L}. \end{aligned}} \tag{4.174}$$

As usual, the exogenous *rate of profits* r has to be chosen arbitrarily in the range $0 \le r < R$, $r = R$ being excluded, because $\mathbf{L} \ge \mathbf{o}$, requiring $w \ge 0$.

Then, we have to calibrate the system.

There are two possibilities.

Either we choose a *numéraire* $i \in \{1, 2, \ldots, n\}$ and set the corresponding price $p_i = p_0$, where p_0 is a given exogenous variable (see PCMC, Par. 3), or we set for the national income $Y = Y_0$, as a given exogenous variable Y_0 (see PCMC, Par. 11).

So now, the *Sraffa price model* (4.174) has $n + 2$ equations and $n + 2$ variables and is solvable under the usual conditions of linear algebra. We further calculate

$$\begin{aligned} X &= \mathbf{q}'\mathbf{p} = \mathbf{e}'\mathbf{x}, \\ K &= \mathbf{e}'\mathbf{S}'\mathbf{p} = (\mathbf{Se})'\mathbf{p}, \\ P &= K \cdot r = (\mathbf{Se})'\mathbf{p} \cdot r, \\ w &= \frac{\tilde{w} \cdot Y}{L}, \\ \tilde{r} &= \frac{P}{Y}, \\ W &= Y - P = \mathbf{d}'\mathbf{p} - (\mathbf{Se})'\mathbf{p} \cdot r = (\mathbf{d}' - (\mathbf{Se})'r)\mathbf{p}. \end{aligned} \tag{4.175}$$

47 For calculation purposes, we have taken here the same commodity flow matrix $\mathbf{S}$ as above, although to be quite consistent this matrix should be modified to account for the fact that all labour is now regrouped within the second left hands terms (see Sraffa ([108], Par. 8, p. 10)).

Aggregating the first equation of (4.174) by left multiplication with the summation vector $\mathbf{e}$, we immediately obtain as expected the economic variables and identities:

$$\mathbf{e}'\left(\mathbf{S}'\mathbf{p}(1+r)+\mathbf{L}\frac{\tilde{w}\cdot Y}{L}\right)=(\mathbf{e}'\mathbf{S}')\mathbf{p}(1+r)+(\mathbf{e}'\mathbf{L})\frac{\tilde{w}\cdot Y}{L})=\mathbf{e}'\hat{\mathbf{q}}\mathbf{p},$$

$$X=(\mathbf{S}\mathbf{e})'\mathbf{p}(1+r)+(L)\frac{\tilde{w}\cdot Y}{L})=K+Kr+\tilde{w}Y=\mathbf{q}'\mathbf{p}, \tag{4.176}$$

and accordingly,

$$X=K+Kr+\tilde{w}Y=K+(P+W)=K+Y. \tag{4.177}$$

As an illustration, see again **Example 4.1.7.**

Now we convert the *single-product Sraffa system* (4.171) into units per commodities, multiplying equation (4.171) from the left by the diagonal matrix $\hat{\mathbf{q}}^{-1}$, obtaining, in analogy to equations (4.56) with the vector $\boldsymbol{\pi}$ of labor per *unit of commodities* (4.58) the *complete single-product Sraffa system*, expressed in *commodity units*, together with equations for national income Y and labour L. We get again $n+2$ equations for the $n+4$ variables r, $\tilde{w}$, L, p_1, p_2, ..., p_n and Y,

$$\boxed{\begin{aligned}\mathbf{C}'\mathbf{p}(1+r)+\boldsymbol{\pi}\cdot\frac{\tilde{w}\cdot Y}{L}&=\mathbf{p},\\ Y&=(\mathbf{q}'-(\mathbf{S}\mathbf{e})')\mathbf{p}=:\mathbf{d}'\mathbf{p},\\ L&=\mathbf{e}'\mathbf{L}.\end{aligned}} \tag{4.178}$$

We conclude this section with the calculation of the Sraffa prices in the case of *basic* and *non-basic* commodities, entering the economy.

Example 4.9.3. Consider the (5×5) semi-positive and not completely reducible matrix $\mathbf{S}$, the vector of labour $\mathbf{L}=[10,8,5,4,8]'>\mathbf{o}$, measured in appropriated units of labour, the vector of total output $\mathbf{q}=[5,4,6,2,2]'>\mathbf{o}$, constituting the production scheme $(\mathbf{S}',\mathbf{L})\to(\hat{\mathbf{q}})$ of Example 4.8.3. Calculate the input-output coefficients matrix $\mathbf{C}=\mathbf{S}\hat{\mathbf{q}}^{-1}$, the Frobenius number λ_C, $0<\lambda_C<1$, the productiveness $R>0$, choose an appropriate rate of profits r, $0<r<R$. Calculate the vector of relative prices $\mathbf{p}_0=[p_1,\ldots,p_5]'$, setting $p_1=1$, and show that $\mathbf{p}_0>\mathbf{o}$.

Solution to Example 4.9.3:

$$\mathbf{S}=\begin{bmatrix}3&2&0&0&0\\2&0&2&0&0\\0&2&0&4&0\\0&0&0&1&1\\0&0&0&0&0\end{bmatrix},\quad \mathbf{C}=\mathbf{S}\hat{\mathbf{q}}^{-1}=\begin{bmatrix}\frac{3}{5}&\frac{1}{2}&0&0&0\\\frac{2}{5}&0&\frac{1}{3}&0&0\\0&\frac{1}{2}&0&2&0\\0&0&0&\frac{1}{2}&\frac{1}{2}\\0&0&0&0&0\end{bmatrix}. \tag{4.179}$$

The (3×3) submatrix in the upper-left corner of matrix $\mathbf{C}$ is irreducible,

$$\mathbf{C}_{11} = \begin{bmatrix} \frac{3}{5} & \frac{1}{2} & 0 \\ \frac{2}{5} & 0 & \frac{1}{3} \\ 0 & \frac{1}{2} & 0 \end{bmatrix}, \quad (\mathbf{I}_3 + \mathbf{C}_{11})^2 = \begin{bmatrix} \frac{69}{25} & \frac{13}{10} & \frac{1}{6} \\ \frac{26}{25} & \frac{41}{30} & \frac{2}{3} \\ \frac{1}{5} & 1 & \frac{7}{6} \end{bmatrix} > \mathbf{0}. \tag{4.180}$$

The characteristic polynomial is

$$\begin{aligned} P_5(\lambda) &= -\frac{1}{600}\lambda(-30 + 170\lambda - 40\lambda^2 - 660\lambda^3 + 600\lambda^4) \\ &= \lambda(\lambda - 0.88639)(\lambda - 0.5)(\lambda - 22194)(\lambda + 0.50832). \end{aligned} \tag{4.181}$$

There exists a real, maximal, non-negative, eigenvalue $\lambda_C = 0.88639 < 1$ of the reducible matrix $\mathbf{C}$ (see Theorem A.10.1), which is equal to the Frobenius number of the irreducible (3×3) upper-left submatrix $\mathbf{C}_{11}$ (4.180) of matrix $\mathbf{C}$ (see Lemma A.10.1), $\lambda_{C_{11}} = 0.88639 < 1$. One then computes the *productiveness* $R = (1/0.88639) - 1 = 0.128176$. We choose as *the rate of profits* $r = 0.1 < R$. The Leontief Inverse exists, Lemma 2.4.1 and we calculate the inverse matrix, which also exists with $\lambda = 1/(1 + r) = 10/11 > \lambda_C = 0.88639$, Theorem A.10.2 and is morewover indeed a *semi-positive* matrix,

$$\left[\mathbf{I} - \frac{1}{\lambda}\mathbf{C}'\right]^{-1} = \begin{bmatrix} 27.123 & 14.494 & 8.222 & 0 & 0 \\ 18.686 & 11.552 & 6.353 & 0 & 0 \\ 6.852 & 4.236 & 3.330 & 0 & 0 \\ 33.497 & 20.707 & 16.278 & 2.222 & 0 \\ 18.423 & 11.389 & 8.953 & 1.222 & 1 \end{bmatrix} \geq \mathbf{0}. \tag{4.182}$$

We calculate the vector of prices $\mathbf{p}$ and the *wage per unit of labour* w with the *complete single-product Sraffa system* (4.57). With the positive vector of labour $\mathbf{L} > \mathbf{o}$, and the positive vector of labour per units of produced commodities, $\boldsymbol{\pi} = \hat{\mathbf{q}}^{-1}\mathbf{L} > \mathbf{o}$. Then due to Lemma 4.1.2 (b), the vector of prices $\mathbf{p}$ also is positive. Indeed,

$$\mathbf{C}'\mathbf{p}(1 + r) + \boldsymbol{\pi} \cdot w = \mathbf{p} \Rightarrow \mathbf{p} = \left[\mathbf{I} - (1 + 0.1)\mathbf{C}'\right]^{-1}\boldsymbol{\pi} \cdot w > \mathbf{o}. \tag{4.183}$$

We set $\mathbf{p}_0 = [1, p_2, p_3, p_4, p_5]'$. Consequently, there are *five* equations and *five* variables p_2, p_3, p_4, p_5, w in (4.183). We then obtain $w = 0.01099$ and the vector of relative prices $\mathbf{p}_0 = [1, 0.7228, 0.2742, 1.3893, 0.8080]'$. ▲

Throughout this section, we have recognised that positive price vectors for the *complete single-product Sraffa system* are guaranteed with *semi-positive reducible* or *irreducible input-output coefficients* matrices $\mathbf{C}$ under Lemma 4.1.2.

The labour value

See also Kurz and Salvadori [52], pp. 110–111, Adam Smith [106] and Schefold [102], p. 75. This is the right place to remember the historical notion of *labour value*. By *labour value*, one understands generally the "quantity of labour" embodied in a commodity. Let $\mathbf{u}$ be the vector of the quantities of labour embodied in the different units of commodities of the production economy. Consider the labour $\boldsymbol{\pi}$ of the present period, the labour $\mathbf{C}'\boldsymbol{\pi}$ of the period before, the labour $\mathbf{C}'^2\boldsymbol{\pi}$ of the preceding period and so on. As the Frobenius number is smaller than 1, $\lambda_C < 1$ because there is a surplus, there is convergence of the series,

$$\mathbf{u} = \boldsymbol{\pi} + \mathbf{C}'\boldsymbol{\pi} + \mathbf{C}'^2\boldsymbol{\pi} + \cdots = (\mathbf{I} - \mathbf{C}')^{-1}\boldsymbol{\pi} \Rightarrow \mathbf{u} = (\mathbf{I} - \mathbf{C}')^{-1}\boldsymbol{\pi}. \tag{4.184}$$

The notion *labour value* involves the idea that all the surplus goes to labour, therefore one sets $r = 0$ in (4.183) and compares the vectors $\mathbf{p} = (\mathbf{I} - \mathbf{C}')^{-1}\boldsymbol{\pi}$ and $\mathbf{u}$ (4.184). We obtain the parallelism $\mathbf{p} \parallel \mathbf{u} \Leftrightarrow \mathbf{p} = w \cdot \mathbf{u}$. This is the solution of Sraffa's *transformation problem*, see Bortis [7], pp. 67–68, namely the transformation of the vector of *labour values* $\mathbf{u}$ into the vector of prices $\mathbf{p}$, precisely valid for the case that all the surplus goes into labour.

For the Example 4.9.3 we then get the vector of labour values (4.184) in units of labour per unit of produced commodities, $\mathbf{u} = [19.95, 14.75, 5.75, 27, 17.5]$, where there is no rate of profits. There is actually non-parallelism, $\mathbf{u} \nparallel \mathbf{p}_0$, with the price vector $\mathbf{p}_0$ because we have computed the price vector $\mathbf{p}_o$, using in (4.183) the positive rate of profits $r = 0.1 > 0$, instead of $r = 0$ as would have been required to get (4.184).

4.10 A digression on growth, constant rate of growth

In this section, we are dealing with an extension of Sraffa's approach which assumes no change over time in the structure of the means of production, however with the possibility of a uniform change in scale over the medium term. The growth of an economy is not mentioned in PCMC. As we have seen, on the one hand, PCMC concerns one reporting period. On the other hand, the centre piece of Sraffa's price model are the conditions of production (Definition 3.1.1, presented in Subsection 3.1.1). Production systems subject to those conditions are in a self-replacing state, having produced a surplus absorbed by demand on the *interindustrial market*. At the end of the period, the system has all available commodities required for interindustry transactions in the next following period. This opens the way to consider successive production periods evolving over time with the possibility of envisaging *growth* in economic output.

Growth is either understood in the present context as an increase in total output $\mathbf{q}$ (including GDP or national income) sector by sector, or, growth operates through a uniform *rate of growth*, acting on the means of production, the surplus, the output and the national income, but keeping the whole structure unchanged.

We assume that labour is incorporated in the matrix **S** with subsistence wages, $w = 0$. Thus, the two growth situations can then be envisaged in this context:

(a) Consider the total output q_j per industry S_j and surplus d_j expressed in quantity terms. The *sectorial surplus ratios* are defined as follows:

$$R_j = \frac{d_j}{(q_j - d_j)}, \quad g = \min_j\{R_j, j = 1, \ldots, n\} \geq 0. \tag{4.185}$$

In general, the *sectorial surplus ratios* vary from industry to industry, but over the medium term there exists the minimum g with the result that for $\Delta R_j = R_j - g$ at least one ΔR_j will be equal to 0.

(b) Now, we consider the obtained term $g \geq 0$ (4.185) as the maximal uniform medium-term *rate of growth* of the production process from one period to the next. The rate g has to increase the means of production, meaning that the components of the matrix $\mathbf{S}'$ are multiplied by the same factor,

$$\mathbf{S}' \to (1 + g)\mathbf{S}', \tag{4.186}$$

the surplus, the national income Y and some other economic variables (K, X, ...). The structure of the technology is unchanged. The commodity flow matrix **S** is just multiplied by a factor, and the proportions remain unchanged. The operating uniform medium term *rate of growth* may be smaller than g.

In the case of growth, a uniform proportion of surplus, up to a maximum g, is allotted to the means of production for increased production in the following periods, and the rest of current surplus corresponding to residuals ΔR_j is going to exogenous demand. The following examples illustrate this concept, considering at first closed economies.

Example 4.10.1. Assume an economy composed of $n = 3$ industries with following sectorial rates of growth: $R_1 = 16\,\%$, $R_2 = 12\,\%$, $R_3 = 20\,\%$. Compute the maximal rate of growth g. Compute the contribution of the sectors to the growth, relative to the sectorial rate of growth.

Solution to Example 4.10.1:
We obtain: $g = \min\{R_1, R_2, R_3\} = 0.12$. Moreover: $\Delta R_1 = R_1 - g = 0.04$, $\Delta R_2 = R_2 - g = 0$, $\Delta R_3 = R_3 - g = 0.08$, and further, $g_1 = (R_1 - \Delta R_1)/R_1 = 0.75$, $g_2 = (R_2 - \Delta R_2)/R_2 = 1$, $g_3 = (R_3 - \Delta R_3)/R_3 = 0.6$. ▲

Example 4.10.2. Medium term rate of growth. Consider Sraffa's third example (PCMC, Par. 5), Example 3.1.3, (3.41), with two industrial sectors, modified by an additional production of 2t of iron, only with subsistence wages, $w = 0$:

The numéraire is wheat, so its price is $p_1 = 1$. Compute the input-output coefficients matrix $\mathbf{C} = \mathbf{S}\hat{\mathbf{q}}^{-1}$, its Frobenius number λ_C and the productiveness $R = (1/\lambda_C) - 1$. Establish the production scheme of the economy.

Compute the price vector $\mathbf{p}$, the sectorial surplus ratios (4.185) R_1 and R_2. Establish the maximal rate of growth g. Compute the vector of additional surplus $\Delta\mathbf{d} = \mathbf{d}(1+g) - \mathbf{d} = g\mathbf{d}$, as well as the national incomes $Y_1 = Y$, $Y_2 = Y_1(1+g)$ and the additional national income $\Delta Y = Y_2 - Y_1 = g \cdot Y_1$. Analyse the sequences $(1+g)^{(k-1)}\mathbf{S}$, $(1+g)^{(k-1)}\mathbf{Se}$, $(1+g)^{(k-1)}\mathbf{d}$ and $(1+g)^{(k-1)}\mathbf{q}$, $k \in \mathbb{N}$.

Solution to Example 4.10.2:
We can now identify

$$\mathbf{S} = \begin{bmatrix} 280 & 120 \\ 12 & 8 \end{bmatrix}, \quad \mathbf{Se} = \begin{bmatrix} 280 & 120 \\ 12 & 8 \end{bmatrix} \begin{bmatrix} 1 \\ 1 \end{bmatrix} = \begin{bmatrix} 400 \\ 20 \end{bmatrix},$$
$$\mathbf{q} = \begin{bmatrix} 575 \\ 22 \end{bmatrix}, \quad \mathbf{d} = \mathbf{q} - \mathbf{Se} = \begin{bmatrix} 575 \\ 22 \end{bmatrix} - \begin{bmatrix} 400 \\ 20 \end{bmatrix} = \begin{bmatrix} 175 \\ 2 \end{bmatrix}. \tag{4.187}$$

As there is no explicit labour, which is at subsistence level included in $\mathbf{S}'$, the production scheme is as follows: $(\mathbf{S}', \mathbf{o}) \to (\hat{\mathbf{q}})$. We calculate the *input-output coefficients* matrix,

$$\mathbf{C} = \mathbf{S}\hat{\mathbf{q}}^{-1} = \begin{bmatrix} 280 & 120 \\ 12 & 8 \end{bmatrix} \begin{bmatrix} \frac{1}{575} & 0 \\ 0 & \frac{1}{22} \end{bmatrix} = \begin{bmatrix} \frac{56}{115} & \frac{60}{11} \\ \frac{12}{575} & \frac{4}{11} \end{bmatrix}, \tag{4.188}$$

and calculate the characteristic polynomial,

$$P_2(\lambda) = \det(\mathbf{C} - \lambda\mathbf{I}) = \lambda^2 - \frac{1076}{1265}\lambda + \frac{16}{253} = (\lambda - 0.082316)(\lambda - 0.768277). \tag{4.189}$$

Thus, the Frobenius number is $\lambda_C = 0.768277 =: 1/(1+R) < 1$, and the *productiveness* $R = (1/\lambda_C) - 1 = (1/0.768277) - 1 = 0.3016 > 0$. The Sraffa price model is set up, leading to an eigenvalue equation, as all the surplus goes to profits and there are no wages, $w = 0$. Wheat is the numéraire, giving the price vector $\mathbf{p} = [1, p_2]'$,

$$\boxed{\begin{aligned} &\mathbf{S}'\mathbf{p}(1+R) = \hat{\mathbf{q}}\mathbf{p} \Rightarrow \mathbf{C}'\mathbf{p} = \frac{1}{1+R}\mathbf{p} = \lambda_C\mathbf{p}, \\ &\begin{bmatrix} \frac{56}{115} & \frac{60}{11} \\ \frac{12}{575} & \frac{4}{11} \end{bmatrix} \begin{bmatrix} 1 \\ p_2 \end{bmatrix} = 0.768277 \begin{bmatrix} 1 \\ p_2 \end{bmatrix}. \end{aligned}} \tag{4.190}$$

We get the vector of relative prices $\mathbf{p} = [1, 13.48]'$. With the vector of surplus $\mathbf{d} = [175, 2]'$, we obtain the *national income* of period 1,

$$Y_1 := Y = \mathbf{d}'\mathbf{p} = [175, 2]' \begin{bmatrix} 1 \\ 13.48 \end{bmatrix} = 201.96. \tag{4.191}$$

The sectorial surplus ratios (4.185) are: $R_1 = d_1/(q_1 - d_1) = 175/400 = 0.4375$ and $R_2 = d_2/(q_2 - d_2) = 2/20 = 0.1$, so the over all *maximal rate of growth* is here $g =$

$\min\{R_1, R_2\} = 0.1$, giving the opportunity of a uniformly growing production system. We present therefore the model of a growing economy with the corresponding maximal multiplier $(1+g) = 1.1$, starting from the actual period, getting number 1,

$$((1+g)^{(k-1)}\mathbf{S}', \mathbf{o}) \to ((1+g)^{(k-1)}k\hat{\mathbf{q}})', \quad k \in \mathbb{N},$$
$$\mathbf{S}'(1+g)^{(k-1)}\mathbf{p}(1+R) = \hat{\mathbf{q}}(1+g)^{(k-1)}\mathbf{p} \Rightarrow \mathbf{C}'\mathbf{p} = \lambda_C\mathbf{p}. \tag{4.192}$$

For this uniformly growing system, we get exactly the upper eigenvalue equation (4.190), leading to the same price vector $\mathbf{p} = [1, 13.48]'$. We present just the national income Y_1 of the second period,

$$Y_2 = (1+g) \cdot Y_1 = 1.1 \cdot 201.96 = 222.15. \tag{4.193}$$

We are here in presence of a closed economy, the factors $(1+g)^{(k-1)}$ being the uniform *dilatation factors* leading from period 1 to the k-th period, all sectors growing at the same rate $(1+g)$ from one period to the next. We remember, all profits going to increase the *means of production*, the national income Y, the circulating capital K, salaries kept at subsistence level.

We compute for the next period the surplus difference $\Delta\mathbf{d} = \mathbf{d}(1+g) - \mathbf{d} = g\mathbf{d}$, the national incomes Y_2 and the national income difference $\Delta Y = Y_2 - Y_1$,

$$\Delta\mathbf{d} = \mathbf{d}(1+0.1) - \mathbf{d} = \begin{bmatrix} 192.5 \\ 2.2 \end{bmatrix} - \begin{bmatrix} 175 \\ 2 \end{bmatrix} = \begin{bmatrix} 17.5 \\ 0.2 \end{bmatrix},$$
$$\Delta Y = g \cdot 201.96 = 0.1 \cdot Y_1 = 20.196. \tag{4.194}$$

Finally, we have an outlook on the development over the sequence of the periods with the *rate of growth* $g = 0.1$, starting with the actual period 1. Let's for illustration present the commodity flow matrices for the periods $k = 0, 1, 2$,

$$\mathbf{S} = \begin{bmatrix} 280 & 120 \\ 12 & 8 \end{bmatrix}, \quad 1.1\mathbf{S} = \begin{bmatrix} 308 & 132 \\ 13.2 & 8.8 \end{bmatrix}, \quad (1.1)^2\mathbf{S} = \begin{bmatrix} 338.8 & 145.2 \\ 14.52 & 9.68 \end{bmatrix}. \tag{4.195}$$

Then, the total amount of the commodities, necessary as the means of production also augment,

$$\mathbf{Se} = \begin{bmatrix} 400 \\ 20 \end{bmatrix}, \quad 1.1\mathbf{Se} = \begin{bmatrix} 440 \\ 22 \end{bmatrix}, \quad (1.1)^2\mathbf{Se} = \begin{bmatrix} 484 \\ 24.2 \end{bmatrix}, \tag{4.196}$$

as well as the surplus of each period k and the vectors of total output,

$$\mathbf{d} = \begin{bmatrix} 175 \\ 2 \end{bmatrix}, \quad 1.1\mathbf{d} = \begin{bmatrix} 192.5 \\ 2.2 \end{bmatrix}, \quad (1.1)^2\mathbf{d} = \begin{bmatrix} 221.75 \\ 2.42 \end{bmatrix},$$
$$\mathbf{q} = \begin{bmatrix} 575 \\ 22 \end{bmatrix}, \quad 1.1\mathbf{q} = \begin{bmatrix} 632.5 \\ 24.2 \end{bmatrix}, \quad (1.1)^2\mathbf{q} = \begin{bmatrix} 695.75 \\ 26.62 \end{bmatrix}. \tag{4.197}$$

Of course, there are the sums, remaining,

$$(1.1)^{(k-1)}\mathbf{q} = (1.1)^{(k-1)}\mathbf{Se} + (1.1)^{(k-1)}\mathbf{d}, \quad k \in \mathbf{N}. \quad \blacktriangle \tag{4.198}$$

We have up to now considered a closed economic. We now consider an open system with exports, in this case namely gold, where the surplus is exclusively constituted by the exports. We then analyse the possible *rates of growth*.

Example 4.10.3. Consider the three sector economy described by Table 4.6, producing wheat, iron and gold. There is a semi-positive commodity flow matrix $\mathbf{S} \geq \mathbf{0}$. The surplus is generated by the third gold-producing sector. Therefore the vector of surplus is $\mathbf{d} = [0, 0, 50]$. Wheat is the numéraire.

Table 4.6: An economy with surplus exclusively in exports.

An open economy				
sectors	**1**	**2**	**3**	**total output**
wheat *in tons*	280	180	115	575
iron *in tons*	12	12	6	30
gold *in kg*	0	0	0	50

Compute the positive input-output coefficients matrix $\mathbf{C} = \mathbf{S}\hat{\mathbf{q}}^{-1}$, the Frobenius number λ_C and the vector of prices $\mathbf{p}$ and the productiveness $R = (1/\lambda_C) - 1$. Determine the possible rates of growth and discuss productiveness and growth.

Solution to Example 4.10.3:
We identify the matrix $\mathbf{S} \geq \mathbf{0}$ and compute the vector of total output $\mathbf{q} = \mathbf{Se} + \mathbf{d}$:

$$\mathbf{S} = \begin{bmatrix} 280 & 180 & 115 \\ 12 & 12 & 6 \\ 0 & 0 & 0 \end{bmatrix}, \quad \mathbf{q} = \begin{bmatrix} 280 & 180 & 115 \\ 12 & 12 & 6 \\ 0 & 0 & 0 \end{bmatrix}\begin{bmatrix} 1 \\ 1 \\ 1 \end{bmatrix} + \begin{bmatrix} 0 \\ 0 \\ 50 \end{bmatrix} = \begin{bmatrix} 575 \\ 30 \\ 50 \end{bmatrix}. \tag{4.199}$$

We start calculating the *input-output coefficients* matrix $\mathbf{C} = \mathbf{S}\hat{\mathbf{q}}^{-1}$:

$$\mathbf{C} = \begin{bmatrix} 280 & 180 & 115 \\ 12 & 12 & 6 \\ 0 & 0 & 0 \end{bmatrix}\begin{bmatrix} \frac{1}{575} & 0 & 0 \\ 0 & \frac{1}{30} & 0 \\ 0 & 0 & \frac{1}{50} \end{bmatrix} = \begin{bmatrix} \frac{56}{115} & 6 & \frac{23}{10} \\ \frac{12}{575} & \frac{2}{5} & \frac{3}{25} \\ 0 & 0 & 0 \end{bmatrix}, \tag{4.200}$$

giving the characteristic equation,

$$P_2(\lambda) = \det(\mathbf{C} - \lambda\mathbf{I}) = \lambda^3 - \frac{102}{115}\lambda^2 + \frac{8}{115}\lambda = \lambda\left(\lambda - \frac{2}{23}\right)\left(\lambda - \frac{4}{5}\right). \tag{4.201}$$

Thus, the Frobenius number is $\lambda_C = 0.8$. We derive then the following *Sraffa price model*,

$$\mathbf{S}'(1+R)\mathbf{p} = \hat{\mathbf{q}}\mathbf{p} \Rightarrow \hat{\mathbf{q}}^{-1}\mathbf{S}'(1+R)\mathbf{p} \Rightarrow \mathbf{C}'\mathbf{p} = \frac{1}{1+R}\mathbf{p} = \lambda_C\mathbf{p}. \quad (4.202)$$

The productiveness is $R = (1/0.8) - 1 = 0.25$. The price vector is $\mathbf{p} = [1, 15, 4.125]$. The prices are expressed in *tons of* wheat per unit of commodities. The sectorial surplus ratios $R_1 = 0/(575-0) = 0$, $R_2 = 0/(30-0) = 0$, $R_3 = 50/(50-50) = \infty$ are computed. There is $g = \min\{R_1, R_2, R_3\} = 0$. As the only possible rate of growth is $g = 0$, we are in presence of a *productive sustainable economy* without growth.

We observe that both first sectors of wheat and iron production do not contribute to surplus, only the export sector 3 contributes to the production of surplus and generates the *productiveness* $R = 0.25$, which benefits all the three sectors. Nevertheless, no positive uniform rate of growth is possible. There is $g = R_1 = R_2 = 0$. ▲

Finally, we have been considering *single-product industries*. The foregoing approach may be extended to *joint production Sraffa systems*, Chapter 6.

Some growth properties of the Sraffa price model.
In general, the sectorial surplus ratios $R_j = d_j/(q_j - d_j) \geq 0, j = 1, \ldots, n$, are different one from each other. If there is a positive minimum $g = \min\{R_1, \ldots, R_n\} > 0$, it is possible to formulate a Sraffa model with *economic growth*. Its effective rate of growth lies in the interval $[0, g]$.

4.11 The ancestor: Quesnay and the economy as a circular process

An item not considered up to now appears in this historical construct: *land*. We ask the reader to take *land* as it is presented here, while it will be examined in more detail in Section 6.7. Contrary to Sraffa's assertion, *productive land* is considered as an indispensable constituent for agriculture and as such directly attached, like the *labour force*, to every production entity. It does not normally belong to the category of commodities, and thus cannot be classified as *non-basic* in this context and obviously in what follows in connection with Quesnay's economic model. The physician and economist F. Quesnay (1694–1774) already clearly understood in the 18th century that no economy of production can exist without these two basic pillars of production.[48]

(1) Historical aspects. Figure 4.17 presents one of the various *Tableaux Economiques* elaborated by F. Quesnay.[49] Here the tableau presents agricultural production and is called "Tableau de la Philosophie Rurale".

48 Productive land for food and energy and equipped land as prerequisites for building purposes and industrial infrastructures.

49 Source: https://www.google.ch/search?q=tableaux+économiques+Quesnay&biw

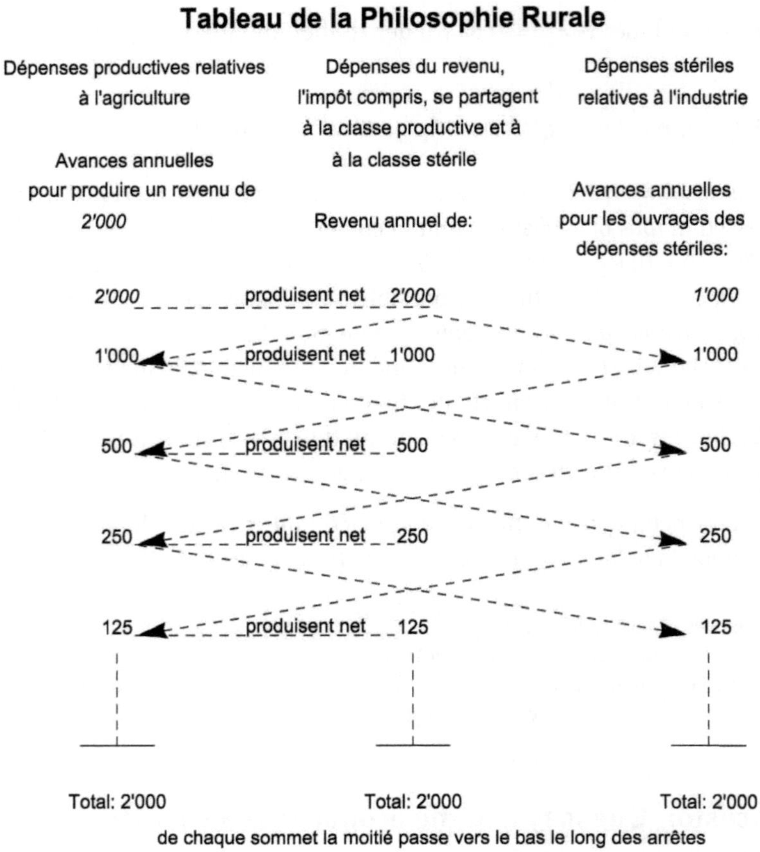

Figure 4.17: One of Quesnay's Tables on agricultural production.

(2) Sraffa Network. The following example is inspired by the foregoing "Tableau de la Philosophie Rurale", where we start by considering *commodity flows*, because we want to construct a *Sraffa Network*. Contrary to the introductory remarks, land, in the hands of landowners, is considered in Quesnay's scheme as a separate commodity put at the disposal of farmers by these landowners against the rent paid. Land thus figures here exceptionally as a commodity in the corresponding Sraffa Network. This reflects prevailing social structures in the 18th century.

The idea behind Quesnay's tableau is twofold:

(a) A circular economic process involving farmers ($\mathcal{F}$), manufacturers and merchants ($\mathcal{M}$) and landowners ($\mathcal{L}$).

(b) A cumulative growth process in total output fuelled by successive autonomous expenditures by landowners.

We shall start by treating (a).

Before proceeding, let us summarise what Quesnay's Tableau illustrates using his figures which are expressed in value terms. Landowners are supposed to enter the initial accounting period with an agricultural income of 2,000 built up in previous periods (say years) which they use to purchase goods from manufactures and merchants (1,000) and to purchase goods from farmers (1,000).

In the new accounting period, farmers have a total output of 5,000 used as follows:

- 2,000 retained in their sector to ensure sustainability;
- 1,000 spent on purchases from manufactures and merchants;
- 2,000 paid as rents due to landowners.

Landowners have autonomous incomes, part of which are constituted by these agricultural rents. To boost agricultural production, considered at the time as the driving component of production, they can decide to distribute part of their total income to the two other sectors, expressed here as 50 % of their initial agricultural income, i. e., 1,000 (see paragraph *(6) Innovations of the Quesnay model* below), thus increasing output. They continue similarly with their incremental agricultural incomes in successive periods.

Example 4.11.1. Quesnay's economic model involving farmers $\mathcal{F}$, manufacturers $\mathcal{M}$ and landowners $\mathcal{L}$, may be represented as follows by a Sraffa Network, representing the **structure of the commodity flows**: The causal links should be read as follows:

Farmers $\mathcal{F}$: produce corn "C" which is required by

- the farmers themselves for agricultural purposes
- manufacturers for their living
- landowners for various uses

Manufacturers and merchants $\mathcal{M}$: produce goods and services "m" required by

- farmers for their activities
- landowners for various uses (buildings, gardens, weapons, works of art, ...)

Landowners $\mathcal{L}$: provide land "l" for farmers

By simplification, as $\mathcal{F}$, $\mathcal{M}$, $\mathcal{L}$ provide only one type of commodity each, the Sraffa Network, Figure 4.18, can be visually compressed into a simpler connected digraph, Figure 4.19, with the corresponding adjacency matrix $\mathbf{W}_3$. Identifying node-pairs of the Sraffa Network with one node as follows: $(\mathrm{C}, \mathcal{F}) \to \mathcal{F}$, $(m, \mathcal{M}) \to \mathcal{M}$, $(l, \mathcal{L}) \to \mathcal{L}$, the three arrows $(\mathcal{F} \to \mathrm{C})$, $(\mathcal{M} \to \mathrm{m})$, $(\mathcal{L} \to \mathrm{l})$ disappear by compression, maintaining the same connectivity, giving the digraph, representing the structure of the commodity flows.

Quesnay was a physiocrat, and in his view agriculture was the motor of the economy: landowners (in his time mainly the aristocracy and the clergy), accumulating land and playing a distributive role, manufacturers just providing the necessaries for

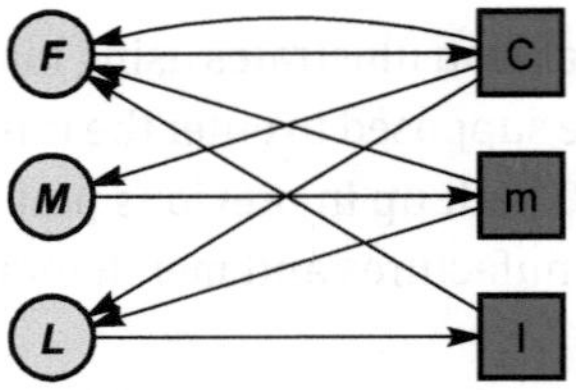

Figure 4.18: Quesnay's economic model (logical structure).

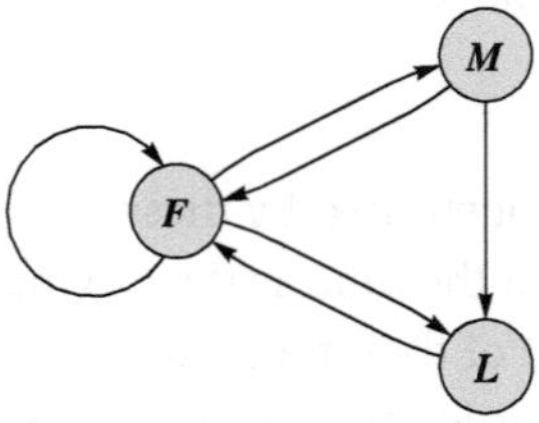

$$\mathbf{W}_3 = \begin{matrix} & F & M & L \\ F & 1 & 1 & 1 \\ M & 1 & 0 & 1 \\ L & 1 & 0 & 0 \end{matrix} . \tag{4.203}$$

Figure 4.19: Associated digraph of the *commodity connection Sraffa Network*, Figure 4.18, and the adjacency matrix $\mathbf{W}_3$ of Quesnay's model 4.11.1.

agriculture and other non-manufacturing activities. This changed radically with the industrial revolution and free trade, the manufacturing and mercantile class becoming the motor of the economy, hence the *dashed arrow* in Figure 4.20 is not considered by Quesnay.

Now starting from the above representation (4.203) and Figure 4.19, Quesnay's original circular process of *commodity flow* is converted into a new directed graph, depicting **payment flows** in relative terms, involving the three sectors: *farmers*, *landowners* and *manufacturers*, Figure 4.20, in a closed economy, starting with an initial accounting period. The matrix of the *payment flows* is the transpose of the *commodity flow* matrix in monetary terms $\mathbf{Z}_3 = (z_{ij})$, $i, j = 1, \ldots, 3$, composed of the *intersectorial transfers* z_{ij} by sector S_i to sector S_j, using Quesnay's values. Its *adjacency matrix* appears here, represented by $\mathbf{W}_3$ (4.203). Matrix $\mathbf{Z}'_3 = (z_{ji})$ then represents the *payment flows* by sector S_j to sector S_i in Figure 4.20.[50]

The understanding of Figure 4.17 is not as trivial as that. We follow the interpretation given by G. Gilibert [36]. Quesnay[51] thus defined a sustainable reallocation process, represented here by the flow commodity matrix $\mathbf{Z}'_3$, Figure 4.20.

Sector 1: Farmers ($\mathcal{F}$) reallocate part of their production of *corn* "C" to sustain the productive capacities ($z_{11} = 2{,}000$, in value terms) of their sector, thus generating

50 This digraph and the numbers indicated are taken form Gilibert [36]. Quesnay's original figures for the initial period of activity, Figure 4.17, are expressed in thousands of units. In Figure 4.20, the figures have been scaled down by 10^{-3} in order to avoid overloading the presentation; the initial relative proportions are maintained.

51 The economy's initial accounting relationships read as follows, noted in $k = 1{,}000$:

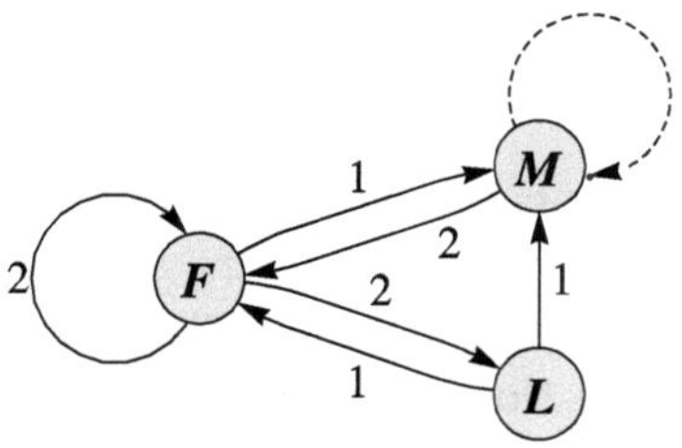

$$\mathbf{Z}_3' = \begin{array}{c} \\ F \\ M \\ L \end{array} \begin{array}{c} \begin{array}{ccc} F & M & L \end{array} \\ \begin{bmatrix} 2{,}000 & 1{,}000 & 2{,}000 \\ 2{,}000 & 0 & 0 \\ 1{,}000 & 1{,}000 & 0 \end{bmatrix} \end{array} \tag{4.204}$$

Figure 4.20: Digraph of the *payment flows* in Figure 4.19, and *payments flow matrix* of Example 4.11.1.

inner-sectorial payments and incomes. The remaining production, in value terms, is used to pay rents due to landowners ($\mathcal{L}$) ($z_{13} = 2{,}000$) and to pay for purchases of *manufactured goods* "m" from manufacturers ($\mathcal{M}$) ($z_{12} = 1{,}000$). The total production of the "productive" sector is thus valued at ($z_{11} + z_{12} + z_{13} = 5{,}000$).

Sector 2: Manufacturers ($\mathcal{M}$) of the "idle" sector use their proceeds from payments by farmers ($\mathcal{F}$) ($z_{21} = 1{,}000$) and landowners ($\mathcal{L}$) ($z_{23} = 1{,}000$) to pay for agricultural products ($z_{12} = 2{,}000$) for their living. The total production of this sector is thus valued at ($z_{21} + z_{23} = 2{,}000$).

Sector 3: Landowners ($\mathcal{L}$) of the "distributive" sector use all their income, from rents due for providing *land* "l" to farmers ($z_{31} = 2{,}000$), to pay for the purchase of agricultural commodities, *corn* "C" ($z_{13} = 1{,}000$) and to pay for *manufactured goods* "m" ($z_{23} = 1{,}000$) to ensure their standard of living. The total value of disposable capital of this sector is thus equal to ($z_{13} = 2{,}000$) to cover those payments.

$$\mathbf{Z}_3 = \begin{bmatrix} 2{,}000 & 2{,}000 & 1{,}000 \\ 1{,}000 & 0 & 1{,}000 \\ 2{,}000 & 0 & 0 \end{bmatrix}, \quad \mathbf{Z}_3\mathbf{e} = \begin{bmatrix} 5{,}000 \\ 2{,}000 \\ 2{,}000 \end{bmatrix}. \tag{4.205}$$

Thus we get the digraph, Figure 4.20, the arrows indicating the payment flow, and the transposed *commodity matrix* $\mathbf{Z}_3' = (z_{ji})$, also indicating the payment from sector S_j to sector S_i.

(3) The Quesnay model following Sraffa. Let us now proceed to represent Quesnay's model following Sraffa's approach, using the *payments flow matrix* $\mathbf{Z}_3' = (z_{ji})$, where

Total Output ($\mathcal{F}$) $= 5k$	total income ($\mathcal{F}$) $= 5k$;	($\mathcal{F}$) requires $2k$ for stainability; Surplus $= 3k$
Total Output ($\mathcal{M}$) $= 2k$	total income ($\mathcal{M}$) $= 2k$;	No surplus
Total Output ($\mathcal{L}$) $= 2k$	total income ($\mathcal{L}$) $= 2k$;	No surplus, but redistribution of income
$9k$	$9k$	

Landowners play the role of redistribution of income that governments are supposed to play today.

the coefficients z_{ji} indicate the value of payments by sector S_j for the obtention of commodities from sector S_i.

For example, $z_{11} = 2{,}000$ are inner sectorial payments/incomes received in the "productive" agricultural sector to ensure sustainability of that sector; z_{32} are payments received by the "idle" class of manufacturers from the "distributive" sector of landowners for purchases of manufactured commodities. Now, the coefficients z_{ij} are composite entities: $z_{ij} = p_i s_{ij}$ (2.18) where $p_i > 0$ is the normally *positive* price of commodity i (expressed in a given numéraire not specified here) and s_{ij} the quantity of commodity i delivered to sector S_j.

A further example: $z_{13} = 2{,}000$ is the value of the commodities received by the "distributive" sector of landowners as rent from the "productive" agricultural sector for providing land to the latter, with quantity and price unspecified, the two only subject to the constraint that their product equals 2,000.

(4) Sraffa's conditions of production. As the Quesnay model works in *monetary terms* and Sraffa in *physical terms*, we introduce an artificial vector of prices $\mathbf{p} = \mathbf{e}$, to enable the passage to a Sraffa system. Now it can be shown that Quesnay's model fulfills Sraffa's *conditions of production*, Definition 3.1.2. There is no surplus and the Frobenius number is $\lambda_C = 1$. We resort to the known matrix relations (2.113) for matrix $\mathbf{Z}_3$ (4.204), using here $\hat{\mathbf{p}} = \hat{\mathbf{e}} = \mathbf{I}$,

$$\mathbf{Z}_3 = \hat{\mathbf{p}}\mathbf{S} = \hat{\mathbf{e}}\mathbf{S} = \mathbf{S}; \quad \mathbf{q} = \mathbf{S}\mathbf{e}; \quad \mathbf{C} = \mathbf{S}\hat{\mathbf{q}}^{-1}, \tag{4.206}$$

and proceed to

Example 4.11.2. Compute matrix $\mathbf{C}$ (4.206) and show that Sraffa's conditions of production are fulfilled. (All the entries of the matrix $\mathbf{S}$ are noted in $k = 1{,}000$!)

Solution to Example 4.11.2:
Matrix $\mathbf{S}$ is *semi-positive* and *irreducible*, in application of Lemma A.8.2,

$$\mathbf{S} = \begin{bmatrix} 2 & 2 & 1 \\ 1 & 0 & 1 \\ 2 & 0 & 0 \end{bmatrix}; \quad (\mathbf{I} + \mathbf{S})^2 = \begin{bmatrix} 13 & 8 & 6 \\ 6 & 3 & 3 \\ 8 & 4 & 3 \end{bmatrix} > 0.$$

$$\mathbf{q} = \mathbf{S}\mathbf{e} = \begin{bmatrix} 2 & 2 & 1 \\ 1 & 0 & 1 \\ 2 & 0 & 0 \end{bmatrix} \begin{bmatrix} 1 \\ 1 \\ 1 \end{bmatrix} = \begin{bmatrix} 5 \\ 2 \\ 2 \end{bmatrix} > \mathbf{o}. \tag{4.207}$$

Then, we compute matrix $\mathbf{C}$

$$\mathbf{C} = \mathbf{S}\hat{\mathbf{q}}^{-1} = \begin{bmatrix} 2 & 2 & 1 \\ 1 & 0 & 1 \\ 2 & 1 & 0 \end{bmatrix} \begin{bmatrix} \frac{1}{5} & 0 & 0 \\ 0 & \frac{1}{2} & 0 \\ 0 & 0 & \frac{1}{2} \end{bmatrix} = \begin{bmatrix} \frac{2}{5} & 1 & \frac{1}{2} \\ \frac{1}{5} & 0 & \frac{1}{2} \\ \frac{2}{5} & 0 & 0 \end{bmatrix} \geq \mathbf{O}, \tag{4.208}$$

its characteristic polynomial and the Frobenius number,

$$P_3(\lambda) = \det(\mathbf{C} - \lambda\mathbf{I}) = \frac{1}{5} + \frac{2}{5}\lambda + \frac{2}{5}\lambda^2 - \lambda^3 = (\lambda - 1)(\lambda - \lambda_2)(\lambda - \lambda_3). \tag{4.209}$$

The Frobenius number is $\lambda_C = 1$. This means that the *Sraffa conditions of production*, Definition 3.1.2, are fulfilled, as may also be seen by summation of the columns components of **S**. ▲

(5) The Quesnay economic model contains only basic commodities. The commodities corn "C", manufactured goods "m" and land "l" are basics because the corresponding Sraffa Network, Figure 4.18, is strongly connected, Definition A.14.9, respectively matrix **S** is *irreducible*, Lemma A.15.3.

This establishes a historical bridge between Quesnay and Sraffa.

(6) Innovations of the Quesnay model.

(a) Quesnay's first innovation. The system is, in a first reference period, in a self-replacing state with constant levels of production and a fixed value of required means of production, set in this example at 2. In the original "Tableau", the corresponding number is 2,000, see Figure 4.17 and Quesnay [88].

(b) Quesnay's second innovation is to incorporate the effect of growth. His reasoning in our modern notation can be summarised as follows, based on his original figures:

The landowner sector starts at some chosen point of time with income valued at $G = 2{,}000$ built up in the past, and farmers with an endowment, also built up in the past, which also ensures the sustainability of their agricultural production. This endowment also equals $G = 2{,}000$. Now landowners have autonomous incomes including their land rents. They reallocate an additional part $g \cdot G$ of these incomes to the other sectors which then operate in the next period with means of production valued at $G + G \cdot g$. Landowners then reallocate in the following period $(G \cdot g)g = G \cdot g^2$ and the sectors operate in the next period with cumulative means of production, increasing to the level at $G + G \cdot g + G \cdot g^2$, etc.

Period after period, the total means of production of the "Agriculture and Manufacture/Merchandise" sector will in fact accordingly increase proportionally according to the following scheme of cumulative autonomous expenditures by landowners:

$$\begin{aligned}
&\text{Period } 1: \quad G,\\
&\text{Period } 2: \quad G + G \cdot g,\\
&\text{Period } 3: \quad G + G \cdot g + G \cdot g^2,\\
&\dots\\
&\text{Period } n: \quad G + G \cdot g + G \cdot g^2 + \dots \to \frac{1}{(1-g)} \cdot G,
\end{aligned} \tag{4.210}$$

the whole economy following in the same proportions.

We have here a geometric series which converges to $\frac{G}{(1-g)}$, $0 \leq g < 1$.

$$S = G + G \cdot g + G \cdot g^2 + \cdots = G \cdot \frac{1}{1-g} = 2{,}000 \cdot \frac{1}{1-0.5} = 4{,}000. \qquad (4.211)$$

As we observe, Quesnay in his "Tableau Economique", Figure 4.17, having set $G = 2{,}000$ and $g = 0.5$, obtains the series, $2{,}000 + 1{,}000 + 500 + 250 + 125 + \cdots$ converging to 4,000, for landowners and farmers, respectively to 2,000 for the idle class. So in the end total agricultural output, measured in payment terms, rises from 5,000 (2,000 + 1,000 + 2,000) to 10,000.

(7) Completing the Quesnay economic model. With the manufacturing class **M** becoming the center piece of the economy, involving interindustry requirements, a new positive term in the diagonal would then enter the adjacency matrix completing Quesnay's scheme: **M** → **M** (dashed arrow) in Figure 4.20. The resulting adjacency matrix

$$\overline{\mathbf{W}}_3 = \begin{bmatrix} 1 & 1 & 1 \\ 1 & 1 & 1 \\ 1 & 0 & 0 \end{bmatrix} \qquad (4.212)$$

is an extension of the adjacency matrix $\mathbf{W}_3$ (4.203) and represents therefore also only *basic* commodities.

Finally, in this connection one also observes by modern standards that in the event landowners do not reinvest in the real productive economy or are not themselves entrepreneurs, they then constitute, contrary to what Quesnay stated in the 18th century context, an idle class living on the land. Economically their land provides their basic income, without labour on their part, and they invest for example in building works of art and the financial markets (remember John Law de Lauriston (1671–1729)!).

5 Sraffa's Standard system and the Standard commodity

What are the *Standard system* and *Standard commodity* all about?

Sraffa explains in PCMC, Chapter IV, Par. 23 and 24, the reasons for seeking a so-called *Standard commodity* and subsequently shows how to devise a system of production resulting in such a commodity for single-product industries. He introduces this chapter by observing:

> *"The necessity of having to express the price of one commodity in terms of another which is arbitrarily chosen as standard, complicates the study of price movements which accompany a change in distribution. It is impossible to tell of any particular price fluctuation whether it arises from the peculiarities of the commodity which is being measured or from those of the measuring standard".*

By "distribution", Sraffa refers to the distribution of surplus between wages and profits based on a given technology, i. e., in the parlance of PCMC, for a given set of means of production. One seeks therefore a measuring standard for which one knows for certain that price fluctuations would originate exclusively in the peculiarities of production of the commodities under scrutiny.[1] Sraffa then continues to explain in PCMC, Par. 24:

> *"It is not likely that an individual commodity could be found which possesses even approximately the necessary requisites. A mixture of commodities, however, or a "composite commodity", would do equally well; ...".*

Such a composite commodity is called a *Standard commodity,* and Sraffa then devises a system of production, the *Standard system*, from which he constructs that *Standard commodity*. The algebraic procedure to obtain a *Standard system* from a non *Standard system* is presented hereafter.

In this chapter, we undertake a complete analysis of the mathematical properties of Sraffa's construct for *Standard systems* of production composed of single-product industries, extending by far the 16 pages he himself devoted to this subject. Kurz and Salvadori ([52], p. 121) argue that Sraffa considered primarily the *Standard commodity* as a useful, although not a necessary, tool of analysis. For our part, we limit ourselves to the technical aspects of the model and leave the economic application to future work.

1 This is Ricardo's problem of "an invariable or standard measure of value", assuming no change in technology. See Roncaglia ([96], p. 36) and also Kurz and Salvadori ([52], p. 121), for comments.

https://doi.org/10.1515/9783110635096-005

5.1 Sraffa's Standard system and the Standard commodity

5.1.1 The general idea and the definitions

Consider a closed economic system of production consisting of single-product industries.

In PCMC, Par. 25–26, Sraffa defines the notion of *Standard system*. He starts by considering a *system of production*, called by him the *actual economic system*, represented by a production scheme, including basic and non-basic commodities.

Within the *actual economic system*, he determines all the basic commodities and considers only equations that link exclusively the basic commodities together, discarding non-basics. These equations form a "miniature" system embedded in the *actual economic system*. In this chapter, we consider only the case where the *actual economic system* produces initially only *basic commodities*.

Using algebraic methods, the obtained miniature *production scheme* is then transformed in a way that ensures that the various commodities, represented in its aggregate means of production, are in the same numerical proportions as they are in the totals produced industry by industry. It is called a *Standard system*.

We shall now proceed to develop the concepts mentioned and their properties in full detail. Sraffa's original description of a *Standard system* will be presented in Example 5.1.2 (see also Emmenegger [27]).

The output obtained in a *Standard system* is called a *Standard composite commodity*, or *Standard commodity* for short (PCMC, Par. 26). Consider an 'actual economic system', an initial system of production, represented by a production scheme that includes basic and non-basic commodities, from which a miniature system of only basic industries, called *a 'Standard system'*, is segregated in such segments as will together form a complete miniature system of equations, endowed with the property that the various commodities are represented among the means of production in the same proportions as they are in its outputs. This leads to the formulation of the following definition, using the notion of parallel vectors[2]:

Definition 5.1.1 (The Standard commodity). The vector of 'aggregate means of production' $\mathbf{Se}$, the vector of total output $\mathbf{q}$, and the vector of surplus $\mathbf{d}$ are parallel, $\mathbf{Se} \parallel \mathbf{q} \parallel \mathbf{d}$, in a Standard system. A set of all basic commodities, issued from such as 'Standard system', constituting a basket of commodities, where the proportions between the quantities of the commodities are the same as those between the components of the three parallel vectors, is called for short a "Standard commodity".

2 The condition for two vectors to be parallel to each other is that if one is a scalar multiple of another, i. e., $\mathbf{a} = [a_1, \dots, a_n]'$, $\mathbf{b} = [b_1, \dots, b_n]'$ be the two vectors, then they are parallel to each other if and only if: $\exists k \in \mathbb{R}, \mathbf{b} = k \cdot \mathbf{a} \Leftrightarrow k = b_i/a_i, i \in \{1, \dots, n\}$, noted as $\mathbf{a} \parallel \mathbf{b}$.

We are familiar with the notion that represents the *surplus* or the *net product* produced in an economy, collected together in the vector of *surplus* $\mathbf{d}$. As by Assumption 2.2.1, the vector of total output is positive, $\mathbf{q} > \mathbf{o}$, according to the required proportionalities between the quantities of commodities of a *Standard commodity*, if the vector of surplus is either *positive*, $\mathbf{d} > \mathbf{o}$, or a *null vector*,[3] $\mathbf{d} = \mathbf{o}$. We consider here only *Standard systems* with *positive* surplus vectors.

Sraffa's aim is to consider the *Standard commodity*, an aggregate of commodities, as a calibrated measurement unit. Sraffa proposes further a calibrated unit. He calls a peculiar *Standard commodity*, where the aggregate employed labour is equal to the labour employed in the initial *actual economic system*, a *Standard net product* (PCMC, Par. 26). We will see later that the proportions of quantities of commodities in a *Standard commodity* also indicate the proportions of their values. This means that we will be able to compare values of commodities by their quantities without resorting to prices, see (5.7), hereafter.

Summing up the rows of matrix $\mathbf{S}$, we get the *total amount of commodities* $\sum_{j=1}^{n} s_{ij}$, the aggregate means of production, necessary for the production of each industry $j \in \{1, \ldots, n\}$. The vector $\mathbf{q}_I = \mathbf{Se}$ contains these sums.

A *Standard system*, composed by definition only of basic commodities produced by *basic* industries, is represented by a *production scheme* $(\mathbf{S}', \mathbf{L}) \rightarrow (\hat{\mathbf{q}})$. For this reason, the *commodity flow* matrices $\mathbf{S}$ of *Standard systems* are *semi-positive* and necessarily *irreducible*, Lemma A.15.3 (i), (v).

Now, according to Definition 5.1.1, we have parallel vectors $\mathbf{q} \parallel \mathbf{d}$. We set $\mathbf{q} = k \cdot \mathbf{d}$, $k \in \mathbb{R}$. We write down the proportions of the components of the vectors $\mathbf{q}$ and $\mathbf{d}$:

$$\frac{q_1}{d_1} = \frac{q_2}{d_2} = \cdots = \frac{q_n}{d_n}. \tag{5.1}$$

For the moment, we do not know the numerical value of the proportionality factor equal of the ratio of the *net product* d_j to the *means of production* $q_j - d_j > 0$, $j = 1, \ldots, n$, a dimensionless constant R', designated as *Standard ratio*.[4] Remember that in a *Standard system* a *semi-positive* vector of surplus $\mathbf{d} \geq \mathbf{o}$ is not possible,

$$\boxed{\frac{d_j}{q_j - d_j} = R' > 0, \quad \Leftrightarrow \frac{d_j}{q_j} = \frac{R'}{1 + R'} \quad j = 1, \ldots, n.} \tag{5.2}$$

Sraffa's intention in setting up a *Standard system* of production is to create a system in which any change in the proportions of the total quantities of commodities entering

---

3 The *conditions of production*, presented by Definition 3.1.1 and Definition 3.1.2, define a special type of a *Standard system*, where the vector of surplus disappears, $\mathbf{d} = \mathbf{o}$.

4 There is: $d_j/(q_j - d_j) = R' \Leftrightarrow d_j = R'q_j - R'd_j \Leftrightarrow d_j(1 + R') = R'q_j \Leftrightarrow d_j/q_j = R'/(1 + R')$.

the production process automatically entails an identical change of proportions in the total outputs of the commodities produced, *independent of prices*.

Then, the proportionality (5.2) in a *Standard system* leads to

Lemma 5.1.1. *The positive vectors* $\mathbf{q} > \mathbf{o}, \mathbf{d} > \mathbf{o}, \mathbf{q} - \mathbf{d} > \mathbf{o}$, *representing a Standard system, being parallel to each other,* $\mathbf{q} \parallel \mathbf{q} - \mathbf{d} \parallel \mathbf{d}$, *the proportionalities between these vectors are given by:*

$$\mathbf{d} = R'\mathbf{Se} = R'(\mathbf{q} - \mathbf{d}) = \frac{R'}{1 + R'}\mathbf{q} \Leftrightarrow \mathbf{q} = (1 + R')\mathbf{Se}, \quad R' > 0. \tag{5.3}$$

Proof. One applies the definition of parallelism to the vectors $\mathbf{d}$ and $\mathbf{q} - \mathbf{d}$, taking the proportionalities $d_j/(q_j - d_j) = R' > 0, j \in \{1, \ldots, n\}$, equation (5.2), we get with $\mathbf{Se} = \mathbf{q} - \mathbf{d}$ (2.15), $\mathbf{d} = R'(\mathbf{q} - \mathbf{d}) = R'\mathbf{Se}$, leading to $\mathbf{d} \parallel \mathbf{q} - \mathbf{d}$. One then has the proportionalities $d_j/q_j = R'/(1 + R')$, giving $\mathbf{d} = (R'/(1 + R'))\mathbf{q} = R'\mathbf{Se}$ and $\mathbf{d} \parallel \mathbf{q}$. Then one concludes, finding $\mathbf{q} = (1 + R')\mathbf{Se}$. □

Now, we have to determine what value the factor R' takes on. We find

Lemma 5.1.2. *Consider a production scheme* $(\mathbf{S}', \mathbf{L}) \to (\hat{\mathbf{q}})$ *representing a Standard system with semi-positive and irreducible or positive commodity flow matrix* $\mathbf{S} \geq \mathbf{0}$. *Then, for every sector* $j \in \{1, \ldots, n\}$, *the Standard ratio* R' *of the net product* d_j *to the means of production* $(q_j - d_j)$ *is equal to the productiveness* R,

$$R' = \frac{d_j}{q_j - d_j} = \frac{d_j p_j}{(q_j - d_j)p_j} = R = \frac{1}{\lambda_C} - 1, \quad j = 1, \ldots, n, \tag{5.4}$$

where λ_C *is the Frobenius number of* $\mathbf{C} = \mathbf{S}\hat{\mathbf{q}}^{-1} \geq \mathbf{0}$. *One obtains,*

$$\mathbf{d} = R\mathbf{Se} = \frac{R}{1 + R}\mathbf{q} > \mathbf{o} \Leftrightarrow \mathbf{q} = (1 + R)\mathbf{Se} > \mathbf{0}; \quad R > 0, \tag{5.5}$$

$$d_j p_j = R(q_j - d_j)p_j \Rightarrow \sum_{j=1}^{n} d_j p_j = R \sum_{j=1}^{n} (q_j - d_j)p_j \Rightarrow$$

$$\mathbf{d}'\mathbf{p} := Y = R(\mathbf{q}' - \mathbf{d}')\mathbf{p} := R \cdot (X - Y) \Leftrightarrow R = \frac{Y}{X - Y} = \frac{Y}{K} =: \tilde{R}. \tag{5.6}$$

Proof. Having excluded $\mathbf{d} = \mathbf{o}$, there remains $\mathbf{d} > \mathbf{o}$ (5.2) for a *Standard system*. As the *input-output coefficients* matrix $\mathbf{C}$ is *semi-positive* and *irreducible* or *positive*, we solve the eigenvalue equation $\mathbf{Cq} = \lambda\mathbf{q}$. Lemma 4.1.1 (b) applies with Frobenius number λ_C, $0 < \lambda_C < 1$. We set $\lambda_C = 1/(1 + R)$ and get $\mathbf{Cq} = \mathbf{S}\hat{\mathbf{q}}^{-1}\mathbf{q} = \mathbf{Se} = \lambda_C\mathbf{q} = (1/(1 + R))\mathbf{q}$. Therefore we have $\mathbf{q} = \mathbf{Se}(1+R)$. With (5.3) we get $R' = R > 0$. The proportions between components of parallel vectors (5.6) lead to $R = \tilde{R}$. □

Remark. Consider any prices $p_j > 0$, $j = 1, \ldots, n$, expressed in any freely chosen numéraire which need not be in monetary terms. Then from (5.4) follows immediately that the proportions of the value $p_j d_j$ to the value $p_j(q_j - d_j)$ are equal to the *net product* d_j to the means of production $(q_j - d_j)$ for any of the sectors. The proportion is evidently equal to the *productiveness* R:

$$\boxed{R' = \frac{p_j d_j}{p_j q_j - p_j d_j} = \frac{p_j d_j}{p_j(q_j - d_j)} = \frac{d_j}{(q_j - d_j)} = R = \frac{Y}{K} = \tilde{R}, \quad j = 1, \ldots, n.} \tag{5.7}$$

The limit case of no surplus

A *semi-positive* and *irreducible* or *positive* input coefficients matrix $\mathbf{C}$ and a *null* vector of surplus $\mathbf{d} = \mathbf{0}$ determine a particular *Standard system* with Frobenius number $\lambda_C = 1$ and no *productiveness* $R = 0$, (PCMC, Par. 1), so we fall back to *interindustrial economies* and stochastic matrices, see Lemma 4.1.1 (a).

The case of surplus

The *Standard ratio* R' of *net product* d_j to the *means of production* $(q_j - d_j)$ expresses the part of the *surplus* with respect to one unit of *means of production* of commodity *j*. In this sense, the *Standard ratio* $R' > 0$ is a productivity rate. It is the productiveness, $R = R'$, as we have shown.[5] As mentioned earlier, Sraffa's intention (PCMC, Par. 28) in setting up a *Standard system*, respectively a *Standard commodity*, Definition 5.1.1, is to obtain a composition of *basic commodities*, as a *homogenous physical measure* of value independent of prices.[6]

The *Standard system* is a valuable theoretical tool in a Ricardian perspective. In addition, it automatically fulfills Sraffa's required assumption for obtention of a uniform rate of profits mentioned in PCMC, Par. 4.

Example 5.1.1. Given the productiveness $R = Y/(X - Y)$ (5.6) of a Standard system, determine the corresponding Frobenius number.

---

5 Seemingly, Sraffa did not apply the theorem of Perron–Frobenius A.9.3. He wrote the equation $R' = r(1-\tilde{w})$ (4.36) noted R' as Standard ratio (PCMC, Par. 29, Par. 30), and said that it is equal to the maximal profit rate R. This means that Sraffa postulated correctly the equality $R = R'$ for *Standard systems*.

6 Roncaglia ([96], p. 36) says: "*Sraffa also constructs a particular analytical tool, namely the 'standard commodity', thanks to which he is able to solve (part of) the Ricardian problem of an invariable measure of value. Ricardo had in fact attributed two meanings to the notion of a 'standard measure of value', which must not be confused: that of having invariable value (...) when changes occur in the distribution of income between wages and profits, the technology remaining unaltered; and that of having invariable value in relation to the changes the technology goes through in the course of time ...Sraffa goes on to show how the former can only be solved in terms of a particular analytic construction, the 'standard commodity'. This is a composite commodity (...) so determined that the aggregate means of production also correspond to a certain quantity of 'standard commodity'*".
And indeed "unaltered technology" is the case with equation (5.1), Lemma 5.1.1 and Lemma 5.1.2.

Solution to Example 5.1.1:
We obtain for the *Frobenius number*

$$\lambda_C = \frac{1}{1+R} = \frac{1}{1+\frac{Y}{X-Y}} = \frac{X-Y}{X}. \quad \blacktriangle \tag{5.8}$$

5.1.2 The notion of the non-Standard actual economic system

Sraffa starts with a *non-Standard* actual economic system, (PCMC, Par. 25): "*The problem is one that concerns industries rather than commodities and is best approached from that angle.*" He continues:

Example 5.1.2. "(PCMC, Par. 25) Suppose we segregate from the actual economic system such fractions of the individual basic industries as will together form a complete miniature system endowed with the property that the various commodities are represented among its aggregate means of production in the same proportions as they are among its products. As an example, let us assume that the actual system from which we start includes only basic industries and that these produce respectively iron, coal and wheat in the following way:"

$$\left(90\text{ t. iron}, 120\text{ t. coal}, 60\text{ qr. wheat}, \frac{3}{16}\text{ labour}\right) \to (180\text{ t. iron}, 0, 0),$$

$$\left(50\text{ t. iron}, 125\text{ t. coal}, 150\text{ qr. wheat}, \frac{5}{16}\text{ labour}\right) \to (0, 450\text{ t. coal}, 0),$$

$$\left(40\text{ t. iron}, 40\text{ t. coal}, 200\text{ qr. wheat}, \frac{8}{16}\text{ labour}\right) \to (0, 0, 480\text{ t. wheat}). \tag{5.9}$$

Solution to Example 5.1.2:
The process of production is presented as a production scheme, including the parts of necessary labour, in analogy to the scheme (3.41).

The *node-complete digraph* of the production scheme just presented, with iron (commodity: C1), coal (commodity: C2) and wheat (commodity: C3) assumes the following aspect, see Figure 5.1.

We apply the matrix algebra developed until now and note that Sraffa uses the *normalised* vector of *labour*. The *commodity flow* matrix **S** and the *semi-positive* vector of *surplus* **d** are easily recognized:

$$\mathbf{S} = \begin{bmatrix} s_{11} & s_{12} & s_{13} \\ s_{21} & s_{22} & s_{23} \\ s_{31} & s_{32} & s_{33} \end{bmatrix} = \begin{bmatrix} 90 & 50 & 40 \\ 120 & 125 & 40 \\ 60 & 150 & 200 \end{bmatrix}, \quad \mathbf{d} = \begin{bmatrix} d_1 \\ d_2 \\ d_3 \end{bmatrix} = \begin{bmatrix} 0 \\ 165 \\ 70 \end{bmatrix}. \tag{5.10}$$

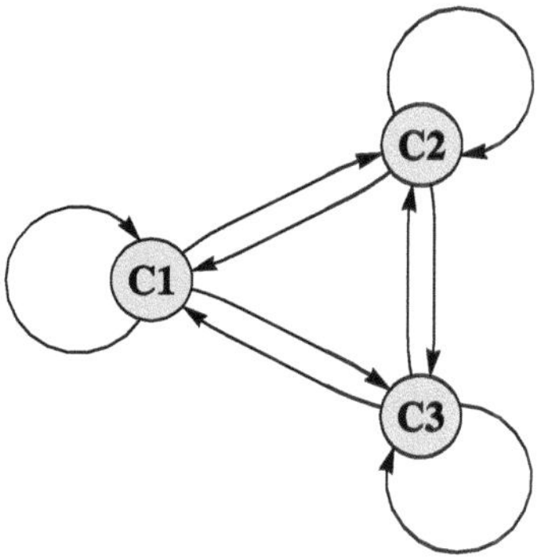

Figure 5.1: *Node-complete digraph* of Sraffa's Example 5.1.2 (PCMC, Par. 25).

Then, we calculate the *means of production* **Se** and the vector of *total output* **q**,

$$\mathbf{Se} = \begin{bmatrix} 90 & 50 & 40 \\ 120 & 125 & 40 \\ 60 & 150 & 200 \end{bmatrix} \begin{bmatrix} 1 \\ 1 \\ 1 \end{bmatrix} = \begin{bmatrix} 180 \\ 285 \\ 410 \end{bmatrix},$$

$$\mathbf{q} = \mathbf{Se} + \mathbf{d} = \begin{bmatrix} 90 & 50 & 40 \\ 120 & 125 & 40 \\ 60 & 150 & 200 \end{bmatrix} \begin{bmatrix} 1 \\ 1 \\ 1 \end{bmatrix} + \begin{bmatrix} 0 \\ 165 \\ 70 \end{bmatrix} = \begin{bmatrix} 180 \\ 450 \\ 480 \end{bmatrix}. \tag{5.11}$$

One has the *normalised* vector of *labour* and the *normalised total quantity of labour*:

$$\mathbf{L} = \begin{bmatrix} L_1 \\ L_2 \\ L_3 \end{bmatrix} = \begin{bmatrix} \frac{3}{16} \\ \frac{5}{16} \\ \frac{8}{16} \end{bmatrix}, \quad L = \mathbf{L}' \cdot \mathbf{e} = \left[\frac{3}{16}, \frac{5}{16}, \frac{8}{16}\right] \cdot \begin{bmatrix} 1 \\ 1 \\ 1 \end{bmatrix} = 1. \tag{5.12}$$

Clearly, the vectors **Se**, **d**, **q**, (5.10) (5.11) are not parallel: **Se** $\nparallel$ **d** $\nparallel$ **q**. So Example 5.1.2 describes a non-Standard system, an *actual economic system* from which a *Standard system*, in accordance with Lemma 5.1.1 and Sraffa's description, must be constructed. ▲

5.2 Application of an orthogonal Euler map to a non-Standard system*

In this section, we develop a **geometric interpretation** of Sraffa's *Standard systems*. We will use familiar notions as *"parallelism of vectors"* and a specific *"linear map"* of the Euclidean space $\mathbb{R}^n \to \mathbb{R}^n$ onto itself. Thus, we will be able to tell how a *non-Standard system* is transformed into a *Standard system*!

5.2.1 Setting the problem and preliminaries

We introduce at first the notion of the top-front-right octant $\mathcal{C}^n \subset \mathbb{R}^n$, called the *commodity space*, because it is used to represent the quantities of produced commodities or products.

Definition 5.2.1 (The commodity space or product space). Consider an n – dim top-front-right octant[7] $\mathcal{C}^n$, called the commodity space (or product space), a subset of the Euclidean space $\mathbb{R}^n$. Consider further in this commodity space $\mathcal{C}^n$ the unit vectors $\vec{e}_i = [0, \dots, 1, \dots, 0]'$ (a 1 on the i-th place), corresponding uniquely to commodities $i \in \{1, \dots, n\}$, giving an orthonormal basis $\mathcal{I}_n = \{\vec{e}_1, \dots, \vec{e}_i, \dots, \vec{e}_n\}$.

> We start here the geometric interpretation of Sraffa's *Standard system*. Sraffa's construction of the *Standard system* (Definition 5.1.1), consists in finding a set of n suitable multipliers $y_1, y_2, \dots, y_n$, forming the diagonal of a matrix $\hat{\mathbf{y}}$, constituting the transformation matrix of an *orthogonal Euler map*.

We consider now such an *orthogonal Euler map*, also called an *orthogonal Euler affinity* (see Definition A.7.1), of the Euclidean space $\mathbb{R}^n \to \mathbb{R}^n$ onto itself.[8] The vectors are supposed to be given relative to the orthonormal basis $\mathcal{I}_n$.

Express now the transposed $n \times n$ *commodity flow* matrix $\mathbf{S}'$ as a row vector of column vectors (Definition A.4.3). We get $\mathbf{S}' = [\mathbf{s}_{1.}, \dots, \mathbf{s}_{n.}]$. The column vectors $\mathbf{s}_{i.} = [s_{i1}, \dots s_{in}]'$ are composed of the quantities s_{ij} of commodity i, necessary as inputs for the production in each sector $j \in \{1, \dots, n\}$. For this reason the column vectors $\mathbf{s}_{i.}$ are called i-th commodity vectors, $i \in \{1, \dots, n\}$.

The *orthogonal Euler map* transforms all the vectors $\mathbf{s}_{i.}$, $i \in \{1, \dots, n\}$. All the *Euler mapped* vectors are noted from now on with a *tilde-sign*, becoming, e. g., $\tilde{\mathbf{s}}_{i.}$ for commodity i and $\tilde{\mathbf{q}}$ for the *total output*:

$$\tilde{\mathbf{s}}_{i.} := \hat{\mathbf{y}}\mathbf{s}_{i.}; \quad \tilde{\mathbf{q}} = \hat{\mathbf{y}}\mathbf{q}. \tag{5.13}$$

This leads to the transposed mapped *commodity flow* matrix $\tilde{\mathbf{S}}' = [\tilde{\mathbf{s}}_{1.}, \dots, \tilde{\mathbf{s}}_{n.}]$ by application of Definition A.4.3. In this context, let us not forget the crucial question of dimensionality. Indeed, for an economic system *with surplus*, one always has $(\hat{\mathbf{q}}-\mathbf{S})\mathbf{e} \geq \mathbf{o}$ and $\det(\hat{\mathbf{q}} - \mathbf{S}) \neq 0$, according to Proposition 3.1.4 meaning the linear independence of the n production processes $(\mathbf{s}_{.j}, q_j)$, implying a full rank: $\operatorname{rank}(\hat{\mathbf{q}} - \mathbf{S}') = n$.

---

7 Any commodity i enters a production process only as a positive quantity, therefore only positive axes need to be considered.

8 The term is translated from the German notion *Orthogonale Eulersche Affinität*. The Swiss mathematician and physicist Leonard Euler (1707–1783) introduced in 1748 the term *affine* (Latin affinis, "related") in his book *Introductio in analysin in infinitorum*, source: http://en.wikipedia.org/wiki/Affine_geometry, 5.12.2013. See also Appendix A.

We then calculate the mapped *commodity flow matrix* $\tilde{\mathbf{S}}$, starting from the transposed matrix $\mathbf{S}'$ and using the elementary identity for the transposed product of matrices, $(\mathbf{AB})' = \mathbf{B}'\mathbf{A}'$, fully explained for this special case with a diagonal matrix in Appendix A, equation (A.36).

$$\tilde{\mathbf{S}}' := \hat{\mathbf{y}}\mathbf{S}' \Leftrightarrow \tilde{\mathbf{S}} = \mathbf{S}\hat{\mathbf{y}}. \tag{5.14}$$

Note also the rule of symmetry (A.37), for the mapped vector of *total output*

$$\tilde{\mathbf{q}} := \hat{\mathbf{y}}\mathbf{q} = \hat{\mathbf{q}}\mathbf{y}. \tag{5.15}$$

Then we can calculate the mapped *means of production*, exemplified for $n = 3$,

$$\begin{aligned}\tilde{\mathbf{S}}\mathbf{e} &= \begin{bmatrix} s_{11}y_1 & s_{12}y_2 & s_{13}y_3 \\ s_{21}y_1 & s_{22}y_2 & s_{23}y_3 \\ s_{31}y_1 & s_{32}y_2 & s_{33}y_3 \end{bmatrix}\begin{bmatrix} 1 \\ 1 \\ 1 \end{bmatrix} = \begin{bmatrix} s_{11}y_1 + s_{12}y_2 + s_{13}y_3 \\ s_{21}y_1 + s_{22}y_2 + s_{23}y_3 \\ s_{31}y_1 + s_{32}y_2 + s_{33}y_3 \end{bmatrix} \\ &= \begin{bmatrix} s_{11} & s_{12} & s_{13} \\ s_{21} & s_{22} & s_{23} \\ s_{31} & s_{32} & s_{33} \end{bmatrix}\begin{bmatrix} y_1 \\ y_2 \\ y_3 \end{bmatrix} = \mathbf{S}\mathbf{y} \Rightarrow \tilde{\mathbf{S}}\mathbf{e} = \mathbf{S}\mathbf{y},\end{aligned} \tag{5.16}$$

which leads us to the following:

Definition 5.2.2 (Euler transformation of a Sraffa system of production). We recall the required matrix and vectors in $\mathbb{R}^n$,

$$\mathbf{S}' = [\mathbf{s}_{1\cdot}, \dots, \mathbf{s}_{n\cdot}] = \begin{bmatrix} s_{11} & s_{21} & \dots & s_{n1} \\ s_{12} & s_{22} & \dots & s_{n2} \\ \dots & \dots & \dots & \dots \\ s_{1n} & s_{2n} & \dots & s_{nn} \end{bmatrix}, \quad \mathbf{L} = \begin{bmatrix} L_1 \\ L_2 \\ \dots \\ L_n \end{bmatrix}, \quad \mathbf{q} = \begin{bmatrix} q_1 \\ q_2 \\ \dots \\ q_n \end{bmatrix}. \tag{5.17}$$

The terminal points of the n vectors $\mathbf{s}_{i\cdot}$, $i = 1, \dots, n$ generate a *polygon* P_n of n vertices in $\mathbb{R}^n$. Now the orthogonal Euler affinity matrix,

$$\hat{\mathbf{y}} = \begin{bmatrix} y_1 & 0 & \dots & 0 \\ 0 & y_2 & \dots & 0 \\ \dots & \dots & \dots & \dots \\ 0 & 0 & \dots & y_n \end{bmatrix}, \tag{5.18}$$

transforms the vectors $\mathbf{s}_{1\cdot}, \dots, \mathbf{s}_{n\cdot}$, $\mathbf{q}$, $\mathbf{L}$, generating mapped vectors and the matrix $\tilde{\mathbf{S}}$

$$\begin{aligned} \tilde{\mathbf{s}}_{i\cdot} &= \hat{\mathbf{y}}\mathbf{s}_{i\cdot}, \quad i = 1, \dots, n, \quad \tilde{\mathbf{L}} = \hat{\mathbf{y}}\mathbf{L}, \\ \tilde{\mathbf{S}} &= [\tilde{\mathbf{s}}_{1\cdot}, \dots, \tilde{\mathbf{s}}_{n\cdot}] = \mathbf{S}\hat{\mathbf{y}}, \quad \tilde{\mathbf{q}} = \hat{\mathbf{y}}\mathbf{q}. \end{aligned} \tag{5.19}$$

The terminal points of the n vectors $\tilde{\mathbf{s}}_{i\cdot}$, $i = 1, \dots, n$ generate the mapped *polygon* $\tilde{P}_n$ of n vertices in $\mathbb{R}^n$.

Notation 5.2.1. In the case of *single-product industries,* a Sraffa system of production', described by the production scheme $(\mathbf{S}', \mathbf{L}) \to (\hat{\mathbf{q}})$, is concisely described as the triple $(\mathbf{S}', \mathbf{q}, \mathbf{L})$ of matrices and vectors.

Geometrically speaking, the multipliers (5.18) in the most general form define an orthogonal Euler map of the Euclidean space $\mathbb{R}^n \to \mathbb{R}^n$, onto itself, operating on the vectors $\mathbf{s}_{i.} = [s_{i1}, \ldots, s_{in}]'$, $i = 1, \ldots, n$, $\mathbf{q}$ and $\mathbf{L}$. In Sraffian terms, an orthogonal Euler map is applied to the *actual economic system* of production $(\mathbf{S}', \mathbf{q}, \mathbf{L})$ resorting to a transformation matrix $\hat{\boldsymbol{\gamma}}$ generating an Euler class $\mathcal{E}$ of systems of production denoted by $(\tilde{\mathbf{S}}', \tilde{\mathbf{q}}, \tilde{\mathbf{L}})$ in application of the transformation (5.19) see Definition 5.2.3.

These systems are in general not *Standard systems*. The specific transformation guaranteeing the parallelism required to obtain a *Standard system*, termed $(\breve{\mathbf{S}}', \breve{\mathbf{q}}, \breve{\mathbf{L}})$, will be determined later. The *breve-sign* $(\breve{\mathbf{q}})$ is used to note an element $(\mathbf{q})$, a vector or a matrix, of a *Standard system*. The system $(\breve{\mathbf{S}}', \breve{\mathbf{q}}, \breve{\mathbf{L}})$ constitute a subclass of the Euler class $\mathcal{E}$. We shall not apply the *breve-sign* to the scalars, like *national income Y* or *total quantity of labour L* of that *Standard system*. We can now introduce

Notation 5.2.2. The 'Standard system', a special system of production, is presented concisely as a triple $(\breve{\mathbf{S}}', \breve{\mathbf{q}}, \breve{\mathbf{L}})$, giving parallel vectors $\breve{\mathbf{S}}\mathbf{e} \parallel \breve{\mathbf{q}} \parallel \breve{\mathbf{d}}$. These three vectors thus have components mutually in constant proportions and constitute examples of Standard composite commodities, or Standard commodities for short (PCMC, Par. 26). The corresponding production scheme is noted $(\breve{\mathbf{S}}', \breve{\mathbf{L}}) \to (\hat{\breve{\mathbf{q}}})$, Definition 5.1.1.

Now, we present the numerical procedures to get a *Standard system* from a non-standard *actual economic system*. For this purpose, we further extend Example 5.1.2, treating it as Sraffa did, multiplying the lines of the *production scheme* (5.9) with appropriate numbers:

Example 5.2.1. Sraffa (PCMC, Par. 25) multiplies the production scheme (5.9) of the non-standard actual economic system, in the following way: the iron row with $\gamma_1 = 1$, the *coal* row with $\gamma_2 = \frac{3}{5}$, the wheat row with $\gamma_3 = \frac{3}{4}$, giving the production scheme of a *Standard system*:

$$\begin{aligned}
&\left(90 \text{ t. iron}, 120 \text{ t. coal}, 60 \text{ qr. wheat}, \frac{3}{16} \text{ labour}\right) \to (180 \text{ t. iron}, 0, 0),\\
&\left(30 \text{ t. iron}, 75 \text{ t. coal}, 90 \text{ qr. wheat}, \frac{3}{16} \text{ labour}\right) \to (0, 270 \text{ t. coal}, 0),\\
&\left(30 \text{ t. iron}, 30 \text{ t. coal}, 150 \text{ qr. wheat}, \frac{6}{16} \text{ labour}\right) \to (0, 0, 360 \text{ t. wheat}). \qquad (5.20)
\end{aligned}$$

We must now develop a general procedure to obtain a Standard system from a non-Standard system!

Solution to Example 5.2.1:
In the present case, the *orthogonal Euler map* with $n = 3$ is described by a 3×3 diagonal matrix and the corresponding mapping equation for each point $\mathbf{x}$,

$$\hat{\mathbf{y}} = \begin{bmatrix} y_1 & 0 & 0 \\ 0 & y_2 & 0 \\ 0 & 0 & y_3 \end{bmatrix} = \begin{bmatrix} 1 & 0 & 0 \\ 0 & \frac{3}{5} & 0 \\ 0 & 0 & \frac{3}{4} \end{bmatrix}; \quad \tilde{\mathbf{x}} = \hat{\mathbf{y}}\mathbf{x}. \tag{5.21}$$

This defines the *orthogonal Euler map* (orthogonal Euler affinity) of the Euclidean space $\mathbb{R}^3 \to \mathbb{R}^3$ into itself. The vectors are given relative to the orthonormal basis $\mathcal{I}_3$ of the vector space $\mathbb{R}^3$.

The multiplication of the equations of the production scheme (5.9) proposed by Sraffa are reproduced here geometrically by the application of that *orthogonal Euler map* on the matrix **S** and the vector of *total output* **q**,[9] giving a *Standard system*:

$$\begin{aligned} \check{\mathbf{S}} = \mathbf{S}\hat{\mathbf{y}} &= \begin{bmatrix} 90 & 50 & 40 \\ 120 & 125 & 40 \\ 60 & 150 & 200 \end{bmatrix} \begin{bmatrix} 1 & 0 & 0 \\ 0 & \frac{3}{5} & 0 \\ 0 & 0 & \frac{3}{4} \end{bmatrix} = \begin{bmatrix} 90 & 30 & 30 \\ 120 & 75 & 30 \\ 60 & 90 & 150 \end{bmatrix}, \\ \check{\mathbf{S}}\mathbf{e} &= \begin{bmatrix} 90 & 30 & 30 \\ 120 & 75 & 30 \\ 60 & 90 & 150 \end{bmatrix} \begin{bmatrix} 1 \\ 1 \\ 1 \end{bmatrix} = \begin{bmatrix} 150 \\ 225 \\ 300 \end{bmatrix}, \\ \check{\mathbf{q}} = \hat{\mathbf{y}}\mathbf{q} &= \begin{bmatrix} 1 & 0 & 0 \\ 0 & \frac{3}{5} & 0 \\ 0 & 0 & \frac{3}{4} \end{bmatrix} \begin{bmatrix} 180 \\ 450 \\ 480 \end{bmatrix} = \begin{bmatrix} 180 \\ 270 \\ 360 \end{bmatrix}. \end{aligned} \tag{5.22}$$

Then we calculate the mapped vector of the new *surplus* and present the obtained results in Figure 5.2:

$$\check{\mathbf{d}} = \check{\mathbf{q}} - \check{\mathbf{S}}\mathbf{e} = \begin{bmatrix} 180 \\ 270 \\ 360 \end{bmatrix} - \begin{bmatrix} 150 \\ 225 \\ 300 \end{bmatrix} = \begin{bmatrix} 30 \\ 45 \\ 60 \end{bmatrix}. \tag{5.23}$$

With Lemma 5.1.1 and Lemma 5.1.2, we obtain the *productiveness* R, e. g., with $i = 1$, giving

$$R = \frac{\check{d}_1}{\check{q}_1 - \check{d}_1} = \frac{30}{180 - 30} = 0.2 = 20\,\%. \tag{5.24}$$

With the *productiveness* R and equation (5.5), and also (2.15), we get the important dilatation equations:

$$\check{\mathbf{q}} = \check{\mathbf{S}}\mathbf{e} + \check{\mathbf{d}} = \check{\mathbf{S}}\mathbf{e}(1 + R) = \check{\mathbf{S}}\mathbf{e} + R\check{\mathbf{S}}\mathbf{e}, \quad \check{\mathbf{d}} = R\check{\mathbf{S}}\mathbf{e}. \tag{5.25}$$

One finds the *mapped* vector of *labour*, applying the rule of symmetry (A.37). The total quantity of labour is no longer normalized as we are at present in a reduced system:

9 The *orthogonal Euler map* with the diagonal matrix $\hat{\mathbf{y}}$ is also applied on the labour vector **L**, further in the present Example (5.26).

$$\breve{\mathbf{L}} = \hat{\mathbf{y}}\mathbf{L} = \begin{bmatrix} L_1 \\ L_2 \\ L_3 \end{bmatrix} = \begin{bmatrix} y_1 & 0 & 0 \\ 0 & y_2 & 0 \\ 0 & 0 & y_3 \end{bmatrix} = \begin{bmatrix} y_1 L_1 \\ y_2 L_2 \\ y_3 L_3 \end{bmatrix},$$

$$\breve{\mathbf{L}} = \begin{bmatrix} 1 \cdot \frac{3}{16} \\ \frac{3}{5} \cdot \frac{5}{16} \\ \frac{3}{4} \cdot \frac{8}{16} \end{bmatrix} = \begin{bmatrix} \frac{3}{16} \\ \frac{3}{16} \\ \frac{6}{16} \end{bmatrix}, \quad L = \breve{\mathbf{L}}' \cdot \mathbf{e} = \left[\frac{3}{16}, \frac{3}{16}, \frac{6}{16}\right] \cdot \begin{bmatrix} 1 \\ 1 \\ 1 \end{bmatrix} = \frac{12}{16}. \quad \blacktriangle \tag{5.26}$$

Recapitulation 5.2.1. There exist orthogonal Euler maps, transforming the non-parallel vectors $\mathbf{d}$, $\mathbf{q}$, $\mathbf{Se}$ (5.9) of a non-standard actual economic system $(\mathbf{S}', \mathbf{q}, \mathbf{L})$ into parallel vectors $\breve{\mathbf{d}}$, $\breve{\mathbf{q}}$, $\breve{\mathbf{S}}\mathbf{e}$ of a Standard system $(\breve{\mathbf{S}}', \breve{\mathbf{q}}, \breve{\mathbf{L}})$. Each vector $\breve{\mathbf{a}}$ parallel to one of the vectors $\breve{\mathbf{d}} \parallel \breve{\mathbf{q}} \parallel \breve{\mathbf{S}}\mathbf{e}$, visualized in Figure 5.2 (right), constitutes a *Standard (composite) commodity* or, shortly speaking, a *Standard commodity*, Definition 5.1.1.

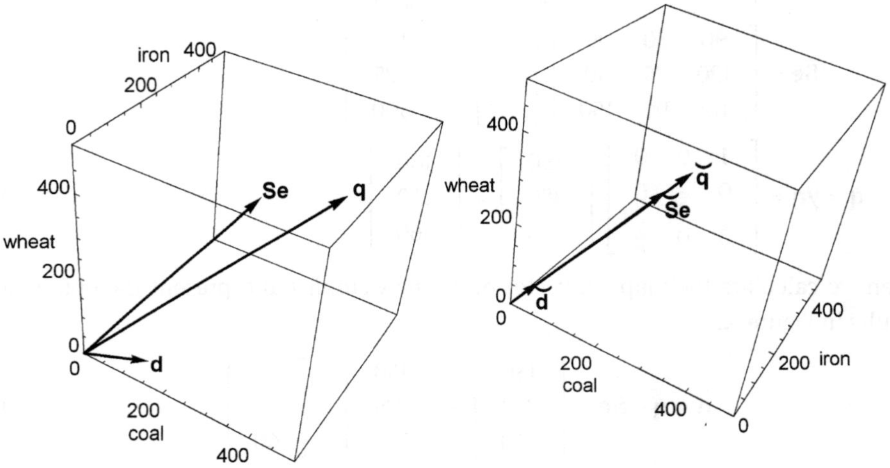

Figure 5.2: *Actual economic system* (left (5.9)) and *Standard system* (right (5.20)).

In the next step, we shall show that an eigenvalue equation is hidden behind the constitution of a *Standard system*, and we will develop the algebra required to calculate Sraffa's multipliers from this perspective.

5.2.2 Calculation of the standard multipliers

At present, we have no general method showing how to compute the multipliers of the diagonal matrix $\hat{\mathbf{y}}$ (5.21), giving in the present case the *Standard system* of Example 5.2.1, generated by the non-standard *actual economic system* of Example 5.1.2.

But preparatory work has been done. We are now ready to propose a mathematical method to compute the *multipliers*; for this purpose, we establish, as those presented by Sraffa in PCMC, Par. 25, Example 5.1.2, a general equation, at first without taking into account the part of *labour* **L** (PCMC, Par. 33).

Let's start with the fundamental equation for the vector of *total output* (2.54), $\mathbf{q} = \mathbf{Se} + \mathbf{d}$. We use the identities $\tilde{\mathbf{q}} = \hat{\mathbf{q}}\mathbf{y}$ and $\tilde{\mathbf{S}} = \mathbf{S}\hat{\mathbf{y}}$, $\tilde{\mathbf{S}}\mathbf{e} = \mathbf{S}\mathbf{y}$, $\tilde{\mathbf{d}} = \tilde{\mathbf{S}}\mathbf{e} - \tilde{\mathbf{q}}$, establishing the Euler mapped *total output* and the Euler mapped *total means of production*. But then, we are finally only interested in the *orthogonal Euler map*, leading to the constituting elements $\breve{\mathbf{S}}$ and $\breve{\mathbf{q}}$ of a *Standard system* (PCMC, Par. 33). We then apply the dilatation equation (5.25) to these elements, here now using the *breve-sign* for the *Standard system*, see Notation 5.2.2, and we obtain:

$$\boxed{\breve{\mathbf{S}}\mathbf{e} + \breve{\mathbf{d}} = \breve{\mathbf{S}}\mathbf{e}(1+R) = \mathbf{S}\mathbf{y}(1+R) = \breve{\mathbf{q}} = \hat{\mathbf{q}}\mathbf{y}.} \tag{5.27}$$

We will then be able to transform (5.27) into an eigenvalue equation enabling us to calculate the multipliers $y_1, y_2, \ldots, y_n$, components of vector $\mathbf{y}$ and diagonal elements of the transformation matrix $\hat{\mathbf{y}}$ (A.36).

We continue to have a look at the set of all *Euler mapped actual economic systems* $(\mathbf{S}', \mathbf{q}, \mathbf{L})$ and will discover some of its properties. Consider the case of n sectors, producing exclusively *basic* products with a surplus. Then, the $n \times n$ *commodity flow* matrix $\mathbf{S}$ is *irreducible*, with Lemma A.15.3 and the corresponding *Standard system* has dimension n (see Section 5.1.1). The vector of *surplus* is *semi-positive*, $\mathbf{d} \geq \mathbf{o}$, and the vector of *total output* is *positive*, $\mathbf{q} = \mathbf{Se} + \mathbf{d} > \mathbf{o}$. We also have a vector of labour $\mathbf{L}$.

$$\mathbf{S} = \begin{bmatrix} s_{11} & s_{12} & \cdots & s_{1n} \\ s_{21} & s_{22} & \cdots & s_{2n} \\ \cdots & \cdots & \cdots & \cdots \\ s_{n1} & s_{n2} & \cdots & s_{nn} \end{bmatrix}, \quad \mathbf{q} = \begin{bmatrix} q_1 \\ q_2 \\ \cdots \\ q_n \end{bmatrix}, \quad \mathbf{L} = \begin{bmatrix} L_1 \\ L_2 \\ \cdots \\ L_n \end{bmatrix}. \tag{5.28}$$

Definition 5.2.3. Take the actual economic system $(\mathbf{S}', \mathbf{q}, \mathbf{L})$ and multiply its three constituents by a diagonal $n \times n$ matrix $\hat{\mathbf{y}}$, whose diagonal elements are the positive multipliers, $y_i > 0$, (5.18) as described above, constituting themself the positive real vector $\mathbf{y} > \mathbf{o}$.[10] One obtains a mapped *system of production* $(\tilde{\mathbf{S}}', \tilde{\mathbf{q}}, \tilde{\mathbf{L}})$ that is an element of the Euler class of *systems of production* $\mathcal{E} = \{(\mathbf{S}', \mathbf{q}, \mathbf{L}) \mid \hat{\mathbf{y}}\}$.

The actual economic system, a system of production $(\mathbf{S}', \mathbf{q}, \mathbf{L})$, (5.28) is called a generator of the Euler class of systems of production $\mathcal{E}$.

Proposition 5.2.1 (Properties of the Euler class of systems of production).

(1) *The actual economic system* $(\mathbf{S}', \mathbf{q}, \mathbf{L})$ *belongs to the Euler class of systems of production with the identity matrix* $\mathbf{y} = \mathbf{I}$, $(\mathbf{S}', \mathbf{q}, \mathbf{L}) \in \mathcal{E}$.

(2) *There is a unique input-output coefficients matrix* $\mathbf{C} = \mathbf{S}\hat{\mathbf{q}}^{-1}$ *for the whole Euler class of systems of production* $\mathcal{E}$*, as we shall now show.*

---

10 For economic purposes, setting $y_i > 0$ is necessary because the signs of the matrix and vector elements have to remain the same. There are no zero equations.

Indeed, applying an orthogonal Euler map on the input-output coefficients matrix $\mathbf{C} = \mathbf{S}\hat{\mathbf{q}}^{-1}$ *(2.16), one defines* $\tilde{\mathbf{C}} := \tilde{\mathbf{S}}\hat{\tilde{\mathbf{q}}}^{-1}$*, leading to:*

$$\tilde{\mathbf{C}} := \tilde{\mathbf{S}}\hat{\tilde{\mathbf{q}}}^{-1} = \begin{bmatrix} s_{11}y_1 & s_{12}y_2 & \dots & s_{1n}y_n \\ s_{21}y_1 & s_{22}y_2 & \dots & s_{2n}y_n \\ \dots & \dots & \dots & \dots \\ s_{n1}y_1 & s_{n2}y_2 & \dots & s_{nn}y_n \end{bmatrix} \begin{bmatrix} \frac{1}{q_1y_1} & 0 & \dots & 0 \\ 0 & \frac{1}{q_2y_2} & \dots & 0 \\ \dots & \dots & \dots & \dots \\ 0 & 0 & \dots & \frac{1}{q_ny_n} \end{bmatrix}$$

$$= \begin{bmatrix} s_{11} & s_{12} & \dots & s_{1n} \\ s_{21} & s_{22} & \dots & s_{2n} \\ \dots & \dots & \dots & \dots \\ s_{n1} & s_{n2} & \dots & s_{nn} \end{bmatrix} \begin{bmatrix} \frac{1}{q_1} & 0 & \dots & 0 \\ 0 & \frac{1}{q_2} & \dots & 0 \\ \dots & \dots & \dots & \dots \\ 0 & 0 & \dots & \frac{1}{q_n} \end{bmatrix} = \mathbf{S}\hat{\mathbf{q}}^{-1} = \mathbf{C}. \quad (5.29)$$

We recognize that with each generator $(\mathbf{S}', \mathbf{q}, \mathbf{L})$ *is associated a unique input-output coefficients matrix* $\tilde{\mathbf{C}} = \mathbf{C} = \mathbf{S}\hat{\mathbf{q}}^{-1}$.

(3) *A similar reduction for the distribution coefficients matrix* $\mathbf{D} = \hat{\mathbf{q}}^{-1}\mathbf{S}$ *(2.20) is obtained as follows:*

$$\tilde{\mathbf{D}} := \hat{\tilde{\mathbf{q}}}^{-1}\tilde{\mathbf{S}} = \begin{bmatrix} \frac{1}{q_1y_1} & 0 & \dots & 0 \\ 0 & \frac{1}{q_2y_2} & \dots & 0 \\ \dots & \dots & \dots & \dots \\ 0 & 0 & \dots & \frac{1}{q_ny_n} \end{bmatrix} \begin{bmatrix} s_{11}y_1 & s_{12}y_2 & \dots & s_{13}y_3 \\ s_{21}y_1 & s_{22}y_2 & \dots & s_{2n}y_n \\ \dots & \dots & \dots & \dots \\ s_{n1}y_1 & s_{n2}y_2 & \dots & s_{nn}y_k \end{bmatrix}$$

$$= \begin{bmatrix} \frac{1}{q_1} & 0 & \dots & 0 \\ 0 & \frac{1}{q_2} & \dots & 0 \\ \dots & \dots & \dots & \dots \\ 0 & 0 & \dots & \frac{1}{q_n} \end{bmatrix} \begin{bmatrix} s_{11} & s_{12} & \dots & s_{13} \\ s_{21} & s_{22} & \dots & s_{23} \\ \dots & \dots & \dots & \dots \\ s_{n1} & s_{n2} & \dots & s_{nn} \end{bmatrix} = \hat{\mathbf{q}}^{-1}\mathbf{S} = \mathbf{D}. \quad (5.30)$$

We recognize that to each generator $(\mathbf{S}', \mathbf{q}, \mathbf{L})$ *is associated a unique distribution coefficients matrix* $\tilde{\mathbf{D}} = \hat{\mathbf{q}}^{-1}\mathbf{S} = \mathbf{D}$*. Clearly, like matrix* $\mathbf{S}$*, matrices* $\mathbf{C}$ *and* $\mathbf{D}$ *are positive or semi-positive and irreducible.*

Returning to Expression (5.27), we can now determine the sought-for eigenvalue equation. We calculate the *distribution coefficients* matrix $\mathbf{D} = \hat{\mathbf{q}}^{-1}\mathbf{S}$ (5.30), common to all elements of the Euler class $\mathcal{E} = \{(\mathbf{S}', \mathbf{q}, \mathbf{L}) \mid \hat{y}\}$ and to the *Standard system* $(\check{\mathbf{S}}', \check{\mathbf{q}}, \check{\mathbf{L}})$ we are looking for. We multiply (5.27) from the left by $\hat{\mathbf{q}}^{-1}$, obtaining:

$$\begin{aligned} \hat{\mathbf{q}}^{-1}(\mathbf{S}y)(1+R) &= (\hat{\mathbf{q}}^{-1}\mathbf{S})y(1+R) = \mathbf{D}y(1+R) \\ &= \hat{\mathbf{q}}^{-1}(\hat{\mathbf{q}}y) = (\hat{\mathbf{q}}^{-1}\hat{\mathbf{q}})y = y, \end{aligned} \quad (5.31)$$

i. e., the *right eigenvector* equation of matrix $\mathbf{D}$,

$$\mathbf{D}y(1+R) = y \Rightarrow \mathbf{D}y = \frac{1}{1+R}y := \lambda_D y. \quad (5.32)$$

The vector of *multipliers* $\mathbf{y}$ is a *positive right eigenvector* of the right eigenvector equation (5.32).

Continuing with the general explicit calculation of the *distribution coefficients* matrix $\mathbf{D}$, taking equation (2.20), we set up,

$$\mathbf{D} = \hat{\mathbf{q}}^{-1}\mathbf{S} = \begin{bmatrix} \frac{1}{q_1} & 0 & 0 \\ 0 & \frac{1}{q_2} & 0 \\ 0 & 0 & \frac{1}{q_3} \end{bmatrix} \begin{bmatrix} s_{11} & s_{12} & s_{13} \\ s_{21} & s_{22} & s_{23} \\ s_{31} & s_{32} & s_{33} \end{bmatrix} = \begin{bmatrix} \frac{s_{11}}{q_1} & \frac{s_{12}}{q_1} & \frac{s_{13}}{q_1} \\ \frac{s_{21}}{q_2} & \frac{s_{22}}{q_2} & \frac{s_{23}}{q_2} \\ \frac{s_{31}}{q_3} & \frac{s_{32}}{q_3} & \frac{s_{33}}{q_3} \end{bmatrix}, \tag{5.33}$$

and redefine the *distribution coefficients*, treating the general case,

$$d_{ij} = \frac{s_{ij}}{q_i}; \quad i,j = 1,\dots,n; \quad \begin{array}{ll} i: & \textit{Input} \text{ index} \\ j: & \textit{Output} \text{ index.} \end{array} \tag{5.34}$$

We recall their economic meaning:

d_{ij}: The *distribution coefficient* determines the fraction (part) of commodity i (Input) required for the production of one unit of commodity j (Output).

Each row $[d_{i1},\dots,d_{ij},\dots,d_{in}]$ of the *distribution coefficients* matrix $\mathbf{D}$ gives accordingly the part of commodity i, necessary for the production of a part of a unit of each commodity $j \in \{1,\dots,n\}$.

Now let us undertake the numerical work for Example 5.1.2.

Example 5.2.2. Consider the following entries, the *positive* matrix $\mathbf{S}$ (5.10), the *semi-positive* vector of surplus $\mathbf{d} \geq \mathbf{o}$ (5.10), and the positive vector of total output $\mathbf{q} = \mathbf{Se} + \mathbf{d} > \mathbf{o}$ (5.11) of Example 5.1.2.

Compute the *distribution coefficients* matrix $\mathbf{D} = \hat{\mathbf{q}}^{-1}\mathbf{S}$ and interpret each of its rows. Compute the characteristic polynomial of matrix $\mathbf{D}$, its Frobenius number λ_D and the associated right eigenvectors. Find the calibration to obtain the multipliers of PCMC, Par. 25.

Solution to Example 5.2.2:

Start with matrix $\mathbf{S}$ (5.10) and vector $\mathbf{q}$ (5.11) and compute the matrix,

$$\mathbf{D} = \hat{\mathbf{q}}^{-1}\mathbf{S} = \begin{bmatrix} \frac{1}{180} & 0 & 0 \\ 0 & \frac{1}{450} & 0 \\ 0 & 0 & \frac{1}{480} \end{bmatrix} \begin{bmatrix} 90 & 50 & 40 \\ 120 & 125 & 40 \\ 60 & 150 & 200 \end{bmatrix} = \begin{bmatrix} \frac{1}{2} & \frac{5}{18} & \frac{2}{9} \\ \frac{4}{15} & \frac{5}{18} & \frac{4}{45} \\ \frac{1}{8} & \frac{5}{16} & \frac{5}{12} \end{bmatrix}. \tag{5.35}$$

The numerator and denominator of the fraction d_{ij} represent quantities of the same commodity in *physical terms*, like *tons of* iron. For this reason, all *distribution coefficients* are dimensionless, $[d_{ij}] = 1$, $i,j = 1,\dots,n$, varying: $0 < d_{ij} < 1$. In equation (5.35), for example, the ratio $d_{12} = 50/180 = 5/18$ determines that 5/18 of commodity 1 (iron) is used in the production of a part of a unit of commodity 2 (coal).

Referring to (5.10), we note:

In the first row of $\mathbf{D}$ (5.35), we have $[\frac{90}{180}, \frac{50}{180}, \frac{40}{180}]$, the sum of these fractions gives $\frac{90}{180} + \frac{50}{180} + \frac{40}{180} = 1$, because there is no surplus of iron production, $d_1 = 0$.

The second row of $\mathbf{D}$ (5.35) is $[\frac{120}{450}, \frac{125}{450}, \frac{40}{450}]$. The sum of these fractions gives $\frac{120}{450} + \frac{125}{450} + \frac{40}{450} = \frac{285}{450}$, indicating that the surplus represents here $\frac{165}{450}$ of one unit of *coal* production, $d_2 = 165$.

For the third row of $\mathbf{D}$ (5.35), we have $[\frac{60}{480}, \frac{150}{480}, \frac{200}{480}]$. The sum of these fractions gives $\frac{60}{480} + \frac{150}{480} + \frac{200}{480} = \frac{410}{480}$, indicating that the surplus represents here $\frac{70}{480}$ of one unit of wheat production, $d_3 = 70$.

As the vector of surplus $\mathbf{d}$ (5.10) is *semi-positive*, we get for the row sums $\sum_{j=1}^{n} d_{ij} \leq 1$, $i = 1, \ldots, n$. If the i-th row sum is smaller than 1, the production process generates a *net product* (surplus) for commodity i.

The eigenvalues of matrix $\mathbf{D}$ are then computed. Set up the characteristic polynomial,

$$P_3(\lambda) = \det(\mathbf{D} - \lambda\mathbf{I}) = -\lambda^3 + \frac{43}{36}\lambda^2 + -\frac{1}{3}\lambda + \frac{35}{1296}. \tag{5.36}$$

The polynomial $P_3(\lambda)$ is then factorised, obtaining,

$$P_3(\lambda) = \left(\lambda - \frac{5}{6}\right)\left(\lambda - \frac{1}{6}\right)\left(\lambda - \frac{7}{36}\right). \tag{5.37}$$

Remember that the present *Standard system* has a *positive* surplus. Consequently Lemma 4.1.1 (b) applies and the *Frobenius number* of matrix $\mathbf{D} = \hat{\mathbf{q}}^{-1}\mathbf{S}$ is smaller than 1, $0 < \lambda_D = 5/6 = 1/(1 + R) < 1$.

We then get the positive *productiveness* $R = 0.2$.

Finally, we obtain the positive eigenvectors $\boldsymbol{\gamma} = k[\frac{4}{3}, \frac{4}{5}, 1]' > 0$, $k \in \mathbb{R}^+$ from the eigenvalue equation, $\mathbf{D}\boldsymbol{\gamma} = \lambda_D\boldsymbol{\gamma}$ (5.32). Setting $k = 3/4$, Sraffa's multipliers $\boldsymbol{\gamma} = [1, \frac{3}{5}, \frac{3}{4}]'$ (PCMC, Par. 25) appear. ▲

Recapitulation 5.2.2. Considering the actual economic system (5.9), which is not a Standard system, the multipliers from the range of Example 5.1.2 generate the Standard system (5.20). Multipliers are the multitude of the positive eigenvectors $\boldsymbol{\gamma} = k[\frac{4}{3}, \frac{4}{5}, 1]'$, $k > 0$, of the positive distribution coefficients matrix $\mathbf{D} > \mathbf{0}$, associated with the Frobenius number less than 1, $\lambda = 5/6 < 1$, as this economy produces a surplus. With the calibration value $k = 3/4$, we get the multipliers $\gamma_1 = 1$, $\gamma_2 = \frac{3}{5}$, $\gamma_3 = 3/4$ of Sraffa (PCMC, Par. 25). The *productiveness* is positive, $R = 0.2 > 0$, (5.24).

5.2.3 Classes of Standard systems and their Standard commodities

In the preceding Subsection 5.2.2, we have shown that Sraffa's multipliers (PCMC, Par. 25) are essentially the components of eigenvectors of the *semi-positive* and irreducible or *positive distribution coefficients matrix* $\mathbf{D}$, associated with its Frobenius number.

In the present subsection, we shall discover a second eigenvalue equation. We continue to consider n sectors and return to the question of generating a *Standard system*. Having a *semi-positive* and *irreducible* or *positive* commodity flow matrix **S** (5.28), consider a generally non-standard *actual economic system* $(\mathbf{S}', \mathbf{q}, \mathbf{L})$. Compute the Euler mapped transposed matrix $\tilde{\mathbf{S}}' = \boldsymbol{y}\mathbf{S}'$, equivalent to $\tilde{\mathbf{S}} = \mathbf{S}\boldsymbol{y}$. The *orthogonal Euler map* is also applied to the vectors **q** and **L**, giving the mapped vector $\tilde{\mathbf{q}} = \boldsymbol{y}\mathbf{q}$ and the mapped vector $\tilde{\mathbf{L}} = \boldsymbol{y}\mathbf{L}$.

We are aiming at a *Standard system*; we accordingly have to calculate $\breve{\mathbf{S}}\mathbf{e}$, a vector that has to be parallel to the vector of *total output* $\breve{\mathbf{q}}$. We apply the dilatation equation (5.25), which we reproduce here now using the *breve-sign* for a *Standard system*, see Notation 5.2.2:

$$\breve{\mathbf{S}}\mathbf{e}(1+R) = \breve{\mathbf{q}}. \tag{5.38}$$

Then, we apply the rule (A.35),

$$\hat{\breve{\mathbf{q}}}^{-1}\breve{\mathbf{q}} = \begin{bmatrix} \frac{1}{\breve{q}_1 y_1} & 0 & \dots & 0 \\ 0 & \frac{1}{\breve{q}_2 y_2} & \dots & 0 \\ \dots & \dots & \dots & \dots \\ 0 & 0 & \dots & \frac{1}{\breve{q}_n y_n} \end{bmatrix} \begin{bmatrix} y_1\breve{q}_1 \\ y_2\breve{q}_2 \\ \dots \\ y_n\breve{q}_n \end{bmatrix} = \begin{bmatrix} 1 \\ 1 \\ \dots \\ 1 \end{bmatrix} = \mathbf{e}, \tag{5.39}$$

to replace vector **e** in (5.38). We then find with the reduction equation (5.29), valid for *Standard systems*, i. e., $\mathbf{C} = \tilde{\mathbf{S}}\hat{\tilde{\mathbf{q}}}^{-1} = \breve{\mathbf{S}}\hat{\breve{\mathbf{q}}}^{-1}$ the eigenvalue equation:

$$\boxed{\begin{aligned} \breve{\mathbf{S}}\mathbf{e}(1+R) &= \breve{\mathbf{S}}(\hat{\breve{\mathbf{q}}}^{-1}\breve{\mathbf{q}})(1+R) = (\breve{\mathbf{S}}\hat{\breve{\mathbf{q}}}^{-1})\breve{\mathbf{q}}(1+R) \\ &= \mathbf{C}(1+R)\breve{\mathbf{q}} = \breve{\mathbf{q}} \\ &= \mathbf{C}\breve{\mathbf{q}} = \frac{1}{1+R}\breve{\mathbf{q}} = \lambda_C\breve{\mathbf{q}}. \end{aligned}} \tag{5.40}$$

The mapped vector of *total output* $\breve{\mathbf{q}}$ is a *right eigenvector* of matrix **C**, associated with the Frobenius number $\lambda_C = 1/(1+R)$. These observations lead to

Lemma 5.2.1. *For a semi-positive and irreducible or positive commodity flow matrix* **S**, *a semi-positive vector of surplus* $\mathbf{d} \geq \mathbf{o}$, *the vector of total output* $\mathbf{q} = \mathbf{S}\mathbf{e} + \mathbf{d} > \mathbf{o}$ *and the vector of labour* **L**, *the Euler class of systems of production* $\mathcal{E} = \{(\mathbf{S}', \mathbf{q}, \mathbf{L}) \mid \hat{y}\}$ *is associated with a unique input-output coefficients matrix* $\mathbf{C} = \mathbf{S}\hat{\mathbf{q}}^{-1}$, *respectively a unique similar distribution coefficients matrix* $\mathbf{D} = \hat{\mathbf{q}}^{-1}\mathbf{S}$, *both are semi-positive and irreducible or positive, with equal Frobenius numbers* $\lambda_D = \lambda_C = 1/(1+R) < 1$ *and equal positive productiveness* $R > 0$. ▲

Lemma 5.2.1 states an invariance property of the elements of the Euler class of systems of production $\mathcal{E}$ under *orthogonal Euler maps* and is based on a property of similar matrices, described by Lemma A.6.2. All the elements of $\mathcal{E}$ resort to analogous technologies in putting to use the available commodities. They have in common an invariant *input-output coefficients* matrix **C** and an invariant similar *distribution coefficients* matrix **D**.

In the next example, we illustrate the generation of a general *Euler mapped system of production* $(\tilde{\mathbf{S}}', \tilde{\mathbf{q}}, \tilde{\mathbf{L}}) \in \mathcal{E}$, which will lead to the required *Standard system.* We take the entries of the non-standard actual *system of production* $(\mathbf{S}', \mathbf{q}, \mathbf{L})$ of Example 5.1.2, (PCMC, Par. 25).

Example 5.2.3. Consider the *Euler class of systems of production* $\mathcal{E} = \{(\mathbf{S}', \mathbf{q}, \mathbf{L}) \mid \hat{\mathbf{y}}\}$, constituted with the matrix **S** (5.10), the vectors **q** (5.11) and **L**, (5.12), generated by $(\mathbf{S}', \mathbf{q}, \mathbf{L})$. Take the general diagonal matrix $\hat{\mathbf{y}}$ (5.21) of positive diagonal elements $y_i > 0$, $i = 1, \ldots, 3$. Write down the general *system of production* $(\tilde{\mathbf{S}}', \tilde{\mathbf{q}}, \tilde{\mathbf{L}}) \in \mathcal{E}$.

Solution to Example 5.2.3:
Set up the matrix **S** (5.10) and vectors **q** (5.11), **L** (5.12) and a general diagonal matrix $\hat{\mathbf{y}}$:

$$\mathbf{S} = \begin{bmatrix} 90 & 50 & 40 \\ 120 & 125 & 40 \\ 60 & 150 & 200 \end{bmatrix}, \quad \mathbf{q} = \begin{bmatrix} 180 \\ 450 \\ 480 \end{bmatrix}, \quad \mathbf{L} = \begin{bmatrix} \frac{3}{16} \\ \frac{5}{16} \\ \frac{8}{16} \end{bmatrix}, \quad \hat{\mathbf{y}} = \begin{bmatrix} y_1 & 0 & 0 \\ 0 & y_2 & 0 \\ 0 & 0 & y_3 \end{bmatrix}. \tag{5.41}$$

Determine the elements $\tilde{\mathbf{S}}$, $\tilde{\mathbf{q}}$, $\tilde{\mathbf{L}}$ of the Euler class $\mathcal{E} = \{(\mathbf{S}', \mathbf{q}, \mathbf{L}) \mid \hat{\mathbf{y}}\}$:

$$\tilde{\mathbf{S}} = \mathbf{S}\hat{\mathbf{y}} = \begin{bmatrix} 90 & 50 & 40 \\ 120 & 125 & 40 \\ 60 & 150 & 200 \end{bmatrix} \begin{bmatrix} y_1 & 0 & 0 \\ 0 & y_2 & 0 \\ 0 & 0 & y_3 \end{bmatrix} = \begin{bmatrix} 90y_1 & 50y_2 & 40y_3 \\ 120y_1 & 125y_2 & 40y_3 \\ 60y_1 & 90y_2 & 150y_3 \end{bmatrix}, \tag{5.42}$$

$$\tilde{\mathbf{q}} = \hat{\mathbf{y}}\mathbf{q} = \begin{bmatrix} y_1 & 0 & 0 \\ 0 & y_2 & 0 \\ 0 & 0 & y_3 \end{bmatrix} \begin{bmatrix} 180 \\ 450 \\ 480 \end{bmatrix} = \begin{bmatrix} y_1 180 \\ y_2 450 \\ y_3 480 \end{bmatrix},$$

$$\tilde{\mathbf{L}} = \hat{\mathbf{y}}\mathbf{L} = \begin{bmatrix} y_1 & 0 & 0 \\ 0 & y_2 & 0 \\ 0 & 0 & y_3 \end{bmatrix} \begin{bmatrix} \frac{3}{16} \\ \frac{5}{16} \\ \frac{8}{16} \end{bmatrix} = \begin{bmatrix} y_1 \frac{3}{16} \\ y_2 \frac{5}{16} \\ y_3 \frac{8}{16} \end{bmatrix}. \tag{5.43}$$

We have now calculated the general *systems of production* $(\tilde{\mathbf{S}}', \tilde{\mathbf{q}}, \tilde{\mathbf{L}})$ (5.20) belonging to the Euler class of systems of production $\mathcal{E} = \{(\mathbf{S}', \mathbf{q}, \mathbf{L}) \mid \hat{\mathbf{y}}\}$. ▲

Now we go on, extracting the *Standard systems* and the corresponding *Standard commodities* (Definition 5.1.1) from $\mathcal{E}$. We then also calculate the proportions between the elements of the *Standard commodities.*

Consider for this purpose again the generator $(\mathbf{S}', \mathbf{q}, \mathbf{L}) \in \mathcal{E}$. Due to Lemma 5.2.1, the *input-output coefficients matrix* $\mathbf{C} = \mathbf{S}\hat{\mathbf{q}}^{-1}$ and the *distribution coefficients matrix* $\mathbf{D} = \hat{\mathbf{q}}^{-1}\mathbf{S}$ are common to all *systems of production* of $\mathcal{E}$. We set up the eigenvalue equations (5.40) and (5.32),[11] with $\lambda_C = \lambda_D$:

11 This means that the determination of a *Standard system* $(\check{\mathbf{S}}', \check{\mathbf{q}}, \check{\mathbf{L}}) \in \mathcal{E}$, obtained by an *orthogonal Euler map* from the *actual economic system* $(\mathbf{S}', \mathbf{q}, \mathbf{L}) \in \mathcal{E}$, the generator, requires the solution of both *eigenvalue equations* (5.44). From this *Standard system* one then may further generate *Standard commodities.*

$$\mathbf{C}\breve{\mathbf{q}} = \lambda_C \breve{\mathbf{q}} = \frac{1}{1+R}\breve{\mathbf{q}}, \quad \mathbf{D}\boldsymbol{y} = \lambda_D \boldsymbol{y} = \frac{1}{1+R}\boldsymbol{y}. \tag{5.44}$$

The *multipliers* of the *orthogonal Euler map* are the components of a *right eigenvector* $\boldsymbol{y}$ of the matrix **D**, corresponding to the Frobenius number λ_D, obtained by an appropriated calibration. They generate the vector of *total output* $\breve{\mathbf{q}}$ (5.22), respectively the vector representing the *means of production* (5.22) $\breve{\mathbf{S}}\mathbf{e}$ and the vector of surplus $\breve{\mathbf{d}}$ (5.23), see Example 5.2.1. These three vectors are parallel to the *right eigenvectors* $\breve{\mathbf{q}}$ of the matrix **C**, corresponding to the Frobenius number λ_C. They are the elements constituting the *Standard systems* and the corresponding *Standard commodities.*

The next example illustrates the relationships between the chosen *total output* eigenvector $\breve{\mathbf{q}}$ and the resulting total quantity of labour L in detail:

Example 5.2.4. Start again with the entries of Example 5.1.2. Compute the *input-output coefficients* matrix $\mathbf{C} = \mathbf{S}\hat{\mathbf{q}}^{-1}$ (5.29) and the *distribution coefficients* matrix $\mathbf{D} = \hat{\mathbf{q}}^{-1}\mathbf{S}$ (5.30), common to all *systems of production* of $\mathcal{E}$, Lemma 5.2.1.

Compute the characteristic polynomial of matrices **C** and **D**, its common Frobenius numbers $\lambda_C = \lambda_D = 1/(1+R)$ (Lemma A.6.1) and the common *productiveness* R.

Compute the right eigenvectors $k \cdot \boldsymbol{y}$ of the positive matrix **D**, associated with the Frobenius number λ_D. Choose the right eigenvector $k \cdot \boldsymbol{y}$, so that the total output of iron becomes $\breve{q}_1 = 1080$. Compute the dilatation coefficient k and the corresponding eigenvector $k \cdot \boldsymbol{y}$ of multipliers. Compute the remaining elements of the *Standard system*. Show the parallelism of the three elements $\breve{\mathbf{S}}_1\mathbf{e}$, $\breve{\mathbf{q}}_1$, $\breve{\mathbf{d}}_1$ of the obtained *Standard system* $(\breve{\mathbf{S}}_1', \breve{\mathbf{q}}_1, \breve{\mathbf{L}}_1)$, described in Recapitulation 5.2.1. Choose the vector of total output $\breve{\mathbf{q}}_1$ as Standard commodity. Compute the total quantity of labour L_1.

Compute the right eigenvectors $k \cdot \breve{\mathbf{q}}$ of the positive matrix **C**, associated to the Frobenius number λ_C. First, calibrate the right eigenvectors $k \cdot \breve{\mathbf{q}}$ to obtain the above vector of total output $\breve{\mathbf{q}}_1$. Second, calibrate the right eigenvectors $k \cdot \breve{\mathbf{q}}$ to get the *Standard system* of Example 5.2.1, equation (5.22), (PCMC, Par. 26). Show again the parallelism of the vectors of the obtained Standard system. Compute Sraffa's multipliers of PCMC, Par. 25.

Solution to Example 5.2.4:
Start with matrix **S** (5.10),[12] and vectors **q** (5.11), **L** (5.12) of Example 5.1.2,

$$\mathbf{S} = \begin{bmatrix} 90 & 50 & 40 \\ 120 & 125 & 40 \\ 60 & 150 & 200 \end{bmatrix}, \quad \mathbf{q} = \begin{bmatrix} 180 \\ 450 \\ 480 \end{bmatrix}, \quad \mathbf{L} = \begin{bmatrix} \frac{3}{16} \\ \frac{5}{16} \\ \frac{8}{16} \end{bmatrix}, \quad \hat{\boldsymbol{y}} = \begin{bmatrix} y_1 & 0 & 0 \\ 0 & y_2 & 0 \\ 0 & 0 & y_3 \end{bmatrix}. \tag{5.45}$$

We compute the *input-output coefficients* matrix $\mathbf{C} = \mathbf{S}\hat{\mathbf{q}}^{-1}$ (5.29) and the *distribution coefficients matrix* $\mathbf{D} = \hat{\mathbf{q}}^{-1}\mathbf{S}$ (5.35), common to all *systems of production* of $\mathcal{E}$

12 *Positivity* of matrices is more restrictive than *semi-positivity* and *irreducibility* of matrices, see Gantmacher ([34], Theorem 1 (Perron) and Theorem 2 (Frobenius), p. 398).

(Lemma 5.2.1) giving

$$\mathbf{C} = \begin{bmatrix} 90 & 50 & 40 \\ 120 & 125 & 40 \\ 60 & 150 & 200 \end{bmatrix} \begin{bmatrix} \frac{1}{180} & 0 & 0 \\ 0 & \frac{1}{450} & 0 \\ 0 & 0 & \frac{1}{480} \end{bmatrix} = \begin{bmatrix} \frac{1}{2} & \frac{1}{9} & \frac{1}{12} \\ \frac{2}{3} & \frac{5}{18} & \frac{1}{12} \\ \frac{1}{3} & \frac{1}{3} & \frac{5}{12} \end{bmatrix} > \mathbf{0},$$

$$\mathbf{D} = \begin{bmatrix} \frac{1}{180} & 0 & 0 \\ 0 & \frac{1}{450} & 0 \\ 0 & 0 & \frac{1}{480} \end{bmatrix} \begin{bmatrix} 90 & 50 & 40 \\ 120 & 125 & 40 \\ 60 & 150 & 200 \end{bmatrix} = \begin{bmatrix} \frac{1}{2} & \frac{5}{18} & \frac{2}{9} \\ \frac{4}{15} & \frac{5}{18} & \frac{4}{45} \\ \frac{1}{8} & \frac{5}{16} & \frac{5}{12} \end{bmatrix} > \mathbf{0}. \tag{5.46}$$

Compute the characteristic polynomial of matrices **C** and **D**

$$P_3(\lambda) = \det(\mathbf{D} - \lambda \mathbf{I}_3) = \det(\mathbf{C} - \lambda \mathbf{I}_3) = -\lambda^3 + \frac{43}{36}\lambda^2 + -\frac{1}{3}\lambda + \frac{35}{1296}. \tag{5.47}$$

Factorise the polynomial, obtaining $P_3(\lambda) = (\lambda - \frac{5}{6})(\lambda - \frac{1}{6})(\lambda - \frac{7}{36})$. As the matrices **D** and **C** are positive and there is a surplus, Lemma 4.1.1 (b) applies and the Frobenius number is $0 < \lambda_C = 1/(1+R) = 5/6 = 1/(1+\frac{1}{5}) < 1$, corresponding to the *productiveness* $R = 0.2 > 0$.

(1). The right eigenvectors of the *positive* matrix **D**, associated with the Frobenius number $\lambda_C = 5/6$, are calculated, resulting in $\mathbf{y} = k[4/3, 4/5, 1]' > 0$, $k \in \mathbb{R}^+$ (see equation (5.32)). Then determine the diagonal matrix of the *orthogonal Euler map*, which will lead to the targeted *Standard system* and *Standard commodity*:

$$\hat{\mathbf{y}} = k \begin{bmatrix} \frac{4}{3} & 0 & 0 \\ 0 & \frac{4}{5} & 0 \\ 0 & 0 & 1 \end{bmatrix}. \tag{5.48}$$

The Euler mapped matrix and Euler mapped vectors are determined in analogy to equations (5.22) and (5.26) in, in order to give the required quantity of iron output:

$$\breve{\mathbf{S}} = \mathbf{S}\hat{\mathbf{y}}, \quad \breve{\mathbf{q}} = \hat{\mathbf{y}}\mathbf{q} = k \begin{bmatrix} \frac{4}{3} & 0 & 0 \\ 0 & \frac{4}{5} & 0 \\ 0 & 0 & 1 \end{bmatrix} \begin{bmatrix} 180 \\ 450 \\ 480 \end{bmatrix} = \begin{bmatrix} 1{,}080 \\ \breve{q}_{12} \\ \breve{q}_{13} \end{bmatrix}, \quad \breve{\mathbf{L}} = \hat{\mathbf{y}}\mathbf{L}. \tag{5.49}$$

We get from (5.49) the linear equation $1{,}080 = k \cdot \frac{4}{3} \cdot 180$ to determine k, and immediately find $k = 4.5$. This gives the required *Standard commodity*, described by the following matrix and vectors:

$$\hat{\mathbf{y}}_1 = 4.5 \begin{bmatrix} \frac{4}{3} & 0 & 0 \\ 0 & \frac{4}{5} & 0 \\ 0 & 0 & 1 \end{bmatrix} = \begin{bmatrix} 6 & 0 & 0 \\ 0 & 3.6 & 0 \\ 0 & 0 & 4.5 \end{bmatrix},$$

$$\breve{\mathbf{S}}_1 = \mathbf{S}\hat{\mathbf{y}}_1 = \begin{bmatrix} 90 & 50 & 40 \\ 120 & 125 & 40 \\ 60 & 150 & 200 \end{bmatrix} \begin{bmatrix} 6 & 0 & 0 \\ 0 & 3.6 & 0 \\ 0 & 0 & 4.5 \end{bmatrix} = \begin{bmatrix} 540 & 180 & 180 \\ 720 & 450 & 180 \\ 360 & 540 & 900 \end{bmatrix},$$

$$\breve{\mathbf{q}}_1 = \hat{\gamma}_1 \mathbf{q} = \begin{bmatrix} 6 & 0 & 0 \\ 0 & 3.6 & 0 \\ 0 & 0 & 4.5 \end{bmatrix} \begin{bmatrix} 180 \\ 450 \\ 480 \end{bmatrix} = \begin{bmatrix} 1{,}080 \\ 1{,}620 \\ 2{,}160 \end{bmatrix},$$

$$\breve{\mathbf{L}}_1 = \hat{\gamma}_1 \mathbf{L} = \begin{bmatrix} 6 & 0 & 0 \\ 0 & 3.6 & 0 \\ 0 & 0 & 4.5 \end{bmatrix} \begin{bmatrix} \frac{3}{16} \\ \frac{5}{16} \\ \frac{8}{16} \end{bmatrix} = \begin{bmatrix} \frac{18}{16} \\ \frac{18}{16} \\ \frac{36}{16} \end{bmatrix},$$

$$\breve{\mathbf{S}}_1 \mathbf{e} = \begin{bmatrix} 900 \\ 1{,}350 \\ 1{,}800 \end{bmatrix}; \quad \boxed{\breve{\mathbf{d}}_1 = \breve{\mathbf{q}}_1 - \breve{\mathbf{S}}_1 \mathbf{e} = \begin{bmatrix} 1{,}080 \\ 1{,}620 \\ 2{,}160 \end{bmatrix} - \begin{bmatrix} 900 \\ 1{,}350 \\ 1{,}800 \end{bmatrix} = \begin{bmatrix} 180 \\ 270 \\ 360 \end{bmatrix}}. \tag{5.50}$$

We clearly recognize the required parallelism of the vectors $\breve{\mathbf{d}}_1 \parallel \breve{\mathbf{q}}_1 \parallel \breve{\mathbf{S}}_1\mathbf{e}$. For the *productiveness* $R = 0.2$, the dilatation equation of *Recapitulation* 5.2.1 *(5.25)* gives

$$\breve{\mathbf{q}}_1 = (1 + R)\breve{\mathbf{S}}_1 \mathbf{e} = 1.2 \cdot \begin{bmatrix} 900 \\ 1{,}350 \\ 1{,}800 \end{bmatrix} = \begin{bmatrix} 1{,}080 \\ 1{,}620 \\ 2{,}160 \end{bmatrix}. \tag{5.51}$$

We have set up the *Standard commodity* $\{1{,}080, 1{,}620, 2{,}160\}$ with $q_1 = 1{,}080$. Moreover, in this example, the *total quantity of labour L* expressed in initial TAL's is:

$$L_1 = \breve{\mathbf{L}}_1' \mathbf{e} = \begin{bmatrix} \frac{18}{16} & \frac{18}{16} & \frac{36}{16} \end{bmatrix} \begin{bmatrix} 1 \\ 1 \\ 1 \end{bmatrix} = 4.5 \text{ TAL}. \tag{5.52}$$

(2). The right eigenvectors of the *positive* matrix **C**, associated with the Frobenius number $\lambda_C = 5/6$, are calculated (see (5.44)) giving $\breve{\mathbf{q}} = k[1/2, 3/4, 1]' > 0$, $k \in \mathbb{R}^+$. To obtain the vector of *total output* of Example 5.2.4, (5.51), we calibrate and set $k = 2{,}160$, giving $\breve{\mathbf{q}}_1 = k[1/2, 3/4, 1]' = [1{,}080, 1{,}620, 2{,}160]'$, again the required *Standard commodity.*

In conclusion, the *right eigenvectors* $k[\frac{1}{2}, \frac{3}{4}, 1]'$ of matrix **C**, equation (5.44), expressed in quantity terms are of course parallel to the mapped vector of *means of production* $\breve{\mathbf{S}}\mathbf{e} = [150, 225, 300]'$, to the mapped vector of *total output* $\breve{\mathbf{q}} = [180, 270, 360]'$ (5.22) and to the mapped vector of *surplus* $\breve{\mathbf{d}} = \breve{\mathbf{S}}\mathbf{e} - \breve{\mathbf{q}} = [30, 45, 60]'$, (5.23) expressed in quantity terms. They represent *Standard commodities* (PCMC, Par. 25).

We also have $\breve{\mathbf{L}} = \hat{\boldsymbol{\gamma}}\mathbf{L} = [3/16, 3/16, 6/16]'$. The obtained system of production $(\breve{\mathbf{S}}', \breve{\mathbf{q}}, \breve{\mathbf{L}}) \in \mathcal{E}$ is the *Standard system* of Example 5.2.1 (second part of PCMC Par. 25).

Setting $k = 3/4$ in the diagonal matrix $\hat{\boldsymbol{\gamma}}$ (5.48), we get exactly Sraffa's multipliers (PCMC, Par. 25), i. e., $\boldsymbol{\gamma} = [1, \frac{3}{5}, \frac{3}{4}]'$ of the *Standard system*! ▲

We note that we are generalizing the procedure of PCMC, Par. 25, generating *Standard systems* and *Standard commodities*. To this effect, we have had to solve two different eigenvalue problems, giving

(1) the *right eigenvectors* $\mathbf{y}$ of the *semi-positive* and *irreducible* matrix $\mathbf{D}$, associated with its Frobenius number λ_D, for the determination of the *multipliers*;
(2) the *right eigenvectors* $\breve{\mathbf{q}}$ of the *semi-positive* and *irreducible* matrix $\mathbf{C}$, associated with its Frobenius number $\lambda_C = \lambda_D$, for the determination of parallel vectors fixing *Standard commodities.*

Let us summarise from different angles the results obtained from this example. The first treats the process of generation of $\mathcal{E}$ and its subsystem composed of *Standard systems*.

Recapitulation 5.2.3. Consider a *semi-positive* and *irreducible* or *positive commodity flow matrix* $\mathbf{S} \geq \mathbf{0}$, a *semi-positive* vector of surplus $\mathbf{d} \geq \mathbf{o}$ and a vector of quantities of *labour* $\mathbf{L}$. Then the non-standard *system of production*, called the actual economic system, $(\mathbf{S}', \mathbf{q}, \mathbf{L})$, and the diagonal matrix $\hat{\mathbf{y}}$ (5.18) of positive multipliers $y_i > 0$ generate the *Euler class* $\mathcal{E} = \{(\mathbf{S}', \mathbf{q}, \mathbf{L}) \mid \hat{\mathbf{y}}\}$ of systems of production.

In PCMC, Par. 25, a construction of a Standard system $(\breve{\mathbf{S}}', \breve{\mathbf{q}}, \breve{\mathbf{L}}) \in \mathcal{E}$ is generated. $(\breve{\mathbf{S}}', \breve{\mathbf{q}}, \breve{\mathbf{L}}) \in \mathcal{E}$. The proportions of the components of the vectors $\breve{\mathbf{S}}\mathbf{e}$, $\breve{\mathbf{q}}$, (5.22), and $\breve{\mathbf{d}}$, (5.23) are determined by the components of the (right) eigenvectors $\breve{\mathbf{q}}$ of matrix $\mathbf{C} = \breve{\mathbf{S}}\hat{\breve{\mathbf{q}}}^{-1} = \mathbf{S}\hat{\mathbf{q}}^{-1}$, (5.46) and (5.42) associated with the Frobenius number λ_C:

$$a : b : c = \frac{1}{2} : \frac{3}{4} : 1 = 180 : 270 : 360 = 150 : 225 : 300 = 30 : 45 : 60.$$

The three vectors are parallel: $\breve{\mathbf{S}}\mathbf{e} \parallel \breve{\mathbf{q}} \parallel \breve{\mathbf{d}}$. Each one represents therefore a *Standard commodity*, and each one is a (right) eigenvector $\breve{\mathbf{a}}$ of matrix $\mathbf{C}$, associated to the Frobenius number λ_C, in this case $\breve{\mathbf{a}} = k[\frac{1}{2}, \frac{3}{4}, 1]' > 0$, $k \in \mathbb{R}^+$.

The second recapitulation treats properties of the Euler class $\mathcal{E}$.

Recapitulation 5.2.4. The *input-output coefficients* matrix $\mathbf{C} = \mathbf{S}\hat{\mathbf{q}}^{-1}$ and the distribution coefficients matrix $\mathbf{D} = \hat{\mathbf{q}}^{-1}\mathbf{S}$, both semi-positive and irreducible or positive, are associated with the Euler class $\mathcal{E} = \{(\mathbf{S}', \mathbf{q}, \mathbf{L}) \mid \hat{\mathbf{y}}\}$ of systems of production, common to each Standard system $(\breve{\mathbf{S}}', \breve{\mathbf{q}}, \breve{\mathbf{L}}) \in \mathcal{E}$.

The components of the positive right eigenvectors $\mathbf{y}$, associated with the Frobenius number $\lambda_D = \lambda_C = 1/(1+R)$ of matrix $\mathbf{D}$ determine the positive multipliers, generating the *Standard systems* in $\mathcal{E}$.

The diagonalization of the eigenvectors $\mathbf{y}$ gives the matrices $\hat{\mathbf{y}}$.

The matrix $\breve{\mathbf{S}} = \mathbf{S}\hat{\mathbf{y}}$, the vectors $\breve{\mathbf{q}} = \hat{\mathbf{y}}\mathbf{q}$, $\breve{\mathbf{L}} = \hat{\mathbf{y}}\mathbf{L}$ and $\breve{\mathbf{d}} = \hat{\mathbf{y}}\mathbf{d}$ determine three parallel vectors $\breve{\mathbf{S}}\mathbf{e} \parallel \breve{\mathbf{q}} \parallel \breve{\mathbf{d}}$ and the resulting *Standard systems* $(\breve{\mathbf{S}}', \breve{\mathbf{q}}, \breve{\mathbf{L}})$.

The proportions between the components of the right eigenvectors $\breve{\mathbf{q}}$ of matrix $\mathbf{C} = \mathbf{S}\hat{\mathbf{q}}^{-1}$ are also the proportions between the elements of any chosen *Standard commodity*, associated with the generated Standard system.

The vectors $\breve{\mathbf{S}}\mathbf{e} \parallel \breve{\mathbf{q}} \parallel \breve{\mathbf{d}}$ are parallel and the components of each one of these vectors taken as a set constitute a *Standard commodity*, Definition 5.1.1.

These recapitulations finalize the results. In the next section, we treat the question of a *Standard commodity* belonging to a *Standard system* where the *total quantity of labour* is dilated to $L_1 = 1$ TAL.

5.3 Constitution of a Standard net product*

In this section, we concentrate our presentation on the concept of the *Standard net product*. Sraffa writes (PCMC, Par. 26):

> "We shall as a rule find it convenient to take as unit of a *Standard commodity* the quantity of it that would form the net product of a *Standard system* employing the whole annual labour of the actual [economic] system."

> " ...Such a unit we shall call the *Standard net product* or *Standard national income*.[13]"

This means, a *Standard net product* is a *Standard commodity*, where the *amount of labour* is equal to the unit of 1 TAL. We have seen in Example 5.1.2 that the quantity of labour of the *actual economic system* is $L = 1$ TAL (5.12). Then, in Example 5.2.1 an *orthogonal Euler map* is applied to the corresponding *actual economic system*, generating the *Standard system* $(\check{\mathbf{S}}, \check{\mathbf{q}}, \check{\mathbf{L}})$, contained in the Euler class $\mathcal{E}$ with the quantity of labour $\check{\mathbf{L}}'\mathbf{e} = 3/4$ TAL (5.26). The result is presented in Recapitulation 5.2.1. Later on, a dilatation by factor $y = 4/3$ will be applied to all the elements of the above *Standard system* $(\check{\mathbf{S}}, \check{\mathbf{q}}, \check{\mathbf{L}})$ in order to attain a *Standard net product* with the quantity of labour $L_1 = 1$ TAL (see below (5.55)).

Sraffa continues (PCMC, Par. 29), indicating a proportion:

> *"Now suppose the Standard net product to be divided between wages and profits, taking care that the share of each consists always, as the whole does, of a Standard commodity: the resulting rate of profits r would be in the same proportion to the productiveness R of the system as the share allotted to profits was to the whole of the Standard net product."*

How do we express this proportion? Sraffa says that the proportion of the value of the profits, represented by a *Standard commodity*, i. e., a basket of commodities, to the value of the *Standard net product* as a whole is equal to r/R. The *rate of profits* is r, situated as usual in the interval, $r \in [0, R]$. Both cited values only make sense and can be calculated, if a price vector $\mathbf{p}$ is introduced (see next section). If the *Standard net product* is represented by the surplus vector $\mathbf{d}$, then its value is the national income: $Y = \mathbf{d}'\mathbf{p}$ (4.32). Let P designate total profits and W total wages. Then, Sraffa's statement reads:

$$Y = P + W, \quad \frac{r}{R} = \frac{P}{Y} \Rightarrow P = \frac{r}{R} \cdot Y. \tag{5.53}$$

13 We abstain here in using the term *Standard national income* (a scalar) whereas Sraffa's term *Standard net product* is a basket of commodities (represented by a vector).

Sraffa (PCMC, Par 30) also resorts to the notion of the *share of total wages* $\tilde{w} = W/Y$, $0 \leq \tilde{w} \leq 1$ (our notations, see Chapter 4) in order to distinguish this variable from w in W, $W = w \cdot L$ with $w \geq 0$). Finally we get[14]

$$Y = P + W = \frac{r}{R} \cdot Y + \tilde{w} \cdot Y \Rightarrow \frac{r}{R} + \tilde{w} = 1 \Rightarrow r = R(1 - \tilde{w}). \tag{5.54}$$

Now we concentrate on Sraffa's *Standard system*, presented in PCMC, Par. 25, from which Sraffa gets by dilatation a *Standard system* with a *Standard net product*. In PCMC, Par. 33, Sraffa sets equation (5.27) to calculate the multipliers and then gives the clue for normalization of labour:

"*...and since we wish the quantity of labour employed in the Standard system to be the same as in the actual system (PCMC, Par. 26) (namely normalized: the authors), we define the unit by an additional equation which embodies that condition, namely:*

$$\breve{\mathbf{S}}\mathbf{y}(1 + R) = \hat{\breve{\mathbf{q}}}\mathbf{y}, \quad L_1 = \breve{\mathbf{L}}_1'\mathbf{e} = y\breve{\mathbf{L}}'\mathbf{e} = y\hat{\mathbf{y}}\mathbf{L}'\mathbf{e} = 1\,\text{TAL.''}^{15} \tag{5.55}$$

Then, from the first equation of (5.55), the eigenvalue equation $\mathbf{D}(1+R)\mathbf{y} = \mathbf{y}$ is obtained to get the multipliers $\mathbf{y}$, using the *distribution coefficients matrix* $\mathbf{D} = \hat{\breve{\mathbf{q}}}^{-1}\breve{\mathbf{S}} = \hat{\mathbf{q}}^{-1}\mathbf{S}$ of the Euler class $\mathcal{E}$ (see Lemma 5.2.1). With the second equation of (5.55) the dilatation coefficient y is calculated. We continue with an illustration of this dilatation process.

Example 5.3.1. Consider the *Standard system* obtained in Example 5.2.1, $(\breve{\mathbf{S}}', \breve{\mathbf{q}}, \breve{\mathbf{L}})$ (PCMC, Par. 25) and proceed with a dilatation (5.55) on $(\breve{\mathbf{S}}', \breve{\mathbf{q}}, \breve{\mathbf{L}})$ to obtain the *Standard net product* with an amount of labour of $L_1 = 1$ TAL (PCMC, Par. 26).

Solution to Example 5.3.1:
We begin with the vector of labour $\breve{\mathbf{L}}$ (5.26). We set $y = 4/3$ and calculate with a dilatation the unit of $L_1 = 1$ TAL:

$$L_1 = \breve{\mathbf{L}}_1'\mathbf{e} = y\breve{\mathbf{L}}'\mathbf{e} = \frac{4}{3} \cdot \left[\frac{3}{16}, \frac{3}{16}, \frac{6}{16}\right] \cdot \begin{bmatrix} 1 \\ 1 \\ 1 \end{bmatrix} = 1\,\text{TAL}, \tag{5.56}$$

so the coefficient of *dilatation* is $y = 4/3$. We accordingly get the vector of labour,

$$\breve{\mathbf{L}}_1' = \left[\begin{array}{ccc} \frac{4}{16} & \frac{4}{16} & \frac{8}{16} \end{array}\right] = \left[\begin{array}{ccc} \frac{1}{4} & \frac{1}{4} & \frac{1}{2} \end{array}\right]. \tag{5.57}$$

---

14 Sraffa normalises throughout his analysis labour $L = 1$ TAL and national income $Y = 1$ GDP. For this reason, he gets $W = w \cdot 1\,\text{TAL} = \tilde{w} \cdot 1\,\text{GDP} \Rightarrow |w| = |\tilde{w}| = |W|$. Therefore Sraffa presents the equation $r = R(1 - w)$ (see Sraffa (PCMC, Par. 30)). Note also that $P = \tilde{r}Y$, so $r = \tilde{r}\tilde{R}$ and of course $\tilde{r} + \tilde{w} = 1$.

15 Equation (5.55) is written in our *notation* and not in Srafa's notation of PCMC, Par. 33.

Then we proceed to determine the *Standard net product*, resorting to the diagonal matrix $\hat{\mathbf{y}}_d$, generated by the dilatation vector $\mathbf{y}_d = [y \;\; y \;\; y]' = [\frac{4}{3} \;\; \frac{4}{3} \;\; \frac{4}{3}]'$:

$$\hat{\mathbf{y}}_d = \begin{bmatrix} y & 0 & 0 \\ 0 & y & 0 \\ 0 & 0 & y \end{bmatrix} = \begin{bmatrix} \frac{4}{3} & 0 & 0 \\ 0 & \frac{4}{3} & 0 \\ 0 & 0 & \frac{4}{3} \end{bmatrix}. \tag{5.58}$$

We apply the dilatation matrix (5.58) to the vector of *labour* $\check{\mathbf{L}}$ (5.26), to the *commodity flow* matrix $\check{\mathbf{S}}$ (5.22) and to the vector of *total output* $\check{\mathbf{q}}$ (5.22). This gives in detail:

$$\check{\mathbf{L}}_1 = \hat{\mathbf{y}}_d\check{\mathbf{L}} = \begin{bmatrix} y & 0 & 0 \\ 0 & y & 0 \\ 0 & 0 & y \end{bmatrix}\begin{bmatrix} \check{L}_1 \\ \check{L}_2 \\ \check{L}_3 \end{bmatrix} = \begin{bmatrix} \frac{4}{3} & 0 & 0 \\ 0 & \frac{4}{3} & 0 \\ 0 & 0 & \frac{4}{3} \end{bmatrix}\begin{bmatrix} \frac{3}{16} \\ \frac{3}{16} \\ \frac{6}{16} \end{bmatrix} = \begin{bmatrix} \frac{1}{4} \\ \frac{1}{4} \\ \frac{1}{2} \end{bmatrix}, \tag{5.59}$$

$$\check{\mathbf{S}}_1 = \check{\mathbf{S}}\hat{\mathbf{y}}_d = \begin{bmatrix} 90 & 30 & 30 \\ 120 & 75 & 30 \\ 60 & 90 & 150 \end{bmatrix}\begin{bmatrix} \frac{4}{3} & 0 & 0 \\ 0 & \frac{4}{3} & 0 \\ 0 & 0 & \frac{4}{3} \end{bmatrix} = \begin{bmatrix} 120 & 40 & 40 \\ 160 & 100 & 40 \\ 80 & 120 & 200 \end{bmatrix},$$

$$\check{\mathbf{q}}_1 = \hat{\mathbf{y}}_d\check{\mathbf{q}} = \begin{bmatrix} \frac{4}{3} & 0 & 0 \\ 0 & \frac{4}{3} & 0 \\ 0 & 0 & \frac{4}{3} \end{bmatrix}\begin{bmatrix} 180 \\ 270 \\ 360 \end{bmatrix} = \begin{bmatrix} 240 \\ 360 \\ 480 \end{bmatrix}. \tag{5.60}$$

And thus we arrive at the production scheme:

$$\begin{aligned} &\left(120 \text{ t. iron}, 160 \text{ t. coal}, 80 \text{ qr. wheat}, \frac{1}{4} \text{ labour}\right) \to (240 \text{ t. iron}, 0, 0), \\ &\left(40 \text{ t. iron}, 100 \text{ t. coal}, 120 \text{ qr. wheat}, \frac{1}{4} \text{ labour}\right) \to (0, 360 \text{ t. coal}, 0), \\ &\left(40 \text{ t. iron}, 40 \text{ t. coal}, 200 \text{ qr. wheat}, \frac{1}{2} \text{ labour}\right) \to (0, 0, 480 \text{ t. wheat}), \end{aligned} \tag{5.61}$$

with the *total means of production*

$$\check{\mathbf{S}}_1\mathbf{e} = \begin{bmatrix} 120 & 40 & 40 \\ 160 & 100 & 40 \\ 80 & 120 & 200 \end{bmatrix}\begin{bmatrix} 1 \\ 1 \\ 1 \end{bmatrix} = \begin{bmatrix} 200 \\ 300 \\ 400 \end{bmatrix}. \tag{5.62}$$

The vector of surplus $\check{\mathbf{d}}_1 = \check{\mathbf{q}}_1 - \check{\mathbf{S}}_1\mathbf{e} = [40, 60, 80]'$, Figure 5.3 (right), (5.60), (5.62), representing the *Standard net product* {40 t. of iron, 60 t. of coal, 80 qr. of wheat} is illustrated in Figure 5.3 (right). ▲

Recapitulation 5.3.1. The proportions $40 : 60 : 80 = 200 : 300 : 400 = 240 : 360 : 480$ follow from the parallelism $\check{\mathbf{d}}_1 \parallel \check{\mathbf{q}}_1 \parallel \check{\mathbf{S}}_1\mathbf{e}$. Aggregate labour is dilated to $L_1 = \check{\mathbf{L}}_1\mathbf{e} = 1$,

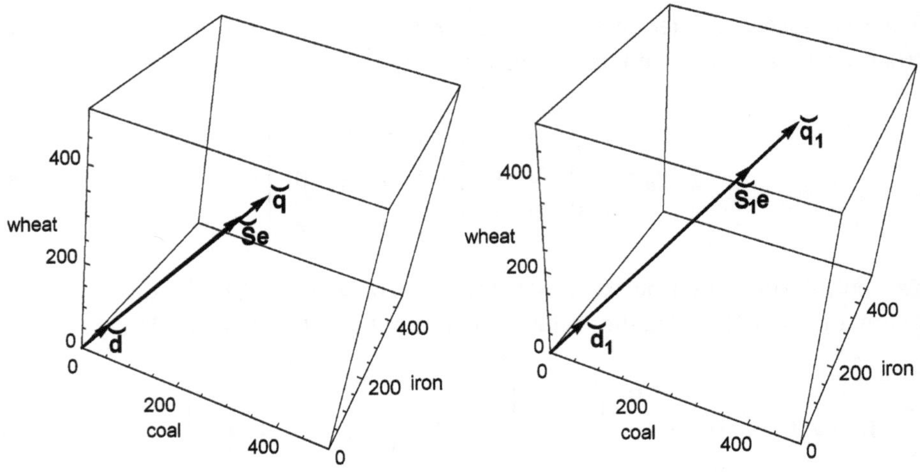

Figure 5.3: *Standard system* (5.20), with total labour $\frac{12}{16}$, with *Standard net product* (5.61), and total labour $L_1 = \frac{16}{16} = 1$.

and the Standard system $(\breve{\mathbf{S}}_1', \breve{\mathbf{q}}_1, \breve{\mathbf{L}}_1)$, obtained by dilatation with the diagonal matrix $\hat{\mathbf{y}}_d$ (5.58), leads to Standard commodities, represented by each of the vectors $\breve{\mathbf{d}}_1$, $\breve{\mathbf{q}}_1$, $\breve{\mathbf{S}}_1\mathbf{e}$. The Standard net product {40 t. of iron, 60 t. of coal, 80 qr. of wheat} is represented by the vector of surplus $\breve{\mathbf{d}}_1 = \breve{\mathbf{q}}_1 - \breve{\mathbf{S}}_1\mathbf{e} = [40, 60, 80]'$, see Notation 5.2.2.

Example 5.3.2. Compute directly the Standard system $(\breve{\mathbf{S}}_1', \breve{\mathbf{q}}_1, \breve{\mathbf{L}}_1)$ (5.59), (5.60), (5.61) out of the constituting elements of the non-standard *actual economic system* (5.9), (PCMC, Par. 25), using the diagonal matrix $\hat{\mathbf{y}}$ (5.21) of the multipliers, obtained in Example 5.2.4 by equation (5.48) and the dilatation parameter $y = 4/3$. Confirm the quantity of labour $L_1 = 1$ corresponding to the obtained *Standard net product* represented by the resulting surplus.

Solution to Example 5.3.2:
Consider the matrices of the non-standard *actual economic system* $(\mathbf{S}', \mathbf{q}, \mathbf{L}) \in \mathcal{E}$, Example 5.1.2, (5.9). Apply at first the *orthogonal Euler map*, represented by the diagonal matrix $\hat{\mathbf{y}}$ (5.21) multiplied by the diagonal *dilatation* matrix $\hat{\mathbf{y}}_d$ (5.58), to the matrix $\mathbf{S}$ and vectors $\mathbf{q}$, $\mathbf{L}$. This will generate directly the *Standard system* $(\breve{\mathbf{S}}_1', \breve{\mathbf{q}}_1, \breve{\mathbf{L}}_1)$. We calculate the transformation matrix,

$$\hat{\mathbf{y}}_d\hat{\mathbf{y}} = \begin{bmatrix} \frac{4}{3} & 0 & 0 \\ 0 & \frac{4}{3} & 0 \\ 0 & 0 & \frac{4}{3} \end{bmatrix} \begin{bmatrix} 1 & 0 & 0 \\ 0 & \frac{3}{5} & 0 \\ 0 & 0 & \frac{3}{4} \end{bmatrix} = \begin{bmatrix} \frac{4}{3} & 0 & 0 \\ 0 & \frac{4}{5} & 0 \\ 0 & 0 & 1 \end{bmatrix}, \tag{5.63}$$

of the composed *orthogonal Euler map*. Then, calculate the matrix $\breve{\mathbf{S}}_1$ and the vector $\breve{\mathbf{q}}_1$ of this *Standard system*, described by the equations (5.9), (5.49) $\breve{\mathbf{S}} = \mathbf{S}\hat{\mathbf{y}}$ and (5.60) $\breve{\mathbf{S}}_1 = \breve{\mathbf{S}}\hat{\mathbf{y}}_d$, giving $\breve{\mathbf{S}}_1 = (\mathbf{S}\hat{\mathbf{y}})\hat{\mathbf{y}}_d = \mathbf{S}(\hat{\mathbf{y}}\hat{\mathbf{y}}_d) = \mathbf{S}(\hat{\mathbf{y}}_d\hat{\mathbf{y}})$, and $\breve{\mathbf{q}}_1 = (\hat{\mathbf{y}}_d\hat{\mathbf{y}})\mathbf{q}$ so,

$$\breve{\mathbf{S}}_1 = \mathbf{S}(\hat{\mathbf{y}}_d\hat{\mathbf{y}}) = \begin{bmatrix} 90 & 50 & 40 \\ 120 & 125 & 40 \\ 60 & 150 & 200 \end{bmatrix} \begin{bmatrix} \frac{4}{3} & 0 & 0 \\ 0 & \frac{4}{5} & 0 \\ 0 & 0 & 1 \end{bmatrix} = \begin{bmatrix} 120 & 40 & 40 \\ 160 & 100 & 40 \\ 80 & 120 & 200 \end{bmatrix}, \tag{5.64}$$

$$\breve{\mathbf{q}}_1 = (\hat{\mathbf{y}}_d\hat{\mathbf{y}})\mathbf{q} = \begin{bmatrix} \frac{4}{3} & 0 & 0 \\ 0 & \frac{4}{5} & 0 \\ 0 & 0 & 1 \end{bmatrix} \begin{bmatrix} 180 \\ 450 \\ 480 \end{bmatrix} = \begin{bmatrix} 240 \\ 360 \\ 480 \end{bmatrix}. \tag{5.65}$$

Of course, we can also compute the mapped *Labour* vector $\breve{\mathbf{L}}_1$ out of the initial *Labour* vector $\mathbf{L}$ with the composed diagonal matrix $\hat{\mathbf{y}}_d\mathbf{y}$. Indeed,

$$\breve{\mathbf{L}}_1 = (\hat{\mathbf{y}}_d\hat{\mathbf{y}})\mathbf{L} = \begin{bmatrix} \frac{4}{3} & 0 & 0 \\ 0 & \frac{4}{5} & 0 \\ 0 & 0 & 1 \end{bmatrix} \begin{bmatrix} \frac{3}{16} \\ \frac{5}{16} \\ \frac{8}{16} \end{bmatrix} = \begin{bmatrix} \frac{4}{16} \\ \frac{4}{16} \\ \frac{8}{16} \end{bmatrix} = \begin{bmatrix} \frac{1}{4} \\ \frac{1}{4} \\ \frac{1}{2} \end{bmatrix}. \tag{5.66}$$

Geometrically this means that the *Standard system* $(\breve{\mathbf{S}}_1', \breve{\mathbf{q}}_1, \breve{\mathbf{L}}_1)$ has been constructed from the initial *actual economic system* (5.9) by composition of an *orthogonal Euler map* and a *dilatation*, giving again an *orthogonal Euler map*, represented by the diagonal matrix $\hat{\mathbf{y}}_d\hat{\mathbf{y}}$ (5.63).

Figure 5.4 presents the scheme to generate a particular *Standard system* from a non-standard *actual economic system*, consisting only of *basic products*, followed by its transformation to a first *Standard system*, then by dilatation to a further *Standard system* exhibiting a *Standard net product*. As has been shown, the triangle of vectors $(\mathbf{Se}, \mathbf{d}, \mathbf{q})$ characterises a *non-standard system of production*, whereas the systems of parallel vectors, $\breve{\mathbf{S}}\mathbf{e} \parallel \breve{\mathbf{d}} \parallel \breve{\mathbf{q}}$ and $\breve{\mathbf{S}}_1\mathbf{e} \parallel \breve{\mathbf{d}}_1 \parallel \breve{\mathbf{q}}_1$ characterise *Standard systems*. The vector $\breve{\mathbf{d}}_1 = [\breve{d}_{11}, \breve{d}_{12}, \breve{d}_{13}]' = \breve{\mathbf{q}}_1 - \breve{\mathbf{S}}_1\mathbf{e}$ represents the *Standard net product* corresponding to $L_1 = \mathbf{e}'\breve{\mathbf{L}}_1 = 1$ TAL. ▲

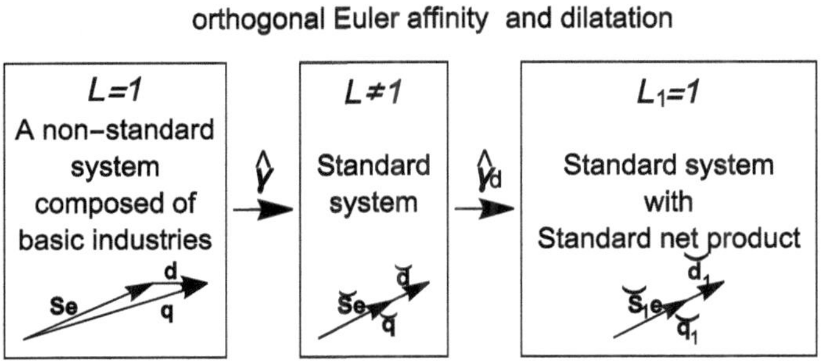

Figure 5.4: Generation of a *Standard system* with a *Standard net product* produced from a non-standard *actual economic system* in two stages, with the resulting *total quantity of labour*.

In the next section, we present the construction of a *Standard system* with prices and the associated economic variables.

5.4 Construction of a Standard system with prices*

As we have seen, Sraffa starts from a non-standard *actual economic system* with surplus, calculating multipliers to attain a *Standard system*, characterised by a positive *productiveness* R, attains by dilatation a further *Standard system*, called a *Standard net product* (PCMC, Par. 26), exhibiting the total *quantity of labour* $L = 1$. Then he introduces a constant rate of profits r, from which depends the *constant wage* w *per unit of labour* (PCMC, Par. 10). He then calculates the prices and thus presents in PCMC, Par. 33, 34, the complete construction of the *price model*, cumulating in the *Standard national income* (PCMC, Par. 34).

5.4.1 From a non-Standard system to a Standard system

Let's now review in detail the various steps of the construction of the *Sraffa price model* for *Standard systems*:

(1) *Setting up a non-standard system.* We start with a *semi-positive* and *irreducible* or *positive* $n \times n$ *commodity flow* matrix $\mathbf{S}$ and a *semi-positive* vector of *surplus* $\mathbf{d} \geq \mathbf{o}$, together with a *normalized* vector of *labour* $\mathbf{L}$, $\mathbf{e}'\mathbf{L} = 1$, and the vector of *total output*, $\mathbf{q} = \mathbf{Se} + \mathbf{d} > \mathbf{0}$. The non-standard system of production $(\mathbf{S}', \mathbf{q}, \mathbf{L}) \in \mathcal{E}$, see Lemma 5.2.1 is constituted. The quantity vectors $\mathbf{Se}$, $\mathbf{d}$, $\mathbf{q}$ are not parallel, $\mathbf{Se} \nparallel \mathbf{d} \nparallel \mathbf{q}$.

(2) *Construction of a Standard system.* Take matrix $\mathbf{S}$ and vector $\mathbf{q}$ of the aforementioned system of production $(\mathbf{S}', \mathbf{q}, \mathbf{L}) \in \mathcal{E}$ and compute the unique *non-negative* and *irreducible* or *positive distribution coefficients matrix* $\mathbf{D} = \hat{\mathbf{q}}^{-1}\mathbf{S}$, appearing in the *right eigenvector* equation (5.32) reproduced here,

$$\mathbf{D}\mathbf{y}(1 + R) = \mathbf{y}. \tag{5.67}$$

Compute the Frobenius number $0 < \lambda_D = 1/(1 + R) < 1$, Lemma 4.1.1 (b), getting the positive *productiveness* $R = (1/\lambda_D) - 1 > 0$. Now we can compute the positive *right eigenvector vectors* determining the *multipliers* $\mathbf{y}$, with positive components are $y_1, y_2, \ldots, y_n$, associated with the Frobenius number λ_D. From vector $\mathbf{y}$ is set up the diagonal matrix $\hat{\mathbf{y}}$ (A.60), generating an *orthogonal Euler map* applied to the vector of *aggregate means of production* $\mathbf{Se}$, the vector of *surplus* $\mathbf{d}$ and the vector of *total output* $\mathbf{q}$.

Generally, one chooses by an appropriated calibration a unique *eigenvector* $\mathbf{y}$ from the calculated eigenspace of the (right) eigenvectors of matrix $\mathbf{D}$ (5.67). Consider that this is now done.

At this stage, we set up the elements of a *Standard system*, the mapped *commodity flow* matrix $\check{\mathbf{S}} = \mathbf{S}\hat{\mathbf{y}}$ (5.14), the mapped *total output* vector $\check{\mathbf{q}} = \hat{\mathbf{y}}\mathbf{q}$, the mapped *aggregate means of production* $\check{\mathbf{S}}\mathbf{e}$ (5.22), the positive mapped vector of *surplus* $\check{\mathbf{d}} = \check{\mathbf{q}} - \check{\mathbf{S}}\mathbf{e} > \mathbf{o}$ (5.23) and the mapped vector of *labour* $\check{\mathbf{L}} = \hat{\mathbf{y}}\mathbf{L}$, no longer normalised.

This obtained *system of production* $(\check{\mathbf{S}}', \check{\mathbf{q}}, \check{\mathbf{L}})$ with $\check{\mathbf{S}}\mathbf{e} \parallel \check{\mathbf{q}} \parallel \check{\mathbf{d}} \nparallel \check{\mathbf{L}}$ is a *Standard system*, see Recapitulation 5.2.3.

(3) *Calculation of the input-output coefficients matrix* **C**. Consider again the actual economic system and the above eigenvector γ. Constitute the *Euler class* $\mathcal{E} = \{\mathbf{S}', \mathbf{q}, \mathbf{L}|\gamma\}$ and compute the *right eigenvectors* $\check{\mathbf{q}}$ of the unique *input coefficients* matrix $\mathbf{C} = \mathbf{S}\hat{\mathbf{q}}^{-1}$ (5.29), (Lemma 5.2.1) of the present *Euler class* $\mathcal{E}$.
One can accordingly concentrate on the eigenvalue equation (5.44), left side, of the *semi-positive* and *irreducible* or *positive input-output coefficients* matrix **C** which we also reproduce here, knowing that there are equal Frobenius numbers $\lambda_C = \lambda_D$,

$$\mathbf{C}\check{\mathbf{q}} = \lambda_C \check{\mathbf{q}} = \frac{1}{1+R}\check{\mathbf{q}}. \tag{5.68}$$

This means that the determination of the vector of *total output* $\check{\mathbf{q}}$ in the *Standard system* $(\check{\mathbf{S}}', \check{\mathbf{q}}, \check{\mathbf{L}})$ is an *eigenvector problem* with Frobenius number λ_C of the above *input-output coefficients* matrix **C**.
The *non-negative* and *irreducible* or *positive* matrices $\mathbf{C} = \mathbf{S}\hat{\mathbf{q}}^{-1}$ and $\mathbf{D} = \hat{\mathbf{q}}^{-1}\mathbf{S}$ have identical eigenvalues λ, see Lemma A.6.1. The Frobenius number $0 < \lambda_C = \lambda_D = 1(/1+R) < 1$, see Lemma 4.1.1 (b), then determines the *productiveness* $R > 0$.

(4) *Construction of the Standard net product*. Only at this stage of the calculations does Sraffa discuss the role of the initial vector of *labour* $\mathbf{L} = [L_1, L_2, \ldots, L_n]'$ (PCMC, Par. 25, 26). The components $L_i, i = 1, \ldots, n$ of the vector of *labour* indicate the part of the *labour* that is used to produce the quantity q_i, relative to the total amount of labour that is normalised to $L = 1$ TAL and used as such by the *actual economic system* $(\mathbf{S}', \mathbf{q}, \mathbf{L})$. In general, having determined the *Standard system*, the vector of labour $\check{\mathbf{L}} = \hat{\gamma}\mathbf{L}$ is no longer normalized, $\check{\mathbf{L}}'\mathbf{e} \neq 1$. For this reason, at this stage, the vector of *labour* $\check{\mathbf{L}}$ is calibrated to the unity of 1 TAL, with a *dilatation* factor γ, generating the dilatation vector $\gamma_d = [\gamma_1, \gamma_2, ..., \gamma_n]'$, resulting in the diagonal matrix $\hat{\gamma}_d$, operating the *dilatation* on the *Standard system* $(\check{\mathbf{S}}', \check{\mathbf{q}}, \check{\mathbf{L}})$, see Example 5.3.2,

$$\begin{aligned} &\check{\mathbf{L}}_1 = (\hat{\gamma}_d\hat{\gamma})\mathbf{L}, \quad L_1 = \check{\mathbf{L}}_1'\mathbf{e} = (\hat{\gamma}_d\hat{\gamma})\mathbf{L}'\mathbf{e} = \hat{\gamma}_d(\hat{\gamma}\mathbf{L})'\mathbf{e} = 1, \\ &\check{\mathbf{S}}_1 = \mathbf{S}(\hat{\gamma}_d\hat{\gamma}), \quad \check{\mathbf{q}}_1 = (\hat{\gamma}_d\hat{\gamma})\mathbf{q}, \end{aligned} \tag{5.69}$$

in order to obtain a further *Standard system* $(\check{\mathbf{S}}_1', \check{\mathbf{q}}_1, \check{\mathbf{L}}_1)$ with the *Standard net product* $\check{\mathbf{d}}_1 = \check{\mathbf{S}}_1\mathbf{e} - \check{\mathbf{q}}_1$.

Sraffa then introduces *relative prices* for the commodities and a *wage rate* with a view to defining a price model and to determine the *national income* Y. The *means of payment* for commodities and wages is a *numéraire*, i. e., any physical unit of a produced commodity (like *qr. of* wheat) for wheat, contained in the present *Standard system* $(\check{\mathbf{S}}', \check{\mathbf{q}}, \check{\mathbf{L}})$.[16]

16 In the present analysis, we will consider enlarging the means of payment. The *numéraire* can be replaced by any currency (CHF, EURO, USD, ...) as the *means of payment*. The ongoing considerations remain valid.

(**5**) *Transformation of the Sraffa price model (4.55) from the non-standard actual economic system onto a Standard system.*
To realize this step, we have to go back to item **(1)** of this review. We have to choose a rate of profits r, $0 \leq r \leq R$. Sraffa (PCMC, Par. 25) starts from the non-standard *actual economic system* $(\mathbf{S}', \mathbf{q}, \mathbf{L}) \in \mathcal{E}$. Then Sraffa (PCMC, Par. 33, Par. 34) considers the positive vector $\mathbf{y}$ of *multipliers*, $y_1 > 0, \ldots, y_n > 0$, which are computed as components of the right eigenvectors of the *semi-positive* and *irreducible* or *positive distribution coefficients* matrix $\mathbf{D}$ (5.67). Vector $\mathbf{y}$ is diagonalised to get the matrix $\hat{\mathbf{y}}$ (A.60), Definition A.7.1.
The *Sraffa price model* (4.55), which determines the prices of the non-standard *system of production* $(\mathbf{S}', \mathbf{q}, \mathbf{L}) \in \mathcal{E}$ are multiplied from the left by the diagonal matrix of multipliers $\hat{\mathbf{y}}$ (5.41),

$$\hat{\mathbf{y}}\left(\mathbf{S}'\mathbf{p}(1+r) + \mathbf{L} \cdot \frac{\tilde{w} \cdot Y}{L}\right) = \hat{\mathbf{y}}(\hat{\mathbf{q}}\mathbf{p}) = (\hat{\mathbf{y}}\hat{\mathbf{q}})\mathbf{p}, \tag{5.70}$$

operating an *orthogonal Euler affinity* on vector $\mathbf{q}$, matrix $\mathbf{S}$ and the *labour* vector $\mathbf{L}$. We find with equations (5.70) and (A.38),

$$(\hat{\mathbf{y}}\mathbf{S}')\mathbf{p}(1+r) + (\hat{\mathbf{y}}\mathbf{L}) \cdot \frac{\tilde{w} \cdot Y}{L} = (\widehat{\hat{\mathbf{y}}\mathbf{q}})\mathbf{p}. \tag{5.71}$$

Remember that we have set $\breve{\mathbf{S}} = \mathbf{S}\hat{\mathbf{y}}$ (5.22), $\breve{\mathbf{q}} = \hat{\mathbf{y}}\mathbf{q}$ (5.22) and $\breve{\mathbf{L}} = \hat{\mathbf{y}}\mathbf{L}$ (5.26) for the matrices and vectors, resulting from this *orthogonal Euler map*. The *system of production* $(\breve{\mathbf{S}}', \breve{\mathbf{q}}, \breve{\mathbf{L}})$ with $\breve{\mathbf{S}}\mathbf{e} \parallel \breve{\mathbf{q}} \parallel \breve{\mathbf{d}}$ now constitutes a *Standard system* (see Recapitulation 5.2.3).

(**6**) *Determination of the ratio of national income to circulating capital.* Now we consider the obtained *Standard system* $(\breve{\mathbf{S}}', \breve{\mathbf{q}}, \breve{\mathbf{L}})$, together with the positive vector $\mathbf{p} = [p_1, \ldots, p_n]'$ of *relative prices*.[17] Taking Lemma 5.1.2, we have for a *Standard system* the equation $\breve{\mathbf{d}} = R \cdot \breve{\mathbf{S}}\mathbf{e}$ (here written with the breve-sign) and the ratio $\tilde{R}$ of *national income* to *circulating capital* is equal to the *productiveness* R. We get

$$Y = R \cdot (\breve{\mathbf{S}}\mathbf{e})'\mathbf{p}, \quad \tilde{R} = \frac{Y}{K} = \frac{\breve{\mathbf{d}}'\mathbf{p}}{(\breve{\mathbf{S}}\mathbf{e})'\mathbf{p}} = \frac{R \cdot (\breve{\mathbf{S}}\mathbf{e})'\mathbf{p}}{(\breve{\mathbf{S}}\mathbf{e})'\mathbf{p}} = R \Rightarrow Y = R \cdot K, \tag{5.72}$$

where Y is the *national income*.[18] National income is the R-th part of the *circulating capital* K. Then the general equation (4.36), connecting $\tilde{R}$, r and $\tilde{w}$, valid

---

17 The physical units of the prices have to be understood as in (3.2). Consider a system of production with $n = 2$ commodities, wheat and iron, where the arbitrary chosen *numéraire* is *quarters of* wheat. The physical units are *quarters of* wheat and *tons of* iron. Then the price p_1 of wheat and the price p_2 of iron are expressed in this numéraire, the units being $[p_1] = (\frac{\text{qr. wheat}}{\text{qr. wheat}}) = 1$ and $[p_2] = (\frac{\text{qr. wheat}}{\text{t. iron}})$.
18 Sraffa (PCMC, Par. 12, Par. 34) normalises national income, $Y = 1$. He calls it the *Standard national income*.

for *general systems of production*, becomes for *Standard systems* specifically with $\tilde{R} = R$,

$$r = R(1 - \tilde{w}). \tag{5.73}$$

- Replacing $\mathbf{S}$ by $\check{\mathbf{S}} = \mathbf{S}\hat{\mathbf{y}}$ (5.14), $\mathbf{q}$ by $\check{\mathbf{q}} = \hat{\mathbf{y}}\mathbf{q}$ (5.15) and $\mathbf{L}$ by $\check{\mathbf{L}} = \hat{\mathbf{y}}\mathbf{L}$ (5.26) in Sraffa's *price model* for a non-standard *system of production* (4.55) gives the *price model* for a *Standard system*

$$\check{\mathbf{S}}'\mathbf{p}(1 + r) + \check{\mathbf{L}} \cdot \left(\frac{\tilde{w} \cdot Y}{L}\right) = \hat{\check{\mathbf{q}}}\mathbf{p}. \tag{5.74}$$

- In Section 4.9 we presented Sraffa's complete *price model* for non-standard *systems of production* (4.174). We now want to obtain Sraffa's complete *price model* for *Standard systems* with the calculation of the *total quantity of labour* L. The former expression for the national income is replaced by equation (5.72), $Y = R \cdot (\check{\mathbf{S}}\mathbf{e})'\mathbf{p}$.

We then have $n + 2$ equations with $n + 3$ variables $p_1, p_2, \ldots, p_n, \tilde{w}, L, Y$ and the known positive parameter $R > 0$, calculated from the Frobenius number $0 < \lambda_D = (1/(1 + R)) < 1$ of matrix $\mathbf{D}$ (point **2**) and the chosen rate of profits $r \in [0, R]$, thus

$$\boxed{\begin{aligned} \check{\mathbf{S}}'\mathbf{p}(1 + r) + \check{\mathbf{L}}\frac{\tilde{w} \cdot Y}{L} &= \hat{\check{\mathbf{q}}}\mathbf{p}, \\ L &= \mathbf{e}'\check{\mathbf{L}}, \\ Y &= R \cdot (\check{\mathbf{S}}\mathbf{e})'\mathbf{p}. \end{aligned}} \tag{5.75}$$

(7) *Normalisation of labour.* Sraffa (PCMC, Par. 33) normalises labour, setting the physical unit, so results $L = 1$ TAL, giving for (5.75) the *price system*

$$\boxed{\begin{aligned} \check{\mathbf{S}}'\mathbf{p}(1 + r) + \check{\mathbf{L}}(\tilde{w} \cdot Y) &= \hat{\check{\mathbf{q}}}\mathbf{p}, \\ Y &= R \cdot (\check{\mathbf{S}}\mathbf{e})'\mathbf{p}. \end{aligned}} \tag{5.76}$$

There are now $n + 1$ equations and $n + 2$ variables $p_1, p_2, \ldots, p_n, \tilde{w}, Y$ and both parameters R, r, just explained above.

(8) *Standard system with normalised national income.* Following Sraffa and normalizing *Standard national income* to $Y = 1$, one then writes, bearing in mind that in this case there is the equality $\tilde{w} = w$,

$$\begin{aligned} \check{\mathbf{S}}'\mathbf{p}(1 + r) + \check{\mathbf{L}}\tilde{w} &= \hat{\check{\mathbf{q}}}\mathbf{p}, \\ Y = R \cdot (\check{\mathbf{S}}\mathbf{e})'\mathbf{p} &= 1. \end{aligned} \tag{5.77}$$

Sraffa proposes to use this normalised *national income*, which he calls the *Standard national income*, PCMC, Par. 34, as measuring unit for the numéraire.

Now, we work with Sraffa's *price model* (5.76), where the total labour is normalised, $L = 1$ TAL. We have to choose a variable, working out the calibration.

Either we choose a *commodity* $i \in \{1, 2, \ldots, n\}$ and set the corresponding price $p_i = p_0$, where p_0 is a given exogenous variable, setting as *numéraire* the price for one unit of that commodity (see PCMC, Par. 3) or we set for the national income $Y = Y_0$, a given exogenous variable Y_0 (see PCMC, Par. 11).

So now, Sraffa's price model (5.76) has $n+1$ equations and $n+1$ endogenous variables, namely either $Y, \tilde{w}, p_1, \ldots, p_{i-1}, p_{i+1}, \ldots, p_n$ or $\tilde{w}, p_1, \ldots, p_n$ and is solvable under the usual conditions known from linear algebra. We have therefore all the elements to calculate the further economic variables, the *total output* X, the total *circulating capital* K, the *total profit* P and the *total wages* W:

$$\begin{aligned} X &= \check{\mathbf{q}}'\mathbf{p}, \quad K = (\check{\mathbf{S}}\mathbf{e})'\mathbf{p}, \quad P = (\check{\mathbf{S}}\mathbf{e})'\mathbf{p} \cdot r = K \cdot r, \\ W &= Y - P = (\check{\mathbf{S}}\mathbf{e})'\mathbf{p} \cdot R - (\check{\mathbf{S}}\mathbf{e})'\mathbf{p} \cdot r = K \cdot (R - r). \end{aligned} \tag{5.78}$$

We recognise that it is easy to calculate or to verify the different ratios, just from the definitions and with (5.72)

$$r = \frac{P}{K}; \quad R = \tilde{R} = \frac{Y}{K}; \quad \tilde{w} = \frac{W}{Y}; \quad w = \frac{W}{L}. \tag{5.79}$$

In contrast, we proceeded, in addition to *the normalisation* of *labour*, to normalise *national income*;[19] we shall for our further calculations keep the *non-normalisation* of *national income*.

5.4.2 Sraffa's price equations expressed in commodity units

Equation (5.74) can be expressed in terms of *commodity units* in analogy to equation (4.59). This is carried out by passing from the *commodity flow matrix* $\mathbf{S}$ to the *input-output coefficients* matrix $\mathbf{C}$ in Sraffa's price equations.

We multiply equation (5.74) from the left with the diagonal matrix $\hat{\check{\mathbf{q}}}^{-1}$. With (5.29) $\check{\mathbf{C}} := \check{\mathbf{S}}\hat{\check{\mathbf{q}}}^{-1} = \mathbf{S}\hat{\mathbf{q}}^{-1} = \mathbf{C} \Rightarrow \mathbf{C}' = \hat{\mathbf{q}}^{-1}\mathbf{S}' = \hat{\check{\mathbf{q}}}^{-1}\check{\mathbf{S}}'$, using elementary rules of matrix algebra, like the rule of multiplication of transposed matrices (A.36) and especially the properties for transposed diagonal matrices, one finds

$$\begin{aligned} \hat{\check{\mathbf{q}}}^{-1}(\check{\mathbf{S}}'\mathbf{p})(1+r) + \hat{\check{\mathbf{q}}}^{-1}\left(\check{\mathbf{L}} \cdot \frac{\tilde{w} \cdot Y}{L}\right) &= (\hat{\check{\mathbf{q}}}^{-1}\check{\mathbf{S}}')\mathbf{p}(1+r) + (\hat{\check{\mathbf{q}}}^{-1}\check{\mathbf{L}}) \cdot \left(\frac{\tilde{w} \cdot Y}{L}\right) \\ &= \hat{\check{\mathbf{q}}}^{-1}(\hat{\check{\mathbf{q}}}\mathbf{p}) = (\hat{\check{\mathbf{q}}}^{-1}\hat{\check{\mathbf{q}}})\mathbf{p} = \mathbf{p} \\ &= \mathbf{C}'\mathbf{p}(1+r) + (\hat{\check{\mathbf{q}}}^{-1}\check{\mathbf{L}}) \cdot \left(\frac{\tilde{w} \cdot Y}{L}\right) = \mathbf{p}. \end{aligned} \tag{5.80}$$

19 According to Sraffa, setting $L = 1$ TAL, $Y = 1$ NI (national income), gives for the *share of total wages* $\tilde{w} = |\frac{w \cdot L}{Y}| = |w|$, the *wage per unit of labour* (this is a numerical identity only) and (5.73) becomes $r = R(1 - |w|)$, Sraffa's famous linear relationship (PCMC, Par. 30).

Now, by applying equation (4.58), we obtain by comparison a unique vector of *labour* $\breve{\boldsymbol{\pi}}$ *per unit of commodities* with the definitions (5.22) of $\breve{\mathbf{q}}$ and (5.26) of $\breve{\mathbf{L}}$, respectively the definitions of $\breve{\mathbf{q}}_1$ (5.65) and of $\breve{\mathbf{L}}_1$ (5.66),

$$\hat{\breve{\mathbf{q}}}^{-1}\breve{\mathbf{L}} = \begin{bmatrix} \frac{1}{\breve{q}_1 y_1} & \dots & \dots & 0 \\ 0 & \frac{1}{\breve{q}_2 y_2} & \dots & 0 \\ \dots & \dots & \dots & \dots \\ 0 & \dots & \dots & \frac{1}{\breve{q}_n y_n} \end{bmatrix} \begin{bmatrix} y_1 \breve{L}_1 \\ y_2 \breve{L}_2 \\ \dots \\ y_n \breve{L}_n \end{bmatrix} = \begin{bmatrix} \frac{\breve{L}_1}{\breve{q}_1} \\ \frac{\breve{L}_2}{\breve{q}_2} \\ \dots \\ \frac{\breve{L}_n}{\breve{q}_n} \end{bmatrix} =: \begin{bmatrix} \breve{\pi}_1 \\ \breve{\pi}_2 \\ \dots \\ \breve{\pi}_n \end{bmatrix} = \breve{\boldsymbol{\pi}}. \tag{5.81}$$

The ratio $\breve{\pi}_i = \breve{L}_i/\breve{q}_i$, $i = 1, \dots, n$ is the labour required for the production of one unit of each commodity, including replacements to satisfy the conditions of production, $\breve{\pi}_i$ and $\breve{\boldsymbol{\pi}}$ are explicitely written *with breve-sign.*

$$\hat{\breve{\mathbf{q}}}_1^{-1}\breve{\mathbf{L}}_1 = \begin{bmatrix} \frac{1}{\breve{q}_1 y y_1} & \dots & \dots & 0 \\ 0 & \frac{1}{\breve{q}_2 y y_2} & \dots & 0 \\ \dots & \dots & \dots & \dots \\ 0 & \dots & \dots & \frac{1}{\breve{q}_n y y_n} \end{bmatrix} \begin{bmatrix} y y_1 \breve{L}_1 \\ y y_2 \breve{L}_2 \\ \dots \\ y y_n \breve{L}_n \end{bmatrix} = \begin{bmatrix} \frac{\breve{L}_1}{\breve{q}_1} \\ \frac{\breve{L}_2}{\breve{q}_2} \\ \dots \\ \frac{\breve{L}_n}{\breve{q}_n} \end{bmatrix} =: \begin{bmatrix} \breve{\pi}_1 \\ \breve{\pi}_2 \\ \dots \\ \breve{\pi}_n \end{bmatrix} = \breve{\boldsymbol{\pi}}. \tag{5.82}$$

Referring to the *Sraffa model* (5.74), of relative prices, we have now the normalized *Sraffa model* for relative prices, expressed for the vector of labour $\breve{\boldsymbol{\pi}}$ per unit of commodities:

$$\mathbf{C}'\mathbf{p}(1+r) + \breve{\boldsymbol{\pi}} \cdot \left(\frac{\tilde{w} \cdot Y}{L}\right) = \mathbf{p} \quad \text{or} \quad [\mathbf{I} - (1+r)\mathbf{C}']\mathbf{p} = \breve{\boldsymbol{\pi}} \cdot \left(\frac{\tilde{w} \cdot Y}{L}\right), \tag{5.83}$$

by analogy to (4.59). The following Lemma highlights an interesting property of *Standard systems.*

Lemma 5.4.1. *Consider a vector* $\mathbf{y}^*$ *of multipliers belonging to an arbitrary orthogonal Euler map and constitute the diagonal matrix* $\hat{\mathbf{y}}^*$. *Let* $(\breve{\mathbf{S}}', \breve{\mathbf{q}}, \breve{\mathbf{L}})$ *be a Standard system, then compute by analogy* $\breve{\mathbf{S}}^{*'} = \hat{\mathbf{y}}^*\breve{\mathbf{S}}'$ *(5.14),* $\breve{\mathbf{q}}^* = \hat{\mathbf{y}}^*\breve{\mathbf{q}}$ *(5.14),* $\breve{\mathbf{d}}^* = \hat{\mathbf{y}}^*\breve{\mathbf{d}}$ *and* $\breve{\mathbf{L}}^* = \hat{\mathbf{y}}^*\breve{\mathbf{L}}$ *(5.26). As* $(\breve{\mathbf{S}}', \breve{\mathbf{q}}, \breve{\mathbf{L}})$ *is a Standard system, the process of production* $(\breve{\mathbf{S}}^{*'}, \breve{\mathbf{q}}^*, \breve{\mathbf{L}}^*)$ *is also a Standard system with* $\breve{\mathbf{q}}^* \parallel \breve{\mathbf{d}}^* \parallel \breve{\mathbf{S}}^*\mathbf{e}$.

Proof. Multiply the parallel vectors $\breve{\mathbf{q}} \parallel \breve{\mathbf{d}} \parallel \breve{\mathbf{S}}\mathbf{e}$ by the diagonal matrix $\hat{\mathbf{y}}^*$ getting with $\hat{\mathbf{y}}^*(\breve{\mathbf{S}}\mathbf{e}) = (\hat{\mathbf{y}}^*\breve{\mathbf{S}})\mathbf{e} = \breve{\mathbf{S}}^*\mathbf{e}$ the parallel vectors $\breve{\mathbf{q}}^* = \hat{\mathbf{y}}^*\breve{\mathbf{q}} \parallel \breve{\mathbf{d}}^* = \hat{\mathbf{y}}^*\breve{\mathbf{d}} \parallel \breve{\mathbf{S}}^*\mathbf{e}$. □

Returning to the *price model* (5.75), we accordingly can also write the matrix equations in developed form with a system of $n + 2$ equations, (PCMC, Par. 33, 34):

$$\begin{aligned}
&y_1\left[(\check{s}_{11}p_1+\cdots+\check{s}_{n1}p_n)(1+r)+L_1\left(\frac{\tilde{w}\cdot Y}{L}\right)\right]=y_1\check{q}_1p_1,\\
&y_2\left[(\check{s}_{12}p_1+\cdots+\check{s}_{n2}p_n)(1+r)+L_2\left(\frac{\tilde{w}\cdot Y}{L}\right)\right]=y_2\check{q}_2p_2,\\
&\ldots,\\
&\ldots,\\
&y_n\left[(\check{s}_{1n}p_1+\cdots+\check{s}_{nn}p_n)(1+r)+L_n\left(\frac{\tilde{w}\cdot Y}{L}\right)\right]=y_n\check{q}_np_n\\
&L=\sum_{i=1}^{n}L_i,\\
&Y=R\cdot\left(\sum_{i=1}\left(\sum_{j=1}\check{s}_{ij}\right)p_i\right).
\end{aligned}\tag{5.84}$$

Then, as we have seen, we have to calibrate the *price system* equation (5.84) which has actually $n+3$ variables, namely $Y, L, \tilde{w}, p_1, \ldots, p_n$, $n+2$ equations and 2 parameters R, r. The positive *productiveness* $R > 0$, being determined by the Frobenius number of matrix $\mathbf{D}$, and the chosen rate of profits $r \in [0, R]$.

Again, either we choose a *commodity* $i \in \{1, 2, \ldots, n\}$ and set the corresponding price $p_i = p_0$, where p_0 is a given exogenous variable, setting as *numéraire* the price for one unit of that commodity (Sraffa PCMC, Par. 3), or we set for the national income $Y = Y_0$, a given exogenous variable Y_0 (PCMC, Par. 11).

So now, Sraffa's price model (5.84) has $n+2$ equations and $n+2$ endogenous variables, namely either $Y, L, \tilde{w}, p_1, \ldots, p_{i-1}, p_{i+1}, \ldots, p_n$ or $\tilde{w}, L, p_1, \ldots, p_n$ and is solvable under the usual conditions known from linear algebra.

We have at present all the elements to calculate *total output* X, the *circulating capital* K, *total profit* P, and *total wages* W.

$$X = Y\cdot\frac{1+R}{R};\quad K=\frac{Y}{R};\quad P=\frac{r}{R}\cdot Y;\quad W=Y-P=\frac{Y}{R}\cdot(R-r).\tag{5.85}$$

It is also easy to calculate or to verify the different ratios,

$$r=\frac{P}{K};\quad \tilde{R}=R=\frac{Y}{K};\quad \tilde{w}=\frac{W}{Y};\quad w=\frac{W}{L}.\tag{5.86}$$

5.4.3 Complete calculation of a Standard system with prices and wages based on Sraffa's elementary example (PCMC, Par. 5, 11)

The aim of this section is to transform *Example* 4.1.7, (PCMC, Par. 5) into a *Standard system* and to calculate the *Standard net product* and Sraffa prices.

Example 5.4.1. We take the same labour vector $\mathbf{L} = [152, 152]'$ as in *Example* 4.1.7 (4.85) with the *total quantity of labour* $L = 304$. We already know the Frobenius eigenvalue $\lambda = 4/5$ of the non-negative, irreducible *input coefficients* matrix $\mathbf{C}$ of the *process of production* in this example, giving the *productiveness* $R = 0.25$.

We take the complete single product Sraffa system, presented as equation (5.84).[20] Before solving it, we identify all matrices of *Example* 4.1.7 (4.85).

- The *commodity flow* matrix $\mathbf{S} = \left[\begin{smallmatrix} 280 & 120 \\ 12 & 8 \end{smallmatrix}\right]$ in *physical terms* is measured in qr. wheat; the second row is measured in t. iron.
- The vector of *labour* $\mathbf{L} = [152 \;\; 152]'$ is normalised and measured in TAL.
- The vector of *surplus* $\mathbf{d} = [d_1 \;\; d_2]' = [175 \;\; 0]'$ in *physical terms* has the mixed units: $[d_1]$ = qr. wheat, $[d_2]$ = t. iron.
- The vector of *total output* $\mathbf{q} = \mathbf{Se} + \mathbf{d} = [575 \;\; 20]'$ in *physical terms*, where the units are mixed: $[q_1]$ = qr. wheat, $[q_2]$ = t. iron.

Perform an *orthogonal Euler map*, computing the right eigenvectors $\mathbf{y}$ of the distribution coefficients matrix $\mathbf{D} = \hat{\mathbf{q}}^{-1}\mathbf{S}$, to obtain the *Standard system*, leading by *dilatation* to the *Standard net product*. Verify that vector $\hat{\mathbf{q}}\mathbf{y}$ is an eigenvector of the input-output coefficients matrix $\mathbf{C} = \mathbf{S}\hat{\mathbf{q}}^{-1}$, see Lemma A.6.2.

Finally, choosing as rate of profits $r = 0.15 < R = 0.25$, calculate the Sraffa *price model* for that *Standard system* with the three following calibrations: (a) national income $Y = 1$ NI (NI stands for an arbitrary unit of *national income*), (b) the price of wheat is set $p_1 = 1$ as *numéraire*, (c) the price of wheat is set $p_1 = 24.60$ CHF/qr. of wheat.

Solution to Example 5.4.1:
We compute the vector of *total output* in physical terms, as

$$\mathbf{q} = \mathbf{Se} + \mathbf{d} = \begin{bmatrix} q_1 \\ q_2 \end{bmatrix} = \begin{bmatrix} 280 & 120 \\ 12 & 8 \end{bmatrix}\begin{bmatrix} 1 \\ 1 \end{bmatrix} + \begin{bmatrix} 175 \\ 0 \end{bmatrix} = \begin{bmatrix} 575 \\ 20 \end{bmatrix} > \mathbf{o}, \tag{5.87}$$

and normalise the vector of *labour* $\mathbf{L}$, presenting labour in the physical unit of *total amount of labour* (TAL),

$$L = \mathbf{L}'\mathbf{e} = \begin{bmatrix} L_1 & L_2 \end{bmatrix}\begin{bmatrix} 1 \\ 1 \end{bmatrix} = \begin{bmatrix} \frac{152}{304} & \frac{152}{304} \end{bmatrix}\begin{bmatrix} 1 \\ 1 \end{bmatrix} = 1\,\text{TAL}. \tag{5.88}$$

These preliminaries are summarised in Table 5.1.

The multipliers and the orthogonal Euler map

We determine the *distribution coefficients* matrix:

$$\mathbf{D} = \hat{\mathbf{q}}^{-1}\mathbf{S} = \begin{bmatrix} \frac{1}{575} & 0 \\ 0 & \frac{1}{20} \end{bmatrix}\begin{bmatrix} 280 & 120 \\ 12 & 8 \end{bmatrix} = \begin{bmatrix} \frac{56}{115} & \frac{24}{115} \\ \frac{3}{5} & \frac{2}{5} \end{bmatrix}. \tag{5.89}$$

Next, the *right eigenvector* equation (5.67) is solved, $\mathbf{D}\mathbf{y} = \lambda\mathbf{y}$. The eigenvalues of the characteristic polynomial $P_2(\lambda) = \det(\mathbf{D} - \lambda\mathbf{I}_2) = \lambda^2 - \frac{102}{115}\lambda + \frac{8}{115} = 0$ are already

20 It was then treated with *normalised labour* (4.85) as a non-standard *system of production*.

Table 5.1: Input-Output Table of Sraffa's model in PCMC, Par. 5, (with labour).

	Buying sectors		Final	Total
	wheat	**iron**	**demand**	**output**
wheat (*qr.* wheat)	$s_{11} = 280$	$s_{12} = 120$	$d_1 = 175$	$q_1 = 575$
iron (*t.* iron)	$s_{21} = 12$	$s_{21} = 8$	$d_2 = 0$	$q_2 = 20$
labour (TAL)	$L_1 = \frac{1}{2}$	$L_2 = \frac{1}{2}$		$L = 1$
	↓	↓		
production (*unit*)	$q_1 = 575$	$q_2 = 20$		

known; the Frobenius number is $\lambda_D = 4/5$, associated with the positive eigenvectors $\boldsymbol{\gamma} = [\frac{2}{3}k, k]'$ with $k \in \mathbb{R}^+$, giving the *productiveness* $R = 0.25$.

With factor $k = 1$ we get a vector of arbitrary multipliers $\boldsymbol{\gamma} = [\frac{2}{3}, 1]'$. We diagonalise this vector, defining the diagonal matrix of an *orthogonal Euler map*:

$$\hat{\boldsymbol{\gamma}} = \begin{bmatrix} \gamma_1 & 0 \\ 0 & \gamma_2 \end{bmatrix} = \begin{bmatrix} \frac{2}{3} & 0 \\ 0 & 1 \end{bmatrix}. \tag{5.90}$$

Then, we get a *Standard system* by computing the mapped vector of *labour* and the total *labour*,

$$\breve{\mathbf{L}} = \hat{\boldsymbol{\gamma}}\mathbf{L} = \begin{bmatrix} \frac{2}{3} & 0 \\ 0 & 1 \end{bmatrix}\begin{bmatrix} \frac{1}{2} \\ \frac{1}{2} \end{bmatrix} = \begin{bmatrix} \frac{1}{3} \\ \frac{1}{2} \end{bmatrix}, \quad L = \breve{\mathbf{L}}'\mathbf{e} = \begin{bmatrix} \frac{1}{3}, \frac{1}{2} \end{bmatrix}\begin{bmatrix} 1 \\ 1 \end{bmatrix} = \frac{5}{6}, \tag{5.91}$$

the mapped *commodity flow* matrix $\breve{\mathbf{S}}$ (5.14),

$$\breve{\mathbf{S}}' = \hat{\boldsymbol{\gamma}}\mathbf{S}' = \begin{bmatrix} \frac{2}{3} & 0 \\ 0 & 1 \end{bmatrix}\begin{bmatrix} 280 & 12 \\ 120 & 8 \end{bmatrix} = \begin{bmatrix} \frac{560}{3} & 8 \\ 120 & 8 \end{bmatrix}, \tag{5.92}$$

the mapped vector of *total output*,

$$\breve{\mathbf{q}} = \hat{\boldsymbol{\gamma}}\mathbf{q} = \begin{bmatrix} \frac{2}{3} & 0 \\ 0 & 1 \end{bmatrix}\begin{bmatrix} 575 \\ 20 \end{bmatrix} = \begin{bmatrix} \frac{1150}{3} \\ 20 \end{bmatrix}, \tag{5.93}$$

together with the mapped *means of production* $\breve{\mathbf{S}}\mathbf{e}$ and the mapped vector of *surplus* $\breve{\mathbf{d}}$ (5.23),

$$\begin{aligned} \breve{\mathbf{S}}\mathbf{e} &= \begin{bmatrix} \frac{560}{3} & 120 \\ 8 & 8 \end{bmatrix}\begin{bmatrix} 1 \\ 1 \end{bmatrix} = \begin{bmatrix} \frac{920}{3} \\ 16 \end{bmatrix}, \\ \breve{\mathbf{d}} &= \breve{\mathbf{q}} - \breve{\mathbf{S}}\mathbf{e} = \begin{bmatrix} \frac{1150}{3} \\ 20 \end{bmatrix} - \begin{bmatrix} \frac{920}{3} \\ 16 \end{bmatrix} = \begin{bmatrix} \frac{230}{3} \\ 4 \end{bmatrix}, \end{aligned} \tag{5.94}$$

and we verify the parallelism,

$$\breve{\mathbf{S}}\mathbf{e} = \begin{bmatrix} \frac{920}{3} \\ 16 \end{bmatrix} \parallel \breve{\mathbf{q}} = \begin{bmatrix} \frac{1150}{3} \\ 20 \end{bmatrix} \parallel \breve{\mathbf{d}} = \begin{bmatrix} \frac{230}{3} \\ 4 \end{bmatrix}. \tag{5.95}$$

Based on Recapitulation 5.2.1, this means that the present *system of production* $(\breve{\mathbf{S}}, \breve{\mathbf{q}}, \breve{\mathbf{L}})$ (5.91), (5.92), (5.94) constitutes a *Standard system*, see Figure 5.5.

Eigenvectors of C as elements of the Standard system

We use the property (5.29) $\breve{\mathbf{C}} := \breve{\mathbf{S}}\hat{\breve{\mathbf{q}}}^{-1} = \mathbf{S}\hat{\mathbf{q}}^{-1} = \mathbf{C}$ to calculate the *input-output coefficients* matrix $\mathbf{C}$ of the mapped system and we find:

$$\mathbf{C} = \mathbf{S}\hat{\mathbf{q}}^{-1} = \begin{bmatrix} 280 & 120 \\ 12 & 8 \end{bmatrix} \begin{bmatrix} \frac{1}{575} & 0 \\ 0 & \frac{1}{20} \end{bmatrix} = \begin{bmatrix} \frac{56}{115} & 6 \\ \frac{12}{575} & \frac{2}{5} \end{bmatrix}. \tag{5.96}$$

The characteristic polynomial is $P_2(\lambda) = \det(\mathbf{C} - \lambda \mathbf{I}_2) = \lambda^2 - \frac{102}{115}\lambda + \frac{8}{115} = 0$. We compute the positive *right eigenvectors* of matrix $\mathbf{C}$, associated with the Frobenius number $\lambda_C = 4/5$, and find $k[115/6, 1]', k \in \mathbb{R}^+$. We then verify that these eigenvectors can also be obtained by calculating,

$$\hat{\mathbf{q}}y = \begin{bmatrix} 575 & 0 \\ 0 & 20 \end{bmatrix} \begin{bmatrix} \frac{2}{3} \\ 1 \end{bmatrix} = \begin{bmatrix} \frac{1{,}150}{3} \\ 20 \end{bmatrix} = \breve{\mathbf{q}} = 20 \begin{bmatrix} \frac{115}{6} \\ 1 \end{bmatrix}, \tag{5.97}$$

according to Lemma A.6.2, giving directly the vector of *total output* of the *Standard system* (5.95), parallel to the vector of *means of production* $\breve{\mathbf{S}}\mathbf{e}$ and the vector of *surplus* $\breve{\mathbf{d}}$.

From the Standard system to the Standard net product

With a view to obtain finally the *Standard national income* corresponding to the *Standard net product*, one has to start by restoring the original quantity of labour, by setting $\breve{\mathbf{L}}_1 = y\breve{\mathbf{L}}$ (5.91), operating a *dilatation* by the factor y on the above obtained *Standard system*:

$$\breve{\mathbf{L}}_1 = y\breve{\mathbf{L}} = y \begin{bmatrix} \frac{1}{3} \\ \frac{1}{2} \end{bmatrix}, \quad L_1 = \breve{\mathbf{L}}_1'\mathbf{e} = y\breve{\mathbf{L}}' \cdot \mathbf{e} = y \begin{bmatrix} \frac{1}{3} & \frac{1}{2} \end{bmatrix} \cdot \begin{bmatrix} 1 \\ 1 \end{bmatrix} = y\frac{5}{6} = 1 \Rightarrow y = \frac{6}{5}, \tag{5.98}$$

giving the normalised vector of *labour* $\mathbf{L}_1 = [2/5, 3/5]'$. Then, the dilated *commodity flow* matrix $\breve{\mathbf{S}}_1$ (5.69) and the dilated *total output* vector $\mathbf{q}_1$ (5.69) are calculated:

$$\begin{aligned} \breve{\mathbf{S}}_1 &= y\breve{\mathbf{S}} = \frac{6}{5} \begin{bmatrix} \frac{560}{3} & 120 \\ 8 & 8 \end{bmatrix} = \begin{bmatrix} 224 & 144 \\ \frac{48}{5} & \frac{48}{5} \end{bmatrix} = \begin{bmatrix} 224 & 144 \\ 9.6 & 9.6 \end{bmatrix}, \\ \breve{\mathbf{q}}_1 &= y\breve{\mathbf{q}} = \frac{6}{5} \begin{bmatrix} \frac{1150}{3} \\ 20 \end{bmatrix} = \begin{bmatrix} 460 \\ 24 \end{bmatrix}; \quad \breve{\mathbf{L}}_1 = \begin{bmatrix} \frac{2}{5} \\ \frac{3}{5} \end{bmatrix}, \end{aligned} \tag{5.99}$$

with the *means of production* of the dilated system,

$$\breve{\mathbf{S}}_1\mathbf{e} = \begin{bmatrix} 224 & 144 \\ 9.6 & 9.6 \end{bmatrix} \begin{bmatrix} 1 \\ 1 \end{bmatrix} = \begin{bmatrix} 368 \\ 19.2 \end{bmatrix}, \tag{5.100}$$

and also the surplus of the dilated system,

$$\mathbf{d}_1 = \mathbf{q}_1 - \mathbf{S}_1\mathbf{e} = \begin{bmatrix} 460 \\ 24 \end{bmatrix} - \begin{bmatrix} 368 \\ 19.2 \end{bmatrix} = \begin{bmatrix} 92 \\ 4.8 \end{bmatrix}. \tag{5.101}$$

We confirm the parallelism:

$$\breve{\mathbf{S}}_1\mathbf{e} = \begin{bmatrix} 368 \\ 19.2 \end{bmatrix} \parallel \breve{\mathbf{q}}_1 = \begin{bmatrix} 460 \\ 24 \end{bmatrix} \parallel \breve{\mathbf{d}}_1 = \begin{bmatrix} 92 \\ 4.8 \end{bmatrix}. \tag{5.102}$$

Finally, with the coefficient of dilatation and the *productiveness* $R = 0.25$, together with Lemma 5.1.2, we confirm: $\breve{\mathbf{q}}_1 = \frac{1+R}{R} \cdot \breve{\mathbf{d}}_1 = (1+R) \cdot \breve{\mathbf{S}}_1\mathbf{e}$.

We then get the production scheme of the *Standard system*, giving the *Standard net product*:

$$\begin{aligned} &\left(224 \text{ t. wheat}, 9.6 \text{ t. iron}, \frac{2}{5} \text{ labour}\right) \to (460 \text{ t. wheat}, 0), \\ &\left(144 \text{ t. wheat}, 9.6 \text{ t. iron}, \frac{3}{5} \text{ labour}\right) \to (0, 24 \text{ t. iron}) \end{aligned} \tag{5.103}$$

The components of the vector of surplus $\breve{\mathbf{d}}_1 = [92, 4.8]'$ form the actual *Standard net product* $\{92, 4.8\}$, presented as a set.

Calculation of prices, wages, the Standard national income and other economic variables

We establish finally the Sraffa *price model* of the obtained *Standard system*, corresponding to equation (5.84), exhibiting the *Standard net product* calculated above, using matrices (5.99). Remember that the *productiveness* is $R = 0.25$ from the Frobenius eigenvalue $\lambda = 4/5$ of the *input-output coefficients* matrix $\mathbf{C}$ (5.29), and that we have set $r = 0.15$. The labour is normalised, $L = 1$, and the *share of total wages to national income* (5.73) is $\tilde{w} = 1 - r/R = 1 - (0.15/0.25) = 0.4$, so that the factor of the labour parts, which are also the *total wages*, in (5.104) is $W = (\tilde{w} \cdot Y)/L = \tilde{w} \cdot Y = 0.4 \cdot Y$. We get the *price model*,

$$\boxed{\begin{aligned} &(224p_1 + 9.6p_2)(1 + 0.15) + \frac{2}{5}(\tilde{w} \cdot Y) = 460p_1, \\ &(144p_1 + 9.6p_2)(1 + 0.15) + \frac{3}{5}(\tilde{w} \cdot Y) = 24p_2, \\ &\qquad\qquad Y = 0.25 \cdot (368p_1 + 19.2p_2). \end{aligned}} \tag{5.104}$$

We set either

(a) as exogenous variable the *national income* $Y = Y_0$ and as endogenous variables the *relative prices* $p_1, p_2, \tilde{w}$ or;

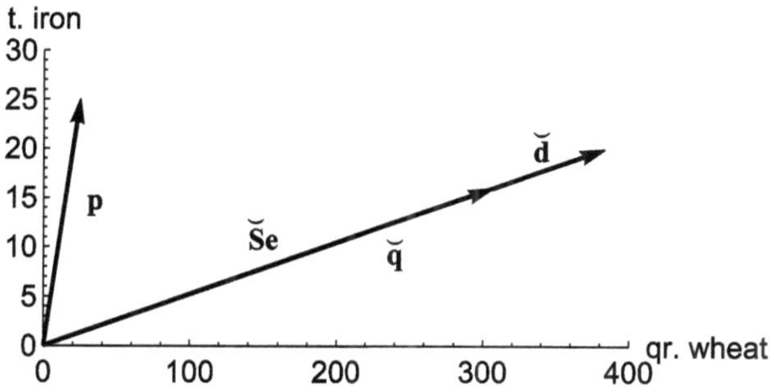

Figure 5.5: Parallelism of the vectors $\breve{\mathbf{S}}\mathbf{e} \parallel \breve{\mathbf{q}} \parallel \breve{\mathbf{d}}$ (5.95) is similar to that of the vectors $\breve{\mathbf{S}}_1\mathbf{e} \parallel \breve{\mathbf{q}}_1 \parallel \breve{\mathbf{d}}_1$ (5.102); the price vector $\mathbf{p}$ is shown.

(b) as exogenous variable one of the *relative price* $p_i = p_0$ and as endogenous variables the other *relative price* p_j, $\tilde{w}$ and the *national income* Y.

We will treat three cases of price calculations in relation to the *remark* following Lemma 5.1.2. Indeed, we can freely choose any physical unit, appearing within this *Standard system*, as a numéraire. Or, we choose any monetary unit to express the prices within the *Standard system*. Clearly, the chosen payment units have no effect on the proportions (5.7) between surplus and circulating capital, equal to the *productiveness R*.

(1) Sraffa's normalization of the national income: the *Standard national income*. (The measurement unit is *national income* = NI). We set $Y = Y_0 = 1$ NI and also have normalised labour size, $L_1 = 1$ TAL. Solving the system (5.104) we have to find the values for p_1 and p_2 and $\tilde{w}$. The physical units are *qr.* wheat and *t.* iron. Then the price p_1 of wheat and the price p_2 of iron are expressed by the units $[p_1] = (\text{NI/qr. wheat})$ and $[p_2] = (\text{NI/t. iron})$. We know the vector of surplus $\mathbf{d}_1 = [92, 4.8]'$. We then set,

$$Y = Y_0 = \mathbf{d}_1'\mathbf{p} = [92, 4.8]' \begin{bmatrix} p_1 \\ p_2 \end{bmatrix} = 1\,\text{NI}, \tag{5.105}$$

having obtained the prices $p_1 = (41/6{,}900)\,\frac{\text{NI}}{\text{qr. wheat}}$, $p_2 = (17/180)\,\frac{\text{NI}}{\text{t. iron}}$, and the wage share of total income $\tilde{w} = 0.4$, one calculates the total wages,

$$W = w \cdot Y_0 = 0.4 \cdot 1\,\text{NI} = 0.4\,\text{NI}. \tag{5.106}$$

The total profit is

$$P = Y_0 - W = 0.6\,\text{NI}, \tag{5.107}$$

and the *circulating capital* is

$$K = \mathbf{e}'\breve{\mathbf{S}}_1'\mathbf{p} = [1, 1] \begin{bmatrix} 224 & 9.6 \\ 144 & 9.6 \end{bmatrix} \begin{bmatrix} \frac{41}{6{,}900} \\ \frac{17}{180} \end{bmatrix} = 4\,\text{NI}. \tag{5.108}$$

Table 5.2: Calculated variables and invariant terms of *Case (1)*.

Variable terms		Invariant terms	
notion	**value**	**notion**	**value**
total output	$X = 5$ NI	productiveness	$R = 0.25$
total capital	$K = 4$ NI	rate of profits	$r = 0.15$
total profit	$P = 0.6$ NI	share of total profits	$\tilde{r} = 0.6$
total wages	$W = 0.4$ NI	surplus ratio	$\tilde{R} = 0.25$
national income	$Y = 1$ NI	share of total wages	$\tilde{w} = 0.4$
average NI per labour	$U = 1 \frac{\text{NI}}{\text{TAL}}$		
wage per unit of labour	$w = 0.4 \frac{\text{NI}}{\text{TAL}}$		

Total output is

$$X = \breve{\mathbf{q}}_1'\mathbf{p} = [460, 24]\begin{bmatrix} \frac{41}{6,900} \\ \frac{17}{180} \end{bmatrix} = K + Y = 5\,\text{NI}. \tag{5.109}$$

The various ratios are calculated, *NI* may appear, Lemma 5.1.2 is applied, $\tilde{R} = R$,

$$\begin{aligned} r &= \frac{P}{K} = \frac{0.6}{4} = 0.15, \quad R = \tilde{R} = \frac{Y}{K} = \frac{1}{4} = 0.25, \\ \tilde{w} &= |w| = \frac{W}{Y} = \left|\frac{W}{L_1}\right| = 0.4, \quad \tilde{r} = \frac{P}{Y} = \frac{0.6}{1} = 0.6, \\ U &= \frac{Y}{L} = \frac{1}{1} = 1\,\frac{\text{NI}}{\text{TAL}}. \end{aligned} \tag{5.110}$$

We also confirm the proportionality (4.36) with the *share of total wages to national income* $\tilde{w} = 0.4$ and the *wage rate* $w = 0.4$, see Table 5.2:

$$r = \tilde{R}(1 - \tilde{w}) = R(1 - w) = 0.25 \cdot (1 - 0.4) = 0.15. \tag{5.111}$$

(2) We choose as *numéraire* the unit qr. wheat and set $p_1 = 1 \frac{\text{qr. NI}}{\text{qr. wheat}}$. We have the endogenous variables Y, p_2, $\tilde{w}$ in the linear system (5.104). The physical units are qr. wheat and t. iron. Then the price p_2 of iron and the *national income* Y are expressed in the units $[p_2] = \frac{\text{qr. wheat}}{\text{t. iron}}$ and $[Y]$ = qr. wheat. We get the price $p_2 = 1{,}955/123 \frac{\text{qr. wheat}}{\text{t. iron}} = 15.89 \frac{\text{qr. wheat}}{\text{t. iron}}$. We then obtain for *national income* Y, total wages W and total profits P,

$$\begin{aligned} Y &= 6{,}900/41 \text{ qr. wheat} = 168.29 \text{ qr. wheat}, \\ W &= w \cdot Y = 0.4 \cdot 168.29 \text{ qr. wheat} = 67.32 \text{ qr. wheat}, \\ P &= Y - W = 100.97 \text{ qr. wheat}. \end{aligned} \tag{5.112}$$

The *circulating capital* is

$$K = \mathbf{e}'\breve{\mathbf{S}}_1'\mathbf{p} = \begin{bmatrix} 1 & 1 \end{bmatrix}\begin{bmatrix} 224 & 9.6 \\ 144 & 9.6 \end{bmatrix}\begin{bmatrix} 1 \\ \frac{1,955}{123} \end{bmatrix} = 673.17 \text{ qr. wheat}, \tag{5.113}$$

Table 5.3: Calculated variables and invariant terms of *Case (2)*.

Variable terms		**Invariant terms**	
notion	**value**	**notion**	**value**
total output	$X = 841.46$ qr. wheat	productiveness	$R = 0.25$
total capital	$K = 673.17$ qr. wheat	rate of profits	$r = 0.15$
total profit	$P = 100.97$ qr. wheat	share of total profits	$\tilde{r} = 0.6$
total wages	$W = 67.32$ qr. wheat	surplus ratio	$\tilde{R} = 0.25$
national income	$Y = 168.29$ qr. wheat	share of total wages	$\tilde{w} = 0.4$
average NI per labour	$U = 168.29 \frac{\text{qr. wheat}}{\text{TAL}}$		
wage per unit of labour	$w = 67.32 \frac{\text{qr. wheat}}{\text{TAL}}$		

the *total output*,

$$X = K + Y = \breve{\mathbf{q}}_1'\mathbf{p} = [460, 24]\begin{bmatrix} 1 \\ \frac{1{,}955}{123} \end{bmatrix} = 841.46 \text{ qr. wheat.} \tag{5.114}$$

The various ratios of entities, expressed in qr. wheat are computed, Lemma 5.1.2 is applied, $\tilde{R} = R$,

$$\begin{aligned} r &= R(1 - \tilde{w}) = \frac{P}{K} = \frac{4{,}140/41}{27{,}600/41} = 0.15, \\ R &= \tilde{R} = \frac{Y}{K} = \frac{6{,}900/41}{27.600/41} = 0.25, \quad |w| = \left|\frac{W}{L}\right| = \frac{2{,}760/41}{1} = 67.32, \\ \tilde{r} &= \frac{P}{Y} = \frac{4{,}140/41}{6{,}900/41} = 0.6, \quad U = \frac{Y}{L} = \frac{6{,}900/41}{1} = 168.29 \frac{\text{qr. wheat}}{\text{TAL}}. \end{aligned} \tag{5.115}$$

We also confirm the key result (4.36), and the *share of total wages to national income*, see Table 5.3,

$$r = \tilde{R}(1 - \tilde{w}) = 0.25 \cdot (1 - 0.4) = 0.15, \quad \tilde{w} = \frac{W}{Y} = \frac{2{,}760/41}{6{,}900/41} = 0.4. \tag{5.116}$$

(3) We choose as *monetary unit* the *Swiss franc* CHF, setting the arbitrary price $p_1 = 24.60 \frac{\text{CHF}}{\text{qr. wheat}}$. We have then the endogenous variables Y, p_2, $\tilde{w}$ in the linear system (5.104). The physical units are qr. wheat and t. iron. The price p_2 of *iron* and the *national income* Y are expressed in the units $[p_2] = \frac{\text{CHF}}{\text{t. iron}}$ and $[Y] = \text{CHF}$. We obtain the price $p_2 = 391$ CHF for iron. For the *national income*, we get $Y = 4{,}140$ CHF, while for the total wages and for the total profit we get

$$\begin{aligned} W &= w \cdot Y = 0.4 \cdot 4{,}140 \text{ CHF} = 1{,}656 \text{ CHF}, \\ P &= Y - W = 4{,}140 - 1{,}656 = 2{,}484 \text{ CHF}. \end{aligned} \tag{5.117}$$

Table 5.4: Calculated variable and invariant terms of *Case (3)*.

Variable terms		Invariant terms	
notion	**value**	**notion**	**value**
total output	$X = 20{,}700$ CHF	productiveness	$R = 0.25$
total capital	$K = 16{,}560$ CHF	rate of profits	$r = 0.15$
total profit	$P = 2{,}484$ CHF	share of total profits	$\tilde{r} = 0.6$
total wages	$W = 1{,}656$ CHF	surplus ratio	$\tilde{R} = 0.25$
national income	$Y = 4{,}140$ CHF	share of total wages	$\tilde{w} = 0.4$
average NI per labour	$U = 4{,}140\ \frac{\text{CHF}}{\text{TAL}}$		
wage per unit of labour	$w = 1{,}656\ \frac{\text{CHF}}{\text{TAL}}$		

For the *circulating capital* and *total output*, we get

$$K = \mathbf{e}'\breve{\mathbf{S}}_1'\mathbf{p} = [1,1]\begin{bmatrix} 224 & 9.6 \\ 144 & 9.6 \end{bmatrix}\begin{bmatrix} 24.6 \\ 391 \end{bmatrix} = 16{,}560 \text{ CHF},$$

$$X = K + Y = \breve{\mathbf{q}}_1'\mathbf{p} = [460, 24]'\begin{bmatrix} 24.6 \\ 391 \end{bmatrix} = 20{,}700 \text{ CHF}. \tag{5.118}$$

To summarize, the various ratios of financial entities calculated in CHF computed, Lemma 5.1.2 is applied, $\tilde{R} = R$,

$$r = \frac{P}{K} = \frac{2{,}484}{16{,}560} = 0.15, \quad R = \tilde{R} = \frac{Y}{K} = \frac{4{,}140}{16{,}560} = 0.25,$$
$$|w| = \left|\frac{W}{Y}\right| = \frac{1{,}656}{1} = 1{,}656, \quad \tilde{r} = \frac{P}{Y} = \frac{2{,}484}{4{,}140} = 0.6,$$
$$U = \frac{Y}{L} = \frac{4{,}140}{1} = 4{,}140\ \frac{\text{CHF}}{\text{TAL}}. \tag{5.119}$$

We again confirm the key result (4.36) and the calculation of the *share to total wages to national income*, see Table 5.4:

$$r = R(1 - \tilde{w}) = \tilde{R}(1 - \tilde{w}) = 0.25 \cdot (1 - 0.4) = 0.15,$$
$$\tilde{w} = \frac{W}{Y} = \frac{1{,}656}{4{,}140} = 0.4 \tag{5.120}$$

Concluding we see that in all the cases $r = 0.15$, $\tilde{R} = R = 0.25$, $\tilde{w} = 0.4$ and $\tilde{r} = 0.6$ are invariants with $\tilde{r} + \tilde{w} = 1$. ▲

5.5 Price fluctuations and the Standard systems*

We continue to deal with single product industries. We analyse here what Sraffa means by stating that *price fluctuations* are independent of the distribution of surplus between profits and wages if an appropriate *composite commodity*, i. e., the Standard commodity, respectively the *Standard net product* (PCMC, Par. 23 and Par. 34) is constituted. This analysis requires examining in what context this assertion is valid, i. e., by looking at the nature of prices in *Standard systems*.

We continue to consider for this purpose n industries and as *system of production* a *Standard system* $(\breve{\mathbf{S}}', \breve{\mathbf{q}}, \breve{\mathbf{L}}) \in \mathcal{E}$. The $n \times n$ matrix $\breve{\mathbf{S}}$ is of course *semi-positive* and *irreducible* or *positive*. The starting points are the proportions (5.7) characterising *Standard systems*, reproduced here (with *breve-sign*):

$$\frac{p_i\breve{d}_i}{p_i\breve{q}_i - p_i\breve{d}_i} = \frac{p_i\breve{d}_i}{p_i(\breve{q}_i - \breve{d}_i)} = \frac{\breve{d}_i}{(\breve{q}_i - \breve{d}_i)} = R, \quad i = 1, \ldots, n, \tag{5.121}$$

where $R > 0$ is the *productiveness* that governs this proportion. Now the explicit knowledge of the *positive* vector of *surplus* $\breve{\mathbf{d}} > \mathbf{o}$ is no longer needed to express the vector of *total output* (5.25); we only need the *productiveness* R and calculate (see also Assumption 2.2.1):

$$\breve{\mathbf{q}} = \breve{\mathbf{S}}\mathbf{e} + \breve{\mathbf{d}} = \breve{\mathbf{S}}\mathbf{e}(1 + R) > \mathbf{o}. \tag{5.122}$$

The crucial point is then to analyse Sraffa's price model (4.55) setting $r = R$. This means there are only profits, without wages. We use directly the defining property (5.122) of a *Standard system*:

$$\breve{\mathbf{S}}'\mathbf{p}(1 + R) = \hat{\breve{\mathbf{q}}}\mathbf{p} = \widehat{\breve{\mathbf{S}}\mathbf{e}}(1 + R)\mathbf{p} \Rightarrow \breve{\mathbf{S}}'\mathbf{p} = (\widehat{\breve{\mathbf{S}}\mathbf{e}})\mathbf{p} = \hat{\breve{\mathbf{q}}}_I\mathbf{p} \Rightarrow \breve{\mathbf{q}}_I = \breve{\mathbf{S}}\mathbf{e} > \mathbf{o}, \tag{5.123}$$

giving the known notation (2.15), again with Assumption 2.2.1, where there is no longer a vector of *surplus*. We are in presence of the *conditions of production*, Definition 3.1.2, (PCMC, Par. 3). The vector of surplus has disappeared, and we just have the means of production of the *Standard system* $(\breve{\mathbf{S}}', \breve{\mathbf{q}}, \breve{\mathbf{L}}) \in \mathcal{E}$. But clearly, if we have a *productiveness* $R = (1/\lambda_C) - 1 > 0$, determined by the Frobenius number λ_C, $0 < \lambda_C < 1$, of matrix $\mathbf{C} = \breve{\mathbf{S}}\hat{\breve{\mathbf{q}}}^{-1} = \mathbf{S}\hat{\mathbf{q}}^{-1}$(see Lemma 5.2.1) and (5.29), the surplus vector $\breve{\mathbf{d}}$ is itself fully determined by R in

$$\breve{\mathbf{q}} = \breve{\mathbf{S}}\mathbf{e}(1 + R) = \breve{\mathbf{q}}_I(1 + R) = \breve{\mathbf{q}}_I + \breve{\mathbf{d}} > \mathbf{o} \Rightarrow \breve{\mathbf{d}} = R(\breve{\mathbf{S}}\mathbf{e}). \tag{5.124}$$

We multiply the concluding equation (without productiveness R!) (5.123) by the diagonal matrix $\hat{\breve{\mathbf{q}}}_I^{-1}$ from the left and get, setting $\mathbf{C} := \breve{\mathbf{S}}\hat{\breve{\mathbf{q}}}_I^{-1}$ an eigenvalue equation,

$$\begin{aligned}\hat{\breve{\mathbf{q}}}_I^{-1}(\breve{\mathbf{S}}'\mathbf{p}) &= (\hat{\breve{\mathbf{q}}}_I^{-1}\breve{\mathbf{S}}')\mathbf{p} = (\breve{\mathbf{S}}\hat{\breve{\mathbf{q}}}_I^{-1})'\mathbf{p} = \mathbf{C}_I'\mathbf{p} = \hat{\breve{\mathbf{q}}}_I^{-1}(\hat{\breve{\mathbf{q}}}_I\mathbf{p}) \\ &= (\hat{\breve{\mathbf{q}}}_I^{-1}\hat{\breve{\mathbf{q}}}_I)\mathbf{p} = \mathbf{p} \Rightarrow \mathbf{C}_I'\mathbf{p} = \mathbf{p}.\end{aligned} \tag{5.125}$$

So, the *semi-positive* and *irreducible* or *positive* transposed matrix $\mathbf{C}_I' = (\breve{\mathbf{S}}\hat{\breve{\mathbf{q}}}_I^{-1})'$ has, due to Lema 4.1.1 (a) positive price eigenvectors $\mathbf{p} > \mathbf{0}$, associated with the Frobenius number $\lambda_{C_I} = 1$.

Taking the *Standard system* $(\breve{\mathbf{S}}', \breve{\mathbf{q}}, \breve{\mathbf{L}}) \in \mathcal{E}$, we now show that the obtained positive eigenvectors $\mathbf{p}$ of matrix $\mathbf{C}_I$ are also positive eigenvectors of matrix $\mathbf{C} = \breve{\mathbf{S}}\hat{\breve{\mathbf{q}}}^{-1}$, associated with the Frobenius number $\lambda_C = 1/(1+R)$.

In fact, from equation (5.124), we get the expression $\breve{\mathbf{q}}_I = \lambda_C \mathbf{q}$, and then find easily the equation $\hat{\breve{\mathbf{q}}}_I^{-1} = (1+R)\hat{\breve{\mathbf{q}}}^{-1}$, which summarising we can accordingly write:

$$\boxed{\begin{aligned} \mathbf{C}_I'\mathbf{p} &= (\breve{\mathbf{S}}\hat{\breve{\mathbf{q}}}_I^{-1})'\mathbf{p} = (\breve{\mathbf{S}}(1+R)\hat{\breve{\mathbf{q}}}^{-1})'\mathbf{p} \\ &= (1+R)(\breve{\mathbf{S}}\hat{\breve{\mathbf{q}}}^{-1})'\mathbf{p} = (1+R)\mathbf{C}'\mathbf{p} = \mathbf{p} \\ &\Rightarrow \mathbf{C}_I'\mathbf{p} = \mathbf{p} \Rightarrow \mathbf{C}'\mathbf{p} = \frac{1}{1+R}\mathbf{p} = \lambda_C \mathbf{p}. \end{aligned}} \tag{5.126}$$

Proposition 5.5.1. *Assume a Standard system* $(\breve{\mathbf{S}}', \breve{\mathbf{q}}, \breve{\mathbf{L}}) \in \mathcal{E}$, *with a semi-positive and irreducible or positive* $n \times n$ *matrix* $\breve{\mathbf{S}}$. *First consider the aggregate means of production* $\breve{\mathbf{q}}_I = \breve{\mathbf{S}}\mathbf{e} > \mathbf{o}$ *without surplus. Matrix* $\mathbf{C}_I = \breve{\mathbf{S}}\hat{\breve{\mathbf{q}}}_I^{-1}$ *(5.29) has then the Frobenius number* $\lambda_{C_I} = 1$. *Second, consider the case with surplus* $\breve{\mathbf{d}} = R\breve{\mathbf{q}}_I > \mathbf{o}$, *implying* $\breve{\mathbf{q}} = \breve{\mathbf{q}}_I(1+R) > \mathbf{o}$, *where* $\mathbf{C} = \breve{\mathbf{S}}\hat{\breve{\mathbf{q}}}^{-1}$ *(5.29) has Frobenius number* $\lambda_C = 1/(1+R) < 1$ *(see further Lemma 4.1.1 (a), (b)).*

The positive eigenvectors $\mathbf{p} > \mathbf{o}$ *of the transposed matrix* $\mathbf{C}_I'$, *associated with the Frobenius number* $\lambda_{C_I} = 1$, *are also positive eigenvectors of the transposed matrix* $\mathbf{C}'$, *associated with the Frobenius number* $\lambda_C = 1/(1+R)$ *(5.126).* ▲

If there is no surplus, then the vector of *total output* is $\breve{\mathbf{q}}_I = \breve{\mathbf{S}}\mathbf{e} > \mathbf{o}$. There is no *national income*, $Y = 0$. Consequently, circulating capital and total output are equal, $K = X = \breve{\mathbf{q}}_I'\mathbf{p}$.

If there is a surplus $\breve{\mathbf{d}} = R\breve{\mathbf{q}}_I > \mathbf{o}$, $\breve{\mathbf{q}} = \breve{\mathbf{q}}_I(1+R) > \mathbf{o}$, then matrix $\mathbf{C} = \breve{\mathbf{S}}\hat{\breve{\mathbf{q}}}^{-1}$ has Frobenius number $\lambda_C = 1/(1+R) < 1$ with productiveness $R > 0$. We then calculate the *total output*, the *circulating capital* and the *national income*, as follows:

$$X = \breve{\mathbf{q}}'\mathbf{p} = (1+R)(\breve{\mathbf{S}}\mathbf{e})'\mathbf{p} = K + Y, \quad K = (\breve{\mathbf{S}}\mathbf{e})'\mathbf{p}, \quad Y = R \cdot K. \tag{5.127}$$

Returning to the complete Sraffa price model (5.75) formulated for a *Standard system*, we present that Sraffa price model expressed in commodity units (5.83), using matrix $\mathbf{C}$ and vector $\breve{\boldsymbol{\pi}}$ (now with *breve-sign*).

$$\boxed{\begin{aligned} \mathbf{C}'\mathbf{p}(1+r) + \breve{\boldsymbol{\pi}} \cdot \left(\frac{\tilde{w} \cdot Y}{L}\right) &= \mathbf{p}, \\ L &= \breve{\mathbf{L}}'\mathbf{e}, \\ Y &= R \cdot (\breve{\mathbf{S}}\mathbf{e})'\mathbf{p}. \end{aligned}} \tag{5.128}$$

We recognise at this stage that the *semi-positive* and *irreducible* or *positive commodity flow* matrix $\breve{\mathbf{S}}$ of that *Standard system* and the vector of *labour* $\breve{\mathbf{L}}$ determine the *conditions of production*. Consequently, the *prices* and all the other economic variables and ratios are determined by the Sraffa price model (5.128).

What is the influence of the vector of labour $\breve{\mathbf{L}}$ on the prices? We focus on the special case of a *Standard system*, where there is the parallelism $\breve{\mathbf{S}}'\mathbf{p} \parallel \breve{\mathbf{L}} \parallel \hat{\breve{\mathbf{q}}}\mathbf{p}$ or equivalently $\mathbf{C}'\mathbf{p} \parallel \breve{\boldsymbol{\pi}} \parallel \mathbf{p}$. In this case, there exists a factor $a \in \mathbb{R}$, such that

$$\breve{\mathbf{L}} \parallel= \hat{\breve{\mathbf{q}}}\mathbf{p} \Leftrightarrow \hat{\breve{\mathbf{q}}}^{-1}\breve{\mathbf{L}} =: \breve{\boldsymbol{\pi}} \parallel \mathbf{p} \Leftrightarrow \breve{\boldsymbol{\pi}} = a \cdot \mathbf{p}. \tag{5.129}$$

Setting $w = (\tilde{w} \cdot Y)/L$ in equation (5.128) and further using the equality $\breve{\boldsymbol{\pi}} = a \cdot \mathbf{p}$ (5.81), we determine the factor a:

$$\begin{aligned}
&\mathbf{C}'\mathbf{p}(1+r) + \left(\frac{\tilde{w} \cdot Y}{L}\right)\breve{\boldsymbol{\pi}} = \mathbf{C}'\mathbf{p}(1+r) + w(a \cdot \mathbf{p}) = \mathbf{p} \Leftrightarrow \\
&\mathbf{C}'\mathbf{p}(1+r) = (1 - a \cdot w)\mathbf{p} \Leftrightarrow \mathbf{C}'\mathbf{p} = \frac{1 - a \cdot w}{1+r}\mathbf{p} =: \frac{1}{1+R}\mathbf{p} \Leftrightarrow \\
&a = \frac{R-r}{w(1+R)} = \frac{L \cdot (R-r)}{\tilde{w} \cdot Y \cdot (1+R)} = \frac{L \cdot (R-r)}{\tilde{w} \cdot Y}\lambda_C \Leftrightarrow \mathbf{C}'\mathbf{p} = \lambda_C \mathbf{p}.
\end{aligned} \tag{5.130}$$

We can therefore state:

Proposition 5.5.2. *Set the variable $a = (L \cdot (R-r))/(\tilde{w} \cdot Y \cdot (1+R))$, determined by the given parameters r, R and the calculated values of $\tilde{w}$, L, Y, obtained from the complete Sraffa price model (5.128), expressed for a standard system. Consider the parallelism $\mathbf{C}'\mathbf{p} \parallel \breve{\boldsymbol{\pi}} \parallel \mathbf{p}$, then the price vector $\mathbf{p}$ is an eigenvector of matrix $\mathbf{C}'$, associated with the Frobenius number $\lambda_C = 1/(1+R) < 1$. That eigenvector $\mathbf{p}$ solves the complete Sraffa price model (5.128) for any profit rate r, $0 \le r \le R$, including the borderline cases $r = 0$ and $r = R$.*

Let us illustrate this subsection by the two following numerical examples:

Example 5.5.1. There are $n = 2$ sectors. Consider the matrices

$$\breve{\mathbf{S}} = \begin{bmatrix} 250 & 110 \\ 30 & 60 \end{bmatrix}, \quad \breve{\mathbf{q}}_I = \breve{\mathbf{S}}\mathbf{e} = \begin{bmatrix} 360 \\ 90 \end{bmatrix} \parallel \breve{\mathbf{q}} = \begin{bmatrix} 432 \\ 108 \end{bmatrix}, \quad \breve{\mathbf{L}} = \begin{bmatrix} 72 \\ 66 \end{bmatrix}, \tag{5.131}$$

constituting a Standard system $(\breve{\mathbf{S}}, \breve{\mathbf{q}}, \breve{\mathbf{L}})$.

(1) Compute the total amount of the means of production $\breve{\mathbf{q}}_I = \breve{\mathbf{S}}\mathbf{e}$, the *input-output coefficients matrix* $\mathbf{C}_I = \breve{\mathbf{S}}\hat{\breve{\mathbf{q}}}_I^{-1}$ and its Frobenius number λ_{C_I}, the *input-output coefficients matrix* $\mathbf{C} = \breve{\mathbf{S}}\hat{\breve{\mathbf{q}}}^{-1}$, its Frobenius number λ_C and the productiveness R. Verify that the price (left) eigenvectors $\mathbf{p} > \mathbf{0}$ of matrices $\mathbf{C}$ and $\mathbf{C}_I$, associated with the respective Frobenius numbers, are equal.

(2) Choose as numéraire commodity 1 and set $p_1 = p_0 = 1$. Choose the rate of profits $r = 0.1$, $0 \le r \le R$, the given vector of labour $\breve{\boldsymbol{\pi}}$ per units of commodities, set up Sraffa's price model (5.128) and solve it. Compute the wage rate w, the relative price p_2, the factor a (5.130). Confirm that the solution price vector is an eigenvector of matrix $\mathbf{C}$, associated with the Frobenius number λ_C.

(3) Continue with the same model (2). Compute the circulating capital K, the national income $Y = R \cdot K$ (5.127), the total output X, the total profits $P = r \cdot K$ and the total wages $W = w \cdot L$. Compute the share of total wages to national income $\tilde{w} = w \cdot (L/Y)$ and the share of total profits to national income $\tilde{r} = P/Y$.

(4) Repeat step (2) with $r = 0$, keeping all the other entries of the given economy unchanged.

Solution to Example 5.5.1:
(1) Consider the vector $\breve{\mathbf{q}}_I = \breve{\mathbf{S}}\mathbf{e} = [360, 90]'$ and compute matrix

$$\mathbf{C}_I = \breve{\mathbf{S}}\hat{\breve{\mathbf{q}}}_I^{-1} = \begin{bmatrix} 250 & 110 \\ 30 & 60 \end{bmatrix} \begin{bmatrix} \frac{1}{360} & 0 \\ 0 & \frac{1}{90} \end{bmatrix} = \begin{bmatrix} \frac{25}{36} & \frac{11}{9} \\ \frac{1}{12} & \frac{2}{3} \end{bmatrix}. \tag{5.132}$$

Then compute the characteristic polynomial,

$$P_2(\lambda) = \det(\mathbf{C}_I - \lambda \cdot \mathbf{I}) = \lambda^2 - \frac{49}{36}\lambda + \frac{13}{36} = (\lambda - 1)\left(\lambda - \frac{13}{36}\right). \tag{5.133}$$

The Frobenius number is as expected $\lambda_{C_I} = 1$, and we solve the corresponding eigenvalue equation associated with λ_{C_I} to find the positive price eigenvectors $\mathbf{p}$,

$$\mathbf{C}_I'\mathbf{p} = \mathbf{p} \Leftrightarrow \begin{cases} \frac{25}{36}p_1 + \frac{1}{12}p_2 = p_1 \\ \frac{11}{9}p_1 + \frac{2}{3}p_2 = p_2. \end{cases} \tag{5.134}$$

We had fixed $p_1 = p_0 = 1$, so $p_2 = 11/3 = 3.667$ and we get the price (left) eigenvector $\mathbf{p} = [1, 11/3]'$ of matrix $\mathbf{C}_I$. Then compute the matrix

$$\mathbf{C} = \mathbf{S}\hat{\mathbf{q}}^{-1} = \begin{bmatrix} 250 & 110 \\ 30 & 60 \end{bmatrix} \begin{bmatrix} \frac{1}{432} & 0 \\ 0 & \frac{1}{108} \end{bmatrix} = \begin{bmatrix} \frac{125}{216} & \frac{55}{54} \\ \frac{5}{72} & \frac{5}{9} \end{bmatrix}, \tag{5.135}$$

and determine the characteristic polynomial

$$P_2(\lambda) = \det(\mathbf{C} - \lambda \cdot \mathbf{I}) = \lambda^2 - \frac{245}{216}\lambda + \frac{315}{1{,}296} = \left(\lambda - \frac{5}{6}\right)\left(\lambda - \frac{65}{216}\right), \tag{5.136}$$

giving the Frobenius number $\lambda_C = 5/6$ and the *productiveness* $R = (1/\lambda_C) - 1 = 0.2$. Having fixed $p_1 = p_0 = 1$, the (left) eigenvector of matrix $\mathbf{C}$, associated with the Frobenius number $\lambda_C = 5/6$ are $\mathbf{p} = [1, 11/3]'$, equal to the eigenvector of matrix $\mathbf{C}_I$, associated with the Frobenius number $\lambda_{C_I} = 1$. This means that for the borderline case $r = R$ of (5.128) without labour vector $\mathbf{L}$, the solution price vector is as stated the eigenvector of matrix $\mathbf{C}$ and of matrix $\mathbf{C}_I$.

(2) We treat now the case of an intermediate *rate of profits* $r_0 = 0.1 \in [0, 0.2]$. For this purpose, we compute $\breve{\boldsymbol{\pi}} = \hat{\breve{\mathbf{q}}}^{-1}\breve{\mathbf{L}} = [(72/432), (66/108)]' = (1/6)[1, (11/3)]'$, the

vector of labour per units of commodities. Accordingly, with the price $p_0 = 1$ for the numéraire, we set up the *Sraffa price model* (5.128) for the given entries,

$$\mathbf{C}'\mathbf{p}(1+0.1) + \left(\frac{\tilde{w}\cdot Y}{L}\right)\breve{\boldsymbol{\pi}} = \mathbf{p}, \quad p_1 = 1, \quad L = \breve{\mathbf{L}}'\mathbf{e}, \quad Y = 0.2\cdot\breve{\mathbf{q}}\mathbf{p}. \tag{5.137}$$

We get with the known values $R = 0.2$, $r = 0.1$ four equations for the four variables $\tilde{w}$, p_2, L, Y,

$$\begin{cases} (1+0.1)(\frac{125}{216}\cdot 1 + \frac{5}{72}\cdot p_2) + (\frac{\tilde{w}\cdot Y}{L})\cdot\frac{1}{6} = 1, \\ (1+0.1)(\frac{55}{54}\cdot 1 + \frac{5}{9}\cdot p_2) + (\frac{\tilde{w}\cdot Y}{L})\cdot\frac{11}{18} = p_2, \\ Y = 0.2\cdot(360\cdot 1 + 90\cdot p_2), \\ L = 72 + 66, \end{cases} \tag{5.138}$$

and we obtain $p_2 = (11/3)$, $L = 138$, $Y = 138$ and $\tilde{w} = 0.5$. We verify the proportionality $r = R(1-\tilde{w}) = 0.2(1-0.5) = 0.1$. The price vector solving the present Sraffa price model (5.138) is $\mathbf{p} = [1, 11/3]' \parallel \breve{\boldsymbol{\pi}} = [1/6, 11/18]'$.

(3) We now compute the circulating capital $K = \mathbf{q}_I\mathbf{p}' = 360\cdot 1 + 90\cdot(11/3) = 690$, the total profits $P = r\cdot K = 0.1\cdot 690 = 69$, the total wages $W = Y - P = 138 - 69 = 69$, the ratio of *total wages to national income* $\tilde{w} = w\cdot(L/Y) = 0.5\cdot(138/138) = 0.5$ and also the ratio of *total profits to national income* $\tilde{r} = (P/Y) = 69/138 = 0.5$, and the *share of national income to circulating capital* $\tilde{R} = Y/K = 138/690 = 0.2$. We confirm the equality $R = \tilde{R} = 0.2$.

(4) Then we establish the *Sraffa price model* (5.128), namely with the *rate of profits* $r = 0$ for the other borderline case. With $r = R(1-\tilde{w}) = 0$, the proportionality, the *share of total wages to national income* and $\tilde{w} = w = 1$, we obtain the model

$$\mathbf{C}'\mathbf{p} + \breve{\boldsymbol{\pi}} = \mathbf{p}. \tag{5.139}$$

We get as solution the price vector $\mathbf{p} = [1, (11/3)]'$. It is easily seen that the three vectors that we set up are parallel:

$$\mathbf{C}'\mathbf{p} = \begin{bmatrix} \frac{125}{216} & \frac{5}{72} \\ \frac{55}{54} & \frac{5}{9} \end{bmatrix}\begin{bmatrix} 1 \\ \frac{11}{3} \end{bmatrix} = \begin{bmatrix} \frac{5}{6} \\ \frac{55}{18} \end{bmatrix} \parallel \breve{\boldsymbol{\pi}} = \begin{bmatrix} \frac{1}{6} \\ \frac{11}{18} \end{bmatrix} \parallel \mathbf{p} = \begin{bmatrix} 1 \\ \frac{11}{3} \end{bmatrix}, \tag{5.140}$$

as presented in Fig. 5.6 (left). For this borderline case $r = 0$, we have obtained as solution the same price vector $\mathbf{p} = [1, (11/3)]'$, as eigenvector of matrix $\mathbf{C}$. ▲

Example 5.5.2. Take the matrices $\breve{\mathbf{S}}$ and the vector $\breve{\mathbf{q}}$ of the *Standard system* in Example 5.5.1 and replace the vector for labour, by the vector $\breve{\mathbf{L}} = [66, 33]$. Set $p_1 = 1$ for the numéraire. Solve the complete Sraffa price model (5.128). ▲

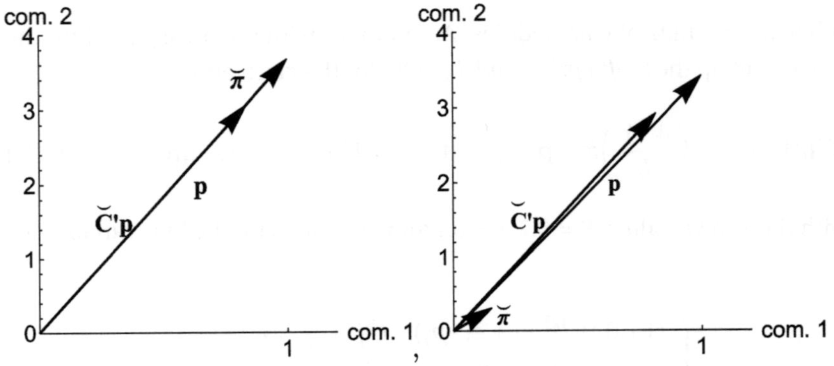

Figure 5.6: Parallelism (left) of the vectors $\breve{\mathbf{C}}'\mathbf{p} \parallel \mathbf{p} \parallel \breve{\boldsymbol{\pi}}$, see Example 5.5.1, and non-parallelism (right) of the vectors $\breve{\mathbf{C}}'\mathbf{p} \nparallel \mathbf{p} \nparallel \breve{\boldsymbol{\pi}}$.

Solution to Example 5.5.2:
We compute $\breve{\boldsymbol{\pi}} = \hat{\breve{\mathbf{q}}}^{-1}\breve{\mathbf{L}} = [(66/432), (33/108)]' = [(11/72), (11/36)]'$, the vector of labour per units of commodities. Accordingly, with the price $p_0 = 1$ for the numéraire, we solve the following *Sraffa price model* (5.128),

$$\begin{cases} (1+0.1)(\frac{125}{216}\cdot 1 + \frac{5}{72}\cdot p_2) + (\frac{\bar{w}\cdot Y}{L})\cdot \frac{11}{72} = 1, \\ (1+0.1)(\frac{55}{54}\cdot 1 + \frac{5}{9}\cdot p_2) + (\frac{\bar{w}\cdot Y}{L})\cdot \frac{11}{36} = p_2, \\ Y = 0.2\cdot(360\cdot 1 + 90\cdot p_2), \\ L = 72 + 66, \end{cases} \tag{5.141}$$

and the price vector $\mathbf{p} = [1, (133/39)]' \nparallel \breve{\boldsymbol{\pi}} = (11/72)[1, 2]'$ which is not an eigenvector of matrix $\mathbf{C}'$ and therefore not parallel to the two vectors $\mathbf{C}'\mathbf{p}$ and $\breve{\boldsymbol{\pi}}$ (see Figure 5.6, right). ▲

We come to the conclusion.

Given a *semi-positive* and *irreducible* or *positive commodity flow* matrix $\breve{\mathbf{S}}$, a numéraire, and a vector of labour $\breve{\mathbf{L}}$, constituting a **Standard system** $(\breve{\mathbf{S}}', \breve{\mathbf{q}}, \breve{\mathbf{L}})$ with surplus, consider the vector of labour per unit of commodities $\breve{\boldsymbol{\pi}} = \hat{\breve{\mathbf{q}}}^{-1}\breve{\mathbf{L}}$, and the input-output coefficients matrix $\mathbf{C} = \breve{\mathbf{S}}\hat{\breve{\mathbf{q}}}^{-1}$. Solve the eigenvalue equation, $\mathbf{C}'\mathbf{p} = \lambda_C\mathbf{p}$, computing the Frobenius number λ_C, the associated eigenvectors $\mathbf{p}$ and the productiveness $R = (1/\lambda_C) - 1 > 0$.

If the parallelism $\mathbf{C}'\mathbf{p} \parallel \breve{\boldsymbol{\pi}} \parallel \mathbf{p}$ exists, then for any rate of profits $r \in [0, R]$, there is exactly *one price vector* $\mathbf{p}$ solving the complete *Sraffa price model* (5.128) and for this reason, Sraffa says that there are no *price fluctuations*, what is confirmerd here. On the other hand, if $\mathbf{p} \nparallel \breve{\boldsymbol{\pi}}$, then there are price variations.

We have here the quantitative explanation for Sraffa's *price fluctuations* in a **Standard system** $(\breve{\mathbf{S}}', \breve{\mathbf{q}}, \breve{\mathbf{L}})$ with a surplus. Summarising, price fluctuations depend on the distribution of the wages among the industrial branches, determined by the vector of labour

$\check{\mathbf{L}}$ and w. Price fluctuations occur if $\mathbf{p} \nparallel \check{\boldsymbol{\pi}}$. Price fluctuations do not occur if $\mathbf{p} \parallel \check{\boldsymbol{\pi}}$. This is the correct solution to Ricardo's problem mentioned in footnote 1 and footnote 6.

For Sraffa (PCMC, Par. 43), the Standard system, endowed with a standard net product, is a purely auxiliary construction. In fact, *Standard systems* can hardly be realised within real economic contexts, unless one resorts to a rigorously centralised planning system imposing the conditions (5.121) on all production sectors.

This being said, Sraffa maintains, having normalised labour, $L = 1$, and the national income, $Y = 1$, that the following reciprocity relation holds:

$$\text{Standard national income} \rightleftarrows r = R(1 - w). \tag{5.142}$$

"This proposition (→, our notation) is reversible, and if we make it a condition of the economic system that w and r should obey the proportionality rule in question, the wage and commodity prices are then ipso facto expressed in Standard net product, without need of defining its composition, since with no other unit can the proportionality rule be fulfilled" (PCMC, Par. 43).

This citation calls for an explanation. Indeed, (5.142) means that $\tilde{R} = R$ (where R the *productiveness* of the present economy and $\tilde{R} = Y/K$ is the *share of national income to circulationg capital*), but this as we know can only be the case if the underlying initial Sraffa price system is already a *Standard system*.

On the background of our investigations, where the national income is no longer normalized, our interpretation is as follows.

Given a *Standard system*, described by a *semi-positive* irreducible technology matrix $\mathbf{S} \geq \mathbf{0}$ with a positive vector of surplus $\mathbf{d} > \mathbf{0}$. One gets $\mathbf{q} = \mathbf{Se} + \mathbf{d} > \mathbf{o}$, then the Frobenius number λ_C of the *input-output coefficients* matrix $\mathbf{C} = \mathbf{S}\hat{\mathbf{q}}^{-1} \geq \mathbf{0}$ is in the interval, $0 < \lambda_C < 1$. The *productiveness* is $R = (1/\lambda_C) - 1 > 0$ and the rate of profits r can be chosen, $r \in]0, R[$. Thus the wage rate is positive $w > 0$, as we assume here a positive vector of labour $\mathbf{L} > \mathbf{o}$. The Sraffa price system is solved, obtaining a positive price vector $\mathbf{p} > \mathbf{o}$. We calculate the *circulating capital* $K = (\mathbf{Se})'\mathbf{p}$. Clearly the *share of national income Y to circulating capital K* and the *productiveness R* are equal, $\tilde{R} := Y/K = R$.

Then the *national income* is equal to the product of *circulating capital* and *productiveness*, i. e., $Y = R \cdot K$.

This means, in the case of a *Standard system*, given by the described production scheme $(\mathbf{S}', \mathbf{L}) \Rightarrow (\mathbf{q})$, the *productiveness R* determines the *national income Y*.

A last remark on Sraffa's extravagant statement (PCMC, Par. 43):

"And it is curious that we should thus be enabled to use a standard without knowing what is consists of." — We may forget it. If $Y_0 = 1$ GDP is used as the standard, we know what it consists of by construction.

6 A new look at joint production analysis

Up to now, we have treated Sraffa's process of production for single-product industries, one industry producing only one product or commodity, presented in the first part of PCMC.[1] In this chapter, we will extend this analysis to the case where n branches or industries produce one or more of an equal number of n commodities. This case is called by Sraffa *joint production* of commodities (PCMC, Chap. VI–IX, XI, specially Par. 50–53), also called economy of *multi-product industries* [23]. We will use throughout this book the term *joint production Sraffa System*, in analogy to B. Schefold's term *single-product Sraffa system* ([103], p. 34). To our knowledge, the first mathematical analysis of this part of PCMC was developed by Manara [61] (in Pasinetti, [83], p. 1–15). Manara [62] published his work at first in Italian in 1968.

We have to mention here the seminal book of Bertram Schefold who treated *"the classical theory of prices, in its significance to the critique of economic theory, and ...special aspects of joint production"* ([103], p. ix). He brought the mathematical analysis of "joint production of commodities" to its completeness 21 years after Manara's presentation of a first mathematical analysis of the subject.[2] We will rely on the algebraic constructions of the present developments on his mathematical inspiration, regarding *joint production.*

Schefold [103], Steedman [114] and Salvadori and Steedman [100] furthermore developed the mathematics to treat the distinction between *basic* and *non-basic* commodities in *joint production analysis* for the calculation of *joint production Sraffa sys-*

1 In order not to overload this introductory text, we shall not enter into the discussion of *fixed capital* considered as joint product (PCMC, Chap. X for further reference), see also Pasinetti (Ed.) ([83], Chap. 1, pp. 1–15). We would just point out that Sraffa's approach to joint production was largely motivated by considering fixed capital (e. g., machinery, data processing equipment, vehicles for transportation) as a commodity entering the process of production. Whilst raw materials for example are usually completely used up in the production process, durable means of production enter production at the beginning of a period, say at the beginning of the year, are then partly worn out and emerge at the end of the production period as a joint commodity with the period's output.
This point of view implies that the same durable means of production, at different periods, should be treated as so many different commodities, each with its own price. So, in order to determine these prices, an equal number of additional equations (and therefore of processes) are required, together with explicit depreciation/amortization/replacement schedules. This complicates significantly Sraffa's model, and, besides land, we have accordingly restricted ourselves to considering joint production in connection with ecological problems (treatment of waste, reduction of greenhouse gases), which is the subject of Chapter 7, and in particular in extending Sraffa's model to account for an open economy, see Section 8.1.

2 Schefold discusses the position of Sraffa's PCMC as a call for a fundamental critique of economic theory and the fact that he is mainly ignored by modern economists, see in [102], p. 315, *"Der Gegensatz zwischen Sraffa's Anspruch, das Fundament für eine Kritik der ökonomischen Theorie schlechthin gelegt zu haben, und seiner Ignorierung durch die Mehrzahl der modernen Ökonomen scheint unüberbrückbar."*

https://doi.org/10.1515/9783110635096-006

tems. Finally, the important contribution of Pasinetti ([83], pp. 36, 45), who introduced the *indirect capital matrix* **H**, will receive due consideration.

We focus on the topics required to attain an operational level. We are inspired by the brilliant analyses of Manara [61], Pasinetti [83], Steedman [114], Steedman and Salvadori [100] and Schefold [103] which are in line with this aim. We continue presenting numerical examples, the complete calculations of their solutions and the resulting economic outlook.

Our contribution consists in introducing some new notions, such as that of the *production space*, which is a tool to better grasp by algebraic treatment the dimensionality of the process hidden in *joint production*. Following the lead of single-product industries, we will also resort to *Sraffa Networks* with their associated *adjacency matrices* to analyse connectivity.

6.1 The production scheme, assumptions and the production space

Single-product industries of n commodities and n industries are enlarged to joint systems of production, where each of the n branches may produce more than one of the n commodities.

Proceeding from self-replacement *single product-industries* to self-replacement *joint production*, we observe that the *output* vector $\mathbf{q} > \mathbf{o}$, presented in the production scheme by the diagonal matrix $\hat{\mathbf{q}}$, is replaced by a *semi-positive* matrix $\mathbf{F} = (F_{ij})$, called the matrix of *outputs* (Schefold [103], p. 49). We will also call it the *output* matrix. Indeed, Manara [61] (in Pasinetti [83], p. 2) called f_{ij} "the quantity of i-th commodity produced by the j-th industry", see Table 6.1.

Table 6.1: Output coefficients matrix $\mathbf{F} = (f_{ij})$.

	Industries *j* producing more than one commodity								**Row sums**
	1	**2**	**3**	**...**	***j***	**...**	**...**	***n***	
1	f_{11}	f_{12}	f_{13}	...	f_{1j}	...	...	f_{1n}	
2	f_{21}	f_{22}	f_{23}	...	f_{2j}	...	...	f_{2n}	
...	...	...	...	...	...	...	...	...	
commodities *i*	f_{i1}	f_{i2}	f_{i3}	...	f_{ij}	...	...	f_{in}	$\sum_{j=1}^{n} f_{ij}$
...	...	...	...	...	...	...	...	...	
...	...	...	...	...	...	...	...	...	
n	f_{n1}	f_{n2}	f_{n3}	...	f_{nj}	...	...	f_{nn}	

The coefficients of matrix **F** are:

f_{ij}: the quantity of commodity i produced by the industry j.

(1) The production scheme. The production scheme of *joint production* needs the *semi-positive* matrices **S** and **F** and the vector of labour **L**, in analogy to that of *single-product industries*, $(\mathbf{S}', \mathbf{L}) \rightarrow (\hat{\mathbf{q}})$ (4.132). Here the diagonal matrix $\hat{\mathbf{q}}$ is replaced by $\mathbf{F}'$, giving a representation, where the commodities appear *vertically* and the industries *horizontally*,

$$\begin{aligned}
&(s_{11}, s_{21}, s_{31}, \ldots, s_{n1}, L_1) \rightarrow (f_{11}, f_{21}, f_{31}, \ldots, f_{n1}),\\
&(s_{12}, s_{22}, s_{32}, \ldots, s_{n2}, L_2) \rightarrow (f_{12}, f_{22}, f_{32}, \ldots, f_{n2}),\\
&(s_{13}, s_{23}, s_{33}, \ldots, s_{n3}, L_3) \rightarrow (f_{13}, f_{23}, f_{33}, \ldots, f_{n3}),\\
&(\ldots, \ldots, \ldots, \ldots, \ldots, \ldots) \rightarrow (\ldots, \ldots, \ldots, \ldots, \ldots),\\
&(s_{1n}, s_{2n}, s_{3n}, \ldots, s_{nn}, L_n) \rightarrow (f_{1n}, f_{2n}, f_{3n}, \ldots, f_{nn}),\\
&\qquad\qquad (\mathbf{S}', \mathbf{L}) \rightarrow (\mathbf{F}').
\end{aligned} \tag{6.1}$$

Observe that each row presents the production of one industry that can no longer be attributed to a sole commodity, as every industry may produce more than one commodity, so we have now a compactly written form of the production process with the matrices **S**, **F** and the vector **L**.

(2) The assumptions.

(a) The **first assumption:** Following the same line of thought as in *single-product industries,* we treat at first the economic assumption of *self-replacement* in the case of *joint production*. The case of *no surplus* corresponds to the conditions of production, Definition 3.1.1, for *single-product industries* expressed by the null vector $\mathbf{d} = \mathbf{o}$. The second case of *self-replacement* guarantees $\mathbf{d} = (\mathbf{F} - \mathbf{S})\mathbf{e} = \mathbf{Fe} - \mathbf{Se} \geq \mathbf{o}$, a *semi-positive* vector of net product. The produced output is only positive for some commodities. The third case of *positive self-replacement* guarantees a *positive* vector of the net product $\mathbf{d} = (\mathbf{F} - \mathbf{S})\mathbf{e} > \mathbf{o}$, a positive output for each product. The last two cases give a surplus, which then is split up into a profit part and a wage part.

Assumption 6.1.1 (Assumption on surplus in joint production). In the case of joint production, the conditions of self-replacement occur in the three different cases, as expressed by the vector of surplus, also termed the vector of net product (see Schefold [103], p. 49):

$$\begin{aligned}
\mathbf{d} &= (\mathbf{F} - \mathbf{S})\mathbf{e} = \mathbf{o}, && \text{no surplus,}\\
\mathbf{d} &= (\mathbf{F} - \mathbf{S})\mathbf{e} \geq \mathbf{o}, && \text{self-replacement,}\\
\mathbf{d} &= (\mathbf{F} - \mathbf{S})\mathbf{e} > \mathbf{o}, && \text{positive self-replacement.}
\end{aligned} \tag{6.2}$$

(b) The **second assumption:** We apply equation (6.2), using the transposed form for the case of *no surplus*, $\mathbf{d}' = \mathbf{e}'(\mathbf{F}' - \mathbf{S}') = \mathbf{o}$, for which the following property applies: $n - \text{rank}(\mathbf{F}' - \mathbf{S}') = \dim(\mathbf{e}') = 1$ (see Nef [69], Theorem 1, p. 121), which has been used for *single-product industries* (3.12) and now reads for *joint production*:

$$\mathbf{e}'(\mathbf{F}' - \mathbf{S}') = \mathbf{o} \Rightarrow \operatorname{rank}(\mathbf{F}' - \mathbf{S}') = n - 1 \Rightarrow \det(\mathbf{F}' - \mathbf{S}') = 0. \tag{6.3}$$

This means: If there is *no surplus*, then matrix $(\mathbf{F}' - \mathbf{S}')$ is singular.

Treating then the case of a *surplus*, we have to describe the production processes. For this purpose, we apply Definition A.4.3 to the *commodity flow matrix* **S**, respectively to the *output* matrix **F**. Consider the $n \times n$ matrix $\mathbf{S} = (s_{ij})$, $i, j = 1, \dots, n$, and its column vectors $\mathbf{s}_{\cdot j} = [s_{1j}, s_{2j}, \dots, s_{nj}]'$, $j = 1, \dots, n$, respectively its row vectors $\mathbf{s}'_{i\cdot} = [s_{i1}, s_{i2}, \dots, s_{in}]$, $i = 1, \dots, n$. Then matrix **S** can be written as a matrix of column vectors or of row vectors as follows

$$\mathbf{S} = \begin{bmatrix} s_{11} & s_{12} & \dots & \dots & s_{1n} \\ s_{21} & s_{22} & \dots & \dots & s_{2n} \\ \dots & \dots & \dots & \dots & \dots \\ \dots & \dots & \dots & \dots & \dots \\ s_{n1} & s_{n2} & \dots & \dots & s_{nn} \end{bmatrix} = [\mathbf{s}_{\cdot 1}, \mathbf{s}_{\cdot 2}, \dots, \dots, \mathbf{s}_{\cdot n}] = \begin{bmatrix} \mathbf{s}'_{1\cdot} \\ \mathbf{s}'_{2\cdot} \\ \dots \\ \dots \\ \mathbf{s}'_{n\cdot} \end{bmatrix}. \tag{6.4}$$

The row vectors $\mathbf{s}'_{i\cdot} = [s_{i1}, s_{i2}, \dots, s_{in}]$, $i = 1, \dots, n$ are composed of the quantities s_{ij} of commodity i, necessary as inputs for the production in each of the different sectors j, $j = 1, \dots, n$. For this reason, the row vectors $\mathbf{s}'_{i\cdot}$ are called *output vectors of the i-th commodity*. The column vectors $\mathbf{s}_{\cdot j} = [s_{1j}, s_{2j}, \dots, s_{nj}]'$, $j = 1, \dots, n$, are composed of the quantities s_{ij} of each commodity i, $i = 1, \dots, n$, necessary as inputs for the production of the specific sector j. The column vectors $\mathbf{s}_{\cdot j}$ are called *input vectors to the j-th production process*.

We set up the output matrix **F** and the column vectors $\mathbf{f}_{\cdot j}$, which are called *output vectors of the j-th production process*, obtaining the analogue decomposition,

$$\mathbf{F} = \begin{bmatrix} f_{11} & f_{12} & \dots & \dots & f_{1n} \\ f_{21} & f_{22} & \dots & \dots & f_{2n} \\ \dots & \dots & \dots & \dots & \dots \\ \dots & \dots & \dots & \dots & \dots \\ f_{n1} & f_{n2} & \dots & \dots & f_{nn} \end{bmatrix} = [\mathbf{f}_{\cdot 1}, \mathbf{f}_{\cdot 2}, \dots, \dots, \mathbf{f}_{\cdot n}] = \begin{bmatrix} \mathbf{f}'_{1\cdot} \\ \mathbf{f}'_{2\cdot} \\ \dots \\ \dots \\ \mathbf{f}'_{n\cdot} \end{bmatrix}. \tag{6.5}$$

If there is a *surplus*, then we have at least $\mathbf{d} = (\mathbf{F} - \mathbf{S})\mathbf{e} \geq \mathbf{o}$. We also require, inspired by observations in the real economy, that the production processes are *different* from one another. In other words: The production processes are *"defined to be different...if they are linearly independent"* (see Schefold [103], p. 50). The required linear independence of the processes results in the regularity of matrix $(\mathbf{F}' - \mathbf{S}')$, and we obtain that the rank[3] of matrix $[\mathbf{S}', \mathbf{F}']$ is equal to n. This is the mathematical expression that the processes are different from each another.

---

3 With $j \neq k$, consider two linear dependent processes $(\mathbf{s}_{\cdot j}, L_j) \to (\mathbf{f}_{\cdot j})$ and $(\mathbf{s}_{\cdot k}, L_k) \to (\mathbf{f}_{\cdot k})$, abbreviated to the pairs $(\mathbf{s}_{\cdot j}, \mathbf{f}_{\cdot j})$ and $(\mathbf{s}_{\cdot k}, \mathbf{f}_{\cdot k})$, labour being irrelevant here. With the real numbers $a, b \neq 0$, the linear combination is equal to the null vector, $a(\mathbf{s}_{\cdot j}, \mathbf{f}_{\cdot j}) + b(\mathbf{s}_{\cdot k}, \mathbf{f}_{\cdot k}) = \mathbf{o}$.

Assumption 6.1.2 (Assumption on linear independence of the processes). In a joint production economy with surplus, the n production processes $(\mathbf{s}_{.j}, L_j) \rightarrow (\mathbf{f}_{.j}), j = 1, \ldots, n$, (6.1) are defined to be different if they are linearly independent. Consequently, the matrix $(\mathbf{F}' - \mathbf{S}')$ is regular,

$$\det\left(\mathbf{F}' - \mathbf{S}'\right) \neq 0 \Leftrightarrow \det(\mathbf{F} - \mathbf{S}) \neq 0. \tag{6.6}$$

Furthermore, we obtain:

$$\operatorname{rank}\left(\left[\mathbf{S}', \mathbf{F}'\right]\right) = \operatorname{rank}\left(\left[\mathbf{S}', \mathbf{F}' - \mathbf{S}'\right]\right) = n. \tag{6.7}$$

(c) The **third assumption**: Assumption 2.2.1 requires a positive vector of *output* $\mathbf{q} > \mathbf{o} \Rightarrow \operatorname{rank}(\hat{\mathbf{q}}) = n$ in *single-product industries*. In *joint production*, there is a similar requirement[4]:

Assumption 6.1.3 (Input-output assumption). Every production process must have at least an *input* and at least an *output* besides labour, $\mathbf{s}_{.j} \geq \mathbf{o}$, $\mathbf{f}_{.j} \geq \mathbf{o}$ (see Schefold ([103], p. 49)). For these economic reasons, the matrices are *semi-positive* $\mathbf{S} \geq \mathbf{0}$, $\mathbf{F} \geq \mathbf{0}$[a]

a In the authors' view, Schefold ([103], p. 49) was the first economist to express this claim clearly and concisely. He said: ...([E]very process has an input besides labour and an output).... The condition $\det(\mathbf{F}) \neq 0$ is a sufficient but not necessary condition for the semi-positivity of the vectors $\mathbf{f}_{.j} \geq \mathbf{o}$. But $\det(\mathbf{F}) \neq 0$ is a necessary condition for systems, which are then called *gross integrated industries*, see Schefold [103], p. 56, and see ensuing content.

Finally, if matrix $\mathbf{F}$ has full rank,

$$\operatorname{rank}(\mathbf{F}) = n \Leftrightarrow \det(\mathbf{F}) \neq 0, \tag{6.8}$$

then one is in presence of a *joint production processes* $(\mathbf{F}'^{-1}\mathbf{S}', \mathbf{F}'^{-1}\mathbf{L}) \Rightarrow (\mathbf{I})$, called *gross integrated industries*, having been transformed into a process of the appearance of a *single-product system*, see Schefold ([103], p. 56).

(3) The commodity space (or product space).

The regularity of matrix $(\mathbf{F}' - \mathbf{S}')$ entails the existence of n – dim subsets in the Euclidean vector space $\mathbb{R}^n$. We use the *commodity space* (or *product space*) $\mathcal{C}^n$, Definition 5.2.1. This vector subset $\mathcal{C}_n$ allows one to represent either the n *means of production*, the produced *commodities* or the *total output* and to discuss the question of dimensionality.

Consider now a *joint production process* $(\mathbf{S}', \mathbf{L}) \Rightarrow (\mathbf{F}')$. We can construct three polyhedrons contained within the *commodity space* $\mathcal{C}^n$, generated by the orthonormal basis $\mathcal{I}_n = \{\vec{e}_1, \ldots, \vec{e}_j, \ldots, \vec{e}_n\}$.

We get $(\mathbf{s}_{.j}, \mathbf{f}_{.j}) = -\frac{b}{a}(\mathbf{s}_{.k}, \mathbf{f}_{.k})$, what is not possible in the case of independence of the processes! Consequently, one process multiplied by factor $-\frac{b}{a}$ gives the other process. These both processes can then be merged into one process so that the two processes become *one process*. There is one process less! This reasoning belongs to linear algebra. Evidently, linear combinations can be set up for more than two processes.

4 Schefold introduced the notion of *gross integrated systems* $(\mathbf{C}_T' = \mathbf{F}'^{-1}\mathbf{S}', \mathbf{F}'^{-1}\mathbf{L}) \Rightarrow (\mathbf{I})$ that can only be defined if $\det(\mathbf{F}) \neq 0$. See Proposition 6.2.1, and for an application look at Example 6.2.1.

Definition 6.1.1 (Commodity, production process and output polyhedrons).

(1) The produced quantities of each commodities in the case of joint production are contained in the vector $\mathbf{q} = \mathbf{Fe} = \mathbf{Se}+\mathbf{d} = [q_1, \dots, q_j, \dots q_n]'$, leading to the diagonal matrix $\hat{\mathbf{q}}$. Let be $\hat{\mathbf{q}} = [\mathbf{q}_{\cdot 1}, \mathbf{q}_{\cdot 2}, \dots, \mathbf{q}_{\cdot n}]$. The endpoints C_j of the geometric vectors $\vec{c}_j := \overrightarrow{OC}_j = [0, \dots 0, q_j, 0, \dots 0]' \Leftrightarrow \mathbf{q}_{\cdot j} = [0, \dots 0, q_j, 0, \dots, 0]'$[5] are points on the axis j, generated by the unit vector $\vec{e}_j$ of the orthonormal basis $\mathcal{I}_n = \{\vec{e}_1, \dots, \vec{e}_j, \dots, \vec{e}_n\}$ with a dilatation by factor q_j. The set of points (vertices) $\mathcal{P}_1 = \{O, C_1, C_2, \dots, C_n\}$ constitute the *commodity polyhedron* in $\mathcal{C}^n$, representing the total production with $\mathbf{q}$ as output vector.

(2) Consider further the production process column vectors $\mathbf{s}_{\cdot j} = [s_{1j}, s_{2j}, \dots, s_{nj}]'$, $j = 1, \dots, n$, relative to the basis $\mathcal{I}_n$ of the *commodity space* $\mathcal{C}^n$. The endpoint S_j of each geometric vector $\vec{s}_j := \overrightarrow{OS}_j = [s_{1j}, s_{2j}, \dots, s_{nj}]' \Leftrightarrow \mathbf{s}_{\cdot j}$ belongs to a polyhedron with n vertices S_j (corners) together with the origin O, the so-called *production process polyhedron* $\mathcal{P}_2 = \{O, S_1, S_2, \dots, S_n\}$ in $\mathcal{C}^n$, representing the n production processes of the present production economy.

(3) Consider then the total output with the column vectors $\mathbf{f}_{\cdot j} = [f_{1j}, f_{2j}, \dots, f_{nj}]'$, $j = 1, \dots, n$, relative to the basis $\mathcal{I}_n$ of the *commodity space* $\mathcal{C}^n$. The endpoint F_j of each geometric vector $\vec{f}_j := \overrightarrow{OF}_j = [f_{1j}, f_{2j}, \dots, f_{nj}]' \Leftrightarrow \mathbf{f}_{\cdot j}$ belongs to a polyhedron with n vertices F_j (corners) together with the origin O, leading finally to the so-called *output polyhedron* $\mathcal{P}_3 = \{O, F_1, F_2, \dots, F_n\}$ in $\mathcal{C}^n$, representing the n outputs of each industry of the present production economy.

We now present the vector sets of these three polyhedrons, see Figure 6.1:

$$\mathcal{P}_1 : \begin{bmatrix} \vec{c}_1 \\ \vec{c}_2 \\ \dots \\ \dots \\ \vec{c}_n \end{bmatrix} = \hat{\mathbf{q}} \begin{bmatrix} \vec{e}_1 \\ \vec{e}_2 \\ \dots \\ \dots \\ \vec{e}_n \end{bmatrix}, \quad \mathcal{P}_2 : \begin{bmatrix} \vec{s}_1 \\ \vec{s}_2 \\ \dots \\ \dots \\ \vec{s}_n \end{bmatrix} = \mathbf{S}' \begin{bmatrix} \vec{e}_1 \\ \vec{e}_2 \\ \dots \\ \dots \\ \vec{e}_n \end{bmatrix}, \quad \mathcal{P}_3 : \begin{bmatrix} \vec{f}_1 \\ \vec{f}_2 \\ \dots \\ \dots \\ \vec{f}_n \end{bmatrix} = \mathbf{F}' \begin{bmatrix} \vec{e}_1 \\ \vec{e}_2 \\ \dots \\ \dots \\ \vec{e}_n \end{bmatrix}. \tag{6.9}$$

Moreover, matrices **S** and **F** realise a *change of basis* in $\mathbb{R}^n$ if they are regular. We also recognise that the distribution coefficients matrix $\mathbf{D} = \hat{\mathbf{q}}^{-1}\mathbf{S}$ transforms the *commodity polyhedron* $\mathcal{P}_1$ into the *production process polyhedron* $\mathcal{P}_2$:

$$\mathcal{P}_2 : \begin{bmatrix} \vec{s}_1 \\ \vec{s}_2 \\ \dots \\ \dots \\ \vec{s}_n \end{bmatrix} = \mathbf{D}' \begin{bmatrix} \vec{c}_1 \\ \vec{c}_2 \\ \dots \\ \dots \\ \vec{c}_n \end{bmatrix} = \mathbf{D}'\hat{\mathbf{q}} \begin{bmatrix} \vec{e}_1 \\ \vec{e}_2 \\ \dots \\ \dots \\ \vec{e}_n \end{bmatrix} = \mathbf{S}'\hat{\mathbf{q}}^{-1}\hat{\mathbf{q}} \begin{bmatrix} \vec{e}_1 \\ \vec{e}_2 \\ \dots \\ \dots \\ \vec{e}_n \end{bmatrix} = \mathbf{S}' \begin{bmatrix} \vec{e}_1 \\ \vec{e}_2 \\ \dots \\ \dots \\ \vec{e}_n \end{bmatrix}. \tag{6.10}$$

5 A geometric vector $\vec{c}_j$ in a vector space corresponds one to one to a column or a row of the diagonal matrix $\hat{\mathbf{q}}$.

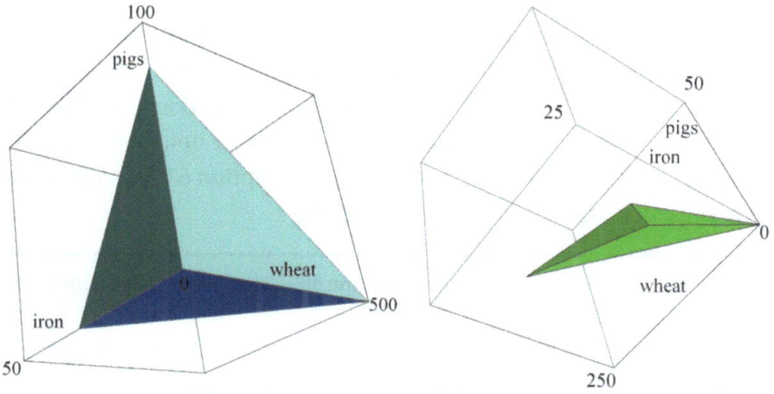

Figure 6.1: Example 6.1.1, $\mathcal{P}_1$ polyhedron (left) and $\mathcal{P}_2$ polyhedron (right).

If matrix **S** is regular, then matrix **D** is also regular. We will now illustrate these notions and refer to Sraffa's second numerical Example 3.1.2, (PCMC, Par. 2), extending it to a case of *joint production*.

Example 6.1.1. We take an arbitrary vector of labour **L** and set up the production scheme $(\mathbf{S}', \mathbf{L}) \rightarrow (\mathbf{F}')$:

$$\begin{aligned}(240 \text{ qr. wheat}, 12 \text{ t. iron}, 18 \text{ pigs}, L_1) &\rightarrow (300 \text{ qr. wheat}, 10 \text{ t. iron}, 0),\\ (90 \text{ qr. wheat}, 6 \text{ t. iron}, 12 \text{ pigs}, L_2) &\rightarrow (200 \text{ qr. wheat}, 20 \text{ t. iron}, 40 \text{ pigs}),\\ (120 \text{ qr. wheat}, 3 \text{ t. iron}, 30 \text{ pigs}, L_3) &\rightarrow (0, 0, 40 \text{ pigs}).\end{aligned} \tag{6.11}$$

Identify the matrices **F**, **S**. Calculate the total output $\mathbf{q} = \mathbf{Fe} > \mathbf{o}$, the distribution coefficients matrix **D** and the determinants $\det(\mathbf{F}' - \mathbf{S}')$, $\det(\mathbf{F})$, $\det(\mathbf{S})$, $\det(\hat{\mathbf{q}})$. Verify the linear independence of the processes, $\operatorname{rank}([\mathbf{S}', \mathbf{F}']) = \operatorname{rank}([\mathbf{S}', \mathbf{F}' - \mathbf{S}']) = 3$. Argue over the three economic assumptions of Schefold. Calculate the vector of *net product*, $\mathbf{d} = (\mathbf{F} - \mathbf{S})\mathbf{e}$. Argue over following the three polyhedrons and their dimensions, Definition 6.1.1.

Solution to Example 6.1.1:
We first identify the matrices **S** and **F** and calculate $\mathbf{q} = \mathbf{Fe}$,

$$\mathbf{S} = \begin{bmatrix} 240 & 90 & 120 \\ 12 & 6 & 3 \\ 18 & 12 & 30 \end{bmatrix}, \quad \mathbf{F} = \begin{bmatrix} 300 & 200 & 0 \\ 10 & 20 & 0 \\ 0 & 40 & 40 \end{bmatrix}, \quad \mathbf{q} = \mathbf{Fe} = \begin{bmatrix} 500 \\ 30 \\ 80 \end{bmatrix}. \tag{6.12}$$

Then, we calculate the distribution coefficients matrix:

$$\mathbf{D} = \hat{\mathbf{q}}^{-1}\mathbf{S} = \begin{bmatrix} \frac{1}{500} & 0 & 0 \\ 0 & \frac{1}{30} & 0 \\ 0 & 0 & \frac{1}{80} \end{bmatrix} \begin{bmatrix} 240 & 90 & 120 \\ 12 & 6 & 3 \\ 18 & 12 & 30 \end{bmatrix} = \begin{bmatrix} \frac{12}{25} & \frac{9}{50} & \frac{6}{25} \\ \frac{2}{5} & \frac{1}{5} & \frac{1}{10} \\ \frac{9}{40} & \frac{3}{20} & \frac{3}{8} \end{bmatrix}. \tag{6.13}$$

Analysing the vectors $\mathbf{s}_{.j}$ and $\mathbf{f}_{.j}$, we see that Assumption 6.1.3, is fulfilled. The column vectors of matrices $\mathbf{S}$ and $\mathbf{F}$ are *semi-positive*. The total output is $\mathbf{q} = \mathbf{Fe} = [500, 30, 80]' > \mathbf{o}$. There is positive self – replacement, Assumption 6.1.1. We calculate the determinants of the matrices: $\det(\mathbf{F}' - \mathbf{S}') = -1{,}940$, $\det(\mathbf{F}) = 160{,}000$, $\det(\mathbf{S}) = 11{,}340$, $\det(\hat{\mathbf{q}}) = 1{,}200{,}000$, Assumption 6.1.2, Assumption 6.1.3 and (6.8) are fulfilled. We are in presence of *gross integrated industries*.

In this example, the dimension of the *commodity polyhedron*, the *production process polyhedron*, the *output polyhedron* are all maximal with $n = 3$.

We then verify that $\text{rank}([\mathbf{S}', \mathbf{F}']) = \text{rank}([\mathbf{S}', \mathbf{F}' - \mathbf{S}']) = 3$. Consequently, the processes $(\mathbf{s}_{.j}, L_j) \to (\mathbf{f}_{.j})$, $j = 1, \ldots, 3$ are linearly independent, as required by Assumption 6.1.2! Then we determine the polyhedrons (6.9) and the *net product* vector

$$\mathcal{P}_1 : \begin{bmatrix} \vec{c}_1 \\ \vec{c}_2 \\ \vec{c}_3 \end{bmatrix} = \begin{bmatrix} 500 & 0 & 0 \\ 0 & 30 & 0 \\ 0 & 0 & 80 \end{bmatrix} \begin{bmatrix} \vec{e}_1 \\ \vec{e}_2 \\ \vec{e}_3 \end{bmatrix}, \quad \mathcal{P}_2 : \begin{bmatrix} \vec{s}_1 \\ \vec{s}_2 \\ \vec{s}_3 \end{bmatrix} = \begin{bmatrix} 240 & 12 & 18 \\ 90 & 6 & 12 \\ 120 & 3 & 30 \end{bmatrix} \begin{bmatrix} \vec{e}_1 \\ \vec{e}_2 \\ \vec{e}_3 \end{bmatrix}, \tag{6.14}$$

$$\mathcal{P}_3 : \begin{bmatrix} \vec{f}_1 \\ \vec{f}_2 \\ \vec{f}_3 \end{bmatrix} = \begin{bmatrix} 300 & 10 & 0 \\ 200 & 20 & 40 \\ 0 & 0 & 40 \end{bmatrix} \begin{bmatrix} \vec{e}_1 \\ \vec{e}_2 \\ \vec{e}_3 \end{bmatrix}, \quad \mathbf{d} = (\mathbf{F} - \mathbf{S})\mathbf{e} = \begin{bmatrix} 50 \\ 9 \\ 20 \end{bmatrix}. \tag{6.15}$$

The reader is invited to apply equation (6.10). We present the vertices of

the *commodity polyhedron* (the polyhedrons $\mathcal{P}_1$ and $\mathcal{P}_2$ are presented in Figure 6.1):

$$\mathcal{P}_1 = \{O(0,0,0), C_1(500,0,0), C_2(0,30,0), C_3(0,0,80)\}, \quad \text{rank}(\mathcal{P}_1) = 3,$$

the *production process polyhedron*:

$$\mathcal{P}_2 = \{O(0,0,0), S_1(240,12,18), S_2(90,6,12), S_3(120,3,30)\}, \quad \text{rank}(\mathcal{P}_2) = 3,$$

the *output polyhedron* (the polyhedrons $\mathcal{P}_2$ and $\mathcal{P}_3$ are presented in Figure 6.2):

$$\mathcal{P}_3 = \{O(0,0,0), F_1(300,10,0), F_2(200,20,40), F_3(0,0,40)\}, \quad \text{rank}(\mathcal{P}_3) = 3. \quad \blacktriangle$$

Example 6.1.1 illustrates Assumption 6.1.2 on the regularity of matrix $(\mathbf{F}' - \mathbf{S}')$ (6.6) and on linear independence of processes, as well as Assumption 6.1.3 on the existence of at least one input and one output to each process and finally on *gross integrated industries*, as matrix $\mathbf{F}$ has full rank.

Now, we transform the entries of Example 4.4.3 to get a *joint production economy*, keeping the same three commodities: the first industry 1 producing wheat and iron, the second industry 2 producing iron and the third industry 3 gold.

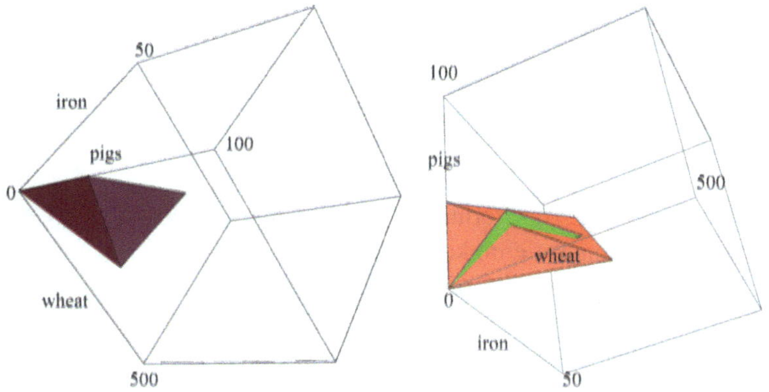

Figure 6.2: Example 6.1.1, $\mathcal{P}_3$ polyhedron (left) and $\mathcal{P}_2$ together with $\mathcal{P}_3$ polyhedron (right).

Example 6.1.2. With the matrices **S** and **F** and an arbitrary vector of labour **L**,

$$\mathbf{S} = \begin{bmatrix} 280 & 180 & 115 \\ 240 & 240 & 120 \\ 0 & 0 & 0 \end{bmatrix}, \quad \mathbf{F} = \begin{bmatrix} 500 & 75 & 0 \\ 0 & 600 & 0 \\ 0 & 0 & \frac{3}{2} \end{bmatrix}, \quad \mathbf{L} = \begin{bmatrix} L_1 \\ L_2 \\ L_3 \end{bmatrix}, \tag{6.16}$$

we consider the production scheme $(\mathbf{S}', \mathbf{L}) \to (\mathbf{F}')$ of this joint production process. Compute the vector of total output $\mathbf{q} = \mathbf{Fe} > \mathbf{o}$, the vector of *net product* $\mathbf{d} = (\mathbf{F} - \mathbf{S})\mathbf{e}$ and the determinants $\det(\mathbf{F}' - \mathbf{S}')$, $\det(\mathbf{F})$, $\det(\mathbf{S})$. Verify the linear independence of the processes, $\operatorname{rank}([\mathbf{S}', \mathbf{F}']) = \operatorname{rank}([\mathbf{S}', \mathbf{F}' - \mathbf{S}']) = 3$. Verify Schefold's three economic assumptions and analyse the dimensions of the three polyhedrons $\mathcal{P}_1$, $\mathcal{P}_2$, $\mathcal{P}_3$.

Solution to Example 6.1.2:
We determine the polyhedrons with (6.9) and the *net product* vector

$$\mathcal{P}_1: \begin{bmatrix} \vec{c}_1 \\ \vec{c}_2 \\ \vec{c}_3 \end{bmatrix} = \begin{bmatrix} 500 & 0 & 0 \\ 0 & 30 & 0 \\ 0 & 0 & 80 \end{bmatrix} \begin{bmatrix} \vec{e}_1 \\ \vec{e}_2 \\ \vec{e}_3 \end{bmatrix}, \quad \mathcal{P}_2: \begin{bmatrix} \vec{s}_1 \\ \vec{s}_2 \\ \vec{s}_3 \end{bmatrix} = \begin{bmatrix} 280 & 240 & 0 \\ 180 & 240 & 0 \\ 115 & 120 & 0 \end{bmatrix} \begin{bmatrix} \vec{e}_1 \\ \vec{e}_2 \\ \vec{e}_3 \end{bmatrix}, \tag{6.17}$$

$$\mathcal{P}_3: \begin{bmatrix} \vec{f}_1 \\ \vec{f}_2 \\ \vec{f}_3 \end{bmatrix} = \begin{bmatrix} 500 & 0 & 0 \\ 75 & 600 & 0 \\ 0 & 0 & \frac{3}{2} \end{bmatrix} \begin{bmatrix} \vec{e}_1 \\ \vec{e}_2 \\ \vec{e}_3 \end{bmatrix}, \; \mathbf{q} = \mathbf{Fe} = \begin{bmatrix} 575 \\ 600 \\ \frac{3}{2} \end{bmatrix}, \; \mathbf{d} = (\mathbf{F} - \mathbf{S})\mathbf{e} = \begin{bmatrix} 0 \\ 0 \\ \frac{3}{2} \end{bmatrix}. \tag{6.18}$$

The vertices are, for

the *commodity polyhedron*:

$$\mathcal{P}_1 = \{O(0,0,0), C_1(500,0,0), C_2(0,30,0), C_3(0,0,80)\},$$

the *production process polyhedron*:

$$\mathcal{P}_2 = \{O(0,0,0), S_1(280,240,0), S_2(180,240,0), S_3(115,120,0)\},$$

the *output polyhedron*:

$$\mathcal{P}_3 = \{O(0,0,0), F_1(500,0,0), F_2(75,600,0), F_3(0,0,1.5)\}.$$

We calculate the ranks of the transformation matrices (6.9), determining the dimension of the polyhedrons: $\dim(\mathcal{P}_1) = \text{rank}(\hat{\mathbf{q}}) = 3$, $\dim(\mathcal{P}_2) = \text{rank}(\mathbf{S}) = 2$, $\dim(\mathcal{P}_3) = \text{rank}(\mathbf{F}) = 3$. It is not necessary for all of them to have dimension $n = 3$(!). We also verify Assumption 6.1.2: $\text{rank}[\mathbf{S}', \mathbf{F}'] = \text{rank}[\mathbf{S}', \mathbf{F}' - \mathbf{S}'] = 3$, and we have consequently *the linear independent* processes $(\mathbf{s}_{.j}, L_j) \to (\mathbf{f}_{.j})$, $j = 1, \ldots, 3$! ▲

6.2 The price model for joint production*

Sraffa's construction of the price model for *single-product industries* was presented in Section 4.9 in four steps. We apply the same methodology for *joint production*, developing **Step I** to **Step IV**. Assumption 6.1.1, to Assumption 6.1.3 hold in the present section.

We will further treat specific *joint production processes* and the system of *gross integrated industries*, where matrices $\mathbf{F}$ are regular, $\det(\mathbf{F}) \neq 0$.

Step I (an economy with no surplus)

We ensure *sustainability*, meaning that the sum of quantities of commodity produced as output by the various industries j must equal the sum of quantities of commodity i required by each industry for production, PCMC, Par. 63, and Manara [61], p. 2, so we must have therefore the equality for the total output

$$\mathbf{q} = \mathbf{F}\mathbf{e} = \mathbf{S}\mathbf{e} \Leftrightarrow \sum_{j=1}^{n} f_{ij} = \sum_{j=1}^{n} s_{ij} \quad \text{(non-diagonal)}. \tag{6.19}$$

Equation (6.19) describes *Sraffa's conditions of joint production*[6] replacing the former formulation, Definition 3.1.1 for *single-product industries*.

In other words, we consider the entries of the column vectors $\mathbf{s}_{.j} = [s_{1j}, \ldots, s_{nj}]'$ of all commodities $i \in \{1, \ldots, n\}$ as inputs, necessarily required for the production by each industry S_j, $j \in \{1, \ldots, n\}$, and the column vectors $\mathbf{f}_{.j} = [f_{1j}, \ldots, f_{nj}]$ as outputs of commodities $i \in \{1, \ldots, n\}$, produced by each of the industries S_j, $j \in \{1, \ldots, n\}$.

6 This is the analogue to Sraffa's *conditions of production* for single-product industries, Definition 3.1.1, included in (6.19); we then return to:

$$\widehat{\mathbf{S}\mathbf{e}} = \hat{\mathbf{q}} \Leftrightarrow \sum_{j=1}^{n} s_{ij} = f_{jj} =: q_j \quad \text{(diagonal } i = j\text{)}, \tag{6.20}$$

and $\mathbf{F}$ indeed reduces to the diagonal form $\hat{\mathbf{q}} = \widehat{\mathbf{S}\mathbf{e}}$, issued from equation (3.6).

Sraffa's *conditions of production in value terms* (PCMC, Par. 3), Definition 3.1.2, $\mathbf{S}'\mathbf{p} = \hat{\mathbf{q}}\mathbf{p}$, become in matrix form for *joint production* with a price vector $\mathbf{p}$:

$$\mathbf{S}'\mathbf{p} = \mathbf{F}'\mathbf{p}. \tag{6.21}$$

If we are in presence of a system of *gross integrated industries* of a *joint production process*, matrix $\mathbf{F}$ is regular 6.8, $\det(\mathbf{F}) \neq 0$, and we are able to define an *input-output coefficients* matrix. This subject is summarised in

Proposition 6.2.1 (Sraffa's conditions of joint production for gross integrated industries). *There is no vector of labour. For gross integrated industries of a joint production process,* $\det(\mathbf{F}) \neq 0$, *an input-output coefficients matrix,*

$$\mathbf{C}_T := \mathbf{S}\mathbf{F}^{-1} \Leftrightarrow \mathbf{C}_T' := \mathbf{F}'^{-1}\mathbf{S}', \tag{6.22}$$

is defined. Sraffa's conditions of joint production, $\mathbf{q} = \mathbf{F}\mathbf{e} = \mathbf{S}\mathbf{e}$, *hold for an economy without surplus, and the inequality,* $\mathbf{q} = \mathbf{F}\mathbf{e} \geq \mathbf{S}\mathbf{e}$, *holds for an economy with surplus. We then obtain an eigenvalue equation for the price vector* $\mathbf{p}$, *multiplying (6.21) from the left by* $\mathbf{F}'^{-1}$, *obtaining,*

$$\mathbf{F}'^{-1}(\mathbf{S}'\mathbf{p}) = (\mathbf{F}'^{-1}\mathbf{S}')\mathbf{p} := \mathbf{C}_T'\mathbf{p} = \mathbf{F}'^{-1}(\mathbf{F}'\mathbf{p}) = (\mathbf{F}'^{-1}\mathbf{F}')\mathbf{p} = \mathbf{p} \Rightarrow$$

$$\mathbf{C}_T'\mathbf{p} = \mathbf{p}. \quad \blacktriangle \tag{6.23}$$

Matrix $\mathbf{C}_T'$ is *not* necessarily *semi-positive* and *irreducible*, and the **Perron–Frobenius theorem A.9.3**, then does not hold. Consequently, negative prices may appear in the price eigenvector $\mathbf{p}$ (6.23), as will be illustrated by Example 6.2.1.

The preliminary *production scheme* without labour, expressed by matrices $\mathbf{S}$ and $\mathbf{F}$, giving the quantities in physical terms, acquires further the aspect of a system of *single-product industries*, presented by the *input-output coefficients* matrix $\mathbf{C}_T$ and the *identity* matrix $\mathbf{I}$,

$$(\mathbf{S}', \mathbf{o}) \to (\mathbf{F}') \Leftrightarrow (\mathbf{C}_T' = \mathbf{F}'^{-1}\mathbf{S}', \mathbf{o}) \to (\mathbf{I}). \tag{6.24}$$

The above explanations lead to the following:

Example 6.2.1. Consider a type of traditional farming in which the means of production and the products are cattle and wheat. There is sustainability, workers are paid with beef and wheat. There are two farms with different methods of production, and the productive processes for one year are represented by the following production scheme:

$$\text{farm 1:} \quad (3 \text{ heads of cattle}, 7 \text{ t. wheat}, 0) \to (5 \text{ heads of cattle}, 8 \text{ t. wheat}),$$

$$\text{farm 2:} \quad (5 \text{ heads of cattle}, 5 \text{ t. wheat}, 0) \to (3 \text{ heads of cattle}, 4 \text{ t. wheat}). \tag{6.25}$$

Present the commodity flow matrix $\mathbf{S}$ and the *output* matrix $\mathbf{F}$. Verify $\det(\mathbf{F}) \neq 0$ and Sraffa's conditions of joint production, Definition 6.1.1. Compute the total output $\mathbf{q}$, the input-output coefficients matrix $\mathbf{C}_T$ and the price eigenvectors $\mathbf{p}$ (6.23).

Solution to Example 6.2.1:
We identify the commodity flow matrix **S** and the *output* matrix **F** and verify the first part of *Sraffa's conditions of joint production.*

$$\mathbf{S} = \begin{bmatrix} 3 & 5 \\ 7 & 5 \end{bmatrix}; \quad \mathbf{F} = \begin{bmatrix} 5 & 3 \\ 8 & 4 \end{bmatrix}, \quad \det(\mathbf{F}) = -4,$$

$$\mathbf{q} = \mathbf{Fe} = \begin{bmatrix} 5 & 3 \\ 8 & 4 \end{bmatrix} \begin{bmatrix} 1 \\ 1 \end{bmatrix} = \mathbf{Se} = \begin{bmatrix} 3 & 5 \\ 7 & 5 \end{bmatrix} \begin{bmatrix} 1 \\ 1 \end{bmatrix} = \begin{bmatrix} 8 \\ 12 \end{bmatrix}. \tag{6.26}$$

We compute the input-output coefficients matrix $\mathbf{C}_T$,

$$\mathbf{F}^{-1} = \begin{bmatrix} -1 & \frac{3}{4} \\ 2 & -\frac{5}{4} \end{bmatrix}, \quad \mathbf{C}_T = \mathbf{SF}^{-1} = \begin{bmatrix} 3 & 5 \\ 7 & 5 \end{bmatrix} \begin{bmatrix} -1 & \frac{3}{4} \\ 2 & -\frac{5}{4} \end{bmatrix} = \begin{bmatrix} 7 & -4 \\ 3 & -1 \end{bmatrix}, \tag{6.27}$$

recognising that matrix $\mathbf{C}_T$ is not *semi-positive*. Now we compute the eigenvalues of matrix $\mathbf{C}_T$,

$$P_2(\lambda) = \det(\mathbf{C}_T - \lambda \mathbf{I}) = \lambda^2 - 6\lambda + 5 = (\lambda - 1)(\lambda - 5) = 0, \tag{6.28}$$

and obtain for the eigenvalue $\lambda = 1$ (no surplus) the eigenvectors $\mathbf{p} = k[-1 \;\; 2]'$ of (6.23). Indeed, we have

$$\mathbf{C}_T'\mathbf{p} = \begin{bmatrix} 7 & 3 \\ -4 & -1 \end{bmatrix} \begin{bmatrix} -k \\ 2k \end{bmatrix} = \begin{bmatrix} -7k + 6k \\ 4k - 2k \end{bmatrix} = \begin{bmatrix} -k \\ 2k \end{bmatrix}. \tag{6.29}$$

Concluding, the eigenvectors $\mathbf{p} = k[-1 \;\; 2]'$ are not *positive*. Thus, the question arises: What conditions must be fulfilled to obtain positive prices? This question will be answered later in Section 6.6. ▲

Step II (economy with a surplus going only into profits)
We now consider Sraffa's *production economy with a surplus* and assume that all the surplus goes into *profits* and that there are no direct wages paid by the surplus, $w = 0$. There are only "subsistence wages".[7]

In the case of *single-product industries* with a surplus, remember that the vector of total output $\mathbf{q} = \mathbf{Se} + \mathbf{d}$ (2.15) leads to the diagonal matrix $\hat{\mathbf{q}}$. At present, the diagonal matrix $\hat{\mathbf{q}}$ in the Sraffa price model (4.168) is replaced by the *output* matrix **F**,

7 In fact, we may consider at this stage that wages are included in "gross profits", i.e., earnings of the productive entity before payment of wages. Wages having been paid out, the remaining profits correspond, as mentioned elsewhere, to EBITDA (earnings before interest, taxes, depreciation and amortization).

containing now the surplus, in order to obtain a *joint production*. The newly constituted *joint production Sraffa system* then takes the form of a *general eigenvalue problem*,[8]

$$\mathbf{S}'\mathbf{p} = \lambda_C \mathbf{F}'\mathbf{p}, \qquad \lambda_C = \frac{1}{1+R}. \tag{6.30}$$

Accordingly, as in the initial price model of *single-product industries*, the *maximal rate of profits*, respectively the *productiveness R*, is also present in the price model of *joint production analysis*. The conditions for the existence of a positive *productiveness* $R > 0$ are treated in Section 6.6. We continue multiplying equation (6.30) from the left by matrix $\mathbf{F}'^{-1}$, giving with $\mathbf{C}_T = \mathbf{S}\mathbf{F}^{-1}$,

$$\mathbf{F}'^{-1}\mathbf{S}'\mathbf{p} = \mathbf{F}'^{-1}(\lambda_C \mathbf{F}'\mathbf{p}) = (\mathbf{F}'^{-1}\mathbf{F}')\lambda_C \mathbf{p} = \mathbf{C}_T'\mathbf{p} = \lambda_C \mathbf{p}. \tag{6.31}$$

The price vector **p**, as in the foregoing remark for **Step I**, may contain negative prices, because matrix $\mathbf{C}_T$ may no longer be *semi-positive* and *irreducible*, as we illustrate in the next

Example 6.2.2. Consider again a type of traditional farming in which the means of production and the commodities produced are cattle and wheat. There are two farming sectors with different methods of production, and the production processes for one year are now represented by the following production scheme:

$$\begin{aligned} f1: &\quad (30 \text{ heads of cattle}, 70 \text{ t. wheat}, 0) \to (40 \text{ heads of cattle}, 80 \text{ t. wheat}) \\ f2: &\quad (50 \text{ heads of cattle}, 50 \text{ t. wheat}, 0) \to (60 \text{ heads of cattle}, 70 \text{ t. wheat}) \end{aligned} \tag{6.32}$$

Write down the commodity flow matrix **S** and the *output* matrix **F**. Compute the vector of surplus **d**. Compute det(**F**), the input-output coefficients matrix $\mathbf{C}_T$, their eigenvalues λ and eigenvectors **p**. Discuss the question of the existence of *productiveness R* associated with an appropriate eigenvalue.

Discuss the applicability of the **Perron–Frobenius theorem A.9.3.**

Solution to Example 6.2.2:

There are no money wages because the workers are paid at "subsistence wages" in beef and wheat. We identify first the commodity flow matrix **S** and the regular *output* matrix **F**,

$$\mathbf{S} = \begin{bmatrix} 30 & 50 \\ 70 & 50 \end{bmatrix}, \quad \mathbf{F} = \begin{bmatrix} 40 & 60 \\ 80 & 70 \end{bmatrix}, \quad \det(\mathbf{F}) = -200, \tag{6.33}$$

and compute the vector of surplus **d**, the inverse of the *output* matrix **F** and then the *input-output coefficients* matrix $\mathbf{C}_T$,

8 Steedman ([114], p. 326) also states the existence of this rate of profits R. In Section 6.6 the conditions to obtain positive prices will be discussed, subsuming also positive rates of profits $R > 0$. The case including wages and profit will be discussed in **Step III** of the present section.

$$\mathbf{d} = (\mathbf{F} - \mathbf{S})\mathbf{e} = \begin{bmatrix} 20 \\ 30 \end{bmatrix}, \quad \mathbf{q} = \mathbf{Fe} = \begin{bmatrix} 100 \\ 150 \end{bmatrix}, \quad \mathbf{F}^{-1} = \begin{bmatrix} -\frac{7}{200} & \frac{3}{100} \\ \frac{1}{25} & -\frac{1}{50} \end{bmatrix},$$

$$\mathbf{C}_T = \mathbf{SF}^{-1} = \begin{bmatrix} 30 & 50 \\ 70 & 70 \end{bmatrix} \begin{bmatrix} -\frac{7}{200} & \frac{3}{100} \\ \frac{1}{25} & -\frac{1}{50} \end{bmatrix} = \begin{bmatrix} \frac{19}{20} & -\frac{1}{10} \\ -\frac{9}{20} & \frac{11}{10} \end{bmatrix}. \tag{6.34}$$

We observe that matrix $\mathbf{C}_T$ is not *semi-positive*. The eigenvalues are obtained with the characteristic polynomial,

$$P_2(\lambda) = \det(\mathbf{C}_T - \lambda\mathbf{I}) = \lambda^2 - \frac{41}{20}\lambda + 1 = \left(\lambda - \frac{4}{5}\right)\left(\lambda - \frac{5}{4}\right) = 0, \tag{6.35}$$

getting $\lambda_1 = 4/5 < 1$ and $\lambda_2 = 5/4 > 1$. Both eigenvalues are positive real numbers. Lemma A.9.3 (b) applies, as well as Lemma A.9.2. We choose $h = 4$, the matrix $\mathbf{C}_{T4} := 4\mathbf{I} - \mathbf{C}_T > \mathbf{O}$ is positive and the **Perron–Frobenius theorem** applies. The transposed matrix $\mathbf{C}'_{T4}$ has with Lemma A.9.2 the same eigenvectors as matrix $\mathbf{C}'_T$. We find positive eigenvectors $\mathbf{p} = k \cdot [3, 1]$, $k \in \mathbb{R}^+$, of the matrix $\mathbf{C}'_T$, corresponding to the smaller eigenvalue $\lambda_1 = 4/5$ of $\mathbf{C}'_T$,

$$\mathbf{C}'_T\mathbf{p} = \begin{bmatrix} \frac{19}{20} & -\frac{9}{20} \\ -\frac{1}{10} & \frac{11}{10} \end{bmatrix} \begin{bmatrix} 3k \\ k \end{bmatrix} = \begin{bmatrix} \frac{48}{20}k \\ \frac{8}{10}k \end{bmatrix} = \frac{4}{5} \begin{bmatrix} 3k \\ k \end{bmatrix} = \lambda_1\mathbf{p}. \tag{6.36}$$

Indeed, the vector $\mathbf{p} = k \cdot [3, 1]' > \mathbf{o}$ is the *positive* price vector of matrix $\mathbf{C}'_T$, associated with the eigenvalue $\lambda_1 = (4/5)$. The vector $\mathbf{p} = k \cdot [3, 1]' > \mathbf{o}$ is also the positive eigenvector of matrix $\mathbf{C}'_{T4}$ associated to $\lambda_{C4} := 4 - \lambda_1 = 16/5 > 0$, which is the Frobenius number of matrix $\mathbf{C}'_{T4}$, as one verifies. Therefore, the eigenvalue $0 < \lambda_1 = (4/5) < 1$ leads to the *productiveness* $R = (1/\lambda_C) - 1 = (5/4) - 1 = 0.25$. Clearly, the price vector $\mathbf{p} = k \cdot [3, 1]' > \mathbf{o}$ solves also model (6.30). ▲

For the further development, we continue to assume a positive price vector $\mathbf{p} > \mathbf{o}$.

We propose now another method to obtain the *productiveness* $R > 0$, given $w = 0$. One multiplies equation (6.30) from the left side with vector $\mathbf{e}'$, resulting in the *total output* $X = (\mathbf{Fe})'\mathbf{p}$. We obtain with (5.78) the *circulating capital* $K = (\mathbf{Se})'\mathbf{p}$ and then the total output

$$X := \mathbf{e}'\mathbf{S}'\mathbf{p}(1 + R) = \mathbf{e}'\mathbf{F}'\mathbf{p} \Rightarrow (\mathbf{Se})'\mathbf{p}(1 + R) = K(1 + R) = (\mathbf{Fe})'\mathbf{p},$$
$$X = K + K \cdot R \Rightarrow K \cdot R = X - K \Rightarrow R = \frac{X - K}{K}, \tag{6.37}$$

remembering that $r = R$ in this case of a *Standard system*, $\mathbf{d} \parallel \mathbf{q}$, equation (6.34), and as $r = \tilde{R}$ with $\tilde{w} = 0$ (4.36) then there is the equality $R = \tilde{R} = (X - K)/K = Y/K$.

We conclude **Step II** with following observation:

Changing from *joint production* to *single-product industries*, the right hand side of the *production scheme* (6.1) contracts to $i = j$ and $f_{ii} = q_i$. The right hand side of the *production scheme* mutes from the full matrix $\mathbf{F}'$ to the diagonal matrix $\hat{\mathbf{q}}$ with $\mathbf{q} := \mathbf{Se} + \mathbf{d}$ (2.15).

$$(\mathbf{S}', \mathbf{o}) \to (\mathbf{F}') \Rightarrow (\mathbf{S}', \mathbf{o}) \to (\hat{\mathbf{q}}). \tag{6.38}$$

Step III (economy with a surplus allotted to profits and wages)

Sraffa's profound intention is to solve the question of the distribution of surplus, which has to be used to pay profits to entrepreneurs and wages to workers. In this section, the *price model* for *single-product industries* (4.171) is transformed into a *price model* for *joint production*.

For this purpose, we take Sraffa's price model (4.171) and apply the procedure used in **Step II**. We introduce a vector of *total labour* $\mathbf{L}$, and directly replace the vector of *total output* $\mathbf{q} = \mathbf{Se} + \mathbf{d}$ by the *output* matrix $\mathbf{F}$. We get the *price model* of the *joint-production Sraffa system*,

$$\mathbf{S}'\mathbf{p}(1+r) + \mathbf{L}\frac{\tilde{w} \cdot Y}{L} = \mathbf{F}'\mathbf{p} = \mathbf{x}. \tag{6.39}$$

Then, we multiply (6.39) from the left by matrix $\mathbf{F}'^{-1}$ and using $\mathbf{C}_T = \mathbf{SF}^{-1}$ and $\mathbf{\Lambda} = \mathbf{F}'^{-1}\mathbf{L}$ we get

$$\mathbf{F}'^{-1}\mathbf{S}'\mathbf{p}(1+r) + \mathbf{F}'^{-1}\mathbf{L}\frac{\tilde{w} \cdot Y}{L} = \mathbf{C}_T'\mathbf{p}(1+r) + \mathbf{\Lambda}\frac{\tilde{w} \cdot Y}{L} = \mathbf{F}'^{-1}\mathbf{F}'\mathbf{p} = \mathbf{p}. \tag{6.40}$$

We get a new equation, where the entries of matrix $\mathbf{C}_T$ and vector $\mathbf{\Lambda}$ are presented per unit of commodities to determine the price vector $\mathbf{p}$,

$$\mathbf{C}_T'\mathbf{p}(1+r) + \mathbf{\Lambda}\frac{\tilde{w} \cdot Y}{L} = \mathbf{p}. \tag{6.41}$$

As we now know, matrix $\mathbf{C}_T$ may no longer be *semi-positive* and *irreducible*.

Under the condition that matrix $\mathbf{G} = (\mathbf{I} - \mathbf{C}_T'(1+r))$ is regular, setting also $w = (\tilde{w} \cdot Y)/L$, we conclude that the price vector $\mathbf{p}$ can be expressed by the other entities of equation (6.41), as follows:

$$\mathbf{p} = (\mathbf{I} - \mathbf{C}_T'(1+r))^{-1}\mathbf{\Lambda}\frac{\tilde{w} \cdot Y}{L} = \mathbf{G}^{-1}\mathbf{\Lambda}w. \tag{6.42}$$

The production scheme for *joint production* with a labour vector is now described in analogy to that of *single-product industries* (4.81) resorting to the matrices $\mathbf{S}$, $\mathbf{F}$, the vector $\mathbf{\Lambda} = \mathbf{F}'^{-1}\mathbf{L}$ and matrix $\mathbf{C}_T = \mathbf{SF}^{-1}$, by

$$(\mathbf{S}', \mathbf{L}) \to (\mathbf{F}') \Leftrightarrow (\mathbf{C}_T', \mathbf{\Lambda}) \to (\mathbf{I}). \tag{6.43}$$

We illustrate **Step III** with the following:

Example 6.2.3. We take Example 6.2.2 and its production scheme (6.32). We add the vector of labour $\mathbf{L} = [L_1, L_2]' = [124, 155]'$, measured in an appropriate *labour unit*, and the rate of profits $r = 0.1$. Compute the determinant of matrix $\mathbf{G} = (\mathbf{I} - \mathbf{C}_T'(1+r))$, its inverse matrix $\mathbf{G}^{-1}$, the vector of total output $\mathbf{q}$ and the vector of surplus $\mathbf{d}$.

Set up the price vector $\mathbf{p} = [p_1, p_2]'$, (6.42), knowing that the price of one head of cattle is fixed at $p_1 = 930$ CHF. Compute then the wage rate w and price p_2.

Solution to Example 6.2.3:
We identify the commodity flow matrix **S**, the *output* matrix **F** and the vector of *labour* **L** and set up the price vector **p**,

$$\mathbf{S} = \begin{bmatrix} 30 & 50 \\ 70 & 50 \end{bmatrix}, \quad \mathbf{F} = \begin{bmatrix} 40 & 60 \\ 80 & 70 \end{bmatrix}, \quad \mathbf{L} = \begin{bmatrix} 124 \\ 155 \end{bmatrix}, \quad \mathbf{p} = \begin{bmatrix} 930 \\ p_2 \end{bmatrix}. \tag{6.44}$$

As matrix **F** is regular, we compute the inverse of the transposed *output* matrix $\mathbf{F}'$, the vector $\mathbf{\Lambda} = \mathbf{F}'^{-1}\mathbf{L}$, the *input-output coefficients* matrix $\mathbf{C}_T$,

$$\mathbf{F}'^{-1} = \begin{bmatrix} -\frac{7}{200} & \frac{1}{25} \\ \frac{3}{100} & -\frac{1}{50} \end{bmatrix}, \quad \mathbf{\Lambda} = \mathbf{F}'^{-1}\mathbf{L} = \begin{bmatrix} -\frac{7}{200} & \frac{1}{25} \\ \frac{3}{100} & -\frac{1}{50} \end{bmatrix} \begin{bmatrix} 124 \\ 155 \end{bmatrix} = \begin{bmatrix} \frac{93}{50} \\ \frac{31}{50} \end{bmatrix}$$

$$\mathbf{C}_T = \mathbf{S}\mathbf{F}^{-1} = \begin{bmatrix} 30 & 50 \\ 70 & 70 \end{bmatrix} \begin{bmatrix} -\frac{7}{200} & \frac{3}{100} \\ \frac{1}{25} & -\frac{1}{50} \end{bmatrix} = \begin{bmatrix} \frac{19}{20} & -\frac{1}{10} \\ -\frac{9}{20} & \frac{11}{10} \end{bmatrix}. \tag{6.45}$$

One recognises that matrix $\mathbf{C}_T$ is not *semi-positive*, but in this case matrix **G** is regular. The inverse of matrix **G** exists and is positive. We compute,

$$\mathbf{G} := (\mathbf{I} - \mathbf{C}_T'(1+r)) = \left(\begin{bmatrix} 1 & 0 \\ 0 & 1 \end{bmatrix} - \begin{bmatrix} \frac{19}{20} & -\frac{1}{10} \\ -\frac{9}{20} & \frac{11}{10} \end{bmatrix} (1+0.1) \right),$$

$$\det(\mathbf{G}) = -\frac{9}{200}, \quad \mathbf{G}^{-1} = \begin{bmatrix} \frac{14}{3} & 11 \\ \frac{22}{9} & 1 \end{bmatrix} > \mathbf{0},$$

$$\mathbf{q} = \mathbf{F}\mathbf{e} = \begin{bmatrix} 100 \\ 150 \end{bmatrix} \parallel \mathbf{d} = (\mathbf{F} - \mathbf{S})\mathbf{e} = \begin{bmatrix} 20 \\ 30 \end{bmatrix} \parallel \mathbf{S}\mathbf{e} = \begin{bmatrix} 80 \\ 120 \end{bmatrix}. \tag{6.46}$$

We observe that we are in presence of a *Standard system*, see Notation 5.2.2. Then, we can compute the price vector with (6.42) and (6.46)

$$\mathbf{p} = \mathbf{G}^{-1}\mathbf{\Lambda}w = \begin{bmatrix} \frac{14}{3} & 11 \\ \frac{22}{9} & 1 \end{bmatrix} \begin{bmatrix} \frac{93}{50} \\ \frac{31}{50} \end{bmatrix} w = \begin{bmatrix} 930 \\ p_2 \end{bmatrix} \Rightarrow w = 60, \quad p_2 = 310,$$

$$\mathbf{p} = \begin{bmatrix} 930 \\ 310 \end{bmatrix} \parallel \mathbf{\Lambda} = \frac{31}{50} \begin{bmatrix} 3 \\ 1 \end{bmatrix} \parallel \mathbf{C}'\mathbf{p} = \begin{bmatrix} 744 \\ 248 \end{bmatrix}. \tag{6.47}$$

We observe moreover that Proposition 5.5.2 applies. Therefore there are no fluctuations of prices, the *Sraffa price models* gives for any rate of profits $r \in [0, 0.25]$ the unvariant price vector $\mathbf{p} = [930, 310]'$ in CHF/*physical term*. The wage rate is $w = 60$ CHF/*labour unit*. We are in presence of a *Standard system* where the prices are independent of the chosen rate of profits $r \in [0, R = 0.25]'$. The reader is invited to think of real economic situations which could be modeled by this *Standard system*. We remind that Sraffa in PCMC was not very explicit on this point. ▲

In Example 6.2.3, a two-sector joint production economy, we have the production scheme $(\mathbf{S}', \mathbf{L}) \to (\mathbf{F}')$. *We also know the uniform rate of profits r and introduced the currency* CHF, *knowing the price* $p_1 = 930$ CHF *for one head of cattle. Then the uniform wage rate w is determined, and the second price* p_2 *can also be computed. Without setting the currency, we would only know the proportion of the prices* p_1/p_2, *and we would have the wage rate w expressed by a factor of* (p_1/p_2). *The economic variables Y, X, K, P, W would be expressed up to a proportionality factor.*

Step IV (The economic variables in Sraffa's joint production system)

Looking at **Step IV**, Section 4.9 of *single-product industries*, the equations fixing the *economic variables* of *national income* Y and of *total output* X have to be modified for *joint production*, using now the output matrix $\mathbf{F}$.

The diagonal matrix $\hat{\mathbf{q}}$ is replaced on the right-hand side of (4.171) by the matrix of *output* $\mathbf{F}'$, where every row j contains the quantities of commodities $i \in \{1, \ldots, n\}$, produced by the sector j.[9] We obtain the Sraffa price model for joint production, which is the centre piece of the resulting *complete joint production Sraffa system*,

$$\boxed{\begin{aligned} \mathbf{S}'\mathbf{p}(1+r) + \mathbf{L}\frac{\tilde{w}\cdot Y}{L} &= \mathbf{F}'\mathbf{p}, \\ Y &= (\mathbf{Fe} - \mathbf{Se})'\mathbf{p} = \mathbf{d}'\mathbf{p}, \\ L &= \mathbf{e}'\mathbf{L}. \end{aligned}} \tag{6.48}$$

The national income Y is obtained by calculating the difference between the summed-up total output per commodity, the vector $(\mathbf{Fe})'$ and the vector of the summed-up total means of production $(\mathbf{Se})'$. This difference vector has to be multiplied with the price vector $\mathbf{p}$, known from the first equation of Sraffa's *complete joint production price model* (6.48). By analogy, we deduce from (4.175) the expressions for the remaining economic variables:

$$\begin{aligned} X &= (\mathbf{Fe})'\mathbf{p}, \\ K &= (\mathbf{Se})'\mathbf{p}, \\ P &= (\mathbf{Se})'\mathbf{p}\cdot r = K\cdot r, \\ W &= Y - P = w\cdot\mathbf{e}'\mathbf{L} = w\cdot L, \\ w &= \frac{\tilde{w}\cdot Y}{L}. \end{aligned} \tag{6.49}$$

Again some of the components of the price vector $\mathbf{p}$ may be negative in the context of *joint production.*

With a positive surplus, the surplus ratio is $\tilde{R} = (X-K)/K = Y/K > 0$. In a *Standard system*, $\mathbf{q} \parallel \mathbf{d}$ (6.46), it is equal to the *productiveness*, $R = \tilde{R} > 0$. The exogenous *rate of*

---

9 The net product is characterised, as we know, by the surplus vector $\mathbf{d} = (\mathbf{F} - \mathbf{S})\mathbf{e}$.

profits r, $0 \leq r \leq R$, or the *share of total wages to national income* $\tilde{w} = W/Y$, $0 \leq \tilde{w} \leq 1$, have to be chosen arbitrarily. The total labour $L > 0$ and the *national income* $Y > 0$ can then be calculated.

For further purposes, it is convenient to introduce the following compact notation, suggested by Schefold ([103], p. 73):

Notation 6.2.1. The constituting elements of a Sraffa *joint production* economy are the *commodity flow* matrix $\mathbf{S}$, the *output* matrix $\mathbf{F}$ and the *labour* vector $\mathbf{L}$, so Schefold ([103], p. 73) symbolised it by the *system* $(\mathbf{S}', \mathbf{F}', \mathbf{L})$. In the case of *single-product industries*, the matrix $\mathbf{F}$ is replaced by the *output* vector $\mathbf{q}$, so we have in analogy the system $(\mathbf{S}', \mathbf{q}, \mathbf{L})$, see Notation 5.2.1.

We illustrate **Step IV** by the following example.

Example 6.2.4. Take the entries of the *Standard system* presented in Example 6.2.3, but change the vector of labour to $\mathbf{L} = [100, 100]'$. Keep the rate of profits $r = 0.1$ and the price $p_1 = 930$ of one head of cattle. Solve the price model (6.48) for this joint production system. Discuss the results, especially the *productiveness* R and the surplus ratio $\tilde{R} = Y/K$.

Compute then the remaining variables $\tilde{w}, w, Y, L, X, K, W, P$.

Solution to Example 6.2.4:
Take up the rate of profits $r = 0.1$, identify the matrices $\mathbf{S}$ and $\mathbf{F}$ and the vector of labour $\mathbf{L}$ and solve the price model (6.48),

$$\mathbf{S} = \begin{bmatrix} 30 & 50 \\ 70 & 50 \end{bmatrix}, \quad \mathbf{F} = \begin{bmatrix} 40 & 60 \\ 80 & 70 \end{bmatrix}, \quad \mathbf{L} = \begin{bmatrix} 100 \\ 100 \end{bmatrix}, \quad \mathbf{d} = (\mathbf{F} - \mathbf{S})\mathbf{e} = \begin{bmatrix} 20 \\ 30 \end{bmatrix}, \tag{6.50}$$

$$\mathbf{S}'\mathbf{p}(1 + r) + \mathbf{L}\frac{\tilde{w} \cdot Y}{L} = \mathbf{F}'\mathbf{p},$$

$$\begin{bmatrix} 30 & 70 \\ 50 & 50 \end{bmatrix}\begin{bmatrix} 930 \\ p_2 \end{bmatrix}(1 + 0.1) + \frac{\tilde{w} \cdot Y}{L}\begin{bmatrix} 100 \\ 100 \end{bmatrix} = \begin{bmatrix} 40 & 80 \\ 60 & 70 \end{bmatrix}\begin{bmatrix} 930 \\ p_2 \end{bmatrix},$$

$$Y = [20, 30]\begin{bmatrix} 930 \\ p_2 \end{bmatrix},$$

$$L = [1, 1]\begin{bmatrix} 100 \\ 100 \end{bmatrix}, \tag{6.51}$$

obtaining: $p_2 = 155$, $L = 200$, $\tilde{w} = 3/5$, $w = 69.75$ and $Y = 23{,}250$, getting therefore the price vector $\mathbf{p} = [930, 155]'$. Then we compute

$$X = (\mathbf{Fe})'\mathbf{p} = \left(\begin{bmatrix} 40 & 60 \\ 80 & 70 \end{bmatrix}\begin{bmatrix} 1 \\ 1 \end{bmatrix}\right)'\begin{bmatrix} 930 \\ 155 \end{bmatrix} = 116{,}250,$$

$$K = (\mathbf{Se})'\mathbf{p} = \left(\begin{bmatrix} 30 & 50 \\ 70 & 50 \end{bmatrix}\begin{bmatrix} 1 \\ 1 \end{bmatrix}\right)'\begin{bmatrix} 930 \\ 155 \end{bmatrix} = 93{,}000. \tag{6.52}$$

With the calculated values Y, K, X, we obtain the other economic variables:

$$P = (\mathbf{Se})'\mathbf{p} \cdot r = K \cdot r = 93{,}000 \cdot 0.1 = 9{,}300,$$

$$W = Y - P = 23{,}230 - 9{,}300 = 13{,}950, \quad \tilde{w} = \frac{w \cdot L}{Y} = \frac{69.75 \cdot 200}{23{,}250} = \frac{3}{5}. \tag{6.53}$$

At last, we compute the *surplus ratio* $\tilde{R} = Y/K = 23, 250/93{,}000 = 0.25$.

The numerical equality $\tilde{R} = R = 0.25$ occurs because we are in presence of a *Standard system*, see Chapter 5, due to the parallelism $\mathbf{d} \parallel \mathbf{q}$ (6.46). The reader is invited to verify that there is no longer the second parallelism $\mathbf{p} = [930, 155]' \nparallel \mathbf{\Lambda} = \mathbf{F}'^{-1}\mathbf{L} = [0.5, 1]'$. Therefore the fluctuation of the prices will reappear, dependent on the chosen rate of profits $r \in [0, 0.25]$. ▲

In Example 6.2.4 of a two-sector joint production economy, we observe that, knowing the flow commodity matrix $\mathbf{S}$*, the output matrix* $\mathbf{F}$*, the vector of labour* $\mathbf{L}$*, the rate of profits* r *and the price* p_1 *for one head of cattle, we get the second price* p_2*, the wage rate* w*, the and the total quantity of labour* L*. The national income* Y *of the economy is determined using the base model of a joint production Sraffa system (6.48).*

Then the remaining economic variables X*,* K*,* P*,* W*, the productiveness* R *and the share of total wages to national income* $\tilde{w}$ *are all computable from this kernel model.*

Remark.

(i) In Sraffa's *joint production* models, the prices composing the price vector are supposedly independent of the industry producing the commodity, and there is one price per commodity irrespective of its provenance. This will be of special importance in the model incorporating land presented in Section 6.7.

(ii) There may appear negative components in the price vector unless certain conditions are met with by the matrices entering the equations. Such negative components have to be interpreted economically, as will be seen in the context of examples relating to ecological economics, presented in Chapter 7.

We have seen that in *joint production analysis* we drop the condition of one industry, respectively one commodity, and we have treated a series of examples. Generally speaking, two situations may arise:

(a) Several separate industries produce one and the same commodity using different technologies and commodities, respectively labour requirements.

Example 6.2.5. Agricultural production. Many production entities, big or small, will grow and sell agricultural products such as wheat, resorting to various technologies from basic to the most advanced, including genetically modified ingredients, on different types of soil in different climates.

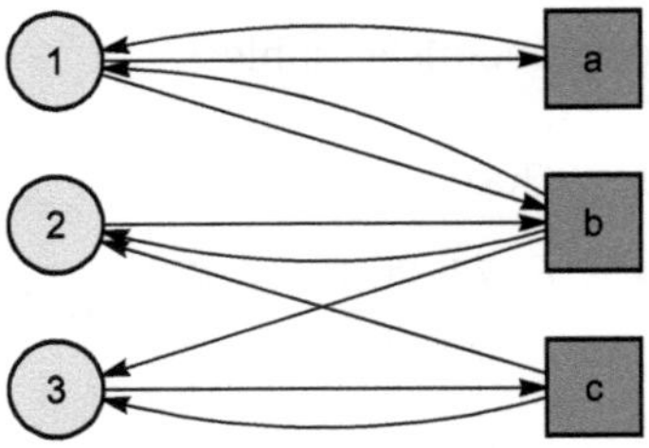

$$\mathbf{W} = \begin{array}{c} \\ 1 \\ 2 \\ 3 \\ a \\ b \\ c \end{array} \begin{array}{c} \begin{array}{ccc|ccc} 1 & 2 & 3 & a & b & c \end{array} \\ \left[\begin{array}{ccc|ccc} 0 & 0 & 0 & 1 & 1 & 0 \\ 0 & 0 & 0 & 0 & 1 & 0 \\ 0 & 0 & 0 & 0 & 0 & 1 \\ \hline 1 & 0 & 0 & 0 & 0 & 0 \\ 1 & 1 & 1 & 0 & 0 & 0 \\ 0 & 1 & 1 & 0 & 0 & 0 \end{array}\right] \end{array} \tag{6.54}$$

Figure 6.3: Elementary Sraffa Network and adjacency matrix.

(b) One and the same industry produces several separate commodities, and the commodities entering the productive entity may generate a set of several new commodities.

Example 6.2.6. The petroleum industry with various by-products and commodities, all derived from crude oil, spanning: gasoline to petrochemical derivatives.

Under *joint production*, Sraffa, Schefold, Pasinetti and other authors usually analyse situation (b). We continue along this line.

6.3 Basics and non-basics in joint production*

Assumption 6.1.1 to Assumption 6.1.3 hold for this section. We will treat the important question of identifying algebraically *basic* and *non-basic* commodities entering the production process, and present the approaches of Schefold [102], Pasinetti [83], Manara [61], Salvadori and Steedman [100].

Before continuing, let us pause to illustrate, by way of a very simple Sraffa Network, why joint production requires a more elaborate approach than hitherto concerning *basic* and *non-basic* commodities.

Example 6.3.1. Consider a three-sector economy (sector 1, 2, 3) producing three commodities (a, b, c) composed of a joint production sector 1 and two single-product sectors 2 and 3. The following network and its adjacency matrix **W** represent this situation, see Figure 6.3.

Solution to Example 6.3.1:

The network is fully connected, and one would accordingly expect all commodities to be basic as would be the case in a process of single-product industries. Now this does not account for the fact that commodity b is produced in two sectors (1 and 2), and this has an incidence on commodity flows not encountered in a process of single-product industries where there is a one-to-one relationship between each sector and the commodity it produces. To be specific, assume just two scenarios chosen amongst others:

Case 1. The production of b in sector 2 is insufficient to ensure the viability of the subsystem (2, 3; b, c) considered separately. Viability requires an input of b from sector

1 which is in surplus. This means there must be a commodity flow from sector 1 to the two other production sectors. Commodity *b* enters then directly into the production of all the commodities; *a* and *c* enter into them directly or indirectly. All commodities and sectors are therefore basic according to Sraffa's original definition (see Section 4.4). However, the status of commodity *c* is ambiguous because sector 1 does not really require this commodity for production.

Case 2. Sector 2 also generates a commodity *b* surplus. Then the flow of *b* produced by sector 1 is diverted to the overall surplus, to which sector 2 also contributes, and does not enter into the production of *b* and *c* in the sectors 2 and 3. Commodity *b* remains basic, and commodities *a* and *c* are *non-basic.* ▲

These are the type of situations commented upon by Sraffa in PCMC, Par. 57, and this simple example shows that in joint production the linear dependence or independence of the quantity flows in the underlying network representation must be taken into account, in addition to connectivity. This was not necessary in single-commodity networks due to the one to one correspondence between an industry and its produced commodity. Furthermore, in *joint production*, contrary to *single-product industries*, a sector producing a basic together with non-basics is not necessarily a basic industry.

In the present situation, the problem of distinguishing between *basics* and *non-basics* accordingly requires a new sophisticated algebraic approach, involving dependencies in the levels of flows, which we will now proceed to present. Note that, economically speaking, the distinction is of fundamental importance also in joint production because, as we have seen in single-product processes, when the basics do not generate a surplus, the sectors producing *non-basics* are not viable if their means of production rely in part or totally, directly or indirectly, on the inputs of basic commodities.

As mentioned in PCMC, Par. 57, the *"criterion previously adopted for distinguishing between basic and non-basic products (namely whether they do, or do not, enter directly or indirectly the means of production of all commodities) now fails, since, each commodity being produced by several industries, it would be uncertain whether a product which entered the means of productions of one of the industries producing a given commodity should or should not be regarded as entering directly the means of production of that commodity."* This rather vague statement of Sraffa has been taken as a starting point by Manara [61], Pasinetti [83] and Schefold [103] to develop precise mathematical tools to investigate the question of *basics* and *non-basics* in *joint production analysis*. In the present section, we treat this question and will add new graphical tools to visualise *basics* and *non-basics*.

This being said, Sraffa defines the notion of *basic* commodity and *non-basic* commodity for *joint production* as follows[10]:

10 In this citation, the designations of the variables are adapted to the notations used in this book, n for the number of industries and commodities, m for the number of *basics*, $1 \leq m < n$.

Definition 6.3.1. (PCMC, Par. 60[11]) "In a system of n productive processes and n commodities (no matter whether produced singly or jointly) we say that a commodity or more generally a group of m linked commodities (where m must be smaller than n and may be equal to 1) are *non-basic* if of the n rows (formed by the $2m$ quantities in which they appear in each process) not more than m rows are independent, the others being linear combinations of these.[12]"

The concept of *basics* and *non-basics* in *joint production* was later further developed and clarified by various economists. We start with Bertram Schefold's [103] masterpiece.

(1) Schefold's definition of basic systems. As the starting point for the various possible characterisations of *basics* and *non-basics* in *joint production analysis,* we will again use the algebraic concept of *reducible* and *irreducible* matrices, masterly presented by Schefold ([103], p. 58). The notions of permutation matrices are presented in Section A.8. We define block partitions of the $n \times n$ matrices $\mathbf{S}$ and $\mathbf{F}$, $(1 \leq m < n)$, noting the $m \times m$ matrices $\mathbf{S}_{22}$ and $\mathbf{F}_{22}$,

$$\mathbf{S} = \begin{bmatrix} \mathbf{S}_{11} & \mathbf{S}_{12} \\ \mathbf{S}_{21} & \mathbf{S}_{22} \end{bmatrix} \Rightarrow \mathbf{S}' = \begin{bmatrix} \mathbf{S}'_{11} & \mathbf{S}'_{21} \\ \mathbf{S}'_{12} & \mathbf{S}'_{22} \end{bmatrix}, \quad \mathbf{F}' = \begin{bmatrix} \mathbf{F}'_{11} & \mathbf{F}'_{21} \\ \mathbf{F}'_{12} & \mathbf{F}'_{22} \end{bmatrix}, \tag{6.55}$$

requiring the transposed matrices $\mathbf{S}'$, $\mathbf{F}'$. Then the $n \times 2m$ matrix $[\mathbf{S}'_2\ \mathbf{F}'_2]$ can be defined,

$$\mathbf{S}'_2 := \begin{bmatrix} \mathbf{S}'_{21} \\ \mathbf{S}'_{22} \end{bmatrix}, \quad \mathbf{F}'_2 := \begin{bmatrix} \mathbf{F}'_{21} \\ \mathbf{F}'_{22} \end{bmatrix} \Rightarrow [\ \mathbf{S}'_2\ \ \mathbf{F}'_2\ ] = \begin{bmatrix} \mathbf{S}'_{21} & \mathbf{F}'_{21} \\ \mathbf{S}'_{22} & \mathbf{F}'_{22} \end{bmatrix}. \tag{6.56}$$

Definition 6.3.2. (adapted essentially from Schefold ([103], p. 58) with our notations) In analogy to the system of production for single-product industries $(\mathbf{S}', \mathbf{q}', \mathbf{L})$, the *system of joint production* $(\mathbf{S}', \mathbf{F}', \mathbf{L})$ is defined. Schefold uses the term "system" for the pair $(\mathbf{S}', \mathbf{F}')$. "A system $(\mathbf{S}', \mathbf{F}')$ is called *non-basic*, if a permutation of the columns [column permutation realised by matrices $\mathbf{Q}$, Definition A.8.3] and a number m exist so that the matrix $[\mathbf{S}'_2\ \mathbf{F}'_2]$ consisting of the last m $(1 \leq m \leq n-1)$ columns of $\mathbf{S}'$ and $\mathbf{F}'$ has at most rank m."

"If the system $(\mathbf{S}', \mathbf{F}')$ is *non-basic*, $n-m$ rows of the $n \times 2m$ matrix $[\mathbf{S}'_2\ \mathbf{F}'_2]$ must by definition be linearly dependent on at most m others."

11 This definition has his roots in notions developed by Lev Semyonovich Pontryagin (1908–1988) in *control theory*, where *optimal control* and *observability* in control systems is investigated (see also Nour Eldin and Heister [74]).

12 Sraffa provided this Definition 6.3.1 in PCMC, Par. 58, 59, illustrated by an example that we reproduce in this text as *Example* 6.5.3.

Proposition 6.3.1 (Matrix rank criterion). *When in an economic system* $(\mathbf{S}', \mathbf{F}')$, $1 \le m \le n-1$, *there is a* $n \times 2m$ *subsystem* $[\mathbf{S}'_2\ \mathbf{F}'_2]$ *defined previously and one gets for the rank of* $[\mathbf{S}'_2\ \mathbf{F}'_2]$, $\operatorname{rank}([\mathbf{S}'_2\ \mathbf{F}'_2]) \le m$, *then there are m non-basics, and the* **matrix rank criterion** *for the non-basics is fulfilled.*

We now illustrate how the number of *non-basics* within an economy can be determined, applying Definition 6.3.2.

Example 6.3.2. Show that the economy, presented in Example 6.2.4 by the pair $(\mathbf{S}', \mathbf{F}')$ is a basic system.

Solution to Example 6.3.2:
There are $n = 2$ sectors, and, evidently, $m = 1$ is the only possible value solving the double inequality $1 \le m < 2$. We make the corresponding block partition of matrices $\mathbf{S}', \mathbf{F}'$,

$$\mathbf{S}' = \begin{bmatrix} \mathbf{S}'_{11} & \mathbf{S}'_{21} \\ \mathbf{S}'_{12} & \mathbf{S}'_{22} \end{bmatrix} = \left[\begin{array}{c|c} 30 & 70 \\ \hline 50 & 50 \end{array}\right], \quad \mathbf{F}' = \begin{bmatrix} \mathbf{F}'_{11} & \mathbf{F}'_{21} \\ \mathbf{F}'_{12} & \mathbf{F}'_{22} \end{bmatrix} = \left[\begin{array}{c|c} 40 & 80 \\ \hline 60 & 70 \end{array}\right]. \tag{6.57}$$

Then we directly constitute with (6.57) the $2m \times n$ matrix $[\mathbf{S}'_2\ \mathbf{F}'_2]$ (6.56),

$$\mathbf{S}'_2 := \begin{bmatrix} 70 \\ 50 \end{bmatrix}, \quad \mathbf{F}'_2 := \begin{bmatrix} 80 \\ 70 \end{bmatrix} \Rightarrow [\ \mathbf{S}'_2\ \ \mathbf{F}'_2\ ] = \begin{bmatrix} 70 & 80 \\ 50 & 70 \end{bmatrix}. \tag{6.58}$$

We determine the $\operatorname{rank}([\mathbf{S}'_2\ \mathbf{F}'_2]) = 2 > m$ and conclude that the system $(\mathbf{S}', \mathbf{F}')$ is basic, see Definition 6.3.2. Both rows are indeed linearly independent. ▲

We introduce the concept of the *direct and indirect capital matrix* **H** introduced by Pasinetti ([83], pp. 20–23) and developed by Steedman ([114], p. 324). We call **H** the *Pasinetti matrix*. We will see that the matrix **H** will give us the algebraic tools to determine directly the number m of *non-basics* in a given production system.

(2) The Pasinetti matrix H. Pasinetti ([85], p. 20, and [83], p. 36) has proposed the concept of matrix **H**:

$$\mathbf{H} = (\mathbf{F}' - \mathbf{S}')^{-1}\mathbf{S}', \quad \det(\mathbf{F}' - \mathbf{S}') \neq 0, \tag{6.59}$$

where its "*j*-th *row*[13] is the vector of capital stock required, directly or indirectly, for the production of one unit of *net* output of commodity *j*" (see Steedman [114], p. 324). Obviously, $(\mathbf{F}' - \mathbf{S}')$ is regular, in accordance with the requirement of Assumption 6.1.2. Thus, the inverse $(\mathbf{F}' - \mathbf{S}')^{-1}$ exists (see Steedman in Pasinetti (Ed.) [83], p. 45) and **H** is defined.

13 Pasinetti said here '*column*'. Manara also adopted Pasinetti's notation ([61], pp. 1-15). But, on the contrary, we follow the notation of B. Schefold ([103], pp. 47–74) who set as matrix **H** Pasinetti's transposed. Therefore we have changed '*column*' to '*row*'.

Suppose additionally regular matrices $\mathbf{F}$, $\det(\mathbf{F}) \neq 0$, (*gross integrated industries*) and the case where all the surplus goes into profits. There are no wages (6.31). We reproduce the corresponding equations here; the matrix $\mathbf{H}$ and $\mathbf{C}_T = \mathbf{S}\mathbf{F}^{-1}$ are obtained through algebraic transformations of the *Sraffa price model for joint production* (6.48),

$$\mathbf{S}'\mathbf{p}(1+R) = \mathbf{F}'\mathbf{p} \Rightarrow \mathbf{F}'^{-1}\mathbf{S}'\mathbf{p} = \frac{1}{1+R}\mathbf{p} \Rightarrow \mathbf{C}_T'\mathbf{p} = \frac{1}{1+R}\mathbf{p},$$

$$\mathbf{S}'\mathbf{p}\cdot R = \mathbf{F}'\mathbf{p} - \mathbf{S}'\mathbf{p} = (\mathbf{F}' - \mathbf{S}')\mathbf{p} \Rightarrow (\mathbf{F}' - \mathbf{S}')^{-1}\mathbf{S}'\mathbf{p} =: \mathbf{H}\mathbf{p} = \frac{1}{R}\cdot\mathbf{p}. \tag{6.60}$$

There appear two eigenvalue equations (6.60) to compute the price vector $\mathbf{p}$. The matrix $\mathbf{F}$ is assumed to be regular, as Assumption 6.1.3 holds. As a consequence of the fact that Assumption 6.1.2 holds, the matrix $\mathbf{I}-\mathbf{C}_T'$ is regular, $\det(\mathbf{I}-\mathbf{C}_T') \neq 0$. We obtain therefore a second definition of matrix $\mathbf{H}$:

$$\mathbf{S}'\mathbf{p}(1+R) = \mathbf{F}'\mathbf{p} \Rightarrow$$

$$\mathbf{F}'^{-1}(\mathbf{S}'\mathbf{p}(1+R)) = (\mathbf{S}\mathbf{F}^{-1})'\mathbf{p}(1+R) = \mathbf{C}_T'\mathbf{p}(1+R) = (\mathbf{F}'^{-1}\mathbf{F}')\mathbf{p} = \mathbf{p} \Rightarrow$$

$$R\cdot\mathbf{C}_T'\mathbf{p} = \mathbf{I}\mathbf{p} - \mathbf{C}_T'\mathbf{p} = (\mathbf{I} - \mathbf{C}_T')\mathbf{p},$$

$$R\cdot(\mathbf{I} - \mathbf{C}_T')^{-1}\mathbf{C}_T'\mathbf{p} = \mathbf{p} \Rightarrow \mathbf{H}_1 := (\mathbf{I} - \mathbf{C}_T')^{-1}\mathbf{C}_T' \Rightarrow \mathbf{H}_1\mathbf{p} = \frac{1}{R}\mathbf{p}. \tag{6.61}$$

Comparing the eigenvalue equations of (6.60) and (6.61), we conclude that both matrices $\mathbf{H}$ and $\mathbf{H}_1$ are identical, indeed,

$$\begin{aligned}\mathbf{H}_1 &= (\mathbf{I} - \mathbf{C}_T')^{-1}\mathbf{C}_T' = (\mathbf{I} - (\mathbf{S}\mathbf{F}^{-1})')^{-1}(\mathbf{S}\mathbf{F}^{-1})' \\ &= (\mathbf{I} - (\mathbf{F}^{-1'})\mathbf{S}')^{-1}(\mathbf{F}^{-1'})\mathbf{S}' = ((\mathbf{F}^{-1'})(\mathbf{F}' - \mathbf{S}'))^{-1}(\mathbf{F}^{-1'})\mathbf{S}' \\ &= (\mathbf{F}' - \mathbf{S}')^{-1}((\mathbf{F}')^{-1})^{-1}(\mathbf{F}')^{-1}\mathbf{S}' = (\mathbf{F}' - \mathbf{S}')^{-1}\mathbf{S}' = \mathbf{H} \Rightarrow \mathbf{H}_1 \equiv \mathbf{H}.\end{aligned} \tag{6.62}$$

We will later show, that matrix $\mathbf{H}$ determines algebraically the number of *basics* and the number of *non-basics*.

(3) Steedman's application of matrix H to single-product industries. Following Steedman's analysis [114], we treat at first the special case of a *single-product industry*. The matrix $\mathbf{F}'$ is again replaced by the diagonal matrix $\mathbf{q} = \mathbf{S}\mathbf{e} + \mathbf{d}$. We set $\mathbf{C} = \mathbf{S}\hat{\mathbf{q}}^{-1}$. We assume a *semi-positive commodity flow* matrix $\mathbf{S}$ and a *semi-positive input-output coefficients* matrix $\mathbf{C}$, which have been transformed into "canonical form" (4.122). They are reproduced here,[14]

$$\mathbf{S}' = \begin{bmatrix} \mathbf{S}_{11}' & \mathbf{0} \\ \mathbf{S}_{12}' & \mathbf{S}_{22}' \end{bmatrix} \geq 0 \Rightarrow \mathbf{C}' = (\mathbf{S}\hat{\mathbf{q}}^{-1})' = \begin{bmatrix} \mathbf{C}_{11}' & \mathbf{0} \\ \mathbf{C}_{12}' & \mathbf{C}_{22}' \end{bmatrix} \geq 0. \tag{6.63}$$

14 We assume that this operation can be done manually, easily feasible in elementary cases. No algorithm to perform this operation on very large matrices is presented in this text.

Steedman ([114], p. 324) then presents the matrix $\mathbf{H} = (\mathbf{I} - \mathbf{C}')^{-1}\mathbf{C}'$, developing it into a convergent geometric series. Indeed, as there is an economy with surplus, $\mathbf{d} \geq \mathbf{o}$, the condition $\lambda_C = \rho(\mathbf{C}) < 1$ for the Frobenius number is fulfilled (see Lemma 4.1.1 (b)), where $\rho(\mathbf{C})$ is the spectral radius of matrix $\mathbf{C}$ (Definition A.9.1). Setting $\lambda = 1 > \lambda_C$, in Theorem A.10.2, equation (A.101), we get the development of $(\mathbf{I} - \mathbf{C}')^{-1}$, and consequently $\mathbf{H}$, into a convergent geometric series

$$\mathbf{H} \equiv (\mathbf{I} - \mathbf{C}')^{-1}\mathbf{C}' \equiv \mathbf{C}' + \mathbf{C}'^2 + \mathbf{C}'^3 + \cdots . \tag{6.64}$$

Steedman then further argues that the following equations hold:

$$\begin{aligned} \mathbf{H}_{11} &= (\mathbf{I} - \mathbf{C}'_{11})^{-1}\mathbf{C}'_{11}, \\ \mathbf{H}_{22} &= (\mathbf{I} - \mathbf{C}'_{22})^{-1}\mathbf{C}'_{22}, \end{aligned} \tag{6.65}$$

and therefore gets a Pasinetti matrix in the form, analogously to (6.63),

$$\mathbf{H} := \begin{bmatrix} \mathbf{H}_{11} & \mathbf{0} \\ \mathbf{H}_{21} & \mathbf{H}_{22} \end{bmatrix} = \begin{bmatrix} (\mathbf{I} - \mathbf{C}'_{11})^{-1}\mathbf{C}'_{11} & \mathbf{0} \\ \mathbf{H}_{21} & (\mathbf{I} - \mathbf{C}'_{22})^{-1}\mathbf{C}'_{22} \end{bmatrix} . \tag{6.66}$$

We illustrate Steedman's concept for *single-product industries* and compute the Pasinetti matrix $\mathbf{H}$, pointing out some of the above properties.

Example 6.3.3. We consider an economy without paid wages producing wheat W, iron I and gold G, described by the following production scheme:

$$\begin{aligned} 1: &\quad (300 \text{ qr. of wheat}, 12 \text{ t. of iron}, 0) \to (600 \text{ qr. of wheat}, 0, 0), \\ 2: &\quad (150 \text{ qr. of wheat}, 9 \text{ t. of iron}, 0) \to (0, 28 \text{ t. of iron}, 0), \\ 3: &\quad (50 \text{ qr. of wheat}, 4 \text{ t. of iron}, 0) \to (0, 0, 5 \text{ kg of gold}). \end{aligned} \tag{6.67}$$

Present the commodity flow matrix $\mathbf{S}$ and the *output* vector $\mathbf{q}$. Present the input-output coefficients matrix $\mathbf{C}$ and the Pasinetti matrix $\mathbf{H}$. Compute also its representation (6.66). Apply Schefold's concept (Definition 6.3.2) to determine the number of non-basics.

Calculate the eigenvalues of matrix $\mathbf{C}$ and determine the development of $\mathbf{H}$ into a geometric series, if such a development exists.

Using matrix $\mathbf{H}$, determine the number $n - m$ of basics and m of non-basics.

Solution to Example 6.3.3:
We identify the *commodity flow* matrix $\mathbf{S}$, the *output* vector $\mathbf{q}$ and the matrix $\hat{\mathbf{q}}^{-1}$ of the $n = 3$ sector economy,

$$\mathbf{S} = \begin{bmatrix} 300 & 150 & 50 \\ 12 & 9 & 4 \\ 0 & 0 & 0 \end{bmatrix}, \quad \mathbf{q} = \begin{bmatrix} 600 \\ 28 \\ 5 \end{bmatrix}, \quad \hat{\mathbf{q}}^{-1} = \begin{bmatrix} \frac{1}{600} & 0 & 0 \\ 0 & \frac{1}{28} & 0 \\ 0 & 0 & \frac{1}{5} \end{bmatrix}, \tag{6.68}$$

and compute the input-output coefficients matrix **C**:

$$\mathbf{C} = \mathbf{S}\hat{\mathbf{q}}^{-1} = \begin{bmatrix} 300 & 150 & 50 \\ 12 & 9 & 4 \\ 0 & 0 & 0 \end{bmatrix} \begin{bmatrix} \frac{1}{600} & 0 & 0 \\ 0 & \frac{1}{28} & 0 \\ 0 & 0 & \frac{1}{5} \end{bmatrix} = \left[\begin{array}{cc|c} \frac{1}{2} & \frac{75}{14} & 10 \\ \frac{1}{50} & \frac{9}{28} & \frac{4}{5} \\ \hline 0 & 0 & 0 \end{array}\right]. \tag{6.69}$$

The matrix **H** is obtained by calculating,

$$\begin{aligned} \mathbf{H} &= (\mathbf{I} - \mathbf{C}')^{-1}\mathbf{C}' \\ &= \begin{bmatrix} \frac{38}{13} & \frac{28}{325} & 0 \\ \frac{300}{13} & \frac{28}{13} & 0 \\ \frac{620}{13} & \frac{168}{65} & 1 \end{bmatrix} \begin{bmatrix} \frac{1}{2} & \frac{1}{50} & 0 \\ \frac{75}{14} & \frac{9}{28} & 0 \\ 10 & \frac{4}{5} & 0 \end{bmatrix} = \begin{bmatrix} \frac{25}{13} & \frac{28}{325} & 0 \\ \frac{300}{13} & \frac{15}{13} & 0 \\ \frac{620}{13} & \frac{168}{65} & 0 \end{bmatrix}. \end{aligned} \tag{6.70}$$

We then set up the sub-matrices of matrix $\mathbf{C}'$, identifying four sub-matrices:

$$\mathbf{C}'_{11} = \begin{bmatrix} \frac{1}{2} & \frac{1}{50} \\ \frac{75}{14} & \frac{9}{28} \end{bmatrix}, \quad \mathbf{C}'_{21} = \begin{bmatrix} 0 \\ 0 \end{bmatrix}, \quad \mathbf{C}'_{12} = \begin{bmatrix} 10 & \frac{4}{5} \end{bmatrix}, \quad \mathbf{C}'_{22} = \begin{bmatrix} 0 \end{bmatrix}. \tag{6.71}$$

Now we compute the sub-matrices of **H**, using the identities (6.65)

$$\begin{aligned} (\mathbf{I} - \mathbf{C}'_{11})^{-1} &= \begin{bmatrix} \frac{38}{13} & \frac{28}{325} \\ \frac{300}{13} & \frac{28}{13} \end{bmatrix}, \quad (\mathbf{I} - \mathbf{C}_{22})^{-1} = \begin{bmatrix} 1 \end{bmatrix}, \\ \mathbf{H}_{11} = (\mathbf{I} - \mathbf{C}'_{11})^{-1}\mathbf{C}'_{11} &= \begin{bmatrix} \frac{38}{13} & \frac{28}{325} \\ \frac{300}{13} & \frac{15}{13} \end{bmatrix} \begin{bmatrix} \frac{1}{2} & \frac{1}{50} \\ \frac{75}{14} & \frac{9}{28} \end{bmatrix} = \begin{bmatrix} \frac{25}{13} & \frac{28}{325} \\ \frac{300}{13} & \frac{15}{13} \end{bmatrix}, \\ \mathbf{H}_{22} = (\mathbf{I} - \mathbf{C}'_{22})^{-1}\mathbf{C}'_{22} &= \begin{bmatrix} 1 \end{bmatrix} \cdot \begin{bmatrix} 0 \end{bmatrix} = \begin{bmatrix} 0 \end{bmatrix}. \end{aligned} \tag{6.72}$$

We recognise that $\mathbf{H}_{11}$, $\mathbf{H}_{22}$ (6.72) are sub-matrices of **H** (6.70) and also $\mathbf{H}_{21} = [\frac{620}{13}\ \frac{168}{65}]$.

$$\mathbf{H} = \begin{bmatrix} \mathbf{H}_{11} & \mathbf{0} \\ \mathbf{H}_{21} & \mathbf{H}_{22} \end{bmatrix} = \begin{bmatrix} (\mathbf{I} - \mathbf{C}'_{11})^{-1}\mathbf{C}'_{11} & \mathbf{0} \\ \mathbf{H}_{21} & (\mathbf{I} - \mathbf{C}'_{22})^{-1}\mathbf{C}'_{22} \end{bmatrix} = \begin{bmatrix} \frac{25}{13} & \frac{28}{325} & 0 \\ \frac{300}{13} & \frac{15}{13} & 0 \\ \frac{620}{13} & \frac{168}{65} & 0 \end{bmatrix}. \tag{6.73}$$

Determine the rank of Schefold's (2×3) matrix $[\mathbf{S}'_2\ \mathbf{F}'_2]$, and we get,

$$\mathbf{S}' = \left[\begin{array}{cc|c} 300 & 12 & 0 \\ 150 & 9 & 0 \\ \hline 50 & 4 & 0 \end{array}\right], \quad \mathbf{F}' := \mathbf{q} = \left[\begin{array}{cc|c} 600 & 0 & 0 \\ 0 & 28 & 0 \\ \hline 0 & 0 & 5 \end{array}\right],$$

$$\begin{bmatrix} \mathbf{S}'_2 & \mathbf{F}'_2 \end{bmatrix} = \begin{bmatrix} 0 & 0 \\ 0 & 0 \\ 0 & 5 \end{bmatrix} \Rightarrow \operatorname{rank}(\begin{bmatrix} \mathbf{S}'_2 & \mathbf{F}'_2 \end{bmatrix}) = m = 1, \tag{6.74}$$

with $m = 1$, a *non-basic* commodity, namely gold, and $n - m = 2$ *basic commodities*, namely wheat and iron.

We next compute the eigenvalues of matrix **C**, obtaining with Definition A.9.1 and the ensemble of the eigenvalues $\sigma(\mathbf{C}) = \{0, \frac{1}{14}, \frac{3}{4}\}$, and determine the spectral radius as $\rho(\mathbf{C}) = \frac{3}{4} < 1$. This means that the geometric series (6.64) is convergent,

$$\mathbf{H} = (\mathbf{I} - \mathbf{C}')^{-1}\mathbf{C}' = \sum_{i=1}^{\infty} \mathbf{C}'^{i}$$

$$= \sum_{i=1}^{\infty} \begin{bmatrix} \frac{1}{2} & \frac{1}{50} & 0 \\ \frac{75}{14} & \frac{9}{28} & 0 \\ 10 & \frac{4}{5} & 0 \end{bmatrix}^{i} = \begin{bmatrix} \frac{25}{13} & \frac{28}{325} & 0 \\ \frac{300}{13} & \frac{15}{13} & 0 \\ \frac{620}{13} & \frac{168}{65} & 0 \end{bmatrix}. \tag{6.75}$$

Now we determine the number of *basics* and *non-basics* in this economy, following the explanations of Steedman ([114], p. 49). Matrix **H** is in “canonical form” and separates the *basics* from the *non-basics*. We determine the rank of submatrix

$$\mathbf{H}_{11} = \begin{bmatrix} \frac{25}{13} & \frac{28}{325} \\ \frac{300}{13} & \frac{15}{13} \end{bmatrix} \Rightarrow \operatorname{rank}(\mathbf{H}_{11}) = 2 = n - m. \tag{6.76}$$

As $\mathbf{H}_{11}$ has full rank, there are $n - m = 2$ *basic* commodities, wheat and iron. The *non-basics* are represented by $\mathbf{H}_{22} = \mathbf{0}$ (Lemma A.15.2 (iii)). This confirms that here there is only one *non-basic* commodity, $m = 1$, namely gold. ▲

The main feature and purpose in the use of the *Pasinetti matrix* **H** is to obtain the “canonical form” and to determine the number of *non-basics* m, respectively the number of *basics* $n - m$. Some invariant properties of matrix transformations will give the solution.

(4) The Schefold transformation matrix T and the Manara transformation matrix M. Remember that the Pasinetti matrix **H** appears re-ordered to a “canonical form”, like in equation (6.66), where *basics* and *non-basics* are separated. The aim of this subsection is precisely to show some invariant properties of matrix transformations, leading to the number m of *non-basics* present in the system. We rely on Schefold’s presentation ([103], pp. 58–60), based on Manara’s ideas (in Pasinetti, Ed. [83], pp. 1–15), as well as on Steedman’s ([114], p. 325) presentation. Let us come back to the matrix partitions (6.55), (6.56),

$$\mathbf{S}' = \begin{bmatrix} \mathbf{S}'_{11} & \mathbf{S}'_{21} \\ \mathbf{S}'_{12} & \mathbf{S}'_{22} \end{bmatrix}, \quad \mathbf{F}' = \begin{bmatrix} \mathbf{F}'_{11} & \mathbf{F}'_{21} \\ \mathbf{F}'_{12} & \mathbf{F}'_{22} \end{bmatrix}, \quad [\ \mathbf{S}'_2 \quad \mathbf{F}'_2\ ] := \begin{bmatrix} \mathbf{S}'_{21} & \mathbf{F}'_{21} \\ \mathbf{S}'_{22} & \mathbf{F}'_{22} \end{bmatrix}, \tag{6.77}$$

and return to the central statements of Definition 6.3.2:

A system $(\mathbf{S}',\mathbf{F}')$ is *non-basic*, if a permutation of the columns and a number m exist, so that with m $(1 \le m \le n-1)$ the $n \times 2m$ matrix $[\mathbf{S}'_2\ \mathbf{F}'_2]$ (6.56) has at most rank m. In this case, the $n-m$ rows of the $n \times 2m$ matrix $[\mathbf{S}'_2\ \mathbf{F}'_2]$ are *non-basic* and must, by definition, be linearly dependent on at most m others.

As the **matrix rank criterion**, Proposition 6.3.1 is fulfilled due to the linear dependence between $\mathbf{S}'_{21}$ and $\mathbf{S}'_{22}$, and also between $\mathbf{F}'_{21}$ and $\mathbf{F}'_{22}$, Sraffa's notion of *non-basics* in joint production implies that there exists an $(n-m) \times n$ matrix $\mathbf{T}$, realising the corresponding transformation. We shall call it *Schefold's transformation matrix* $\mathbf{T}$, according to Schefold's presentation in ([103], p. 58), setting,

$$[\ \mathbf{S}'_{21}\quad \mathbf{F}'_{21}\ ] = \mathbf{T}[\ \mathbf{S}'_{22}\quad \mathbf{F}'_{22}\ ], \tag{6.78}$$

and the *non-basics* are a linear transformation of the last m rows of $[\mathbf{S}'_{22}\ \mathbf{F}'_{22}]$ taken as a basis.

We then construct a matrix, according to Manara, in Pasinetti [83], p. 12. We call it the *Manara transformation matrix* which reads in Schefhold's representation:

$$\mathbf{M} \equiv \begin{bmatrix} \mathbf{I}_{n-m} & -\mathbf{T} \\ \mathbf{0} & \mathbf{I}_m \end{bmatrix}, \quad \det(\mathbf{M}) = \det(\mathbf{I}_{n-m})\det(\mathbf{I}_m) = 1, \tag{6.79}$$

transforming $(\mathbf{S}',\mathbf{F}')$ into a pair of almost triangular matrices,

$$\mathbf{MS}' = \begin{bmatrix} \mathbf{S}'_{11} - \mathbf{TS}'_{12} & \mathbf{0} \\ \mathbf{S}'_{12} & \mathbf{S}'_{22} \end{bmatrix}, \quad \mathbf{MF}' = \begin{bmatrix} \mathbf{F}'_{11} - \mathbf{TF}'_{12} & \mathbf{0} \\ \mathbf{F}'_{12} & \mathbf{F}'_{22} \end{bmatrix}. \tag{6.80}$$

Matrices $\mathbf{MS}'$ and $\mathbf{MF}'$ are new 'canonical forms' of the present system $(\mathbf{F}',\mathbf{S}')$, exploiting the linear dependence between *basics* and *non-basics*. The initial system $(\mathbf{F}',\mathbf{S}')$ undergoes a transformation and we say:

Notation 6.3.1. When the system $(\mathbf{F}',\mathbf{S}')$ is transformed by matrix $\mathbf{M}$ to the system $(\mathbf{MF}',\mathbf{MS}')$, it undergoes an **M-transformation.**

Schefold writes ([103], p. 59): "*The smallest such system (with the largest m)* $[\mathbf{S}'_{11} - \mathbf{TS}_{21}\ \mathbf{F}'_{11} - \mathbf{TF}'_{21}]$ *will be called the basic system. If it is identical with* $(\mathbf{S}',\mathbf{F}')$, $(\mathbf{S}',\mathbf{F}')$ *will be called basic.*" Clearly, when $(\mathbf{S}',\mathbf{F}')$ is basic, then $m = 0$ and $\mathbf{S}'_{22} = \mathbf{F}'_{22} = \emptyset$, exhibiting empty matrices.

The smallest basic system $[\mathbf{S}'_{11} - \mathbf{TS}_{21}\ \mathbf{F}'_{11} - \mathbf{TF}'_{21}]$ contains the $n-m$ *basic* commodities. The matrix $\mathbf{S}_{22}$, respectively $\mathbf{F}_{22}$, contain the *m non-basic* commodities.

We then compute the Pasinetti matrix $\mathbf{H}$ for a system $(\mathbf{MF}',\mathbf{MS}')$ that has undergone an **M-transformation.** Since initially $\mathbf{H} \equiv (\mathbf{F}' - \mathbf{S}')^{-1}\mathbf{S}'$, we further show that $\mathbf{H} \equiv (\mathbf{MF}' - \mathbf{MS}')^{-1}\mathbf{MS}'$, which leads, using the *regularity* of matrix $\mathbf{M}$, the block-matrix inversion rules and $\mathbf{V}_{ij} \equiv (\mathbf{F}_{ij} - \mathbf{S}_{ij})$, $i, j = 1, 2$, easily to

$$
\begin{aligned}
(\mathbf{MF}' - \mathbf{MS}')^{-1}\mathbf{MS}' &= (\mathbf{M}(\mathbf{F}' - \mathbf{S}'))^{-1}\mathbf{MS}' \\
&= (\mathbf{F}' - \mathbf{S}')^{-1}\mathbf{M}^{-1}\mathbf{MS}' = (\mathbf{F}' - \mathbf{S}')^{-1}\mathbf{S}' = \mathbf{H} = \begin{bmatrix} \mathbf{H}_{11} & \mathbf{H}_{12} \\ \mathbf{H}_{21} & \mathbf{H}_{22} \end{bmatrix} \\
&= \begin{bmatrix} (\mathbf{V}'_{11} - \mathbf{T}\mathbf{V}'_{12})^{-1}(\mathbf{S}'_{11} - \mathbf{T}\mathbf{S}'_{12}) & \mathbf{0} \\ \mathbf{V}'^{-1}_{22}[\mathbf{S}'_{12} - \mathbf{V}'_{12}(\mathbf{V}'_{11} - \mathbf{T}\mathbf{V}'_{12})^{-1}(\mathbf{S}'_{11} - \mathbf{T}\mathbf{S}'_{12})] & \mathbf{V}'^{-1}_{22}\mathbf{S}'_{22} \end{bmatrix}.
\end{aligned}
\tag{6.81}
$$

It is immediately clear that $\mathbf{H}$ is invariant under the **M-transformation** of the system $(\mathbf{F}', \mathbf{S}')$. This fact is proved by the development (6.81).

(5) Invariance properties of matrix H. The distinction of *basic* and *non-basic* commodities requires permutations of columns within Schefold's system $(\mathbf{F}', \mathbf{S}')$. We will now investigate invariance properties of the Pasinetti matrix $\mathbf{H}$ subject to row- and column-permutations. We set the following notions:

Notation 6.3.2. The multiplication of the matrix $\mathbf{A}$ by a permutation matrix $\mathbf{Q}$ from the right, $\mathbf{AQ}$, results in a **column-permutation** of matrix $\mathbf{A}$ by matrix $\mathbf{Q}$. The multiplication of the matrix $\mathbf{A}$ by a permutation matrix $\mathbf{P}$ from the left, $\mathbf{PA}$, results in a **row-permutation** of matrix $\mathbf{A}$ by matrix $\mathbf{P}$.

We continue applying an **M-transformation** on a row-permutation of matrix $\mathbf{S}'$, respectively $\mathbf{F}'$, by the same permutation matrix $\mathbf{P}$, realising a permutation σ and compute then the corresponding Pasinetti $\mathbf{H}_\sigma$,

$$
\begin{aligned}
\mathbf{H}_\sigma &\equiv (\mathbf{M}(\mathbf{PF}') - \mathbf{M}(\mathbf{PS}'))^{-1}\mathbf{M}(\mathbf{PS}') = ((\mathbf{MP})(\mathbf{F}' - \mathbf{S}'))^{-1}(\mathbf{MP})\mathbf{S}' \\
&= (\mathbf{F}' - \mathbf{S}')^{-1}(\mathbf{MP})^{-1}(\mathbf{MP})\mathbf{S}' = (\mathbf{F}' - \mathbf{S}')^{-1}\mathbf{S}' = \mathbf{H}.
\end{aligned}
\tag{6.82}
$$

We apply a column-permutation to matrices $\mathbf{S}'$ and $\mathbf{F}'$ by the permutation matrix $\mathbf{Q}$, realising a permutation τ and compute then the Pasinetti matrix,

$$
\begin{aligned}
\mathbf{H}_\tau &\equiv ((\mathbf{F}'\mathbf{Q}) - (\mathbf{S}'\mathbf{Q}))^{-1}(\mathbf{S}'\mathbf{Q}) = ((\mathbf{F}' - \mathbf{S}')\mathbf{Q})^{-1}\mathbf{S}'\mathbf{Q} \\
&= \mathbf{Q}^{-1}((\mathbf{F}' - \mathbf{S}')^{-1}\mathbf{S}')\mathbf{Q} = \mathbf{Q}^{-1}\mathbf{HQ}.
\end{aligned}
\tag{6.83}
$$

Matrices $\mathbf{H}_\tau$ and $\mathbf{H}$ are *similar matrices* and represent the same *linear operator* under two bases, with $\mathbf{Q}$ being the change of basis matrix. We obtain:

Lemma 6.3.1 (Invariance properties of the Pasinetti matrix). *The Pasinetti matrix* $\mathbf{H} = (\mathbf{F}' - \mathbf{S}')^{-1}\mathbf{S}'$ *is invariant under the* **M-transformation.** *The Pasinetti matrix* $\mathbf{H}$ *is invariant under a row-permutation. A column-permutation operated on matrices* $\mathbf{S}'$ *and* $\mathbf{F}'$, *realised by the same permutation matrix* $\mathbf{Q}$, *results in matrix* $\mathbf{Q}^{-1}\mathbf{HQ}$, *similar to the above Pasinetti matrix* $\mathbf{H}$.

The *invariance property* of the Pasinetti matrix $\mathbf{H}$ gives besides the "*matrix rank criterion*" a further tool to determine the number of *non-basics* in a production economy. We formulate it so:

Proposition 6.3.2 (Determination of the number of *non-basics*). *Given a system* $(\mathbf{F}', \mathbf{S}')$, *if the Pasinetti matrix* $\mathbf{H}$, *or a matrix similar to it, appears in 'canonical form',*

$$\mathbf{H} = \left(\mathbf{F}' - \mathbf{S}'\right)^{-1}\mathbf{S}' = \begin{bmatrix} \mathbf{H}_{11} & \mathbf{0} \\ \mathbf{H}_{21} & \mathbf{H}_{22} \end{bmatrix}, \tag{6.84}$$

then the number of columns $m > 0$ *of the* $(n-m) \times m$ *matrix* $\mathbf{H}_{12} = \mathbf{0}$ *is the number of the non-basics. The* $n - m$ *commodities contained in the* $(n-m) \times (n-m)$ *matrix* $\mathbf{H}_{11}$ *are then the basic commodities.* ▲

We now illustrate the key role played by the Pasinetti matrix **H** and its invariance property under the **M-transformation.** We also illustrate the fact that, under the action of the permutation matrix **Q**, its similar matrix $\mathbf{Q}^{-1}\mathbf{H}\mathbf{Q}$ represents the same *linear operator*, able to reveal the number of *non-basics*.

Example 6.3.4. We consider an economy producing wheat, iron and gold, described by following production scheme:

$$\begin{aligned} 1: &\quad (300 \text{ qr. of wheat}, 12 \text{ t.of iron}, 0) \to (600 \text{ qr. of wheat}, 0, 0), \\ 2: &\quad (150 \text{ qr. of wheat}, 9 \text{ t. iron}, 0) \to (0, 30 \text{ t. of iron}, 6 \text{ kg ofgold}), \\ 3: &\quad (50 \text{ qr. of wheat}, 4 \text{ t. of iron}, 0) \to (0, 0, 4 \text{ kg of gold}). \end{aligned} \tag{6.85}$$

There are three products, $n = 3$. Present the commodity flow matrix **S** and the *output* matrix **F**; calculate the Pasinetti matrix **H** (6.59) and determine the non-basics and basics. Set up the Schefold transformation matrix **T** and the Manara transformation matrix **M**.

Carry out the transformation, computing $\mathbf{MS}'$ and $\mathbf{MF}'$, compute the transformed matrix $\mathbf{H} = (\mathbf{MF}' - \mathbf{MS}')^{-1}\mathbf{MS}'$ (6.81), confirming its invariance under the transformation by matrix **M** and determine directly the submatrix $\mathbf{H}_{11} = (\mathbf{V}'_{11} - \mathbf{T}\mathbf{V}'_{12})^{-1}(\mathbf{S}'_{11} - \mathbf{T}\mathbf{S}'_{12})$.

Solution to Example 6.3.4:
We identify the transposed *commodity flow* matrix $\mathbf{S}'$ and the transposed *output* matrix $\mathbf{F}'$,

$$\mathbf{S}' = \begin{bmatrix} 300 & 12 & 0 \\ 150 & 9 & 0 \\ 50 & 4 & 0 \end{bmatrix}, \quad \mathbf{F}' = \begin{bmatrix} 600 & 0 & 0 \\ 0 & 30 & 6 \\ 0 & 0 & 4 \end{bmatrix}. \tag{6.86}$$

The matrix $(\mathbf{F}' - \mathbf{S}')$ is generally not positive,

$$\mathbf{F}' - \mathbf{S}' = \begin{bmatrix} 300 & -12 & 0 \\ -150 & 21 & 6 \\ -50 & -4 & 4 \end{bmatrix}, \quad (\mathbf{F}' - \mathbf{S}')^{-1} = \begin{bmatrix} \frac{3}{800} & \frac{1}{600} & -\frac{1}{400} \\ \frac{1}{96} & \frac{1}{24} & -\frac{1}{16} \\ \frac{11}{192} & \frac{1}{16} & \frac{5}{32} \end{bmatrix}. \tag{6.87}$$

Calculate now the Pasinetti matrix (6.59), appearing directly in "canonical form",

$$\mathbf{H} = (\mathbf{F}' - \mathbf{S}')^{-1}\mathbf{S}'$$
$$= \begin{bmatrix} \frac{3}{800} & \frac{1}{600} & -\frac{1}{400} \\ \frac{1}{96} & \frac{1}{24} & -\frac{1}{16} \\ \frac{11}{192} & \frac{1}{16} & \frac{5}{32} \end{bmatrix} \begin{bmatrix} 300 & 12 & 0 \\ 150 & 9 & 0 \\ 50 & 4 & 0 \end{bmatrix} = \left[\begin{array}{cc|c} \frac{5}{4} & \frac{1}{20} & 0 \\ \frac{25}{4} & \frac{1}{4} & 0 \\ \hline \frac{275}{8} & \frac{15}{8} & 0 \end{array}\right]. \tag{6.88}$$

We consider the (2×1) zero matrix $\mathbf{H}_{12} = [0 \;\; 0]'$ and are led to set $m = 1$, claiming that gold is *non-basic*. Therefore, there are $n - m = 2$ *basics*, namely wheat and iron. Accordingly, we identify now the sub-matrices of $\mathbf{S}'$ and $\mathbf{F}'$ (6.77)

$$\mathbf{S}'_{11} = \begin{bmatrix} 300 & 12 \\ 150 & 9 \end{bmatrix}, \quad \mathbf{S}'_{21} = \begin{bmatrix} 0 \\ 0 \end{bmatrix}, \quad \mathbf{S}'_{12} = [\,50 \;\; 4\,], \quad \mathbf{S}'_{22} = [\,0\,],$$
$$\mathbf{F}'_{11} = \begin{bmatrix} 600 & 0 \\ 0 & 30 \end{bmatrix}, \quad \mathbf{F}'_{21} = \begin{bmatrix} 0 \\ 6 \end{bmatrix}, \quad \mathbf{F}'_{12} = [\,0 \;\; 0\,], \quad \mathbf{F}'_{22} = [\,4\,]. \tag{6.89}$$

With $n = 3$ and $m = 1$, we set up the $((n-m)\times m) = (2\times1)$ Schefold *transformation* matrix $\mathbf{T} = [t_1, t_2]'$ and subsequently the (3×3) Manara *transformation* matrix $\mathbf{M}$, computing,

$$\mathbf{S}'_{21} = \mathbf{T}\mathbf{S}'_{22} \quad \text{and} \quad \mathbf{F}'_{21} = \mathbf{T}\mathbf{F}'_{22} \Leftrightarrow$$
$$\begin{bmatrix} 0 \\ 0 \end{bmatrix} = \begin{bmatrix} t_1 \\ t_2 \end{bmatrix}[0] \quad \text{and} \quad \begin{bmatrix} 0 \\ 6 \end{bmatrix} = \begin{bmatrix} t_1 \\ t_2 \end{bmatrix}[4] \quad \text{obtaining } t_2 = \frac{3}{2}. \tag{6.90}$$

We get the Schefold transformation matrix

$$\mathbf{T} = \begin{bmatrix} 0 \\ \frac{3}{2} \end{bmatrix}, \tag{6.91}$$

and conclude, setting up the Manara transformation matrix,

$$\mathbf{M} \equiv \begin{bmatrix} \mathbf{I}_{n-m} & -\mathbf{T} \\ \mathbf{0} & \mathbf{I}_m \end{bmatrix} = \begin{bmatrix} 1 & 0 & 0 \\ 0 & 1 & -\frac{3}{2} \\ 0 & 0 & 1 \end{bmatrix}. \tag{6.92}$$

We now compute the transformed *commodity flow* and *output* matrices,

$$\mathbf{M}\mathbf{S}' = \begin{bmatrix} \mathbf{I}_{n-m} & -\mathbf{T} \\ \mathbf{0} & \mathbf{I}_m \end{bmatrix}\mathbf{S}' = \begin{bmatrix} 1 & 0 & 0 \\ 0 & 1 & -\frac{3}{2} \\ 0 & 0 & 1 \end{bmatrix}\mathbf{S}' = \begin{bmatrix} 300 & 12 & 0 \\ 75 & 3 & 0 \\ 50 & 4 & 0 \end{bmatrix},$$
$$\mathbf{M}\mathbf{F}' = \begin{bmatrix} \mathbf{I}_{n-m} & -\mathbf{T} \\ \mathbf{0} & \mathbf{I}_m \end{bmatrix}\mathbf{F}' = \begin{bmatrix} 1 & 0 & 0 \\ 0 & 1 & -\frac{3}{2} \\ 0 & 0 & 1 \end{bmatrix}\mathbf{F}' = \begin{bmatrix} 600 & 0 & 0 \\ 0 & 30 & 0 \\ 0 & 0 & 4 \end{bmatrix}. \tag{6.93}$$

Finally, one computes the transformed Pasinetti matrix **H**, illustrating again the *invariance property* of the **M**-transformation,

$$\mathbf{H} \equiv (\mathbf{MF}' - \mathbf{MS}')^{-1}\mathbf{MS}' = \begin{bmatrix} \frac{3}{800} & \frac{1}{600} & 0 \\ \frac{1}{96} & \frac{1}{24} & 0 \\ \frac{11}{192} & \frac{1}{16} & \frac{1}{4} \end{bmatrix} \begin{bmatrix} 300 & 12 & 0 \\ 75 & 3 & 0 \\ 50 & 4 & 0 \end{bmatrix} = \begin{bmatrix} \frac{5}{4} & \frac{1}{20} & 0 \\ \frac{25}{4} & \frac{1}{4} & 0 \\ \frac{275}{8} & \frac{15}{8} & 0 \end{bmatrix}. \tag{6.94}$$

We recognise the sub-matrices $\mathbf{H}_{ij}$ of matrix **H** (6.94) as:

$$\mathbf{H} = \left[\begin{array}{cc|c} \frac{5}{4} & \frac{1}{20} & 0 \\ \frac{25}{4} & \frac{1}{4} & 0 \\ \hline \frac{275}{8} & \frac{15}{8} & 0 \end{array}\right] = \begin{bmatrix} \mathbf{H}_{11} & \mathbf{0} \\ \mathbf{H}_{21} & \mathbf{H}_{22} \end{bmatrix} \Rightarrow \mathbf{H}_{11} = \begin{bmatrix} \frac{5}{4} & \frac{1}{20} \\ \frac{25}{4} & \frac{1}{4} \end{bmatrix},$$
$$\mathbf{H}_{21} = \begin{bmatrix} \frac{275}{8} & \frac{15}{8} \end{bmatrix}, \quad \mathbf{H}_{22} = \begin{bmatrix} 0 \end{bmatrix}, \tag{6.95}$$

and can calculate all the sub-matrices $\mathbf{H}_{ij}$, $i, j = 1, 2$, applying the equalities (6.81); let's do it for $\mathbf{H}_{11}$,

$$\mathbf{S}'_{11} - \mathbf{TS}'_{12} = \begin{bmatrix} 300 & 12 \\ 75 & 3 \end{bmatrix}, \quad \mathbf{V}'_{11} = \mathbf{F}'_{11} - \mathbf{S}'_{11} = \begin{bmatrix} 300 & -12 \\ -150 & 21 \end{bmatrix},$$
$$\mathbf{V}'_{12} = \mathbf{F}'_{12} - \mathbf{S}'_{12} = \begin{bmatrix} -50 & -4 \end{bmatrix}, \quad (\mathbf{V}'_{11} - \mathbf{TV}'_{12})^{-1} = \begin{bmatrix} \frac{3}{800} & \frac{1}{600} \\ \frac{1}{96} & \frac{1}{24} \end{bmatrix}, \tag{6.96}$$

getting finally,

$$\mathbf{H}_{11} = (\mathbf{V}'_{11} - \mathbf{TV}'_{12})^{-1}(\mathbf{S}'_{11} - \mathbf{TS}'_{12}) = \begin{bmatrix} \frac{3}{800} & \frac{1}{600} \\ \frac{1}{96} & \frac{1}{24} \end{bmatrix} \begin{bmatrix} 300 & -150 \\ -12 & 21 \end{bmatrix} = \begin{bmatrix} \frac{5}{4} & \frac{1}{20} \\ \frac{25}{4} & \frac{1}{4} \end{bmatrix}. \tag{6.97}$$

We obtain again the above matrix $\mathbf{H}_{11}$ (6.95) and confirm its positiveness. Matrix $\mathbf{H}_{11}$ is therefore *irreducible*, and we again conclude:

Matrix $\mathbf{H}_{11}$ represents $n - m = 2$ *basics*, namely the commodities wheat and iron, and we have one *non-basic* commodity ($m = 1$), namely gold. We also calculate the rank

$$\operatorname{rank}\left(\begin{bmatrix} \mathbf{S}'_2 & \mathbf{F}'_2 \end{bmatrix}\right) = \operatorname{rank}\left(\begin{bmatrix} 0 & 0 \\ 0 & 6 \\ 0 & 4 \end{bmatrix}\right) = 1 \leq m = 1, \tag{6.98}$$

confirming Sraffa's initial definition (PCMC, Par. 60) or the *matrix rank criterion* Proposition 6.3.1, stating that *"the rank is less than, or equal to, m."* ▲

We consolidate now our understanding of the Pasinetti matrix **H** with an illustration of the Manara transformation matrix **M** and the Schefold transformation matrix **T**. Consider the following

Example 6.3.5. Establish the system $(\mathbf{S}', \mathbf{F}')$ of a joint production process with $n = 5$ commodities, represented by the following matrices,

$$\mathbf{S}' = \begin{bmatrix} 30 & 0 & 15 & 45 & 25 \\ 30 & 35 & 10 & 95 & 45 \\ 10 & 20 & 10 & 90 & 50 \\ 10 & 0 & 0 & 40 & 30 \\ 0 & 0 & 5 & 50 & 20 \end{bmatrix}, \quad \mathbf{F}' = \begin{bmatrix} 30 & 20 & 30 & 55 & 60 \\ 40 & 40 & 20 & 115 & 100 \\ 30 & 30 & 10 & 110 & 120 \\ 10 & 10 & 0 & 50 & 80 \\ 10 & 10 & 10 & 60 & 40 \end{bmatrix}. \tag{6.99}$$

Calculate the Pasinetti matrix **H** and determine the numbers $n - m$ of basics and m of non-basics. Compute the Manara transformation matrix **T** (6.78) and set up matrix **M** (6.79) (6.79).

Compute the matrix products $\mathbf{MS}'$, respectively $\mathbf{MF}'$ and compute the Pasinetti matrix **H** (6.81) of this **M-transformation.**

Solution to Example 6.3.5:
We start by computing the Pasinetti matrix (6.59), $n = 5$,

$$\mathbf{H} = (\mathbf{F}' - \mathbf{S}')^{-1}\mathbf{S}'$$
$$= \left[\begin{array}{ccc|cc} \frac{5}{16} & \frac{13}{16} & \frac{17}{16} & 0 & 0 \\ -\frac{5}{4} & -\frac{9}{4} & -\frac{1}{4} & 0 & 0 \\ \frac{25}{8} & \frac{17}{8} & \frac{13}{8} & 0 & 0 \\ \hline -\frac{61}{24} & -\frac{7}{8} & -\frac{49}{24} & \frac{17}{3} & \frac{4}{3} \\ \frac{23}{24} & \frac{5}{8} & \frac{11}{24} & -\frac{1}{3} & \frac{1}{3} \end{array}\right] = \begin{bmatrix} \mathbf{H}_{11} & \mathbf{0} \\ \mathbf{H}_{21} & \mathbf{H}_{22} \end{bmatrix}. \tag{6.100}$$

As there is a 3×2 zero matrix $\mathbf{H}_{12}$, we conclude with Lemma 6.3.1 that there are $m = 2$ *non-basics* and $n - m = 3$ *basics*. We extract the required sub-matrices from the given matrices $\mathbf{S}'$ and $\mathbf{F}'$ (6.99),

$$\mathbf{S}'_{21} = \begin{bmatrix} 45 & 25 \\ 95 & 45 \\ 90 & 50 \end{bmatrix} =: \begin{bmatrix} \mathbf{y}_1 & \mathbf{y}_2 \end{bmatrix}, \quad \mathbf{S}'_{22} = \begin{bmatrix} 40 & 30 \\ 50 & 20 \end{bmatrix} =: \begin{bmatrix} \mathbf{x}_1 & \mathbf{x}_2 \end{bmatrix},$$
$$\mathbf{F}'_{21} = \begin{bmatrix} 55 & 60 \\ 115 & 100 \\ 110 & 120 \end{bmatrix} =: \begin{bmatrix} \mathbf{y}_3 & \mathbf{y}_4 \end{bmatrix}, \quad \mathbf{F}'_{22} = \begin{bmatrix} 50 & 80 \\ 60 & 40 \end{bmatrix} =: \begin{bmatrix} \mathbf{x}_3 & \mathbf{x}_4 \end{bmatrix}, \tag{6.101}$$

and applying the *matrix rank criterion*, we find,

$$\begin{bmatrix} \mathbf{S}'_2 & \mathbf{F}'_2 \end{bmatrix} = \begin{bmatrix} \mathbf{S}'_{21} & \mathbf{F}'_{21} \\ \mathbf{S}'_{22} & \mathbf{F}'_{22} \end{bmatrix} \Rightarrow \operatorname{rank}(\begin{bmatrix} \mathbf{S}'_2 & \mathbf{F}'_2 \end{bmatrix}) = 2 = m. \tag{6.102}$$

Then Manara's transformation matrix **T** (6.78) can be calculated, either with the *commodity flow* or the *output* matrix, necessarily giving the same resulting matrix **T**,

$$\begin{aligned} \mathbf{S}'_{21} &= \mathbf{T}_1\mathbf{S}'_{22} \Rightarrow \mathbf{T}_1 = \mathbf{S}'_{21}(\mathbf{S}_{22}{}')^{-1} \\ \mathbf{F}'_{21} &= \mathbf{T}_2\mathbf{F}'_{22} \Rightarrow \mathbf{T}_2 = \mathbf{F}'_{21}(\mathbf{F}_{22}{}')^{-1}. \end{aligned} \tag{6.103}$$

Numerically, we obtain,

$$\mathbf{T}_1 = \begin{bmatrix} 45 & 25 \\ 95 & 45 \\ 90 & 50 \end{bmatrix} \begin{bmatrix} -\frac{1}{35} & \frac{3}{70} \\ \frac{1}{14} & -\frac{2}{35} \end{bmatrix} = \begin{bmatrix} \frac{1}{2} & \frac{1}{2} \\ \frac{1}{2} & \frac{3}{2} \\ 1 & 1 \end{bmatrix},$$

$$\mathbf{T}_2 = \begin{bmatrix} 55 & 60 \\ 115 & 100 \\ 110 & 120 \end{bmatrix} \begin{bmatrix} -\frac{1}{70} & \frac{1}{35} \\ \frac{3}{140} & -\frac{1}{56} \end{bmatrix} = \begin{bmatrix} \frac{1}{2} & \frac{1}{2} \\ \frac{1}{2} & \frac{3}{2} \\ 1 & 1 \end{bmatrix}. \tag{6.104}$$

The resulting identity of both Manara transformation matrices, $\mathbf{T} = \mathbf{T}_1 = \mathbf{T}_2$ confirms the fact that there are two *non-basics*, due to the linear dependency between the matrices $\mathbf{S}'_{21}$ and $\mathbf{S}'_{22}$, respectively matrices $\mathbf{F}'_{21}$ and $\mathbf{F}'_{22}$.

For the commodity flow matrix **S**, the Schefold transformation matrix **T** maps with the application $\mathbf{S}'_{21} = \mathbf{T}\mathbf{S}'_{22}$ the vector $\mathbf{x}_1 = [40, 50]'$ on the corresponding vector $\mathbf{y}_1 = [45, 95, 90]'$ and the vector $\mathbf{x}_2 = [30, 20]'$ on the corresponding vector $\mathbf{y}_2 = [25, 45, 50]'$. For the output matrix **S**, the Schefold transformation matrix **T** maps with the application $\mathbf{F}'_{21} = \mathbf{T}\mathbf{F}'_{22}$ the vector $\mathbf{x}_3 = [50, 60]'$ on the corresponding vector $\mathbf{y}_3 = [55, 115, 110]'$ and the vector $\mathbf{x}_4 = [80, 40]'$ on the corresponding vector $\mathbf{y}_4 = [60, 100, 120]'$.

Using the Schefold transformation matrix **T**, we define the Manara transformation matrix **M**,

$$\mathbf{M} = \begin{bmatrix} \mathbf{I}_{n-m} & -\mathbf{T} \\ \mathbf{0} & \mathbf{I}_m \end{bmatrix} = \begin{bmatrix} 1 & 0 & 0 & -\frac{1}{2} & -\frac{1}{2} \\ 0 & 1 & 0 & -\frac{1}{2} & -\frac{3}{2} \\ 0 & 0 & 1 & -1 & -1 \\ 0 & 0 & 0 & 1 & 0 \\ 0 & 0 & 0 & 0 & 1 \end{bmatrix}, \tag{6.105}$$

and calculate the matrix products $\mathbf{MS}'$, respectively $\mathbf{MF}'$. In the right upper corner of the matrix product $\mathbf{MS}'$, respectively $\mathbf{MF}'$, are generated as expected the zero matrix $\mathbf{S}'_{21} - \mathbf{T}\mathbf{S}'_{22}$, respectively the zero matrix $\mathbf{F}'_{21} - \mathbf{T}\mathbf{F}'_{22}$, getting altogether,

$$\mathbf{MS}' = \begin{bmatrix} 25 & 0 & 12.5 & 0 & 0 \\ 25 & 35 & 2.5 & 0 & 0 \\ 0 & 20 & 5 & 0 & 0 \\ 10 & 0 & 0 & 40 & 30 \\ 0 & 0 & 5 & 50 & 20 \end{bmatrix}, \quad \mathbf{MF}' = \begin{bmatrix} 20 & 10 & 25 & 0 & 0 \\ 20 & 20 & 5 & 0 & 0 \\ 10 & 10 & 0 & 0 & 0 \\ 10 & 10 & 0 & 50 & 80 \\ 10 & 10 & 10 & 60 & 40 \end{bmatrix}. \tag{6.106}$$

We now calculate the inverse matrix,

$$(\mathbf{MF}' - \mathbf{MS}')^{-1} = \begin{bmatrix} \frac{1}{20} & -\frac{3}{80} & \frac{17}{160} & 0 & 0 \\ 0 & -\frac{1}{20} & -\frac{1}{40} & 0 & 0 \\ \frac{1}{10} & \frac{1}{40} & \frac{1}{16} & 0 & 0 \\ -\frac{1}{6} & \frac{11}{120} & -\frac{49}{240} & -\frac{1}{15} & \frac{1}{6} \\ \frac{1}{30} & -\frac{1}{120} & \frac{11}{240} & \frac{1}{30} & -\frac{1}{30} \end{bmatrix}, \tag{6.107}$$

and the Pasinetti matrix, according to (6.81),

$$\mathbf{H} = (\mathbf{MF}' - \mathbf{MS}')^{-1}\mathbf{MS}'$$
$$= \left[\begin{array}{ccc|cc} \frac{5}{16} & \frac{13}{16} & \frac{17}{16} & 0 & 0 \\ -\frac{5}{4} & -\frac{9}{4} & -\frac{1}{4} & 0 & 0 \\ \frac{25}{8} & \frac{17}{8} & \frac{13}{8} & 0 & 0 \\ \hline -\frac{61}{24} & -\frac{7}{8} & -\frac{49}{24} & \frac{17}{3} & \frac{4}{3} \\ \frac{23}{24} & \frac{5}{8} & \frac{11}{24} & -\frac{1}{3} & \frac{1}{3} \end{array}\right] = \begin{bmatrix} \mathbf{H}_{11} & \mathbf{0} \\ \mathbf{H}_{21} & \mathbf{H}_{22} \end{bmatrix}. \tag{6.108}$$

The computation confirms the invariance of matrix **H** under the **M-transformation** of the economic system and the pertinence of determining the number of *non-basics* from the Pasinetti matrix **H** in “canonical form”. ▲

These examples have shown the practical importance of the invariance property, Lemma 6.3.1. We have also highlighted the similarity between the “canonical forms” of matrix **S** in the case of *single-product industries* and matrix **H** in the case of *joint production.*

The following example illustrates the *row-permutation*, meaning interchanging the individual sectors of a production process.

Example 6.3.6. Based on Example 3.1.2, (PCMC, Par. 2), we set up the following joint production economy with three sectors: wheat, iron, pigs. We present the system of production with a dummy vector of labour **L**:

$$\begin{aligned} &(240 \text{ qr. of wheat}, 12 \text{ t. of iron}, 18 \text{ pigs}, L_1) \to (400 \text{ qr. of wheat}, 10 \text{ t. of iron}, 0) \\ &(90 \text{ qr. of wheat}, 6 \text{ t. of iron}, 12 \text{ pigs}, L_2) \to (100 \text{ qr. of wheat}, 8 \text{ t. of iron}, 40 \text{ pigs}) \\ &(120 \text{ qr. of wheat}, 3 \text{ t. of iron}, 30 \text{ pigs}, L_3) \to (0,\ 10 \text{ t. of iron}, 40 \text{ pigs}) \end{aligned} \tag{6.109}$$

Consider the permutation matrix and its inverse,

$$\mathbf{P} = \begin{bmatrix} 1 & 0 & 0 \\ 0 & 0 & 1 \\ 0 & 1 & 0 \end{bmatrix}, \quad \mathbf{P}^{-1} = \begin{bmatrix} 1 & 0 & 0 \\ 0 & 0 & 1 \\ 0 & 1 & 0 \end{bmatrix}. \tag{6.110}$$

Set up the commodity flow matrix $\mathbf{S}'$, the output matrix $\mathbf{F}'$ and the dummy vector of labour $\mathbf{L}$. Realise a row-permutation of matrices $\mathbf{S}'$ and $\mathbf{F}'$, computing the products $\mathbf{PS}'$, $\mathbf{PF}'$ and the vector $\mathbf{PL}$. One establishes with these matrices the transformed system of production $(\mathbf{PS}', \mathbf{PL}) \to (\mathbf{PF}')$. Compute the corresponding Pasinetti matrix $\mathbf{H}$. Determine the number of basics and the number of non-basics.

Solution to Example 6.3.6:
Identify the *commodity flow* matrix $\mathbf{S}'$, the *output* matrix $\mathbf{F}'$ and finally the vector of labour $\mathbf{L}$,

$$\mathbf{S}' = \begin{bmatrix} 240 & 12 & 18 \\ 90 & 6 & 12 \\ 120 & 3 & 30 \end{bmatrix}, \quad \mathbf{F}' = \begin{bmatrix} 400 & 10 & 0 \\ 100 & 8 & 40 \\ 0 & 10 & 40 \end{bmatrix} \mathbf{L} = \begin{bmatrix} L_1 \\ L_2 \\ L_3 \end{bmatrix}, \tag{6.111}$$

and compute the matrix products,

$$\mathbf{PS}' = \begin{bmatrix} 240 & 12 & 18 \\ 120 & 3 & 30 \\ 90 & 6 & 12 \end{bmatrix}, \quad \mathbf{PF}' = \begin{bmatrix} 400 & 10 & 0 \\ 0 & 10 & 40 \\ 100 & 8 & 40 \end{bmatrix},$$

$$\mathbf{PL} = \begin{bmatrix} 1 & 0 & 0 \\ 0 & 0 & 1 \\ 0 & 1 & 0 \end{bmatrix} \begin{bmatrix} L_1 \\ L_2 \\ L_3 \end{bmatrix} = \begin{bmatrix} L_1 \\ L_3 \\ L_2 \end{bmatrix}, \tag{6.112}$$

obtaining the corresponding reordered system of production, corresponding to the economic process:

$$\begin{gathered}(240 \text{ qr. of wheat}, 12 \text{ t. of iron}, 18 \text{ pigs}, L_1) \to (400 \text{ qr. of wheat}, 10 \text{ t. of iron}, 0), \\ (120 \text{ qr. of wheat}, 3 \text{ t. of iron}, 30 \text{ pigs}, L_3) \to (0,\ 10 \text{ t. of iron}, 40 \text{ pigs}), \\ (90 \text{ qr. of wheat}, 6 \text{ t. of iron}, 12 \text{ pigs}, L_2) \to (100 \text{ qr. of wheat}, 8 \text{ t. of iron}, 40 \text{ pigs}).\end{gathered} \tag{6.113}$$

Then, compute matrix $\mathbf{H} = (\mathbf{PF}' - \mathbf{PS}')^{-1}(\mathbf{PS}') = (\mathbf{F}' - \mathbf{S}')^{-1}\mathbf{S}'$,

$$\mathbf{H} = \frac{1}{149} \begin{bmatrix} 301 & 15.6 & 28 \\ 8{,}000 & 327 & 1{,}160 \\ -200 & 3 & -29 \end{bmatrix}, \tag{6.114}$$

The Pasinetti matrix does not contain a zero component. Therefore, there is an empty submatrix, $\mathbf{H}_{22} = \emptyset$. There is no *non-basic* commodity. We have $m = 0$.

We observe that the sectors appear re-ordered. The *production* is not altered in itself. The overall *production structure* is unchanged. The Pasinetti matrix **H** is invariant under *row permutation* (6.82). ▲

6.4 Joint production, Sraffa Networks and adjacency matrices

A formal presentation of the *Sraffa Network* is given in Section A.14. We go on, considering the structure of the *Sraffa system* for *joint production* industries (6.48),

$$\boxed{\begin{aligned} \mathbf{S}'\mathbf{p}(1+r) + \mathbf{L}\frac{\tilde{w}\cdot Y}{L} &= \mathbf{F}'\mathbf{p}, \\ Y &= \mathbf{e}'(\mathbf{F}-\mathbf{S})'\mathbf{p} = \mathbf{d}'\mathbf{p}, \\ L &= \mathbf{e}'\mathbf{L}. \end{aligned}} \tag{6.115}$$

We consider the *commodity flow* matrix $\mathbf{S}$, describing the technology, and the *output* matrix $\mathbf{F}$ of this *joint production economy*, describing the production during the considered period.

We construct the following square $2n \times 2n$ matrix

$$\mathbf{\Sigma} = (\sigma_{kl}) = \begin{bmatrix} \mathbf{0} & \mathbf{F}' \\ \mathbf{S} & \mathbf{0} \end{bmatrix}; \quad k,l = 1,\dots,2n. \tag{6.116}$$

Then, for matrix $\mathbf{\Sigma}$, we set up the *adjacency matrix*, where the output matrix $\mathbf{F}$ is replaced by the matrix $\mathbf{Q} = (q_{ij})$ and matrix $\mathbf{S}$ is replaced by a matrix $\mathbf{V} = (\upsilon_{ij})$, consisting both of '0' and '1', in accordance to Definition A.8.5 getting,

$$\begin{aligned} \upsilon_{ij} = w_{n+i,j} &= \begin{cases} 1 & \text{if } s_{ij} \neq 0 \\ 0 & \text{if } s_{ij} = 0, \end{cases} \qquad i,j \in \{1,\dots,n\}, \\ q_{ij} = w_{i,n+j} &= \begin{cases} 1 & \text{if } f_{ij} \neq 0 \\ 0 & \text{if } f_{ij} = 0, \end{cases} \qquad i,j \in \{1,\dots,n\}. \end{aligned} \tag{6.117}$$

We have now got the adjacency matrix corresponding to matrix $\mathbf{\Sigma}$:

$$\mathbf{W} = (w_{kl}) = \begin{bmatrix} \mathbf{0} & \mathbf{Q}' \\ \mathbf{V} & \mathbf{0} \end{bmatrix}; \quad k,l = 1,\dots,2n, \tag{6.118}$$

and we have a very useful

Lemma 6.4.1. *The rank of matrix* $\mathbf{\Sigma}$ *in* (6.116)*, respectively in* (4.145) *is equal to the sum: rank of matrix* $\mathbf{S}$ *plus rank of matrix* $\mathbf{F}'$*, respectively rank of matrix* $\hat{\mathbf{q}}$*. Indeed, in the case of* (4.145)*,* $\operatorname{rank}(\mathbf{F})$ *is replaced by* $\operatorname{rank}(\hat{\mathbf{q}})$*,*

$$\operatorname{rank}(\mathbf{\Sigma}) = \operatorname{rank}(\mathbf{S}) + \operatorname{rank}(\mathbf{F}). \tag{6.119}$$

Proof. Take a $2n \times 2n$ permutation matrix $\mathbf{R}$, whose rank is by definition equal to $2n$, $\operatorname{rank}(\mathbf{R}) = 2n$, by applying Lemma A.8.1 permuting the rows of matrix $\mathbf{\Sigma}$ in order to get a block diagonal matrix, in such a way that $\mathbf{F}'$ and $\mathbf{S}$ are symmetrically interchanged as is shown by the next calculation (6.120). Then, apply the calculation rule for ranks of block diagonal matrices, see Horn and Johnson [43], p. 31, and as matrix **R** is regular, there is $\operatorname{rank}(\mathbf{R}^{-1}\mathbf{\Sigma}\mathbf{R}) = \operatorname{rank}(\mathbf{\Sigma})$,

$$\mathbf{R}^{-1}\mathbf{\Sigma}\mathbf{R} = \mathbf{R}^{-1}\begin{bmatrix} \mathbf{0} & \mathbf{F}' \\ \mathbf{S} & \mathbf{0} \end{bmatrix}\mathbf{R} = \begin{bmatrix} \mathbf{F}' & \mathbf{0} \\ \mathbf{0} & \mathbf{S} \end{bmatrix} \Rightarrow$$

$$\operatorname{rank}(\mathbf{R}^{-1}\mathbf{\Sigma}\mathbf{R}) = \operatorname{rank}(\mathbf{\Sigma}) = \operatorname{rank}\left(\begin{bmatrix} \mathbf{F}' & \mathbf{0} \\ \mathbf{0} & \mathbf{S} \end{bmatrix}\right) = \operatorname{rank}(\mathbf{F}) + \operatorname{rank}(\mathbf{S}). \tag{6.120}$$

□

Example 6.4.1. Consider the following example, inspired by PCMC, Par. 73, with the industries: grazing, shearing, wool manufacturing, manufacturing meat, manufacturing iron, service industries and the commodities: sheep, shorn sheep, mutton, wool, iron, services. The connections are freely created. The Sraffa Network of this economy is given. We seek the adjacency matrix **W**.

Solution to Example 6.4.1:
Limiting to mutton and wool, one sees that in a first approach the number of processes is n and the number of commodities is $m = n+1$. If one seeks to have $m = n$, one can define mutton and wool as a composite commodity, but this brings an artificial constraint into the model. The correct approach is to analyse the production process in more detail, and this entails distinguishing sheep and shorn sheep and introducing service industries providing services. Service industries will require mutton, wool, iron for their labour forces, in other words: food, clothing, manufactured goods.

$$\mathbf{W} = \begin{array}{c} \\ 1 \\ 2 \\ 3 \\ 4 \\ 5 \\ 6 \\ \mathrm{S} \\ \mathrm{SS} \\ \mathrm{M} \\ \mathrm{W} \\ \mathrm{I} \\ \mathrm{SV} \end{array} \begin{array}{c} \begin{array}{cccccc|cccccc} 1 & 2 & 3 & 4 & 5 & 6 & \mathrm{S} & \mathrm{SS} & \mathrm{M} & \mathrm{W} & \mathrm{I} & \mathrm{SV} \end{array} \\ \left[\begin{array}{cccccc|cccccc} 0 & 0 & 0 & 0 & 0 & 0 & 1 & 0 & 0 & 0 & 0 & 0 \\ 0 & 0 & 0 & 0 & 0 & 0 & 0 & 1 & 0 & 1 & 0 & 0 \\ 0 & 0 & 0 & 0 & 0 & 0 & 0 & 0 & 1 & 0 & 0 & 0 \\ 0 & 0 & 0 & 0 & 0 & 0 & 0 & 0 & 0 & 1 & 0 & 0 \\ 0 & 0 & 0 & 0 & 0 & 0 & 0 & 0 & 0 & 0 & 1 & 0 \\ 0 & 0 & 0 & 0 & 0 & 0 & 0 & 0 & 0 & 0 & 0 & 1 \\ \hline 1 & 1 & 0 & 0 & 0 & 0 & 0 & 0 & 0 & 0 & 0 & 0 \\ 1 & 0 & 1 & 0 & 0 & 0 & 0 & 0 & 0 & 0 & 0 & 0 \\ 0 & 0 & 0 & 0 & 0 & 1 & 0 & 0 & 0 & 0 & 0 & 0 \\ 0 & 0 & 0 & 1 & 0 & 1 & 0 & 0 & 0 & 0 & 0 & 0 \\ 0 & 0 & 1 & 1 & 1 & 1 & 0 & 0 & 0 & 0 & 0 & 0 \\ 1 & 1 & 1 & 1 & 1 & 0 & 0 & 0 & 0 & 0 & 0 & 0 \end{array}\right] \\ \mathbf{V} \end{array} \begin{array}{c} \mathbf{Q}' \\ \\ \\ \\ \\ \\ \\ \\ \\ \\ \\ \end{array} \tag{6.121}$$

The corresponding *Sraffa Network* is presented in Figure 6.4. One of the industrial nodes, namely 2 shearing, has an outdegree $y^{+}(2) > 1$. This digraph, Figure 6.4, therefore represents a *joint production economy* (Lemma A.14.1) with adjacency matrix (6.121). The submatrix **Q** is no longer diagonal (6.121). ▲

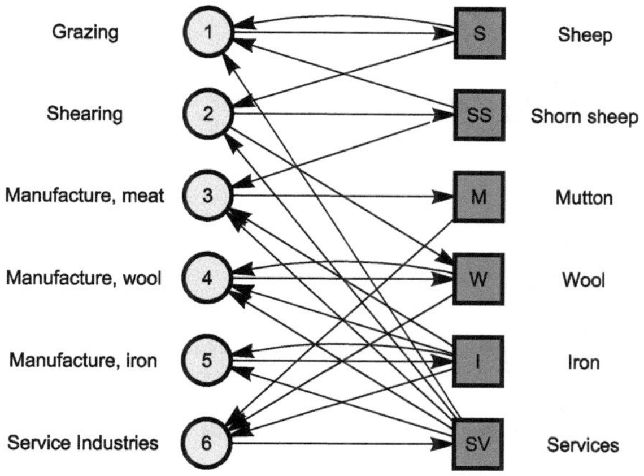

Figure 6.4: Joint production, Sraffa ([108], Chap. VI–IX,XI).

Example 6.4.2. Consider the following Sraffa Network, assumed reasonably realistic for illustrative purposes, that is embedded in a larger network. This modest example already shows the complexities of modern economies of production and exchange.

Dashed arrows indicate incoming flows from or outgoing flows of commodities to the production entities of the larger network, designated hereafter as "other".[15]

The reader should note the clear separation and the anti-symmetry between the two categories of commodities in a production economy:

- **Category 1: basic commodities**

The *input* to basic industries is constituted exclusively of *basic* commodities and *basic* commodities are composed of *basic* commodities only.

The *interindustry* output of basic industries is composed of basic and non-basic commodities.

- **Category 2: non-basic commodities**

The input to non-basic industries can include basic and *non-basic* commodities. *Non-basic* commodities can thus be composed of *basic* and/or *non-basic* commodities.

The *interindustry output* of *non-basic* industries however can only be composed of *non-basic* commodities.

Solution to Example 6.4.2:
The following list indicates the meaning of the various production entities and commodities entering the network.

15 For example, outgoing water W in the network is again tapped and will in part reappear as input for cooling in the nuclear power production ⑥.

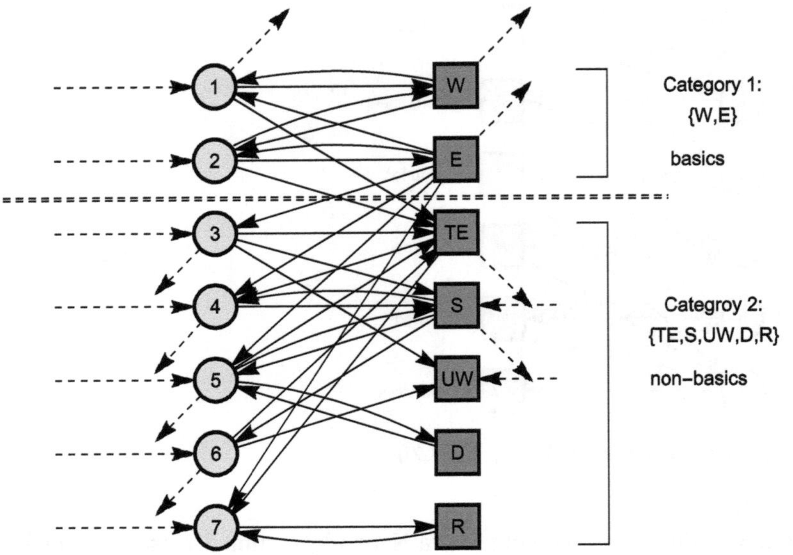

Figure 6.5: Sraffa Network of *basics* and *non-basics*.

Round nodes:

① Water collection
② Hydroelectricity production
③ Transport facilities | all activities assumed here
④ Telecommunication facilities | restricted to the non-basic subsystem
⑤ Data-processing facilities | under consideration
⑥ Nuclear power production
⑦ Horse breeding and racing

Square nodes:

[W] Water
[E] Electricity
[TE] Thermal Energy
[S] Specialised services
[UW] Unrecycled waste
[D] Stored data
[R] Race horses[16]

16 We have taken Sraffa's exotic example, PCMC, Par. 6. Note that horse racing is a cultural and speculative activity. At our present time, we could just as well replace horse breeding and trading by financial markets and race horses by speculative financial products, both clearly *non-basic* by nature and in many circumstances detrimental to sustainable economic growth.

It has to be noted that the present Sraffa Network contains *single-product industries* and *joint production industries*. Schefold's Definition 6.3.2 is also used in such cases to distinguish *basic* from *non-basic* commodities because such examples as a whole are cases of *joint production*.

Referring to what Sraffa (PCMC, Par. 35 and Par. 58) had in mind, the non- basic commodities of this example illustrate:

(i) UW, Non-basics excluded from the means of production

(ii) TE, S, D, Non-basics involved directly or indirectly in the production of one or more other non-basics

(iii) R, Non-basics involved only directly in their own production

The production processes (incoming/outgoing arrows) associated with each industry are as follows:

① Water Collection: Water + Electricity + Other → Water + Thermal Energy + Other

② Hydroelectricity production: Water + Electricity + Other → Electricity + Thermal energy

③ Transport facilities XXX: Electricity + Other → Specialised Services + Thermal Energy + Other + Waste

④ Telecommunication facilities YYY: Electricity + Thermal Energy + Specialised Services + Other → Specialised Services + Other

⑤ Data-processing facilities ZZZ: Electricity + Specialised Services + Stored Data + Other → Specialised Services + Data Storage + Thermal Energy + Other

⑥ Nuclear power production: Specialised Services + Other → Thermal Energy + Other + Waste

⑦ Horse breeding: Racehorses + Electricity + Thermal energy + Other → Racehorses

We invite the reader to devise corresponding approximate real world scenarios.

The reader must be aware that the distinction between *basic* and *non-basic* commodities is not clear-cut. This is a source of debate in connection with sustainable economies.

Indeed, assume a Sraffa model incorporating electric power as the energy provider entering directly or indirectly all industries and therefore corresponding to a basic commodity.

Now if this model of a production economy is refined to more detail, i. e., is expanded to include more commodities than the original model, electric power being then earmarked according to the method of production used:
- nuclear generated electricity;
- hydroelectric power;

- solar generated electricity;
- wind generated electricity;

these more detailed commodities are *basic* or *non-basic*, depending on their utilisation. This property is determined by application of the fundamental definitions presented in the case of *single-product* industries in Section 4.4 and in the case of *joint production* in Section 6.3.

Furthermore, commodities entering a subsystem G' embedded in a global system of production G may be *non-basic* in G but *basic* in the restricted system G' taken separately.

Accordingly, because the distinction between basic commodities and non-basic commodities is conditional, it will be model-dependent.

This fact may be seen by considering the *sub-digraph* G', spanned by the nodes ④, ⑤, [S], [D] of *digraph* G, Figure 6.6, $G' \preceq G$, *non-basics* in G are *basic* in G'. The *sub-digraph* G' is composed only of *basics* because G' is here indeed clearly strongly connected (Lemma A.15.3).

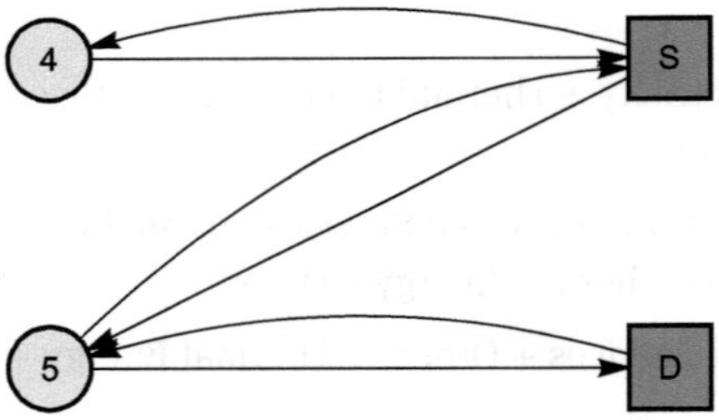

Figure 6.6: Fully connected *sub-digraph* G' of the *Sraffa Network* G, Figure 6.5.

The *sub-digraph* G', $G' \preceq G$, of the *Sraffa Network* G is here clearly strongly connected (Definition A.14.9), and [S], [D] are *basics*:

Finally, one should bear in mind that commodities may be composite commodities, like machinery, which enter into *joint production*, see Chapter 6 and that *non-basics* may with time often become *basic* commodities. For example, it appears fairly certain that with time portable smart phones, lap-tops and robots will become indispensable basic commodities in all sectors of production. ▲

Non-basic commodities furthermore call for the following comments:

(1) Assuming that all *non-basics* have been identified and earmarked, the dashed incoming arrows illustrated in Figure 6.5 are necessarily carriers of mixes of basic products entering the subsystems depicted. All commodities of the *non-basic* subsystems must then require linear combinations of quantities of *basics* for their

production, and this explains PCMC (Par. 60) where the concept of *non-basics* is outlined.

(2) If the *non-basics* are totally separated from a given set of *basics* (no incoming dashed arrows), they fall out of the linear combinations, as their factors are equal to 0. They then form a separate process.

(3) In this latter case, the price of *non-basics* has of course no impact on the prices of *basic* commodities of the former process. But this is not the case for all *non-basics*, as the discussion of Example 4.4.3 has shown.

6.5 National accounting and joint production

In the next example, variables used in the context of national economic accounts are calculated.

Example 6.5.1. Consider three industries 1, 2, 3 and three groups of commodities: agricultural products A, services B and finally housing accommodation C. These industries also produce a surplus which corresponds to national income Y.

Industry 1 produces primarily housing as output and also generates in the process agricultural by-products and residual services, This industry requires agricultural products, services and housing as inputs.

Industry 2 produces mainly agricultural products and provides certain services, requiring as inputs a certain amount of agricultural products, specific services and housing.

Industry 3 is a pure service industry requiring here only certain specific services to operate.

Finally in the economic process national income is sustained here through the surplus generated by agricultural products and housing facilities.

Consider 3 cases:

(a) National income Y is allotted exclusively to profits.

(b) National income Y is allotted completely to wages for the labour forces attached to the industries. For this we assume that all surplus goes to labour, so $r = 0$, and treat the special case, where there is no profit.[17]

(c) Assume a rate of profits $r = 0.2$, so the surplus then goes to profits and wages.

Now we assume that

---

17 Remark: We are in presence of the *joint production Sraffa system* $(\mathbf{S}', \mathbf{F}', \mathbf{L})$ where all the profits goes into labour. Schefold ([103], p. 75) computes the vector of prices $\mathbf{p} = (\mathbf{F}' - \mathbf{S}')^{-1} w \cdot \mathbf{L}$, using the *vector of labour* $\mathbf{L}$, which is here positive, and terms the vector $\mathbf{u} = \mathbf{p}/w$ the *vector of labour values* (see further also (7.5)), which is therefore necessarily positive for a *semi-positve* inverse matrix $(\mathbf{F}' - \mathbf{S}')^{-1}$.

Industry 1 proceeds with 300 units of commodity A plus 120 units of commodity B plus 160 units of commodity C and $\frac{3}{16}$ TAL (total amount of labour), resulting in 260 units of commodity A plus 20 units of commodity B plus 450 units of commodity C.

Industry 2 proceeds with 110 units of commodity A plus 40 units of commodity B plus 125 units of commodity C and $\frac{5}{16}$ TAL, resulting in 220 units of commodity A plus 100 units of product B.

Industry 3 proceeds with 20 units of commodity B plus $\frac{8}{16}$ TAL, resulting in 60 units of commodity B, see Figure 6.7.

Commodity C operates as numéraire.

The underlying production process reads as follows:

$$(\mathbf{S}', \mathbf{L}) \rightarrow (\mathbf{F}')$$

$$\begin{aligned}
&\left(300A, \quad 120B, \quad 160C, \quad \frac{3}{16}w\right) \rightarrow (260A, \quad 20B, \quad 450C),\\
&\left(110A, \quad 40B, \quad 125C, \quad \frac{5}{16}w\right) \rightarrow (220A, \quad 100B, \quad 0),\\
&\left(0, \quad 20B, \quad 0, \quad \frac{8}{16}w\right) \rightarrow (0, \quad 60B, \quad 0).
\end{aligned} \tag{6.122}$$

Identify the *commodity flow* matrix **S**, the *output* matrix **F** and the vector of *labour* **L**, defining the system of production $(\mathbf{S}', \mathbf{F}', \mathbf{L})$, using Notation 6.2.1. Establish the *Sraffa Network* and the corresponding *adjacency matrix* **W**.

We follow the scheme of the three last steps considered in Section 4.9,

(a) **Step II** with surplus going exclusively into profits: Verify that matrix **F** is regular. Calculate the *input-output coefficients* matrix $\mathbf{C}_T$, the *productiveness* R and the corresponding price vector **p**;

(b) **Step III** with surplus going only into labour: Calculate the price vector, here called "labour values";

(c) **Step IV** with surplus going into profits and wages: Consider the rate of profits $r = 0.2$. Calculate the price vector, the national income Y, the total output X, the circulating capital K, the total profit P, the total wages W and the share of total wages to national income $\tilde{w}$ (6.49).

Solution to Example 6.5.1:

We start identifying the *commodity flow* matrix **S**, describing the technology, and the *output matrix* **F** of this *joint production economy*, giving the result of the production during the considered period, and also the labour vector **L**,

$$\mathbf{S} = \begin{bmatrix} 300 & 110 & 0 \\ 120 & 40 & 20 \\ 160 & 125 & 0 \end{bmatrix}, \quad \mathbf{F} = \begin{bmatrix} 260 & 220 & 0 \\ 20 & 100 & 60 \\ 450 & 0 & 0 \end{bmatrix}, \quad \mathbf{L} = \begin{bmatrix} \frac{3}{16} \\ \frac{5}{16} \\ \frac{8}{16} \end{bmatrix}. \tag{6.123}$$

We set up the square $(2n \times 2n)$ matrix (6.116),

$$\mathbf{\Sigma} = \begin{bmatrix} \mathbf{0} & \mathbf{F}' \\ \mathbf{S} & \mathbf{0} \end{bmatrix} = \left[\begin{array}{ccc|ccc} 0 & 0 & 0 & 260 & 20 & 450 \\ 0 & 0 & 0 & 220 & 100 & 0 \\ 0 & 0 & 0 & 0 & 60 & 0 \\ \hline 300 & 110 & 0 & 0 & 0 & 0 \\ 120 & 40 & 20 & 0 & 0 & 0 \\ 160 & 125 & 0 & 0 & 0 & 0 \end{array}\right]; \quad k, l = 1, \ldots, 2n. \qquad (6.124)$$

We can now set up the corresponding Sraffa Network and the adjacency matrix $\mathbf{W}$ (6.118) on the basis of the definition (6.116) of matrix $\mathbf{\Sigma}$,

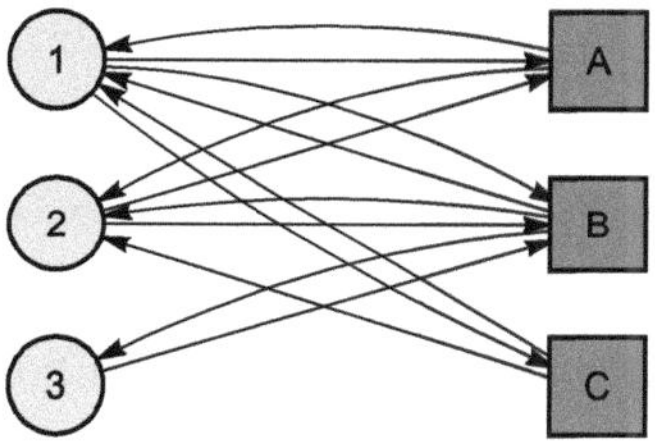

$$\mathbf{W} = \begin{array}{c} \\ 1 \\ 2 \\ 3 \\ A \\ B \\ C \end{array} \begin{array}{c} \begin{array}{ccc|ccc} 1 & 2 & 3 & A & B & C \end{array} \\ \left[\begin{array}{ccc|ccc} 0 & 0 & 0 & 1 & 1 & 1 \\ 0 & 0 & 0 & 1 & 1 & 0 \\ 0 & 0 & 0 & 0 & 1 & 0 \\ \hline 1 & 1 & 0 & 0 & 0 & 0 \\ 1 & 1 & 1 & 0 & 0 & 0 \\ 1 & 1 & 0 & 0 & 0 & 0 \end{array}\right] \end{array}, \qquad (6.125)$$

Figure 6.7: Sraffa Network and adjacency matrix of Example 6.5.1.

calculating,

$$\mathbf{We} = \begin{bmatrix} 0 & 0 & 0 & 1 & 1 & 1 \\ 0 & 0 & 0 & 1 & 1 & 0 \\ 0 & 0 & 0 & 0 & 1 & 0 \\ 1 & 1 & 0 & 0 & 0 & 0 \\ 1 & 1 & 1 & 0 & 0 & 0 \\ 1 & 1 & 0 & 0 & 0 & 0 \end{bmatrix} \begin{bmatrix} 1 \\ 1 \\ 1 \\ 1 \\ 1 \\ 1 \end{bmatrix} = \begin{bmatrix} 3 \\ 2 \\ 1 \\ 2 \\ 3 \\ 2 \end{bmatrix} = \begin{bmatrix} y^+(1) \\ y^+(2) \\ y^+(3) \\ y^+(A) \\ y^+(B) \\ y^+(C) \end{bmatrix}, \qquad (6.126)$$

one obtains exactly the outdegrees of the nodes of the *Sraffa Network*.

The labour vector $\mathbf{L} = [3/16, 5/16, 8/16]'$ does not figure in the *Sraffa Network* of the *joint production process* $(\mathbf{S}', \mathbf{F}', \mathbf{L})$, Definition 6.3.2, because labour is directly attached to each industry node.

Now we are ready to follow the three last steps of the proposed methodology to calculate the *joint production Sraffa system* with different *profit rates* r and *wage rates* w.

Starting from the structure of the *joint production Sraffa system* (6.48),

$$\boxed{\begin{aligned} \mathbf{S}'\mathbf{p}(1+r)+\mathbf{L}\frac{\tilde{w}\cdot Y}{L} &= \mathbf{F}'\mathbf{p},\\ Y &= (\mathbf{Fe}-\mathbf{Se})'\mathbf{p},\\ L &= \mathbf{e}'\mathbf{L},\\ \text{setting } p_C &= 1, \end{aligned}} \tag{6.127}$$

clearly, the labour is normalised, $L = \mathbf{e}'\mathbf{L} = \frac{3}{16} + \frac{5}{16} + \frac{8}{16} = 1$. Sraffa's conditions of joint production $\mathbf{Se} = \mathbf{Fe}$ (6.19) are exceeded,

$$\begin{aligned} \mathbf{Se} &= \begin{bmatrix} 300 & 110 & 0 \\ 120 & 40 & 20 \\ 160 & 125 & 0 \end{bmatrix}\begin{bmatrix} 1 \\ 1 \\ 1 \end{bmatrix} = \begin{bmatrix} 410 \\ 180 \\ 285 \end{bmatrix} \\ \leq \mathbf{Fe} &= \begin{bmatrix} 260 & 220 & 0 \\ 20 & 100 & 60 \\ 450 & 0 & 0 \end{bmatrix}\begin{bmatrix} 1 \\ 1 \\ 1 \end{bmatrix} = \begin{bmatrix} 480 \\ 180 \\ 450 \end{bmatrix}, \end{aligned} \tag{6.128}$$

indicating the existence of a *semi-positive* surplus $\mathbf{d} = (\mathbf{F}-\mathbf{S})\mathbf{e} = [70, 0, 165]' \geq 0$. Therefore we directly apply to case (a)

(Step II) of the method presented in Section 4.9 (all the surplus goes into profits), the *share of total wages to national income* reduces to $\tilde{w} = 0$.

We verify $\det(\mathbf{F}) = 5{,}940{,}000$ and then start computing the inverse matrix of $\mathbf{F}$ with (6.123), getting,

$$\mathbf{F}^{-1} = \begin{bmatrix} 260 & 220 & 0 \\ 20 & 100 & 60 \\ 450 & 0 & 0 \end{bmatrix}^{-1} = \begin{bmatrix} 0 & 0 & \frac{1}{450} \\ \frac{1}{220} & 0 & -\frac{13}{4{,}950} \\ -\frac{1}{132} & \frac{1}{60} & \frac{1}{275} \end{bmatrix}. \tag{6.129}$$

Then we calculate the new *input-output coefficients* matrix $\mathbf{C}_T = \mathbf{SF}^{-1}$, using the matrix $\mathbf{S}$ (6.123) with (6.129) and see that we obtain a *semi-positive* matrix,

$$\begin{aligned} \mathbf{C}_T &= \mathbf{SF}^{-1} \\ &= \begin{bmatrix} 300 & 110 & 0 \\ 120 & 40 & 20 \\ 160 & 125 & 0 \end{bmatrix}\begin{bmatrix} 0 & 0 & \frac{1}{450} \\ \frac{1}{220} & 0 & -\frac{13}{4{,}950} \\ -\frac{1}{132} & \frac{1}{60} & \frac{1}{275} \end{bmatrix} = \begin{bmatrix} \frac{1}{2} & 0 & \frac{17}{45} \\ \frac{1}{33} & \frac{1}{3} & \frac{116}{495} \\ \frac{25}{44} & 0 & \frac{3}{110} \end{bmatrix} \geq \mathbf{0}. \end{aligned} \tag{6.130}$$

The *joint production Sraffa system* (6.127) becomes:

$$\mathbf{S}'\mathbf{p}(1+R) = \mathbf{F}'\mathbf{p} \Rightarrow \mathbf{C}_T'\mathbf{p}(1+R) = \mathbf{p}. \tag{6.131}$$

We multiply the above eigenvalue equation (6.131) with $\lambda_C = \frac{1}{1+R}$ and get for the price vector,

$$\mathbf{C}_T'\mathbf{p} = \lambda_C\mathbf{p}. \tag{6.132}$$

At present, consider the characteristic polynomial of matrix $\mathbf{C}_T$,

$$\begin{aligned} P_3(\lambda) &= \det(\mathbf{C}_T - \lambda\mathbf{I}_3) = -\lambda^3 + \frac{142}{165}\lambda^2 + \frac{5}{198}\lambda - \frac{199}{2{,}970} \\ &= (\lambda - 0.33333)(\lambda + 0.256473)(\lambda - 0.783746). \end{aligned} \tag{6.133}$$

We take matrix $\mathbf{C_T}$ and compute,

$$(\mathbf{I} + \mathbf{C}_T)^2 = \begin{bmatrix} \frac{244}{99} & 0 & \frac{2{,}363}{2{,}475} \\ \frac{53}{242} & \frac{16}{9} & \frac{46{,}117}{81{,}675} \\ \frac{695}{484} & 0 & \frac{34{,}574}{27{,}225} \end{bmatrix} \geq 0, \tag{6.134}$$

and conclude with Lemma A.8.2 that $\mathbf{C}_T \geq \mathbf{0}$ is *reducible*. Lemma 4.1.1 (b) applies. We get a positive Frobenius number less than one, $1 > \lambda_C = 1/(1 + R) = 0.7837 > 0$, $1 > \lambda_C > 0$, revealing the *productiveness* $R = 27.59\,\% > 0$. Theorem A.10.1, guarantees *non-negative* eigenvectors associated to the Frobenius number λ_C.

In fact, in this case, the *right eigenvector* equation (6.132) gives directly positive eigenvectors $\mathbf{p} = k[1.332, 0.610, 1]' > 0$. Finally, we compute the national income, using (6.128) and the fact that commodity C has been chosen as numéraire, $p_C = 1$, thus, we have $k = 1$, so:

$$Y = (\mathbf{Fe} - \mathbf{Se})' \cdot \mathbf{p} = [70, 0, 165] \begin{bmatrix} 1.332 \\ 0.610 \\ 1 \end{bmatrix} = 258.198, \tag{6.135}$$

presented in units of commodity C.

Now we proceed to handle case (b)

(Step III) (Sraffa's *price equations with no profit*: Labour values)

We consider Sraffa's price equation with *normalised* labour and no profit, $r = 0$ and $P = 0$. As mentioned, the resulting prices are often called *labour values* (see Schefold [103], p. 75). We get the system

$$\boxed{\begin{aligned} \mathbf{S}'\mathbf{p} + \mathbf{L}\frac{\tilde{w} \cdot Y}{L} &= \mathbf{F}'\mathbf{p}, \\ (\mathbf{Fe} - \mathbf{Se})'\mathbf{p} &= Y, \\ L = \mathbf{e}'\mathbf{L} &= 1, \\ p_C &= 1, \end{aligned}} \tag{6.136}$$

then, computing the *wage rate* $w = \frac{\tilde{w} \cdot Y}{L}$, we take the first matrix equation of (6.136) and modify it algebraically,

$$\mathbf{S}'\mathbf{p} + \mathbf{L}w = \mathbf{F}'\mathbf{p} \Rightarrow \mathbf{S}'\mathbf{p} - \mathbf{F}'\mathbf{p} + \mathbf{L}w = \mathbf{0}. \tag{6.137}$$

So we get finally,

$$\mathbf{p} = w \cdot [\mathbf{F}' - \mathbf{S}']^{-1}\mathbf{L}, \tag{6.138}$$

and numerically for the inverse,

$$(\mathbf{F}' - \mathbf{S}')^{-1} = \begin{bmatrix} -40 & -100 & 290 \\ 110 & 60 & -125 \\ 0 & 40 & 0 \end{bmatrix}^{-1} = \begin{bmatrix} \frac{5}{1{,}076} & \frac{29}{2{,}690} & -\frac{49}{10{,}760} \\ 0 & 0 & \frac{1}{40} \\ \frac{11}{2{,}690} & \frac{2}{1{,}345} & \frac{43}{5{,}380} \end{bmatrix}. \tag{6.139}$$

We now calculate the vector $\mathbf{p}' = [p_A, p_B, p_C]$, and obtain from (6.138) and (6.139)

$$\begin{aligned} \mathbf{p} &= w \cdot \begin{bmatrix} \frac{5}{1{,}076} & \frac{29}{2{,}690} & -\frac{49}{10{,}760} \\ 0 & 0 & \frac{1}{40} \\ \frac{11}{2{,}690} & \frac{2}{1{,}345} & \frac{43}{5{,}380} \end{bmatrix} \begin{bmatrix} \frac{3}{16} \\ \frac{5}{16} \\ \frac{8}{16} \end{bmatrix} \\ &= w \cdot \begin{bmatrix} \frac{169}{86{,}080} \\ \frac{1}{80} \\ \frac{45}{8{,}608} \end{bmatrix} = w \cdot \mathbf{u} = \begin{bmatrix} p_A \\ p_B \\ 1 \end{bmatrix} = \begin{bmatrix} 0.376 \\ 2.391 \\ 1 \end{bmatrix}, \end{aligned} \tag{6.140}$$

due to the fact that commodity C is the numéraire, we set $p_C = 1 = w \cdot \frac{45}{8{,}608}$ in (6.140). This gives the wage rate $w = 191.289$. We compute with (6.140) $p_A = 0.376$ an $p_B = 2.391$ and get the national income with (6.128)

$$Y = (\mathbf{Fe} - \mathbf{Se})' \cdot \mathbf{p} = [70, 0, 165] \begin{bmatrix} 0.376 \\ 2.391 \\ 1 \end{bmatrix} = 191.289, \tag{6.141}$$

presented in units of commodity C. Clearly, $L = 1$, the total wages are $W = w \cdot L = w$ and $Y = W + P = W$, therefore $Y = W = w = 191.289$ and we also obtain the vector of labour values $\mathbf{u} = [\frac{169}{86{,}080}, \frac{1}{80}, \frac{45}{8{,}608}]' = [0.00196, 0.01250, 0.00522]'$ in units of labour per unit of produced commodities, each component representing the total "quantity of labour" necessary over all preceding periods for the production of the units of each of the commodities.

Finally, we treat case (c), (**Step IV**), and establish the joint production Sraffa system with a *profit rate* $r = 0.2$ and the labour vector $\mathbf{L} = [\frac{3}{16}, \frac{5}{16}, \frac{8}{16}]'$ and the same numéraire C, $p_C = 1$,

$$\boxed{\begin{aligned} \mathbf{S}'\mathbf{p}(1 + 0.2) + \mathbf{L}\frac{\tilde{w} \cdot Y}{L} &= \mathbf{F}'\mathbf{p}, \\ (Y &= \mathbf{Fe} - \mathbf{Se})'\mathbf{p}, \\ L &= \mathbf{e}'\mathbf{L} = 1, \\ p_C &= 1. \end{aligned}} \tag{6.142}$$

We have of course to compute $w = \frac{\tilde{w} \cdot Y}{L}$.

(A) If matrix **F** is regular, we can multiply the first equation of (6.142) by the matrix $\mathbf{F}'^{-1}$ from the left, and we get a system of equations for the price vector **p**,

$$\mathbf{S}'\mathbf{p}(1+r)+\mathbf{L}w=\mathbf{F}'\mathbf{p}\Rightarrow \mathbf{F}'^{-1}\mathbf{S}'\mathbf{p}(1+r)+w\cdot\mathbf{F}'^{-1}\mathbf{L}=\mathbf{F}'^{-1}\mathbf{F}'\mathbf{p}=\mathbf{p}. \tag{6.143}$$

We have now the joint production Sraffa system, using the *input-output coefficients* matrix $\mathbf{C}_T=\mathbf{SF}^{-1}$ and the labour parts $w\cdot(\mathbf{F}'^{-1}\mathbf{L})=:w\cdot\boldsymbol{\Lambda}$. So, we have an equation for the prices p_i, $i\in\{A,B,C\}$ of each commodity i:

$$\mathbf{C}_T'\mathbf{p}(1+r)+w\cdot\boldsymbol{\Lambda}=\mathbf{p}. \tag{6.144}$$

We know the numerical expression of matrix $\mathbf{C}_T$ (6.130), and we now also need the numerics of the product

$$\boldsymbol{\Lambda}:=\mathbf{F}'^{-1}\mathbf{L}=\begin{bmatrix}0 & \frac{1}{220} & -\frac{1}{132}\\ 0 & 0 & \frac{1}{60}\\ \frac{1}{450} & -\frac{13}{4{,}950} & \frac{1}{275}\end{bmatrix}\begin{bmatrix}\frac{3}{16}\\ \frac{5}{16}\\ \frac{8}{16}\end{bmatrix}=\begin{bmatrix}-\frac{5}{2{,}112}\\ \frac{1}{120}\\ \frac{7}{4{,}950}\end{bmatrix}. \tag{6.145}$$

We thus obtain from (6.144) the price vector $\mathbf{p}=[1.422,0.842,1]'$ expressed in the *numéraire C*. We observe that the vector $\boldsymbol{\Lambda}$ contains a negative value. This is an artefact coming from the inverse matrix $\mathbf{F}'^{-1}$, which is not *non-negative*. We observe that in *joint production analysis* the vector $\boldsymbol{\Lambda}$ needs an extended interpretation of the term "quantity of labour per unit of commodity", which can be negative. This interpretation needs further work not treated in this book.

(B) If matrix $\mathbf{F}$ is not regular, what is not the case here, we can take the price model with the original *commodity flow* matrix and the matrix of *outputs*. We have in this case

$$\mathbf{S}'\mathbf{p}(1+0.2)+w\cdot\mathbf{L}=\mathbf{F}'\mathbf{p}\Rightarrow$$

$$\begin{bmatrix}300 & 120 & 160\\ 110 & 40 & 125\\ 0 & 20 & 0\end{bmatrix}\begin{bmatrix}p_1\\ p_2\\ 1\end{bmatrix}1.2+w\begin{bmatrix}\frac{3}{16}\\ \frac{5}{16}\\ \frac{8}{16}\end{bmatrix}=\begin{bmatrix}260 & 20 & 450\\ 220 & 100 & 0\\ 0 & 60 & 0\end{bmatrix}\begin{bmatrix}p_1\\ p_2\\ 1\end{bmatrix}. \tag{6.146}$$

We then obtain, from this system (6.146) of unknown variables $p_1, p_2\; w$, again the price vector $\mathbf{p}=[1.422,0.842,1]'$, expressed in the *numéraire* C, and we observe the *wage per unit of labour* $w=60.622$ has changed with respect to **Step III.**

Finally, we compute the national income, using (6.128) and the requested economic variables, according to (6.49).

$$Y=(\mathbf{Fe}-\mathbf{Se})'\cdot\mathbf{p}=[70,0,165]\begin{bmatrix}1.422\\ 0.842\\ 1\end{bmatrix}=264.561,$$

$$X=(\mathbf{Fe})'\mathbf{p}=[480,180,450]\begin{bmatrix}1.422\\ 0.842\\ 1\end{bmatrix}=1{,}284.26,$$

$$K = (\mathbf{Se})'\mathbf{p} = [410, 180, 285] \begin{bmatrix} 1.422 \\ 0.842 \\ 1 \end{bmatrix} = 1{,}019.7, \tag{6.147}$$

$$\begin{aligned} P &= (\mathbf{Se})'\mathbf{p} \cdot r = K \cdot r = 1{,}284.26 \cdot 0.2 = 203.939, \\ \tilde{w} &= w \cdot \frac{L}{Y} = 60.622 \cdot \frac{1}{264.561} = 0.2291, \\ W &= Y - P = 60.622 = w \cdot L = 60.622 \cdot 1 = 60.622, \end{aligned} \tag{6.148}$$

presented in units of commodity C. We can now easily derive the *accountable balance*, multiplying the complete Sraffa price model (6.146) for joint production by vector $\mathbf{e}'$

$$\begin{aligned} \mathbf{e}'\mathbf{S}'\mathbf{p}(1+r) + \mathbf{e}'w \cdot \mathbf{L} = \mathbf{e}'\mathbf{F}'\mathbf{p} &= (\mathbf{Se})'\mathbf{p}(1+r) + w \cdot (\mathbf{e}'\mathbf{L}) = (\mathbf{Fe})'\mathbf{p} \\ &= K + P + w \cdot L = K + (P + W) = K + Y = X. \end{aligned} \tag{6.149}$$

We have indeed the principal *accountable identity* of the *National accounting system*, which even can be brought to $X = K + P + W + (E - M)$ with $Y = P + W + (E - M)$, if Sraffa's model is extended. ▲

Note that in *joint production* the conditions for the application of the **Perron–Frobenius theorem A.9.3** are in general not fulfilled, as in **Example 6.5.1**, even if by chance all the price vectors here are positive.

So again, the question of the conditions required to obtain positive price vectors, $\mathbf{p} > 0$, has to be treated. This will be undertaken in Section 6.6.

We continue to apply Sraffa Networks to *joint production* processes and see how they present a useful complement to the algebraic treatment of such processes. As a further example of a Sraffa production system $(\mathbf{S}', \mathbf{F}', \mathbf{L})$, Definition 6.3.2, let us modify the system of production handled in Example 6.5.1:

Example 6.5.2. Consider the production scheme

$$\begin{gathered} (\mathbf{S}', \mathbf{L}) \rightarrow (\mathbf{F}') \\ \left(300\text{A}, 120\text{B}, 160\text{C}, \frac{3}{16}w\right) \rightarrow (260\text{A}, 20\text{B}, 450\text{C}), \\ \left(110\text{A}, 40\text{B}, 125\text{C}, \frac{5}{16}w\right) \rightarrow (220\text{A}, 100\text{B}, 0), \\ \left(20\text{A}, 20B, 0, \frac{8}{16}w\right) \rightarrow (0, 60\text{B}, 0), \end{gathered} \tag{6.150}$$

derived from Example 6.5.1, where the production process of Industry 3 has been modified. Identify the *commodity flow* matrix **S** and the *output* matrix **F**. Determine the Sraffa Network, the $2n \times 2n$ adjacency matrix **W**. Determine the basic products and the non-basic products.

Solution to Example 6.5.2:

We start identifying the *commodity flow* matrix **S**, describing the technology, and the *output matrix* $\mathbf{F}'$ of this *joint production economy*, giving the result of the production during the considered period,

$$\mathbf{S} = \begin{bmatrix} 300 & 110 & 20 \\ 120 & 40 & 20 \\ 160 & 125 & 0 \end{bmatrix}, \quad \mathbf{F}' = \begin{bmatrix} 260 & 20 & 450 \\ 220 & 100 & 0 \\ 0 & 60 & 0 \end{bmatrix}, \tag{6.151}$$

We set up the square $(2n \times 2n)$ matrix (6.116),

$$\boldsymbol{\Sigma} = \begin{bmatrix} \mathbf{0} & \mathbf{F}' \\ \mathbf{S} & \mathbf{0} \end{bmatrix} = \left[\begin{array}{ccc|ccc} 0 & 0 & 0 & 260 & 20 & 450 \\ 0 & 0 & 0 & 220 & 100 & 0 \\ 0 & 0 & 0 & 0 & 60 & 0 \\ \hline 300 & 110 & 20 & 0 & 0 & 0 \\ 120 & 40 & 20 & 0 & 0 & 0 \\ 160 & 125 & 0 & 0 & 0 & 0 \end{array}\right]; \quad k, l = 1, \ldots, 2n. \tag{6.152}$$

We can now get the adjacency matrix $\mathbf{W}$ (6.118) on the basis of the definition (6.117) of its coefficients. The *Sraffa Network*, Figure 6.8, corresponding to the adjacency matrix follows immediately,

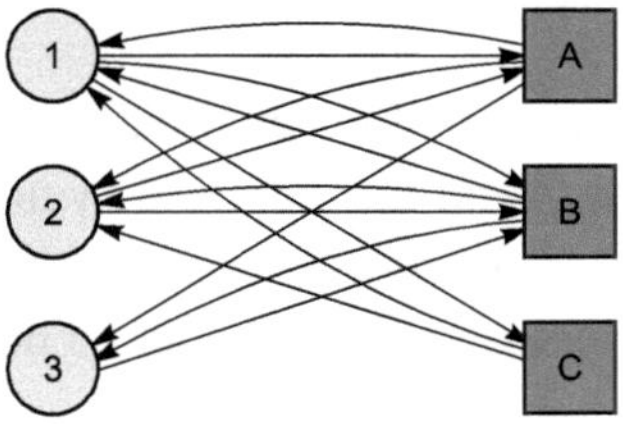

$$\mathbf{W} = \begin{array}{c} \\ 1 \\ 2 \\ 3 \\ A \\ B \\ C \end{array} \begin{array}{c} \begin{array}{ccc|ccc} 1 & 2 & 3 & A & B & C \end{array} \\ \left[\begin{array}{ccc|ccc} 0 & 0 & 0 & 1 & 1 & 1 \\ 0 & 0 & 0 & 1 & 1 & 0 \\ 0 & 0 & 0 & 0 & 1 & 0 \\ \hline 1 & 1 & 1 & 0 & 0 & 0 \\ 1 & 1 & 1 & 0 & 0 & 0 \\ 1 & 1 & 0 & 0 & 0 & 0 \end{array}\right] \end{array}. \tag{6.153}$$

Figure 6.8: Sraffa Network and adjacency matrix of Example 6.5.2.

We continue analysing the *irreducibility* of matrices $\mathbf{S}$ and $\mathbf{F}$,

$$(\mathbf{S} + \mathbf{I})^2 = \begin{bmatrix} 301 & 110 & 20 \\ 120 & 41 & 20 \\ 160 & 125 & 1 \end{bmatrix}^2 = \begin{bmatrix} 107{,}001 & 40{,}120 & 8{,}240 \\ 44{,}240 & 17{,}381 & 3{,}240 \\ 63{,}320 & 22{,}850 & 5{,}701 \end{bmatrix} > 0, \tag{6.154}$$

$$(\mathbf{F} + \mathbf{I})^2 = \begin{bmatrix} 261 & 220 & 0 \\ 20 & 101 & 60 \\ 450 & 0 & 1 \end{bmatrix}^2 = \begin{bmatrix} 72{,}521 & 79{,}640 & 13{,}200 \\ 34{,}240 & 14{,}601 & 6{,}120 \\ 117{,}900 & 99{,}000 & 1 \end{bmatrix} > 0. \tag{6.155}$$

We see that neither **S** nor **F** are reducible, Lemma A.8.2. So, we can take the next step and compute the Pasinetti **H**, first computing the inverse matrix,

$$(\mathbf{F}' - \mathbf{S}') = \begin{bmatrix} -40 & -110 & 290 \\ 110 & 60 & -125 \\ -20 & 40 & 0 \end{bmatrix}, \quad (\mathbf{F}' - \mathbf{S}')^{-1} = \begin{bmatrix} \frac{5}{1{,}174} & \frac{29}{2{,}935} & \frac{-49}{11{,}740} \\ \frac{5}{2{,}348} & \frac{29}{5{,}870} & \frac{269}{11{,}740} \\ \frac{14}{2{,}935} & \frac{9}{2{,}935} & \frac{43}{5{,}870} \end{bmatrix}, \tag{6.156}$$

thus obtaining,

$$\mathbf{H} = (\mathbf{F}' - \mathbf{S}')^{-1}\mathbf{S}' = \begin{bmatrix} 2.281 & 0.823 & 1.917 \\ 1.641 & 0.911 & 0.958 \\ 1.915 & 0.842 & 1.147 \end{bmatrix} > 0, \tag{6.157}$$

which is positive and for this reason also *irreducible*. So we conclude that the three commodities A, B, C are *basic*. ▲

We continue with a historical example where there are *basics* and *non-basics*. It is the abstract example of Sraffa in PCMC, Par. 59. We start presenting the production scheme, having in mind the structure $(\mathbf{S}', \mathbf{F}', \mathbf{L})$. We differ from the original notation, by designating the processes {**1, 2, 3, 4**} and the products {1, 2, 3, 4}:

Example 6.5.3. "Suppose that, in a system of four processes and four products, two commodities, 2 and 3 are jointly produced by one process,[18] and are produced by no other; but while 2 does not enter the means of production of any process, 3 enters the means of all four processes. Further suppose that the process producing 2 and 3 be represented by equation

$$(s_{11}p_1 + s_{31}p_3 + s_{41}p_4)(1 + r) + w \cdot L_1 = f_{11}p_1 + f_{21}p_2 + f_{31}p_3 + f_{41}p_4.\text{"} \tag{6.158}$$

The present production economy of four processes and products may be embedded into a greater economy.

Sraffa's original formulation of the example is in fact incomplete: there is no indication of what the processes (i. e., industries) **2**, **3** and **4** produce.

Nevertheless, we complete Sraffa's production scheme by inserting into the corresponding Sraffa Network inputs of commodities 1 and 4 by industries **2, 3, 4**. Furthermore, we present a maximal scheme by adding *quantities of labour*. Being a system in itself, necessarily the three Assumptions 6.1.1–6.1.3 hold.

The resulting production scheme then presents itself as follows:

18 Sraffa has chosen the process **1** in PCMC, Par. 59. The authors respect this choice and keep it.

$$\begin{array}{lllllllll} & 1 & 2 & 3 & 4 & 1 & 2 & 3 & 4 \\ \mathbf{1} & (s_{11}, & 0, & s_{31}, & s_{41}, L_1) \to (f_{11}, & f_{21}, & f_{31}, & f_{41}) \\ \mathbf{2} & (s_{12}, & 0, & s_{32}, & s_{42}, L_2) \to (f_{12}, & 0, & 0, & f_{42}) \\ \mathbf{3} & (s_{13}, & 0, & s_{33}, & s_{43}, L_3) \to (f_{13}, & 0, & 0, & f_{43}) \\ \mathbf{4} & (s_{14}, & 0, & s_{34}, & s_{44}, L_4) \to (f_{14}, & 0, & 0, & f_{44}) \\ & & & & (\mathbf{S}', \mathbf{L}) \to (\mathbf{F}'). \end{array} \tag{6.159}$$

Determine the rank of matrix $\mathbf{F}'$.

Apply the matrix rank criterion, Proposition 6.3.1 to prove that the commodities 2 and 3 are non-basics.

Solution to Example 6.5.3:
We identify the *commodity flow* matrix $\mathbf{S}$ and the *output* matrix $\mathbf{F}$,

$$\mathbf{S} = \begin{array}{c} \\ 1 \\ 2 \\ 3 \\ 4 \end{array} \overset{\begin{array}{cccc} \mathbf{1} & \mathbf{2} & \mathbf{3} & \mathbf{4} \end{array}}{\begin{bmatrix} s_{11} & s_{12} & s_{13} & s_{14} \\ 0 & 0 & 0 & 0 \\ s_{31} & s_{32} & s_{33} & s_{34} \\ s_{41} & s_{42} & s_{43} & s_{44} \end{bmatrix}}, \quad \mathbf{F} = \begin{array}{c} \\ 1 \\ 2 \\ 3 \\ 4 \end{array} \overset{\begin{array}{cccc} \mathbf{1} & \mathbf{2} & \mathbf{3} & \mathbf{4} \end{array}}{\begin{bmatrix} f_{11} & f_{12} & f_{13} & f_{14} \\ f_{21} & 0 & 0 & 0 \\ f_{31} & 0 & 0 & 0 \\ f_{41} & f_{42} & f_{43} & f_{44} \end{bmatrix}}, \tag{6.160}$$

and find rank($\mathbf{F}$) $=$ 3. This means that matrix $\mathbf{F}$ is *not regular* and the *input-output coefficients* matrix $\mathbf{C}_T$, (6.22), Proposition 6.2.1, does not exist.

Then we set up the rectangular matrix, according to Definition 6.3.1,

$$[\mathbf{S}', \mathbf{F}'] = \overset{\begin{array}{cccccccc} 1 & 2 & 3 & 4 & 1 & 2 & 3 & 4 \end{array}}{\begin{bmatrix} s_{11} & 0 & s_{31} & s_{41} & f_{11} & f_{21} & f_{31} & f_{41} \\ s_{12} & 0 & s_{32} & s_{42} & f_{12} & 0 & 0 & f_{42} \\ s_{13} & 0 & s_{33} & s_{43} & f_{13} & 0 & 0 & f_{43} \\ s_{14} & 0 & s_{34} & s_{44} & f_{14} & 0 & 0 & f_{44} \end{bmatrix}}, \tag{6.161}$$

and extract the columns corresponding to commodities 2 and 3, calculating the rank of the resulting submatrix of (6.161),

$$\operatorname{Rank}\left(\overset{\begin{array}{cccc} 2 & 3 & 2 & 3 \end{array}}{\begin{bmatrix} 0 & s_{31} & f_{21} & f_{31} \\ 0 & s_{32} & 0 & 0 \\ 0 & s_{33} & 0 & 0 \\ 0 & s_{34} & 0 & 0 \end{bmatrix}} \right) = 2. \tag{6.162}$$

This means: commodities 2 and 3 are *non-basics*, consequently, one of the commodities 1 or 4 is *basic*, the other is *basic* or *non-basic*, depending on the matrix coefficients. ▲

Now, we develop Example 6.5.3 to obtain a numerical version of the Sraffa problem (PCMC, Par. 59) and to calculate the Manara transformation matrix $\mathbf{T}$, matrix $\mathbf{M}$ and the Pasinetti matrix $\mathbf{H}$, to identify the number of *non-basic* commodities.

Example 6.5.4. Consider Example 6.5.3 and replace all variables s_{ij}, f_{ij} by arbitrary chosen non-negative numbers in order to create a numerical example. Apply Proposition 6.3.2 to determine the number of non-basic commodities.

Then compute the determinant $\det(\mathbf{F}' - \mathbf{S}')$, the vector of surplus $\mathbf{d} = (\mathbf{F} - \mathbf{S})\mathbf{e}$ and the rank($\mathbf{F}$).

Compute the Schefold transformation matrix $\mathbf{T}$ (6.78), the Manara transformation matrix $\mathbf{M}$ (6.79), and the Pasinetti matrix $\mathbf{H}$ (6.59).

The chosen production scheme is as follows:

$$\begin{array}{cllllcllll} & 1 & 2 & 3 & 4 & & 1 & 2 & 3 & 4 \\ \mathbf{1} & (20, & 0, & 50, & 30, L_1) & \to & (100, & 150, & 200, & 100) \\ \mathbf{2} & (10, & 0, & 50, & 60, L_2) & \to & (200, & 0, & 0, & 150) \\ \mathbf{3} & (30, & 0, & 10, & 10, L_3) & \to & (100, & 0, & 0, & 200) \\ \mathbf{4} & (40, & 0, & 40, & 50, L_4) & \to & (50, & 0, & 0, & 100) \\ & & & & (\mathbf{S}', \mathbf{L}) & \to & (\mathbf{F}'). & & & \end{array} \tag{6.163}$$

Present the *Sraffa Network* of the present sub-system of $n = 4$ products and discuss the aspects that become visible in it.

Solution to Example 6.5.4:
We identify the *commodity flow* matrix $\mathbf{S}$ and the *output* matrix $\mathbf{F}$ of the just-described economy, fulfilling the principle of production, described by Assumption 2.2.1 and by Assumption 2.2.2,

$$\mathbf{S} = \begin{array}{c} \\ 1 \\ 2 \\ 3 \\ 4 \end{array} \begin{array}{c} \begin{array}{cccc} \mathbf{1} & \mathbf{2} & \mathbf{3} & \mathbf{4} \end{array} \\ \begin{bmatrix} 20 & 10 & 30 & 40 \\ 0 & 0 & 0 & 0 \\ 50 & 50 & 10 & 40 \\ 30 & 60 & 10 & 50 \end{bmatrix} \end{array}, \quad \mathbf{F} = \begin{array}{c} \\ 1 \\ 2 \\ 3 \\ 4 \end{array} \begin{array}{c} \begin{array}{cccc} \mathbf{1} & \mathbf{2} & \mathbf{3} & \mathbf{4} \end{array} \\ \begin{bmatrix} 100 & 200 & 100 & 50 \\ 150 & 0 & 0 & 0 \\ 200 & 0 & 0 & 0 \\ 100 & 150 & 200 & 100 \end{bmatrix} \end{array}. \tag{6.164}$$

We then compute the rank($\mathbf{F}$) $= 2$ and conclude that we are not in presence of a system of *gross integrated industries*.

We compute $\det(\mathbf{F}' - \mathbf{S}') = -177{,}900{,}000$, confirming Assumption 6.1.2. We then compute the vector of surplus $\mathbf{d} = (\mathbf{F} - \mathbf{S})\mathbf{e} = [350, 150, 50, 400]' > \mathbf{0}$. Now permute the rows of matrices $\mathbf{S}$ and $\mathbf{F}$, corresponding to the commodities, attaining a maximal number of 0 in the lower parts of the matrices,

$$\tilde{\mathbf{S}} = \begin{array}{c} \\ 1 \\ 4 \\ 3 \\ 2 \end{array} \begin{array}{c} \begin{array}{cccc} \mathbf{1} & \mathbf{2} & \mathbf{3} & \mathbf{4} \end{array} \\ \begin{bmatrix} 20 & 10 & 30 & 40 \\ 30 & 60 & 10 & 50 \\ 50 & 50 & 10 & 40 \\ 0 & 0 & 0 & 0 \end{bmatrix} \end{array}, \quad \tilde{\mathbf{F}} = \begin{array}{c} \\ 1 \\ 4 \\ 3 \\ 2 \end{array} \begin{array}{c} \begin{array}{cccc} \mathbf{1} & \mathbf{2} & \mathbf{3} & \mathbf{4} \end{array} \\ \begin{bmatrix} 100 & 200 & 100 & 50 \\ 100 & 150 & 200 & 100 \\ 200 & 0 & 0 & 0 \\ 150 & 0 & 0 & 0 \end{bmatrix} \end{array}. \tag{6.165}$$

One identifies the sub-matrices of $\tilde{\mathbf{S}}$ and $\tilde{\mathbf{F}}$, according to (6.77) as:

$$\tilde{\mathbf{S}}'_{11} = \begin{bmatrix} 20 & 30 \\ 10 & 60 \end{bmatrix}, \quad \tilde{\mathbf{S}}'_{12} = \begin{bmatrix} 30 & 10 \\ 40 & 50 \end{bmatrix}, \quad \tilde{\mathbf{S}}'_{21} = \begin{bmatrix} 50 & 0 \\ 50 & 0 \end{bmatrix}, \quad \tilde{\mathbf{S}}'_{22} = \begin{bmatrix} 10 & 0 \\ 40 & 0 \end{bmatrix},$$
$$\tilde{\mathbf{F}}'_{11} = \begin{bmatrix} 100 & 100 \\ 200 & 150 \end{bmatrix}, \quad \tilde{\mathbf{F}}'_{12} = \begin{bmatrix} 100 & 200 \\ 50 & 100 \end{bmatrix}, \quad \tilde{\mathbf{F}}'_{21} = \begin{bmatrix} 200 & 150 \\ 0 & 0 \end{bmatrix},$$
$$\tilde{\mathbf{F}}'_{22} = \begin{bmatrix} 0 & 0 \\ 0 & 0 \end{bmatrix}. \tag{6.166}$$

We have to confirm that the 2×2 Schefold transformation matrix $\mathbf{T}$ exists. Thus, we can compute the 4×4 Manara transformation matrix $\mathbf{M}$,

$$\mathbf{T} = \begin{bmatrix} t_{11} & t_{12} \\ t_{21} & t_{22} \end{bmatrix}, \quad \tilde{\mathbf{S}}'_{21} = \mathbf{T}\tilde{\mathbf{S}}'_{22}, \quad \tilde{\mathbf{F}}'_{21} = \mathbf{T}\tilde{\mathbf{F}}'_{22}, \tag{6.167}$$

$$\begin{bmatrix} 50 & 0 \\ 50 & 0 \end{bmatrix} = \begin{bmatrix} t_{11} & t_{12} \\ t_{21} & t_{22} \end{bmatrix} \begin{bmatrix} 10 & 0 \\ 40 & 0 \end{bmatrix} \Rightarrow \text{Calculate matrix } \mathbf{T},$$
$$\begin{bmatrix} 200 & 150 \\ 0 & 0 \end{bmatrix} = \begin{bmatrix} t_{21} & t_{21} \\ t_{21} & t_{22} \end{bmatrix} \begin{bmatrix} 0 & 0 \\ 0 & 0 \end{bmatrix} \Rightarrow \text{leading to any } \mathbf{T}. \tag{6.168}$$

We obtain the Schefold transformation matrix $\mathbf{T}$ from the upper matrix equation of (6.168), the second equation providing any matrix $\mathbf{T}$. Realising the corresponding matrix products, we come to parametric equations in the coefficients $t_{11}, t_{12}, t_{21}, t_{22}$ with parameters $a, b \in \mathbb{R}$, chosen for any two of the four just mentioned variables t_{ij}. $i, j = 1, 2$, as here, i. e., $t_{21} = a$ and $t_{12} = b$,

$$\begin{bmatrix} 10t_{11} + 40t_{12} = 50 \\ 10t_{21} + 40t_{22} = 50 \\ t_{21} = a \\ t_{12} = b \end{bmatrix} \Rightarrow \begin{bmatrix} 10 & 40 & 0 & 0 \\ 0 & 0 & 10 & 40 \\ 0 & 0 & 1 & 0 \\ 0 & 1 & 0 & 0 \end{bmatrix} \begin{bmatrix} t_{11} \\ t_{12} \\ t_{21} \\ t_{22} \end{bmatrix} = \begin{bmatrix} 50 \\ 50 \\ a \\ b \end{bmatrix}. \tag{6.169}$$

$$\mathbf{T} = \begin{bmatrix} t_{11} & t_{12} \\ t_{21} & t_{22} \end{bmatrix} = \begin{bmatrix} 5 - 4b & b \\ a & \frac{5-a}{4} \end{bmatrix}, \quad \operatorname{rank}(\mathbf{T}) \le 2. \tag{6.170}$$

With matrix $\mathbf{T}$, we attain the Manara transformation matrix $\mathbf{M}$ and its determinant:

$$\mathbf{M} \equiv \begin{bmatrix} \mathbf{I}_1 & -\mathbf{T} \\ \mathbf{0} & \mathbf{I}_3 \end{bmatrix} = \begin{bmatrix} 1 & 0 & -5 + 4b & -b \\ 0 & 1 & -a & \frac{-5+a}{4} \\ 0 & 0 & 1 & 0 \\ 0 & 0 & 0 & 1 \end{bmatrix}, \quad \det(\mathbf{M}) = 1. \tag{6.171}$$

We compute the Pasinetti matrix $\mathbf{H}$, applying Lemma 6.3.1, observing that the result is independent of the variables $a, b \in \mathbb{R}$,

$$\mathbf{H} = (\mathbf{MF}' - \mathbf{MS}')^{-1}\mathbf{MS}'$$
$$\equiv (\mathbf{F}' - \mathbf{S}')^{-1}\mathbf{S}' = \frac{1}{593}\begin{bmatrix} -153 & -\frac{15}{2} & 0 & 0 \\ 125 & -\frac{11}{2} & 0 & 0 \\ -475 & -750 & -593 & 0 \\ \frac{1{,}732}{3} & \frac{5{,}251}{6} & \frac{2{,}372}{3} & 0 \end{bmatrix}. \quad (6.172)$$

Concluding, with Proposition 6.3.2 we have shown that there are $m = 2$ *non-basics* because there is a 2×2 zero matrix $\mathbf{H}_{22}$ in the upper right corner of the Pasinetti matrix **H** (6.172).

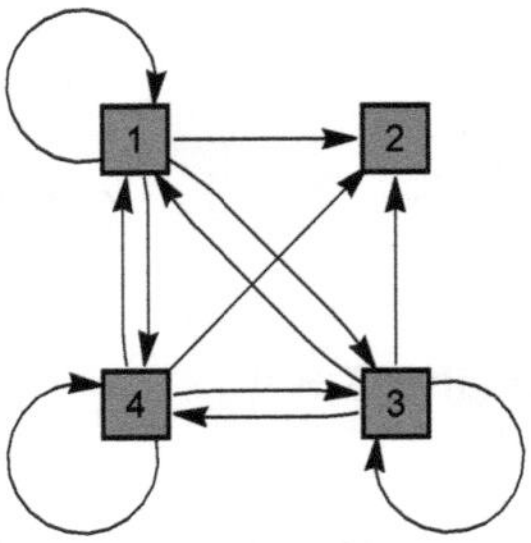

$$\mathbf{W}_C = \begin{matrix} & \mathbf{1} & \mathbf{4} & \mathbf{3} & \mathbf{2} \\ 1 & 1 & 1 & 1 & 1 \\ 4 & 1 & 1 & 1 & 1 \\ 3 & 1 & 1 & 1 & 1 \\ 2 & 0 & 0 & 0 & 0 \end{matrix} \quad (6.173)$$

Figure 6.9: Commodity digraph $G(\mathbf{W}_C)$ and corresponding adjacency matrix $\mathbf{W}_C$, Example 6.5.4.

Now we come back to the permuted matrices $\tilde{\mathbf{S}}$ and $\tilde{\mathbf{F}}'$, for which we define the corresponding adjacency matrices,

$$\tilde{\mathbf{V}} = \begin{bmatrix} 1 & 1 & 1 & 1 \\ 1 & 1 & 1 & 1 \\ 1 & 1 & 1 & 1 \\ 0 & 0 & 0 & 0 \end{bmatrix}, \quad \tilde{\mathbf{Q}}' = \begin{bmatrix} 1 & 1 & 1 & 1 \\ 1 & 1 & 0 & 0 \\ 1 & 1 & 0 & 0 \\ 1 & 1 & 0 & 0 \end{bmatrix}, \quad (6.174)$$

and compose with them the adjacency matrix $\tilde{\mathbf{W}}$ of the *Sraffa Network*, Figure 6.10,

$$\tilde{\mathbf{W}} = \begin{bmatrix} \mathbf{0} & \tilde{\mathbf{Q}}' \\ \tilde{\mathbf{V}} & \mathbf{0} \end{bmatrix} = \begin{matrix} & \mathbf{1} & \mathbf{4} & \mathbf{3} & \mathbf{2} & 1 & 4 & 3 & 2 \\ \mathbf{1} & 0 & 0 & 0 & 0 & 1 & 1 & 1 & 1 \\ \mathbf{4} & 0 & 0 & 0 & 0 & 1 & 1 & 0 & 0 \\ \mathbf{3} & 0 & 0 & 0 & 0 & 1 & 1 & 0 & 0 \\ \mathbf{2} & 0 & 0 & 0 & 0 & 1 & 1 & 0 & 0 \\ 1 & 1 & 1 & 1 & 1 & 0 & 0 & 0 & 0 \\ 4 & 1 & 1 & 1 & 1 & 0 & 0 & 0 & 0 \\ 3 & 1 & 1 & 1 & 1 & 0 & 0 & 0 & 0 \\ 2 & 0 & 0 & 0 & 0 & 0 & 0 & 0 & 0 \end{matrix}. \quad (6.175)$$

The associated *Sraffa Network*[19] is weakly connected (Definition A.14.10).

19 Note that the number of connections of the *Sraffa Network* is correct: 12 arcs entering the four industries and ten arcs entering the four commodities, giving 10 + 12 = 22, the total number of entries "1" of the adjacency matrix $\tilde{\mathbf{W}}$.

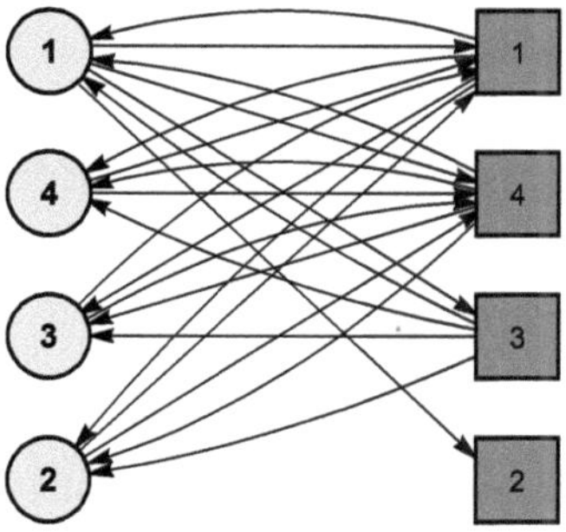

Figure 6.10: The Sraffa Network of Example 6.5.3.

We compute now the matrix product $\tilde{\mathbf{V}}\tilde{\mathbf{Q}}' \sim \mathbf{W}_C$, presented in Figure 6.9. See also Example A.14.6 for further illustrations.

$$\tilde{\mathbf{V}}\tilde{\mathbf{Q}}' = \begin{bmatrix} 1 & 1 & 1 & 1 \\ 1 & 1 & 1 & 1 \\ 1 & 1 & 1 & 1 \\ 0 & 0 & 0 & 0 \end{bmatrix} \begin{bmatrix} 1 & 1 & 1 & 1 \\ 1 & 1 & 0 & 0 \\ 1 & 1 & 0 & 0 \\ 1 & 1 & 0 & 0 \end{bmatrix} = \begin{bmatrix} 4 & 4 & 1 & 1 \\ 4 & 4 & 1 & 1 \\ 4 & 4 & 1 & 1 \\ 0 & 0 & 0 & 0 \end{bmatrix} \sim \mathbf{W}_C, \tag{6.176}$$

in order to define the associated adjacency matrix $\mathbf{W}_C$ (6.173), which directly leads to the commodity digraph $G(\mathbf{W}_C)$ representing this production process, Definition A.14.8, Figure 6.9. ▲

6.6 Conditions for positive price vectors in viable economies

It is important to note that in present day *economic systems* positive price vectors are no longer the only requested solution of *joint production systems* (6.115). It is known that in *ecological economy*, zero prices or negative prices may appear for produced waste, if such sectors are included in the investigated *economic system*. Indeed, to get to grips, e. g., with the immense 'Pacific garbage patch', the contemporary accumulation of plastic in the oceans, economic models, considering the *economic recycling* and utilisation of waste, have to be developed. The *joint production Sraffa system* becomes a useful tool in this problem area. We will develop such ideas later, see Chapter 7.

But for now, we are concerned with conditions for positive price vectors $\mathbf{p} > \mathbf{o}$.

Pasinetti (Ed.) [83] (Chap. II, p. 17) writes in the line of PCMC: "*An economic system will be considered in which all commodities are produced by means of commodities, used as capital goods. Commodities enter the process of production at the beginning of each 'year' as inputs, jointly with labour services, and commodities come out at the end of the year as outputs. The economic system is supposed to be viable in the sense that it is capable of producing larger quantities of commodities than those required to replace used-up capital goods.*" See Definition 2.2.1.

Manara investigated conditions to obtain positive prices, see Pasinetti (Ed.) [83] (Chap. I, p. 4, original paper published in 1968). He states:

> *"It is quite obvious that such prices must constitute the components of a positive vector."*

Then, Manara makes the following overall condition, treating the simplified situation:

> *"Let us suppose for the sake of simplicity all the commodities under consideration are basic commodities ..."*

In a first step, we will present these *sufficient but not necessary conditions*. They are *stability conditions* for the economic system. Manara stated his conditions about 50 years ago. We will incorporate them in an up-to-date strengthen form. We start considering the *Sraffa price model of joint production*, the first equation in (6.39) in its most general form for *semi-positive* matrices $\mathbf{S} \geq 0$, $\mathbf{F} \geq 0$, $w = \frac{\bar{w} \cdot Y}{L}$,

$$\mathbf{S}'\mathbf{p}(1+r) + \mathbf{L}w = \mathbf{F}'\mathbf{p}, \tag{6.177}$$

remaining on the level of the *commodity flows*. We do not go down to the *input-output coefficients* and do not require *gross integrated industries*, see Assumption 6.1.3. Therefore we cannot require the regularity of matrix $\mathbf{F}$, $\det(\mathbf{F}) \neq 0$ (6.8). Consequently, we will not obtain eigenvalue equations for *non-negative* and *irreducible* matrices, as is illustrated in Example 6.2.2, equation (6.34). Therefore, the **Perron–Frobenius theorem A.9.3** is far out of reach.

We introduce at first the notion of *convex polyhedral cones*.

Definition 6.6.1 (Convex polyhedral cone).
(1) A subset $\mathcal{C}$ of the vector space $\mathbb{R}^n$ is a cone if, for every $\mathbf{x} \in \mathcal{C}$ and positive scalars $\alpha > 0$, the product $\alpha\mathbf{x}$ is in $\mathcal{C}$.
(2) A cone $\mathcal{C}$ is convex if $\alpha\mathbf{x} + \beta\mathbf{y} \in \mathcal{C}$ for any possible positive scalars α, β and any $\mathbf{x}, \mathbf{y} \in \mathcal{C}$.
(3) A cone $\mathcal{C}$ is called polyhedral if there is some square $n \times n$ matrix $\mathbf{J}$ such that $\mathcal{C} = \{\mathbf{x} \mid \mathbf{J}\mathbf{x} \geq \mathbf{o}\}$. This is the inequality description. There is a second description by *weighted sums* of a polyhedral cone. The cone $\mathcal{C}$ is given by a finite set of "generating vectors" $\mathbf{\upsilon}_1, \ldots, \mathbf{\upsilon}_k$ such that $\mathcal{C} = \{\alpha_1\mathbf{\upsilon}_1 + \cdots + \alpha_k\mathbf{\upsilon}_k \mid \alpha_1 \geq 0, \ldots, \alpha_k \geq 0\}$ is the *weighted sum* of the vector set $\{\mathbf{\upsilon}_1, \ldots, \mathbf{\upsilon}_k\}$, acting as a "generator" of the cone.

Notation 6.6.1. We also observe that the rank($\mathbf{J}$) $= k$ is the dimension of the vector space spanned by the vectors $\{\mathbf{\upsilon}_1, \ldots, \mathbf{\upsilon}_k\}$. We say that the corresponding cone has dimension $\dim(\mathcal{C}) = k$, in the way we speak of a 3–dim cube in the 3–dim Euclidian space. If $\mathbf{J}\mathbf{x} > \mathbf{o}$, the vectors $\mathbf{x}$ are inside the cone; if $\mathbf{J}\mathbf{x} = \mathbf{o}$, the vectors $\mathbf{x}$ are on the point set surface of the cone.

Manara transforms *Sraffa's joint production price model* (6.177) to the following form with $r \geq 0$, $w > 0$, requiring positive vectors of sectorial wages $\mathbf{w} > \mathbf{o}$ and therefore also positive vectors of labour $\mathbf{L} > \mathbf{o}$:

$$\mathbf{w} = \mathbf{L} \cdot w = [\mathbf{F}' - \mathbf{S}'(1+r)]\mathbf{p} > \mathbf{o}. \tag{6.178}$$

We follow with Manara's assumptions on positive prices. He labeled them as 'unstated assumptions', **UA1** to **UA4**, see [61] (in Pasinetti (Ed.) [83], Chap. I).[20] We will see that some of the expressed conditions are covered by statements contained in Assumption 6.1.1 and Assumption 6.1.2.

- (**UA1**) (*The quantity produced exceeds the means of production*). Manara [61] says: *"The overall quantity of every commodity used as means of production is less than the total quantity of the same commodity produced in the whole economic system."* This condition may be expressed by the positive vector of surplus:

$$\mathbf{d} = (\mathbf{F} - \mathbf{S})\mathbf{e} > \mathbf{o}. \tag{6.179}$$

  This *self-replacement* condition requests a positive surplus for every commodity and is part of Assumption 6.1.1.
- (**UA2**) (*Condition for the existence of positive price vectors*). Manara postulates: *"There exists at least one positive vector* $\bar{\mathbf{p}}$*, such that the value of the commodities used as means of production by every individual industry, evaluated at those prices, is smaller than the value of the products, also evaluated at those prices."* The condition is formulated as follows,

$$\exists \bar{\mathbf{p}} \text{ such that } \{\bar{\mathbf{p}} > \mathbf{o} \wedge (\mathbf{F}' - \mathbf{S}')\bar{\mathbf{p}} > \mathbf{o}\}. \tag{6.180}$$

  Then Manara sets up an octant of the vector space $\mathbb{R}^n$, made up of the ensemble of the potential *non-negative* price vectors $\mathbf{p}$, defined as[21]

$$\mathcal{P} = \{\mathbf{p} \mid \mathbf{p} \geqq \mathbf{o}\} \subset \mathbb{R}^n. \tag{6.181}$$

  The purpose is to define subsets of *convex polyhedral cones*, contained in the defined octant (6.181), $V(r) \subseteq \mathcal{P}$. The *convex polyhedral cones* $V(r)$ will be made up of positive price vectors $\mathbf{p} > \mathbf{o}$, when used in equation (6.178) will lead to *positive* vectors of sectorial wages $\mathbf{w} > \mathbf{o}$, as a result, i. e.,

$$V(r) = \{\mathbf{p} \mid \mathbf{p} \in \mathcal{P} \wedge \mathbf{w} = (\mathbf{F}' - \mathbf{S}'(1 + r))\mathbf{p} > \mathbf{o}\}. \tag{6.182}$$

  The non-empty *convex polyhedral cones* $V(r)$ are defined for all admitted rate of profits r. The Example 6.6.1 and Example 6.6.2 illustrate such *polyhedral cones*. At first, a positive price vector $\bar{\mathbf{p}} \in \mathcal{P}$ has to be found, showing that $V(0)$ is not an empty set, $V(0) \neq \emptyset$ (6.182) for $r = 0$, i. e., $\bar{\mathbf{p}} \in V(0)$.
- (**UA3**) (*The linear independence of the processes*). This condition is identical to Assumption 6.1.2 and is valid for all the *joint production* processes. Manara formulates the assumption (in Pasinetti (Ed.) [83], p. 6) in the same way,

20 We do not treat **UA5**, which relates to *Standard systems* of *joint production economies*, leading to the general eigenvalue equation $\mathbf{S}(1 + R)\mathbf{y} = \mathbf{F}\mathbf{y}$.

21 We are especially interested in positve prices, $\mathbf{p} > \mathbf{o}$, i. e., the open set $\dot{\mathcal{P}} = \{\mathbf{p} \mid \mathbf{p} > \mathbf{o}\} \subset \mathbb{R}^n$.

$$\det(\mathbf{F}' - \mathbf{S}') \neq 0 \Leftrightarrow \det(\mathbf{F} - \mathbf{S}) \neq 0, \tag{6.183}$$

and says: *"This ensures, that for at least one value of r (the value r = 0), the vectors forming the rows of the matrix* $(\mathbf{F} - \mathbf{S}(1 + r))$ *are linearly independent."*. Then the real polynomial $f(r)$ of the variable r is defined as

$$f(r) = \det(\mathbf{F} - \mathbf{S}(1 + r)), \tag{6.184}$$

which is evidently continuous. Manara further defines a subset on the half-line $[0, \infty[$, defined by non-negative rate of profits, $r \geq 0$, such that,

$$\Phi = \{r \mid f(r) = \det(\mathbf{F} - \mathbf{S}(1 + r)) \neq 0\}. \tag{6.185}$$

The set Φ *"is closed on the left-hand side, having r = 0 as its minimum"*. Concluding, we will now refer to the interval $\Phi \subset [0, \infty[$, for which the cones $V(r)$ (6.182) are not empty. Defining $\mathbf{J}(r) := \mathbf{F}' - \mathbf{S}'(1+r)$, we recognise that for $r \in \Phi$ the cones $V(r)$ have as dimension the number of sectors of the economic system $(\mathbf{F}', \mathbf{S}')$, $\dim(V(r)) = n$.

Manara considers again the positive vector of *sectorial wages* $\mathbf{w} := w \cdot \mathbf{L} > \mathbf{o}$, giving the total wages $W = \mathbf{e}'\mathbf{w} > 0$, and he computes

$$\mathbf{w} := w \cdot \mathbf{L} = (\mathbf{F}' - \mathbf{S}'(1 + r))\mathbf{p} > \mathbf{o}. \tag{6.186}$$

He observes that (6.186) *"does not possess as a solution a price vector* $\mathbf{p}$ *which is positive for any positive vector* $\mathbf{L} > \mathbf{o}$ *of the quantity of labour absorbed by the industries of the system"* (in Pasinetti (Ed.) [83], p. 7). For this reason, it is necessary to postulate a further condition (in Pasinetti (Ed.) [83], p. 8) yielding a positive price vector $\mathbf{p} \in V(r)$ for $r \in \Phi$. We calculate the vector $\mathbf{w}$ (6.186) of sectorial wages, and according to Manara we define $V'(r)$ as the image of $V(r)$ as follows,

$$V'(r) = \{\mathbf{w} \mid \{\mathbf{w} = (\mathbf{F}' - \mathbf{S}'(1 + r))\mathbf{p} > \mathbf{o}\} \wedge \mathbf{p} \in V(r)\}. \tag{6.187}$$

The intention is that the non-empty set $V(r) \neq \emptyset$ of the positive price vectors $\mathbf{p}$ has to guarantee that the vectors of *sectorial wages* $\mathbf{w} = (\mathbf{F}' - \mathbf{S}'(1 + r))\mathbf{p} > \mathbf{o}$ are *positive*! By construction the matrix $\mathbf{F}' - \mathbf{S}'(1 + r)$ is regular and its inverse matrix exists. For this reason one obtains the equivalence relation

$$\mathbf{w} = (\mathbf{F}' - \mathbf{S}'(1 + r))\mathbf{p} > \mathbf{o} \Leftrightarrow \mathbf{p} = (\mathbf{F}' - \mathbf{S}'(1 + r))^{-1}\mathbf{w} > \mathbf{o}, \tag{6.188}$$

leading to the condition[22]:

- (**UA4**) (*Positive vectors of sectorial wages associated with positive price vectors*). We formulate this condition as an inclusion, a binary relation leading to an equivalence relation:

$$r \in \Phi \Rightarrow (\mathbf{p} \in V(r) \wedge \mathbf{w} = \mathbf{J}(r)\mathbf{p} \Leftrightarrow \mathbf{w} \in V'(r) \wedge \mathbf{p} = \mathbf{J}^{-1}(r)\mathbf{w}). \tag{6.189}$$

We give now two illustrations of Manara's conditions:

---

22 Manara originally formulated it as follows: $r \in \Phi \Rightarrow \mathbf{w} \in V'(r)$.

Example 6.6.1. Given a joint production system $(\mathbf{S}', \mathbf{F}')$ of $n = 2$ sectors, a general rate of profits $r \geq 0$, a specific rate of profits $r_1 = 0.5$, the wage rate $w = 9$, the vector of labour $\mathbf{L}$, Assumption 2.5.1 holds,

$$\mathbf{S} = \begin{bmatrix} 1 & 1 \\ 1 & 1 \end{bmatrix}, \quad \mathbf{F} = \begin{bmatrix} 0 & 3 \\ 5 & 1 \end{bmatrix}, \quad \mathbf{L} = \begin{bmatrix} 2 \\ 3 \end{bmatrix}. \tag{6.190}$$

Verify the Assumption 6.1.1, to Assumption 6.1.2 and the conditions **UA1** to **UA4**. Set up the square matrix $\mathbf{J}(r) = \mathbf{F}' - \mathbf{S}'(1 + r)$, defining a *convex polyhedral cone*,

$$V(r) = \{\mathbf{y} \mid \mathbf{J}(r)\mathbf{y} \geq \mathbf{o} \wedge \mathbf{y} \in P\}, \tag{6.191}$$

and transform it into the weighted-sum description. Choose $r_1 = 0.5$ and present the cones $V(0.5)$ and $V'(0.5)$ geometrically in Euclidean planes. Determine the polynomial $f(r) = \det(\mathbf{J}(r))$, the interval Φ (6.185) and discuss the positivity of the price vector $\mathbf{p} > \mathbf{o}$ and of the sectorial wage vector $\mathbf{w} > \mathbf{o}$. Compute the maximal profit rate R and discuss the solvability of the price model (6.177).

Solution to Example 6.6.1:
We verify condition **UA1**, previously Assumption 6.1.1, computing $\mathbf{d} = (\mathbf{F} - \mathbf{S})\mathbf{e} = [1, 4]' > \mathbf{o}$.

We determine then the validity of condition **UA2**, choosing the vector $\overline{\mathbf{p}} = [1, 2]'$ and compute the image

$$\overline{\mathbf{w}} = (\mathbf{F}' - \mathbf{S}')\overline{\mathbf{p}} = \begin{bmatrix} -1 & 4 \\ 2 & 0 \end{bmatrix}\begin{bmatrix} 1 \\ 2 \end{bmatrix} = \begin{bmatrix} 7 \\ 2 \end{bmatrix} > \mathbf{o} \Rightarrow \overline{\mathbf{p}} \in V(0). \tag{6.192}$$

We verify condition **UA3**, previously Assumption 6.1.2, computing $\det(\mathbf{F}' - \mathbf{S}') = -8$. For any positive vector $\mathbf{p} > \mathbf{o}$, we set up the convex polyhedral cone $V'(r)$ and matrices $\mathbf{J}(\mathbf{r}), \mathbf{J}^{-1}(r)$,

$$\begin{aligned} \mathbf{w} = \mathbf{J}(r)\mathbf{p} &:= (\mathbf{F}' - \mathbf{S}'(1 + r))\mathbf{p} = \left(\begin{bmatrix} 0 & 5 \\ 3 & 1 \end{bmatrix} - (1 + r)\begin{bmatrix} 1 & 1 \\ 1 & 1 \end{bmatrix}\right)\begin{bmatrix} p_1 \\ p_2 \end{bmatrix} \\ &= \begin{bmatrix} -1 - r & 4 - r \\ 2 - r & -r \end{bmatrix}\begin{bmatrix} p_1 \\ p_2 \end{bmatrix} = \left\{ \begin{matrix} (-1 - r)p_1 + (4 - r)p_2 \geq 0 \\ (2 - r)p_1 - rp_2 \geq 0 \end{matrix} \right\}, \end{aligned}$$

$$\mathbf{J}(\mathbf{r}) = \begin{bmatrix} -1 - r & 4 - r \\ 2 - r & -r \end{bmatrix}, \quad \mathbf{J}^{-1}(r) = \begin{bmatrix} \frac{r}{8-7r} & \frac{4-r}{8-7r} \\ \frac{2-r}{8-7r} & \frac{1+r}{8-7r} \end{bmatrix} \geq \mathbf{o}. \tag{6.193}$$

We choose now $r = 0.5$ and evaluate (6.193) for this rate of profits, in order to set up the convex polyhedral $V'(0.5)$,

$$\mathbf{J}(0.5)\mathbf{p} = \begin{bmatrix} -\frac{3}{2} & \frac{7}{2} \\ \frac{3}{2} & -\frac{1}{2} \end{bmatrix}\begin{bmatrix} p_1 \\ p_2 \end{bmatrix} = \left\{ \begin{matrix} (-\frac{3}{2})p_1 + \frac{7}{2}p_2 \geq 0 \\ \frac{3}{2}p_1 - \frac{1}{2}p_2 \geq 0 \end{matrix} \right\}, \tag{6.194}$$

and determine the straight lines $-1.5p_1 + 3.5p_2 = 0$ and $1.5p_1 - 0.5p_2 = 0$, and the point-set surface of the *convex polyhedral cone*, on which we determine the direction vectors which are, $\mathbf{\upsilon}_1(0.5) = [3.5, 1.5]'$ and $\mathbf{\upsilon}_2(0.5) = [0.5, 1.5]'$, generators of the cone $V(0.5)$,

$$V(0.5) = \left\{\mathbf{p} \mid \mathbf{p} = \alpha_1 \begin{bmatrix} 3.5 \\ 1.5 \end{bmatrix} + \alpha_2 \begin{bmatrix} 0.5 \\ 1.5 \end{bmatrix} \mid \alpha_1 \geq 0, \alpha_2 \geq 0\right\}. \tag{6.195}$$

We present now the *convex polyhedral cone* $V'(0.5)$, image of $V(0.5)$. For this purpose, we determine the images of the generators $\mathbf{v}_1(0.5)$ and $\mathbf{v}_2(0.5)$ of the *convex cone* $V(0.5)$, which are $\mathbf{w}_k = \mathbf{J}(0.5)\mathbf{v}_k$, $k = 1, 2$,

$$\mathbf{w}_1 = \begin{bmatrix} -\frac{3}{2} & \frac{7}{2} \\ \frac{3}{2} & -\frac{1}{2} \end{bmatrix} \begin{bmatrix} \frac{7}{2} \\ \frac{3}{2} \end{bmatrix} = \begin{bmatrix} 0 \\ \frac{9}{2} \end{bmatrix}, \quad \mathbf{w}_2 = \begin{bmatrix} -\frac{3}{2} & \frac{7}{2} \\ \frac{3}{2} & -\frac{1}{2} \end{bmatrix} \begin{bmatrix} \frac{1}{2} \\ \frac{3}{2} \end{bmatrix} = \begin{bmatrix} \frac{9}{2} \\ 0 \end{bmatrix}. \tag{6.196}$$

We now have obtained the system of generators $\{\mathbf{w}_1, \mathbf{w}_2\}$ of the *convex polyhedral cone* $V'(0.5) = \{\mathbf{w} \mid \mathbf{w} = \beta_1\mathbf{w}_1 + \beta_2\mathbf{w}_2 \mid \beta_1 \geq 0, \beta_2 \geq 0\}$. The corresponding *convex polyhedral cones* $V(0.5)$ and $V'(0.5)$ are represented in Fig. 6.11.

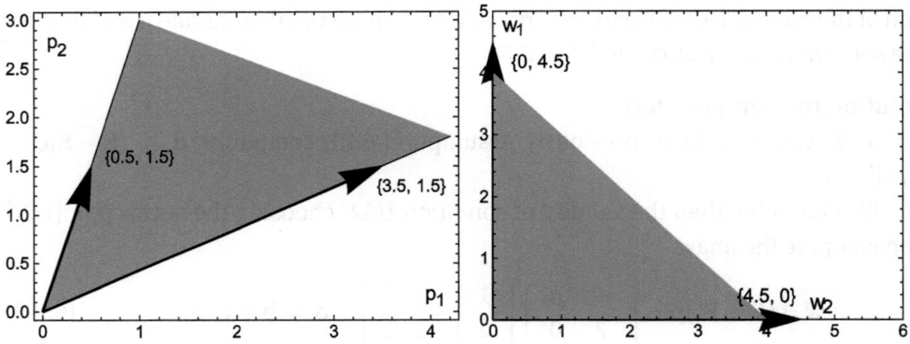

Figure 6.11: Convex polyhedral cones $V(0.5)$ (left) and $V'(0.5)$ (right), Example 6.6.1.

We continue exploring whether there are positive price vectors within the *convex polyhedral cone* $V(0.5)$. We choose the vectors $\mathbf{p}_1 = [4, 1]'$ and $\mathbf{p}_2 = [1, 1]'$, calculating the images $\mathbf{w}_1 = \mathbf{J}(0.5)\mathbf{p}_1 = [-2.5, 5.5]' \notin V'(0.5)$ and $\mathbf{w}_2 = \mathbf{J}(0.5)\mathbf{p}_2 = [2, 1]' \in V'(0.5)$. We recognise finally that $\mathbf{p}_2 \in V(0.5) \neq \emptyset$, whereas $\mathbf{p}_1 \notin V(0.5)$.

Then, we verify condition[23] **UA4**, determining the polynomial $f(r)$ and its roots, which is of order 1 in this example, see Figure 6.12

$$f(r) = \det(\mathbf{F}' - \mathbf{S}'(1 + r)) = -8 + 7r = 0 \Rightarrow r_1 = \frac{8}{7} > 0. \tag{6.197}$$

Moreover, in this example we have *gross integrated industries*, $\det(\mathbf{F}) = -15$. We compute the *input-output coefficients* matrix $\mathbf{C}' = \mathbf{F}'^{-1}\mathbf{S}'$. We consider the vector of labour $\mathbf{L}$ in the price model (6.177) and tend to the limit $\mathbf{L} \Rightarrow \mathbf{o}$, $L = 0$, we obtain the eigenvalue equation, where the rate of profits r has to be replaced by the maximal rate of profits (productiveness) R.

23 One shows that when for $n \times n$ matrices $\mathbf{S}$, $\mathbf{F}$, there is $\det(\mathbf{S}) = 0$, and then the degree of the polynomial $f(r) = \det(\mathbf{F}' - (1 + r)\mathbf{S}')$ is smaller than n.

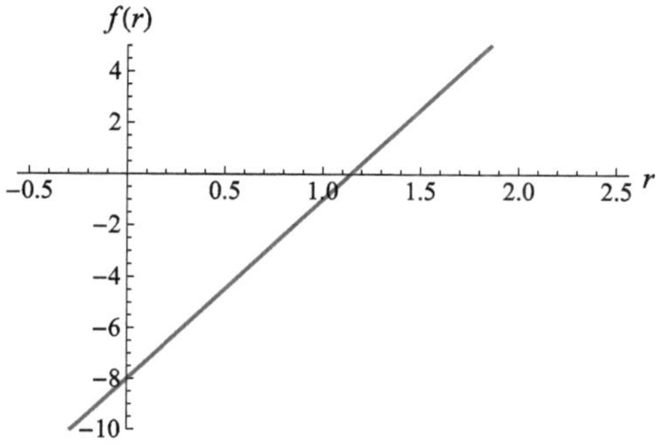

Figure 6.12: The polynomial $f(r)$ (6.197).

$$\mathbf{C}'\mathbf{p}(1+R) = \mathbf{p} \Rightarrow \mathbf{C}'\mathbf{p} = \frac{1}{15}\begin{bmatrix} 4 & 4 \\ 3 & 3 \end{bmatrix}\mathbf{p} = \lambda_C \mathbf{p}, \quad \lambda_C = \frac{1}{1+R} \tag{6.198}$$

The Frobenius number of the positive matrix $\mathbf{C}$ is $\lambda_C = 7/15$ and the maximal rate of profits is $R = 8/7 = r_1$ equal the zero of the polynomial $f(r)$ (6.197). We have now the interval $\Phi = [0, \frac{8}{7}[$. The upper limit $R = 8/7$ is excluded, because condition (6.185) requests the existence of the inverse matrix $\mathbf{J}^{-1}$, i. e., the regularity of $\mathbf{J}$ (6.197).

As for $r \in \Phi$, we have $\det(\mathbf{J}(r)) \neq 0$, and this means, $\dim(V(r)) = \dim(V'(r)) = 2$, see Notation 6.6.1. We need now to show that for all $r \in \Phi$ the price vectors are positive, $\mathbf{p} > \mathbf{o}$, and the vectors of sectorial wages are *semi-positive*, $\mathbf{w} > \mathbf{o}$. We go back to the general convex polyhedral (6.193) to determine the vectors of prices and sectorial wages. We have

$$\mathbf{J}(r)\mathbf{p} = \begin{bmatrix} -1-r & 4-r \\ 2-r & -r \end{bmatrix}\begin{bmatrix} p_1 \\ p_2 \end{bmatrix} = \begin{Bmatrix} (-1-r)p_1 + (4-r)p_2 \geq 0 \\ (2-r)p_1 - rp_2 \geq 0 \end{Bmatrix}, \tag{6.199}$$

and determine the straight lines $(-1-r)p_1 + (4-r)p_2 = 0$ and $(2-r)p_1 - rp_2 = 0$, the *convex polyhedral cone* point-set surface, on which we determine the *direction vectors* which are, $\mathbf{v}_1(r) = [4-r, 1+r]'$ and $\mathbf{v}_2(r) = [r, 2-r]'$, generators of the cone $V(r)$. To get positive price vectors in the *first quadrant* of the Euclidian plane $\mathbb{R}^2$, the condition between tangents of elevation angles of the vectors $\mathbf{v}_1(r)$ (α) and $\mathbf{v}_2(r)$ (β) must be fulfilled.

$$\tan(\alpha) = \frac{1+r}{4-r} \leq \tan(\beta) = \frac{2-r}{r}. \tag{6.200}$$

In the limit, inequality (6.200) leads to equation $\frac{1+r}{4-r} = \frac{2-r}{r}$ or to $r_1 = \frac{8}{7}$. This is the third way to compute the Frobenius number of matrix $\mathbf{C}$. We now compute the generators of the *convex polyhedral cone* $V'(r)$, which are: $\mathbf{w}_k = \mathbf{J}(r)\mathbf{v}_k$, $k = 1, 2$,

$$\begin{aligned} \mathbf{w}_1 &= \begin{bmatrix} -1-r & 4-r \\ 2-r & -r \end{bmatrix}\begin{bmatrix} 4-r \\ 1+r \end{bmatrix} = \begin{bmatrix} 0 \\ 8-7r \end{bmatrix}, \\ \mathbf{w}_2 &= \begin{bmatrix} -1-r & 4-r \\ 2-r & -r \end{bmatrix}\begin{bmatrix} r \\ 2-r \end{bmatrix} = \begin{bmatrix} 8-7r \\ 0 \end{bmatrix}. \end{aligned} \tag{6.201}$$

Thus, we have the inclusion (6.189). As the inverse matrix $\mathbf{J}^{-1}(r) \geq \mathbf{o}$ (6.193) is *semi-positive* and the vector of labour $\mathbf{L} > \mathbf{o}$ is assumed *positive* by Assumption 2.5.1, one only has to consider positive vectors of sectorial wages $\mathbf{w} = w \cdot \mathbf{L} > \mathbf{o}$,

$$r \in \Phi \Rightarrow (\mathbf{p} \in V(r) \wedge \mathbf{w} = \mathbf{J}(r)\mathbf{p} > \mathbf{o} \Leftrightarrow \mathbf{w} \in V'(r) \wedge \mathbf{p} = \mathbf{J}^{-1}(r)\mathbf{w} > \mathbf{o}). \tag{6.202}$$

We recognize that for every $r \in \Phi = [0, 8/7[$ the economic condition for positive price vectors $\mathbf{p}$ and positive vectors of sectorial wages $\mathbf{w}$ means that these vectors are elements of the open cone sets: $\mathbf{p} \in \dot{V}(r)$, $\mathbf{w} \in \dot{V}'(r)$.

The cone $\dot{V}(r)$ of positive prices $\mathbf{p} > \mathbf{o}$ is a subset of the first quadrant, and the cone $\dot{V}'(r)$ of sectorial wages $\mathbf{w} > \mathbf{o}$ is in this case equal to the open first quadrant of the Euclidean plane. It is easy to verify that for all $r > 8/7$ neither $V(r)$ nor $V'(r)$ contain only positive vectors, and for this reason admissible rates of profits r are only in Φ, $r \in \Phi$. The reader may verify that for $r = 8/7$ there are no wages.

How do we get in this example positive price vectors $\mathbf{p} > \mathbf{o}$ as a solution of the price model (6.177) for *joint production*? It is easy! Taking the given wage rate $w = 9$, we have to choose a rate of profits $r \in \Phi$ and a vector of sectorial wages $\mathbf{w} = w \cdot \mathbf{L} = 9 \cdot [2, 3]' = [18, 27]' \in \dot{V}'(r)$, and then compute $\mathbf{p} = \mathbf{J}^{-1}(r)\mathbf{w} = [\frac{9(12-r)}{8-7r}, \frac{9(7+r)}{8-7r}]' \in \dot{V}(r)$ which is positive. For $r = 0.5$, we obtain a positive price vector $\mathbf{p} = [23, 15]' \in \dot{V}(0.5)$. ▲

From Example 6.6.1, we learn that for *joint production Sraffa systems* of the type of *gross integrated industries*, where there is a positive *input-output coefficients* matrix $\mathbf{C} = \mathbf{S}\mathbf{F}^{-1} > \mathbf{0}$, obtaining the Frobenius number $\lambda_C > 0$, we compute the *productiveness* $R = (1/\lambda_C) - 1$ and conclude for the rate of profits: $r \in \Phi = [0, R[$ (6.202).

Example 6.6.2. Consider for $n = 3$ sectors a system $(\mathbf{S}', \mathbf{F}')$ of joint production with matrices,

$$\mathbf{S} = \begin{bmatrix} 3 & 2 & 1 \\ 2 & 3 & 1 \\ 1 & 1 & 2 \end{bmatrix}, \quad \mathbf{F} = \begin{bmatrix} 6 & 1 & 0 \\ 0 & 6 & 1 \\ 1 & 0 & 6 \end{bmatrix}. \tag{6.203}$$

Show that we are in presence of gross integrated industries and compute the input-output coefficients matrix. If it is positive, compute the Frobenius number λ_C and the maximal rate of profits R.

Show that all the three conditions **UA1** to **UA3** of Manara are fulfilled.

Compute the polynomial $f(r) = \det(\mathbf{F}' - \mathbf{S}'(1 + r))$ and its roots and present the graph. Determine the set Φ (6.185).

Present the inequality description $\mathbf{J}(r)\mathbf{p} \geq \mathbf{0}$ of the convex polyhedral cone $V(r)$ (6.182), evaluate $\mathbf{J}(0.1)\mathbf{p}$ and compute the generating vectors of $V(0.1)$.

Set up the convex polyhedral cone $V(0.1)$ in weighted-sum description, and compute its image $V'(0.1)$. Argue over the positivity of the vector of labour $\mathbf{w} \in V'(0.1)$ and the price vector $\mathbf{p} \in V(0.1)$ as solutions of the Sraffa price model (6.177).

Compute for $r \in \Phi$ the matrices $\mathbf{J}(r)$ and $\mathbf{J}(r)^{-1}$, compute the spanning vectors of $V(r)$ and $V'(r)$ and show that the Manara's condition **UA4** is fulfilled.

Solution to Example 6.6.2:
We start computing det(**F**) = 217 and are therefore in presence of a system of *gross integrated industries*. We further compute

$$\mathbf{F}^{-1} = \frac{1}{217}\begin{bmatrix} 36 & -6 & 1 \\ 1 & 36 & -6 \\ -6 & 1 & 36 \end{bmatrix}, \quad \mathbf{C} = \mathbf{S}\mathbf{F}^{-1} = \frac{1}{217}\begin{bmatrix} 104 & 55 & 27 \\ 69 & 97 & 20 \\ 25 & 32 & 67 \end{bmatrix} > \mathbf{0}. \tag{6.204}$$

As matrix **C** is positive, we compute the Frobenius number, getting $\lambda_C = 0.8056$ and the *maximal rate of profits* $R = (1/0.8056) - 1 = 0.2413$. Then, we verify step by step Manara's four conditions:

(1) We note that the vector of surplus $\mathbf{d} = (\mathbf{F} - \mathbf{S})\mathbf{e} = [1, 1, 3]' > \mathbf{o}$ is positive. Therefore condition **UA1** is fulfilled.

(2) We now present the general matrix $\mathbf{J}(r)$ to obtain the convex polyhedral cone $V(r)$ (6.182) for $r \in \mathbb{R}_0^+$, as well as the specific matrix $\mathbf{J}(0)$,

$$\begin{aligned} \mathbf{J}(r) &= (\mathbf{F}' - \mathbf{S}'(1+r)) \\ &= \begin{bmatrix} 3-3r & -2-2r & -r \\ -1-2r & 3-3r & -1-r \\ -1-r & -r & 4-2r \end{bmatrix}, \quad \mathbf{J}(0) = \begin{bmatrix} 3 & -2 & 0 \\ -1 & 3 & -1 \\ -1 & 0 & 4 \end{bmatrix}. \end{aligned} \tag{6.205}$$

We set up the *inequality description* $\mathbf{J}(0.1)\mathbf{p} \geq \mathbf{o}$ of the convex polyhedral cone $V(0.1)$, obtaining

$$\mathbf{J}(0.1)\mathbf{p} = \frac{1}{10}\begin{bmatrix} 27 & -22 & -1 \\ -12 & 27 & -11 \\ -11 & -1 & 38 \end{bmatrix}\begin{bmatrix} p_1 \\ p_2 \\ p_3 \end{bmatrix} = \left\{ \begin{array}{c} 27p_1 - 22p_2 - p_3 \geq 0 \\ -12p_1 + 27p_2 - 11p_3 \geq 0 \\ -11p_1 - p_2 + 38p_3 \geq 0 \end{array} \right\}. \tag{6.206}$$

Choosing the price vector $\bar{\mathbf{p}} = [1, 1, 1]'$, one evaluates $\bar{\mathbf{w}} = \mathbf{J}(0)\bar{\mathbf{p}} = [1, 1, 3]' > \mathbf{o}$. For this reason, there is $\bar{\mathbf{p}} \in V(0) \neq \emptyset$, a non empty set. The condition **UA2** is fulfilled.

(3) We continue, calculating

$$f(0) = \det(\mathbf{F}' - \mathbf{S}') = 26 \neq 0, \tag{6.207}$$

and conclude with Assumption 6.1.2 that we have the requested *linear independence of the processes*.

We compute now the three vectors spanning the convex polyhedral cone $V(0.1)$. For this purpose, we now need normal vectors $\mathbf{n}_1 = [27, -22, -1]'$, $\mathbf{n}_2 = [-12, 27, -11]'$, $\mathbf{n}_3 = [-11, -1, 38]'$ of the planes describing the three inequalities (6.206), computing the following *cross product*, giving

$$\begin{aligned} \mathfrak{v}_1 &= \mathbf{n}_2 \times \mathbf{n}_3 = \begin{bmatrix} 1{,}015 \\ 577 \\ 309 \end{bmatrix}, \quad \mathfrak{v}_2 = \mathbf{n}_3 \times \mathbf{n}_1 = \begin{bmatrix} 837 \\ 1{,}015 \\ 269 \end{bmatrix}, \\ \mathfrak{v}_3 &= \mathbf{n}_1 \times \mathbf{n}_2 = \begin{bmatrix} 269 \\ 309 \\ 465 \end{bmatrix}. \end{aligned} \tag{6.208}$$

Finally, we compute the image $V'(0.1)$ of the polyhedral cone $V(0.1)$. For this purpose we have to establish $\mathbf{w}_1 = \mathbf{J}(0.1)\boldsymbol{\upsilon}_1 = [1{,}440.2, 0, 0]'$, $\mathbf{w}_2 = \mathbf{J}(0.1)\boldsymbol{\upsilon}_2 = [0, 1{,}440.2, 0]'$, $\mathbf{w}_3 = \mathbf{J}(0.1)\boldsymbol{\upsilon}_3 = [0, 0, 1{,}440.2]'$, giving an orthogonal basis of the convex polyhedral cone $V(0.1)'$, which is in this case identical to the first octant of $\mathbb{R}^3$.

$$V'(0.1) = \left\{ \mathbf{w} | \mathbf{w} = \alpha_1 \begin{bmatrix} 1 \\ 0 \\ 0 \end{bmatrix} + \alpha_2 \begin{bmatrix} 0 \\ 1 \\ 0 \end{bmatrix} + \alpha_3 \begin{bmatrix} 0 \\ 0 \\ 1 \end{bmatrix} \mid \alpha_1 \geq 0, \alpha_2 \geq 0 \alpha_3 \geq 0 \right\}. \quad (6.209)$$

As we consider the specific case $r = 0.1 > 0$, one can choose any *semi-positive* vector $\mathbf{w} \geq \mathbf{o}$, one always gets a *positive* price vector $\mathbf{p} = \mathbf{J}(0.1)^{-1}\mathbf{w} \in V(0.1)$, which then is a solution of the price model (6.177). Consider for example: $\mathbf{p} = [837, 1{,}015, 269]' = \mathbf{J}(0.1)^{-1}[0, 1{,}440.2, 0]'$. We go on extending the domain of rates of profits to get the general solution. For this purpose, we consider the polynomial,

$$f(r) = \det(\mathbf{F}' - \mathbf{S}'(1+r)) = 26 - 122r + 61r^2 - 8r^3. \quad (6.210)$$

The polynomial $f(r)$ is presented in Fig. 6.13. We compute its roots, take the minimal positive root $r_1 = 0.2413$ and see that it corresponds to the Frobenius number $\lambda_C = 0.8056 = 1/(1+R) = 1/(1+0.2413)$, determining the above defined set $\Phi = [0, 0.2413[$. As for $r \in \Phi$, we have $f(r) > 0$, so the cone spanned by a maximal number of linearly independent vectors of matrix $(\mathbf{F}' - \mathbf{S}'(1+r))$ has dimensions equal to $\dim(V(r)) = n = 3$. Therefore $V(r) \neq \emptyset$ for every r for which $f(r) \neq 0$. Moreover $f(r)$ is continuous. This means that the condition **UA3** is fulfilled.

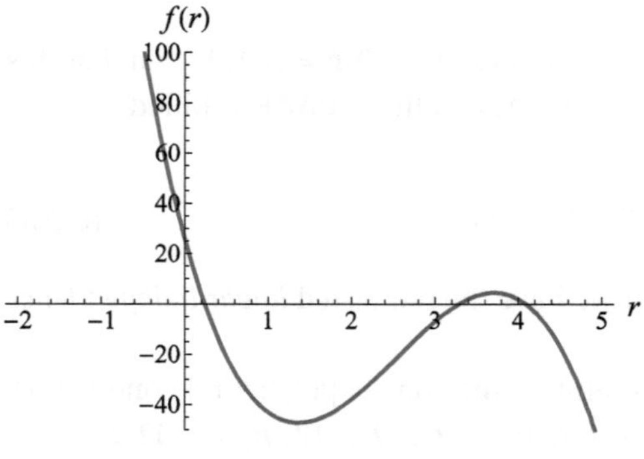

Figure 6.13: The polynomial $f(r)$ (6.210).

(4) Consequently, we compute again the matrix $\mathbf{J}(r)$ and also its inverse $\mathbf{J}(r)^{-1}$, existing for all $r \in \Phi$,

$$\mathbf{J}(r) = \begin{bmatrix} 3-3r & -2-2r & -r \\ -1-2r & 3-3r & -1-r \\ -1-r & -r & 4-2r \end{bmatrix},$$

$$\mathbf{J}(r)^{-1} = \frac{1}{f(r)} \begin{bmatrix} 12-19r+5r^2 & 8+4r-3r^2 & 2+7r-r^2 \\ 5+8r-3r^2 & 12-19r+5r^2 & 3+r-r^2 \\ 3+r-r^2 & 2+7r-r^2 & 7-24r+5r^2 \end{bmatrix} > \mathbf{0}, r \in \Phi. \quad (6.211)$$

It is moreover easy to show that all the numerators of the entries of matrix $\mathbf{J}(r)^{-1}$ are positive for $r \in \Phi$, because their roots lie outside the set $\Phi = [0, 0.2413[$. One also verifies that the numerators and the denominator $f(r)$ are all positive for $r \in \Phi, f(r) > 0$. For this reason, the existing inverse matrix $\mathbf{J}(r)^{-1} > \mathbf{0}$ is positive.

The rows of matrix $\mathbf{J}(r)$ are normal vectors to the 3 planes edging the convex polyhedral cone $V(r)$. They are: $\mathbf{n}_1 = [3-3r, -2-2r, -r]'$, $\mathbf{n}_2 = [-1-2r, 3-3r, -1-r]'$, $\mathbf{n}_3 = [-1-r, -r, 4-2r]'$. We then calculate the cross vectors (6.208), $\boldsymbol{\upsilon}_1 = \mathbf{n}_2 \times \mathbf{n}_3 = [12-19r+5r^2, 5+8r-3r^2, 3+r-r^2]'$ $\boldsymbol{\upsilon}_2 = \mathbf{n}_3 \times \mathbf{n}_1 = [8+4r-3r^2, 12-19r+5r^2, 2+7r-r^2]'$, $\boldsymbol{\upsilon}_3 = \mathbf{n}_1 \times \mathbf{n}_2 = [2+7r-r^2, 3+r-r^2, 7-24r+5r^2]'$, spanning the convex polyhedral cone $V(r)$. Having set $f(r) = 26 - 122r + 61r^2 - 8r^3$, we calculate $\mathbf{w}_1 = \mathbf{J}(r)\boldsymbol{\upsilon}_1 = [f(r), 0, 0]'$, $\mathbf{w}_2 = \mathbf{J}(r)\boldsymbol{\upsilon}_2 = [0, f(r), 0]'$, $\mathbf{w}_3 = \mathbf{J}(r)\boldsymbol{\upsilon}_3 = [0, 0, f(r)]'$, spanning the convex polyhedral cone $V'(r)$ (6.209), which is the whole first octant of $\mathbb{R}^3$. We conclude that Manara's condition **UA4** (6.189) is fulfilled,

$$r \in \Phi \Rightarrow (\mathbf{w} = \mathbf{J}(r)^{-1}\mathbf{p} \in V'(r) \Leftrightarrow \mathbf{p} = \mathbf{J}(r)\mathbf{w} \in V(r)). \quad \blacktriangle \quad (6.212)$$

The techniques proposed in this section for vector spaces of dimension $n = 2$ and $n = 3$ can be extended to dimensions $n > 3$, applying the methods of analytical geometry of the vector space $\mathbf{R}^n$.

6.7 Accounting for land and natural resources

We start this section by taking Example 3.1.2, which we shall require later on in the extended Example 6.7.2 including land. Consider the *Garden of Eden economy* with the system of production for wheat, iron and pigs:

$$\begin{aligned} &(240 \text{ qr. wheat}, 12 \text{ t. iron}, 18 \text{ pigs}) \to (450 \text{ qr. wheat}, 0, 0), \\ &(90 \text{ qr. wheat}, 6 \text{ t. iron}, 12 \text{ pigs}) \to (0, 21 \text{ t. iron}, 0), \\ &(120 \text{ qr. wheat}, 3 \text{ t. iron}, 30 \text{ pigs}) \to (0, 0, 60 \text{ pigs}). \end{aligned} \quad (6.213)$$

There is no surplus. Computing the *input-output coefficients* matrix $\mathbf{C} = \mathbf{S}\hat{\mathbf{q}}^{-1}$ (3.33), we obtain the *left eigenvector* equation (3.34), $\mathbf{p}'\mathbf{C} = \lambda\mathbf{p}'$ and the price vector $\mathbf{p} = [1, 10, 5]'$ for this *Garden of Eden* economy, giving Sraffa's (PCMC, Par. 2), here with wheat as the *numéraire*. The Frobenius number is $\lambda_C = 1$. The production scheme (6.213) gives us also the vector of total output $\mathbf{q}_I = [450, 21, 60]'$. We compute the *circulating capital* $K = \mathbf{q}_I'\mathbf{p} = 960$ which is evidently equal to *total output* X.

From now on, we develop Example 3.1.2 adding a surplus of production and a labour force (measured in *man-years*). This gives

Example 6.7.1. Extend the system of production (6.213) to an economy producing surplus,

$$\begin{aligned}(240 \text{ qr. wheat}, 12 \text{ t. iron}, 18 \text{ pigs}, L_1 = 84) &\to (600 \text{ qr. wheat}, 0, 0),\\ (90 \text{ qr. wheat}, 6 \text{ t. iron}, 12 \text{ pigs}, L_2 = 42) &\to (0, 30 \text{ t. iron}, 0),\\ (120 \text{ qr. wheat}, 3 \text{ t. iron}, 30 \text{ pigs}, L_3 = 42) &\to (0, 0, 90 \text{ pigs}).\end{aligned} \tag{6.214}$$

Then consider the three cases:

(a) Extreme case: "exploitation of labour economy", where all the surplus goes to the entrepreneurs and no wages are paid to the workers, the wage rate is $w = 0$.

(b) Extreme case: "domination of labour economy", where all the surplus goes to the workers and no profit is paid to the entrepreneurs, the rate of profits is $r = 0$.

(c) A Sraffa-type case of "uniform distributive economy", all the workers have the same *wage rate* $w > 0$ and all the entrepreneurs have the same rate of profits $r = 0.10$.

Set up the single-product Sraffa system (4.174) with the *numéraire* wheat and calculate the price vector $\mathbf{p} = [p_1 = 1, p_2, p_3]'$, and calculate the *total output* X, the *circulating capital* K, *the national income* Y, the *total profit* P, the *total wages* W, the *share of total profits* $\tilde{r}$ and the *share of total wages* $\tilde{w}$ (4.175).

Solution to Example 6.7.1:

We reproduce the Sraffa price model (4.174)

$$\boxed{\begin{aligned}\mathbf{S}'\mathbf{p}(1+r) + \mathbf{L}w &= \hat{\mathbf{q}}\mathbf{p},\\ Y &= (\mathbf{q}' - (\mathbf{Se})')\mathbf{p} =: \mathbf{d}'\mathbf{p},\\ L &= \mathbf{e}'\mathbf{L},\\ w &= \frac{\tilde{w} \cdot Y}{L},\end{aligned}} \tag{6.215}$$

and the definitions of the various economic variables (4.175)

$$\begin{aligned} X = \mathbf{q}'\mathbf{p}, \quad & K = (\mathbf{Se})'\mathbf{p},\\ P = (\mathbf{Se})'\mathbf{p} \cdot r = K \cdot r, \quad & W = Y - P,\\ w = \frac{\tilde{w} \cdot Y}{L}, \quad & \tilde{r} = \frac{P}{Y}. \end{aligned} \tag{6.216}$$

At first, we have to identify the matrices and vectors

$$\mathbf{S} = \begin{bmatrix} 240 & 90 & 120 \\ 12 & 6 & 3 \\ 18 & 12 & 30 \end{bmatrix}, \quad \mathbf{q} = \begin{bmatrix} 600 \\ 30 \\ 90 \end{bmatrix}, \quad \mathbf{L} = \begin{bmatrix} 84 \\ 42 \\ 42 \end{bmatrix}. \tag{6.217}$$

(a) As the *wage rate* disappears, $w = 0$, the Sraffa price equation (6.215) becomes the eigenvalue equation $\mathbf{C}'\mathbf{p}(1+R) = \mathbf{p}$. Therefore, we have to compute the matrices

$$\hat{\mathbf{q}}^{-1} = \begin{bmatrix} \frac{1}{600} & 0 & 0 \\ 0 & \frac{1}{30} & 0 \\ 0 & 0 & \frac{1}{90} \end{bmatrix}, \quad \mathbf{C} = \mathbf{S}\hat{\mathbf{q}}^{-1} = \begin{bmatrix} \frac{2}{5} & 3 & \frac{4}{3} \\ \frac{1}{50} & \frac{1}{5} & \frac{1}{30} \\ \frac{3}{100} & \frac{2}{5} & \frac{1}{3} \end{bmatrix}. \tag{6.218}$$

Then, we calculate the vector of prices $\mathbf{p}$, using the left eigenvector of matrix $\mathbf{C}$ with $p_1 = 1$. We get $\mathbf{p} = [1, 9.203, 4.314]'$. Then, for this "exploitation of labour economy", we calculate *total output* $X = 1{,}264.38$, the *circulating capital* $K = 902.12$, and *national income* $Y = 362.26$ which is equal to *total profits* $P = 362.26$ because there are no *wages* $W = 0$, only subsistence wages incorporated in the process of production. Clearly, $\tilde{r} = 1$, $\tilde{w} = 0$.

(b) As the *rate of profits* disappears, $r = 0$, the Sraffa price equations (6.215) become

$$\begin{aligned}\mathbf{S}'\mathbf{p} + \mathbf{L}w &= \hat{\mathbf{q}}\mathbf{p},\\ Y &= (\mathbf{q}' - (\mathbf{S}\mathbf{e})')\mathbf{p},\\ L &= \mathbf{e}'\mathbf{L},\\ w &= \frac{\tilde{w} \cdot L}{L}.\end{aligned} \tag{6.219}$$

With $p_1 = 1$, we calculate the vector of prices $\mathbf{p} = [1, 9.394, 3.939]'$. Then we calculate for this "domination of labour economy" the *total output* $X = 1{,}236.36$, the *circulating capital* $K = 883.64$, and the *national income* $Y = 352.73$ which is equal to *total wages* $W = 352.73$ because there is no *profit* $P = 0$. Clearly, $\tilde{r} = 0$, $\tilde{w} = 1$.

(c) Finally, we set $r = 0.1$ for the *rate of profits* and solve the Sraffa price equations (6.215), again with $p_1 = 1$. We calculate the vector of prices $\mathbf{p} = [1, 9.343, 4.024]'$. Then we calculate for this "uniform distributive economy" *total output* $X = 1{,}242.39$, the *circulating capital* $K = 887.60$, the *national income* $Y = 354.79$, the *total profits* $P = 88.76$, the *total wages* $W = 266.03$, the *wage share of total income* $\tilde{w} = 0.7498$ and the *profit share of total income* $\tilde{r} = 0.2502$.

We observe that the variations in prices are very small between these three regimes. ▲

We are at present ready to introduce the use of land and natural resources, (PCMC, Chap. XI). The two must be distinguished: natural resources such as sunlight, water, minerals, crude oil and natural rubber (latex), must be exploited and conditioned to become commodities entering into the means of production. As is obvious from the few foregoing examples, some of these natural resources require land acting as the carrier for the exploitation.

Now in PCMC, Par. 85, Sraffa writes:

> *"Natural resources which are used in production, such as land and mineral deposits, and which being in short supply enable their owners to obtain a rent, can be said to occupy among means of production a position equivalent to that of "non-basics" among products. Being employed in production, but not themselves produced, they are the converse of commodities which, although produced, are not used in production. They are in fact already included under the wider definition of non-basics given in Par 60."*

and adds a few lines later:

"...(natural resources) only appear on one side of the production process."

This apodeictic statement calls for a *caveat lecture*: the situation is far more subtle than implied by Sraffa and depends in particular on how land and natural resources are incorporated in the price model of a system of production.

Let us first consider land. Every productive entity is situated in one way or another on a land site that is attached to it, just as a labour force[24] is attached to each productive entity, as already mentioned earlier in this text.

Either

(a) the land is owned by the productive entity and is part of its fixed capital (not addressed in this text), or
(b) the productive entity has obtained a concession to exploit the land and pays a royalty to the landowner (terms defined case by case), or
(c) the land is rented and a rent is paid by the productive entity to the landowner.

In cases (b) and (c), the rents or royalties are then explicitly accounted for in the price equations on the left-hand side of the price equations, like wages paid. Land accounted for in this manner then does not have the status of a commodity, just as the labour force is not a commodity. Here, to be specific, we follow the approach (c).

However this does not preclude land from being considered as a commodity in certain processes. For example, a landowner of fallow land will invest in it, and then offer equipped land as a commodity for real-estate building purposes, or sell forest land to the lumber industry.

Sraffa considers in PCMC (Par. 86) a specific *system of joint production* of n processes and n commodities where there is a *single-product* agricultural industry producing "corn" on arable land. We will now develop the production scheme (6.1), corresponding to this economy, and we assume that the agricultural industry is labeled as the first industry $j = 1$. We further assume that this industry does not produce any other commodities. This gives for the agricultural industry $j = 1$ a production scheme with a total "corn" production, $i = 1$, as output f_{11},

$$(s_{11}, s_{21}, s_{31}, \ldots, s_{n1}, L_1) \rightarrow (f_{11}, 0, 0, \ldots, 0). \tag{6.220}$$

We then explicitly assume that the other remaining industries $j \in \{2, \ldots, n\}$ produce the remaining $n - 1$ commodities $i \in \{2, \ldots, n\}$ but not "corn". Therefore, the resulting production scheme has 0 elements in the first column of $\mathbf{F}'$ and looks as follows,

24 In present-day parlance, one speaks of human resources. This term is not harmless, one step further in ultra-liberal economic theory, and human resources, for which there is a market with demand and supply, and a price (wages), become a commodity. A commodity which in the future will be replaced by another commodity, robots, in many economic sectors.

$$\begin{aligned}
&(s_{12}, s_{22}, s_{32}, \ldots, s_{n2}, L_2) \to (0, f_{22}, f_{32}, \ldots, f_{n2}),\\
&(s_{13}, s_{23}, s_{33}, \ldots, s_{n3}, L_3) \to (0, f_{23}, f_{33}, \ldots, f_{n3}),\\
&(\ldots, \ldots, \ldots, \ldots, \ldots, \ldots) \to (0, \ldots, \ldots, \ldots, \ldots),\\
&(s_{1n}, s_{2n}, s_{3n}, \ldots, s_{nn}, L_n) \to (0, f_{2n}, f_{3n}, \ldots, f_{nn}).
\end{aligned} \tag{6.221}$$

Ricardo analysed the question of the land rent, and considered an economic situation where the production of "corn", commodity $i = 1$, had to be increased and new land be explored. He considered that the profitability of the new land is lower than that of the old land and that the new land needed more labour. Sraffa took up Ricardo's idea and considered that the initial "corn" industry is developed in m agricultural sectors, each sector working on different qualities and quantities of land. As there is an extension of the initial single agricultural industry, it is assumed that the m new sectors together produce a greater amount of "corn" than the initial agricultural industry (6.220). Thus, each of the m sectors produces "corn" on a area of G_k, $k = 1, \ldots, m$, acres of arable land. The input quantities s_{i1} are replaced by m non-negative parts $u_{ik} \geq 0$, as also the output quantity f_{11} by m $v_{1k} \geq 0$ and the total labour L_1 of the initial agricultural industry $j = 1$ by m $l_k \geq 0$. According to the problem set by Ricardo, we obtain the inequalities $\sum_{k=1}^{m} u_{ik} \geq s_{i1}$, $i = 1, \ldots, n$ for the inputs, the inequalities $\sum_{k=1}^{m} v_{1k} \geq f_{11}$ for the output and the inequality $\sum_{k=1}^{m} l_k \geq L_1$ for the total labour. The m sector-subsystem thus obtained presents itself now as follows.[25] The areas of arable land G_k are newly introduced in the production scheme:

$$\begin{aligned}
&(u_{11}, u_{21}, u_{31}, \ldots, u_{n1}, l_1, G_1) \to (v_{11}, 0, 0, 0, 0),\\
&(\ldots, \ldots, \ldots, \ldots, \ldots, \ldots, \ldots\ldots) \to (\ldots, 0, 0, 0, 0),\\
&(u_{1m}, u_{2m}, u_{3m}, \ldots, u_{mn}, l_m, G_m) \to (v_{1m}, 0, 0, 0, 0).
\end{aligned} \tag{6.222}$$

On the basis of the present production scheme constituted by *multi-product industries*[26] (6.221) and the set of *single-product industry* sectors (6.222), we are able to define the elements of production of this specific production process. On the right side, there is a $(m + n - 1) \times n$ matrix reflecting the fact that the price of "corn" is the same for all *m agricultural sectors*:

$$\begin{aligned}
&(u_{11}, u_{21}, u_{31}, \ldots, u_{n1}, l_1, G_1) \to (v_{11}, 0, 0, 0, 0),\\
&(\ldots, \ldots, \ldots, \ldots, \ldots, \ldots, \ldots) \to (\ldots, 0, 0, 0, 0),\\
&(u_{1m}, u_{2m}, u_{3m}, \ldots, u_{mn}, l_m, G_m) \to (v_{1m}, 0, 0, 0, 0),\\
&(s_{12}, s_{22}, s_{32}, \ldots, s_{n2}, L_2, 0) \to (0, f_{22}, f_{32}, \ldots, f_{n2}),\\
&(s_{13}, s_{23}, s_{33}, \ldots, s_{n3}, L_3, 0) \to (0, f_{23}, f_{33}, \ldots, f_{n3}),\\
&(\ldots, \ldots, \ldots, \ldots, \ldots, \ldots, 0) \to (0, \ldots, \ldots, \ldots, \ldots),\\
&(s_{1n}, s_{2n}, s_{3n}, \ldots, s_{nn}, L_n, 0) \to (0, f_{2n}, f_{3n}, \ldots, f_{nn}).
\end{aligned} \tag{6.223}$$

25 Further investigations on this topic are treated by Kurz and Salvadori ([52], Chap. 10).

26 The terms *joint production* and *multi-product industries* are used synonymously by F. Duchin and A. E. Stenge [23].

We can identify following matrices:

- $\mathbf{U} = \mathbf{U}_{n\times m} = (u_{ij})$ is the *commodity flow* $n \times m$ submatrix limited to the agricultural sectors, the extension of the initial agricultural industry $j = 1$;
- $\underline{\mathbf{S}} = \underline{\mathbf{S}}_{n\times(n-1)} = (s_{ij})$, $i = 1, \ldots, n$, is the technological $n \times (n-1)$ submatrix of the initial *commodity flow* $n \times n$ matrix $\mathbf{S}$, containing the $n-1$ industries $j\, j = 2, \ldots, n$;
- $\mathbf{l} = \mathbf{l}_{m\times 1} = [l_1, l_2, \ldots, l_m]'$ is the labour vector attached to each of the m agricultural sectors (6.222). l_k is the labour needed to work for agricultural activities on land k, $k = 1, \ldots, m$;
- $\underline{\mathbf{L}} = \underline{\mathbf{L}}_{(n-1)\times 1} = [L_2, L_3, \ldots, L_n]'$ is the vector of labour attached to the $n-1$ remaining industries j, $j = 2, \ldots, n$ (6.221);
- $\mathbf{G} = \mathbf{G}_{(m+n-1)\times 1} = [G_1, G_2, \ldots, G_m, 0, 0, \ldots, 0]'$ is the vector of m productive lands based on an appropriate surface measure, such as m^2 or acres, attached to each agricultural sector k, $k = 1, \ldots m$, extended with 0 to a $n + m - 1$ vector. Then we define the corresponding diagonal matrix $\hat{\mathbf{G}} = \text{diag}(\mathbf{G})$ (A.17);
- in each agricultural sector k, the landowner gets a land rent, $R_k = G_k \rho_k$, where ρ_k expresses the *land-based profit rate* per m^2 or acres, $k = 1, \ldots, m$. Without limitation of generality, we assume that the rents R_k are arranged in descending order. According to the concept and assumption of Sraffa, the land of lowest arable quality gives no profit. The corresponding *land-based profit* (land rent) vanishes, $\rho_m = 0$.
  The vector of *land based profit rates* is

$$\boldsymbol{\rho} = \boldsymbol{\rho}_{(n+m-1)\times 1} = [\rho_1, \rho_2, \ldots, \rho_{m-1}, 0, 0, \ldots, 0]';$$

- then, we define the $n \times m$ output matrix of the m agricultural sectors, the output coefficients matrix $\mathbf{V}$, and the $n \times (n-1)$ output matrix of the $n-1$ initial industry sectors, the output coefficients matrix $\underline{\mathbf{F}}$,

$$\mathbf{V} = \begin{bmatrix} v_{11} & v_{12} & \ldots & v_{1m} \\ 0 & 0 & 0 & 0 \\ 0 & 0 & 0 & 0 \\ \ldots & \ldots & \ldots & \ldots \\ 0 & 0 & 0 & 0 \end{bmatrix}; \quad \underline{\mathbf{F}} = \begin{bmatrix} 0 & 0 & 0 & 0 \\ f_{22} & f_{23} & \ldots & f_{2n} \\ f_{32} & f_{33} & \ldots & f_{3n} \\ \ldots & \ldots & \ldots & \ldots \\ f_{n2} & f_{n3} & \ldots & f_{nn} \end{bmatrix}; \tag{6.224}$$

- w is the wage per unit of labour;
- r is the uniform rate of profits defined by the global system of production in which the subsystem is imbedded;
- with all these items, $p_C = p_1$ is the price of the numéraire corn, we calculate the $n \times 1$ price vector: $\mathbf{p} = [p_C, p_2, \ldots, p_n]'$, the vector of n prices of the n commodities. The price p_C of corn is assumed the same for everyone of the m considered *agricultural sectors*.

The production scheme (6.223) can now be written with the just-defined matrices and vectors as follows:

$$\left(\left[\begin{array}{c}\mathbf{U}'\\ \underline{\mathbf{S}}'\end{array}\right],\left[\begin{array}{c}\mathbf{l}\\ \underline{\mathbf{L}}\end{array}\right],\hat{\mathbf{G}}\right)\rightarrow\left(\left[\begin{array}{c}\mathbf{V}'\\ \underline{\mathbf{F}}'\end{array}\right]\right). \tag{6.225}$$

Now, we establish the Sraffa price model in analogy to (6.48). We adapt the calculation of the national income Y by including land revenue. We have to add the *total land rent* $R_L = \sum_{k=1}^{m} R_k$. This *total land rent* is the summed up product of amounts of acres multiplied by the *land-based profit rate* per acre, $R_L = \mathbf{e}'(\hat{\mathbf{G}}\boldsymbol{\rho})$. We also include the total labour L, necessary to calculate the wage rate $w = \frac{\tilde{w}\cdot Y}{L}$ (4.175),

$$\boxed{\begin{array}{c}\left[\begin{array}{c}\mathbf{U}'\\ \underline{\mathbf{S}}'\end{array}\right]\mathbf{p}(1+r)+\left[\begin{array}{c}\mathbf{l}\\ \underline{\mathbf{L}}\end{array}\right]\dfrac{\tilde{w}\cdot Y}{L}+\hat{\mathbf{G}}\boldsymbol{\rho}=\left[\begin{array}{c}\mathbf{V}'\\ \underline{\mathbf{F}}'\end{array}\right]\mathbf{p},\\ Y=\mathbf{e}'\left[\begin{array}{c}\mathbf{V}'-\mathbf{U}'\\ \underline{\mathbf{F}}'-\underline{\mathbf{S}}'\end{array}\right]\mathbf{p},\\ L=\mathbf{e}'\left[\begin{array}{c}\mathbf{l}\\ \underline{\mathbf{L}}\end{array}\right].\end{array}} \tag{6.226}$$

Having obtained the prices, we then have to adapt the expressions to calculate total output X, the *circulating capital* K, total profits P, total wages W and the wage rate w of the system, as in equation (6.49). We get:

$$\begin{array}{l}X=\left(\mathbf{e}'\left[\begin{array}{c}\mathbf{V}'\\ \underline{\mathbf{F}}'\end{array}\right]\right)\mathbf{p},\quad K=\left(\mathbf{e}'\left[\begin{array}{c}\mathbf{U}'\\ \underline{\mathbf{S}}'\end{array}\right]\right)\mathbf{p},\quad R_L=\mathbf{e}'(\hat{\mathbf{G}}\boldsymbol{\rho}),\\ P=\left(\mathbf{e}'\left[\begin{array}{c}\mathbf{U}'\\ \underline{\mathbf{S}}'\end{array}\right]\right)\mathbf{p}\cdot r=K\cdot r,\quad W=\left(\mathbf{e}'\left[\begin{array}{c}\mathbf{l}\\ \underline{\mathbf{L}}\end{array}\right]\right)\cdot w,\\ w=\dfrac{\tilde{w}\cdot Y}{L},\quad Y=X-K=P+W+R_L.\end{array} \tag{6.227}$$

As has been mentioned, Sraffa sets arbitrarily the *land-based profit rate* of the least productive land number m equal to zero, $\rho_m = 0$. The other $k-1$ land rents must be strictly positive. This just indicates that the landowner of the least productive land receives no rent, but the entrepreneur exploiting this land, employing labour and machinery, can still make a profit at the common rate r.

Some of the components of the price vector $\mathbf{p}$ may be negative in the present context of joint production.

A further assumption is that the m agricultural sectors of land of different quality all produce "corn" at the same price p_C. This additional assumption is expressed on the right side of equation (6.226) by the product $\mathbf{V}'\mathbf{p}$.

Let us now count the number of variables that have to be calculated in the equations (6.226). There are the *n prices* $p_C, p_2, \ldots, p_n$ of price vector $\mathbf{p}$, the *rate of profits* r, the $m-1$ *land profit rates* $\rho_1, \rho_2, \ldots, \rho_{m-1}$, the *national income* Y and the *total labour* L. This gives $n+m+2$ unknown variables in a system of $n+m+1$ equations. If all the equations are linearly independent, then we have to determine exogenously one of the variables.

Here, the rate of profits $r_0 = r$ is exogenously determined.

We now illustrate this land rent system starting from Example 6.7.1.

We follow the approach of Ricardo and Sraffa. In a first step, we replace the agricultural industry by $m = 2$ agricultural sectors, such that both sectors need the same amounts of commodities as means of production and produce together the same amount of wheat as the initial agricultural industry. The first sector farms $G_1 = 3{,}000$ *acres* of land with a quantity of labour of $l_1 = 56$ *man-year*; the second sector farms $G_2 = 1{,}000$ *acres* of new arable land of lower quality with a quantity of labour of $l_2 = 28$ *man-years*. The second and the third industries remain unchanged. The capacity of the entire economy remains unchanged. This leads us to

Example 6.7.2. Without specifying the units of the quantity of labour and of the arable land in the present production scheme:

$$\begin{aligned}
(140, 7, 10, l_1 = 56, G_1 = 3{,}000, 0, 0, 0) &\to (400, 0, 0),\\
(100, 5, 8, l_2 = 28, 0, G_2 = 1{,}000, 0, 0) &\to (200, 0, 0),\\
(90, 6, 12, L_2 = 42, 0, 0, 0, 0) &\to (0, 30, 0),\\
(120, 3, 30, L_3 = 42, 0, 0, 0, 0) &\to (0, 0, 90).
\end{aligned} \tag{6.228}$$

- The profit rate is $r = 0.1$. The prices of both "corn" qualities are equal, $p_C = p_{C_1} = p_{C_2}$.
- The land profit rate (land rent of the second agricultural sector) is $G_2 p_2 = 0$, in accordance with the views of Ricardo and Sraffa on this point.

Set up the single product Sraffa system (6.226) with the *numéraire* wheat and calculate the price vector $\mathbf{p} = [p_C = 1, p_2, p_3]'$, the *total output* X, the *circulating capital* K, the *national income* Y, the *total profits* P, the *total wages* W, the *total land rent* R_L, the *share of total profits* $\tilde{r}$, the *share of total wages* $\tilde{w}$ (6.227) and the *wage rate* w.

Solution to Example 6.7.2:

We start by identifying the matrices of the system (6.228)

$$\mathbf{U} = \begin{bmatrix} 140 & 100 \\ 7 & 5 \\ 10 & 8 \end{bmatrix}, \quad \underline{\mathbf{S}} = \begin{bmatrix} 90 & 120 \\ 6 & 3 \\ 12 & 30 \end{bmatrix}, \quad \mathbf{V} = \begin{bmatrix} 400 & 200 \\ 0 & 0 \\ 0 & 0 \end{bmatrix},$$

$$\underline{\mathbf{F}} = \begin{bmatrix} 0 & 0 \\ 30 & 0 \\ 0 & 90 \end{bmatrix}, \quad \mathbf{l} = \begin{bmatrix} 56 \\ 28 \end{bmatrix}, \quad \underline{\mathbf{L}} = \begin{bmatrix} 42 \\ 42 \end{bmatrix}, \quad \mathbf{G} = \begin{bmatrix} 3{,}000 \\ 1{,}000 \\ 0 \\ 0 \end{bmatrix}. \tag{6.229}$$

With $r = 0.1$ and $p_C = 1$ we solve the Sraffa price equations (6.226),

$$\begin{bmatrix} 140 & 7 & 10 \\ 100 & 5 & 8 \\ 90 & 6 & 12 \\ 120 & 3 & 30 \end{bmatrix} \begin{bmatrix} p_C \\ p_2 \\ p_3 \end{bmatrix} (1+r) + \frac{\tilde{w} \cdot Y}{L} \begin{bmatrix} 56 \\ 28 \\ 42 \\ 42 \end{bmatrix}$$

$$+ \begin{bmatrix} 3{,}000 & 0 & 0 & 0 \\ 0 & 1{,}000 & 0 & 0 \\ 0 & 0 & 0 & 0 \\ 0 & 0 & 0 & 0 \end{bmatrix} \begin{bmatrix} \rho_1 \\ \rho_2 \\ 0 \\ 0 \end{bmatrix} = \begin{bmatrix} 400 & 0 & 0 \\ 200 & 0 & 0 \\ 0 & 30 & 0 \\ 0 & 0 & 90 \end{bmatrix} \begin{bmatrix} p_C \\ p_2 \\ p_3 \end{bmatrix},$$

$$Y = [1,1,1,1] \begin{bmatrix} 260 & -7 & -10 \\ 100 & -5 & -8 \\ -90 & 24 & -12 \\ -120 & -3 & 60 \end{bmatrix} \begin{bmatrix} p_C \\ p_2 \\ p_3 \end{bmatrix},$$

$$L = [1,1,1,1] \cdot [56,28,42,42]' = 168. \tag{6.230}$$

We obtain the vector of prices $\mathbf{p} = [1, 7.39, 3.28]'$ and the *land-based profit rates* $\boldsymbol{\rho} = [0.03735, 0, 0, 0]'$. Then, we calculate for the present "uniform distributive economy" with a split-up agricultural industry the *total output* $X = 1{,}117.19$, the *circulating capital* $K = 802.19$, the *national income* $Y = 315.00$, the *total profits* $P = 80.22$, the *total wages* $W = 122.73$, the *total land rent* $R_L = 112.06$, the *wage share of total income* $\tilde{w} = 0.3896$, the *profit share of total income* $\tilde{r} = 0.2547$, the *rent share to total income* $\tilde{s} = 0.3557$, verifying the sum $\tilde{r} + \tilde{w} + \tilde{s} = 1$, and finally the *wage rate* $w = 0.7305$.

The total surplus is now distributed (see Bortis ([8], p. 144) or Lipsey ([58], p. 494)). The national income $Y = P + W + R_L = 80.22 + 122.73 + 112.06 = 315.00$ is distributed as indicated in the proceeding explanations. ▲

We now proceed to analyse how a variation of the arable area acts on the prices and the various economic variables. For this purpose, we take *Example* 6.7.2 and double the capacity of the second agricultural sector of arable land of lower quality.

Example 6.7.3. Consider the production scheme (6.228) and replace the second agricultural sector by:

$$(200, 10, 16, l_2 = 56, G_2 = 2{,}000) \to (400, 0, 0) \tag{6.231}$$

The rate of profits remains $r = 0.1$. The prices of both "corn" qualities are equal, $p_C = p_{C_1} = p_{C_2}$.

Set up the single-product Sraffa system (6.226) with the *numéraire* wheat and calculate the price vector $\mathbf{p} = [p_C = 1, p_2, p_3]'$, the *total output* X, the *circulating capital* K, the *national income* Y, the *total profit* P, the *total wages* W, the *total land rent* R_L, the *share of total profits* $\tilde{r}$, the *share of total wages* $\tilde{w}$ (6.227) and finally the *wage rate* w.

Solution to Example 6.7.3:
We have to identify the modified matrices of the system (6.228), getting

$$\mathbf{U} = \begin{bmatrix} 140 & 200 \\ 7 & 10 \\ 10 & 16 \end{bmatrix}, \quad \mathbf{V} = \begin{bmatrix} 400 & 400 \\ 0 & 0 \\ 0 & 0 \end{bmatrix}, \quad \begin{bmatrix} \mathbf{l} \\ \underline{\mathbf{L}} \end{bmatrix} = \begin{bmatrix} 56 \\ 56 \\ 42 \\ 42 \end{bmatrix}, \quad \mathbf{G} = \begin{bmatrix} 3{,}000 \\ 2{,}000 \\ 0 \\ 0 \end{bmatrix}. \tag{6.232}$$

We set for the *rate of profits* $r = 0.1$ and solve the Sraffa price equations (6.226) with $p_C = 1$, because wheat remains the *numéraire*. We obtain the vector of positive prices $\mathbf{p} = [1, 7.39, 3.28]'$ and the *land-based profit rates* $\boldsymbol{\rho} = [0.03735, 0, 0, 0]'$. Then, we calculate for this "uniform distributive economy" with a split-up agricultural industry the *total output* $X = 1{,}317.19$, the *circulating capital* $K = 965.41$, the *national income* $Y = 351.78$, the *total profits* $P = 96.54$, the *total wages* $W = 143.18$, the *total land rent* $R_L = 112.06$, the *wage share of total income* $\tilde{w} = 0.4070$, the *profit share of total income* $\tilde{r} = 0.2744$, the *rent share to total income* $\tilde{s} = 0.3186$, verifying $\tilde{r} + \tilde{w} + \tilde{s} = 1$, and the *wage of profit* $w = 0.7305$.

As before, the total surplus is now distributed between three players.[27] The workers and entrepreneurs having foregone a potential further participation in surplus to the advantage of the landowner, and this will accordingly also appear in the national accounts for the whole economy, i. e., in addition to profits and wages, (land) rents enter the national accounts. ▲

We now analyse the evolution of the (relative) prices and the economic processes issuing from Example 6.7.1 to 6.7.3.

Interpreting the results summarised in Table 6.2, taking into account that wheat is the *numéraire* and therefore the variables are measured in one unit of wheat, we observe:

Table 6.2: Comparison of the calculated examples (rounded numbers).

Examples	Relative prices			Economic variables					
	wheat	iron	pigs	X	K	Y	P	W	R_L
3.1.2	1	10	5	960	960	0	0	0	0
6.7.1(a)	1	9.20	4.31	1,264	902	362	362	0	0
6.7.1(b)	1	9.39	3.40	1,236	884	353	0	353	0
6.7.1(c)	1	9.34	4.02	1,242	888	355	89	266	0
6.7.2	1	7.39	3.28	1,117	802	315	80	123	112
6.7.3	1	7.39	3.28	1,317	965	352	97	143	112

27 This example nicely illustrates the "Three Rents" process in an agricultural economy, first described by Richard Cantillon (1680–1734) in the early 18th century, see Murphy ([67], Chap. 4).

(a) The introduction of *land rents* and different qualities of land makes the price of wheat increase with respect to iron and pigs. On the other hand, one needs more wheat in Examples 6.7.1 (a), (b) (c), Example 6.7.2 and Example 6.7.3 than in the preceding Example 3.1.2, represented by the system of production (6.213).

(b) In Example 6.7.3 we just have doubled the capacity of the agricultural sector 2, compared to Example 6.7.3. This did not alter the structure of agricultural sector 2. Therefore the relative prices remained unchanged, as well as the *land rent* R_L issued form the agricultural sector 1.

7 Sraffa's theory of joint production as a tool in ecological economics

7.1 Introduction

Up to now, we have abundantly treated *single-product industries* and *joint production processes* and have developed algebraic and graphical methods of analysis. We come back to this important characterisation of production processes and will apply the model of *joint production* processes to ecological economics.[1]

Sraffa's book PCMC was written at a time when ecological problems received little attention. But in 1983, when the ecological crisis had become evident, Schefold ([102], p. 323) proposed to apply Sraffa's theory of *joint production*. He evokes here the question of the "stability of the world climate" that can be increased if the *carbon dioxide* (CO_2) in Earth's atmosphere can be reduced. For this reason, the use of *fossil fuels* must be reduced. This implies: (1). Enhancement of the energy production efficiency (like compound gas-steam power stations), (2). The use of natural gas and wind energy, (3). Electro drives for transportation. In fact, Schefold says ([103], p. 30): *"This is 'the problem of environment', but it is also 'the problem of joint production'. Whereas all materials are recycled in the biosphere, so that the excretions of one species are food to others, human production results in both goods to be sold as commodities and wastes"*.

Section 7.6 to 7.5 of the present chapter are specifically devoted to this subject. But before proceeding, we will begin with four sections concerning fundamental technical aspects, some of them already addressed in previous chapters, which repeatedly appear when analysing ecological and related topics.

7.2 From Sraffa to ecological economics: technical preliminaries

We shall illustrate this by looking at Sraffa's approach: He considers (PCMC, Par. 58) a system of n processes and a group of $m = 3$ commodities ($n > m$), labelled $i \in \{1, \ldots, 3\}$. For each process, he arranges the *"quantities in which these commodities enter any one process, as means of production, and as products, in a row."* He thus obtains the following matrix with n rows and $2m = 6$ columns

$$[\ \mathbf{S}' \quad \mathbf{F}'\ ] = \begin{bmatrix} s_{11} & s_{21} & s_{31} & f_{11} & f_{21} & f_{31} \\ s_{12} & s_{22} & s_{32} & f_{12} & f_{22} & f_{32} \\ \cdots & \cdots & \cdots & \cdots & \cdots & \cdots \\ s_{1n} & s_{2n} & s_{3n} & f_{1n} & f_{2n} & f_{3n} \end{bmatrix}. \tag{7.1}$$

Then, if *basic* and *non-basic* commodities are present, Definition 6.3.1 can be applied to determine the number m, $(1 \le m \le n-1)$ of *non-basics*.

1 This chapter with the examples is principally due to H. Knolle.

https://doi.org/10.1515/9783110635096-007

Example 7.2.1. As an illustration of this definition for the case $n = 3$, $m = 1$, let us consider the following example, an extension of the wheat–iron model (4.1) with the breeding of race horses added to the production of wheat and iron:

$$\begin{gathered}(\mathbf{S}', \mathbf{0}) \to (\mathbf{F}'),\\ (280 \text{ qr. wheat}, 12 \text{ t. iron}, 0) \to (575 \text{ qr. wheat}, 0, 0),\\ (120 \text{ qr. wheat}, 8 \text{ t. iron}, 0) \to (0, 20 \text{ t. iron}, 0),\\ (175 \text{ qr. wheat}, 0, 9 \text{ horses}) \to (0, 0, 10 \text{ horses}).\end{gathered} \tag{7.2}$$

Solution to Example 7.2.1:
In order to decide whether horses are *basic* or *non-basic*, we have to compute matrices $\mathbf{S}_2'$ and $\mathbf{F}_2'$ (6.56) as well as the rank of the following matrix:

$$\operatorname{rank}([\ \mathbf{S}_2' \quad \mathbf{F}_2'\ ]) = \operatorname{rank}\left(\begin{bmatrix} 0 & 0 \\ 0 & 0 \\ 9 & 10 \end{bmatrix}\right) = 1 = m. \tag{7.3}$$

So, horses[2] are *non-basic*. We confirm this statement with the calculation of the Pasinetti matrix, see also Lemma 6.3.1,

$$\mathbf{H} = (\mathbf{F}' - \mathbf{S}')^{-1}\mathbf{S}' = \begin{bmatrix} \frac{16}{7} & \frac{4}{35} & 0 \\ \frac{230}{7} & \frac{38}{21} & 0 \\ 575 & 20 & 9 \end{bmatrix} = \begin{bmatrix} \mathbf{H}_{11} & \mathbf{0} \\ \mathbf{H}_{21} & \mathbf{H}_{22} \end{bmatrix}. \tag{7.4}$$

Indeed, we have found the 2×1 zero matrix $\mathbf{H}_{12} = [0, 0]'$, wherein the number $m = 1$ of columns is the number of *non-basics*. ▲

In this chapter, we assume linear independence of the n production processes, $\det(\mathbf{F} - \mathbf{S}) \neq 0$, and in some cases also to be in presence of *gross integrated industries* (6.8), $\det(\mathbf{F}) \neq 0$, see Proposition 6.2.1. As we have here linear independence, one can transform the Sraffa price model of *joint production* with labour and without profits (6.138) as follows:

2 The inclusion of racehorses, a *non-basic* luxury product mentioned in PCMC, may seem somewhat out of place here. In fact, they could be replaced by the semi-aquatic rodent nutria, or coypus (French: ragondin; German: Wasserratte). This animal used to be in high demand because of its fur being included in fashion wear. In the early 1930s, nutrias were imported from Argentina, in particular to the USA, for breading purposes to cover increasing demand for this fur. By the end of the 1930s and early 1940s, the fashion faded and demand for nutria fur decreased sharply. Various breeders let loose their stocks of this rodent. With a very high reproduction rate, nutrias proliferated out of control in some southeastern parts of the USA, becoming actually a very acute ecological problem, especially in Louisiana, wreaking havoc in wetlands and road networks due to their herbivorous feeding habits and extensive burrowing activities.

$$\mathbf{u} = [u_1, \ldots, u_n]' := \frac{\mathbf{p}}{w} = (\mathbf{F}' - \mathbf{S}')^{-1}\mathbf{L}. \tag{7.5}$$

The vector **u** is called the vector of *labour values* (see Schefold [103], p. 75).[3]

Consider the inverse matrix $(h_{ij}) = (\mathbf{F}' - \mathbf{S}')^{-1}$. We then constitute the n vectors $\mathbf{h}_{i.} = \mathbf{e}_i(\mathbf{F}' - \mathbf{S}')^{-1}$, using the unit vectors $\mathbf{e}_i$, $i \in \{1, \ldots, n\}$, Section A.2, point **(2)**, one vector for every commodity i. The coefficients h_{ij} represent the *values per unit of labour* allotted by commodity i to each commodity j, $j \in \{1, \ldots, n\}$ during the production process, leading to the labour value u_i. Thus the quadratic form

$$u_i = \mathbf{e}_i(\mathbf{F}' - \mathbf{S}')^{-1}\mathbf{L} = \sum_{j=1}^{n} h_{ij} \cdot L_j = \mathbf{h}_{i.}\mathbf{L} \tag{7.6}$$

is a scalar product, summing up the contributions $h_{ij} \cdot L_j$, resulting in the labour value u_i of commodity i.

Definition 7.2.1. (waste good) The commodity i is called a waste good, if at least one of the components of the corresponding row vector $\mathbf{h}_{i.} = \mathbf{e}_i(\mathbf{F}' - \mathbf{S}')^{-1}$ is negative. It is used with the vector of labour **L** to compute its labour value $u_i = \mathbf{h}_{i.}\mathbf{L}$.

In this connection one should distinguish between
- waste as such that it must be eliminated, e. g., nuclear wastes;
- recycled waste that enters as a commodity into the means of production of certain industries, e. g., paper, certain plastics, aluminium cans, etc. The classic example is (or was) dung used as fertilizer in agriculture.

Example 7.2.2. Consider Example 6.5.4. We suppose a given vector of labour $\mathbf{L} = [100, 200, 300, 100]'$, a wage rate $w = 593/100$. Determine the waste goods of this economy and the vector **u** of labour values.

Solution to Example 7.2.2:
We present the *commodity flow* matrix **S** and the *output* matrix **F** (6.164) of the present economy

$$\mathbf{S} = \begin{bmatrix} 20 & 10 & 30 & 40 \\ 0 & 0 & 0 & 0 \\ 50 & 50 & 10 & 40 \\ 30 & 60 & 10 & 50 \end{bmatrix}, \quad \mathbf{F} = \begin{bmatrix} 100 & 200 & 100 & 50 \\ 150 & 0 & 0 & 0 \\ 200 & 0 & 0 & 0 \\ 100 & 150 & 200 & 100 \end{bmatrix}, \tag{7.7}$$

3 The vector of *labour values* **u** is a mathematical expression belonging to a much wider concept of *labour theory of value (LTV)*. The notion *value* is used by *earlier liberal economists* such as Adam Smith and David Ricardo. The *earlier liberal economists* argued that the economic value of a good or service is determined by the total amount of socially necessary labour required to produce it. The vector of *labour values* **u** here defined is a way to express this value in the framework of Sraffa's joint production economies.

and then compute

$$\mathbf{F}' - \mathbf{S}' = \begin{bmatrix} 80 & 150 & 150 & 70 \\ 190 & 0 & -50 & 90 \\ 70 & 0 & -10 & 190 \\ 10 & 0 & -40 & 50 \end{bmatrix},$$

$$(\mathbf{F}' - \mathbf{S}')^{-1} = \begin{bmatrix} 0 & \frac{71}{11{,}860} & \frac{-11}{11{,}860} & \frac{-43}{5{,}930} \\ \frac{1}{150} & \frac{-139}{177{,}900} & \frac{-1{,}699}{177{,}900} & \frac{841}{29{,}650} \\ 0 & \frac{-4}{2{,}965} & \frac{43}{5{,}930} & \frac{-149}{5{,}930} \\ 0 & \frac{-27}{11{,}860} & \frac{71}{11{,}860} & \frac{-4}{2{,}965} \end{bmatrix}. \tag{7.8}$$

We recognise that all the commodities are *waste goods*, as every row of the matrix $(\mathbf{F}'-\mathbf{S}')^{-1}$ has at least one negative componenent. We then compute the vector of *labour values*, taking the *wage rate* $w = 593/100$,

$$\mathbf{u} = \frac{1}{w}(\mathbf{F}' - \mathbf{S}')^{-1}\mathbf{L} = \frac{100}{593}(\mathbf{F}' - \mathbf{S}')^{-1} \begin{bmatrix} 100 \\ 200 \\ 300 \\ 100 \end{bmatrix} = \begin{bmatrix} \frac{115}{593} \\ \frac{857}{1{,}779} \\ -\frac{360}{593} \\ \frac{875}{593} \end{bmatrix}, \tag{7.9}$$

where positivity of the *labour value* are not guaranteed because matrix $(\mathbf{F}' - \mathbf{S}')^{-1}$ is not *semi-positive*. We see with (7.9) that *waste goods* can induce negative *labour values* and therefore negative prices with $\mathbf{p} = w \cdot \mathbf{u} = [\frac{23}{20}, \frac{857}{300}, -\frac{18}{5}, \frac{35}{4}]'$. ▲

It makes sense to study the economics of release and abatement or recycling of undesired by-products within the framework of a theory of joint production. So, far from finding Sraffa's development (PCMC, Chapter VII) on *joint production* "too abstract", see Definition 6.3.1, we are going to apply it to very concrete problems. For example, we will extend Sraffa's wheat–iron model to include the recycling of scrap, and we will consider an alternative CO_2-emissions trading scheme.

As we know, the theory of joint production becomes difficult because new problems arise:

(a) negative prices may occur;
(b) basics and non-basics must be defined in another way than in the case of single product industries, as explained by Sraffa (PCMC, Par. 60) and in the present text by Definition 6.3.1.

In Section 4.4, we presented Sraffa's (PCMC, Par. 6) definition of basic and non-basic products for *single-product industries*.

Non-basic products are the generalisation of what Adam Smith and David Ricardo called luxuries. A classic example is gold. This case is illustrated by an extension of Sraffa's wheat–iron model (PCMC, Par. 7), see Example 4.4.3.

We cite here again Sraffa's abstract definition of *basics* and *non-basics* for the case of *joint production*:

> *"In a system of n productive processes and n commodities (no matter whether produced singly or jointly) we say that a commodity or more generally a group of m linked commodities (where m must be smaller than n and may be equal to 1) are non-basic if of the n rows (formed by the 2m quantities in which they appear in each process) not more than m rows are independent, the others being linear combinations of these."*

Inspired by Schefold ([103], p. 58), we have reformulated this definition in Proposition 6.3.1 as a *matrix rank criterion* for *non-basic* systems $(\mathbf{S}', \mathbf{F}')$.

It is not easy to understand why Sraffa's definition (PCMC, Par 60) of *basics* and *non-basics* in *joint production* is a generalisation of Sraffa's definition (Sraffa, Par. 6) for the corresponding notion for *single-product industries*. We will therefore proceed step by step following the developments presented by Pasinetti [83], Steedman [114] and especially Schefold [103].

7.3 Two commodities and two industries

The comparison of parallel industries using and producing the same commodities has its importance in ecological economics because such industries, differing in the methods of production, may produce differing pollution rates and may thus incur different penalties.

In this section, we consider an economy generating a surplus with two processes of different technologies, each of which produces jointly two commodities A and B. Sraffa explains why it is not unreasonable to assume the availability of two different methods of production, capable of being employed side by side. Against the possible objection that the less productive method would not be employed at all, he argues: *"No such condition as to equal productiveness is implied, nor would it have a definite meaning before the prices were determined; and, with different proportions of products, a set of prices can generally be found at which different methods are equally profitable"* (PCMC, Par. 50). In what follows, it will be shown that one of the prices may be negative (Example 7.7.1 below).

In Schefold's terminology [103] the production scheme is presented as follows with the labour vector $\mathbf{L} = [L_1, L_2]'$ in scalar or matrix form,

$$\begin{aligned} &\text{process 1} \quad (s_{11}, s_{12}, L_1) \to (f_{11}, f_{12}), \\ &\text{process 2} \quad (s_{21}, s_{22}, L_2) \to (f_{21}, f_{22}), \\ &\qquad\qquad (\mathbf{S}', \mathbf{L}) \to (\mathbf{F}'). \end{aligned} \tag{7.10}$$

If we define the *semi-positive* matrices,

$$\mathbf{S} = \begin{bmatrix} s_{11} & s_{12} \\ s_{21} & s_{22} \end{bmatrix} \quad \text{and} \quad \mathbf{F} = \begin{bmatrix} f_{11} & f_{12} \\ f_{21} & f_{22} \end{bmatrix}, \tag{7.11}$$

then the positive, maximal rate of profits $R > 0$, and the pertinent price vector $\mathbf{p}$ satisfy the price system $(1 + R)\mathbf{S}'\mathbf{p} = \mathbf{F}'\mathbf{p}$. Then, we assume *gross integrated industries*, i. e., rank($\mathbf{F}$) = 2, (6.8). Therefore, the transposed *input-output coefficients* matrix $\mathbf{C}_T' = \mathbf{F}'^{-1}\mathbf{S}'$ must be calculated. We obtain,

$$\det(\mathbf{F}')\mathbf{C}_T' = \det(\mathbf{F}')\mathbf{F}'^{-1}\mathbf{S}' = \begin{bmatrix} s_{11}f_{22} - s_{12}f_{21} & s_{21}f_{22} - f_{21}s_{22} \\ s_{12}f_{11} - s_{11}f_{22} & s_{22}f_{11} - s_{21}f_{12} \end{bmatrix}. \tag{7.12}$$

A more practical way of calculating this matrix works as follows. Define the rectangular matrix,

$$\mathbf{M} = [\ \mathbf{S}' \quad \mathbf{F}'\ ] = \begin{bmatrix} s_{11} & s_{21} & f_{11} & f_{21} \\ s_{12} & s_{22} & f_{12} & f_{22} \end{bmatrix}. \tag{7.13}$$

As matrix $\mathbf{F}$ is assumed to be regular, $\det(\mathbf{F}) = 2$, matrix $\mathbf{M}$ has rank 2 and the two processes are really different. If we denote the columns of $\mathbf{M}$ with 1, 2, 3, 4 and the determinant of the matrix with columns i, k with $\det(i, k)$, then we have,

$$\det(\mathbf{F}')\mathbf{F}'^{-1}\mathbf{S}' = \begin{bmatrix} \det(1,4) & \det(2,4) \\ -\det(1,3) & -\det(2,3) \end{bmatrix}. \tag{7.14}$$

Although this matrix may have one or more negative elements, Frobenius' theory can still be applied in certain cases. These cases are:

- negative elements occur only in the main diagonal;
- both elements in the upper-right and the lower-left corner are negative.

This follows from Lemma A.9.2 as will be seen in the proof of the next theorem.

In what follows, we will use the concept of "partial productivity".

Definition 7.3.1. Consider processes with $n = 2$ sectors. The partial productivity of process i with respect to product 1 resp. 2 is the quotient f_{1i}/s_{1i} resp. f_{2i}/s_{2i}.

Theorem 7.3.1. *The productive system (7.10) admits a uniform profit rate and positive prices if process* **1** *has greater partial productivity than process* **2** *with respect to one product and process* **2** *has greater partial productivity with respect to the other product.*

Proof. Assume at first the inequalities $f_{11}/s_{11} > f_{12}/s_{12}$ and $f_{21}/s_{21} < f_{22}/s_{22}$. Then $\det(2,4) > 0$ and $\det(1,3) < 0$. Therefore both elements in the upper-right and the lower-left corner of $\mathbf{F}'^{-1}\mathbf{S}'$ are positive, and for some positive h the matrix $\mathbf{F}'^{-1}\mathbf{S}' + h\mathbf{I}$ is positive and irreducible. In the opposite case, both elements in the upper-right and the lower-left corner of $\mathbf{F}'^{-1}\mathbf{S}'$ are negative. Therefore, the matrix $h\mathbf{I} - \mathbf{F}'^{-1}\mathbf{S}'$ is positive for some positive h. In both cases, the Lemma A.9.2 and the theorem of Perron–Frobenius, Theorem A.9.3, complete the proof. □

Example 7.3.1. We consider two farms that produce cattle and wheat, described by the following production scheme (without labour **L**):

$$(\mathbf{S}',\mathbf{0}) \to (\mathbf{F}'),$$
$$\begin{aligned}
&\text{farm 1:} \quad (3 \text{ heads of cattle}, 7 \text{ t. wheat}, 0) \to (4 \text{ heads of cattle}, 8 \text{ t. wheat}),\\
&\text{farm 2:} \quad (5 \text{ heads of cattle}, 5 \text{ t. wheat}, 0) \to (6 \text{ heads of cattle}, 7 \text{ t. wheat}).
\end{aligned} \tag{7.15}$$

Identify the matrices **S**, **F**, compute the vector of surplus **d**, the matrix $\mathbf{C}_T$ (or $\mathbf{C}_T'$) and its eigenvalues and the eigenvectors of matrix $\mathbf{C}_T'$ and interpret the results.

Solution to Example 7.3.1:
There are no money wages because the workers are paid at "subsistence wages" in beef and wheat. In this case, we have:

$$\mathbf{S} = \begin{bmatrix} 3 & 5 \\ 7 & 5 \end{bmatrix}, \quad \mathbf{F} = \begin{bmatrix} 4 & 6 \\ 8 & 7 \end{bmatrix}, \quad \mathbf{d} = (\mathbf{F} - \mathbf{S})\mathbf{e} = \begin{bmatrix} 1 & 1 \\ 1 & 2 \end{bmatrix} \begin{bmatrix} 1 \\ 1 \end{bmatrix} = \begin{bmatrix} 2 \\ 3 \end{bmatrix}. \tag{7.16}$$

The (2×4) rectangular matrix $[\mathbf{S}',\mathbf{F}']$ is therefore,

$$\begin{bmatrix} \mathbf{S}' & \mathbf{F}' \end{bmatrix} = \begin{bmatrix} 3 & 7 & 4 & 8 \\ 5 & 5 & 6 & 7 \end{bmatrix}, \tag{7.17}$$

with determinants: $\det(\mathbf{F}') = -20$, $\det(1,4) - 19$, $\det(2,4) = 9$, $\det(1,3) = -2$, $\det(2,3) = 22$. With equation (7.12), one obtains:

$$\det(\mathbf{F}')\mathbf{C}_T' = (-20)\mathbf{C}_T' = \begin{bmatrix} -19 & 9 \\ 2 & -22 \end{bmatrix} \Rightarrow \mathbf{C}_T' = \begin{bmatrix} 0.95 & -0.45 \\ -0.1 & 1.1 \end{bmatrix}. \tag{7.18}$$

Computing the rank$([\mathbf{S}',\mathbf{F}'])$, we recognise that the *matrix rank criterion*, Proposition 6.3.1 is fulfilled. Therefore, we conclude that both commodities, wheat and cattle, are basic.

One verifies that matrices **S** and **F** (7.16) satisfy the conditions of Theorem 7.3.1,

$$\begin{aligned}
\det(2,4) &= \begin{vmatrix} s_{21} & f_{21} \\ s_{22} & f_{22} \end{vmatrix} = s_{21}f_{22} - f_{21}s_{22} > 0 \Leftrightarrow \frac{f_{22}}{s_{22}} = \frac{7}{5} > \frac{f_{21}}{s_{21}} = \frac{8}{7},\\
\det(1,3) &= \begin{vmatrix} s_{11} & f_{11} \\ s_{12} & f_{12} \end{vmatrix} = s_{11}f_{12} - f_{11}s_{12} < 0 \Leftrightarrow \frac{f_{12}}{s_{12}} = \frac{6}{5} < \frac{f_{11}}{s_{11}} = \frac{4}{3}.
\end{aligned} \tag{7.19}$$

The characteristic polynomial is then calculated to obtain the eigenvalues:

$$P_2(\lambda) = \det(\mathbf{C}_T - \lambda\mathbf{I}) = \lambda^2 - \frac{41}{20}\lambda + 1 = \left(\lambda - \frac{4}{5}\right)\left(\lambda - \frac{5}{4}\right) = 0. \tag{7.20}$$

The eigenvalues are $\lambda_1 = 4/5 < 1$ and $\lambda_2 = 5/4 > 1$ with associated eigenvectors $\mathbf{p}_1 = k[3,1]'$, $k \in \mathbb{R}^+$ and $\mathbf{p}_2 = k[-1.5,1]'$, $k \in \mathbb{R}^+$. In application of Theorem 7.3.1

or its generalisation, Lemma A.9.3, the obtained minimal eigenvalue is the Frobenius number $\lambda_1 := \lambda_{\min} = (4/5) < 1$ associated with the positive eigenvector $\mathbf{p}_1' = [3,1]'$ of matrix $\mathbf{C}_T'$.

$$\mathbf{C}_T'\mathbf{p} = \begin{bmatrix} \frac{19}{20} & -\frac{9}{20} \\ -\frac{1}{10} & \frac{11}{10} \end{bmatrix} \begin{bmatrix} 3k \\ k \end{bmatrix} = \begin{bmatrix} \frac{48}{20}k \\ \frac{8}{10}k \end{bmatrix} = \frac{4}{5} \begin{bmatrix} 3k \\ k \end{bmatrix} = \lambda_1 \mathbf{p}. \tag{7.21}$$

For this reason, we obtain the *maximal rate of profits* or, equivalently, the productiveness $R = (1/\lambda_{\min}) - 1 = (1/0.8) - 1 = 0.25$. We can further choose any uniform positive rate of profits $r \in [0, 0.25]$ and any vector of labor $\mathbf{L} \geq \mathbf{o}$ to solve a complete Sraffa price model for joint production. ▲

We continue with the following example:

Example 7.3.2. This is a modification of Example 7.2.1. Consider the model (again without labour $\mathbf{L}$):

$$(\mathbf{S}', \mathbf{0}) \to (\mathbf{F}'),$$

$$\begin{aligned} &\text{farm 1:} \quad (3 \text{ heads of cattle}, 5 \text{ t. wheat}, 0) \to (4 \text{ heads of cattle}, 6 \text{ t. wheat}, 0) \\ &\text{farm 2:} \quad (5 \text{ heads of cattle}, 10 \text{ t. wheat}, 0, 0) \to (7 \text{ heads of cattle}, 12 \text{ t. wheat}, 0). \end{aligned} \tag{7.22}$$

Identify the matrices **S**, **F**, and compute matrix $\mathbf{C}_T$ and its eigenvalues and eigenvectors.

Solution to Example 7.3.2:
So we have:

$$\mathbf{S} = \begin{bmatrix} 3 & 5 \\ 5 & 10 \end{bmatrix} \quad \text{and} \quad \mathbf{F} = \begin{bmatrix} 4 & 7 \\ 6 & 12 \end{bmatrix}, \tag{7.23}$$

and, following Schefold, establish the block partitions of matrices $\mathbf{S}'$ and $\mathbf{F}'$, in application of PCMC (Par. 60) and Definition 6.3.1,

$$\mathbf{S}' = \begin{bmatrix} \mathbf{S}_{11}' & \mathbf{S}_{21}' \\ \mathbf{S}_{12}' & \mathbf{S}_{22}' \end{bmatrix} = \left[\begin{array}{c|c} 3 & 5 \\ \hline 5 & 10 \end{array}\right], \quad \mathbf{F}' = \begin{bmatrix} \mathbf{F}_{11}' & \mathbf{F}_{21}' \\ \mathbf{F}_{12}' & \mathbf{F}_{22}' \end{bmatrix} = \left[\begin{array}{c|c} 4 & 6 \\ \hline 7 & 12 \end{array}\right]. \tag{7.24}$$

Then we directly constitute with (7.24) the $(2m \times n)$ matrix $[\mathbf{S}_2' \ \mathbf{F}_2']$ (6.56),

$$\mathbf{S}_2' := \begin{bmatrix} 5 \\ 10 \end{bmatrix}, \quad \mathbf{F}_2' := \begin{bmatrix} 6 \\ 12 \end{bmatrix} \Rightarrow [\ \mathbf{S}_2' \ \ \mathbf{F}_2'\ ] = \begin{bmatrix} 5 & 6 \\ 10 & 12 \end{bmatrix} \Rightarrow \operatorname{rank}([\ \mathbf{S}_2' \ \ \mathbf{F}_2'\ ]) = 1, \tag{7.25}$$

concluding that commodity 2 (wheat) is *non-basic*, Proposition 6.3.1. So the classification of *non-basics* as luxuries only is no longer acceptable. We have

$$\mathbf{C}_T' = \frac{1}{6} \begin{bmatrix} 6 & 0 \\ -1 & 5 \end{bmatrix} \Rightarrow f(\lambda) = \lambda^2 - \frac{11}{6}\lambda + \frac{5}{6} = (\lambda - 1)\left(\lambda - \frac{5}{6}\right). \tag{7.26}$$

This matrix is reducible. It has the eigenvalues $\lambda_1 = 1$ with eigenvector of prices $\mathbf{p}_1' = [1, -1]'$ and $\lambda_2 = \frac{5}{6}$ with eigenvector of prices $\mathbf{p}_2' = [0, 1]'$. The eigenvalue $\lambda_2 = \frac{5}{6}$ would imply the maximal profit rate $R = (6/5) - 1 = 0.2$, which can be attained by both industries selling product 2 (wheat) alone.

This result is strange because, with a positive price of 1 for cattle, both industries could attain profit rates r that are different, but in any case higher than $R = 0.2$. This suggests that the assumption of equal profit rates may be inadequate in the presence of *joint production*. See also Section 8.3.

It is not known whether the condition of Theorem 7.3.1 for positive prices is necessary. Therefore, a different approach to this issue is given.

Taking the *joint production Sraffa system* (6.30), we set with $\mathbf{r} = [r_1, r_2]'$ in analogy to (8.16),

$$\mathbf{S}'\mathbf{p}(1 + R) = \mathbf{F}'\mathbf{p} \Rightarrow \mathbf{S}'\mathbf{p}(1 + \mathbf{r}) = \mathbf{F}'\mathbf{p}, \tag{7.27}$$

separating both obtained equations leads to the following proportions

$$1 + r_j = \frac{f_{1j}p_1 + f_{2j}p_2}{s_{1j}p_1 + s_{2j}p_2}, \quad j = 1, 2. \tag{7.28}$$

We divide numerator and denominator by p_1 and set $x = \frac{p_1}{p_2}$. This gives us:

$$1 + r_j = \frac{f_{1j} + f_{2j}x}{s_{1j} + s_{2j}x} = h_j(x), \quad j = 1, 2. \tag{7.29}$$

When both prices are positive, then x is positive. This is assumed here. By calculating derivatives we see that for positive x each of these rational functions are monotonically decreasing.

This means that we have the maxima f_{1j}/s_{1j} for $x = 0$ and the minima f_{2j}/s_{2j} for $x = \infty$. Therefore, with positive prices, the equality of profit rates is possible if and only if the intervals $I_1 = [f_{21}/s_{21}, f_{11}/s_{11}]$ and $I_1 = [f_{22}/s_{22}, f_{12}/s_{12}]$ have a non-empty intersection. In our example, the profit rates r_1, r_2 are in the intersection $[1.2, 1.3333] = [1.2, 1.3333] \cap [1.2, 1.4]$. The rate of profits $R = 0.2$ found here is in fact minimal in the present case.

Now, we consider a case in which positive prices are incompatible with a common profit rate. ▲

Example 7.3.3. Consider a model with the matrices $\mathbf{S}$, $\mathbf{F}$, and compute $\mathbf{C}_T$ and its eigenvalues and eigenvectors.

$$\mathbf{S} = \begin{bmatrix} 5 & 3 \\ 7 & 5 \end{bmatrix} \quad \text{and} \quad \mathbf{F} = \begin{bmatrix} 6 & 4 \\ 8 & 7 \end{bmatrix}. \tag{7.30}$$

Discuss the obtained prices.

Solution to Example 7.3.3:
The intervals $[f_{21}/s_{21} = 8/7, f_{11}/s_{11} = 6/5]$ and $[f_{12}/s_{12} = 4/3, f_{22}/s_{22} = 7/5]$ of real numbers have no common points. Therefore, for arbitrary positive prices the profit rates must be different. We have with (7.14) and $\mathrm{Det}(\mathbf{F}') = 10$:

$$\mathbf{C}_T' = \mathbf{F}'^{-1}\mathbf{S}' = \frac{1}{10}\begin{bmatrix} 11 & 9 \\ -2 & 2 \end{bmatrix}. \tag{7.31}$$

The elements in the upper-right and the lower-left corners have opposite signs, therefore Theorem 7.3.1 cannot be applied. The matrix has eigenvalue $\lambda = 0.8$ with eigenvector $\mathbf{p}' = [3, -1]'$. There are negative prices. ▲

These three examples cover the whole range of relevant cases of systems with 2 industries and 2 commodities.

7.4 Three commodities and three industries

This is also a preparatory section concerning profits and prices. It can be modified to incorporate recycled waste and its effects on profits, as will be shown in Section 7.7. We continue the discussion, now for simplified systems with three industries and three commodities, illustrated by two examples. Their general form without labour is:

$$\begin{aligned} (\mathbf{S}', \mathbf{0}) &\rightarrow (\mathbf{F}'), \\ (s_{11}, s_{21}, s_{31}, 0) &\rightarrow (f_{11}, f_{21}, f_{31}), \\ (s_{12}, s_{22}, s_{32}, 0) &\rightarrow (f_{12}, f_{22}, f_{32}), \\ (s_{13}, s_{23}, s_{33}, 0) &\rightarrow (f_{13}, f_{23}, f_{33}). \end{aligned} \tag{7.32}$$

We consider the special case $s_{31} = f_{31} = 0$. In the sense of Sraffa's definition, Definition 6.3.1, product 3 is *non-basic*, if the matrix

$$\begin{bmatrix} 0 & 0 \\ s_{32} & f_{32} \\ s_{33} & f_{33} \end{bmatrix} \tag{7.33}$$

has rank 1. This condition can be written as $s_{32}f_{33} = s_{33}f_{32}$. We will show that the profit rate and the prices of 1 and 2 are independent of f_{32} and f_{33} if 3 is *non-basic*. The equations for the profit rate and prices are,

$$\begin{aligned} (1+r)(s_{11}p_1 + s_{21}p_2) &= f_{11}p_1 + f_{21}p_2, \\ (1+r)(s_{12}p_1 + s_{22}p_2 + s_{32}p_3) &= f_{12}p_1 + f_{22}p_2 + f_{32}p_3, \\ (1+r)(s_{13}p_1 + s_{23}p_2 + s_{33}p_3) &= f_{13}p_1 + f_{23}p_2 + f_{33}p_3. \end{aligned} \tag{7.34}$$

After multiplying the second equation of (7.34) with s_{33} and the third equation of (7.34) with s_{32} and subtracting, we obtain, since $s_{32}f_{33} - s_{33}f_{32} = 0$,

$$(1 + r)\big[(s_{12}s_{33} - f_{13}s_{32})p_1 + (s_{22}s_{33} - f_{23}s_{32})p_2\big]$$
$$= (f_{12}s_{33} - f_{13}s_{32})p_1 + (f_{22}s_{33} - f_{23}s_{32})p_2 \qquad (7.35)$$

Now, we have two equations, the first equation of (7.34) and equation (7.35) for the profit rate and the prices of products 1 and 2, in which the coefficients f_{32} and f_{33} are eliminated. Furthermore, if s_{32} and s_{33} are multiplied by the same factor, the whole equation (7.35) is multiplied by this factor. This implies that the *partial productivity* with respect to 3 has no influence on the profit rate and the prices of 1 and 2.

Example 7.4.1. We start again with Sraffa's wheat–iron model, but now we add a branch that produces diamonds. If diamonds were used only as luxuries, we would have the same model as with gold. But diamonds have, due to their extreme hardness, a number of important industrial applications. Therefore, we may assume that diamonds are used in the mining of iron, as well as of diamonds, and that they are not consumed in the process of mining. We consider the production system again without the labour vector **L**,

$$(\mathbf{S}', \mathbf{0}) \to (\mathbf{F}'),$$
$$(280 \text{ qr. wheat}, 12 \text{ t. iron}, 0, 0) \to (575 \text{ qr. wheat}, 0, 0),$$
$$(120 \text{ qr. wheat}, 8 \text{ t. iron}, 2 \text{ kg diamonds}, 0) \to (0, 30 \text{ t. iron}, 2 \text{ kg diamonds}),$$
$$(60 \text{ qr. wheat}, 4 \text{ t. iron}, 2 \text{ kg diamonds}, 0) \to (0, 0, 3 \text{ kg diamonds}). \qquad (7.36)$$

Identify the matrices **S**, **F**, and compute matrix $\mathbf{C}_T$ and its eigenvalues and eigenvectors and the rate of profits r.

Solution to Example 7.4.1:
We have the following input and output matrices:

$$\mathbf{S} = \begin{bmatrix} 280 & 120 & 60 \\ 12 & 8 & 4 \\ 0 & 2 & 2 \end{bmatrix} \quad \text{and} \quad \mathbf{F} = \begin{bmatrix} 575 & 0 & 0 \\ 0 & 30 & 0 \\ 0 & 2 & 3 \end{bmatrix}. \qquad (7.37)$$

As in the proceeding sections, we combine all inputs and outputs in the rectangular matrix

$$\begin{bmatrix} \mathbf{S}' & \mathbf{F}' \end{bmatrix} = \begin{bmatrix} 280 & 12 & 0 & 575 & 0 & 0 \\ 120 & 8 & 2 & 0 & 30 & 2 \\ 60 & 4 & 2 & 0 & 0 & 3 \end{bmatrix}, \quad \det(\mathbf{F}) = 51{,}750 \neq 0. \qquad (7.38)$$

According to Sraffa's definition, diamonds are a *basic* product in this model because the matrix consisting of the 3rd and 6th column of $[\mathbf{S}' \ \ \mathbf{F}']$ has

$$\operatorname{rank}\left(\begin{bmatrix} 0 & 0 \\ 2 & 2 \\ 2 & 3 \end{bmatrix}\right) = 2. \qquad (7.39)$$

For prices and the rate of profits, we have the equations

$$\begin{aligned}(1+R)(280p_1+12p_2) &= 575p_1,\\ (1+R)(120p_1+8p_2+2p_3) &= 30p_2+2p_3,\\ (1+R)(60p_1+4p_2+2p_3) &= 3p_3.\end{aligned} \tag{7.40}$$

Therefore (with rounded numbers):

$$\mathbf{C}_T' = (\mathbf{SF}^{-1})' = \begin{bmatrix} 0.487 & 0.021 & 0 \\ 2.667 & 0.178 & 0.022 \\ 20 & 1.333 & 0.667 \end{bmatrix}. \tag{7.41}$$

Here again, Frobenius' theorem applies, and we have the eigenvalue $\lambda = 0.8$ with eigenvector $\mathbf{p} = [1, 15, 300]'$. So the profit rate is $R = 0.25$, and one kg of diamonds costs as much as 300 quarters of wheat or 20 tons of iron. If the technology of mining diamonds is improved, the profit rate must rise, and the price of iron is also affected because diamonds are a *basic* product. For example, if the output of the third industry is four kg of diamonds instead of three kg, the profit rate becomes $R = 0.343$, and the price of iron falls to $p_2 = 12.35$ quarters of wheat and that of diamonds to $p_3 = 111.8$ quarters of wheat. ▲

Example 7.4.2. In the preceding example, the diamonds used in the iron industry play the role of fixed capital which leaves the production process in the same quantity as it has entered it. Now we assume that the mines of the iron industry produce, in addition to iron, also new diamonds. If we replace in the iron industry the $f_{32} = 2$ kg of diamonds on the right hand of (7.36) by a quantity greater than two kg, but different from three kg, we have again a system with three basic products. But diamonds become *non-basic* if the iron industry produces exactly three kg of them, resulting in a surplus of one kg. In Example 7.4.1, the price of one kg of diamonds is equal to the price of 20 t. of iron. Therefore, the productivity of the iron industry, measured in the same prices, does not change, if we go to the following system:

$$\begin{aligned}(\mathbf{S}', \mathbf{0}) &\rightarrow (\mathbf{F}'),\\ (280\text{ qr. wheat}, 12\text{ t. iron}, 0, 0) &\rightarrow (575\text{ qr. wheat}, 0, 0),\\ (120\text{ qr. wheat}, 8\text{ t. iron}, 2\text{ kg diamonds}, 0) &\rightarrow (0, 10\text{ t. iron}, 3\text{ kg diamonds}),\\ (60\text{ qr. wheat}, 4\text{ t. iron}, 2\text{ kg diamonds}, 0) &\rightarrow (0, 0, 3\text{ kg diamonds}).\end{aligned} \tag{7.42}$$

Identify the matrices **S**, **F**, establish the matrix pair $[\mathbf{S}' \;\; \mathbf{F}']$, determine the number of non-basics and calculate the prices of the Sraffa price model without labour vector.

Solution to Example 7.4.2:
We have the following input and output matrices:

$$\mathbf{S} = \begin{bmatrix} 280 & 120 & 60 \\ 12 & 8 & 4 \\ 0 & 2 & 2 \end{bmatrix} \quad \text{and} \quad \mathbf{F} = \begin{bmatrix} 575 & 0 & 0 \\ 0 & 10 & 0 \\ 0 & 3 & 3 \end{bmatrix}. \tag{7.43}$$

As before, we combine all inputs and outputs in the rectangular matrix

$$\begin{bmatrix} \mathbf{S}' & \mathbf{F}' \end{bmatrix} = \begin{bmatrix} 280 & 12 & 0 & 575 & 0 & 0 \\ 120 & 8 & 2 & 0 & 10 & 3 \\ 60 & 4 & 2 & 0 & 0 & 3 \end{bmatrix}. \tag{7.44}$$

According to Sraffa's definition, applying the *matrix rank criterion*, Proposition 6.3.1, diamonds are a *non-basic* product in this model because the matrix consisting of the 3rd and 6th column of $[\mathbf{S}' \ \ \mathbf{F}']$ has

$$\operatorname{rank}\left(\begin{bmatrix} 0 & 0 \\ 2 & 3 \\ 2 & 3 \end{bmatrix}\right) = 1. \tag{7.45}$$

The usual equations in the present case are

$$\begin{aligned} (1+R)(280p_1 + 12p_2) &= 575p_1, \\ (1+R)(120p_1 + 8p_2 + 2p_3) &= 10p_2 + 3p_3, \\ (1+R)(60p_1 + 4p_2 + 2p_3) &= 3p_3. \end{aligned} \tag{7.46}$$

Following Sraffa, *non-basics* can be eliminated from a system of equations of *joint production* by a suitable linear transformation PCMC (Par. 61). In the present case, we have to subtract the third equation of (7.46) from the second equation. The result is

$$(1+R)(60p_1 + 4p_2) = 10p_2. \tag{7.47}$$

The system of equations composed of the first equation of (7.46) and equation (7.47) is essentially the same as the system (4.13) (see also equations (3.42)). We know already the solution $R = 0.25$, $p_1 = 1$, $p_2 = 15$; inserting these values into the third equation of (7.46) and solving for p_3 gives $p_3 = 300$. ▲

Now we can check the general law that a technological change in the production of a *non-basic* good does not change the profit rate and the prices of *basics*. Indeed, if the coefficients of p_3 on the right-hand side of the second and third equation of (7.46) are changed to the same extent, the same algebraic procedure as before leads again to equation (7.47), which together with the first equation of (7.46) determines the rate of profits and the prices of wheat and iron.

7.5 The impact of taxation and technological innovations

The application of penalties by way of specific taxes on polluting products is a central theme in ecological economics. This section, following the approach of the preceding sections, is devoted to illustrating algebraically the incidence of taxation on profits and prices within Sraffa's models.

Taxes on specific commodities are a tool of economic policies that aim at the reduction of environmental damages. In some cases, such taxes have an impact on the whole economy, while in other cases they have not. Taxation of fossil energy or of tourist helicopter flights are examples for the first and the second type of taxes, respectively. Sraffa considers the impact of taxes in PCMC, Par 65. He writes: "*A tax on a basic product then will affect all prices and cause a fall in the rate of profits that corresponds to a given wage, while if imposed on a non-basic it will have no effect beyond the price of the taxed commodity and those of such other non-basics as may be linked with it.*" In the following, we will consider examples that confirm Sraffa's statement about the impact of taxes. But it is also shown that a tax on non-basics that are jointly produced with basics may affect the rate of profits, contrary to Sraffa's statement in PCMC, Par 6. Following the same line of argument, we will consider the impact of technical innovations.

A cornerstone of the following considerations is:

Theorem 7.5.1. *The Frobenius eigenvalue of a non-negative, irreducible matrix is a strictly monotone increasing function of each of its elements.*

Proof. Pasinetti [80], Theorem 3, p. 272.

We illustrate these concepts with two examples:

Example 7.5.1. Again we consider Sraffa's simplified wheat–iron model, without the labour vector **L**. If taxes must be paid for the production of both wheat and iron, and the corresponding tax ratios are t_1 $(0 \leq t_1 < 1)$ and t_2 $(0 \leq t_2 < 1)$, then the equations (3.42) become:

$$\begin{aligned}(1+R)(280p_1+12p_2) &= 575p_1(1-t_1),\\ (1+R)(120p_1+8p_2) &= 20p_2(1-t_2).\end{aligned} \tag{7.48}$$

Solution to Example 7.5.1:

In this example, the impact of taxes is very clear. As we know, see Section 3.1.3, without taxes and if wages are included in the means of production, the Frobenius number is $\lambda_C = 0.8$, the profit rate is $R = (1/\lambda_C) - 1 = (1/0.8) - 1 = 0.25$, and the ratio of prices is 1:15.

If only wheat is taxed and $t_1 = 0.181$, then $R = 0.111$ and the price ratio is 1:12. If only iron is taxed and $t_2 = 0.225$, then $R = 0.106$ and the price ratio is 1:20. If both products are taxed and $t_1 = t_2 = t$, then the price ratio remains unchanged, and the Frobenius eigenvalue is divided by $1-t$, with a corresponding change of the profit rate.

For example, if $t = 0.1$, then $R = ((1-t)/\lambda_C) - 1 = (0.9/0.8) - 1 = 0.125$. In this special case, a tax of 10 % on both products reduces the profit rate from $R = 0.25$ to $R = 0.125$. □

Example 7.5.2. In the following, gold is an example of an arbitrary non-basic product. We extend the model considered before to the case in which gold is produced in two different ways: as a single commodity like in Example 6.3.3 and as a joint product in connection with the extraction of iron, as in Example 6.3.4. We may consider the following system:

$$\begin{aligned}(\mathbf{S}', \mathbf{0}) &\to (\mathbf{F}'),\\ (300\text{ qr. wheat}, 12\text{ t. iron}, 0) &\to (600\text{ qr. wheat}, 0, 0),\\ (150\text{ qr. wheat}, 9\text{ t. iron}, 0) &\to (0, 30\text{ t. iron}, 6\text{ kg gold}),\\ (50\text{ qr. wheat}, 4\text{ t. iron}, 0) &\to (0, 0, 4\text{ kg gold}).\end{aligned} \tag{7.49}$$

Solution to Example 7.5.2:
The sum of all inputs is 500 qr. wheat and 25 t. iron. Therefore, if all the surplus of wheat and iron is reinvested, the rate of growth is $\rho = 20\,\%$.[4] Since gold is not used as a means of production, the rate of growth is not affected by the abundance of gold. So, the maximum rate of growth of the system is $\rho = 20\,\%$.

Now we consider the case in which gold is taxed with ratio t. Then the equations for prices and the profit rate are:

$$\begin{aligned}(1+R)(300p_1 + 12p_2) &= 600p_1,\\ (1+R)(150p_1 + 9p_2) &= 30p_2 + 6p_3(1-t),\\ (1+R)(50p_1 + 4p_2) &= 4p_3(1-t).\end{aligned} \tag{7.50}$$

This system is equivalent to the following:

$$\begin{aligned}(1+R)(0.5p_1 + 0.02p_2) &= p_1,\\ (1+R)(5p_1 + 0.3p_2) &= p_2 + 0.2p_3(1-t),\\ (1+R)\left(\frac{12.5}{1-t}p_1 + \frac{1}{1-t}p_2\right) &= p_3,\end{aligned} \tag{7.51}$$

and may be written in the form

$$(1+R)\mathbf{S}'\mathbf{p} = \mathbf{F}'\mathbf{p}, \tag{7.52}$$

where $\mathbf{p} = [p_1, p_2, p_3]'$ is the price vector and $\mathbf{S} = (s_{ij})$, $\mathbf{F} = (f_{ij})$, $i, j = 1, \ldots, 3$, are matrices of the general transposed form

$$\mathbf{S}' = \begin{bmatrix} s_{11} & s_{21} & 0\\ s_{12} & s_{22} & 0\\ s_{13} & s_{23} & 0\end{bmatrix}, \quad \mathbf{F}' = \begin{bmatrix} 1 & 0 & 0\\ 0 & 1 & f_{32}\\ 0 & 0 & 1\end{bmatrix}. \tag{7.53}$$

4 Calculation of the rate of growth: $\rho = (600-500)/500 = (30-25)/25 = 0.2$.

Note that, in our example,

$$f_{32} = 0.2(1-t), \quad s_{13} = \frac{12.5}{1-t} \quad s_{23} = \frac{1}{1-t}. \tag{7.54}$$

The maximal rate of profit R is related to the Frobenius number of the matrix $\mathbf{C}_T = \mathbf{SF}^{-1}$. In this case,

$$\mathbf{C}'_T = \begin{bmatrix} s_{11} & s_{21} & 0 \\ s_{12} - f_{32}s_{13} & s_{22} - f_{32}s_{23} & 0 \\ s_{13} & s_{23} & 0 \end{bmatrix}. \tag{7.55}$$

The nontrivial eigenvalues of this matrix are the two eigenvalues of the 2×2 matrix in the upper left corner. Only the elements in the second row are related to the production of gold via the parameters s_{13}, s_{23} and f_{32}, but the taxation factor $1 - t$ cancels out. Therefore taxation of gold does not affect the rate of profits in this case.

However, it is evident that the technology of extraction of gold, accounted for by the parameters f_{32}, s_{13} and s_{23}, has an impact on the eigenvalues of $\mathbf{C}_T = \mathbf{SF}^{-1}$ and hence on the maximal rate of profits. An improvement in the technology of joint production of gold, i. e., greater f_{32}, leads to a smaller Frobenius eigenvalue, and therefore to a greater profit rate (this follows from the fact that the Frobenius eigenvalue is a monotone increasing function of every element of the matrix). But an improvement in the single-production of gold, i. e., smaller s_{13} and s_{23}, leads to a smaller profit rate, a counter-intuitive result! If gold is classified as non-basic, both results differ from Sraffa's statement made for the case of single-product industries.

With the numerical values of our example and $t = 0$, we have:

$$\mathbf{S}' = \begin{bmatrix} 0.5 & 0.02 & 0 \\ 5 & 0.3 & 0 \\ 12.5 & 1 & 0 \end{bmatrix}, \quad \mathbf{F}' = \begin{bmatrix} 1 & 0 & 0 \\ 0 & 1 & 0.2 \\ 0 & 0 & 1 \end{bmatrix}, \quad \mathbf{C}'_T = \begin{bmatrix} 0.5 & 0.02 & 0 \\ 2.5 & 0.1 & 0 \\ 12.5 & 1 & 0 \end{bmatrix}, \tag{7.56}$$

With these matrices, we can calculate $\mathbf{p}$ and λ_C from the equation

$$\mathbf{S}'\mathbf{p} = \lambda \mathbf{F}'\mathbf{p} \Rightarrow \mathbf{C}'_T\mathbf{p} = \lambda \mathbf{p}. \tag{7.57}$$

We obtain the Frobenius number $\lambda_C = 0.6$. Then we compute the productiveness $R_1 = (10/6) - 1 = 0.6667$ and the prices $p_1 = 1$, $p_2 = 5.0$, $p_3 = 29.2$. If the output of gold in the second sector of the joint production of gold is doubled by technological improvement, without increasing the inputs ($f_{32} = 0.4$), the profit rate rises to $R_2 = 1$. If this technological improvement is obtained in the single production of gold (s_{13} and s_{23} are halved), then the profit rate falls to $R_3 = 0.51$.

Now we look at the definition of basics and non-basics given by Sraffa for the case of joint production. First we consider the "group of two linked commodities" wheat and iron. This is the case $n = 3$, $m = 2$ in Sraffa's definition, Definition 6.3.1 and Proposition 6.3.1. ▲

From the 3×6 matrix $(\mathbf{S}',\mathbf{F}')$, we select the four columns that correspond to wheat and iron. This gives the matrix,

$$[\ \mathbf{S}_2'\quad \mathbf{F}_2'\ ] = \begin{bmatrix} s_{11} & s_{21} & 1 & 0 \\ s_{12} & s_{22} & 0 & 1 \\ s_{13} & s_{23} & 0 & 0 \end{bmatrix} \Rightarrow \begin{bmatrix} 0.5 & 0.02 & 1 & 0 \\ 2.5 & 0.15 & 0 & 1 \\ 12.5 & 1 & 0 & 0 \end{bmatrix}. \tag{7.58}$$

Since the determinant of the matrix formed by the first three columns is,

$$\det\left(\begin{bmatrix} 0.5 & 0.02 & 1 \\ 2.5 & 0.15 & 0 \\ 12.5 & 1 & 0 \end{bmatrix}\right) = 1 \times 2.5 - 0.15 \times 12.5 = 0.625 > 0, \tag{7.59}$$

we get $\operatorname{rank}([\mathbf{S}_2'\ \mathbf{F}_2']) = 3$. Therefore wheat and iron are *basics*.

Then, with respect to gold, we consider the case $n = 3$, $m = 1$, and get,

$$\operatorname{rank}([\ \mathbf{S}_1'\quad \mathbf{F}_1'\ ]) = \operatorname{rank}\left(\begin{bmatrix} 0 & 0 \\ 0 & f_{32} \\ 0 & 1 \end{bmatrix}\right) = 1. \tag{7.60}$$

Since this matrix has rank 1, gold is non-basic. ▲

An important observation from this example is that the joint production of basics (here iron and wheat) and non-basics (here gold) may lead to a situation, in which the rate of profits ($R_1 = 66.7\,\%$) is much higher than the maximal rate of growth ($\rho = 20\,\%$).

The next three sections are devoted to specific environmental issues: treatment of waste, pollution and CO_2 emission.

7.6 A model with recycling of iron scrap

Today, the economic system produces immense quantities of waste and harmful or toxic by-products. It makes sense to *consider these by-products as commodities with a negative price*. This would be an interesting alternative to the concept of "internalizing external costs", which is the cornerstone of neoclassical environmental economics. Sraffa himself was uncomfortable in the presence of negative prices; in PCMC, Par. 50, he wrote: "*..., only those methods of production are practicable which ...do not involve other than positive prices*". However the great advantage of his theory of joint production is precisely that negative prices may occur in a quite natural way. The price of certain by-products may change from negative to positive if an efficient technology of recycling becomes available.

A Sraffa model is justified, when its construction is in relation to any official Input-Output Table (IOT). In Section 2.1 we have presented the *European standard classification of productive economic activities* (NACE [16]) and the *European Classification of Products by Activities* (CAP) which classifies all the products, appearing in an IOT, and are used in an economy as means of production or consumption goods. The products are ordered into the sectors of the IOT according to the nomenclature of CAP. As we will extend the historical iron-wheat model, we rely on the products classified by CAP, relying on the NACE report [16], because the headings of the divisions in NACE lead directly to the *branch headings* of the IOTs, decribing the *palettes* of the products. In the extended "wheat-iron" model we will indicate the number of the divisions of the *Detailed Structure* of the products, which will be used in the extended wheat-iron model.

We continue to use "wheat" which appears in division 1 (NACE, p. 61), treated in the sector of *agriculture, forestry and fishing* and "iron" in division 24 (NACE, p. 66), typically treated in a sector of *manufacture of basic metals*. As a new mean of production we will introduce metallic "scrap" which appears in division 46 (NACE, p. 74), treated typically in a sector of *wholesale and retail*.

Let us outline that waste recycling can be modelled as:

(1). Outlays for by-product waste like "scrap" (Iron/Steel sector).

(2). The recycling industry can be represented as a recycling sector/branch within the *input-output table* (IOT). The sector recycles different items/objects with different technologies of recycling and with different efficiency. Such joint production can be treated within the *input-output tables*, see Section 9.2.1.

We do not describe the *technological process of production* of iron, which lies outside the scope of this book. We will only study in this simplifying model the occurring prices relative to the efficieny η of a theoretical recycling production process.

It could be objected that the economy described by Sraffa's wheat-iron model is not sustainable, because it produces scrap and needs iron, a non-renewable resource. In a green economy scrap would be partly recycled. Therefore, let us modify the model in the following way. We enter now a *recycling industry*. We assume that all the "iron tools" of the *agricultural sector* turns to scrap after a period. We assume that the *iron industry* is divided into a branch which uses iron and a branch which collects and recycles scrap. We denote with η the efficiency of recycling ($\eta < 1$). This means that a ton of scrap yields a fraction η of a ton of iron. We consider the following

Example 7.6.1. The labour force gets wages in kind distributed 1 : 1 among the branches of iron industry. Then we set up the following model, taking the quantities of PCMC, Par. 2, where the first sector is just copied and the quantities of the iron sector are distributed on two sectors, one using iron and the other using scrap as means of production, again abstracting from a separate labour vector **L**:

$$\begin{gathered}(\mathbf{S}', \mathbf{0}) \to (\mathbf{F}'),\\ (280\text{ qr. wheat}, 12\text{ t. iron}, 0) \to (575\text{ qr. wheat}, 0, 12\text{ t. scrap}),\\ (60\text{ qr. wheat}, 4\text{ t. iron}, 0) \to (0, 10\text{ t. iron}, 0),\\ (60\text{ qr. wheat}, 12\text{ t. scrap}, 0) \to (0, 12\eta\text{ t. iron}, 0).\end{gathered} \tag{7.61}$$

Solution to Example 7.6.1:
We note the following input and output matrices, and the price vector $\mathbf{p} = [p_1, p_2, p_3]'$,

$$\mathbf{S} = (s_{ij}) = \begin{bmatrix} 280 & 60 & 60 \\ 12 & 4 & 0 \\ 0 & 0 & 12 \end{bmatrix}, \quad \mathbf{F} = (f_{ij}) = \begin{bmatrix} 575 & 0 & 0 \\ 0 & 10 & 12\eta \\ 12 & 0 & 0 \end{bmatrix}, \quad i,j = 1,\ldots,3. \tag{7.62}$$

When the trade between the three branches is fair, each branch must have the same profit rate. If we denote with p_3 the price of 1 t. of scrap and with R the uniform maximal profit rate, then the prices and R satisfy the equations:

$$\begin{aligned} (1+R)(280p_1 + 12p_2) &= 575p_1 + 12p_3, \\ (1+R)(60p_1 + 4p_2) &= 10p_2, \\ (1+R)(60p_1 + 12p_3) &= 12\eta p_2. \end{aligned} \tag{7.63}$$

In order to have four equations for the four unknowns $p_1(\eta), p_2(\eta), p_3(\eta), R(\eta)$, we may add the convention $p_1(\eta) = 1$. The price system (7.63) can be written in matrix form, with the price vector $\mathbf{p} = [1, p_2, p_3]'$, as a *general eigenvalue problem* in physical terms,

$$\mathbf{S}'\mathbf{p} = \lambda_C \mathbf{F}'\mathbf{p}, \quad \lambda_C = \frac{1}{(1+R)}. \tag{7.64}$$

The values of λ for which a non-zero price vector exists are the solutions of the characteristic polynomial $P_3(\lambda) = \det(\mathbf{S}' - \lambda\mathbf{F}') = 0$. Here, due to the special form of $\mathbf{S}$ and $\mathbf{F}$, the characteristic polynomial is merely quadratic (coefficient: $a_3 = 0$) in λ, namely,

$$P_3(\lambda) = \det(\mathbf{S} - \lambda\mathbf{F}) = a_2\lambda^2 + a_1\lambda + a_0 = (8{,}640\eta + 61{,}800)\lambda^2 - 58{,}320\lambda + 4{,}800. \tag{7.65}$$

One verifies that the characteristic equation $P_3(\lambda) = 0$ has two positive solutions $\lambda_1 > 0$, $\lambda_2 > 0$ for the proposed η, see Table 7.1. The maximal positive eigenvalue $\lambda_C = \max\{\lambda_1, \lambda_2\} > 0$ is related to the *productiveness*, $R = (1/\lambda_C) - 1$, based on (7.64). The Table 7.1 shows how the *productiveness* R and the price of scrap p_3 in this model depend on the efficiency η of recycling. If efficiency is very low, the price of scrap is negative, $p_3 < 0$; for $\eta = (5/12)$, one finds $p_3 = 0$, and it is positive, $p_3 > 0$, for higher efficiencies. In this case, wheat becomes cheaper in relation to iron, due to the fact that farmers can sell scrap in addition to wheat, see Table 7.1. ▲

This example shows how Sraffa's theory can be applied to determine the prices of waste in an economic system with recycling.

7.7 An alternative CO_2-emissions trading scheme

Every productive process by which CO_2 is emitted can be considered as a process of joint production with CO_2 as by-product. On the other hand, CO_2 is necessary for photosynthesis in plants, i. e., the conversion of solar energy into biomass. Therefore,

Table 7.1: Recycling of scrap.

Efficiency of recycling η	Profit rate R	Price of iron p_2	Price of scrap p_3	Price of wheat p_1
0.25	0.219	14.3	−2.07	1
$\frac{5}{12}$	0.25	15.0	0	1
0.5	0.266	15.4	1.08	1
0.8	0.322	16.8	5.19	1

forestry and traditional agriculture that uses no fossil fuels are absorbers of CO_2. This leads to the idea to treat CO_2 as a "commodity" that could be "traded" between emitters and absorbers of CO_2.

Neoclassical economists have already proposed market-based instruments for attaining a reduction of emissions of greenhouse gases in developed countries. A well-known example is the Emissions Trading Scheme (ETS) of the European Union, operational since 2005. In this trading scheme, governments create *emission permits* as artificial "commodities" that can be traded between EU-based firms. But the Stern Review admits that "it has been difficult to ensure scarcity in the EU ETS market" ([115], p. 374). Therefore, the price of *emission permits* in the EU trading scheme was very low, oscillating between 10 and 25 EURO per ton of CO_2 ([115], p. 372), whereas in Switzerland subsidies for CO_2 reduction are between 56 and 70 CHF per ton CO_2 (NZZ, 18.9.2013).

The EU ETS deals with greenhouse-gas emissions, but the role of forests in carbon sequestration is not considered. The Kyoto Protocol failed to mention climate forestry in its Article 12 on the CDM (clean development mechanism), but at the conference held at Bonn in 2000, afforestation and reforestation were made eligible under the CDM (see Michaelova [64]). Still later, the topic was addressed in a decision taken by the Copenhagen Conference on Climate Change (2009), which highlights *"the importance of reducing emissions from deforestation and forest degradation, and the role of conservation, sustainable management of forests and enhancement of forest carbon stocks in developing countries"* (see Bottazzi [10]). But hitherto, there is no economic theory which links CO_2 emission and absorption.

During the ten years of its operation, the ETS of the EU has had only a small impact on prices, while it has triggered the appearance of many products with the label "CO_2-neutral", but the meaning of this label is not clear. Therefore, the following definition is proposed:

> The products of a firm which emits CO_2 are CO_2 – neutral if the firm has made a contract with one or more absorbers of CO_2 to receive all CO_2 emitted by the firm. The payment for this service must entail equal profit rates for emitters and absorbers.

Such a trading scheme requires a framework of strong international institutions with the task of measuring the quantities of CO_2 emitted and absorbed and of controlling the flow of payments from emitters to absorbers. This topic is not to be discussed here. It is simply assumed that such a trading scheme does exist. It will be shown that Sraffa's theory of joint production can be applied in order to calculate a negative "price" of CO_2 which would determine the money transfer per unit of CO_2 from emitters to absorbers, in such a way that an equal rate of profit can be obtained by all parties. With an abstract example the influence of CO_2 – trading on the choice of technology and on prices will be demonstrated.

Example 7.7.1. Let us consider a circular system of production consisting of two commodities W_1 and W_2 and three processes with constant returns to scale. Process 1 absorbs CO_2 and produces W_1. This may be traditional agriculture or forestry. The processes *2a* and *2b* produce W_2, the first with a low-carbon, the other with a high-carbon technology. W_1 is a necessary consumption good whereas W_2 can be used both in production and in luxury consumption, as is the case of many products of modern technology. Wages are regarded as consisting of the necessary subsistence goods for the workers, so they are not mentioned explicitly. It is assumed that a year's operations can be tabulated as follows, see Table 7.2:

Table 7.2: A model of emission and absorption of CO_2.

process 1	$(60\ W_1, 30\ W_2, 80\ CO_2)$	$\rightarrow (100\ W_1, 0, 0)$
process 2a	$(15\ W_1, 25\ W_2, 0)$	$\rightarrow (0, 50\ W_2, 30\ CO_2)$
process 2b	$(10\ W_1, 25\ W_2, 0)$	$\rightarrow (0, 50\ W_2, 50\ CO_2)$

Solution to Example 7.7.1:
At these levels of activity, a final demand of 15 W_1 and 20 W_2 can be satisfied, and all CO_2 emitted is absorbed by process 1, so W_2 is CO_2 – neutral in the sense of our definition. But if CO_2 is not regarded as a harmful by-product, then obviously process *2b* is more productive than process *2a* which would not be activated. In order to satisfy the final demand for W_2, the level of activity of process *2b* would be doubled and emission of CO_2 would *increase* to 100. The equations for the prices p_i of W_i, $(i = 1, 2)$ and the uniform profit rate R would be:

$$\begin{aligned}(1+R)(60p_1+30p_2) &= 100p_1,\\ (1+R)(20p_1+50p_2) &= 100p_2.\end{aligned} \tag{7.66}$$

As W_1 is the numéraire, $p_1 = 1$, the positive solutions are $R = 0.25$ and $p_2 = (2/3)$.

Now we assume that a trading scheme as described above has been installed. Formally, we treat CO_2 in the same way as the commodities W_1 and W_2. Whether p_3, the "price" of CO_2, will be positive or negative is not known in advance. We assume that the payments for CO_2 are made at the beginning of each year. Then the equations for prices and profit rate are:

$$\begin{aligned}(1+R)(60p_1+30p_2+80p_3) &= 100p_1,\\ (1+R)(15p_1+25p_2) &= 50p_2+30p_3,\\ (1+R)(10p_1+25p_2) &= 50p_2+50p_3.\end{aligned} \tag{7.67}$$

This system can be written in matrix form, after identifying the matrices,

$$\mathbf{S}=\begin{bmatrix}60 & 15 & 10\\ 30 & 25 & 25\\ 80 & 0 & 0\end{bmatrix},\quad \mathbf{F}=\begin{bmatrix}100 & 0 & 0\\ 0 & 50 & 50\\ 0 & 30 & 50\end{bmatrix},\quad \det(\mathbf{F})=100{,}000, \tag{7.68}$$

giving,

$$(1+R)\mathbf{S}'\mathbf{p}=\mathbf{F}'\mathbf{p}. \tag{7.69}$$

In the present case, **F** has an inverse, and we have a *gross integrated industries*. Therefore we calculate

$$\mathbf{F}^{-1}=\begin{bmatrix}\frac{1}{100} & 0 & 0\\ 0 & \frac{1}{20} & -\frac{1}{20}\\ 0 & -\frac{3}{100} & \frac{1}{20}\end{bmatrix},\quad \mathbf{C}_T=\mathbf{S}\mathbf{F}^{-1}=\begin{bmatrix}\frac{3}{5} & \frac{9}{20} & -\frac{1}{4}\\ \frac{3}{10} & \frac{1}{2} & 0\\ \frac{4}{5} & 0 & 0\end{bmatrix}. \tag{7.70}$$

The calculation of R and the price vector amounts to finding an eigenvalue and an eigenvector of the matrix $\mathbf{C}_T = \mathbf{SF}^{-1}$. The most appropriate method for doing this is the method of successive approximations. With the choice $p_1 = 1$, we find: $R = 0.25$, $p_1 = 1$, $p_2 = 1.5$ and $p_3 = -0.3125$. ▲

This result shows that an emissions trading scheme that includes CO_2 absorbing processes can make a low-carbon technology as profitable as a high-carbon technology. But this requires a substantial change of the price system. Those commodities that are produced with high CO_2 emissions will be much more expensive than in an economy without trading of emission permits.

7.8 Prelude for the Sraffian approach in ecological economics

The biosphere, which is the home of millions of animal and plant species and of mankind, owes its stability to the cycles of water, carbon, nitrogen and other vital material cycles. Most agrarian civilizations have taken advantage of these cycles without interfering with them. Unless metals were used, all residues of production and consumption could be recycled by humans or nature. The Industrial Revolution shifted human activity to production processes that could not be integrated into a natural cycle. In the millennia before, the production of CO_2 by human and animal respiration and the combustion of wood was compensated for by the consumption of CO_2 in the metabolism of plants. But the combustion of fossil fuels releases more CO_2 into the atmosphere worldwide than can be absorbed globally, by vegetation only. And the chemical industry produces large quantities of environmental toxins and plastics that cannot be decomposed into harmless substances by any natural decomposition process.

The characteristic expression of this mode of production is the neoclassical production function that indicates the quantities of "production factors" labour (L) and capital (K) used in the manufacture of a quantity Y of an undescribed product: $Y = f(L, K)$. Production is seen here as a one-way street from L and K to Y. What unwanted by-products are produced, where the workforce and capital come from, whether and how often the production process can be repeated—this is not asked. An ecological economy in the true sense of the word cannot be content to include natural resources as further production factors in the neoclassical production function, and otherwise leave everything as it is. Rather, it must conceive of an ecological economy as a cycle that repeats itself from year to year or at longer intervals and is capable of recycling all undesirable by-products and waste.

Sraffa's theory does just that. But the representatives of ecological economics do not know this theory, and let Georgescu-Roegen [35] lure them onto the wrong track. This is why they hardly distance themselves from neoclassicism and adopt its basic assumptions without opposition. Just like Paul Samuelson [101] in the neoclassical standard textbook, Herman Daly [21] begins his book "Ecological Economics" with the sentence: "Economics is the study of the use of limited or scarce resources for alternative, competing purposes". But today, there are already many products whose proper disposal is more expensive than their manufacture. Most ecological problems in the industrialised countries do not arise from scarcity, but from abundance.

This also applies to global environmental problems. The combustion of coal, oil, natural gas and their derivatives produces CO_2, which is evenly distributed throughout the Earth's atmosphere. Consequently, the proportion of CO_2 in the air has increased continuously in recent decades, and the global climate is getting warmer. As a result of advances in chemistry, it has become possible to produce all kinds of packaging, containers and other plastic objects extremely cheaply from the waste from oil refineries. So plastic is no longer a scarce commodity, and, on the contrary, there is already far too much of it. A large proportion of empty plastic bags and containers are discharged into rivers and finally into the sea. One part breaks down into tiny toxic particles swallowed by marine animals, while another part resists the ravages of time and pollutes the beaches or forms huge floating carpets in the open sea. Where is the scarce commodity?

The series of examples could be extended at will. They show that an economic theory that concentrates on scarce goods cannot adequately represent global ecological problems.

8 Sraffa and extensions

This chapter follows from the work of Pasinetti's [80] and Schefold's [103] explanations of the Sraffa production economy.[1] In the first three sections, this chapter is devoted to illustrating the potential of Sraffa's price model. It is not a rigid construction limited to producing commodities by means of commodities in a closed economy. The Sraffa price model for *joint production* is extended to *imports* and *exports*, and a discussion of the incidence of including services as commodities is included. The model may even be extended to situations involving profit rates and wages which vary from industry to industry. Section 8.4 reformulates Sraffa's price model by incorporating the notion of a mark-up k which replaces both the wage rate and the rate of profits. The remaining sections position Sraffa's conceptual approach with respect to other economic approaches.

8.1 Sraffa and the open economy

We now develop the Sraffa system of *joint production* to an open economy including *imports* and *exports*. For this purpose, we start from the production scheme (4.143), $(\mathbf{S}', \mathbf{L}) \to (\mathbf{F}')$, and add an $n \times n$ matrix $\mathbf{E}^{\uparrow}$, involving the export volume of this economy, and an $(g \times n)$ matrix $\mathbf{M}^{\downarrow}$, representing the $g \in \mathbb{N}$ imported goods.

$$(\mathbf{S}', \mathbf{M}^{\downarrow\prime}, \mathbf{L}) \to (\mathbf{F}', \mathbf{E}^{\uparrow\prime}). \tag{8.1}$$

We now define the *non-negative* square $n \times n$ *export volume* matrix $\mathbf{E}^{\uparrow} = (e_{ij})$, $i, j = 1, \ldots, n$, and also the *non-negative* rectangular $g \times n$ *import volume* matrix $\mathbf{M}^{\downarrow} = (m_{kj})$, $k = 1, \ldots, g$, $j = 1, \ldots, n$ as follows:

e_{ij}: is the total quantity of commodity i exported by the sector S_j to all foreign economies that acquire this commodity i from the present economy. The coefficient e_{ij} is called *export coefficient*, expressed in physical terms of commodity i.

m_{kj}: is the total quantity of the good $k \in \{1, \ldots, g\}$, imported by the sector S_j, $j \in \{1, \ldots, n\}$. The coefficient m_{kj} is called *import coefficient* of good k to sector S_j, expressed in physical terms of good k.

Now we describe the import and export prices of the various commodities.

$p_{E,i}$: is the *export price* of commodity i. We assume a constant ratio to the production price p_i, expressed as $p_{E,i} = a_i p_i$, leading to the price vector $\mathbf{p}_E = [p_{E,1}, \ldots, p_{E,n}]' = [a_1 p_1, \ldots, a_n p_n]'$. The coefficients $a_1, a_2, \ldots, a_n$ are exogenous parameters fixed by the exporters.

1 This chapter with the examples is principally due to D. Chable and J.-F. Emmenegger.

https://doi.org/10.1515/9783110635096-008

$p_{M,k}$: is the *import price* of good k, acquired by any of the sectors $S_j, j \in \{1, \ldots, n\}$. We assume constant import prices for each imported good $k \in \{1, \ldots, g\}$, leading to the price vector $\mathbf{p}_M = [p_{M,1}, \ldots, p_{M,g}]' = \text{const}$. The import prices are also exogenous parameters depending on external markets.

The matrices are now fully presented:

$$\mathbf{E}^{\uparrow} = \begin{bmatrix} e_{11} & e_{12} & \ldots & e_{1n} \\ e_{21} & e_{22} & \ldots & e_{2n} \\ \ldots & \ldots & \ldots & \ldots \\ e_{n1} & e_{n2} & \ldots & e_{nn} \end{bmatrix}, \quad \mathbf{M}^{\downarrow} = \begin{bmatrix} m_{11} & m_{12} & \ldots & m_{1n} \\ m_{21} & m_{22} & \ldots & m_{2n} \\ \ldots & \ldots & \ldots & \ldots \\ \ldots & \ldots & \ldots & \ldots \\ m_{g1} & m_{g2} & \ldots & m_{gn} \end{bmatrix}. \tag{8.2}$$

The equations of the *joint production* Sraffa price model for an open economy as an extension of (6.48) supplemented by equations for total exports and total imports, are now presented. We need the diagonal matrix $\hat{\mathbf{a}}$, according to definition (A.17) obtained from the vector $\mathbf{a} = [a_1, \ldots, a_2]'$ of the just-defined ratios $a_j, j = 1, \ldots, n$. We set up the system:

$$\boxed{\begin{aligned} \mathbf{S}'\mathbf{p}(1+r) + \mathbf{M}^{\downarrow\prime}\mathbf{p}_M + \mathbf{L}\frac{\tilde{w} \cdot Y}{L} &= (\mathbf{F}' + \mathbf{E}^{\uparrow\prime}\hat{\mathbf{a}})\mathbf{p}, \\ E_X &= (\mathbf{E}^{\uparrow}\mathbf{e})'\mathbf{p}_E, \\ M_I &= (\mathbf{M}^{\downarrow}\mathbf{e})'\mathbf{p}_M, \\ Y &= (\mathbf{Fe} - \mathbf{Se})'\mathbf{p} + E_X - M_I, \\ L &= \mathbf{e}'\mathbf{L}. \end{aligned}} \tag{8.3}$$

The system (8.3) has $n + 4$ equations. Setting $r = r_0$ and $p_1 = p_0$ exogenously, there are also $n + 4$ variables, of which two, M_I and L, can immediately be calculated. The remaining $n+2$ unknowns form a linear system and unique solutions for each variable exist if the system satisfies conditions, similar to Manara's conditions, see Section 6.6, which we will not treat in this text.

We will now present a simplified example for *single-product* industries with exports of two commodities and one commodity imported to all sectors.

Example 8.1.1. Consider an economy producing electric power (E) in *GWh* (standing for energy), iron, measured in *tons of* iron, (standing for manufactured products) and wheat, also measured in *tons*, (standing for agricultural products). Electric power (E) and wheat are exported (↑). A fourth commodity, petroleum (P), measured in *tons*, is imported (↓) ($g = 1$) and directly split between the three production sectors. There furthermore exists a labour vector $\mathbf{L} = [L_1, L_2, L_3]'$. The production scheme reads as follows:

$$(\mathbf{S}', \mathbf{M}^{\downarrow\prime}, \mathbf{L}) \rightarrow (\mathbf{F}', \mathbf{E}^{\uparrow\prime})$$

$$\begin{gathered}(120\text{ GWh E}, 50\text{ t. iron}, 10\text{ t. wheat}, 20\text{ t. P}, L_1) \rightarrow (400\text{ GWh E}, 0, 0, 50\text{ GWh E}, 0, 0)\\ (100\text{ GWh E}, 100\text{ t. iron}, 10\text{ t. wheat}, 30\text{ t. P}, L_2) \rightarrow (0, 250\text{ t. iron}, 0, 0, 0, 0)\\ (20\text{ GWh E}, 30\text{ t. iron}, 50\text{ t. wheat}, 30\text{ t. P}, L_3) \rightarrow (0, 0, 120\text{ t. wheat}, 0, 0, 30\text{ t. wheat})\end{gathered} \tag{8.4}$$

The vector of ratios $\mathbf{a} = [1.2, 0.9, 0.8]'$ and the vector of the import price of petroleum $\mathbf{p}_M = [5]'$ are given. Identify the five matrices, respectively vectors, involved in (8.4) and the diagonal matrix $\hat{\mathbf{a}}$:

$$\mathbf{S} = \begin{bmatrix} 120 & 100 & 20 \\ 50 & 100 & 30 \\ 10 & 10 & 50 \end{bmatrix}, \quad \mathbf{F} = \begin{bmatrix} 400 & 0 & 0 \\ 0 & 250 & 0 \\ 0 & 0 & 120 \end{bmatrix}, \quad \mathbf{L} = \begin{bmatrix} 40 \\ 80 \\ 50 \end{bmatrix},$$
$$\mathbf{E}^{\uparrow} = \begin{bmatrix} 50 & 0 & 0 \\ 0 & 0 & 0 \\ 0 & 0 & 30 \end{bmatrix}, \quad \mathbf{M}^{\downarrow} = [\ 20 \quad 30 \quad 30\ ], \quad \hat{\mathbf{a}} = \begin{bmatrix} 1.2 & 0 & 0 \\ 0 & 0.9 & 0 \\ 0 & 0 & 0.8 \end{bmatrix}. \tag{8.5}$$

Compute the adapted vector of total output $\mathbf{q} = (\mathbf{F} + \hat{\mathbf{a}}\mathbf{E}_X^{\uparrow})\mathbf{e}$ and the vector of surplus $\mathbf{d} = \mathbf{q} - \mathbf{Se}$. Compute then the adapted input-output coefficients matrix $\mathbf{C} = \mathbf{S}\hat{\mathbf{q}}^{-1}$, its Frobenius number λ_C and the productiveness $R = (1/\lambda_C) - 1$. Verify that the rate of profits $r = 0.2$ belongs to the interval $[0, R]$ and solve the Sraffa price system (8.3) with the numéraire E and $p_{\text{E}} = p_1 = 1\,(\text{MWh/MWh}) = 1$. Compute the national income Y, the circulating capital K, the total export volume E_X, the total import volume M_I, the total output X, the total profit P, the total wages W, the wage rate w and the share of total wages to national income $\tilde{w}$. Establish then the Sraffa Network.

Solution to Example 8.1.1:
Compute the adapted vector of total output $\mathbf{q}$ and the vector of surplus $\mathbf{d}$,

$$\mathbf{q} = (\mathbf{F} + \hat{\mathbf{a}}\mathbf{E}_X^{\uparrow})\mathbf{e} = \left[\begin{bmatrix} 400 & 0 & 0 \\ 0 & 250 & 0 \\ 0 & 0 & 120 \end{bmatrix} + \begin{bmatrix} 60 & 0 & 0 \\ 0 & 0 & 0 \\ 0 & 0 & 24 \end{bmatrix} \right] \begin{bmatrix} 1 \\ 1 \\ 1 \end{bmatrix} = \begin{bmatrix} 460 \\ 250 \\ 144 \end{bmatrix},$$
$$\mathbf{d} = \mathbf{q} - \mathbf{Se} = \begin{bmatrix} 460 \\ 250 \\ 144 \end{bmatrix} - \begin{bmatrix} 120 & 100 & 20 \\ 50 & 100 & 30 \\ 10 & 10 & 50 \end{bmatrix} \begin{bmatrix} 1 \\ 1 \\ 1 \end{bmatrix} = \begin{bmatrix} 220 \\ 70 \\ 74 \end{bmatrix}, \tag{8.6}$$

as well as the adapted input-output coefficients matrix $\mathbf{C} = \mathbf{S}\hat{\mathbf{q}}^{-1}$,

$$\mathbf{C} = \begin{bmatrix} 120 & 100 & 20 \\ 50 & 100 & 30 \\ 10 & 10 & 50 \end{bmatrix} \begin{bmatrix} \frac{1}{460} & 0 & 0 \\ 0 & \frac{1}{250} & 0 \\ 0 & 0 & \frac{1}{144} \end{bmatrix} = \begin{bmatrix} \frac{6}{23} & \frac{2}{5} & \frac{5}{36} \\ \frac{5}{46} & \frac{2}{5} & \frac{5}{24} \\ \frac{1}{46} & \frac{1}{25} & \frac{25}{72} \end{bmatrix}, \tag{8.7}$$

and calculate the characteristic polynomial, giving

$$P_3(\lambda) = \det(\mathbf{C} - \lambda\mathbf{I}) = -\lambda^3 + \frac{8{,}347}{8{,}280}\lambda^2 - \frac{77}{276}\lambda + \frac{167}{8{,}280}$$
$$= (\lambda - 0.1138)(\lambda - 0.2964)(\lambda - 0.5979). \tag{8.8}$$

Thus, the Frobenius number is $\lambda_C = 1/(1+R) = 0.5979 < 1$, and we calculate the *maximal rate of profits* $R = (1/\lambda_C) - 1 = (1/0.5979) - 1 = 0.6725 > 0$. The uniform *rate of profits* belongs to the real interval $[0, 0.6725]$, $r = 0.2 \in [0, 0.6725]$. Then we set up the *Sraffa price model* (8.3) with price vector $\mathbf{p} = [1, p_2, p_3]'$ and the adapted equation for *total output* X and *circulating capital* K (6.147). There is the vector $\mathbf{a} = [a_1, a_2, a_3]' = [1.2, 0.9, 0.8]$ and the electric power is the *numéraire*, giving:

$$\begin{aligned}
&\begin{bmatrix} 120 & 50 & 10 \\ 100 & 100 & 10 \\ 20 & 30 & 50 \end{bmatrix}\begin{bmatrix} 1 \\ p_2 \\ p_3 \end{bmatrix}(1+0.2) + \begin{bmatrix} 20 \\ 30 \\ 30 \end{bmatrix}[\,6\,] + \begin{bmatrix} 40 \\ 80 \\ 50 \end{bmatrix}\frac{\tilde{w}\cdot Y}{L} \\
&= (\mathbf{F}' + \mathbf{E}^{\uparrow'}\hat{\mathbf{a}})\mathbf{p} = \left(\begin{bmatrix} 400 & 0 & 0 \\ 0 & 250 & 0 \\ 0 & 0 & 120 \end{bmatrix} + \begin{bmatrix} 50\cdot 1.2 & 0 & 0 \\ 0 & 0 & 0 \\ 0 & 0 & 30\cdot 0.8 \end{bmatrix}\right)\begin{bmatrix} 1 \\ p_2 \\ p_3 \end{bmatrix}, \\
&E_X = \mathbf{e}'(\mathbf{E}^{\uparrow}\hat{\mathbf{a}})'\mathbf{p} = [\,1 \quad 1 \quad 1\,]'\begin{bmatrix} 60 & 0 & 0 \\ 0 & 0 & 0 \\ 0 & 0 & 24 \end{bmatrix}\begin{bmatrix} 1 \\ p_2 \\ p_3 \end{bmatrix}, \\
&M_I = (\mathbf{M}^{\downarrow}\mathbf{e})'\mathbf{p}_M = \left([\,20 \quad 30 \quad 30\,]\begin{bmatrix} 1 \\ 1 \\ 1 \end{bmatrix}\right)[\,5\,], \\
&X = (\mathbf{F}\mathbf{e})'\mathbf{p} + E_X - M_I \\
&= \left(\begin{bmatrix} 400 & 0 & 0 \\ 0 & 250 & 0 \\ 0 & 0 & 120 \end{bmatrix}\begin{bmatrix} 1 \\ 1 \\ 1 \end{bmatrix}\right)'\begin{bmatrix} 1 \\ p_2 \\ p_3 \end{bmatrix} + E_X - M_I, \\
&K = (\mathbf{S}\mathbf{e})'\mathbf{p} \\
&= \left(\begin{bmatrix} 120 & 100 & 20 \\ 50 & 100 & 30 \\ 10 & 10 & 50 \end{bmatrix}\begin{bmatrix} 1 \\ 1 \\ 1 \end{bmatrix}\right)'\begin{bmatrix} 1 \\ p_2 \\ p_3 \end{bmatrix}, \\
&Y = X - K, \\
&L = \mathbf{e}'\mathbf{L} = [\,1 \quad 1 \quad 1\,]\begin{bmatrix} 40 \\ 80 \\ 50 \end{bmatrix}.
\end{aligned} \tag{8.9}$$

As a result we get the vector of relative prices $\mathbf{p} = [1, 2.6427, 3.4438]'$, and we compute the total output $X = 1{,}216.58$, the circulating capital $K = 956.75$, the national income $Y = 259.83$, the value of exports $E_X = 142.65$ and the value of imports $M_I = 400$. The wage rate $w = 0.4028$, the share of total wages to national income $\tilde{w} = 0.2636$, the total wages $W = 68.48$, the total profits $P = 191.35$ and the quantity of labour $L = 170$.

The corresponding *Sraffa Network*, see Figure 8.1, with EL (not to be confused with E for the product *electric power*) for the sector of electricity production, I for the sector or iron production and W for the sector of wheat production is established with the adjacency matrix depicting inland operations:

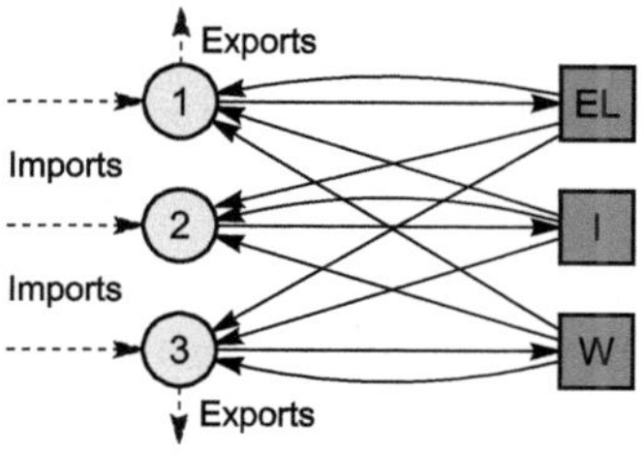

$$\mathbf{W} = \begin{array}{c} \\ 1 \\ 2 \\ 3 \\ E \\ I \\ W \end{array} \begin{array}{c} \begin{array}{cccccc} 1 & 2 & 3 & E & I & W \end{array} \\ \left[\begin{array}{ccc|ccc} 0 & 0 & 0 & 1 & 0 & 0 \\ 0 & 0 & 0 & 0 & 1 & 0 \\ 0 & 0 & 0 & 0 & 0 & 1 \\ \hline 1 & 1 & 1 & 0 & 0 & 0 \\ 1 & 1 & 1 & 0 & 0 & 0 \\ 1 & 1 & 1 & 0 & 0 & 0 \end{array}\right] \end{array} \tag{8.10}$$

Figure 8.1: An open economy.

The imported amount of petroleum (↓), i.e, refined oil, is split for internal use between the three production entities involved and is represented by dashed arrows entering the three sectors. The exports of the sector of wheat production and the sector of electric power production are represented by outgoing dashed arrows.

Exports are what Bortis [8], p. 146, terms an autonomous (exogenous) component of GDP. Referring to the national income equation (2.144) with $E_X = E, M_I = M$, we see how Sraffa's model of an open economy fits in with:

$$\begin{aligned} &Y = X - K = (\mathbf{Fe} - \mathbf{Se})'\mathbf{p} + E_X - M_I = C + I + G + (E - M) = W + P, \\ &C + G + I = Y - (E - M). \end{aligned} \tag{8.11}$$

The Sraffa Network also clearly establishes that we are in presence of a *single-product economy* and that the inland (internal) production process forms a separate entity embedded in an open economy. ▲

The system (8.3) can be used to describe a joint production system. But neither for *single-product* industries nor for *joint production* can positive price vectors $\mathbf{p}$ be guaranteed because, as mentioned, adapted Manara conditions, which we do not treat in this text, must be fulfilled for this purpose. In Chapter 10 we present an example taken from the real world, illustrating the Sraffa model for an open economy.

8.2 Extension of the Sraffa system to services

The present section is just an initial foray into a complex domain which requires much future research but which we have touched upon several times in foregoing chapters: it is the question of Sraffa's PCMC in relation to complete modern monetary production economies, including services, like education, banking, health care, tourism, regional and state administration and tourism.

Undoubtedly, Sraffa deliberately focussed his price system on a part of the economy, namely the production part, following the program presented in the preface to PCMC. He writes in PCMC, Par. 44:

> *"The rate of profits, as a ratio, has a significance which is independent of any prices, and can be 'given' before the prices are fixed. It is accordingly susceptible of being determined from outside the system of production, in particular by the level of the money rate of interest".*

This means in fact that Sraffa, well-versed in banking matters, considers that finance, money and banking lie *outside* the production system he describes. This also entails that the relevant prices obtained by his system are *relative prices* based on an arbitrarily chosen physical numéraire like wheat. As soon as we want ***absolute prices***, see Example 4.4.3 the currency adopted comes from outside Sraffa's production economy.

A classical breakdown of economic activity distinguishes the *primary sector*, involving the retrieval and production of raw materials, such as corn, wood, coal, iron ore, and the *secondary sector*, involving the transformation of raw or intermediate materials into goods, e. g., manufacturing steel into cars. Then the *tertiary sector*, involves the supply of services to consumers and businesses, such as banking, health care, education, research. Obviously Sraffa treats in the practical examples of PCMC only branches from the primary and secondary sectors.

One aim of this text is to embed the Sraffa production system in a *monetary economy of production* and *exchange*, comprising all three economic sectors, and therefore also all essential service branches. For this purpose, we have referred in Chapter 10 the breakdown of the branches of Swiss Input-Output Tables (Swiss IOT 2008 and Swiss IOT 2014), according to the NOGA system (French abbreviation for General Classification of Economic Activities; see Nathani [68], p. 47). The branches of the primary and secondary sectors are in the upper-left corner of the Swiss IOTs 2008 and 2014, as is the case for the IOT of most countries. These up-to-date IOTs moreover are based on the Input-Output Tables of Leontief, where the entries of the tables are expressed in monetary terms.

We have described in Chapter 2 the connection between Sraffa's *input-output coefficients* matrix $\mathbf{S}$ (2.13) and the Leontief IOTs. The branches chosen by Sraffa for his examples of the production systems $(\mathbf{S}', \mathbf{L}) \Rightarrow (\hat{\mathbf{q}})$ (PCMC, Par. 1, Par. 2, Par. 5) are all selected in the upper-left corner of present day national IOTs based on Leontief's concepts. In this connection, if we want to enlarge the Sraffa price system to services (*tertiary sector*), we would have to generate the entries of an *input-output* matrix $\mathbf{A}$ in *monetary terms*; and the performances of the services are then directly expressed in money, even if they no longer correspond to products, like wheat or iron, but, e. g., to education tasks and the like. Clearly, passing from Leontief's IOT to Sraffa's *commodity flow* matrix $\mathbf{S}$, each selling sector comprise then exactly one commodity. There are alltogether n commodities and n industrial sectors.

Returning back, we consider a Sraffa price model with a physical numéraire, as in Example 4.7.1. Then we set up the *Sraffa Network*, respectively the commodity flows, see Figure 8.2 (left). At this stage, there is no flow of money. Payments are expressed in relative prices that can be calculated with the Sraffa price model using the price unit of a physical numéraire. If this Sraffa production system is embedded in a larger monetary production economy with a tertiary sector comprising a *Central Bank* and a *banking system*, then payments will be made in the corresponding national currency, and there are *payment flows*, inverse to the *commodity flows*, see Figure 8.2 (right).

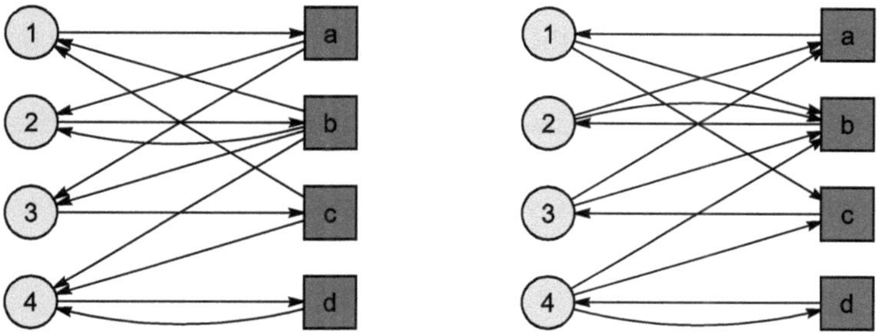

Figure 8.2: Commodity flow (left) and payment flow (right).

The *various flows* are governed by the following equations where the entries may be easily converted from *monetary terms* to *physical terms* and vice versa (see also Miller and Blair [65], p. 48):

$$\begin{cases} x_i = p_i q_i, \\ z_{ij} = p_i s_{ij}, \\ f_i = p_i d_i, \end{cases} \Leftrightarrow \begin{cases} \mathbf{x} = \hat{\mathbf{p}}\mathbf{q} = \hat{\mathbf{q}}\mathbf{p}; \quad \hat{\mathbf{x}} = \hat{\mathbf{p}}\hat{\mathbf{q}}, \\ \mathbf{Z} = \hat{\mathbf{p}}\mathbf{S} \Leftrightarrow \mathbf{S} = \hat{\mathbf{p}}^{-1}\mathbf{Z}, \\ \mathbf{f} = \hat{\mathbf{p}}\mathbf{d} \Leftrightarrow \mathbf{d} = \hat{\mathbf{p}}^{-1}\mathbf{f}. \end{cases} \tag{8.12}$$

These ideas are of course open to further developments.

8.3 Beyond the uniform rates of profits and wages

The assumption of a uniform rate of profits r and a uniform *wage rate* w for all industries is often considered a major weakness of Sraffa's model by economists. As already mentioned, it is not our intention to discuss the economic issues of this assumption in this textbook. We present the mathematics, the operational aspects and the numerical applications.

Nevertheless, in order to illustrate the open-endedness of Sraffa's approach, permitting various extensions, we present here an extension of Sraffa's model which alleviates the critique leveled at the assumption of a uniform *rate of profits* and a uniform *wage rate*, see also Emmenegger [28].

Inspired by an input-output table (IOT) of a country, say the Swiss IOT 2008, containing $n = 44$ sectors of industries and other economic activities, see Chapter 10, we return to a Sraffa system of n industries, each one producing exactly one commodity. Alltogether, there are again n commodities and n industries. Assume further individual rates of profit r_j and wage rates w_j, $j = 1, \dots, n$, for each of the sectors. This gives two distributions, one for the rate of profits and one for the wage rates, which we regroup separately in two vectors, the vector $\mathbf{r} = [r_1, r_2, \dots, r_n]'$ of the *distribution of the rate of profits* and the vector $\mathbf{w} = [w_1, w_2, \dots, w_n]'$ of the *distribution of the wage rates*. We specify again that the distribution $\mathbf{r}$ contains a profit rate r_j for each sector and the distribution $\mathbf{w}$ contains a wage rate w_j for each sector. These sector profits and wage rates can be considered as mean rates of the corresponding sectors, estimated on the basis of available economic wage and profit statistics.

We diagonalise these vectors $\mathbf{r}$ and $\mathbf{w}$ and get $n \times n$ matrices $\hat{\mathbf{r}}$, called the diagonal matrix of the *distribution of the rate of profits* and the matrix $\hat{\mathbf{w}}$, called the diagonal matrix of the *distribution of the wage rates*. We set

$$\hat{\mathbf{r}} = \begin{bmatrix} r_1 & 0 & \dots & \dots & 0 \\ 0 & r_2 & \dots & \dots & 0 \\ \dots & \dots & \dots & \dots & \dots \\ \dots & \dots & \dots & \dots & \dots \\ 0 & 0 & \dots & \dots & r_n \end{bmatrix}; \quad \hat{\mathbf{w}} = \begin{bmatrix} w_1 & 0 & \dots & \dots & 0 \\ 0 & w_2 & \dots & \dots & 0 \\ \dots & \dots & \dots & \dots & \dots \\ \dots & \dots & \dots & \dots & \dots \\ 0 & 0 & \dots & \dots & w_n \end{bmatrix}. \tag{8.13}$$

Remember that in Section 4.9, with equation (4.174) we have presented Sraffa's *price model* of $n+2$ equations and $n+4$ variables, expressed by the *wage rate* $w = (\tilde{w} \cdot Y)/L$, containing the *share of total wages to national income* $\tilde{w}$. We set now the factor w into the *price model* and can therefore drop the equations for national income Y and the total amount of labour (required working time) L. We get

$$\mathbf{S}'\mathbf{p}(1 + r) + w \cdot \mathbf{L} = \hat{\mathbf{q}}\mathbf{p} = \mathbf{x}. \tag{8.14}$$

The Sraffa *price model* has now n equations and $n+2$ variables, namely r, w and the n prices of the price vector $\mathbf{p}$. We have to discuss the question of the choice of two variables to solve the system (8.14). Remember that, to get a solvable system of equations, we selected the *rate of profits* r, determining the wage rate w, and we set the price of a commodity, giving the link to a *numéraire* or to a currency coming from outside.

At this stage, we want to remind the reader that solving the Sraffa *price model* (8.14), as described, we also formally solve the *distribution problem* of Ricardo in the sense of Sraffa. Namely, given the production scheme $(\mathbf{S}', \mathbf{L}) \to (\hat{\mathbf{q}})$ of an economy producing a surplus, based on a semi-positive matrix $\mathbf{S}$. The *input-output coefficients* matrix $\mathbf{C} = \mathbf{S}\hat{\mathbf{q}}^{-1}$ has a positive Frobenius number less than one, $0 < \lambda_C < 1$, and a maximal profit rate $R = (1/\lambda_C) - 1 > 0$. The national income is calculated with $Y = R \cdot (\mathbf{Se})'\mathbf{p} = R \cdot K > 0$, and is then split up into *total profits* P and *total wages* W, $Y = P + W$. A value for the *rate of profits* r is chosen, $r \in \,]0, R[$, giving $0 < P = r \cdot K < Y$ and $W = Y - P > 0$. We define the positive *wage to profit ratio*, $w_p := W/P > 0$, which expresses the repartition of wages and profits between entrepreneurs and workers. Moreover, we recognise an analogy,

$$(R > 0, 0 < r < R) \sim w_p = \frac{W}{P} > 0. \tag{8.15}$$

Choosing a value for the rate of profits, $r \in]0, R[$, in the case of constant rates of profits and wages is analogue to the fact to set the *wage to profit ratio* $w_p = W/P$ in the case of profit and wage distributions.

With this idea in mind, we at present abandon the concept of *uniformity* of the rates of profits and of the wage rates. We switch to *profit and wage distributions*, given by the diagonal matrices $\hat{\mathbf{r}}$ and $\hat{\mathbf{w}}$ (8.13), in order to extend Sraffa's initial *price system* (6.215), to a system, where each sector S_j has own *profit rates* r_j and own *wage rates* w_j, $j \in \{1, \dots, n\}$.

We have n unknown prices $p_1, p_2, \dots, p_n$, as previously described, collected in the vector $\mathbf{p}$, and we have to choose a *numéraire*, setting the price of one unit of that commodity, in order to get *relative prices*. On the other hand, if we aim at *absolute prices*, we have to choose a currency from *outside* the production economy.

Addressing the *distribution problem* of Ricardo requires necessarily an economy producing a surplus, which is shared as profit and wages between the industrial sectors and the workers. At this instant, there are no longer free parameters to regulate the share of profits and wages. But we will benefit of the already developed equivalence statement (8.15) and propose the choice of the positive *wage to profit ratio* $w_p = w_{p0} > 0$. It is an exogenous parameter to the system, replacing the initial rate of profits $r \in]0, R[$. Then, we need two calibrating factors n_p, n_w, one for each distribution $\hat{\mathbf{r}}$ and $\hat{\mathbf{w}}$, and replace r and w in the proceeding system (8.14) of uniform rates. Finally, we need the summation vector $\mathbf{e}' = [1, \dots, 1]$. Thus, we obtain

$$\boxed{\begin{aligned} &\mathbf{S}'(\mathbf{I} + n_p \cdot \hat{\mathbf{r}})\mathbf{p} + n_w \cdot \hat{\mathbf{w}}\mathbf{L} = \hat{\mathbf{q}}\mathbf{p}, \\ &\qquad\qquad p_i = p_0, \quad i \in \{1, \dots, n\}, \\ &\qquad\qquad w_p = w_{p0}, \\ &\qquad\qquad P = w_{p0} \cdot W, \\ &\qquad\qquad P = n_p \cdot \mathbf{e}'(\mathbf{S}'\hat{\mathbf{r}})\mathbf{p}, \\ &\qquad\qquad W = n_w \cdot \mathbf{e}'(\hat{\mathbf{w}}\mathbf{L}). \end{aligned}} \tag{8.16}$$

Finally, we have to select an exogenously given p_0, the price $p_i = p_0$ of the *numéraire* i. The extended Sraffa *price model* (8.16) has the $n+5$ variables $n_p, n_w, W, P, w_p, p_1, \dots, p_n$ and $n + 5$ equations. The $n + 5$ equations are apparently not linear, due to the product $\dots + n_p \cdot \hat{\mathbf{r}}\mathbf{p} + \dots$ in the first n equations, but is solvable with advanced software packages.[2]

---

2 *MATHEMATICA*, Wolfram Research, Inc., Champaign, IL (2016), or, *MATLAB*, MathWorks, Inc. Natick, Massachusetts, USA.

With the calculated price vector $\mathbf{p}$ (8.16), the distribution matrices of the profit rates $\hat{\mathbf{r}}$ and wage rates $\hat{\mathbf{w}}$, one then obtains the three remaining economic variables:

$$\begin{aligned} Y &= \mathbf{d}'\mathbf{p} = n_p \cdot \mathbf{e}'\mathbf{S}'(\hat{\mathbf{r}}\mathbf{p}) + n_w \cdot \mathbf{e}'(\hat{\mathbf{w}}\mathbf{L}) = P + W, \\ K &= \mathbf{e}'\mathbf{S}'\mathbf{p} = (\mathbf{S}\mathbf{e})'\mathbf{p}, \\ X &= \mathbf{q}'\mathbf{p} = Y + K. \end{aligned} \tag{8.17}$$

To make things clear, we assume that the *commodity flow* matrix $\mathbf{S}$ is *positive* or *irreducible* and *semi-positive;* then these properties are also transferred to the *input-output coefficients* matrix $\mathbf{C}$ and the **Perron–Frobenius theorem A.9.3** applies. The Frobenius number of $\mathbf{C}$ is positive and less than one, $0 < \lambda_C := 1/(1+R) < 1$, with *maximal rate of profits* $R > 0$, which is the productiveness.[3]

Under the described conditions the positive national income Y (8.17) of any Sraffa price model (8.16) is approximately equal to the national income $Y_0 = R \cdot (\mathbf{S}\mathbf{e})'\mathbf{p}$ of an economy only with *subsistence wages* ($\mathbf{L} = \mathbf{o}$) and uniform profit rate R, as many calculations have illustrated.

The *extended Sraffa price model* (8.16) enables us to solve new problems, where profit rates and wage rates are given by distributions $\hat{\mathbf{r}}$ and $\hat{\mathbf{w}}$. For this purpose, we extend Example 3.1.2 (PCMC, Par. 2) where we now introduce the production of a surplus.

Example 8.3.1. Three commodities (wheat, iron, pigs) are produced in an economy of single-product industries. The production process is symbolised as follows, each line corresponding to an industry: The physical units are 'quarter of wheat', 'tons of iron', 'number of pigs' and the quantity of labour is measured in 'man-years' (MY),

$$\begin{aligned} &(240 \text{ qr. wheat}, 12 \text{ t. iron}, 18 \text{ pigs}, 42 \text{ MY}) \to (600 \text{ qr. wheat}, 0, 0), \\ &(90 \text{ qr. wheat}, 6 \text{ t. iron}, 12 \text{ pigs}, 84 \text{ MY}) \to (0, 30 \text{ t. iron}, 0), \\ &(120 \text{ qr. wheat}, 3 \text{ t. iron}, 30 \text{ pigs}, 42 \text{ MY}) \to (0, 0, 90 \text{ pigs}). \end{aligned} \tag{8.18}$$

Begin by identifying the matrices $\mathbf{S}, \mathbf{L}, \mathbf{q}$ of this economy. Set up the system of production in matrix form. Determine the input-output coefficients matrix $\mathbf{C} = \mathbf{S}\hat{\mathbf{q}}^{-1}$. Calculate the productiveness R of matrix $\mathbf{C}$.

Compute the national income Y_0 of the corresponding economy that exhibits only *subsistence wages*, where all the surplus goes into the profit of the sectors.

Then, set up the distribution of the wage rates in arbitrary units that is specified in 'monetary units/man-years': $\mathbf{w}' = [1.2, 1.5, 1.1]$; set up the distribution of the rate of

3 The reader will notice that the condition can be relaxed to *reducibility* and *semi-positivity* of the *input-output coefficients* matrix $\mathbf{C}$. Such matrices can be brought by permutations into 'canonical form' (A.73) where at least the submatrix $\tilde{\mathbf{C}}_{11}$ is *positive* or *irreducible* and *semi-positive*, see Example A.10.3. Then Theorem A.10.1 and Lemma A.10.1 apply and the Frobenius number is positive, $\lambda_C > 0$.

profits by dimensionless components: $\mathbf{r}' = [0.15, 0.2, 0.12]$. Choose $w_{p0} = W/P = 1$ for the wages to profit ratio and treat two cases:
(a) The commodity wheat is the numéraire; therefore its price is $p_1 = p_0 = 1$.
(b) The economy adopts the currency CHF. Thus, the production price of wheat is given, say to $p_1 = p_0 = 55$ CHF.

Compute the prices p_2, p_3, the norm factors n_p and n_w, the effective distributions $n_p \cdot \mathbf{r}$ and $n_w \cdot \mathbf{w}$ and then the values of the economic variables P, W, Y, K and X.

Solution to Example 8.3.1:
We start by identifying the matrices

$$\mathbf{S} = \begin{bmatrix} 240 & 90 & 120 \\ 12 & 6 & 3 \\ 18 & 12 & 30 \end{bmatrix}, \quad \mathbf{q} = \begin{bmatrix} 600 \\ 30 \\ 90 \end{bmatrix}, \quad \mathbf{L} = \begin{bmatrix} 42 \\ 84 \\ 42 \end{bmatrix} \tag{8.19}$$

and set up the system of production

$$(\mathbf{S}', \mathbf{L}) \to (\mathbf{q}). \tag{8.20}$$

Then, we compute the matrix $\mathbf{C} = \mathbf{S}\hat{\mathbf{q}}^{-1}$ (2.16):

$$\mathbf{C} = \begin{bmatrix} 240 & 90 & 120 \\ 12 & 6 & 3 \\ 18 & 12 & 30 \end{bmatrix} \begin{bmatrix} \frac{1}{600} & 0 & 0 \\ 0 & \frac{1}{30} & 0 \\ 0 & 0 & \frac{1}{90} \end{bmatrix} = \begin{bmatrix} \frac{2}{5} & 3 & \frac{4}{3} \\ \frac{1}{50} & \frac{1}{5} & \frac{1}{30} \\ \frac{3}{100} & \frac{2}{5} & \frac{1}{3} \end{bmatrix}. \tag{8.21}$$

The matrix $\mathbf{C}$ is *irreducible* and *non-negative*, see Definition A.8.3, so that the **Perron–Frobenius theorem A.9.3** applies.

We set up the characteristic function

$$P_3(\lambda) = \det(\mathbf{C} - \lambda\mathbf{I}) = \lambda^3 - \frac{14}{15}\lambda^2 + \frac{1}{6}\lambda - \frac{7}{1{,}000}. \tag{8.22}$$

The polynomial is factorized, $P_3(\lambda) = (\lambda - 0.7134)(\lambda - 0.1576)(\lambda - 0.0623)$, the Frobenius number is calculated, $\lambda_C = 0.7173$, and the *productiveness* or maximal rate of profits is then obtained, $R = (1/0.7134) - 1 = 0.4017$.

We first compute the price model with matrix $\mathbf{C}$ for an economy with subsistence wages and maximal uniform rate of profits R,

$$\boxed{\begin{aligned} &\mathbf{C}'(1+R)\mathbf{p} = (1+0.4017) \begin{bmatrix} \frac{2}{5} & \frac{1}{50} & \frac{3}{100} \\ 3 & \frac{1}{5} & \frac{2}{5} \\ \frac{4}{5} & \frac{1}{30} & \frac{1}{3} \end{bmatrix} \begin{bmatrix} p_1 \\ p_2 \\ p_3 \end{bmatrix} = \begin{bmatrix} p_1 \\ p_2 \\ p_3 \end{bmatrix} = \mathbf{p}, \\ &p_1 = 1, \\ &Y = R \cdot (\mathbf{S}\mathbf{e})'\mathbf{p} = 0.4017 \cdot \left(\begin{bmatrix} 240 & 90 & 120 \\ 12 & 6 & 3 \\ 18 & 12 & 30 \end{bmatrix} \begin{bmatrix} 1 \\ 1 \\ 1 \end{bmatrix} \right)' \begin{bmatrix} p_1 \\ p_2 \\ p_3 \end{bmatrix}. \end{aligned}} \tag{8.23}$$

We obtain the price vector $\mathbf{p} = [1, 9.2031, 4.3143]'$, and then we can calculate the national income $Y_0 = 362.26$.

Then, the extended single product Sraffa price model (8.16) with the arbitrary distributions of the *profit rates* $\hat{\mathbf{r}}$ and the *wage rates* $\hat{\mathbf{w}}$ is established. The exogenous price $p_0 = 1$ for wheat and the *wages to profit ratio* $w_p = 1$ are chosen.

$$\begin{aligned}
&\mathbf{S}'(\mathbf{I} + n_p \cdot \hat{\mathbf{r}})\mathbf{p} + n_w \cdot \hat{\mathbf{w}}\mathbf{L} \\
&= \begin{bmatrix} 240 & 12 & 18 \\ 90 & 6 & 12 \\ 120 & 3 & 30 \end{bmatrix} \left(\begin{bmatrix} 1 & 0 & 0 \\ 0 & 1 & 0 \\ 0 & 0 & 1 \end{bmatrix} + n_p \begin{bmatrix} 0.15 & 0 & 0 \\ 0 & 0.2 & 0 \\ 0 & 0 & 0.12 \end{bmatrix} \right) \begin{bmatrix} p_1 \\ p_2 \\ p_3 \end{bmatrix} \\
&\quad + n_w \begin{bmatrix} 1.2 & 0 & 0 \\ 0 & 1.5 & 0 \\ 0 & 0 & 1.1 \end{bmatrix} \begin{bmatrix} 42 \\ 84 \\ 42 \end{bmatrix} = \hat{\mathbf{q}}\mathbf{p} = \begin{bmatrix} 600 & 0 & 0 \\ 0 & 30 & 0 \\ 0 & 0 & 90 \end{bmatrix} \begin{bmatrix} p_1 \\ p_2 \\ p_3 \end{bmatrix}, \\
&p_1 = p_0, \\
&W = P, \\
&W = n_q \cdot \mathbf{e}'(\hat{\mathbf{w}}\mathbf{L}) = n_q \cdot [1, 1, 1] \begin{bmatrix} 1.2 & 0 & 0 \\ 0 & 1.5 & 0 \\ 0 & 0 & 1.1 \end{bmatrix} \begin{bmatrix} 42 \\ 84 \\ 42 \end{bmatrix}, \\
&P = n_p \cdot \mathbf{e}'(\mathbf{S}'\hat{\mathbf{r}})\mathbf{p} \\
&\quad = n_p \cdot [1, 1, 1] \begin{bmatrix} 240 & 12 & 18 \\ 90 & 6 & 12 \\ 120 & 3 & 30 \end{bmatrix} \begin{bmatrix} 0.15 & 0 & 0 \\ 0 & 0.2 & 0 \\ 0 & 0 & 0.12 \end{bmatrix} \begin{bmatrix} p_1 \\ p_2 \\ p_3 \end{bmatrix}.
\end{aligned} \tag{8.24}$$

(a) We solve the *extended Sraffa price model* (8.24). We obtain the prices for iron $p_2 = 12.1808$ qr. wheat/t. iron and for pigs $p_3 = 4.1391$ qr. wheat/pig, giving the price vector $\mathbf{p} = [1, 12.1808, 4.1391]'$. Then, with the price vector we calculate the total wages $W = 191.9$ qr. wheat and the total profits $P = 191.9$ qr. wheat, the national income $Y = 383.8$ qr. wheat ~ $Y_0 = 362.26$ qr. wheat and the *norm factors* $n_p = 1.2926$ and $n_w = 0.8621$. These factors lead to the equilibrated distributions $n_p\mathbf{r}' = [0.1939, 0.2585, 0.1551]$ and $n_w\mathbf{w}' = [1.0345, 1.2931, 0.9483]$ in qr. wheat/man-years.

Then, we calculate the economic variables in physical units qr. wheat:

$$\begin{aligned}
Y &= \mathbf{d}'\mathbf{p} = [150, 9, 30] \begin{bmatrix} 1 \\ 12.1808 \\ 4.1391 \end{bmatrix} = P + W = 383.8, \\
K &= (\mathbf{Se})'\mathbf{p} = [450, 21, 60] \begin{bmatrix} 1 \\ 12.1808 \\ 4.1391 \end{bmatrix} = 954.14, \\
X &= \mathbf{q}'\mathbf{p} = [600, 30, 90] \begin{bmatrix} 1 \\ 12.1808 \\ 4.1391 \end{bmatrix} = K + Y = 1{,}337.94.
\end{aligned} \tag{8.25}$$

(b) We choose the currency CHF, setting the wheat price $p_0 = 55$ CHF/qr. wheat, and solve the *Sraffa extended price model* (8.24). We get the prices $p_2 = 669.95$ CHF/t. iron and $p_3 = 227.65$ CHF/pig. Then, the *norm factor* $n_p = 1.2926$ gives the profit rate distribution $n_p\mathbf{r}' = [0.1939, 0.2585, 0.1551]$ and the *norm factor* $n_w = 47.4146$ gives the wage rate distribution $n_w\mathbf{w}' = [56.90, 71.12, 52.16]$ in CHF/man-years. We calculate with the formulas (8.24), (8.25) the *total profits* $P = 10{,}554.50$ CHF, the *total wages* $W = 10{,}554.50$ CHF, the *national income* $Y = 21{,}109$ CHF, the *circulating capital* $K = 52{,}477.80$ CHF and the *total output* $X = 73{,}586.80$ CHF.[4] ▲

Example 8.3.2. Set up Sraffa's extended price model (8.16) with specific rates of profits and wage rates for each industrial sector. Show that with uniform rates r and a uniform wage rate w that the extended model reduces to Sraffa's initial price model for single product industries.

Solution to Example 8.3.2:
For uniform rates apply the reduction:

$$\hat{\mathbf{r}} = \begin{bmatrix} r & 0 & \dots & 0 \\ 0 & r & \dots & 0 \\ \dots & \dots & \dots & \dots \\ 0 & \dots & \dots & r \end{bmatrix} = r\mathbf{I}; \quad \hat{\mathbf{w}} = \begin{bmatrix} w & 0 & \dots & 0 \\ 0 & w & \dots & 0 \\ \dots & \dots & \dots & \dots \\ 0 & 0 & \dots & w \end{bmatrix} = w\mathbf{I}. \tag{8.26}$$

Introduce the diagonal matrices $\hat{\mathbf{r}}$ and $\hat{\mathbf{w}}$ in (8.16) and write

$$\mathbf{S}'(\mathbf{I} + \hat{\mathbf{r}})\mathbf{p} + \hat{\mathbf{w}}\mathbf{L} = \hat{\mathbf{q}}\mathbf{p} = \mathbf{x}. \tag{8.27}$$

So we have $(\mathbf{I} + \hat{\mathbf{r}}) = \mathbf{I} + r\mathbf{I} = \mathbf{I}(1 + r)$, and the general equations reduce to

$$\mathbf{S}'(\mathbf{I} + \hat{\mathbf{r}})\mathbf{p} + \hat{\mathbf{w}}\mathbf{L} = (\mathbf{S}'\mathbf{I})\mathbf{p}(1 + r) + w(\mathbf{I}\mathbf{L}) = \mathbf{S}'\mathbf{p}(1 + r) + w\mathbf{L} = \hat{\mathbf{q}}\mathbf{p} = \mathbf{x}, \tag{8.28}$$

Sraffa's equations (PCMC) for single product industries (8.14). ▲

Recapitulation. Consider an economy producing surplus with a known *technology* matrix **S** and given vector of labour (required working time) **L**. There are further the distributions of the *rates of profits* **r** and of the *wage rates* **w**.

David Ricardo [91] postulated that the determination of the laws that regulate the distribution of what is produced by the Earth, all what is *"derived from its surface by the united application of labour, machinery, and capital"* is *"the principal problem in Political Economy"*.

The Sraffa price model (8.16) is an extension of Sraffa's formal solution of Ricardo's *distribution problem* with *profit rates* and *wage rates* distributions.

The share between profits and wages, set by the *wage to profit ratio* $w_p = W/P$ and the price p_i for a unit of a *numéraire* i have to be chosen. Then a *factor* n_p calibrating the distribution **r** of *profit rates*,

4 The prices and amounts are unrealistic for today economies because the entries are taken from the example PCMC, Par. 2.

a *factor* n_w calibrating the distribution **w** of *wage rates* and the price vector **p** of the n commodities are calculated by the extended Sraffa price model (8.16). One obtains the profit and wage distributions $n_p \cdot \mathbf{r}$ and $n_w \cdot \mathbf{w}$, which formally solve Ricardo's distribution problem in an extended way.

8.4 The mark-up *k* and the Weintraub representation

In this section, we make use of the mark-up concept. Loosely speaking, in microeconomics following a frequent business practice, a mark-up k is the relationship between the selling price Y of a given quantity of a commodity and the *production cost* P_C of that quantity, i.e, $Y = k \cdot P_C$. The exact definition of k may vary from one enterprise to another depending on what one includes in *"production costs"*.

The mark-up concept has also been introduced in the more general context of macroeconomics, notably by Kalecki ([46], Chapter 2), and the post-Keynesian economist Sidney Weintraub (1914–1983) [120], pp. 44–47. Bortis [6], pp. 436–444, uses this concept in his macroeconomic analysis.

Simply stated, in macroeconomics, using our notation and referring to aggregates:

- Kalecki considers the mark-up k_K as defined by $P + K + W = k_K(K + W) = Y + K$. In this presentation, circulating capital K is considered as composed of ongoing variable costs and fixed overheads.
- Weintraub, for his part, sets more simply $P + W = k_W \cdot W = Y$, an identity fixing a different numerical value for the mark-up.

So also at the macroeconomic level definitions will differ; one readily finds in this case that $k_W \geq k_K \geq 1$.

Here we shall follow Weintraub,[5] using now k instead of k_W, and reformulate the aggregate Y as,

$$Y = k \cdot W = k \cdot (wL) = k \cdot w\mathbf{e}'\mathbf{L} = \mathbf{d}'\mathbf{p} \quad \text{or}$$

$$\left(\frac{W}{Y}\right) \cdot k = \tilde{w} \cdot k = 1. \tag{8.29}$$

In this representation, using the *wage rate* w and *wage* $W = w \cdot L$, entrepreneurial profits no longer appear explicitly as in $P + W = r \cdot K + w \cdot L = Y$, where prices are related to wages W and *circulating capital* K both contributing to "production costs". Profits P now disappear as a separate term, they are tucked up indirectly in k, considered as given, and the equations, replacing Sraffa's price model, relate prices to wages only, which here represent *"production costs"*, assuming given conditions of production.

The *rate of profits* r and the mark-up k are then directly linked through the expression for the profit $P = r \cdot K = w \cdot L(k - 1) = w \cdot L \cdot k - W$ expressed in *physical units*.

5 This multiplicative approach $\tilde{w} \cdot k = 1$ replaces the additive $\tilde{r} + \tilde{w} = 1$ with $\tilde{w} = \frac{1}{k}$, $\tilde{r} = 1 - \frac{1}{k}$.

We come back to the production scheme, containing a *positive* or *semi-positive* and *irreducible commodity flow* matrix $\mathbf{S}$, representing the *interindustrial economy* means of production. The total amount of each commodity of the means of production are then accumulated, $\mathbf{q}_I = \mathbf{Se}$. We assume for convenience a positive vector of surplus $\mathbf{d} > \mathbf{o}$, which we add up to get the vector of total output $\mathbf{q} = \mathbf{Se} + \mathbf{d} > \mathbf{o}$, as well as a positive vector of labour $\mathbf{L} > \mathbf{o}$. Introducing prices, the total value of the means of production is represented by the *circulating capital* K:

$$(\mathbf{S}', \mathbf{L}) \rightarrow (\hat{\mathbf{q}}), \quad \mathbf{q}_I = \mathbf{Se}, \quad K = (\mathbf{Se})'\mathbf{p}, \quad \mathbf{q} = \mathbf{Se} + \mathbf{d}. \tag{8.30}$$

We are now able to present as counterpart to Sraffa's representation (8.14). Throughout this section, we assume positive *rate of profits* $r > 0$ and the positive *wage rate* $w > 0$, i. e.:

$$\mathbf{S}'\mathbf{p}(1 + r) + w\mathbf{L} = \hat{\mathbf{q}}\mathbf{p} = \mathbf{x}, \tag{8.31}$$

the Weintraub representation depending on the *wage rate* w only and the constant mark-up k[6]

$$\mathbf{S}'\mathbf{p} + wk\mathbf{L} = \hat{\mathbf{q}}\mathbf{p} = \mathbf{x}. \tag{8.34}$$

Note on notation: We will now have to distinguish the prices of the *single-product Sraffa system*, which we denote in this section with $\mathbf{p}_S(r, w)$ instead of $\mathbf{p}$ from the prices of the Weintraub price model, which we denote hereafter with, $\mathbf{p}_W(k)$ instead of $\mathbf{p}$.[7]

(1) A first **question** arises:
Consider a Sraffa production scheme $(\mathbf{S}', \mathbf{L}) \rightarrow (\hat{\mathbf{q}})$ with given *means of production* (8.30), the vector of *labour* $\mathbf{L}$, the wage rate w and then either the *rate of profits* r or alternatively a given exogenous *national income* $Y_K = Y_0$. One applies the *Sraffa price*

6 Bortis speaks of the Kalecki–Weintraub price equation, see ([6] (19.7), p. 440) also written in aggregate form as *national income*

$$Y = PQ = w \cdot k \cdot L, \text{or} P = wk\left(\frac{Q}{L}\right)^{-1}, \tag{8.32}$$

$(\frac{Q}{L})$ being the aggregate productivity of labour, a pure numerical ratio. P is the average price level and Q an index of the real value of final expenditures, i.e, of the total quantity of goods and services purchased (see Mankiw ([63], pp. 82–83)).
We compute the national incomes, according to a Sraffa price model with (4.32) and to Weintraub (8.29) or Bortis (8.32):

$$\begin{aligned} Y_W &= \mathbf{d}'\mathbf{p}_W(k) = k \cdot w\mathbf{e}'\mathbf{L} = k \cdot w \cdot L = k \cdot W; \quad \textbf{Weintraub} \\ Y_S &= \mathbf{d}'\mathbf{p}_S(r, w) = r \cdot \mathbf{e}'\mathbf{S}\mathbf{p}_S(r, w) + w \cdot \mathbf{e}'\mathbf{L} = r \cdot K + w \cdot L; \quad \textbf{Sraffa}. \end{aligned} \tag{8.33}$$

7 The index S stands for Sraffa, the index W for Weintraub.

model (8.31) and the *Weintraub price model* (8.33), and one calculates the price vectors of both models. Can we expect that the obtained price vectors are equal, i. e., $\mathbf{p}_S(r,w) = \mathbf{p}_W(k)$, except for the pair $(k=1, r=0)$, where it is evident?

The answer is NO.

We illustrate this statement with an example.

Example 8.4.1. We consider a production scheme with the input-output matrix **S** and vector of labour **L**, the vector of total output **q**, the rate of profits $r_0 = 0.1$ and the wage rate $w_0 = 1$,

$$\begin{aligned}&(250(W), 1{,}100(M), 80) \rightarrow 575(W),\\&(110(W), 720(M), 20) \rightarrow 2{,}000(M).\end{aligned} \tag{8.35}$$

(a) Set up the Sraffa price model (8.31), and compute the price vector $\mathbf{p}_S(r,w)$, the total profit P_S, the total wages W_S, the *national income* Y_S, the operating capital K_S and total output X_S.

(b) Then set up the Weintraub price model (8.33), require equal *national incomes*, $Y_W = Y_S$, and compute the price vector $\mathbf{p}_W(k)$, the mark–up k the total profit P_W and the total wages W_W.

Compare the price vectors $\mathbf{p}_S(r,w)$ and $\mathbf{p}_W(k)$.

Solution to Example 8.4.1:
We identify the matrices:

$$\mathbf{S} = \begin{bmatrix} 250 & 110 \\ 1{,}100 & 720 \end{bmatrix}, \quad \mathbf{Se} = \begin{bmatrix} 250 & 110 \\ 1{,}100 & 720 \end{bmatrix}\begin{bmatrix} 1 \\ 1 \end{bmatrix} = \begin{bmatrix} 360 \\ 1{,}820 \end{bmatrix},$$
$$\mathbf{L} = \begin{bmatrix} 80 \\ 20 \end{bmatrix}, \quad \mathbf{q} = \begin{bmatrix} 575 \\ 2{,}000 \end{bmatrix}, \tag{8.36}$$

and the surplus quantity vector is then calculated

$$\mathbf{d} = \mathbf{q} - \mathbf{Se} = \begin{bmatrix} 575 \\ 2{,}000 \end{bmatrix} - \begin{bmatrix} 360 \\ 1{,}820 \end{bmatrix} = \begin{bmatrix} 215 \\ 180 \end{bmatrix}. \tag{8.37}$$

(a) At first, we solve the Sraffa price model in (8.31) with the exogenous variables r_0 and w_0 to get the price vector $\mathbf{p}_S$, here noted with index S. All the computed values are rounded off. We get

$$\boxed{\begin{aligned}&\begin{bmatrix} 250 & 1{,}100 \\ 110 & 720 \end{bmatrix}\begin{bmatrix} p_1 \\ p_2 \end{bmatrix}(1+r) + w \cdot \begin{bmatrix} 80 \\ 20 \end{bmatrix} = \begin{bmatrix} 575 & 0 \\ 0 & 2{,}000 \end{bmatrix}\begin{bmatrix} p_1 \\ p_2 \end{bmatrix},\\&r = r_0 = 0.1,\\&w = w_0 = 1.\end{aligned}} \tag{8.38}$$

This gives the *vector of prices* $\mathbf{p}_S = [0.5595, 0.0726]'$ (two price variables, two exogenous values and four equations).

We compute now the five essential economic variables P_S, W_S, Y_S, X_S, K_S,

$$P_S = r \cdot (\mathbf{Se})'\mathbf{p} = 0.1 \cdot [360, 1{,}820] \begin{bmatrix} 0.55947 \\ 0.07260 \end{bmatrix} = 33.353,$$
$$W_S = w \cdot \mathbf{e}'\mathbf{L} = 1 \cdot [1, 1] \begin{bmatrix} 80 \\ 20 \end{bmatrix} = 100,$$
$$Y_S = \mathbf{d}'\mathbf{p} = [215, 180] \begin{bmatrix} 0.55947 \\ 0.07260 \end{bmatrix} = P_S + W_S = 133.353,$$
$$K_S = (\mathbf{Se})'\mathbf{p} = [360, 1{,}820] \begin{bmatrix} 0.55947 \\ 0.07260 \end{bmatrix} = 333.534,$$
$$X_S = \mathbf{q}'\mathbf{p} = [575, 2{,}000] \begin{bmatrix} 0.55947 \\ 0.07260 \end{bmatrix} = K_S + Y_S = 466.887. \tag{8.39}$$

(b) Second, we compute the Weintraub price model (8.33) with the exogenous variables $w_0 = 1$, and $Y_W = Y_S = 133.353$, to get the price vector $\mathbf{p}_W$, here noted with index W. We get

$$\boxed{\begin{aligned} &\begin{bmatrix} 250 & 1{,}100 \\ 110 & 720 \end{bmatrix} \begin{bmatrix} p_1 \\ p_2 \end{bmatrix} + k \cdot w \begin{bmatrix} 80 \\ 20 \end{bmatrix} = \begin{bmatrix} 575 & 0 \\ 0 & 2{,}000 \end{bmatrix} \begin{bmatrix} p_1 \\ p_2 \end{bmatrix}, \\ &w = w_0 = 1, \quad Y = Y_W = 133.353, \\ &k = w \cdot \frac{Y}{L}, \quad L = [1, 1] \begin{bmatrix} 80 \\ 20 \end{bmatrix}. \end{aligned}} \tag{8.40}$$

This gives here the *vector of prices* $\mathbf{p}_W = [0.5623, 0.0692]'$, $k = 1.3335$, $L = 100$ (four variables, two exogenous values and six equations). Then, we compute the total wages $W_W = w \cdot \mathbf{e}'\mathbf{L} = 100$ and the total profits $P_W = Y - W = 33.353$. We get the same values as in (a); the reason is that we had required equal national incomes $Y_S = Y_W$.

Clearly the resulting Sraffa prices are different from the Weintraub prices (8.33), $\mathbf{p}_W = [0.56234, 0.06916]' \neq \mathbf{p}_S = [0.55947, 0.07260]'$. ▲

(2) A second **question** arises:
Consider a Sraffa production scheme $(\mathbf{S}', \mathbf{L}) \rightarrow (\hat{\mathbf{q}})$ with given *means of production* (8.30), a given vector of *labour* $\mathbf{L}$, a uniform *rate of profits* r and the wage rate w. The single-product Sraffa system (8.31) is applied. The question arises how to **modify** the Weintraub price model (8.34), to obtain identical price vectors $\mathbf{p}_S(r, w) = \mathbf{p}_W(.)$. We do not note the dependent variable k, because the mark-up k will be modified!

The proposition is to replace the uniform mark-up k in (8.34) by the vector $\mathbf{k} = [k_1, \ldots, k_n]'$ of specific *industry mark-up coefficients* k_j varying from industry to industry, with which we form the diagonal matrix $\hat{\mathbf{k}}$:

$$\hat{\mathbf{k}} = \begin{bmatrix} k_1 & 0 & 0 & \dots & 0 \\ 0 & k_2 & 0 & \dots & 0 \\ 0 & 0 & k_3 & \dots & 0 \\ \dots & \dots & \dots & \dots & \dots \\ 0 & 0 & 0 & \dots & k_n \end{bmatrix}, \quad k_j \geq 1, j = 1, \dots, n, \tag{8.41}$$

the diagonal matrix of industry specific *mark-up coefficients*.

We then get a further Weintraub price model depending on the *wage rate* w and the diagonal matrix $\hat{\mathbf{k}}$ of *industry specific mark-up coefficients* k_j,

$$\boxed{\mathbf{S}'\mathbf{p} + w \cdot \hat{\mathbf{k}}\mathbf{L} = \hat{\mathbf{q}}\mathbf{p}.} \tag{8.42}$$

Before developing further theoretical concepts, we will illustrate the answer to the question on equal price vectors with the next example.

Example 8.4.2. We refer to the production scheme (8.35) of Example 8.4.1 with the same matrices and coefficient $w_0 = 1$.

We require the equality $\mathbf{p}_W(\mathbf{k}) = \mathbf{p}_S(r, w) = [0.55947, 0.072596]$. Compute the Weintraub price model (8.42), where k has been replaced by $\mathbf{k}$,

$$\hat{\mathbf{k}} = \begin{bmatrix} k_1 & 0 \\ 0 & k_2 \end{bmatrix}. \tag{8.43}$$

Then compute the coefficients of the diagonal matrix $\hat{\mathbf{k}}$ of industry specific mark-up and compute the five principal economic variables P_W, W_W, Y_W, K_W, X_W.

Solution to Example 8.4.2:
With the given matrices and elements, we set up the Weintraub price model (8.42) and compute k_1, k_2 (2 variables, 3 exogenous values and 5 equations),

$$\boxed{\begin{aligned} \begin{bmatrix} 250 & 1{,}100 \\ 110 & 720 \end{bmatrix} \begin{bmatrix} p_1 \\ p_2 \end{bmatrix} + w \cdot \begin{bmatrix} k_1 & 0 \\ 0 & k_2 \end{bmatrix} \begin{bmatrix} 80 \\ 20 \end{bmatrix} &= \begin{bmatrix} 575 & 0 \\ 0 & 2{,}000 \end{bmatrix} \begin{bmatrix} p_1 \\ p_2 \end{bmatrix}, \\ p_1 &= 0.55947, \\ p_2 &= 0.072596, \\ w = w_0 &= 1, \end{aligned}} \tag{8.44}$$

giving the *industry mark-up coefficients* matrix

$$\hat{\mathbf{k}} = \begin{bmatrix} 1.2747 & 0 \\ 0 & 1.5691 \end{bmatrix}. \tag{8.45}$$

One observes that the values of the five essential economic variables P_W, W_W, Y_W, X_W, K_W are identical to those obtained in Example 8.4.1, revealing that required equal Sraffa and Weintraub prices result in equal economic variables Y, W, P, K and X.

$$P_W = P_S = \mathbf{e}'(\hat{\mathbf{k}} - w \cdot \mathbf{I})\mathbf{L} = [1,1]\begin{bmatrix} 0.27465 & 0 \\ 0 & 0.56905 \end{bmatrix}\begin{bmatrix} 80 \\ 20 \end{bmatrix} = 33.353,$$

$$W_W = W_S = w \cdot \mathbf{e}'\mathbf{L} = 1 \cdot [1,1]\begin{bmatrix} 80 \\ 20 \end{bmatrix} = 100,$$

$$Y_W = Y_S = \mathbf{e}'\hat{\mathbf{k}}\mathbf{L} = [1,1]\begin{bmatrix} 1.27465 & 0 \\ 0 & 1.56905 \end{bmatrix}\begin{bmatrix} 80 \\ 20 \end{bmatrix} = P_W + W_W = 133.353$$

$$K_W = K_S = (\mathbf{Se})'\mathbf{p} = [360, 1, 820]\begin{bmatrix} 0.55947 \\ 0.07260 \end{bmatrix} = 333.534,$$

$$X_W = X_S = \mathbf{q}'\mathbf{p} = [575, 2, 000]\begin{bmatrix} 0.55947 \\ 0.07260 \end{bmatrix} = K_S + Y_S = 466.887. \quad \blacktriangle \tag{8.46}$$

The *mean macroeconomic mark-up* $\overline{k}$ is then the weighted average obtained from the definition of *national income*

$$Y_W = \overline{k}\mathbf{e}'\mathbf{L} = \overline{k}L = \mathbf{e}'\hat{\mathbf{k}}\mathbf{L} \Rightarrow \overline{k} = \frac{Y_W}{L_W}, \tag{8.47}$$

i. e., the *mean macroeconomic mark-up* $\overline{k}$ is the *national income* per unit of *required working time L*.

Taking again Example 8.4.2, we indeed get

$$\overline{k} = (1.27465 \cdot 80 + 1.56905 \cdot 20)/100 = 1.334. \tag{8.48}$$

Summarising, the general Weintraub price system is given by (8.42). This alternative approach is best analysed in connection with the extended single-product Sraffa system introducing variable rates of profits and wages as presented in Section 8.3, opening up new avenues of research, which we shall not pursue here.

The matrix $\mathbf{S}$ is assumed to be positive or *irreducible* and *semi-positive*. For this reason, the **Perron–Frobenius theorem A.9.3** applies and we get

Proposition 8.4.1. *Given a Sraffa production scheme* $(\mathbf{S}', \mathbf{L}) \rightarrow (\hat{\mathbf{q}})$, *a mark-up k and the wage rate w and a positive vector of total surplus,* $\mathbf{d} > 0$, *one applies the Sraffa price model (4.29) (I) and the Weintraub price models (8.34) and (8.42). Then, the Weintraub price vectors are positive,* $\mathbf{p}_W(k) > \mathbf{0}$ *and* $\mathbf{p}_W(\mathbf{k}) > \mathbf{0}$.

Proof. (a) One computes the positive vector of *total output*, $\mathbf{q} = \mathbf{Se} + \mathbf{d}$ (2.15). The *productiveness* $R > 0$ is positive because there is a positive vector of surplus $\mathbf{d}$. The productiveness is chosen as *rate of profits*. There are no wages. We then know that the Sraffa price model,

$$\mathbf{S}'(1 + R)\mathbf{p}_S(R, 0) = \hat{\mathbf{q}}\mathbf{p}_S(R, 0), \tag{8.49}$$

leads to an eigenvalue problem. Therefore, we compute the *input-output coefficients* matrix $\mathbf{C} = \mathbf{S}\hat{\mathbf{q}}^{-1}$, getting the Frobenius number $\lambda_C = 1/(1+R) < 1$ and here the *productiveness* $R > 0$. Now we set the *left eigenvector equation* of matrix $\mathbf{C}$ to obtain the positive price eigenvector $\mathbf{p}_S(R,0) := \mathbf{p}$, associated with the Frobenius number λ_C.

$$\mathbf{C}'\mathbf{p}_S(R,0) = \lambda_C \cdot \mathbf{p}_S(R,0) = \frac{1}{1+R} \cdot \mathbf{p}_S(R,0). \tag{8.50}$$

We know that the vector $\mathbf{p}_S(R,0) > 0$ is positive. Its uniqueness is guaranteed by the Perron–Frobenius theorem A.9.3 (iii), requiring the sum of the components to be normalised, namely: $\sum_{i=1}^{n} p_i = 1$.

(b) Consider now the *Weintraub price model* (8.34) and (8.42) with constant markup $k > 1$, respectively vector $\mathbf{k}$, a positive *wage rate* $w > 0$ and a positive vector of *required working time* $\mathbf{L} > 0$. Apply the *input-output coefficients* matrix $\mathbf{S} = \mathbf{C}\hat{\mathbf{q}}$. One obtains:

$$\mathbf{S}'\mathbf{p}_W(k) + wk\mathbf{L} = \hat{\mathbf{q}}\mathbf{C}'\mathbf{p}_W(k) + wk\mathbf{L} = \hat{\mathbf{q}}\mathbf{p}_W(k),$$
$$\mathbf{S}'\mathbf{p}_W(\mathbf{k}) + wk\mathbf{L} = \hat{\mathbf{q}}\mathbf{C}'\mathbf{p}_W(\mathbf{k}) + wk\mathbf{L} = \hat{\mathbf{q}}\mathbf{p}_W(\mathbf{k}). \tag{8.51}$$

As the Frobenius number is less than 1, $\lambda_C < 1$, the transposed Leontief Inverse exists, and with $\boldsymbol{\pi} = \hat{\mathbf{q}}^{-1}\mathbf{L}$ we obtain,

$$\mathbf{p}_W(k) = (\mathbf{I} - \mathbf{C}')^{-1} \cdot wk \cdot \hat{\mathbf{q}}^{-1}\mathbf{L} := (\mathbf{I} - \mathbf{C}')^{-1} \cdot wk \cdot \boldsymbol{\pi},$$
$$\mathbf{p}_W(\mathbf{k}) = (\mathbf{I} - \mathbf{C}')^{-1} \cdot w\hat{\mathbf{k}} \cdot \hat{\mathbf{q}}^{-1}\mathbf{L} := (\mathbf{I} - \mathbf{C}')^{-1} w\hat{\mathbf{k}} \cdot \boldsymbol{\pi}. \tag{8.52}$$

Thus, with Theorem A.10.2 the transposed Leontief Inverse is positive, $(\mathbf{I} - \mathbf{C}')^{-1} > \mathbf{0}$ and because the vector of labour is also positive, $\mathbf{L} > \mathbf{o}$, the Weintraub price vectors are consequently positive, $\mathbf{p}_W(k) > \mathbf{o}$ and $\mathbf{p}_W(\mathbf{k}) > \mathbf{o}$. □

Let us illustrate Proposition 8.4.1 by resorting to the following example

Example 8.4.3. We refer to the production scheme (8.35) of Example 8.4.1 with the same matrices and vectors (8.36), (8.37), the wage rate $w = 1$, the mark-up $k = 2$ and the industry mark-up coefficients matrix

$$\hat{\mathbf{k}} = \begin{bmatrix} 2 & 0 \\ 0 & 1.5 \end{bmatrix}. \tag{8.53}$$

Compute the positive input-output coefficients matrix $\mathbf{C} = \mathbf{S}\hat{\mathbf{q}}^{-1}$, the characteristic polynomial of $\mathbf{C}$, the Frobenius number $\lambda_C < 1$ and the productiveness $R > 0$.

Compute the transposed Leontief Inverse $(\mathbf{I} - \mathbf{C}')^{-1}$, Weintraub price vector $\mathbf{p}_K(k)$ and $\mathbf{p}_K(\mathbf{k})$, according to (8.52).

Solution to Example 8.4.3:
We calculate the positive *input-output matrix* **C**

$$\mathbf{C} = \mathbf{S}\hat{\mathbf{q}}^{-1} = \begin{bmatrix} 250 & 110 \\ 1{,}100 & 720 \end{bmatrix} \begin{bmatrix} \frac{1}{575} & 0 \\ 0 & \frac{1}{2{,}000} \end{bmatrix} = \begin{bmatrix} \frac{10}{23} & \frac{11}{200} \\ \frac{44}{23} & \frac{9}{25} \end{bmatrix} > \mathbf{0} \tag{8.54}$$

and compute the characteristic polynomial

$$P_2(\lambda) = \det(\mathbf{C} - \lambda\mathbf{I}) = \det\left(\begin{bmatrix} \frac{10}{23} - \lambda & \frac{11}{200} \\ \frac{44}{23} & \frac{9}{25} - \lambda \end{bmatrix}\right)$$

$$= \frac{59}{1{,}150} - \frac{457}{575} + \lambda^2 = (\lambda - 0.0709)(\lambda - 0.7329). \tag{8.55}$$

The **Perron–Frobenius theorem A.9.3** applies as matrix $\mathbf{C} > \mathbf{0}$ is positive and the Frobenius number is $\lambda_C = 0.7329 < 1$, giving directly the *productiveness* $R = (1/\lambda_C) - 1 = 0.38138 > 0$. Compute now the transposed Leontief Inverse,

$$(\mathbf{I} - \mathbf{C}')^{-1} = \begin{bmatrix} \frac{736}{295} & \frac{440}{59} \\ \frac{253}{1{,}180} & \frac{130}{59} \end{bmatrix} > \mathbf{0}. \tag{8.56}$$

Then, one computes the vector of labour components per units of produced commodities,

$$\boldsymbol{\pi} = \hat{\mathbf{q}}^{-1}\mathbf{L} = \begin{bmatrix} \frac{1}{575} & 0 \\ 0 & \frac{1}{2{,}000} \end{bmatrix} \begin{bmatrix} 80 \\ 20 \end{bmatrix} = \begin{bmatrix} \frac{16}{115} \\ \frac{1}{100} \end{bmatrix} > \mathbf{o}. \tag{8.57}$$

Finally, we compute both Weintraub price vectors (8.52),

$$\mathbf{p}_W(k) = (\mathbf{I} - \mathbf{C}')^{-1} \cdot wk \cdot \boldsymbol{\pi}$$

$$= \begin{bmatrix} \frac{736}{295} & \frac{440}{59} \\ \frac{253}{1{,}180} & \frac{130}{59} \end{bmatrix} \cdot 1 \cdot 2 \cdot \begin{bmatrix} \frac{16}{115} \\ \frac{1}{100} \end{bmatrix} = \begin{bmatrix} 0.8434 \\ 0.1037 \end{bmatrix},$$

$$\mathbf{p}_W(\mathbf{k}) = (\mathbf{I} - \mathbf{C}')^{-1} w\hat{\mathbf{k}} \cdot \boldsymbol{\pi}$$

$$= \begin{bmatrix} \frac{736}{295} & \frac{440}{59} \\ \frac{253}{1{,}180} & \frac{130}{59} \end{bmatrix} \cdot 1 \cdot \begin{bmatrix} 2 & 0 \\ 0 & 1.5 \end{bmatrix} \begin{bmatrix} \frac{16}{115} \\ \frac{1}{100} \end{bmatrix} = \begin{bmatrix} 0.8061 \\ 0.0927 \end{bmatrix}. \tag{8.58}$$

Here, we terminate this subsection, where we have discovered a further price model, based upon the mark–up concept of Sidney Weintraub, called the Weintraub price model.

8.5 Where does Sraffa fit in the strain of economic theories?

8.5.1 Sraffa in context

Having previously examined some relationships between Input-Output Analysis and Sraffa's approach, we must understand the economic context in which the latter is to be conceived. In fact, Sraffa's initial model applies to one period of time, say period t, embedded in an ongoing economic process over successive periods as mentioned in PCMC, Par. 1. This may be illustrated as follows (we resort here to the notation stipulated of our text):

One considers the short period in a *closed economy*:

- Wages are paid ex post at the end of the period.
- Commodities are produced during that period and are sold or exchanged on a "market" (PCMC, Par. 1).
- If a surplus emerges by the end of the period, it is distributed as wages W to workers, the rest reverting to entrepreneurs as profits P.

If this is to make sense, it is implicitly assumed that:

- Industries pay for their operating expenses in period t to sustain production.
- The wage earners of period $t-1$ spend their earnings W in period t (or even later).
- Entrepreneurs spend their profits P similarly.
- Accordingly, this will create a *demand* that exercises its effects in period t (and possibly later).
- This *effective demand* drives production in excess of interindustry production required for sustainability (viability) and generates a surplus, i. e., drives *total output*.
- In period t, a *total output*, ensuring sustainability and a surplus, therefore emerges to respond to demand generated by earnings in period $t-1$ (and possibly in preceding periods) and in period t. At the end of period t, inter-industry purchases and *final demand* $\mathbf{f}$, in value terms, are registered. We have encountered these items in the introduction to Input-Output Analysis, according to Miller and Blair [65][8] presented in Chapter 2.

The ongoing evolution of *effective demand* and *total output* from period to period is graphically presented in Figure 8.3. *Demand* is presented by a green arrow and *output by a red arrow.*[9]

---

8 Incidentally, the foregoing confirms Bailly's result (Bailly [3], pp. 374–378):
During phases of static activity in a closed economy, i. e., $Y_{t-1} = Y_t, W_{t-1} = W_t = W$. If demand from W_{t-1} absorbs all profits P in period t, then in the aggregate $\frac{P}{X} \leq 0.5$ (X is total output in period t). Indeed $\frac{P}{X} = \frac{P}{(P+W+K)}$, so if $W = P$ then $\frac{W}{(W+W+K)} = \frac{1}{(1+1+\frac{K}{W})} \leq \frac{1}{(2+c)} \leq 0.5$, with $\frac{K}{W} = c \geq 0$ (K is the value of circulating capital, i. e., the value of Sraffa's means of production, in period t).

9 Sraffa's basic conditions of production (see Subsection 3.1.1) in quantity terms, expressing interindustry flows of commodities do not vary from period to period, i. e., by assumption, the coeffi-

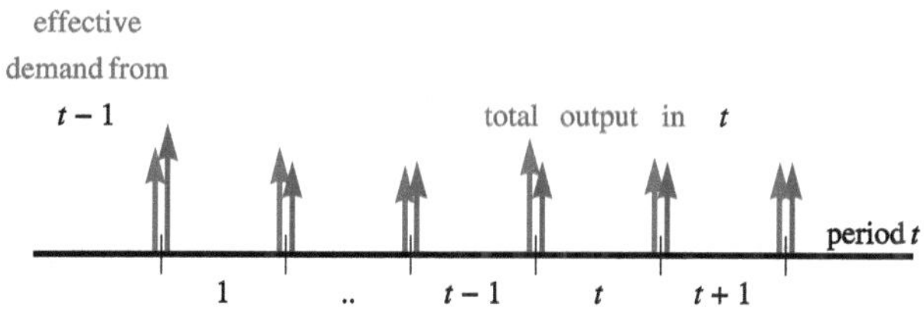

Figure 8.3: Variation of *demand* and *output* from period to period.

We thus have a representative circular process of the type illustrated by Figure 2.2 (in that case without surplus).

More generally, as J. W. Forrester, that forgotten pioneer, showed more than 40 years ago [32], the networks of economies of production and exchange with their many interwoven circuits (of the type addressed in Chapter 6 hereafter) represent circular processes. As such, they generate over successive production periods economic cycles such as the short-term business cycle, the intermediate *Kuznets swing* and the long-term *Kondratiev cycle*.

Total output is the measurable response of the production process to *effective demand*. If total output exceeds effective demand, inventories are allocated to future use; if effective demand is higher, existing inventories are depleted (see Miller and Blair [65], Section 4.5).[10]

As regards final demand, we have $\mathbf{f} = \hat{\mathbf{p}}\mathbf{d}$, equation (2.105) as outcome in period t, and one considers the commodity surplus $\mathbf{d}$ as given, the stance adopted by Sraffa.

We cannot leave this subsection without a word about the central role played by the uniform rate of profits r in Sraffa's scheme. The rate of profits r is a macroeconomic parameter concerning the ensemble of all industries, or sectors, entering a given economy of production and exchange: in fact we know that $r = \tilde{r}\tilde{R} = (P/Y) \cdot (Y/K)$. This independent parameter is not a "long-term competitive equilibrium rate of profits" to speak in mainstream terms, nor is it the discount rate of an expected future income system.

In the Sraffa context, one considers a sustainable circular economic process obeying the conditions of production, and if a surplus is generated, such surplus ensures a positive national income of the corresponding productive economy.

cients of matrix **S** are constants. But expressed in value terms, interindustry transactions may vary because of price fluctuations due for example to changes in the value-added components of the economy or, in other terms, due to changes in final demand.

10 The inner workings of the dynamics between supply and effective demand are essentially inaccessible to direct observation: only the effects are statistically measurable.

This situation is similar to that in quantum physics: atomic and subatomic processes are only measurable via their effects (electromagnetic spectra, trajectories in wire chambers, etc.), see for example Plotnitsky [87], Chapter 2.

This production system furthermore remains stable over time as long as the technology matrix $\mathbf{S}'$ remains unchanged. It is in this sense that the expression "long-term equilibrium", in other words equilibrium over time, must be understood in the present context.

Now, the desired stability can only be attained if all the industries involving at least one basic commodity that participate in the functioning of the economy remain over time in the production process regardless of their effective varying degrees of profitability. This is realised by the redistribution of the surplus at the end of each reference period, as explained in PCMC, Par. 4. In practice, such a redistribution could be realised through appropriate institutions or via taxation, in proportion to the circulating capital (in monetary terms) K of each industry sector. This leads to a uniform rate of profits r, indicating that the system is in stable "equilibrium" in the Sraffian sense.

In a sustainable, viable and stable closed circular monetary economy of production and exchange, creative adjustments are made on the basis of interindustrial solidarity. There is no Schumpeterian "creative competitive destruction".

8.5.2 Leontief and Sraffa: basic aspects compared

The process illustrated in Figure 8.3 can be represented either by the Leontief model or by Sraffa's construct. So let us try to see how the two compare, assuming that $\mathbf{S}$, $\mathbf{q}$, $\mathbf{d}$, $\mathbf{L}$, X and Y are the same in both models. We shall start by referring again to the Leontief price equations ($\mathbf{p}_L$ meaning here prices obtained from the Leontief model), see Chapter 2.

Let us consider the cost-push input-output Leontief price model (2.111) in physical terms, to be in line with PCMC:

$$\mathbf{p}_L = \mathbf{C}'\mathbf{p}_L + \mathbf{v}_c \quad \text{or} \quad \mathbf{p}_L = (\mathbf{I} - \mathbf{C}')^{-1}\mathbf{v}_c. \tag{8.59}$$

Remember that $\mathbf{v}_c$ is a value-added vector, expressed directly as such in value terms, where no distinction is made between profits and labour costs.

We know furthermore from (2.113) that $\mathbf{A}$ (2.9) and $\mathbf{C}$ (2.17) are similar matrices: $\mathbf{A} = \hat{\mathbf{p}}_L\mathbf{C}\hat{\mathbf{p}}_L^{-1}$.

For Sraffa ($\mathbf{p}_S$ denoting Sraffa prices), we refer to equation (4.178)

$$\mathbf{p}_S = \mathbf{C}'\mathbf{p}_S(1 + r) + w \cdot \boldsymbol{\pi}, \tag{8.60}$$

with $w = \tilde{w} \cdot \frac{Y}{L}$, $\mathbf{C}' = \hat{\mathbf{q}}^{-1}\mathbf{S}'$ (2.16), and $\boldsymbol{\pi} = \hat{\mathbf{q}}^{-1}\mathbf{L}$ (4.58) or explicitly:

$$\mathbf{p}_S = [\mathbf{I} - \mathbf{C}'(1 + r)]^{-1} w \cdot \boldsymbol{\pi} \tag{8.61}$$

Again referring to (2.17) and (4.61), setting $\boldsymbol{u}_c := [r\mathbf{C}'\mathbf{p}_S + w \cdot \boldsymbol{\pi}]$, the surplus part in the prices, (8.60) can be rewritten as

$$\mathbf{p}_S = \mathbf{C}'\mathbf{p}_S + [r\mathbf{C}'\mathbf{p}_S + w \cdot \boldsymbol{\pi}] = \mathbf{C}'\mathbf{p}_S + \boldsymbol{u}_c. \tag{8.62}$$

We can explicitly calculate the Sraffa prices,

$$\mathbf{p}_S = (\mathbf{I} - \mathbf{C}')^{-1}\boldsymbol{u}_c, \tag{8.63}$$

and setting,

$$\boldsymbol{u}_c := \hat{\mathbf{q}}^{-1}(\boldsymbol{\Pi} + \boldsymbol{\Omega}), \quad \boldsymbol{\Pi} := r \cdot \hat{\mathbf{q}}\mathbf{C}'\mathbf{p}_S, \quad \boldsymbol{\Omega} := w \cdot \hat{\mathbf{q}}\boldsymbol{\pi}, \tag{8.64}$$

we finally get

$$\boldsymbol{\Pi} := r \cdot (\hat{\mathbf{q}}\mathbf{C}')\mathbf{p}_S = r \cdot \mathbf{S}'\mathbf{p}_S, \quad \boldsymbol{\Omega} := w \cdot (\hat{\mathbf{q}}\boldsymbol{\pi}) = w \cdot \mathbf{L}, \tag{8.65}$$

where $\boldsymbol{\Pi} = r{\cdot}\mathbf{S}'\mathbf{p}_S$ defines the vector of *gross profits* per industry, and $\boldsymbol{\Omega} = w\mathbf{L}$, the vector of *labour costs* per industry. Realising the summation, we get total profits $P = \mathbf{e}'\boldsymbol{\Pi}$ and total wages $W = \mathbf{e}'\boldsymbol{\Omega}$.

The vector $\boldsymbol{u}_c$ is the total *value added* per unit of produced commodities in each industry (sector).

Indeed: $\mathbf{e}'\hat{\mathbf{q}}\mathbf{u}_c = \mathbf{e}'r \cdot (\hat{\mathbf{q}}\mathbf{C}')\mathbf{p}_S + \mathbf{e}'w \cdot (\hat{\mathbf{q}}\boldsymbol{\pi}) = r \cdot \mathbf{e}'\mathbf{S}'\mathbf{p}_S + w \cdot \mathbf{e}'\mathbf{L} = P + W = Y$.

The first observation to be noted is that the Leontief price model (8.59) and the single-product Sraffa system (8.63) have the same formal algebraic structure and the same inverse intervenes in both systems.

But, the comparison ends here because the components of value added in Sraffa are calculated differently than in Leontief. In fact, there still subsists the price vector $\mathbf{p}_S$ in the profit component of $\boldsymbol{u}_c$.

A direct relationship between $\mathbf{p}_L$ and $\mathbf{p}_S$ may be obtained as follows: write $\boldsymbol{v}_c = \hat{\mathbf{v}}_c\mathbf{e}$ and $\boldsymbol{u}_c = \hat{\mathbf{u}}_c\mathbf{e}$ then

$$\begin{aligned} &\hat{\mathbf{v}}_c^{-1}\mathbf{p}_L = (\mathbf{I} - \mathbf{C}')'^{-1}\mathbf{e} = \boldsymbol{u}_c^{-1}\mathbf{p}_S, \quad \text{accordingly} \\ &\mathbf{p}_L = \hat{\mathbf{v}}_c\hat{\boldsymbol{u}}_c^{-1}\mathbf{p}_S \end{aligned} \tag{8.66}$$

Leontief and Sraffa prices are therefore not equal, except in the special case where there are no profits and where the unit wage is equal in all industries. In that case, $\boldsymbol{v}_c = \boldsymbol{u}_c$ and so $\mathbf{p}_L = \mathbf{p}_S$.

We have just considered vectors; the price vectors $\mathbf{p}_L$ and $\mathbf{p}_S$ are not the same but not completely independent. As we are dealing with the same economic process, both models must fulfil the same national accounting identities. This means that the price vectors must also fulfil the following conditions in the aggregate for a closed economy:

$$\begin{aligned} &X = \mathbf{e}'\mathbf{S}'\mathbf{p}_L + \mathbf{e}'\hat{\mathbf{d}}\mathbf{p}_L = \hat{\mathbf{q}}\mathbf{p}_L = \hat{\mathbf{q}}\mathbf{p}_S = \mathbf{e}'\mathbf{S}'\mathbf{p}_S + \mathbf{e}'\hat{\mathbf{d}}\mathbf{p}_S \quad \text{and} \\ &Y = \mathbf{e}'\mathbf{f} = \mathbf{e}'\hat{\mathbf{d}}\mathbf{p}_L = \mathbf{e}'\hat{\mathbf{d}}\mathbf{p}_S, \end{aligned} \tag{8.67}$$

so as corollary for the value of the means of production[11]

$$K = \mathbf{e}'\mathbf{S}'\mathbf{p}_L = \mathbf{e}'\mathbf{S}'\mathbf{p}_S. \tag{8.68}$$

Other common characteristics are:

(1) The production process generates two types of output:
  (a) Commodities covering interindustry demands (Sraffa's conditions of production);
  (b) Commodities entering the surplus which corresponds to final demand, i. e., national income (GDP) in a closed economic system.

  The addition of (a) and (b) gives in both models total output.

(2) National income is the total of *value added* for a closed economy in both models.

(3) In the absence of a surplus, respectively of final demand, the Leontief model is identical to Sraffa's model, if the same *monetary numéraire* is adopted.

  The main differences in this approach are:

  ① Loosley speaking, Leontief's Input-Output model starts from the total output of a production process which is then split between interindustry purchases and a surplus element termed final demand. Whereas Sraffa's starts by saying let us consider a process, composed of means of production, which is self-replacing and is then assumed to produce a surplus.

  ② The choice of the numéraire is left open in Sraffa. If money is chosen, his model is consistent with the identity (simplified) $P + W = Y$. Note however that Sraffa makes no reference to national accounting per se and implicitly assumes a *numéraire* defined in *physical terms.*

  Input-Output Analysis on the contrary resorts systematically to money as numéraire or to *index prices* to fix the value of flows. National accounting identities are central to this approach.

  ③ Sraffa incorporates explicitly labour, wages and profits as separate entities in his model.

  In Input-Output Analysis wages and profits appear indirectly as one whole value-added element. There is no split between profits and wages and no specific assumptions are made regarding these two items.

  ④ In fact, referring to Table 2.1 which is the centrepiece of all that follows, Leontief's approach is horizonal, and his equations are based on the *commodity flow* matrix $\mathbf{Z}$ in *monetary terms.* Whereas Sraffa's approach is vertical, his equations are based on the transposed *commodity flow* matrix $\mathbf{S}'$ in *physical terms* (the starting point of Pasinetti's concept of "vertical integration", see Pasinetti [83], Chapter 2.)

  ⑤ More precisely, the Leontief prices $\mathbf{p}_L$ depend on an empirical distribution of total *value added* throughout the various industries.

11 Geometrically, this just means that the price vectors are not the same but lie in the same hyperplane.

In Sraffa (PCMC) *value added*, i. e., the surplus, is assumed *in fine* to be distributed proportionally to the means of production (circulating capital). Furthermore, in calculating the value added, it is assumed that the rate of profits r and the unit wage rate w are the same in all industries.

⑥ Leontief's Input-Output model also incorporates imports and exports, so it is immediately applicable to open economics, whereas Sraffa's original model refers to a closed economy. The open economy requires an extension of Sraffa's model, see Section 8.1.

⑦ Finally, Leontief's Input-Output methodology and Sraffa's PCMC can both be extended to joint production. This is an important aspect and brings these model closer to applications, see A. E. Steenge [110] and T. L. Saaty [99].

This explains the apparent higher complexity of Sraffa's model and the importance attributed to the allocation of surplus between wages and profits, an aspect not addressed to in Input-Output Analysis.

8.6 Sraffa and neoclassical theory

Piero Sraffa's PCMC opened a new route of economic research because his approach was totally different from marginalism, the dominant paradigm of economic thought in the West since about 1900. We shall comment on Sraffa and Walras in Section 8.8 and pursue here aspects of the discussion.

The *preface* of PCMC presents the economic concept of Sraffa's book. His ideas are then developed in a series of paragraphs, with numerical examples, some equations, but no linear algebra. In our text, we have concentrated on the mathematical background of Sraffa's work; let us now take a short look at the central assumptions of *neoclassical theory*, which today is the current label for marginalism.

In neoclassical theory, production is seen as a one-way avenue that leads from "factors of production" to consumption goods. A "production function" f links the amounts of the factors "capital" (K) and "labour" (L) to the final output Q, i. e., $Q = f(K, L)$. Labour L can be measured in units of time, but it is not clear how K and Q can be measured, unless prices of capital goods and consumption goods are known. The function f must have certain strong mathematical properties: namely for positive values of K and L it must have positive first derivatives and a negative definite Hessian matrix. This condition implies the existence of so-called 'indifference curves': any small amount of labour can be replaced by a small amount of capital, without changing the output Q. The prices of goods depend on subjective factors such as the tastes and preferences of individuals, and market clearing is attained by the mechanism of supply and demand.

Sraffa does not resort to the marginal method because his approach centers on objective laws of production rather than on the individual tastes of consumers and ideological considerations. He writes in the Preface of PCMC:

"The investigation is concerned exclusively with such properties of an economic system as do not depend on changes in the scale of production or in the proportions of factors".

Clearly, in Sraffa's context the number of workers (working hours) L is a constant, as also the total product Q is constant for each industry. Sraffa notes: The marginal cost, or alternatively the marginal product 'just would not be there to be found', (PCMC, page v). Indeed, the production–labor–capital space contains in the Sraffa case only the point (Q, K, L), the three values being constant. There are no continuous functions $Q = f(K)$ or $Q = f(K, L)$ and no applications of differential calculus.

So, the assumption of constant returns in all industries is not a necessary working hypothesis, as Sraffa says in his preface. In fact, this assumption is implicit in his whole approach as mathematical formalisations have shown, in particular for the construction of a *Standard system* and the corresponding *standard commodity* and in questions concerning growth. In all the developments of PCMC, there are otherwise no changes in the proportions in which the various means of production are used by an industry, and this is the standpoint of classical economists from Adam Smith (1723–1790) to David Ricardo (1772–1823) and others. The Cobb-Douglas production function $Q = f(K, L)$ belongs to neoclassical theory and is heavily criticised by Sraffa.

8.7 Sraffa versus the Cobb–Douglas production function

Roncaglia pointed out ([96], pp. 51–60) that Sraffa, following the classical economists, versus the mainstream neoclassical economists, operates in a different representation of an economy of production and exchange. We enter this discussion by the presentation of the Cobb–Douglas production function.

Mankiw ([63], p. 45) states, presenting the *Cobb–Douglas production function*, that the *"available production technology determines how much output is produced from given amounts of capital K and labour L. Neoclassical economists express the available technology using a production function"*.

Letting Q denote the amount of output, the strict *Cobb–Douglas production function* is then specified as follows ([63], p. 71):

$$Q = F(K, L) = AK^{\alpha}L^{\beta} \quad \text{where } \alpha + \beta = 1, \tag{8.69}$$

the parameter α is the *production elasticity of capital and* β the *production elasticity of labour*[12] so they are both dimensionless. The narrative is: when there is *full employment*, unemployment is always voluntary in this approach, then all the production

12 The *production elasticity of capital* is $\alpha = \frac{\frac{\partial Y}{\partial K}}{\frac{Y}{K}}$ and the *production elasticity of labour* is $\beta = \frac{\frac{\partial Y}{\partial L}}{\frac{Y}{L}}$ and are dimensionless numbers. D. G. Champernowne [14] in 1936 introduced the notion of *elasticity* to get a dimensionless economic measure, circumventing the difficult problem of units in measuring economic variables.

is consumed. If Y is the measure of the consumed goods, say *Gross Domestic Product* (GDP) or *National Income* (NI), then one sets $Q = Y$. The technology variable A measures the productivity of the available technology and is a positive dimensionless numerical parameter.[13] The Cobb–Douglas production function is routinely used by neoclassical econmists; it was reintroduced in 1927 in the USA by Paul Douglas and Charles Cobb, based on earlier work by the Swedish economist K. Wicksell, 1851–1926, see Mankiw [63], pp. 71–73. The condition $\alpha + \beta = 1$ means *constant returns to scale*.[14]

The marginalist economists searched for a production function incorporating the property, which produces constant factor shares if the factors always earned their marginal product. This property is realised by the notion of *MPK = marginal product of capital*: an extra amount of output obtained from an extra amount of capital and the *MPL = marginal product of labour*, an extra amount of output obtained from an extra amount of labour (see Mankiw ([63], p. 51)), as follows

$$\frac{\partial F(K,L)}{\partial K} \sim F(K+1,L) - F(K,L), \quad \frac{\partial F(K,L)}{\partial L} \sim F(K,L+1) - F(K,L). \tag{8.72}$$

Paul Douglas and Charles Cobb proposed the strict *production function* with exactly those properties,

$$\text{Capital Income} = \frac{\partial F(K,L)}{\partial K} K = \alpha Y, \tag{8.73}$$

$$\text{Labour Income} = \frac{\partial F(K,L)}{\partial L} L = (1-\alpha) Y. \tag{8.74}$$

This means in other words that the marginalists set the *marginal physical product of capital* (*MPK*) equal to the *rate of profits r*, the profit per unit of capital, and the

13 That is, a parameter which is supposed to indicate the incidence of available technology on productivity, whatever that means. In fact, various sources consulted (Mankiw [63], Branson [11], Chiang and Wainwright [19]) do not give clear definitions of this parameter involving capital, labour and technology. The latter references are even inconsistent in defining productivity and the variables involved. Neither Kurz and Salvadori [52], Miller and Blair [65] nor Takayama [116] mention the technology factor. So the important question is: How to measure the technology parameter A? Solow ([107], p. 312), for example, considers the technology parameter as a function of time t and writes

$$Y = A(t) \cdot F(K,L). \tag{8.70}$$

Solow then calculates the total differential of (8.70) and gets

$$\frac{\dot{Y}}{Y} = \frac{\dot{A}}{A} + A\frac{\partial F}{\partial K}\frac{\dot{K}}{Y} + A\frac{\partial F}{\partial L}\frac{\dot{L}}{Y}, \tag{8.71}$$

where dots indicate the time derivative. Solow then is able to estimate the function $A(t)$.

14 A production function has constant returns to scale if an increase of an equal percentage in the factors of production K, L causes an increase in output Y of the same percentage, $z \cdot Y = F(zK, zL)$.

marginal physical product of labour (*MPL*) equals the *wage rate* w, i. e., the wage per unit of labour. The profound sense of this statement is that given the *production function*, the price of capital, namely the rate of profits $r = P/K$, and the price of labour, namely the wage rate $w = W/L$, are detemined.[15] We have here the fundamental axiom of the *theory of market economy*: With Cobb-Douglas the *price* of capital r is determined as the *marginal product of capital*, the *price* of labour w is determined as the *marginal product of labour*. Here, the capital K and labour L are prior to both prices r and w,

$$0 \leq MPK = \frac{\partial F(K,L)}{\partial K} = r \leq 1; \quad 0 \leq MPL = \frac{\partial F(K,L)}{\partial L} = w. \tag{8.75}$$

The factors K and L are aggregate variables obtained from census data, so they give no indication as to their composition and their valuation; they are determined independently and thus in principle can be substituted for one another.

This methodology has been criticized by various authors. For example, Birner [5] notes pointedly in the first passage of the preface: "*One does not need a degree in economics to know that in reality this is not the case. The economy consists of a bewildering variety of buildings, machines, software, skills and ways of organizing production. And as everyone who has filed a corporate tax declaration knows, it is not even possible to give more than a rather inexact estimation of their value*".

In fact, more prominently, Joan Robinson [92] was one of the first to systematically point out that the constitution of capital, in other words the way it is measured, an indispensable step, is not indicated; furthermore, she noted that capital cannot be a separate independent factor of production and that there is no such thing as the substitution of labour by capital.

Concerning the *theory of distribution*, Cobb–Douglas ([20]. p. 163) reads: "The degree of correspondence is however sufficient to give a considerable degree of corroboration to the *law of production* which has been worked out and to indicate that the *process of distribution* follows in large measure the *process of production* if sufficient time is allowed".

As Utz-Peter Reich writes [89]: "*This kernel statement explicates the opposite role which Cobb-Douglas concedes to "distribution" compared to Sraffa.* With Cobb–Douglas the prices of the commodities are given, as well as the capital K and the size of labour employed L. Together with the process of production, described by the production function $Y = F(K,L)$, the *rate of profits* r and the *wage rate* w are determined by the marginal product, defining ex post the *distribution*.

On the contrary, with Sraffa the *structure* of the *means of production* among the production sectors is described ex ante by the *commodity flow* matrix $\mathbf{S}$ and the distribution of the surplus Y between entrepreneurs and workers by choosing the *rate*

15 Indeed, by definition, the property $MPK \geq 0$, $MPL \geq 0$ is always fulfilled.

of profits r, $0 \leq r \leq R$, which precedes the determination of the prices of the commodities from which the value of capital K is computed. In other word, with Sraffa the distribution is determined by the prior choice of the *rate of profits r*, the price of the capital, which precedes logically the production. We present here again Sraffa's masterpiece of work for *joint production* (8.76), and summarize the procedure to attain the solution,

$$\boxed{\begin{aligned} \mathbf{S}'\mathbf{p}(1+r) + \mathbf{L}\frac{\tilde{w}\cdot Y}{L} &= \mathbf{F}',\mathbf{p} \\ Y &= (\mathbf{Fe}-\mathbf{Se})'\mathbf{p}, \\ L = \mathbf{e}'\mathbf{L} &= 1, \\ p_C &= 1. \end{aligned}} \tag{8.76}$$

The classical economists, such as Quesnay, Leontief, Sraffa, start from the structure of the economy, aiming to present it by a *tableau économique*, an *Input-Output Table* of a *commodity flow matrix*. Then Sraffa, given the labour inputs, solved principally the question of distribution of the surplus between the workers and the entrepreneurs, by choosing the rate of profits, contrary to the neoclassic economists Cobb and Douglas, where the rate of profits is a result. Sraffa then sets up a complete price system, leading to the calculation of the waged rate and wages and the prices for every commodity of the actual economy, bringing the system into equilibrium, and resulting in the calculation of the capital K and the national income Y. We present here again this masterpiece of work for *joint production* (8.76).

On the contrary, Cobb and Douglas start with capital K and labour L as "production factors", giving national income, leading to the prices of capital as *marginal physical product of capital* and the price of labour, as *marginal physical product of labour*.

This presentation shows that the approach of the classical economists like Quesnay, Leontief and Sraffa to the production economy, starting from the structure of the economy is irremediably irreconcilable with the approach of the neoclassic economists like Cobb and Douglas with their approach through a production function. In this sense, the seminal passage of Joan Robinson ([92], p. 81) gains a new actuality: "*Moreover, the production function has been a powerful instrument of miseducation. The student of economic theory is taught to write $Q = F(L,K)$ where L is a quantity of labour, K a quantity of capital and Q a rate of output of commodities. He is instructed to assume all workers alike, and to measure L in man-hours of labour; he is told something about the index-number problem involved in choosing a unit of output; and then he is hurried on to the next question, in the hope that he will forget to ask in what units K is measured. Before ever he does ask, he has become a professor, and so sloppy habits of thought are handed on from one generation to the next.*"[16]

16 The variables in the production function have been adapted to the actual text.

8.8 Sraffa and Walras

At this point, we may also ask, regarding prices, what, if anything, have the prices obtained in Sraffa's model in common with the equilibrium prices obtained in the Walras model? Can the two representations of different economic realities, market prices and production prices be reconciled in one way or another in technical terms?

Léon Walras (1834 – 1910) was a French mathematical economist and taught political economy at the University of Lausanne. He developed the general equilibrium theory and published in 1874 the first version of "Éléments d'économie politique pure". He proved that any individual market was necessarily in equilibrium if all other markets were also in equilibrium. This rule became known as Walras's Law. Walras set up the theory of general equilibrium by beginning with a few equations and then increasing the complexity reaching a general system of n equations. There are many assumptions inside the general equilibrium framework. There are a finite number of goods and a finite number of agents. The notion of utility is introduced. Each agent has a continuous and strictly concave utility function. There is a specified and limited set of *market prices* for the goods in this theoretical economy, essentially determined by the intersection of a supply and demand curve for various goods. General equilibrium analyses the economy as a whole, on the basis of single markets where there is a partial equilibrium and it exists when supply and demand are balanced.

We must emphasise straightaway that, besides algebraic aspects *"Sraffa's model is not a particular case of the Walrasian equilibrium model"*, as explained by Oetsch in [78]. Sraffa, following Leontief, addresses a succession of short-term periods. Starting from a theoretical analysis, we have shown clearly enough in this text that Sraffa's model can be put to use in the analysis of production problems, both for single-product industries and joint production. The Walrasian system on the contrary treated the notion of "utility" and initiated the "marginal revolution", describing the equilibrium and practical equilibrium in the market between demand and supply. He considered a long-term general economic equilibrium.[17]

Consider Walras's model in its simplest general form as presented in Takayama ([116], Chapter 2, Section E):

We suppose all entries given in monetary terms in this optimisation model,

x: the vector of commodities comprising total output of the economy;

p: the price vector of the corresponding commodities $\mathbf{x}$, $\mathbf{p} = f(\mathbf{x})$, is a function expressing the price of each commodity as a function of the quantity demanded;

17 Miller and Blair ([65], p. 730) comment on this as follows: "Some theorists characterise Leontief's model as an approximation of the Walrasian model but with several important simplifications, that allow a theory of general equilibrium to be applied and implemented empirically." We leave it up to readers to each make up their mind on this.

$\mathbf{\upsilon}$: the vector of maximum commodity amounts that can be produced, per industry (single product industries);

$\mathbf{A} = (a_{ij})$ the *productivity coefficients matrix*, indicating the amount of, commodity i, required for the production of one unit of commodity j. It is identical to the *input-output coefficients* matrix.

Let be the feasible set $X = \{\mathbf{x} : \mathbf{x} \geq \mathbf{o}, \mathbf{A}\mathbf{x} \leq \mathbf{\upsilon}\}$ where $\mathbf{A} = (a_{ij})$. One easily shows that X is not empty, but compact and convex. Then consider the linear programming problem:

$$\max_{\mathbf{x}}(\mathbf{p}'\mathbf{x}),$$
$$\mathbf{A}\mathbf{x} \leq \mathbf{\upsilon}, \quad \mathbf{p} \geq 0, \quad \mathbf{x} \geq 0, \quad \mathbf{\upsilon} > 0. \tag{8.77}$$

This is the problem of optimizing an objective function, here $\mathbf{p}'\mathbf{x}$, under linear constraints described by matrix $\mathbf{A}$, the final values of both $\mathbf{p}$ and $\mathbf{x}$ are to be determined.

Chiang [19], p. 51, says: "*This is essentially the way that Leon Walras approached the problem of the existence of a general market equilibrium. In the modern literature, there can be found a number of sophisticated mathematical proofs of the existence of a competitive market equilibrium under cetain postulated economic conditions.* Takayama ([116], Chapter 2, Section E) also describes the solution of this *linear program*.

We see again that no prerequisites, as the *rate of profits* r or the vector of labour $\mathbf{L}$ of any industrial sector, have to be known in before.

As for Sraffa, let us again consider the complete price model for *joint production*, with labour, and typically with the *positive* or *semi-positive* and *irreducible commodity flow* matrix $\mathbf{S}$ and the output matrix $\mathbf{F}$,

$$\boxed{\begin{aligned} \mathbf{S}'\mathbf{p}(1+r) + \mathbf{L}\frac{\tilde{w} \cdot Y}{L} &= \mathbf{F}'\mathbf{p}, \\ Y &= (\mathbf{F}\mathbf{e} - \mathbf{S}\mathbf{e})'\mathbf{p}, \\ L = \mathbf{e}'\mathbf{L} &= 1, \\ p_C &= 1. \end{aligned}} \tag{8.78}$$

Here, on the contrary, with Sraffa the *structure* of the *means of production* among the production sectors, is described ex ante by the *commodity flow* matrix $\mathbf{S}$ and the output matrix $\mathbf{F}$. The distribution of the surplus Y between entrepreneurs and workers, by choosing the *rate of profits* r, $0 \leq r \leq R$, precedes the determination of the prices of the commodities from which the value of capital K is computed. The distribution precedes logically the production. The distribution precedes the determination of prices through the Sraffa's price model based on the constituent means of production.

Concluding, the Walras representation of *market prices* and Sraffa's representation of *production costs*, termed 'prices', cannot be reconciliated. They follow different economic thoughts and their mathematical bases have their roots in different deep mathematical theorems, for Sraffa, it is the group around the *Perron–Frobenius theorem*, belonging to matrix algebra, for Walras, it is the *Brouwer fixed-point theorem*, belonging to topological algebra, see Takayama ([116], Chapter 2, Section E).

8.9 PCMC and cost management

Sraffa's book is not a production-management tool, contrary to what one might expect on reading the title of it: *Production of Commodities by Means of Commodities*. Production as a business activity, comprising the complete chain from purchasing, manufacturing, packaging, distribution onto sales, is an activity of the real economy at the microeconomic level. An important element of such activity is cost management, from the primary economic sector to the tertiary sector.

Now, as explained in the preface to PCMC, Sraffa's model investigates at the macroeconomic level such properties of an economic system that do not depend on changes in the scale of production or in the proportions of the factors of production during the reference period under consideration.

Is this approach compatible with the current business practice of cost management, if we consider that the rows entering Sraffa's price model reflect the means of production of the corresponding productive entity, including labour, in other words, the question is: Are these factors of production? Basically yes, with one caveat that does not however basically put into question Sraffa's approach.

Present day methodology in cost management considers effective costs as constant over a short period, typically on a quarterly (three months) basis. Costs are then recalculated at such regular intervals and constitute, e. g., the basis for:
- cataloging in detail the components entering into the means of production;
- the determination ex post of average production costs for reporting purposes;
- forecasting of costs for reporting, planning and sales purposes;
- optimising production costs by reallocations;
- assistance in establishing best practices in production;
- fixing pricing policy for targeted gross profits (EBITDA) etc.

The aforementioned caveat is based on the fact that production costs are of two types:
- fixed costs (administrative overheads, amortization, rents, local base taxes etc.) and;

- variable costs depending on quantities produced (raw products, energy, distribution and labour costs etc.).

The latter depend linearly (constant marginal cost) or step-wise on the quantities involved; their effect will however normally be felt only in passing from one short period to the next, and in that sense the numerical proportions of certain factors may change, in particular with respect to fixed costs. This only concerns the numerical value of certain production coefficients entering the technology matrix **S**, not its intrinsic structure.

What is however quite clear is that marginalist methods of optimisation, such as calculations of the marginal product of capital and of the marginal product of labour (8.75), following neoclassical economic theory (see, for example, Mankiw [63], Chapter 3.2), are not used as a practical tool in day-to-day production management, as noted by Sraffa in his preface to PCMC.

So, the prices obtained by Sraffa's price equations are not *today* production costs, and they are not determined according to the corresponding microeconomic methodology. Sraffa's prices, more precisely, Sraffa's production costs, on the contrary, define a general "normal" or equilibrium level of prices for a given technology at the macroeconomic level, given total output, i. e., interindustry transactions plus final exogenous demand fuelled by household consumption, private investments, government purchases and exports. These prices will however indirectly reflect accepted business practice in cost management and may serve economic policy makers ex ante for planning and forecasting purposes as well as ex post in analysing reasons of deviations of prices from the norm.

There nevertheless arises the question of a possible relationship between absolute Sraffa prices, expressed in monetary terms, and the pricing policies of individual firms. PCMC is clearly a piece of pure economic theory showing how fundamental economic processes, operating at the macroeconomic level work in principle within a social and circular interindustry framework governing the prices of production and regulating, to a certain degree, the distribution of surplus between producers and workers.

Based on the *Markup Approach* (see Section 8.4 and Bortis [8], p. 468), in which mark-ups differ from industry to industry, are introduced in order to obtain prices that are equal to the prices obtained in PCMC for the single-product industries price model. One can indeed consider the Sraffa prices as an approximation of industry production prices that are based at the outset on standard (normal) costing (see Bortis op. cit. and Bortis [7], pp. 66–67). The same statement is valid for the extended Sraffa price model for single-product industries with profits and wages varying from industry to industry (see Section 8.3). So, following Bortis, we are dealing with what one may refer to as "normal" prices based on economic principles. The markets have then the role to implement these "normal" prices with possible corrections depending on effective demand.

A better approximation is then to incorporate in the technology matrix $\mathbf{S}'$ fixed and variable components (following the logic applied by industries for the determination of fixed and variable costs), an avenue which we leave to further research. At the microeconomic level, industry production prices will nevertheless continue to differ, based on corporate pricing policies which will account for further considerations, such as optimisation of marginal contributions and of course commercial aspects, such as marketing costs.

9 The algebraic structure behind the "Leontief–Sraffa" interindustrial economy*

We have introduced several times in this text the fundamental importance of interindustrial transactions and of the conditions of production in connection with the Leontief and Sraffa models. This deserves special attention. Accordingly, this chapter written by H. A. Nour Eldin presents a complete analysis of the algorithmic properties of these topics for a closed economy of *single-product* industries, accompanied by some illustrative examples.

This chapter exclusively treats the *interindustrial market (Figure 9.1)*, which is just the production process without surplus $\mathbf{d}$. The *interindustrial market* just concerns the quantity $\mathbf{q}_I = \mathbf{Se}$ produced in physical terms or also expressed as $\mathbf{x}_I = \mathbf{Ze}$ in value terms (Figure 9.1).

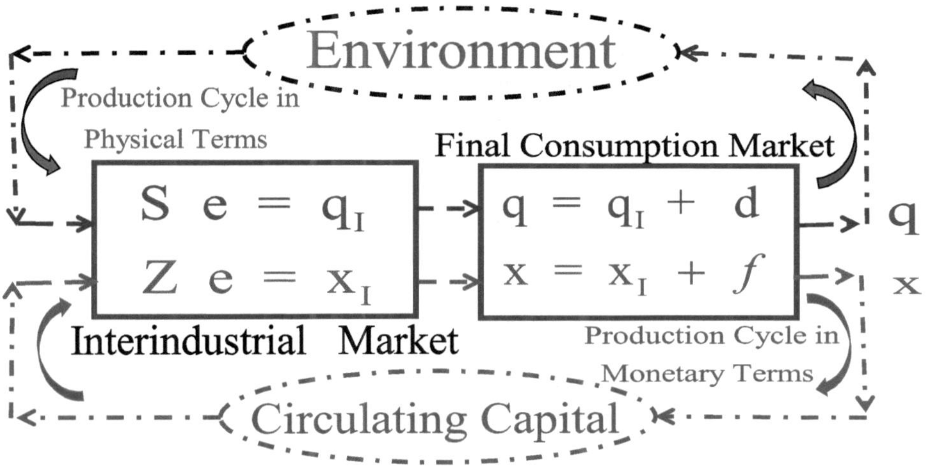

Figure 9.1: Production cycle in monetary and physical terms.

9.1 The input-output matrix Z of an interindustrial economy

The Sraffa commodity flow matrix $\mathbf{S}$ (2.13) in PCMC and the state matrices $\mathbf{A}$ (2.9), $\mathbf{C}$ (2.17), $\mathbf{D}$ (2.12) together with the *Leontief Inverse* $(\mathbf{I} - \mathbf{A})^{-1}$ (2.31) are derived in the previous Chapter 2 using the total production value vector $\mathbf{x} = \mathbf{Ze} + \mathbf{f}$, $\mathbf{x} = \mathbf{Z}'\mathbf{e} + \boldsymbol{\upsilon}$ (2.7). In this chapter, we concentrate the analysis on the interindustrial (intermediate) market as the core of production and technology. Therefore, the existing vector of final demand $\mathbf{f}$ is set to vanish and be replaced by $(\mathbf{f} \equiv \mathbf{0})$. The *interindustrial market* alone will be considered. Such treatment is known as an input-output economy *without surplus* or, an *interindustrial economy*.

https://doi.org/10.1515/9783110635096-009

For the interindustrial economy, the vectors **x**, **p**, **q** are respectively the *value* vector **x**, the *price vector* **p** and the *quantity* vector **q** of the interindustrial production.

The Figures 9.2–9.3 and Figures 9.7–9.8 are *three-dimensional* representations of the (n, n) Input-Output matrix and n is the number of lines of the corresponding square matrices $\mathbf{Z}_{44}$, $\mathbf{Z}_{80}$, $\mathbf{Z}_6$, $\mathbf{D}_6$. The direction of the continuously numbered horizontal axis represents the rows of the matrices. This means, at each line i, the quantities of commodities $i, i \in \{1, \ldots, n\}$ are represented vertically. The second horizontal axis, perpendicular to the first axis, represents the columns j of the matrices or the industrial sectors $j, j \in \{1, \ldots, n\}$. The vertical axis represents the entries or components of the matrix.

Thus, in Figure 9.2 the first line with the peak at number 1, corresponding to the component $(z_{1,1} = 2,568)$ *million* CHF, it is the input of the sector *products of agriculture*[1] in itself. Then, at number 5, corresponding to the component $(z_{1,5} = 9,796)$ *million* CHF, it is the highest input of the first sector to all the 44 sectors and goes into 5th appearing group of activities, the sector *food products, beverages and tobacco products*, see Figure 10.1. Then comes the sector *product of forestry;* then one after the other to the last line, corresponding to the activities: *other services; private households with employed persons*, $(z_{44,j})$ $j \in \{1, \ldots, 44\}$, see Figure 10.2. The highest peak, corresponding to the matrix component $(z_{12,12} = 24,878)$ *million* CHF, is the input of the sector *coke, refined* petroleum *products and nuclear fuel; chemicals and chemical products* by itself.

The Perron–Frobenius number λ_F expresses through the used currency the overall strength of the economy whose structure is captured by the *initial* Input-Output matrix **Z**. For normalisation, we define the λ_F-normalised matrix $\mathbf{Z}_\lambda$,

$$\mathbf{Z}_\lambda = \left(\frac{1}{\lambda_F}\right)\mathbf{Z}. \tag{9.1}$$

As a result of this λ_F-normalization, other vectors, matrices and variables, e. g., the value vector **x**, are consequently also λ_F-normalised, $\mathbf{x}_\lambda = (1/\lambda_F)\mathbf{x}$.

Thus, one presents the λ_F-normalised *Input-Output commodity flow* matrix $\mathbf{Z}_\lambda$ of interindustrial production, see Table 2.1 and Figures 9.2–9.3. We present in detail the *semi-positive* λ_F-normalized Input-Output matrix, the output vector $\mathbf{x}_\lambda$ and the circulating capital K_λ,

$$\mathbf{Z}_\lambda = \begin{bmatrix} z_{\lambda 11} & z_{\lambda 12} & \cdots & z_{\lambda 1j} & \cdots & z_{\lambda 1n} \\ z_{\lambda 21} & z_{\lambda 22} & \cdots & z_{\lambda 2j} & \cdots & z_{\lambda 2n} \\ \cdots & \cdots & \cdots & \cdots & \cdots & \cdots \\ \cdots & \cdots & \cdots & \cdots & \cdots & \cdots \\ z_{\lambda(n-1)1} & z_{\lambda(n-1)2} & \cdots & z_{\lambda(n-1)j} & \cdots & z_{\lambda(n-1)n} \\ z_{\lambda n1} & z_{\lambda n2} & \cdots & z_{\lambda nj} & \cdots & z_{\lambda nn} \end{bmatrix}, \quad \mathbf{x}_\lambda = \begin{bmatrix} x_{\lambda 1} \\ x_{\lambda 2} \\ \cdots \\ \cdots \\ x_{\lambda n-1} \\ x_{\lambda n} \end{bmatrix}$$

$$\mathbf{x}_\lambda = \sum_{j=1}^{n} z_{\lambda ij} = \mathbf{Z}_\lambda \mathbf{e} \quad \text{(row–sum)}, \tag{9.2}$$

1 The NOGA (*Nomenclature Générale des Activités économiques*) classification is used.

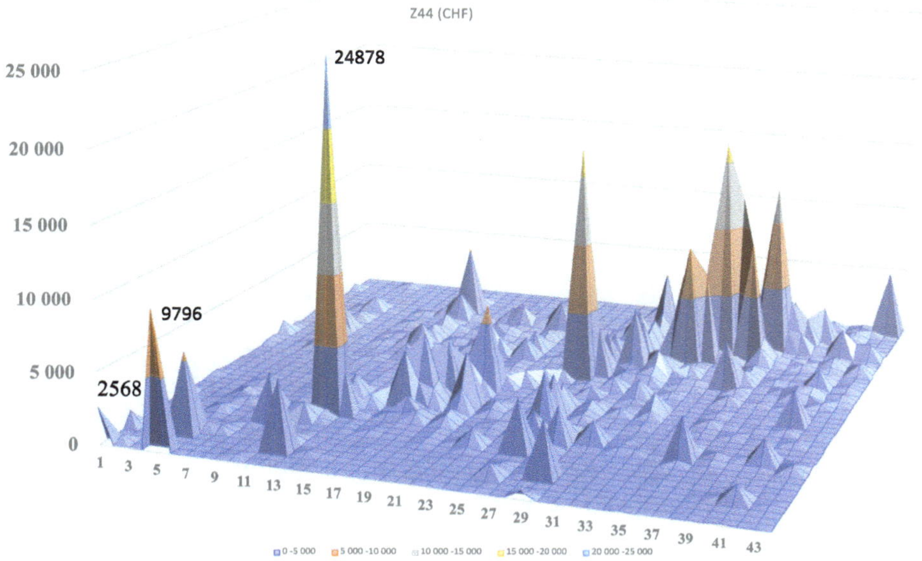

Figure 9.2: The Input-Output matrix $\mathbf{Z}_{44}$ of the Swiss interindustrial production 2008 (510.79 Billion CHF).

$$\mathbf{Z}_\lambda \mathbf{e} = \mathbf{x}_\lambda, \quad \mathbf{x}_\lambda = [x_{\lambda 1}, x_{\lambda 2}, \dots, x_{\lambda n}]', \quad x_{\lambda i} = \sum_{j=1}^{n} z_{\lambda ij}, \quad \sum_{k=1}^{n} x_{\lambda k} = K_\lambda =: K^*. \tag{9.3}$$

In the interindustrial economy, the variables are usually expressed in λ_F units, e. g., the *circulating capital*, is expressed as, $K^* := K_\lambda$.[2] But we will not overcharge the notation, rather we continue to write the matrices, vectors and economic variables without index λ when they are λ_F-normalized, within Sections 9.2 to 9.4. Thus, e. g., matrix $\mathbf{Z}_\lambda$ remains noted as $\mathbf{Z}$, like other matrices, vectors and variables.

It is assumed that a sector (branch) i produces a total quantity $q_i = \sum_{j=1}^{n} s_{ij}$ of a *single* product (object) that is offered to sale at a certain *single* price p_i to all sectors (branches).

As a further illustration, we present the Input-Output Table of Japan 2011, comprising 80 sectors (Figure 9.3).

9.2 The interindustrial production matrices Z, S and D

9.2.1 The "Sraffa" commodity flow matrix S, no surplus

The above type of single-product–*single*-price industries (*single* product per sector/branch) leads to an interindustrial production vector $\mathbf{q} = [q_1, q_2, \dots, q_n]'$, where q_i is the product quantity produced by sector i. It is assumed further that q_i is offered

2 In the examples, we use the asterisk * to note the λ_F-normalized circulating capital: $K^* := K_\lambda$.

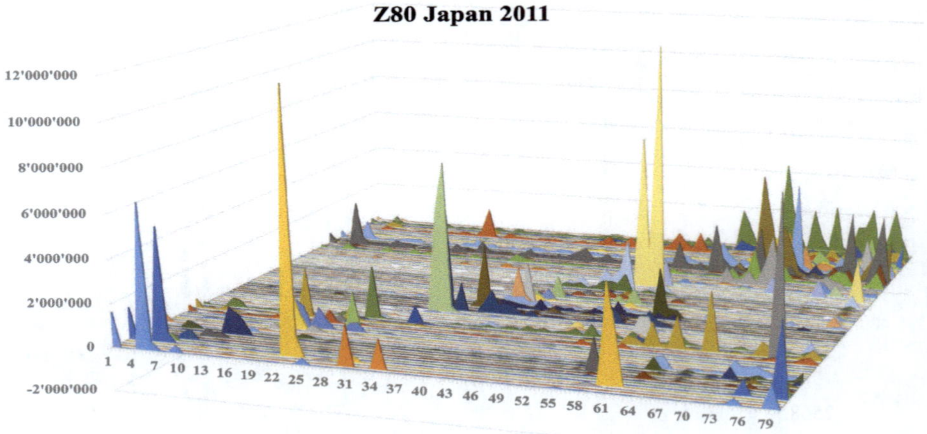

Figure 9.3: The Input-Output matrix $\mathbf{Z}_{80}$ of the interindustrial production of Japan 2011, K = 44,197,405 Mio YEN (2010, at the exchange rate it gives 3,295.96 Billion CHF).

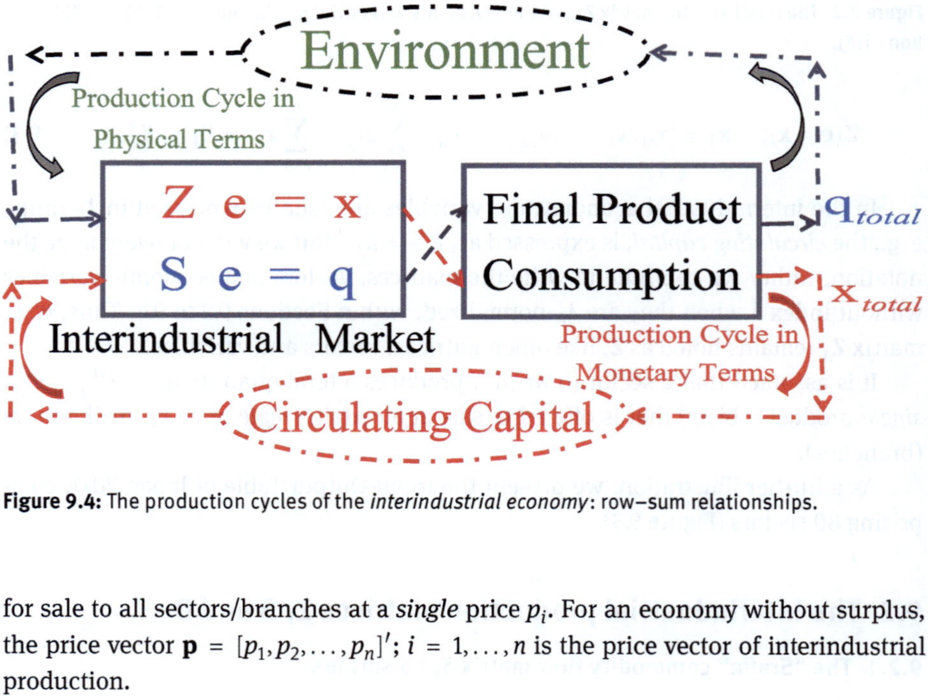

Figure 9.4: The production cycles of the *interindustrial economy*: row–sum relationships.

for sale to all sectors/branches at a *single* price p_i. For an economy without surplus, the price vector $\mathbf{p} = [p_1, p_2, \ldots, p_n]'$; $i = 1, \ldots, n$ is the price vector of interindustrial production.

$$\mathbf{S} = \begin{bmatrix} s_{11} & \ldots & \ldots & s_{1n} \\ s_{21} & \ldots & \ldots & s_{2n} \\ \ldots & \ldots & \ldots & \ldots \\ s_{n1} & \ldots & \ldots & s_{nn} \end{bmatrix}, \quad \mathbf{q} = \begin{bmatrix} q_1 \\ \ldots \\ \ldots \\ q_n \end{bmatrix}; \quad \text{(Table 2.2)} \tag{9.4}$$

$$q_i = \sum_{j=1}^{n} s_{ij}, \quad \mathbf{q} = \mathbf{S}\mathbf{e}, \tag{9.5}$$

$$\mathbf{Z} = \hat{\mathbf{p}}\mathbf{S} = \begin{bmatrix} p_1 s_{11} & p_1 s_{12} & \dots & \dots & p_1 s_{1n} \\ p_2 s_{21} & p_2 s_{22} & \dots & \dots & p_2 s_{2n} \\ \dots & \dots & \dots & \dots & \dots \\ \dots & \dots & \dots & \dots & \dots \\ p_n s_{n1} & p_n s_{n2} & \dots & \dots & p_n s_{nn} \end{bmatrix}, \quad \mathbf{p} = \begin{bmatrix} p_1 \\ p_2 \\ \dots \\ \dots \\ p_n \end{bmatrix}. \tag{9.6}$$

The matrix **S** is Sraffa's *commodity flow* matrix in *physical terms* (branch product), and the vector **q** is the total branch-production quantity of its *single-product industries* (object/item).

For *multi-product industries* or *joint production*, one can consult Chapters 6 and Chapter 7. One can treat joint production in a similar way to the one presented here by just including for every sector a certain number of representative sector products. Ultimately, one can even include hundreds of products for each sector. The methodology introduced in this chapter is independent on the number of variables used. Dealing with hundreds of sectors or thousands of products does not—principally—lead to substantial difficulties. Some properties as the matrix sparseness or the irreducibility problems of non-negative matrices may need to be adequately taken into consideration.

In the interindustrial market, the *single*-product–*single*-price industries sell and buy the quantity vector **q** at a given price vector $\mathbf{p} = [p_1, \dots, p_n]'$, one can apply directly the equation (9.6):

$$\mathbf{S} = \hat{\mathbf{p}}^{-1}\mathbf{Z}, \quad \mathbf{Z} = \hat{\mathbf{p}}\mathbf{S}. \tag{9.7}$$

Since the *interindustrial market* satisfies the price-quantity-value relations, we get, with equation (2.105),

$$p_k q_k = x_k, \quad \hat{\mathbf{p}}\hat{\mathbf{q}} = \hat{\mathbf{x}},$$
$$\sum_{k=1}^{n} x_k = K, \tag{9.8}$$

and one achieves directly: $\mathbf{S}\mathbf{e} = \hat{\mathbf{p}}^{-1}\mathbf{Z}\mathbf{e} = \hat{\mathbf{p}}^{-1}\hat{\mathbf{x}}\mathbf{e} = \hat{\mathbf{q}}\mathbf{e} = \mathbf{q}$.

In summary, the interindustrial matrices **Z** and **S** (Sraffa-*commodity flow* matrix in *physical terms*) satisfy the *interindustrial* row–sum relationships, see Figure 9.4, with equations (2.5) we have now set $\mathbf{x} := \mathbf{x}_I$, and (3.6),

$$\mathbf{Z}\mathbf{e} = \mathbf{x}, \quad \mathbf{S}\mathbf{e} = \mathbf{q}. \tag{9.9}$$

One should mention here again the *main difference* between the value vector **x**, formerly written as $\mathbf{x}_I$, of the interindustrial market as the row sum of the normalised matrix **Z** – whose *component-sum* is the *circulating capital* K ($\mathbf{x} = \mathbf{Z}\mathbf{e}$; $\mathbf{x}'\mathbf{e} = K$) – and the vector $\mathbf{x} = \mathbf{Z}\mathbf{e} + \mathbf{f}$, equation (2.5), $\mathbf{x} = \mathbf{Z}'\mathbf{e} + \boldsymbol{\upsilon}$, equation (2.6).

9.2.2 The stochastic production matrix D of the interindustrial economy

The stochastic row–sum product of Z and S

An arbitrary non-negative matrix **M** is representable by the matrix product

$$\mathbf{M} := (M_{ij}) = \hat{\boldsymbol{\alpha}}\mathbf{M}^*, \quad \mathbf{M}^* = \hat{\boldsymbol{\alpha}}^{-1}\mathbf{M}, \quad \mathbf{Me} = \boldsymbol{\alpha}, \quad \mathbf{M}^*\mathbf{e} = \mathbf{e}, \quad \text{where}$$

$$\boldsymbol{\alpha} = [\alpha_1, \alpha_2, \dots, \alpha_n]', \quad \alpha_i = \sum_{j=1}^{n} M_{ij} = (\mathbf{Me})_i, \quad \text{(row–sum)}. \tag{9.10}$$

The matrix $\mathbf{M}^*$ is a stochastic matrix (Gantmacher ([34], Section 13.6)), and the positive vector $\boldsymbol{\alpha}$ is the row–sum of **M**.

By applying this *stochastic* row–sum product together with the market relationships $\hat{\mathbf{p}}\hat{\mathbf{q}} = \hat{\mathbf{x}}$ on the matrices **Z** and **S**, one obtains the *value*, *quantity* and *object* products of the Interindustrial Economy (Figure 9.6, Figure 9.7, Figure 9.8),

$$\begin{array}{llll} \text{value} & \mathbf{x} & \mathbf{Z} = \hat{\mathbf{x}}\mathbf{D}, & \mathbf{Ze} = \hat{\mathbf{x}}\mathbf{De} = \mathbf{x} \quad \text{row–sum}, \\ \text{quantity} & \mathbf{q} & \mathbf{S} = \hat{\mathbf{q}}\mathbf{D}, & \mathbf{Se} = \hat{\mathbf{q}}\mathbf{De} = \mathbf{q} \quad \text{row–sum}, \\ \text{object} & \mathbf{e} & \mathbf{D} = (\mathbf{I})\mathbf{D}, & \mathbf{De} = \mathbf{e} \quad \text{row–sum} = 1, \\ & & \mathbf{D} = \hat{\mathbf{x}}^{-1}\mathbf{Z} = \hat{\mathbf{q}}^{-1}\mathbf{S}. & \end{array} \tag{9.11}$$

The stochastic production matrix **D** is also called the *distribution coefficients* matrix of the interindustrial production. The *distribution coefficients* of **D** are defined in Section 2.1.2, equations (2.20), (2.21) as:

$$\begin{array}{lll} \mathbf{D} = (d_{ij}); & \mathbf{D} = \hat{\mathbf{q}}^{-1}\mathbf{S}; & \text{equations (2.20), (2.21)} \\ d_{ij} = \dfrac{s_{ij}}{q_i}; & 0 < d_{ij} < 1; & \displaystyle\sum_{j=1}^{n} d_{ij} = 1 \\ i,j = 1,\dots,n; & i: \text{input index}; & j: \text{output index} \end{array} \tag{9.12}$$

The economic significance of the distribution coefficients is that a coefficient d_{ij} determines the fraction (part) of input-commodity i per unit of commodity i required for the production of output-commodity j. All distribution coefficients d_{ij} are dimensionless.

The definitions and equations here coincide exactly with the stochastic row–sum product of the Sraffa *commodity* matrix **S** (equation (9.11)). Thus, the *distribution coefficients matrix* **D** of interindustrial economy is a *stochastic production matrix* **D** (Figure 9.7 and Figure 9.8).

$$\mathbf{De} = \mathbf{e}. \tag{9.13}$$

This *stochastic production matrix* **D** represents a basic algebraic structure matrix of the *Sraffa–Leontief Interindustrial Economy*.

9.3 The value and environmental cycles of "Sraffa" interindustrial economies

The monetary value vector **x** of the interindustrial production satisfies the basic value and quantity equations

$$\begin{array}{llll} \text{value} & \mathbf{x} & \mathbf{Z} = \hat{\mathbf{x}}\mathbf{D}; \quad \mathbf{Ze} = \hat{\mathbf{x}}\mathbf{De} = \mathbf{x}; & \text{row–sum} \\ \text{quantity} & \mathbf{q} & \mathbf{S} = \hat{\mathbf{q}}\mathbf{D}; \quad \mathbf{Se} = \hat{\mathbf{q}}\mathbf{De} = \mathbf{q}; & \text{row–sum} \end{array} \tag{9.14}$$

The monetary value vector **x** and the quantity vector **q** are respectively the row sums of the *Input-Output* matrices **Z** and **S**. The matrix **D** is a right *stochastic production matrix.* Also, for the summation vector **e**, there is:

$$\hat{\mathbf{p}}\mathbf{Se} = \hat{\mathbf{p}}\mathbf{q} = \mathbf{x} = \mathbf{Ze}. \tag{9.15}$$

The role of the vectors of value **x** and quantity **q** within a "Sraffa"-Interindustrial Economy reflects the fact that the quantity vector $\mathbf{q} = \mathbf{Se}$ is the distinct core variable of production in Sraffa's *production of commodities (products/services) by means of commodities (products/services).*

9.3.1 The Environmental Cycle of Sraffa's Production of Commodities by Means of Commodities (PCMC)

In Sraffa's *production of commodities (products/services) by means of commodities (products/services),* the quantity vector **q** is a function of all successive previous products of the prior production stages in the cascaded processes of interindustrial production during a given period, see Figure 9.5:

$$\mathbf{Se} = \mathbf{q}, \quad \mathbf{q} = \mathbf{q}(\mathbf{q}_{-1}(\ldots(\mathbf{q}_{\text{environment}}))). \tag{9.16}$$

In this sense, the interpretation of the interindustrial production as a Sraffa-*production of commodities by means of commodities* leads to a production of products that are cascaded in the interindustrial production chain with the result that: "The almost primary quantity vector $\mathbf{q}_{\text{environment}}$ is always extracted from the environment." Further, and due to the consumption processes (Figures 9.4–9.6): All interindustrial production ultimately returns back to the environment. In a similar argumentation, the *circulating capital* (Figures 9.4) of the interindustrial economy—as a value vector **x** (value added)—is a function of the labor value vector of all the sub-product values of the previous production stages—within every branch—in the cascaded process of interindustrial production (Figure 9.5):

$$\mathbf{Ze} = \mathbf{x}, \quad \mathbf{x} = (\mathbf{x}_L(\mathbf{x}_{L-1}(\ldots(\mathbf{x}_{\text{environment}})))). \tag{9.17}$$

The components of vector $\mathbf{x}_\alpha$, $\alpha = L, L-1, \ldots$, are the *labor costs* at production stage α. At every micro-production stage α, the corresponding value vector $\mathbf{x}_\alpha$ is a decision vector

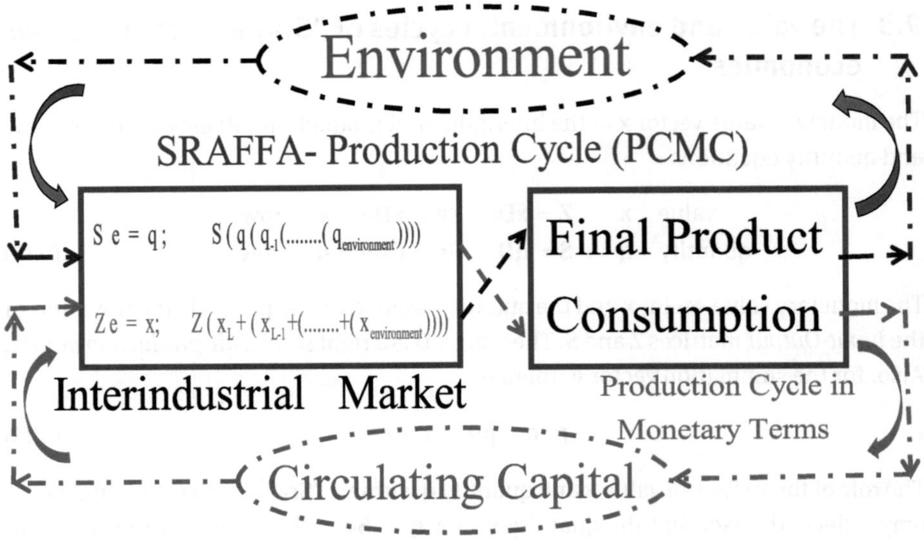

Figure 9.5: The environmental cycle of Sraffa's *Production of Commodities by Means of Commodities* (PCMC).

that allocates the currency value $\mathbf{x}_\alpha$ to the product vector according to the priority weights of decisions taken within this stage of interindustrial production.
As Sraffa's *Production of Commodities by Means of Commodities* (PCMC) is processed, Figure 9.4 and Figure 9.5, the monetary value $\mathbf{x}_\alpha$ is generated and is added up (accumulated) **over the cascade** of the value added out of the various production stages.

Ultimately, at the environmental stage of production, the final environment relation $\mathbf{x}_{\text{environment}} = \hat{\mathbf{p}}_{\text{environment}}\mathbf{q}_{\text{environment}}$ holds, in analogy to $\mathbf{x} = \hat{\mathbf{p}}\mathbf{q}$.

9.3.2 Production recursions of Sraffa interindustrial economies

At every production stage s, the production recursions for the quantity vector $\mathbf{q}$ and its corresponding value vector $\mathbf{x}$ are given by the interindustrial recursions (Figure 9.5):

$$\begin{aligned} \mathbf{q}_{s+1} &= \mathbf{q}_{s+1}(\mathbf{q}_s), \quad \mathbf{q}_0 = \mathbf{q}_{\text{environment}}, \\ \mathbf{x}_{s+1} &= \mathbf{x}_{s+1} + \mathbf{x}_s(\mathbf{q}_s). \end{aligned} \tag{9.18}$$

The *Interindustrial Economy* is therefore governed by the following interindustrial production relationchips (Figure 9.6),

$$\begin{array}{lllll} & & \mathbf{Z} = \hat{\mathbf{x}}\mathbf{D} = \mathbf{A}\hat{\mathbf{x}} = \hat{\mathbf{p}}\mathbf{S}, & & \\ \text{value} & \mathbf{x} & \mathbf{Z} = \hat{\mathbf{x}}\mathbf{D}, & \mathbf{Ze} = \mathbf{x} & \text{row–sum}, \\ \text{quantity} & \mathbf{q} & \mathbf{S} = \hat{\mathbf{q}}\mathbf{D}, & \mathbf{Se} = \mathbf{q} & \text{row–sum}, \\ & & \sum_{k=1}^{n} x_k = \sum_{k=1}^{n} p_k q_k = K. & & \end{array} \tag{9.19}$$

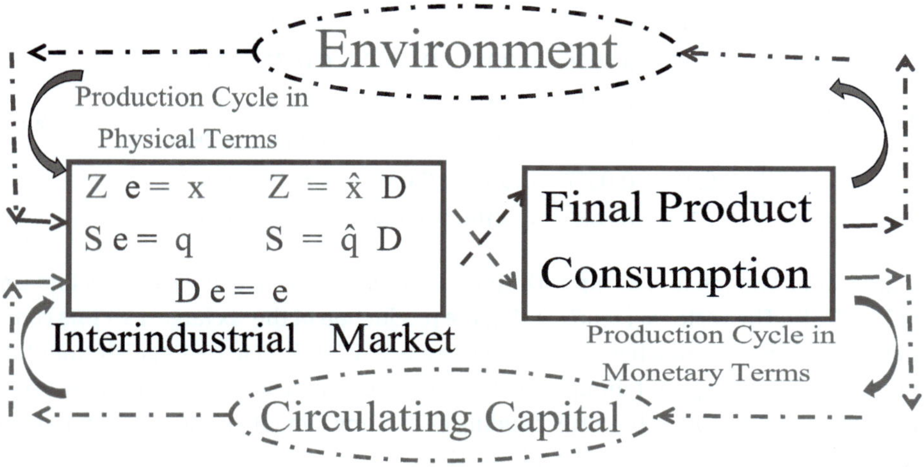

Figure 9.6: The row–sum and the stochastic production matrix relationships of the environmental cycle and of the interindustrial economy.

9.3.3 The input-output matrix $\mathbf{Z}_6$ and the stochastic production matrix $\mathbf{D}_6$

We consider the *interindustrial economy* of Switzerland, see Chapter 10.

Example 9.3.1. The (44×44) input-output matrix $\mathbf{Z}_{44}$ of the interindustrial production of Switzerland 2008 (Figure 9.2) has 44 sectors with a circulating capital, without net commodity taxes, of $K_{44} = 510.79$ Billion CHF. We define the (44×1) sector production value vector $\mathbf{x}_{44} = \mathbf{Z}_{44}\mathbf{e}_{44}$ (Figure10.2) and extract the six largest sectors in descending order, giving $\tilde{\mathbf{x}}_6 = [62.467, 44.069, 42.515, 29.710, 26.178, 21.837]'$.

Compute with $\mathbf{e}_6 = [1,1,1,1,1,1]'$ the circulating capital $\tilde{K}_6 = \mathbf{e}_6'\tilde{\mathbf{x}}_6$ of these six largest sectors. Establish the (6×6) (entries in Billion CHF) input-output matrix $\mathbf{Z}_6$ and the distribution coefficients matrix $\mathbf{D}_6$ and present them in appropriated graphs. Verify that $\mathbf{D}_6$ is stochastic.

Solution to Example 9.3.1:
The circulating capital is $\tilde{K}_6 = \mathbf{e}_6'\tilde{\mathbf{x}}_6 = 226.78$ billion CHF (44.4 per cent of the interindustrial production CH 2008). We calculate the (6×6) input-output matrix $\mathbf{Z}_6$ and the distribution coefficients matrix $\mathbf{D}_6$,

$$\mathbf{Z}_6 = \begin{bmatrix} 12.499 & 2.0108 & 6.0905 & 5.1138 & 0.5045 & 1.1164 \\ 0.8490 & 24.878 & 0.0524 & 0.9406 & 0.4730 & 0.4628 \\ 1.9100 & 1.1924 & 16.241 & 2.3434 & 0.4207 & 0.8568 \\ 0.7710 & 1.8367 & 0.0603 & 4.6234 & 0.0201 & 0.7237 \\ 0.3951 & 1.1311 & 0.1988 & 0.6243 & 16.888 & 0.3582 \\ 0.1066 & 1.4884 & 0.0069 & 0.4554 & 0.0699 & 2.1435 \end{bmatrix}, \quad \mathbf{x}_6 = \begin{bmatrix} 27.335 \\ 27.655 \\ 22.964 \\ 8.036 \\ 19.596 \\ 4.271 \end{bmatrix}, \tag{9.20}$$

$$\mathbf{D}_6 = \hat{\tilde{\mathbf{x}}}_6^{-1}\mathbf{Z}_6 = \begin{bmatrix} 0.4573 & 0.0736 & 0.2228 & 0.1871 & 0.0185 & 0.0408 \\ 0.0307 & 0.8996 & 0.0019 & 0.0340 & 0.0171 & 0.0167 \\ 0.0832 & 0.0519 & 0.7072 & 0.1020 & 0.0183 & 0.0373 \\ 0.0960 & 0.2286 & 0.0075 & 0.5754 & 0.0026 & 0.0901 \\ 0.0202 & 0.0577 & 0.0101 & 0.0319 & 0.8618 & 0.0183 \\ 0.0250 & 0.3485 & 0.0016 & 0.1066 & 0.0164 & 0.5019 \end{bmatrix}. \quad (9.21)$$

Figure 9.7 shows the (6 × 6) Input-Output matrix $\mathbf{Z}_6$ of the first six largest sectors (in billion CHF). The vector $\tilde{\mathbf{x}}_6$, representing the production of the six largest sectors coming from all the 44 branches of the whole Swiss IOT 2008, is to distinguish from the production vector $\mathbf{x}_6$, containing only the production of those selected six largest sectors among themselves. The corresponding circulating capital for the (6 × 6) market is $\mathbf{x}_6 = \mathbf{Z}_6\mathbf{e}_6$ about $K_6 = \mathbf{e}_6'\mathbf{x}_6 = 109.9$ Billion CHF.

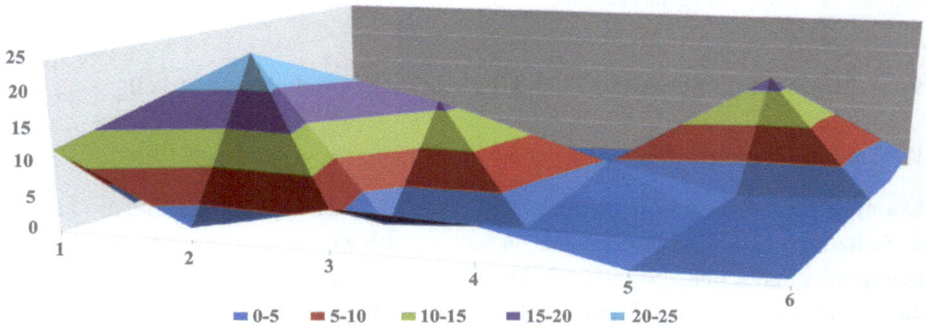

Figure 9.7: The input-output matrix $\mathbf{Z}_6$ of the six largest sectors of Switzerland 2008.

One easily verifies the equation $\mathbf{D}_6\mathbf{e}_6 = \mathbf{e}_6$. The Frobenius eigenvalue (Frobenius number) of $\mathbf{D}_6$ is $\lambda_D = 1$ with corresponding eigenvector $\mathbf{e}_6$ because matrix $\mathbf{D}_6$ is stochastic (Figure 9.8). ▲

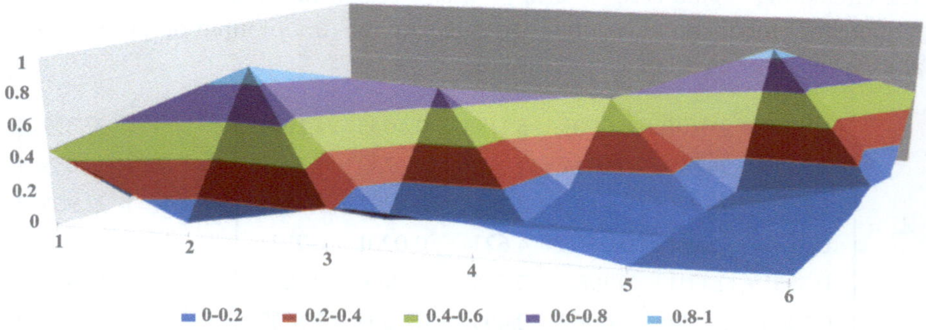

Figure 9.8: The stochastic production matrix $\mathbf{D}_6$ of the six largest sectors of Switzerland CH 2008.

9.4 Main result: the stochastic similarity table of an interindustrial economy (GDP-Table)

Given the *semi-positive Input-Output* matrix **Z** of an interindustrial economy, then with the Frobenius eigenvalue of the *stochastic production matrix* **D**, the matrices **Z, S, A, C, D** satisfy the following matrix and vector relations:

Matrix Relationships:

$$\begin{array}{llll}
(\lambda_{\text{Frobenius}} = 1) & & \text{I/O matrices} & \text{Stochastic Similarity} \\
\text{value} & \mathbf{x} & \mathbf{Z} = \hat{\mathbf{x}}\mathbf{D} & \mathbf{A} = \hat{\mathbf{x}}\mathbf{D}\hat{\mathbf{x}}^{-1} \\
\text{price} & \mathbf{p} & \mathbf{T} = \hat{\mathbf{p}}\mathbf{D} & \mathbf{B} = \hat{\mathbf{p}}\mathbf{D}\hat{\mathbf{p}}^{-1} \\
\text{quantity} & \mathbf{q} & \mathbf{S} = \hat{\mathbf{q}}\mathbf{D} & \mathbf{C} = \hat{\mathbf{q}}\mathbf{D}\hat{\mathbf{q}}^{-1} \\
\text{object} & \mathbf{e} & \mathbf{D} = \hat{\mathbf{e}}\mathbf{D} & \mathbf{D} = \hat{\mathbf{e}}\mathbf{D}\hat{\mathbf{e}}^{-1} = \mathbf{D}
\end{array} \tag{9.22}$$

Vector Relationships: $\mathbf{e}'\mathbf{x} = K$

$$\begin{array}{llll}
(\lambda_{\text{Frobenius}} = 1) & & \text{Row–sum} & \text{FP–Eigenvector} \\
\text{value} & \mathbf{x} & \mathbf{Ze} = \mathbf{x} & \mathbf{Ax} = \mathbf{x} \\
\text{price} & \mathbf{p} & \mathbf{Te} = \mathbf{p} & \mathbf{Bp} = \mathbf{p} \\
\text{quantity} & \mathbf{q} & \mathbf{Se} = \mathbf{q} & \mathbf{Cq} = \mathbf{q} \\
\text{object} & \mathbf{e} & \mathbf{De} = \mathbf{e} & \mathbf{De} = \mathbf{e}
\end{array} \tag{9.23}$$

The interindustrial vector of value **x**, the price vector **p**, the quantity vector **q** and the object (item) vector **e** are respectively the positive Perron–Frobenius eigenvectors of the quadruple of state matrices (**A, B, C, D**). They are simultaneously the row–sums of the quadruple of I/O matrices (**Z, T, S, D**). The *stochastic production matrix* **D** is the stochastic similarity matrix of all state matrices with the similarity transformation matrices of value $\hat{\mathbf{x}}$, of price $\hat{\mathbf{p}}$, of quantity $\hat{\mathbf{q}}$ and of object (item) $\hat{\mathbf{e}} = \mathbf{I}$. The matrix and vector relationships here can be summed up in the

Stochastic similarity table of an interindustrial economy (GDP-Table)

$$\begin{array}{llllll}
(\lambda_{\text{Frobenius}} = 1) & & \text{Row–sum} & \text{PF-Eigenvector} & \text{Stochastic similarity} & \text{I/O matrices} \\
\text{value} & \mathbf{x} & \mathbf{Ze} = \mathbf{x} & \mathbf{Ax} = \mathbf{x} & \mathbf{A} = \hat{\mathbf{x}}\mathbf{D}\hat{\mathbf{x}}^{-1} & \mathbf{Z} = \hat{\mathbf{x}}\mathbf{D} \\
\text{price} & \mathbf{p} & \mathbf{Te} = \mathbf{p} & \mathbf{Bp} = \mathbf{p} & \mathbf{B} = \hat{\mathbf{p}}\mathbf{D}\hat{\mathbf{p}}^{-1} & \mathbf{T} = \hat{\mathbf{p}}\mathbf{D} \\
\text{quantity} & \mathbf{q} & \mathbf{Se} = \mathbf{q} & \mathbf{Cq} = \mathbf{q} & \mathbf{C} = \hat{\mathbf{q}}\mathbf{D}\hat{\mathbf{q}}^{-1} & \mathbf{S} = \hat{\mathbf{q}}\mathbf{D} \\
\text{object} & \mathbf{e} & \mathbf{De} = \mathbf{e} & \mathbf{De} = \mathbf{e} & \mathbf{D} = \hat{\mathbf{e}}\mathbf{D}\hat{\mathbf{e}}^{-1} & \mathbf{D} = \hat{\mathbf{e}}\mathbf{D}
\end{array} \tag{9.24}$$

Beside the GDP-table above with $\mathbf{x} = \mathbf{Ze}$ (production/output table, summation of the rows), there is an adjoint/dual table corresponding to $\mathbf{y} = \mathbf{Z}'\mathbf{e}$ (purchase/input table, summation of the columns). It has the same structure. The two tables are connected through the production–purchase equation $\mathbf{y} = \mathbf{D}'\mathbf{x}$,[3] $k_x = \mathbf{e}'\mathbf{x}$, $k_y = \mathbf{e}'\mathbf{y}$, $k_x \neq k_y$, of the interindustrial market.

The vectors **x**, **p**, **q** and the object vector **e** are the internal state vectors of the interindustrial market model, responsible for the production within the market. The matrices **A**, **B**, **C** and **D** are the corresponding state matrices of production. The monetary state vector **x** is directly measurable through $\mathbf{x} = \mathbf{Ze}$. As the other internal state vectors **p** and **q** are not directly measurable through the matrix **Z**, these vectors should be treated as computed *model state vectors* that can only be identified through model computations. In our case, one refers to the Sraffa price model (4.29) exceptionally to the Oosterhaven price model (2.109) and we are in presence of a *model-based system*.

There is an inherent state identification/observation problem arising from the value-price equation $\mathbf{x} = \hat{\mathbf{p}}\mathbf{q} = \hat{\mathbf{q}}\mathbf{p}$, with units $[\mathbf{x}]$ = *currency*, $[\mathbf{q}]$ = *quantity* and $[\mathbf{p}]$ = *currency/quantity*. Additional information/measurements are necessary to ensure that the units—such as the price unit $[\mathbf{p}]$ = *currency/quantity* – come from statistical market prices, generally through the *statistical offices* of the countries. So long as such information is *unavailable*, the price vectors **p** and quantity vectors **q** of the corresponding tables should be regarded as *computed state vectors* that are compatible with the realisation of row–sums, getting the total output, or column–sums, getting the total outlays, Table 2.1. "Price" or "quantity" design the usual economic notions and appear conventionally in the construction of IOTs. See also the citation from the Eurostat Manual [72], p. 239, and equation (2.18).

The left eigenvector $\mathbf{y}$ of the object matrix **D**, equation $\mathbf{D}'\mathbf{y} = \mathbf{y}$, determines the stationary solution for both total output and total outlays. For an accountably balanced economy, the accounting identity relationship relating the total outlays to the total inputs (2.7) takes the form $\mathbf{x} = \mathbf{x}_I + \mathbf{f} = \mathbf{y}_I + \boldsymbol{\upsilon} = \mathbf{y}$. The circulating capital of such an economy is $K = \mathbf{e}'\mathbf{x}$ (see also Nour Eldin [73]).

9.4.1 The right and the left eigenvectors of the matrices A and Z

1. The internal exchange (state vectors) and the external (boundary) exchange (Input-Output) processes of the interindustrial market are governed by the right and the left eigenvectors of the matrices **A** and **Z**. The solution of the symmetrical boundary value problem of the interindustrial market, described as

Boundary Value Problem

$$\begin{bmatrix} \mathbf{0} & \mathbf{Z} \\ \mathbf{Z}' & \mathbf{0} \end{bmatrix} \begin{bmatrix} \mathbf{e} \\ \mathbf{e} \end{bmatrix} = \begin{bmatrix} \mathbf{x} \\ \mathbf{y} \end{bmatrix}, \quad \sigma = \begin{bmatrix} \mathbf{0} & \mathbf{I} \\ \mathbf{I} & \mathbf{0} \end{bmatrix}, \quad \sigma^2 = \mathbf{I}, \tag{9.25}$$

is determined from the following symmetrical eigenvalue problem of exchange. We are in the presence of the right and the left eigenvector problem.

3 Matrix **D** is stochastic. For this reason, there is: $\mathbf{y} = \mathbf{Z}'\mathbf{e} = \mathbf{D}'\mathbf{x} = \mathbf{D}'\mathbf{Ze} \Rightarrow \mathbf{e} = (\mathbf{Z}')^{-1}\mathbf{D}'\mathbf{Ze} = (\mathbf{Z}')^{-1}(\hat{\mathbf{x}}^{-1}\mathbf{Z})'\mathbf{Ze} = (\mathbf{Z}')^{-1}\mathbf{Z}'\hat{\mathbf{x}}^{-1}\mathbf{Ze} = \hat{\mathbf{x}}^{-1}\mathbf{Ze} = \mathbf{De} = \mathbf{e}$.

Symmetrical Eigenvalue Problem

$$\text{State Eigenvector:} \quad \begin{bmatrix} \mathbf{0} & \mathbf{A} \\ \mathbf{A}' & \mathbf{0} \end{bmatrix} \begin{bmatrix} \boldsymbol{\beta} \\ \mathbf{x} \end{bmatrix} = \begin{bmatrix} \mathbf{0} & \mathbf{I} \\ \mathbf{I} & \mathbf{0} \end{bmatrix} \begin{bmatrix} \boldsymbol{\beta} \\ \mathbf{x} \end{bmatrix},$$

$$\text{Input-Output Eigenvector:} \quad \begin{bmatrix} \mathbf{0} & \mathbf{Z} \\ \mathbf{Z}' & \mathbf{0} \end{bmatrix} \begin{bmatrix} \mathbf{s}_2 \\ \mathbf{s}_1 \end{bmatrix} = \begin{bmatrix} \mathbf{0} & \mathbf{I} \\ \mathbf{I} & \mathbf{0} \end{bmatrix} \begin{bmatrix} \mathbf{s}_2 \\ \mathbf{s}_1 \end{bmatrix}. \tag{9.26}$$

The matrix $\boldsymbol{\sigma}$ is the self-inverse matrix of exchange. For every solution $\{\mathbf{x}, \boldsymbol{\beta}, \lambda_F\}$ with λ_F, there exists an adjoint/dual solution $\{\mathbf{x}, -\boldsymbol{\beta}, -\lambda_F\}$ (payments solution).

2. The eigenvectors $\mathbf{s}_1$ and $\mathbf{s}_2$ lead to the stochastic similarity matrices:

stochastic similarity matrix $\mathbf{Z}_1$ (right/Labour Market Boundary),

$$\mathbf{Z}_1 = \hat{\mathbf{s}}_1^{-1}\mathbf{Z}\hat{\mathbf{s}}_1,$$

$$\mathbf{Z}\mathbf{s}_1 = \mathbf{s}_1, \quad \mathbf{Z} = \hat{\mathbf{s}}_1\mathbf{Z}_1\hat{\mathbf{s}}_1^{-1}, \quad \mathbf{Z}_1\mathbf{e} = \mathbf{e}, \quad \mathbf{Z}_1\hat{\mathbf{s}}_1^{-1} = \hat{\mathbf{s}}_1^{-1}\mathbf{Z},$$

stochastic similarity matrix $\mathbf{Z}_2$ (left/Consumption Market Boundary),

$$\mathbf{Z}_2 = \hat{\mathbf{s}}_2^{-1}\mathbf{Z}'\hat{\mathbf{s}}_2,$$

$$\mathbf{Z}'\mathbf{s}_2 = \mathbf{s}_2, \quad \mathbf{Z}' = \hat{\mathbf{s}}_2\mathbf{Z}_2\hat{\mathbf{s}}_2^{-1}, \quad \mathbf{Z}_2\mathbf{e} = \mathbf{e}, \quad \mathbf{Z}_2\hat{\mathbf{s}}_2^{-1} = \hat{\mathbf{s}}_2^{-1}\mathbf{Z}'. \tag{9.27}$$

In the following sections, the aforementioned matrix and vector relationships will be deduced.

9.4.2 The stochastic product family (Z, T, S, D)

The (**Z**, **T**, **S**, **D**)-family satisfies the row–sum and stochastic row–sum product relationships

Stochastic Product Family $\{\mathbf{Z}, \mathbf{T}, \mathbf{S}, \mathbf{D}\}$

		row–sum	stochastic product
value	$\mathbf{x}$	$\mathbf{Ze} = \mathbf{x}$	$\mathbf{Z} = \hat{\mathbf{x}}\mathbf{D}$
price	$\mathbf{p}$	$\mathbf{Te} = \mathbf{p}$	$\mathbf{T} = \hat{\mathbf{p}}\mathbf{D}$
quantity	$\mathbf{q}$	$\mathbf{Se} = \mathbf{q}$	$\mathbf{S} = \hat{\mathbf{q}}\mathbf{D}$
object	$\mathbf{e}$	$\mathbf{De} = \mathbf{e}$	$\mathbf{D} = \hat{\mathbf{e}}\mathbf{D}$

(9.28)

The relationships of this matrix family result from the basic interindustrial production relations (2.8), (2.11), (2.17), (2.19).

value	$\mathbf{x}$	$\mathbf{Z} = \hat{\mathbf{x}}\mathbf{D}$	$\mathbf{Ze} = \mathbf{x}$	row sum
quantity	$\mathbf{q}$	$\mathbf{S} = \hat{\mathbf{q}}\mathbf{D}$	$\mathbf{Se} = \mathbf{q}$	row sum

(9.29)

Introduction of the price coefficients matrix T

In combination with the matrix form of the interindustrial market $\hat{\mathbf{x}} = \hat{\mathbf{p}}\hat{\mathbf{q}}$, one obtains by substitution $\mathbf{Z} = \hat{\mathbf{x}}\mathbf{D} = \hat{\mathbf{p}}\hat{\mathbf{q}}\mathbf{D} = \hat{\mathbf{q}}(\hat{\mathbf{p}}\mathbf{D}) := \hat{\mathbf{q}}\mathbf{T}$. The stochastic row–sum products for the matrices $\mathbf{Z}, \mathbf{T}, \mathbf{S}$ are then:

$$\mathbf{Z} = \hat{\mathbf{x}}\mathbf{D}, \quad \mathbf{T} = \hat{\mathbf{p}}\mathbf{D}, \quad \mathbf{S} := \hat{\mathbf{q}}\mathbf{D},$$
$$\mathbf{Z} = \hat{\mathbf{p}}\mathbf{S} = \hat{\mathbf{q}}\mathbf{T} = \hat{\mathbf{x}}\mathbf{D}. \tag{9.30}$$

The new price coefficients matrix **T** results as a consequence to the market-price relationships $\hat{\mathbf{x}} = \hat{\mathbf{p}}\hat{\mathbf{q}}$. Its row–sum is determined by the multiplication with the object vector **e**,

$$\mathbf{Te} = \hat{\mathbf{p}}\mathbf{De} = \hat{\mathbf{p}}\mathbf{e} = \mathbf{p} \Rightarrow \mathbf{Te} = \mathbf{p} \tag{9.31}$$

The row–sum of the matrix **T** is therefore the price vector **p**. The row fractions of the matrix **T** indicate how the prices $p_i = \sum_{i=1}^{n} t_{ij}$ are composed, where each matrix entry t_{ij} is the contribution of the commodities i in sector j to the price of commodity i.

Whenever the relative prices are required, the price vector **p** can be normalised through the relation $\mathbf{e}'\mathbf{p} = 1$. With this normalisation, the linear norm of the matrix **T** ($\mathbf{e}'\mathbf{Te} = \mathbf{e}'\mathbf{p} = 1$) will be equal to one.

9.4.3 The stochastic similarity family (A, B, C, D)

Referring to equations (2.8) and (2.9) that define the matrix of technical coefficients of the *input-output coefficients* matrix **A**: The coefficient of production a_{ij}, equation (2.8), is the monetary value z_{ij} of commodity i (input), divided by the monetary value x_j of commodity j (output) produced, equation (2.8)

$$\mathbf{A} = (a_{ij}), \quad a_{ij} = \frac{z_{ij}}{x_j},$$
$$i : \text{Input index}, \quad j : \text{Output index}. \tag{9.32}$$

With this definition, the matrix relationships between the monetary input-output matrix **Z** and input-output coefficient matrix **A** are given by

$$\mathbf{A} = \mathbf{Z}\hat{\mathbf{x}}^{-1}, \quad \mathbf{Z} = \mathbf{A}\hat{\mathbf{x}}. \tag{9.33}$$

(a) The Frobenius–Perron eigenvectors of the (A, B, C, D)-family

Applying the row–sum relationship (branch production value) $\mathbf{Ze} = \mathbf{x}$ to equation (9.33) results in

$$\mathbf{Ze} = \mathbf{A}\hat{\mathbf{x}}\mathbf{e} = \mathbf{Ax} \equiv \mathbf{x} \quad \text{or} \quad \mathbf{Ax} = \mathbf{x}. \tag{9.34}$$

The value vector **x** is therefore the *Perron–Frobenius* eigenvector of matrix **A** at the Frobenius eigenvalue $\lambda_A = 1$.

Introduction of the model-generation matrix B

By further application of the interindustrial market relationships $\hat{\mathbf{x}} = \hat{\mathbf{p}}\hat{\mathbf{q}}$, the following relationships are obtained:

$$\begin{aligned}
\mathbf{Ax} = \mathbf{x} &\Rightarrow (\hat{\mathbf{q}}^{-1}\mathbf{A}\hat{\mathbf{q}})\mathbf{p} = \hat{\mathbf{q}}^{-1}\mathbf{x} = \mathbf{p} && \Rightarrow \mathbf{Bp} = \mathbf{p} \\
&\Rightarrow \mathbf{B} := \hat{\mathbf{q}}^{-1}\mathbf{A}\hat{\mathbf{q}} && \mathbf{A} = \hat{\mathbf{q}}\mathbf{B}\hat{\mathbf{q}}^{-1} \\
\mathbf{Ax} = \mathbf{x} &\Rightarrow (\hat{\mathbf{p}}^{-1}\mathbf{A}\hat{\mathbf{p}})\mathbf{q} = \hat{\mathbf{p}}^{-1}\mathbf{x} = \mathbf{q} && \Rightarrow \mathbf{Cq} = \mathbf{q} \\
&\Rightarrow \mathbf{C} := \hat{\mathbf{p}}^{-1}\mathbf{A}\hat{\mathbf{p}} && \mathbf{A} = \hat{\mathbf{p}}\mathbf{C}\hat{\mathbf{p}}^{-1} \\
\mathbf{Ax} = \mathbf{x} &\Rightarrow (\hat{\mathbf{x}}^{-1}\mathbf{A}\hat{\mathbf{x}})\mathbf{e} = \hat{\mathbf{x}}^{-1}\mathbf{x} = \mathbf{e} && \Rightarrow \mathbf{De} = \mathbf{e} \\
&\Rightarrow \mathbf{D} := \hat{\mathbf{x}}^{-1}\mathbf{A}\hat{\mathbf{x}}
\end{aligned} \tag{9.35}$$

The matrix **C** is the *input coefficients matrix*, see Section 2.4.2. The matrix **B** arises due to the market price conditions. Its *Perron–Frobenius* eigenvector is the price vector **p**. Therefore, we term it the *model-generation matrix* **B**. The *Perron–Frobenius* eigenvectors of the (**A**, **B**, **C**, **D**)-family are respectively the value vector **x**, the price vector **p**, the quantity vector **q** and the object (item) vector **e**:

$$\begin{array}{lll}
\mathbf{Ax} = \mathbf{x} & \text{value} & \mathbf{x}, \\
\mathbf{Bp} = \mathbf{p} & \text{price} & \mathbf{p}, \\
\mathbf{Cq} = \mathbf{q} & \text{quantity} & \mathbf{q}, \\
\mathbf{De} = \mathbf{e} & \text{object} & \mathbf{e}.
\end{array} \tag{9.36}$$

(b) The stochastic similarity relationships for the (A, B, C, D)-family

In equation (9.35), there is a similarity between the matrices **A**, **B**, **C**, **D**. Such similarity is due to a fundamental property of *non-negative* matrices (see Gantmacher [34]). Let the vector $\boldsymbol{\beta}$ be the positive Perron–Frobenius eigenvector of the non-negative matrix **M** with the Frobenius-eigenvalue λ_M, then, the matrix

$$\mathbf{M}_F = \frac{1}{\lambda_F(\mathbf{M})}\hat{\boldsymbol{\beta}}^{-1}\mathbf{M}\hat{\boldsymbol{\beta}} \tag{9.37}$$

is the stochastic similarity matrix of **M**, ($\mathbf{M}_F\mathbf{e} = \mathbf{e}$). The matrix $\hat{\boldsymbol{\beta}}$ is the corresponding similarity transformation matrix.

The similarity relationships in equation (9.35) and the *Perron–Frobenius* eigenvector relations (9.36) lead to the stochastic similarity relationships for the (**A**, **B**, **C**, **D**)-family. Especially:

The right stochastic production matrix **D**, that is $\mathbf{De} = \mathbf{e}$, is the stochastic similarity matrix of all the (**A**, **B**, **C**, **D**)-family matrices with the similarity transformation matrices $\{\hat{\mathbf{x}}, \hat{\mathbf{p}}, \hat{\mathbf{q}}, \mathbf{I} = \hat{\mathbf{e}}\}$ of value **x**, price **p**, quantity **q** and object (item) **e**.

$$\begin{array}{lll}
\mathbf{A} = \hat{\mathbf{x}}\mathbf{D}\hat{\mathbf{x}}^{-1} & \text{value} & \mathbf{x} \\
\mathbf{B} = \hat{\mathbf{p}}\mathbf{D}\hat{\mathbf{p}}^{-1} & \text{price} & \mathbf{p} \\
\mathbf{C} = \hat{\mathbf{q}}\mathbf{D}\hat{\mathbf{q}}^{-1} & \text{quantity} & \mathbf{q} \\
\mathbf{D} = \hat{\mathbf{e}}\mathbf{D}\hat{\mathbf{e}}^{-1} & \text{object} & \mathbf{e}
\end{array}$$

$$\begin{aligned}
\mathbf{A} &= \hat{\mathbf{q}}\mathbf{B}\hat{\mathbf{q}}^{-1} = \hat{\mathbf{p}}\mathbf{C}\hat{\mathbf{p}}^{-1} = \hat{\mathbf{x}}\mathbf{D}\hat{\mathbf{x}}^{-1} \\
\mathbf{D} &= \hat{\mathbf{q}}^{-1}\mathbf{C}\hat{\mathbf{q}} = \hat{\mathbf{p}}^{-1}\mathbf{B}\hat{\mathbf{p}} = \hat{\mathbf{x}}^{-1}\mathbf{A}\hat{\mathbf{x}}
\end{aligned} \tag{9.38}$$

In summary:

The matrices of the (**A**, **B**, **C**, **D**)-family are all similar and have the same *Perron–Frobenius* eigenvalue $\lambda_{\text{Frobenius}} = 1$. Their positive *Perron–Frobenius* eigenvectors are respectively the interindustrial market vectors of value **x**, price **p**, quantity **q** and object (item) **e**.

(c) The stochastic similarity of the Leontief matrix L

The Leontief matrix $\mathbf{L} = (\mathbf{I} - \mathbf{A})$, whose inverse is the Leontief Inverse $(\mathbf{I} - \mathbf{A})^{-1}$ (2.31), satisfies the following similarity relations

$$\mathbf{L} = (\mathbf{I} - \mathbf{A}) = \hat{\mathbf{q}}(\mathbf{I} - \mathbf{B})\hat{\mathbf{q}}^{-1} = \hat{\mathbf{p}}(\mathbf{I} - \mathbf{C})\hat{\mathbf{p}}^{-1} = \hat{\mathbf{x}}(\mathbf{I} - \mathbf{D})\hat{\mathbf{x}}^{-1}. \tag{9.39}$$

As the value vector **x** is the row–sum of **Z**, it is interesting to build the product

$$\mathbf{L}\mathbf{x} = (\mathbf{I} - \mathbf{A})\mathbf{x} = \hat{\mathbf{q}}(\mathbf{I} - \mathbf{B})\hat{\mathbf{q}}^{-1}\mathbf{x} = \hat{\mathbf{p}}(\mathbf{I} - \mathbf{C})\hat{\mathbf{p}}^{-1}\mathbf{x} = \hat{\mathbf{x}}(\mathbf{I} - \mathbf{D})\hat{\mathbf{x}}^{-1}\mathbf{x} \equiv \mathbf{0}. \tag{9.40}$$

One concludes from these identities the following geometrical relationship: The value vector **x** is perpendicular to all rows of matrix **L**.

9.4.4 Relationships between the (Z, T, S, D)- and the (A, B, C, D)-families

Based on the equations (9.28) and (9.36), the main relationships between the matrices of the two families are given by

$$\begin{array}{lll} \text{value} & \mathbf{x} & \mathbf{Z} = \hat{\mathbf{x}}\mathbf{D} = \mathbf{A}\hat{\mathbf{x}} \\ \text{price} & \mathbf{p} & \mathbf{T} = \hat{\mathbf{p}}\mathbf{D} = \mathbf{B}\hat{\mathbf{p}} \\ \text{quantity} & \mathbf{q} & \mathbf{S} = \hat{\mathbf{q}}\mathbf{D} = \mathbf{C}\hat{\mathbf{q}} \\ \text{object} & \mathbf{e} & \mathbf{D} = \hat{\mathbf{e}}\mathbf{D} = \mathbf{D}\hat{\mathbf{e}} \end{array} \tag{9.41}$$

9.4.5 The stochastic similarity table of an interindustrial economy (GDP-Table)

In summary, the matrix and vector relationships given in the last sections are grouped in a Stochastic Similarity Table of an *interindustrial economy* (Table 9.1, Fig. 9.9):

Table 9.1: (GDP-Table).

$\lambda_{\text{Frobenius}} = 1$	Row sum	FP Eigenvector	Stochastic similarity	I/O matrices
value **x**	$\mathbf{Ze} = \mathbf{x}$	$\mathbf{Ax} = \mathbf{x}$	$\mathbf{A} = \hat{\mathbf{x}}\mathbf{D}\hat{\mathbf{x}}^{-1}$	$\mathbf{Z} = \hat{\mathbf{x}}\mathbf{D}$
price **p**	$\mathbf{Te} = \mathbf{p}$	$\mathbf{Bp} = \mathbf{p}$	$\mathbf{B} = \hat{\mathbf{p}}\mathbf{D}\hat{\mathbf{p}}^{-1}$	$\mathbf{T} = \hat{\mathbf{p}}\mathbf{D}$
quantity **q**	$\mathbf{Se} = \mathbf{q}$	$\mathbf{Cq} = \mathbf{q}$	$\mathbf{C} = \hat{\mathbf{q}}\mathbf{D}\hat{\mathbf{q}}^{-1}$	$\mathbf{S} = \hat{\mathbf{q}}\mathbf{D}$
object **e**	$\mathbf{De} = \mathbf{e}$	$\mathbf{De} = \mathbf{e}$	$\mathbf{D} = \hat{\mathbf{e}}\mathbf{D}\hat{\mathbf{e}}^{-1}$	$\mathbf{D} = \hat{\mathbf{e}}\mathbf{D}$

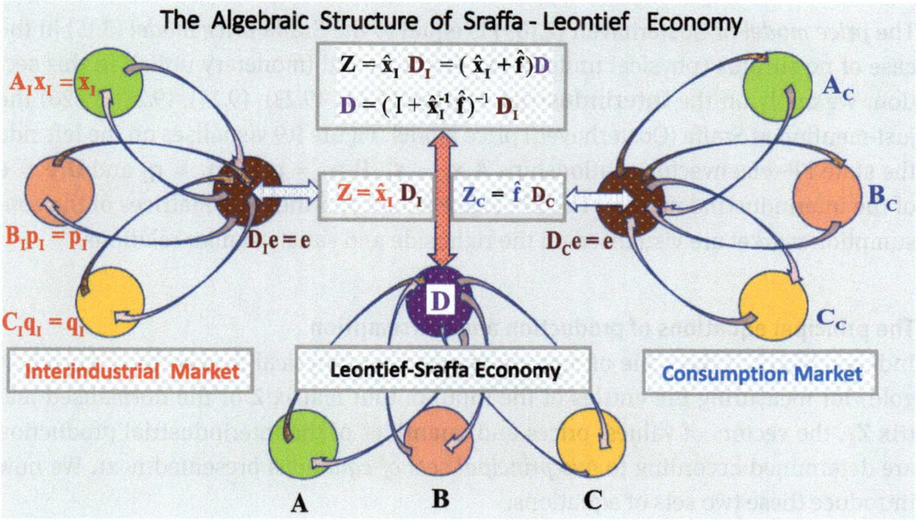

Figure 9.9: The algebraic structure of the Sraffa–Leontief economy with the state matrices and eigenvector relationships of the interindustrial and the consumption markets.

9.5 Productivity and interindustrial production*

Notation 9.5.1. See Figure 9.9. In the next Sections 9.5 and 9.6, the coefficients matrices **A**, **C**, **D**, **S** and vectors **x**, **y** with regard to the interindustrial market are written with an index I, that is as $\mathbf{A}_I$, $\mathbf{C}_I$, $\mathbf{D}_I$, $\mathbf{S}_I$, $\mathbf{x}_I$, $\mathbf{y}_I$ if it is needed to distinguish them from the corresponding complete economy, where surplus **d**, respectively final demand **f** is included. The matrix **S** and the vector **p** of the Leontief–Sraffa economy are used without indices.[4]

The state matrices of the interindustrial market have the identical FP–eigenvalue $\lambda_F(\mathbf{A}_I) = \lambda_F(\mathbf{B}_I) = \lambda_F(\mathbf{C}_I) = \lambda_F(\mathbf{D}_I) = 1$. The state matrices **A, B, C, D** have also identical FP–eigenvalues, but are different from the above FP–eigenvalues of the interindustrial market. Courtesy of Nour Eldin [73], where the consumption market is treated.

In Summary: The right eigenvector $\mathbf{s}_1$ responsible for the value-added/labor market boundary (9.27) together with the interindustrial purchase vector $\mathbf{y}_I = \mathbf{Z}'\mathbf{e}$ and the total input/output vector $\mathbf{y} = \mathbf{y}_I + \mathbf{u} = \mathbf{Ze} + \mathbf{f} = \mathbf{x}$ uniquely determine the FP–eigenvalue $\lambda_F(\mathbf{C}) = 1/(1+R)$, as well as the factor of productiveness R of the complete Leontief–Sraffa economy, including surplus or final demand (Figure 9.9, Figure 9.10, Figure 9.11).

The eigenvalue $\lambda_F(\mathbf{C})$, as well as the productivity R, are analytically determined through either the scaler product relations (Figure 9.10) or the hyperbolic relations (Figure 9.11).

---

4. The consumption market with matrices such as $\mathbf{A}_C$, $\mathbf{B}_C$, $\mathbf{C}_C$ are treated in Nour Eldin [73].

The *price model* of Oosterhaven (2.109) is equal to the Sraffa *price model* (3.15) in the case of no surplus (physical units) or no final demand (monetary units). In this section, we apply on the **interindustrial market** (9.22), (9.23), (9.24), (9.25), (9.26) the just-mentioned Sraffa (Oosterhaven) price model. Figure 9.9 visualises on the left side the state FP–eigenvector relationships $\mathbf{A}_I\mathbf{x}_I = \mathbf{x}_I$, $\mathbf{B}_I\mathbf{p}_I = \mathbf{p}_I$, $\mathbf{C}_I\mathbf{q}_I = \mathbf{q}_I$ and $\mathbf{D}_I\mathbf{e} = \mathbf{e}$ of the interindustrial market. The FP–eigenvectors and the state matrices of the consumption market are visualised on the right side and satisfy similar relations.

The principal equations of production and consumption

Independently of the scale of measurements used (physical numéraire, a currency, gold) for measuring the entries of the input-output matrix $\mathbf{Z}$ or the normalised matrix $\mathbf{Z}_\lambda$, the vectors of values, prices and quantities of the interindustrial production are determined according to *two principal sets of equations* presented next. We now introduce these two sets of equations.

Suppose that an interindustrial (intermediate) market is given. Ex ante, the prices p_i of the commodities i and the quantities of commodities q_i are initially given through statistical information, typically collected by national statistical offices. Through the basic rule $x_i = p_i q_i$ for one commodity i, the values x_i are obtained and the IOT is constructed. This process is formalised in **Part I**, where the price vector $\mathbf{p}$ is supposed to be known (statistical-based price determination) (see also the remark in the Eurostat Manual [72] and equation (2.18)). Ex post, on the basis of available IOTs $\mathbf{Z}$, the *price model-based approach* is possible, which is presented in **Part II** with the Sraffa price model (4.29) where the price vector is calculated through that model (model-based price determination). Figure 9.9 presents an extension of the complete process of the interindustrial and consumption market, which here is not treated in detail:

I. Here are summarised the **accounting balance identity** (2.7) the right eigenvector $\mathbf{s}_1$ of the matrix $\mathbf{Z}$ (9.26) describing the input-output exchange of the interindustrial market and the basic value-price-quantity relationships (2.18), (2.19). Simultaneously are satisfied, the interindustrial market relationships (9.22), (9.23), (9.24), (9.25), the GDP-Table 9.1, the input-output relationships and the determinations of the quantity vectors (2.15),

$$\text{Accounting Identity:} \quad \mathbf{x} = \mathbf{Ze} + \mathbf{f} = \mathbf{x}_I + \mathbf{f} = \mathbf{y}_I + \boldsymbol{\upsilon} = \mathbf{Z}'\mathbf{e} + \boldsymbol{\upsilon} = \mathbf{y}, \tag{9.42}$$

$$\text{Interindustrial Market:} \quad \mathbf{x}_I = \hat{\mathbf{p}}_I\mathbf{q}_I, \quad \mathbf{e}'\mathbf{p}_I = 1,$$

$$\text{Right eigenvector of exchange:} \quad \mathbf{Zs}_1 = \mathbf{s}_1 \quad \mathbf{e}'\mathbf{s}_1 = 1,$$

$$\text{Sraffa's PCMC:} \quad \mathbf{Z} = \hat{\mathbf{p}}\mathbf{S}, \quad \mathbf{q} = \mathbf{q}_I + \mathbf{d} = \mathbf{Se} + \mathbf{d}. \tag{9.43}$$

The normalised vector $\mathbf{s}_1$ (9.26) is the right eigenvector of the input-output matrix $\mathbf{Z}$. The vector $\mathbf{q}$ is the PCMC quantity vector with surplus (see Figures 9.1, 9.4, 9.6 and 9.9). We set the index I when the price vector normed to 1, i. e., $\mathbf{e}'\mathbf{p}_I = 1$, but in the Sraffa price model the price vector $\mathbf{p}$ has no index.

II. The Perron–Frobenius eigenvalue $\lambda_F(\mathbf{C})$ of the Sraffa input-output coefficients matrix $\mathbf{C}$ (3.44) is uniquely determined by the interindustrial production output $\mathbf{y}_I$ and the total production output $\mathbf{y}$ through the scalar products with the eigenvector $\mathbf{s}_1$ (projection on $\mathbf{s}_1$). The normalisation $\mathbf{e}'\mathbf{s}_1 = 1$ of $\mathbf{s}_1$ leads to the normalisation $\mathbf{e}'\mathbf{y}_I = 1$ of the interindustrial outlay/purchase $\mathbf{y}_I$. We get,

$$\begin{aligned}
&\text{Frobenius eigenvalue} \\
&\mathbf{C}'\mathbf{p} = \lambda_F(\mathbf{C})\mathbf{p}, \\
\text{FP–eigenvalue relationship} \quad & \lambda_F^{-1}(\mathbf{C}) = \mathbf{s}_1'\mathbf{y} = (1+R) > 1, \\
\text{normalisation} \quad & \mathbf{s}_1'\mathbf{e} = \mathbf{s}_1'\mathbf{y}_I = 1, \\
\text{productiveness relationship} \quad & \mathbf{s}_1'\mathbf{y} = \mathbf{s}_1'\mathbf{y}_I + \mathbf{s}_1'\boldsymbol{\upsilon} = (1+R), \\
\text{productiveness relationship} \quad & R = \mathbf{s}_1'\boldsymbol{\upsilon}
\end{aligned} \tag{9.44}$$

$$\begin{aligned}
&\alpha\text{-parametrisation relations} \\
\text{FP–eigenvalue:} \quad & \lambda_F^{-1}(\mathbf{C}) = \mathbf{s}_1'\mathbf{y} = \cosh^2(\alpha_v), \\
\text{productiveness :} \quad & R = \mathbf{s}_1'\boldsymbol{\upsilon}_I = \sinh^2(\alpha_v), \\
\text{eigenvalue relationship:} \quad & \lambda_F^{-1}(\mathbf{C}) = \cosh^2(\alpha_v) = 1 + \sinh^2(\alpha_v) = (1+R), \\
\text{productiveness relatioship:} \quad & \cosh^2(\alpha_v) = 1 + \sinh^2(\alpha_v) = (1+R).
\end{aligned} \tag{9.45}$$

The factor R is called the productiveness (see Knolle [49]). The value R characterises the overall productivity of an economy in the sense of Krugman [47] *(see Section 3.1.3).*

The proof of the **principal equations** is based on the **Sraffa price model** (see Section 3.1.3, (3.43) and (3.44)).

I. The Sraffa price model

$$\mathbf{S}'\mathbf{p} = \frac{1}{1+R}\hat{\mathbf{q}}\mathbf{p} = \frac{1}{1+R}\mathbf{y}, \quad \mathbf{C}'\mathbf{p} = \lambda_F(\mathbf{C})\mathbf{p}, \tag{9.46}$$

$$\lambda_F(\mathbf{C}) = \frac{1}{1+R} < 1, \quad \lambda_F^{-1}(\mathbf{C}) = 1 + R > 0. \tag{9.47}$$

These eigenvalue equations can be transformed in terms of the monetary matrix $\mathbf{Z}$ using the following identities:

$$\begin{aligned}
&\text{II. Right eigenvector normalization} \\
&\text{normalisation:} \quad \mathbf{Z}\mathbf{s}_1 = \mathbf{s}_1 \quad \text{or} \quad \mathbf{s}_1' = \mathbf{s}_1'\mathbf{Z}', \quad \mathbf{s}_1'\mathbf{e} = 1, \\
&\mathbf{Z}' = \mathbf{S}'\hat{\mathbf{p}}, \quad \mathbf{C}' = \hat{\mathbf{q}}^{-1}\mathbf{S}', \quad \mathbf{Z}' = \hat{\mathbf{q}}\mathbf{C}'\hat{\mathbf{p}}, \\
&\qquad \mathbf{Z}\mathbf{e} = \mathbf{x}_I, \quad \mathbf{Z}'\mathbf{e} = \mathbf{y}_I, \\
&\text{normalised } \mathbf{y}_I \quad \mathbf{s}_1'\mathbf{e} = \mathbf{s}_1'\mathbf{Z}'\mathbf{e} = \mathbf{s}_1'\mathbf{y}_I = 1.
\end{aligned} \tag{9.48}$$

The following accounting identity is a vector-addition relationship:

III. The accounting identity

$$\mathbf{x} = (\mathbf{Ze} + \mathbf{f}) = (\mathbf{x}_I + \mathbf{f}) = (\mathbf{y}_I + \boldsymbol{\upsilon}) = (\mathbf{Z}'\mathbf{e} + \boldsymbol{\upsilon}) = \mathbf{y}. \tag{9.49}$$

The Sraffa price system (**I.**) and the normalised eigenvector $\mathbf{s}_1$ leading to the normalisation of the interindustrial outlay vector $\mathbf{y}_I$ (**II.**) should satisfy the accounting identity (**III.**). The relationships (**I.**), (**II.**) and (**III.**) are only simultaneously valid when the interindustrial input-output vectors (**II.**) and the total input-output vectors (**I.**) satisfy the accounting identity (**III.**). This leads directly to the input-output relationships (9.50) (see also Figure 9.10 and equation (9.44)).

$$\begin{aligned}
&\text{Interindustrial boundary vector} \qquad \text{Interindustrial Input}\\
&\mathbf{s}_1' = \mathbf{s}_1'\mathbf{Z}', \quad \mathbf{s}_1'\mathbf{e} = 1, \qquad \mathbf{y}_I = \mathbf{Z}'\mathbf{e} = \mathbf{S}'\mathbf{p}.\\
&\text{It follows a scalar addition relatioship:} \quad (2.18), \mathbf{x} = \mathbf{y}\\
&\mathbf{s}_1'\mathbf{y}_I = \mathbf{s}_1'\mathbf{S}'\mathbf{p} = \frac{1}{1+R}(\mathbf{s}_1'\hat{\mathbf{q}}\mathbf{p}) = \frac{1}{1+R}(\mathbf{s}_1'\mathbf{y}) = \frac{1}{1+R}\mathbf{s}_1'(\mathbf{y}_I + \boldsymbol{\upsilon})\\
&\quad = \lambda_F(\mathbf{C})\mathbf{s}_1'\mathbf{x} = \lambda_F(\mathbf{C})\mathbf{s}_1'\mathbf{y} = 1,\\
&\text{FP–eigenvalue:} \quad \lambda_F(\mathbf{C}) = \frac{\mathbf{s}_1'\mathbf{y}_I}{\mathbf{s}_1'\mathbf{y}} = \frac{1}{\mathbf{s}_1'\mathbf{y}} = \frac{1}{\mathbf{s}_1'\mathbf{x}} = \frac{1}{1+R}\\
&\rightarrow \text{productiveness}: \quad R = \mathbf{s}_1'\boldsymbol{\upsilon} \quad \text{(value-added).}
\end{aligned} \tag{9.50}$$

The FP–eigenvalue $\lambda_F(\mathbf{C})$ is the ratio between the scalar products $\mathbf{s}_1'\mathbf{y}_I = 1$ and $\mathbf{s}_1'\mathbf{y}$, giving $1/(1+R)$, where R is the productiveness of the present economy.

The hyperbolic parametrisation (α-parametrisation)

The parametrisation of the scalar products just treated, through the hyperbolic functions $\mathbf{s}_1'\boldsymbol{\upsilon} = \sinh^2(\gamma_\nu) = R$, $\mathbf{s}_1'\mathbf{y} = \cosh^2(\gamma_\nu)$ and $\lambda_F^{-1}(\mathbf{C}) = \mathbf{s}_1'\mathbf{y} = \cosh^2(\gamma_\nu)$ uncovers the hyperbolic behaviour of the markets and the economy as an algebraic exchange process (see Figure 9.11). The parametrisation is the result of the scaler-product projection of the accounting relation $\mathbf{x} = (\mathbf{Ze}+\mathbf{f}) = (\mathbf{x}_I+\mathbf{f}) = (\mathbf{y}_I+\boldsymbol{\upsilon}) = (\mathbf{Z}'\mathbf{e}+\boldsymbol{\upsilon}) = \mathbf{y}$ on the normalised interindustrial boundary vector $\mathbf{s}_1$. That scaler product projection leads directly to the productiveness factor R, as well as the hyperbolic function relationships of the accounting conditions $\cosh^2(\gamma_\nu) - \sinh^2(\gamma_\nu) = 1$ that parameterises the scaler-product relationship $\mathbf{s}_1'\mathbf{y} - \mathbf{s}_1'\boldsymbol{\upsilon} = \mathbf{s}_1'\mathbf{y}_I = 1$. For these relationships, the normalisation $\mathbf{e}_1'\mathbf{s}_1 = 1$ of the interindustrial boundary vector $\mathbf{s}_1$ leads directly to the normalisation of the interindustrial outlay $\mathbf{e}_1'\mathbf{s}_1 = (\mathbf{e}'\mathbf{Z})\mathbf{s}_1 = \mathbf{y}_I'\mathbf{e} = 1$, as well as to the determination of the eigenvalue inverse $\lambda_F^{-1}(\mathbf{C}) = (1 + \mathbf{s}_1'\boldsymbol{\upsilon}) > 1$. Both normalisations play a central role on the derivation, as well as on the parametrisation.

Thus, the scalar-product relationship $\mathbf{s}_1'\mathbf{y} = \mathbf{s}_1'\mathbf{y}_I + \mathbf{s}_1'\boldsymbol{\upsilon}$ is parameterised through the α-variable, such that the scalar-product relationship is transferred to the hyperbolic relationship $\cosh^2(\alpha_\nu) = 1 + \sinh^2(\alpha_\nu)$. The projection of the accounting equation is always satisfied for every value of the variables α_ν and α_f (see Figure 9.11).

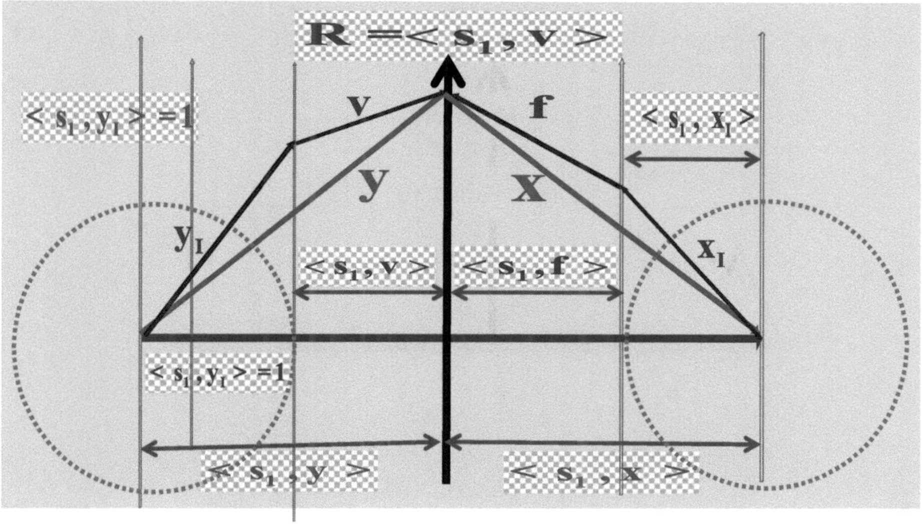

Figure 9.10: Geometrical illustration of the productiveness R with the corresponding scaler products and projection on the right eigenvector, responsible for the external input-output exchange of the interindustrial market (exchange with the value-added/Labor market). The productivity factor R reaches a maximum when the vector of value added u and the right eigenvector $\mathbf{s}_1$ of $\mathbf{Z}$ are parallel, $R = \mathbf{s}_1'\mathbf{u}$ (same direction).

The accounting identity at the boundary

The accounting identity (2.7) relating the total outlays to total inputs has to be split into two boundary (external) accounts. This is achieved by the right and the left eigenvectors of the matrix $\mathbf{Z}$. Figures 9.10 and 9.11 illustrate the projection on $\mathbf{s}_1$. Similar geometric illustrations are valid for $\mathbf{s}_2$. Optimum productivity is achieved when the value-added vector is in the same direction (parallel) as $\mathbf{s}_1$. Similar argumentation is valid at the boundary of the final product-consumption market.

	Account identity	
Right Boundary:	$\mathbf{x} = \mathbf{Z}\mathbf{e} + \mathbf{f} = \mathbf{x}_I + \mathbf{f}$	final consumption market
Left Boundary:	$\mathbf{y} = \mathbf{Z}'\mathbf{e} + \boldsymbol{\upsilon} = \mathbf{y}_I + \boldsymbol{\upsilon}$	value-added/Labour market
	Account identity at the boundary	
Right Boundary:	$\langle \mathbf{s}_2, \mathbf{x}\rangle = \langle \mathbf{s}_2, \mathbf{x}_I\rangle + \langle \mathbf{s}_2, \mathbf{f}\rangle$	final consumption market
Left Boundary:	$\langle \mathbf{s}_1, \mathbf{y}\rangle = \langle \mathbf{s}_1, \mathbf{y}_I\rangle + \langle \mathbf{s}_1, \boldsymbol{\upsilon}\rangle$	value-added/Labour market (9.51)

9.6 Computational examples

We finish this chapter with two examples illustrating the interindustrial market, where the *Sraffa price model* is used to compute the price vectors. In the first production system, there is no surplus, in contrast to the second one that exhibits a vector of surplus.

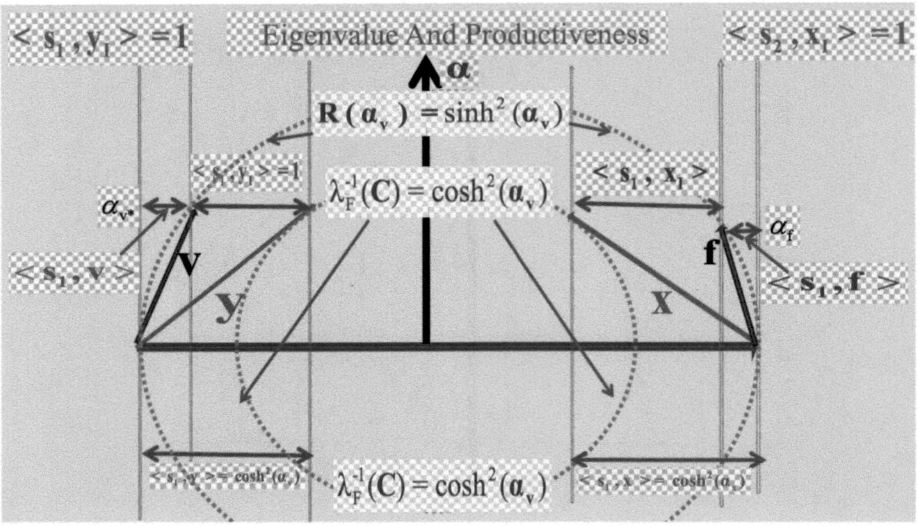

Figure 9.11: Graphical illustration of the locus of productiveness R as the function $\sinh^2(\alpha_v)$. The independent variables α_v and α_f connect the productiveness R to the eigenvalue $\lambda_F^{-1}(\mathbf{C}) = \mathbf{p}_l'\mathbf{y} = \cosh^2(\alpha_v)$.

Example 9.6.1. Consider the interindustrial production of a partial production economy with two sectors, a = wheat and b = iron, represented for a given period of production T. Wheat is the numéraire, $p_1 = 1$. Every producer produces one single product which is transferred to the other producer at a price to be determined through the Sraffa price model (4.29),

$$\begin{aligned}(3\text{ t. wheat}, 1\text{ t. iron}) &\to (7\text{ t. wheat},\ 0),\\ (4\text{ t. wheat}, 2\text{ t. iron}) &\to (0,\ 3\text{ t. iron}).\end{aligned} \tag{9.52}$$

Identify the flow commodity matrix **S**. Compute the vector of total output **q** in physical terms, the input-output coefficients matrix **C**, the Frobenius number λ_C and the price vector **p** as the Perron–Frobenius right eigenvector of matrix $\mathbf{C}'$.

Compute then the matrices **D**, **A**, **B**, **T** (9.22). Compute the commodity flow matrix **Z** in monetary terms (here a numéraire) and the corresponding Frobenius number λ_F of matrix **Z**, the vector of total output **x**, the right and left eigenvectors $\mathbf{s}_1$ and $\mathbf{s}_2$ of matrix **Z**, $\mathbf{Z}\mathbf{s}_1 = \mathbf{s}_1$, $\mathbf{Z}'\mathbf{s}_2 = \mathbf{s}_2$, the circulating capital K, the λ_F-normalised commodity flow matrix $\mathbf{Z}_\lambda$, the λ_F-normalised circulating capital K_λ and the stochastic matrix $\mathbf{Z}_0$, see Figure 9.12.

Verify that the vector of total output $\mathbf{x} = \mathbf{Z}\mathbf{e}$ is a left eigenvector of the distribution coefficients matrix **D**, associated with the Frobenius number λ_D, according to equation (4.21) (2). Apply (9.50) to compute the productiveness of this economy, and check the obtained value with the Frobenius number λ_C.

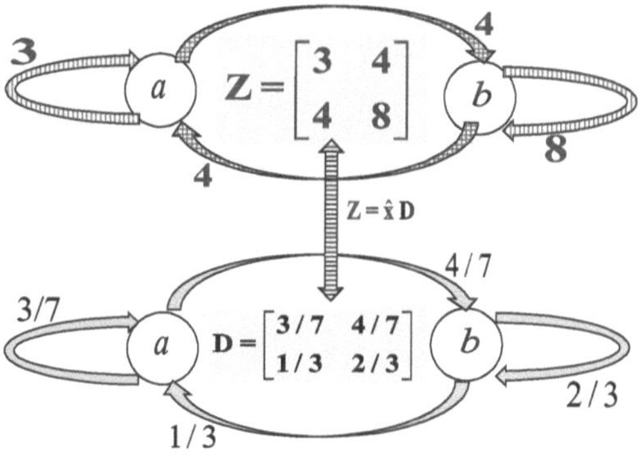

Figure 9.12: The directed graphs associated to the input-output matrix **Z** and the stochastic production matrix **D**.

Solution to Example 9.6.1:
Identify the commodity flow matrix **S**, and compute the vector $\mathbf{q} = \mathbf{Se} + \mathbf{d}$ (with $\mathbf{d} = \mathbf{o}$ for the interindustrial market) and the input-output coefficients matrix $\mathbf{C} = \mathbf{S}\hat{\mathbf{q}}^{-1}$,

$$\mathbf{S} = \begin{bmatrix} 3 & 4 \\ 1 & 2 \end{bmatrix}, \quad \mathbf{q} = \mathbf{Se} = \begin{bmatrix} 7 \\ 3 \end{bmatrix}, \quad \mathbf{C} = \begin{bmatrix} 3 & 4 \\ 1 & 2 \end{bmatrix}\begin{bmatrix} \frac{1}{7} & 0 \\ 0 & \frac{1}{3} \end{bmatrix} = \begin{bmatrix} \frac{3}{7} & \frac{4}{3} \\ \frac{1}{7} & \frac{2}{3} \end{bmatrix}. \tag{9.53}$$

The characteristic polynomial

$$P_2(\lambda) = \det(\mathbf{C} - \lambda\mathbf{I}_2) = (\lambda - 1)\left(\lambda + \frac{2}{21}\right) \tag{9.54}$$

is established. We identify the Frobenius number $\lambda_C = 1$. Having set $p_1 = 1$ for the numéraire, we calculate the requested price vector $\mathbf{p} = [1, 4]'$ of the Sraffa price model $\mathbf{C}'\mathbf{p} = \lambda_C\mathbf{p}$ (equation (4.21), (1)).

Then we compute the distribution coefficients matrix **D** (9.22) and check its stochastic property,

$$\mathbf{D} = \hat{\mathbf{q}}^{-1}\mathbf{S} = \begin{bmatrix} \frac{1}{7} & 0 \\ 0 & \frac{1}{3} \end{bmatrix}\begin{bmatrix} 3 & 4 \\ 1 & 2 \end{bmatrix} = \begin{bmatrix} \frac{3}{7} & \frac{4}{7} \\ \frac{1}{3} & \frac{2}{3} \end{bmatrix}, \quad \mathbf{De} = \begin{bmatrix} 1 \\ 1 \end{bmatrix} = \mathbf{e}, \tag{9.55}$$

where one verifies that $\mathbf{A} = \mathbf{D}'$, the non stochastic *price coefficients* matrix,

$$\mathbf{T} = \hat{\mathbf{p}}\mathbf{D} = \begin{bmatrix} 1 & 0 \\ 0 & 4 \end{bmatrix}\begin{bmatrix} \frac{3}{7} & \frac{4}{7} \\ \frac{1}{3} & \frac{2}{3} \end{bmatrix} = \begin{bmatrix} \frac{3}{7} & \frac{4}{7} \\ \frac{4}{3} & \frac{8}{3} \end{bmatrix}, \quad \mathbf{Te} = \mathbf{p} = \begin{bmatrix} 1 \\ 4 \end{bmatrix}, \tag{9.56}$$

the *model generation matrix*, equal to the matrix $\mathbf{C}'$,

$$\mathbf{B} = \hat{\mathbf{p}}\mathbf{D}\hat{\mathbf{p}}^{-1} = \begin{bmatrix} 1 & 0 \\ 0 & 4 \end{bmatrix}\begin{bmatrix} \frac{3}{7} & \frac{4}{7} \\ \frac{1}{3} & \frac{2}{3} \end{bmatrix}\begin{bmatrix} 1 & 0 \\ 0 & \frac{1}{4} \end{bmatrix} = \begin{bmatrix} \frac{3}{7} & \frac{1}{7} \\ \frac{4}{3} & \frac{2}{3} \end{bmatrix} = \mathbf{C}', \tag{9.57}$$

the flow commodity matrix $\mathbf{Z}$ in monetary terms (here with the numéraire wheat) using equation (9.30),

$$\mathbf{Z} = \hat{\mathbf{p}}\mathbf{S} = \begin{bmatrix} 1 & 0 \\ 0 & 4 \end{bmatrix}\begin{bmatrix} 3 & 4 \\ 1 & 2 \end{bmatrix} = \begin{bmatrix} 3 & 4 \\ 4 & 8 \end{bmatrix}, \quad \mathbf{x} = \mathbf{Z}\mathbf{e} = \begin{bmatrix} 7 \\ 12 \end{bmatrix}. \tag{9.58}$$

We obtain a symmetric matrix $\mathbf{Z}$, which is positive and accountably balanced ($\mathbf{x} = \mathbf{y}$). Its Frobenius-Eigenvalue λ_F is now computed,

$$P_2(\mathbf{Z}, \lambda) = \det(\mathbf{Z} - \lambda\mathbf{I}_2) = \lambda^2 - 11\lambda + 8 = (\lambda - 10.217)(\lambda - 0.783),$$

$$\lambda_F = 10.217, \quad \lambda_2 = 0.783,$$

$$\mathbf{Z}_\lambda = \frac{1}{\lambda_F}\mathbf{Z} = \frac{1}{10.217}\begin{bmatrix} 3 & 4 \\ 4 & 8 \end{bmatrix} = \begin{bmatrix} 0.2936 & 0.3915 \\ 0.3915 & 0.7830 \end{bmatrix},$$

$$K = \mathbf{e}'\mathbf{x} = 19, \quad \mathbf{x}_\lambda = \mathbf{Z}_\lambda\mathbf{e} = \begin{bmatrix} 0.6851 \\ 1.1745 \end{bmatrix},$$

$$\mathbf{y}_I = \mathbf{Z}'\mathbf{e} = \begin{bmatrix} 7 \\ 12 \end{bmatrix} = \mathbf{x}, \quad \mathbf{y} = \mathbf{y}_I + \boldsymbol{\upsilon} = \mathbf{x} \Rightarrow \boldsymbol{\upsilon} = \mathbf{o},$$

$$\text{normalised circulating capital:} \quad K_\lambda = \frac{K}{\lambda_F} = \frac{19}{10.217} = 1.8597. \tag{9.59}$$

Then, we compute the stochastic similarity matrix $\mathbf{Z}_0 = \hat{\mathbf{s}}_1^{-1}\mathbf{Z}_\lambda\hat{\mathbf{s}}_1$. We compute the right and left eigenvectors of matrix $\mathbf{Z}_\lambda$: $\mathbf{s}_1 = \mathbf{s}_2 = [0.5542, 1]$, with $\mathbf{Z}_\lambda\mathbf{s}_1 = \mathbf{s}_1$ and $\mathbf{Z}_\lambda'\mathbf{s}_2 = \mathbf{s}_2$, which in the present case are identical, because matrix $\mathbf{Z}_\lambda$ is symmetric:

$$\mathbf{Z}_0 = \begin{bmatrix} 1.8044 & 0 \\ 0 & 1 \end{bmatrix}\begin{bmatrix} 0.2936 & 0.3915 \\ 0.3915 & 0.7830 \end{bmatrix}\begin{bmatrix} 0.5542 & 0 \\ 0 & 1 \end{bmatrix} = \begin{bmatrix} 0.2936 & 0.7064 \\ 0.2170 & 0.7830 \end{bmatrix}. \tag{9.60}$$

We verify that $\mathbf{Z}_0\mathbf{e} = \mathbf{e}$. Thus, $\mathbf{Z}_0$ is stochastic. Clearly the Frobenius number is $\lambda_{F_0} = 1$. The eigenvector $\mathbf{s}_1$ is economically not interpreted, but it may be understood to indicate a direction of the development of the present economy, weighting the commodities. Further work is necessary to obtain a precise interpretation.

Now, we check that the output vector $\mathbf{x}' = [7, 12]$ is a left eigenvector of the distribution coefficients matrix $\mathbf{D}$. Indeed,

$$\mathbf{D}'\mathbf{x} = \begin{bmatrix} \frac{3}{7} & \frac{1}{3} \\ \frac{4}{7} & \frac{2}{3} \end{bmatrix}\begin{bmatrix} 7 \\ 12 \end{bmatrix} = \begin{bmatrix} 7 \\ 12 \end{bmatrix} = \mathbf{x}. \tag{9.61}$$

Finally, we calculate the productiveness of this economy, either with the Frobenius number or with equation (9.50),

$$R = \frac{1}{\lambda_C} - 1 = \frac{1}{1} - 1 = 0 \quad \text{or}$$

$$\lambda_C = \frac{\mathbf{s}_1' \cdot \mathbf{y}_I}{\mathbf{s}_1' \cdot \mathbf{y}} = \frac{15.9897}{15.8797} = 1 = \frac{1}{1+R} \Rightarrow R = 0. \quad \blacktriangle \tag{9.62}$$

Now we treat an economy with surplus, respectively with final demand.

Example 9.6.2. Consider the production scheme of the partial production economy with two sectors a = wheat and b = iron for a given period of production of T,

$$\begin{aligned}&(3\,\text{t. wheat},\ 1\,\text{t. iron}) \to (9\,\text{t. wheat},\ 0),\\ &(4\,\text{t. wheat}, 2\,\text{t. iron}) \to (0, 5\,\text{t. iron}).\end{aligned} \tag{9.63}$$

Wheat is the numéraire, $p_1 = 1$. Identify the flow commodity matrix $\mathbf{S}$ and the vector of surplus $\mathbf{d}$. Compute the vector of total output $\mathbf{q}$ in physical terms, the input-output coefficients matrix $\mathbf{C}$, the Frobenius number λ_C and the price vector $\mathbf{p}$ of the Sraffa price model as Perron–Frobenius left eigenvector of matrix $\mathbf{C}$. Compute the flow commodity matrix $\mathbf{Z} = \hat{\mathbf{p}}\mathbf{S}$ in monetary terms, the vector of final demand $\mathbf{f} = \hat{\mathbf{p}}\mathbf{d}$ and the vector of total output $\mathbf{x} = \mathbf{Ze} + \mathbf{f}$.

Compute the total outlays $\mathbf{y}_I = \mathbf{Z}'\mathbf{e}$, the vector of value added $\boldsymbol{\upsilon} = \mathbf{x} - \mathbf{y}_I$. Apply equation (9.44) to compute the productiveness R of this economy. Check R using its relation to the Frobenius number λ_C.

Solution to Example 9.6.2:

We identify the commodity flow matrix $\mathbf{S}$ and the vector of surplus $\mathbf{d}$ and compute the vector $\mathbf{q} = \mathbf{Se} + \mathbf{d}$ and input-out coefficients matrix $\mathbf{C} = \mathbf{S}\hat{\mathbf{q}}^{-1}$,

$$\mathbf{S} = \begin{bmatrix} 3 & 4 \\ 1 & 2 \end{bmatrix}, \quad \mathbf{d} = \begin{bmatrix} 2 \\ 2 \end{bmatrix}, \quad \mathbf{q} = \mathbf{Se} + \mathbf{d} = \begin{bmatrix} 9 \\ 5 \end{bmatrix}, \quad \mathbf{C}' = \begin{bmatrix} \frac{1}{3} & \frac{1}{9} \\ \frac{4}{5} & \frac{2}{5} \end{bmatrix}. \tag{9.64}$$

The characteristic polynomial is

$$P_2(\lambda) = \det(\mathbf{C} - \lambda \mathbf{I}_2) = \lambda^2 - \frac{11}{15}\lambda + \frac{2}{45} = \left(\lambda - \frac{2}{3}\right)\left(\lambda - \frac{1}{15}\right) \tag{9.65}$$

is established. We identify the Frobenius number $\lambda_C = 2/3$. Having set $p_1 = 1$ for the numéraire wheat, we calculate the price vector $\mathbf{p} = [1, 3]'$ as left eigenvector of $\mathbf{C}$, solving the Sraffa price model $\mathbf{C}'\mathbf{p} = \lambda_C \mathbf{p}$ in (4.21) (1). We then obtain the productiveness

$$R = \frac{1}{\lambda_C} - 1 = \frac{3}{2} - 1 = \frac{1}{2} \tag{9.66}$$

of this economy. We compute the *input-output commodity flow* matrix $\mathbf{Z}$ in monetary terms and the vector of final consumption $\mathbf{f} = \hat{\mathbf{p}}\mathbf{d}$ in (2.105),

$$\mathbf{Z} = \hat{\mathbf{p}}\mathbf{S} = \begin{bmatrix} 1 & 0 \\ 0 & 3 \end{bmatrix}\begin{bmatrix} 3 & 4 \\ 1 & 2 \end{bmatrix} = \begin{bmatrix} 3 & 4 \\ 3 & 6 \end{bmatrix} \quad \mathbf{f} = \hat{\mathbf{p}}\mathbf{d} = \begin{bmatrix} 1 & 0 \\ 0 & 3 \end{bmatrix}\begin{bmatrix} 2 \\ 2 \end{bmatrix} = \begin{bmatrix} 2 \\ 6 \end{bmatrix}. \tag{9.67}$$

Then, we compute the vector of total output $\mathbf{x} = \mathbf{Ze} + \mathbf{f}$ in monetary terms, the vector of total outlays $\mathbf{y}_I = \mathbf{Z}'\mathbf{e}$ and the vector of value added $\boldsymbol{\upsilon} = \mathbf{x} - \mathbf{y}_I$,

$$\mathbf{x} = \begin{bmatrix} 3 & 4 \\ 3 & 6 \end{bmatrix}\begin{bmatrix} 1 \\ 1 \end{bmatrix} + \begin{bmatrix} 2 \\ 6 \end{bmatrix} = \begin{bmatrix} 9 \\ 15 \end{bmatrix},$$

$$\mathbf{y}_I = \mathbf{Z}'\mathbf{e} = \begin{bmatrix} 3 & 3 \\ 4 & 6 \end{bmatrix}\begin{bmatrix} 1 \\ 1 \end{bmatrix} = \begin{bmatrix} 6 \\ 10 \end{bmatrix}, \quad \boldsymbol{\upsilon} = \mathbf{x} - \mathbf{y}_I = \begin{bmatrix} 9 \\ 15 \end{bmatrix} - \begin{bmatrix} 6 \\ 10 \end{bmatrix} = \begin{bmatrix} 3 \\ 5 \end{bmatrix}. \tag{9.68}$$

Finally, we determine the Perron–Frobenius right eigenvector of matrix $\mathbf{Z}$.

$$\mathbf{Zs}_1 = \lambda_F \mathbf{s}_1 \Rightarrow \mathbf{s}_1 = \begin{bmatrix} 0.7583 \\ 1 \end{bmatrix} \tag{9.69}$$

Then we verify that the above obtained productiveness R and the FP–eigenvalue λ_C can further be calculated (9.44) using the right eigenvector $\mathbf{s}_1$ of matrix $\mathbf{Z}$, the vector of value added $\boldsymbol{\upsilon}$ and the vector of total purchase $\mathbf{y}_I$,

$$R_a = \frac{\mathbf{s}_1' \cdot \boldsymbol{\upsilon}}{\mathbf{s}_1' \cdot \mathbf{y}_I} = \frac{[0.7783, 1][3, 5]'}{[0.7783, 1][6, 10]'} = \frac{1}{2} = R, \quad \lambda_C = \frac{1}{1+R} = \frac{2}{3}. \quad \blacktriangle \tag{9.70}$$

This example illustrates that the FP–eigenvalue $\lambda_C := \lambda_C(\mathbf{C})$ and the productiveness R of the present economy are determined either by the matrix $\mathbf{Z}$ in monetary terms or the matrix $\mathbf{S}$ of the Sraffa price model.

10 Exploration of Input-Output Tables

10.1 The Swiss Input-Output Tables and the productiveness

This chapter examines the structure of the symmetric Swiss Input-Output Tables (IOT) 2008 and 2014. An IOT consists of three rectangular tables: *the homogenous branches*, the *final use* and the part of the *outlay* including the *value added* (see Table 10.1[1]).

Table 10.1: Structure of the Swiss IOTs 2008 and 2014 at basic prices.

Input homogenous branches								**Final use**					
products	**homogenous branches (CPA)**						**total input**	**consumption**	**government**	**investment**	**export**	**tot. final use**	**total output**
(CPA)	S_1	S_2	...	S_j	...	S_n							
S_1	z_{11}	z_{12}	...	z_{1j}	...	z_{1n}	$\sum z_{1j}$	c_1	g_1	i_1	e_1	f_1	x_1
S_2	z_{21}	z_{22}	...	z_{2j}	...	z_{2n}	$\sum z_{2j}$	c_2	g_2	i_2	e_2	f_2	x_2
⋮	⋮	⋮	...	⋮	...	⋮	⋮	⋮	⋮	⋮	⋮	⋮	⋮
S_i	z_{i1}	z_{i2}	...	z_{ij}	...	z_{in}	$\sum z_{ij}$	c_i	g_i	i_i	e_i	f_i	x_i
⋮	⋮	⋮	...	⋮	...	⋮	⋮	⋮	⋮	⋮	⋮	⋮	⋮
S_n	z_{n1}	z_{n2}	...	z_{nj}	...	z_{nn}	$\sum z_{nj}$	c_n	g_n	i_n	e_n	f_n	x_n
TOT	$\sum z_{i1}$	$\sum z_{i2}$		$\sum z_{ij}$	...	$\sum z_{in}$	K	C	G	I	E	F	X
NCT	t_1	t_2	...	t_j	...	t_n	t_Z	t_C	t_G	t_I	t_E	t_F	t_{CT}
TIC	z_1	z_2	...	z_j	...	z_n	K_t	C_t	G_t	I_t	E_t	F_t	X_t
VAP	v_1	v_2	...	v_j	...	v_n	V_Z						
OBP	o_1	o_2	...	o_j	...	o_n	O_Z						
IMP	m_1	m_2	...	m_j	...	m_n	M						
OBP	x_1	x_2	...	x_j	...	x_n	X						

The symmetric IOT are established on the basis of a supply and a use table. *"A supply table shows the supply of goods and services by product and by type of supplier, distinguishing supply by domestic industries and imports from those of other countries"*, according to the Eurostat Manual [72], p. 18, and further: *"The use table is a product by industry based table with products and components of value added in the rows and industries, categories of final uses and imports in the columns. A use table shows the use*

1 CPA = Classification of Products by Activities.

https://doi.org/10.1515/9783110635096-010

of goods and services by product and by type of use, i. e., as intermediate consumption by industry, final consumption, gross capital formation or exports." The Swiss IOT 2014 comprises $n = 49$ branches or sectors, whose exact designations are given in the third columns of Figure 10.1, Figure 10.2 and Figure 10.4 (see Nathani [68] and Carsten [17]). The first columns contain the *total input* of the sectors, and the second columns contain the *sector numbers* and in brackets the *sector codes*. The sector number results from an aggregation of more detailing sector codes, numbered from 1 to 95, appearing in world Input-Output Tables.

13'618	1(01)	Products of agriculture
1'031	2(02)	Products of forestry
46	3(05)	Products of fishing
6'542	4(10-14)	Products of mining and quarrying
13'527	5(15-16)	Food products, beverages and tobacco products
2'876	6(17)	Textiles
891	7(18)	Wearing apparel, furs
531	8(19)	Leather and leather products
10'010	9(20)	Wood and products of wood and cork (except furniture); articles of straw and plaiting materials
5'082	10(21)	Pulp, paper and paper products
9'909	11(22)	Printed matter and recorded media
44'069	12(23-24)	Coke, refined petroleum products and nuclear fuel; chemicals and chemical products
9'050	13(25)	Rubber and plastic products
7'178	14(26)	Other non-metallic mineral products
12'812	15(27)	Basic metals
21'837	16(28)	Fabricated metal products, except machinery and equipment
6'821	17(29)	Machinery and equipment n.e.c.
12'545	18(30-31)	Office machinery, computers and electrical machinery n.e.c.
7'027	19(32)	Radio, television and communication equipment and apparatus
10'467	20(33)	Medical, precision and optical instruments, watches and clocks
2'197	21(34)	Motor vehicles, trailers and semi-trailers
2'352	22(35)	Other transport equipment

Figure 10.1: Designations of the sectors of the SWISS IOT 2008.

4'747	23(36)	Furniture; other manufactured goods n.e.c.
2'007	24(37)	Secondary raw materials
26'178	25(40-41)	Electrical energy, gas, steam, hot water; collected and purified water and distribution services of water
15'495	26(45)	Construction work
6'992	27(50)	Trade, maintenance and repair services of motor vehicles and motorcycles; retail sale of automotive fuel
29'710	28(51-52)	Wholesale trade and commission trade services, except of motor vehicles and motorcycles, Retail trade
3'730	29(55)	Hotel and restaurant services
16'962	30(60-62)	Transport services
16'789	31(63)	Supporting and auxiliary transport services; travel agency services
16'807	32(64)	Post and telecommunication services
42'515	33(65)	Financial intermediation services, except insurance and pension funding services
18'168	34(66)	Insurance and pension funding services, except compulsory social security services
7'001	35(70, 96-97)	Real estate services (incl. private households)
62'467	36(71, 74)	Renting of machinery and equipment without operator and of personal and household goods; other business
8'637	37(72)	Computer and related services
9'462	38(73)	Research and development services
5'591	39(75)	Public administration and defence services; compulsory social security services
1'483	40(80)	Education services
1'010	41(85)	Health and social work services
3'464	42(90)	Sewage and refuse disposal services, sanitation and similar services
9'888	43(91-92)	Membership organisation services n.e.c.; recreational, cultural and sporting services
1'266	44(93-95)	Other services; private households with employed persons
510'789		**Total Production value**

Figure 10.2: Designations of the sectors of the SWISS IOT 2008.

In this section, the Swiss Input-Output Tables 2008 [68] and 2014 are explored. Their productiveness is computed. The variables in use are explained[2]:

z_{ij}: quantity of commodity i (Input), required for the production of commodity j (Output), expressed in *monetary terms* at *basic prices*, also called intermediate transactions from sector i to sector j;
c_i: final consumption of sector i;
g_i: final government expenditure of sector i;
i_i: final investment of sector i;
e_i: final export of sector i;
f_i: final exogenous demand for sales of sector i equal to the total final use at *basic prices* (TFUBP) per sector $f_i = c_i + g_i + i_i + e_i$;
x_i: the total demand of commodity i by sector S_i (total Output) *necessarily equal* to the total production $x_i = \sum_{j=1}^{n} z_{ij} + f_j$;
$\sum z_{ij}$: total production of the sector $j \in \{1, \dots, n\}$, sum of z_{ij} over index i;
t_j: net commodity taxes of sector j;
z_j: total intermediate consumption of sector j;
v_j: total value-added expenditures of sector j;
o_j: output at basic prices of each sector j: $o_j = z_j + v_j$;
m_j: import of sector j;
x_j: total expenditure of sector j (total Input): $x_j = m_j + o_j$;
TOT: total interindustrial production at basic prices;
NCT: net commodity taxes;
TIC: total intermediate consumption/final use at purchasers' prices.

Abbreviations used in the lower outlay part[3] of the Swiss IOTs, Table 10.1:
VAP: value added at basic prices;
OBP: output at basic prices;
IMP: imports cif (cost, insurance and freight);
OBP: outlay at *basic prices* as total use (output) at basic prices.

Economic variables in use in the Swiss IOTs 2008 and 2014, Table 10.1:
K: total operating capital, $K = \sum_{i=1}^{n}(\sum_{j=1}^{n} z_{ij})$;
C: total private consumption;
G: total government purchases or expenditures;
I: total private investments;
E: total exports;

2 The term "basic price" is in German "Herstellungspreis". The term "purchasers' price" is "basic price plus transport and trade margin".

3 In the SIOTs 2008 and 2014, the term "supply at basic prices" is used instead of "outlay at basic prices", as it has been introduced in Chapter 2, Table 2.1.

F: total final demand (use), $F = C + G + I + E$;
X: total *output of the production of the economy*, $X = \sum_{i=1}^{n} x_i$;
t_Z: total net commodity taxes of all sectors t_j, that is $t_Z = \sum_{j=1}^{n} t_j$;
t_C: net commodity taxes on total consumption;
t_G: net commodity taxes on total government expenditure;
t_I: net commodity taxes on total investment;
t_E: net commodity taxes on total export;
t_F: net commodity taxes on total final demand, $t_F = t_C + t_G + t_I + t_E$;
t_{CT}: total net commodity taxes on total output, $t_{CT} = t_F + t_Z$;
K_t: total of intermediate consumption, $K_t = \sum_{j=1}^{n} z_j = K + t_Z$;
C_t: total consumption with additionally net commodity taxes;
G_t: total government expenditure with additionally net commodity taxes;
I_t: total investment with additionally net commodity taxes;
E_t: total export with additionally net commodity taxes;
F_t: total final demand with additionally net commodity taxes, $F_t = C_t + G_t + I_t + E_t = F + t_F$;
X: sum of total outlay (output) x_j at *basic prices*, $X = \sum_{j=1}^{n} x_j = M + O_Z$;
V_Z: sum of total value added v_j *at basic prices*, $v_Z = \sum_{j=1}^{n} v_j$;
O_Z: sum of total output o_j *at basic prices*, $O_Z = \sum_{j=1}^{n} o_j = V_Z + K_t$;
M: sum of imports m_j, $M = \sum_{j=1}^{n} m_j$;
X_t: total *output* (outlay) of production, comprising total net commodity taxes, $X_t = X + t_{CT} = K_t + F_t$.

The irreducibility of the *flow commodity matrix* $\mathbf{Z}$ (or a submatrix of it) contained in the explored annual Swiss or German IOTs gives the possibility to calculate the productiveness R of the corresponding economies.[4] As mentioned, the *productiveness* R is an *economic indicator* of the production power of a national economy.

The *national accounting equations* $Y = C + I + G + (E - M)$ together with the equations for the total final demand (the total final use) $F = C + I + G + E$, for the circulating capital $K = \mathbf{e}'\mathbf{x}_I$ and for the total output (the total use) $X = K + F$ can be obtained by summation of the adequate columns of the IOTs. Thanks to a recent revision, the value of the economic variable Y, obtained from the Swiss IOT 2014, is equal to the value of the Swiss GDP, obtained from the National Accounts and published by the Official Federal Institutions (© SOFS 2017). The next two examples give illustrations.

4 One needs from here on the complete Swiss IOT 2008 and Swiss IOT 2014 (calculation level of May 2018) from the *Federal Statistical Office, Switzerland*, containing the supply, the use-table and the symmetric IOT (siot), https://www.bfs.admin.ch/bfs/en/home/statistics/. Then we analyse the German IOT 2013, see: Statistisches Bundesamt Destatis, – Fachserie 18 Reihe 2 -Volkswirtschaftliche Gesamtrechnungen des Bundes – Input-Output-Rechnung 2013 (Revision 2014, Stand: August 2017). The calculations of the three following examples have been realised with MATLAB.

The *semi-positive* (44×44) Swiss IOT 2008 $\mathbf{Z}$ contains $44 \cdot 44 = 1{,}936$ entries, graphically presented in Figure 9.2. Hundreds of the entries are zero. Thus, one could expect that matrix $\mathbf{Z}$ is *reducible*. Calculations will show that this is not the case. Matrix $\mathbf{Z}$ is *irreducible*.

Example 10.1.1. Consider the *total intermediate consumption/final use at producers's prices (incl. net commodity taxes)* of the Swiss IOT 2008. Compute the vector of output $\mathbf{x}_I = \mathbf{Ze}$, the circulating capital $K = \mathbf{e}'\mathbf{x}_I$, the vector of final use $\mathbf{f} = \mathbf{c} + \mathbf{i} + \mathbf{g} + \mathbf{e}_x$ (the vector designations are chosen in analogy to C: household consumption, G: government expenditure [consumption of government], I: total gross capital formation [investments], E: exports) and the vector of total output $\mathbf{x} = \mathbf{x}_I + \mathbf{f}$. Compute the input-output coefficients matrix $\mathbf{A} = \mathbf{Z}\hat{\mathbf{x}}^{-1}$ with regard to the vector of total output $\mathbf{x}$. Compute its Frobenius numbers λ_Z and λ_A, as well as the productiveness $R = (1/\lambda_A) - 1$ and the ratio of *national income to circulating capital* $\tilde{R} = Y/K$. Verify that the Swiss IOT 2008 $\mathbf{Z} \geq \mathbf{O}$ is irreducible.

Solution to Example 10.1.1:
One establishes the positive vector of interindustrial output $\mathbf{x}_I = \mathbf{Ze} > \mathbf{o}$, the vector of *final use at basic prices* $\mathbf{f} > \mathbf{o}$, which is an aggregation of the columns of the *final use table* the Swiss IOT 2008. Sub-aggregations of the *total final use at basic prices* f_i are *consumption* c_i, *government expenditure* g_i, *investment* i_i and *export* e_i (see Table 10.1). Then, we compute the vector of total output (= vector of total use at basic prices) $\mathbf{x} = \mathbf{x}_I + \mathbf{f} > \mathbf{o}$ and the input-output coefficients matrix $\mathbf{A} = \mathbf{Z}\hat{\mathbf{x}}^{-1}$ with regard to the vector of *total output* $\mathbf{x}$.

Matrix $\mathbf{A}$ shows to be primitive: Indeed, one gets $\mathbf{A}^2 > \mathbf{O}$, Lemma A.9.1. Lemma A.8.2, also applies, $(\mathbf{A} + \mathbf{I})^{43} > \mathbf{O}$.[5] For these reasons, $\mathbf{Z}$ and $\mathbf{A}$ are *irreducible*. This means that the Perron–Frobenius theorem applies and the Frobenius numbers are positive.

We obtain the Frobenius numbers $\lambda_Z = 26{,}425 > 0$, necessarily positive, and furthermore, $0 < \lambda_A := \lambda_F(\mathbf{A}) = 0.47779 < 1$ within the interval $]0,1[$, according to Lemma 4.1.1 (b).[6] Consequently, the Leontief Inverse $(\mathbf{I} - \mathbf{A})^{-1}$ exists because all the eigenvalues λ are smaller or equal to λ_A, $\lambda \leq \lambda_A < 1$. We are in presence of a productive Leontief model. We get the equality

$$\mathbf{x} = (\mathbf{I} - \mathbf{A})^{-1}\mathbf{f}. \tag{10.1}$$

We compute the *total final use at basic prices* $F = \mathbf{e}'\mathbf{f}$ (2.1), obtaining exactly the table value of the Swiss IOT 2008 $F = C + I + G + E = 789{,}676$. We also compute the *total*

5 There is a number of $n = 44$ sectors in the SWISS IOT; the exponent is $(n - 1)$.

6 Choose as fictive positive price vector, the summation vector $\mathbf{p} = \mathbf{e} > \mathbf{0}$ to formally shift to the Sraffa system in physical terms, obtaining with (2.18), the *commodity flow* matrix $\mathbf{S} = \mathbf{Z} > \mathbf{0}$, the quantity output vector $\mathbf{q} = \mathbf{x}$ and the *input-output coefficients* matrix $\mathbf{C} = \mathbf{S}\hat{\mathbf{q}}^{-1}$ and other necessary vectors. Then apply Lemma 4.1.1.

use at basic prices, obtaining exactly the table value $X = F + K = 1{,}308{,}588$. Then we compute the *productiveness* $R = (1/\lambda_A) - 1 = 1.093$ of the input-output coefficients matrix **A**. We compute from various columns of the Swiss IOT 2008 (all terms in *million* CHF) the *circulation capital* $K = \mathbf{e}'\mathbf{x}_I = 518{,}913$ and the imports $M = 245{,}480$. In the following items, the net commodity taxes are included: the *total final consumption by households* $C = 298{,}029$, the overall *government expenditure* $G = 69{,}247$, the *total investments* $I = 114{,}945$, the *exports* $E = 307{,}454$, the *circulating capital* $K = 518{,}913$. We then compute the *Gross domestic product (GDP)*[7] $Y = C + G + I + (E - M) = 544{,}196$. We compute the ratio of *national income* to *circulating capital* $\tilde{R} = Y/K = 544{,}196/518{,}913 = 1.049 < R = 1.093$, which is slightly smaller than the productiveness. ▲

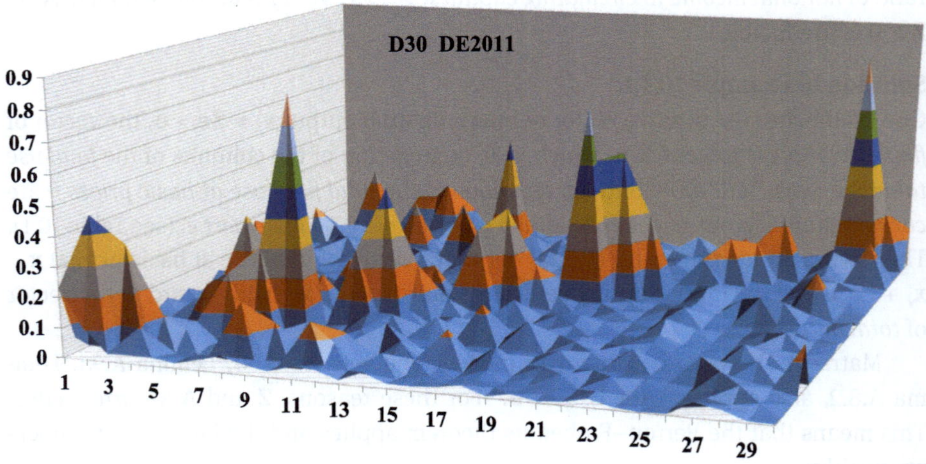

Figure 10.3: The sorted matrix **D** with the 30 largest sectors (Germany IOT 2011).

In analogy to Example 10.1.1, analyse the Swiss IOT 2014.

Example 10.1.2. Consider the *total intermediate consumption/final use at producer's prices (incl. net commodity taxes)* of the *semi-positive* symmetric (49×49) Swiss IOT 2014 table at basic prices **Z** (© SFSO 2017). The sector No. 49, called *Activities of households as employers of domestic personnel* is skiped, because it is a null vector. There remain therefore $48 \cdot 48 = 2{,}304$ entries. One observes that hundreds of the entries are zero. Verify that the obtained (48×48) submatrix which we also denote by **Z** is irreducible.

7 The GDP (noted by the Federal Statistical Office (CH) as "Bruttoinlandprodukt") is designed in our text as Y, also called in PCMC *national income*, a notation we are using.

Compute the vector of *total output* (= vector of total use at basic prices) $\mathbf{x} = \mathbf{Ze} + \mathbf{f} > \mathbf{o}$, the matrix $\mathbf{A} = \mathbf{Z}\hat{\mathbf{x}}^{-1}$ with regard to the vector of *total output* $\mathbf{x}$, the Frobenius numbers λ_Z and λ_A and the productiveness $R = (1/\lambda_A)-1$. Compute the GDP Y, the *total final consumption of households*, the *government expenditures* G, the *total investment* I, the *export* E, the *total output (total use at basic prices)* $X = \mathbf{e}'\mathbf{x}$, and the ratio of *national income* to *circulating capital* $\tilde{R} = Y/K$.

Table 10.2 represents the values of *total outputs*, called the *total use at basic prices* on the left edge of the Swiss IOT 2014.

Table 10.2: *Total use at basic prices* in the Swiss IOT 2014 (in Mio. CHF).

No.	x_i	No.	x_i	No.	x_i	No.	x_i
1	16,049.6	13	24,506.9	25	12,842.2	37	40,658.7
2	7,427.6	14	81,863.3	26	124,674.8	38	74,849.5
3	47,910.4	15	29,369.3	27	28,348.3	39	81,090.3
4	14,624.8	16	44,243.8	28	57,513.1	40	30,724.6
5	10,363.3	17	15,589.0	29	26,969.1	41	8,787.9
6	5,933.1	18	8,768.1	30	6,762.4	42	48,331.4
7	3,784.0	19	6,178.6	31	9,380.7	43	42,303.7
8	51,761.5	20	22,492.7	32	18,113.7	44	35,859.8
9	106,188.0	21	4,834.5	33	11,303.2	45	55,383.9
10	13,571.9	22	42,845.9	34	17,698.8	46	25,802.2
11	9,787.8	23	9,596.5	35	39,638.1	47	13,784.9
12	22,484.8	24	78,730.8	36	66,793.4	48	15,199.7
						49	2,062.6

Solution to Example 10.1.2:
One obtains the input-output coefficients matrix $\mathbf{A} = \mathbf{Z}\hat{\mathbf{x}}^{-1}$ with regard to the vector of *total output* $\mathbf{x}$. Matrix $\mathbf{A}$ is again primitive, $\mathbf{A}^2 > \mathbf{O}$, with Lemma A.9.1. For this reason, $\mathbf{Z}$ and $\mathbf{A}$ are *irreducible*. The Perron–Frobenius theorems applies, and the Frobenius numbers are positive.

We calculate the Frobenius numbers, obtaining $\lambda_Z = 34{,}452 > 0$, necessarily positive and $0 < \lambda_A := \lambda_F(\mathbf{A}) = 0.4647 < 1$, in the interval $]0, 1[$, according to Lemma 4.1.1 (b). The justification is as in footnote 6. Consequently, the Leontief Inverse $(\mathbf{I} - \mathbf{A})^{-1}$ exists because all the eigenvalues λ are smaller or equal than λ_A, $\lambda \leq \lambda_A < 1$, and we are in presence of a productive Leontief model. Then we compute from the Swiss IOT 2014 the following *items*, where the net commodity taxes are included (all terms in *million* CHF): we compute from various columns of the Swiss IOT 2014 the *total final consumption by households* $C = 345{,}035$, the *government expenditures* $G = 77{,}777$, the *total investments* $I = 158{,}682$, the *exports* $E = 351{,}212$, the *total final use at basic prices* $F = C+I+G+E = \mathbf{e}'\mathbf{f} = 932{,}706$, which in this case is the exact table value of the Swiss IOT. Then we compute the *circulating capital* $K = 662{,}275$, the *imports* $M = 282{,}987$

and the *total output* (= total use at basic prices) $X = F + K = 1{,}594{,}980$, which is also the exact table value in the Swiss IOT 2014. We then obtain the *Gross Domestic Product* (GDP) $Y = C + G + I + (E - M) = F - M = 649{,}718$, the official value of Swiss GDP 2014 published by the Swiss Federal Statistical Office (© SFSO 2019).

Finally, we compute the *productiveness* $R = (1/\lambda_A) - 1 = 1.1518$ of matrix **A**, and the ratio of *national income* to *circulating capital* $\tilde{R} = Y/K = 649{,}718/662{,}275 = 0.9810$, smaller than the productiveness $R = 1.1518$. ▲

Example 10.1.3. Consider the *semi-positive* (71×71) German IOT 2013 (revision 2014: state August 2017) **Z** without sector 72: activities of households as employers of domestic personnel.[8] We have therefore $71 \cdot 71 = 5{,}041$ entries. Hundreds of the entries are zero. Verify that this submatrix is reducible.

Consider the vector **x** of *total output (use)*[9] *at basic prices*. Compute the input-output coefficients (71×71) matrix $\mathbf{A} = \mathbf{Z}\hat{\mathbf{x}}^{-1}$ with regard to the vector **x**, then the Frobenius numbers λ_Z and λ_A. Compute the GDP 2013, the ratio of *national income* to *circulating capital* $\tilde{R} = Y/K$. Compute the productiveness $R = (1/\lambda_A) - 1$ and compare it with the ratio of *national income* to *circulating capital* $\tilde{R} = Y/K$.

Solution to Example 10.1.3:
We compute the input-output coefficients matrix $\mathbf{A} = \mathbf{Z}\hat{\mathbf{x}}^{-1}$ of the German IOT 2013 with regard to the vector of total use **x**. Because matrix $(\mathbf{A}+\mathbf{I})^{70} \geq 0$ is *semi-positive*, matrices **Z** and **A** are *reducible* according to Lemma A.8.2. This means that Theorem A.10.1 applies, and therefore the Frobenius numbers are non-negative. We consequently calculate the Frobenius numbers, obtaining $\lambda_F(\mathbf{Z}) = 119{,}596 > 0$, necessarily positive, and $0 < \lambda_A := \lambda_F(\mathbf{A}) = 0.596695 < 1$, in the interval $]0,1[$. The justification is as in footnote 6.

We compute the *productiveness* $R = (1/\lambda_A) - 1 = 0.6771$ of the input-output coefficients matrix **A**. In the following items of *intermediate consumption* the net commodity taxes are included[10] (all terms in *million EURO*): We compute from the various columns of the German IOT 2013 the *total final consumption by households* $C = 1{,}472{,}436$, the *government expenditure* $G = 593{,}728$, the *total investments* $I = 551{,}462$, the *exports* $E = 1{,}257{,}691$ and the *circulating capital* $K = 2{,}831{,}297$. We then compute the *Gross Domestic Product* (GDP), giving $Y = C + G + I + (E - M) = 2{,}826{,}240$, the official number of the German GDP 2013 published by the Federal Statistical Office of Germany in 2018. We compute the ratio of *national income* to *circulating capital*. $\tilde{R} = Y/K = 2{,}826{,}240/2{,}831{,}297 = 0.9982 > R = 0.6771$. ▲

---

8 The German designation is: Input-Output-Tabelle 2013 zu Herstellungspreisen – Inländische Produktion und Importe, in Mill. EURO, die $n = 72$ laufend numerierte Sektoren umfasst.

9 In German, the term *total use of products* is translated as: *Gesamte Verwendung von Gütern.*

10 The German notion for this items is: "Vorleistungen der Produktionsbereiche zu Anschaffungspreisen", Anschaffungspreisen = *purchaser's prices* in English.

Recapitulation. The productiveness $R = (1/\lambda_A) - 1$ is calculated for the analysed symmetric IOTs of Switzerland and of Germany. The Frobenius number λ_A of the *input-output coefficients* matrix **A** is in the range $]0,1[$, $0 < \lambda_A < 1$. The productiveness R is a measure of the production power of an economy.

We observe that R is near the *national income to circulating capital ratio* $\tilde{R}$, $R \sim \tilde{R}$. We remember that for *Standard systems* there is equality, $R = \tilde{R}$.

10.2 Analysis of an aggregated SWISS IOT 2014

In this section, we aggregate the $n = 49$ branches of the Swiss IOT 2014 to six group of branches according to following scheme:
S_1: Food products, grouping initial sectors: $\{1, 2, 3\}$
S_2: Textiles, grouping initial sectors: $\{4, 5, 6, 7\}$
S_3: Machinery, grouping initial sectors: $\{8, \dots, 23\}$
S_4: Construction work, grouping initial sectors: $\{24\}$
S_5: Services, grouping initial sectors: $\{25, \dots, 39\}$
S_6: Social and education, grouping initial sectors: $\{38, \dots, 49\}$

The official IOTs of countries are too large to be presented in a book in their totalities. For this reason, one uses extractions or aggregations of them.

We start with a reduced (6×6) *commodity flow matrix* **Z** obtained by aggregation of branches according to the given scheme of the Swiss IOT 2014 and the corresponding (1×6) vector of *final demand* **f**, obtaining:

$$\mathbf{Z} = \begin{bmatrix} 18{,}121 & 332 & 6{,}818 & 667 & 6{,}814 & 3{,}823 \\ 681 & 5{,}959 & 4{,}761 & 4{,}127 & 3{,}871 & 1{,}859 \\ 4{,}419 & 2{,}295 & 149{,}184 & 15{,}777 & 20{,}690 & 22{,}166 \\ 221 & 84 & 2{,}291 & 10{,}673 & 8{,}080 & 2{,}482 \\ 6{,}904 & 2{,}187 & 47{,}195 & 10{,}452 & 187{,}030 & 34{,}960 \\ 1{,}682 & 373 & 9{,}273 & 2{,}307 & 25{,}168 & 28{,}710 \end{bmatrix}, \quad \mathbf{f} = \begin{bmatrix} 34{,}812 \\ 13{,}446 \\ 279{,}350 \\ 54{,}900 \\ 327{,}909 \\ 210{,}726 \end{bmatrix}. \tag{10.2}$$

There are, in this reduced IOT 2014, Table 10.3, also aggregated values, that have not been modified with regard to the initial IOT 2014. One verifies that following variables are unchanged: *total final consumption by households* $C = 345{,}035$, the *government expenditure* $G = 77{,}777$, the *total investments* $I = 158{,}682$, the *exports* $E = 351{,}212$ and the *circulating capital* $K = 662{,}275$. We therefore can again compute the *Gross Domestic Product (GDP)* $Y = C + G + I + (E - M) = 649{,}718$, getting the value published by the Swiss Federal Statistical Office in 2018. Given these data, we formulate the following

Example 10.2.1 (Frobenius numbers and eigenvectors in relation to the reduced Swiss IOT 2014, of six groups of sectors). Compute the vector of interindustrial production

Table 10.3: Reduced SWISS IOT 2014 with 6 groups of the initial 49 industrial sectors, see Section 10.2.

Symmetric Input-Output Table								Final use					
selling sectors	**Food products**	**Textile products**	**Machinery**	**Construction work**	**Services**	**Social & Education**	**total input**	**Consumption**	**Government expenditure**	**Investment**	**Exports**	**total demand**	**total output**
Prod.	S_1	S_2	S_3	S_4	S_5	S_6		C	G	I	E	F	
S_1	18,121	332	6,818	667	6,814	3,823	36,576	24,484	0	953	9,375	34,812	71,388
S_2	681	5,959	4,761	4,127	3,871	1,859	21,259	7,993	0	125	5,328	13,446	34,705
S_3	4,419	2,295	149,184	15,777	20,690	22,166	214,532	36,189	1,065	47,874	194,221	279,350	493,883
S_4	221	84	2,291	10,673	8,080	2,482	23,831	3,081	0	50,253	1,566	54,900	78,731
S_5	6,904	2,187	47,195	10,452	187,030	34,960	288,727	173,493	102	28,519	125,795	327,909	616,636
S_6	1,682	373	9,273	2,307	25,168	28,710	67,514	93,698	76,357	26,756	13,916	210,726	278,241
TOT	32,029	11,232	219,522	44,002	251,653	94,001	652,440	338,938	77,524	154,480	350,202	921,143	1,573,583
NCT	794	50	645	256	4,310	3,780	9,835	6,097	253	4,202	1,010	11,562	21,397
TIC	32,823	11,282	220',167	44,259	255,962	97,782	662,275	345,035	77,777	158,682	351,212	932,706	1,594,980
VAP	14,537	7,182	110,178	33,901	300,921	161,602	628,321						
OBP	47,360	18,464	330,345	78,160	556,883	259,384	1,290,596						
IMP	24,028	16,241	163,537	571	59,753	18,857	282,987						
SBP	71,388	34,705	493,883	78,731	616,636	278,241	1,573,583						

11'237	1	1-3	Agriculture, forestry and fishing
7'149	2	3-5	Mining and quarrying
18'191	3	9-10	Manufacture of food and tobacco products
3'965	4	13 - 15	Manufacture of textiles and apparel
9'604	5	16	Manufacture of wood and of products of wood and cork, except furniture
4'091	6	17	Manufacture of paper and paper products
3'599	7	18	Printing and reproduction of recorded media
23'046	8	19 - 20	Manufacture of coke, chemicals and chemical products
44'786	9	21	Manufacture of basic pharmaceutical products and Pharmaceutical preparations
7'494	10	22	Manufacture of rubber and plastic products
8'322	11	23	Manufacture of other non-metallic mineral products
11'527	12	24	Manufacture of basic metals
18'047	13	25	Manufacture of fabricated metal products, except machinery And equipment
33'175	14	26	Manufacture of computer, electronic and optical products
12'286	15	27	Manufacture of electrical equipment
7'728	16	28	Manufacture of machinery and equipment n.e.c.
1'895	17	29	Manufacture of motor vehicles, trailers and semi-trailers
2'030	18	30	Manufacture of other transport equipment
650	19	31	Manufacture of furniture
2'042	20	32	Other manufacturing
4'144	21	33	Repair and installation of machinery and equipment
32'470	22	35	Electricity, gas, steam and air-conditioning supply
4'891	23	36 - 39	Water supply, waste management
23'831	24	41 - 43	Construction
5'829	25	45	Wholesale and retail trade and repair of motor vehicles and motorcycles
64'563	26	46	Wholesale trade, except of motor vehicles and motorcycles
3'829	27	47	Retail trade, except of motor vehicles and motorcycles
38'045	28	49 - 51	Land, water and air transport and transport via pipelines
21'767	29	52	Warehousing and support activities for transportation
4'687	30	53	Postal and Courier Aktivitas
1'318	31	55	Accommodation
1'996	32	56	Food and beverage service activities
7'040	33	58 - 60	Publishing, audiovisual and broadcasting activities
8'845	34	61	Telecommunications
13'488	35	62 - 63	IT and other information services
34'117	36	64	Financial service activities
8'652	37	65	Insurance
10'162	38	68	Real estate activities
64'390	39	69 - 71	Legal, accounting, management, architecture, engineering activities
1'567	40	72	Scientific Research and Developern
7'232	41	73 - 75	Other professional, scientific and technical activities
34'633	42	77 - 82	Administrative and support service activities
4'988	43	84	Public administration
6'945	44	85	Education
1'138	45	86	Human health activities
2'664	46	87 - 88	Residential care and social work activities
4'836	47	90 - 93	Arts, entertainment and recreation
3'511	48	94 - 96	Other service activities
0	49	97 - 98	Activities of households as employers of domestic personnel / Undifferentiated goods- and services-producing activities of private households for own use
652'440			TOTAL

Figure 10.4: Designations of the sectors of the SWISS IOT 2014.

$\mathbf{x}_I = \mathbf{Ze}$ and the *vector of total use at basic prices*, $\mathbf{x} = \mathbf{x}_I + \mathbf{f}$ (2.5). Compute the *input-output coefficients* matrix $\mathbf{A} = \mathbf{Z}\hat{\mathbf{x}}^{-1}$ (2.8) with regard to vector $\mathbf{x}$ and the *input-output coefficients* matrix of *interindustrial production* $\mathbf{A}_I = \mathbf{Z}\hat{\mathbf{x}}_I^{-1}$ with regard to vector $\mathbf{x}_I$.

Then compute the characteristic polynomials $P_6(\lambda) = \det(\mathbf{A} - \lambda I_6)$ and $Q_6(\lambda) = \det(\mathbf{A}_I - \lambda I_6)$. Compute the Frobenius numbers of matrices $\mathbf{Z}$, $\mathbf{A}$, $\mathbf{A}_I$. Interpret in an economic context the obtained results.

Solution of Example 10.2.1:
We compute the vector of interindustrial production of the reduced Swiss IOT (6×6) matrix: $\mathbf{x}_I = \mathbf{Ze} = [36{,}575; 21{,}258; 214{,}531; 23{,}831; 288{,}728; 67{,}513]'$ and the vector $\mathbf{x} = \mathbf{x}_I + \mathbf{f}$. Then we compute at first matrix

$$\mathbf{A} = \mathbf{Z}\hat{\mathbf{x}}^{-1} = \begin{bmatrix} 0.2538 & 0.0096 & 0.0138 & 0.0084 & 0.0111 & 0.0137 \\ 0.0095 & 0.1717 & 0.0096 & 0.0524 & 0.0063 & 0.0067 \\ 0.0619 & 0.0661 & 0.3021 & 0.2004 & 0.0336 & 0.0797 \\ 0.0031 & 0.0024 & 0.0046 & 0.1356 & 0.0131 & 0.0089 \\ 0.0967 & 0.0630 & 0.0956 & 0.1328 & 0.3033 & 0.1256 \\ 0.0237 & 0.0107 & 0.0188 & 0.0293 & 0.0408 & 0.1032 \end{bmatrix}, \tag{10.3}$$

followed by the matrix $\mathbf{A}_I$

$$\mathbf{A}_I = \mathbf{Z}\hat{\mathbf{x}}_I^{-1} = \begin{bmatrix} 0.4954 & 0.0156 & 0.0318 & 0.0280 & 0.0236 & 0.0566 \\ 0.0186 & 0.2803 & 0.0222 & 0.1732 & 0.0134 & 0.0275 \\ 0.1208 & 0.1080 & 0.6954 & 0.6620 & 0.0717 & 0.3283 \\ 0.0060 & 0.0040 & 0.0107 & 0.4479 & 0.0280 & 0.0368 \\ 0.1888 & 0.1029 & 0.2200 & 0.4386 & 0.6478 & 0.5178 \\ 0.0460 & 0.0175 & 0.0432 & 0.0968 & 0.0872 & 0.4253 \end{bmatrix}. \tag{10.4}$$

Then, there are the characteristic polynomials of both matrices $\mathbf{A}$ and $\mathbf{A}_I$:

$$\begin{aligned} P_6(\lambda) &= \det(\mathbf{A} - \lambda I_6) = \lambda^6 - 1.26967\lambda^5 + 0.63658\lambda^4 \\ &\quad - 0.16134\lambda^3 + 0.02195\lambda^2 - 0.00151\lambda + 0.00004 = (\lambda - 0.4175)P_5(\lambda), \\ Q_6(\lambda) &= \det(\mathbf{A}_I - \lambda I_6) = \lambda^6 - 2.9921\lambda^5 + 3.5576\lambda^4 - 2.1715\lambda^3 \\ &\quad + 0.7214\lambda^2 - 0.1241\lambda + 0.0087 = (\lambda - 1)Q_5(\lambda). \end{aligned} \tag{10.5}$$

The Frobenius numbers are: $\lambda_{Z_6} = 212{,}342$, $\lambda_A = \lambda_F(\mathbf{A}) = 0.4175$ and $\lambda_{A_I} = 1$; the *productiveness* $R_6 = (1/\lambda_A) - 1 = (1/0.4175) - 1 = 1.3952 > R_{48} = 1.1518$, slightly greater than the productiveness found in Example 10.1.2.

The economical meaning of this result is that *productiveness* is sensitive to the degree of aggregation of the sectors of an economy. Greater aggregation makes the economy to appear stronger.

We treat the following extreme case of aggregation:

Example 10.2.2. Aggregate the Swiss IOT 2014 to one single sector and compute the resulting matrices, Frobenius numbers and productiveness.

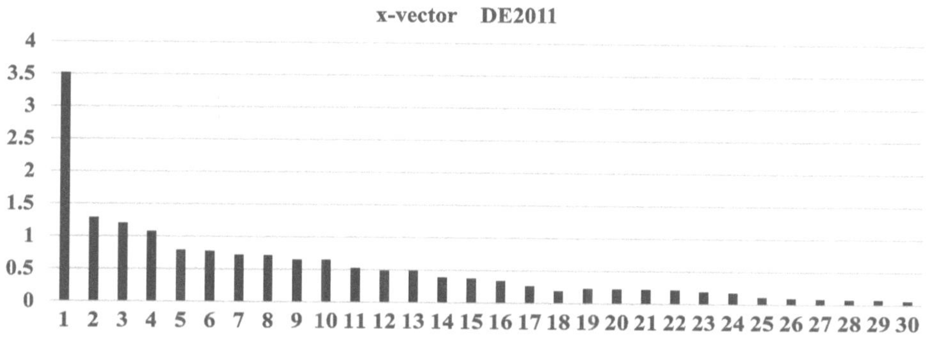

Figure 10.5: IOT 2011 of Germany—the row–sum of the sorted 30 largest sectors.

Solution of Example 10.2.2:
We get the (1×1) *commodity flow* matrix $\mathbf{Z}_1 = [662{,}275]$, the vectors $\mathbf{f}_1 = [932{,}705]$, $\mathbf{x}_{I1} = \mathbf{Z}_1\mathbf{e}_1 = [662{,}275]$, vector $\mathbf{x}_1 = \mathbf{Z}_1\mathbf{e}_1 + \mathbf{f}_1 = [1{,}594{,}980]$. We compute the *input-output coefficients matrix* $\mathbf{A}_1 = \mathbf{Z}_1\hat{\mathbf{x}}_1^{-1} = [0.415225]$. The Frobnius numbers are $\lambda_{Z1} = 662275 > \lambda_{Z_6} = 221{,}342$ and $\lambda_{A_1} = 0.415225$ and the productiveness $R_1 = (1/\lambda_{A_1}) - 1 =$, $(1/0.415225) - 1 = 1.40833 > R_6 > R_{48}$. ▲

Recapitulation. We recognise that, in the three Examples 10.1.1, 10.1.2, 10.1.3, the *commodity flow matrices* **Z** and **A** are *semi-positive,* and the Frobenius numbers are smaller than 1 and positive, $\lambda_A < 1$.

This means that for the German IOT 2013 and Swiss IOTs 2008, 2014 *productive Leontief models* exist, Theorem A.12.1. One verifies that the inverse Leontief matrices $(\mathbf{I} - \mathbf{A})^{-1}$ exist and that in the five cases for any positive vector of *total final use (demand)*, $\mathbf{f} > \mathbf{o}$; the positive vector of *total use (output)* $\mathbf{x} > \mathbf{o}$ is computed as follows:

$$\mathbf{x} = (\mathbf{I} - \mathbf{A})^{-1}\mathbf{f} > \mathbf{o}. \tag{10.6}$$

This statement is also valid for the two further developed aggregated Swiss IOTs of Examples 10.2.1, 10.2.2.

There are several figures that we present here. The 30 largest sectors of the sorted **D**-matrix (German IOT 2011) are presented in Figure 10.3. In Figure 10.5, we present the row–sums of the sorted 30 largest sectors of the German IOT 2011. Figure 10.7 shows the monthly operating capital K^* (in λ_F-units) for the USA, GBR, DE (Germany) and the Euro-Zone.

10.3 Exploration of interindustrial economies

In this section, we start exploring the *interindustrial market* of the economies. We concentrate on the $n \times n$ matrix **Z** and do not consider the surplus or the GDP. We will apply the algebraic structures and calculate the matrices, developed in Chapter 9, concerning the interindustrial market. This will be presented by the next computations and figures.

Table 2 USA	Euro_zone	DE	GRC	PRT	GBR
1.633	1.8545	1.715	1.396	1.6954	1.7022
1.5689	1.8113	1.6444	1.3559	1.566	1.6727
1.499	1.7869	1.616	1.4168	1.3838	1.6865
1.3802	1.7651	1.5881	1.519	1.281	1.6044
1.3146	1.7256	1.521	1.5797	1.297	1.5487
1.25257	1.7303	1.5352	1.696	1.2924	1.4887
1.439	1.6731	1.511	1.8474	1.269	1.4279
1.273	1.6474	1.47405	1.9198	1.1886	1.3625
1.22557	1.6435	1.42778	2.135	1.223	1.3276
1.17646	1.6392	1.4467	2.069	1.1428	1.2926
1.16949	1.5899	1.3883	2.099	1.1017	1.3492
1.12579	1.5513	1.3883	1.792	1.0598	1.3114
1.10208	1.5463	1.5303	1.9785	1.1766	1.2701
1.1722	1.542	1.4575	1.9546	1.1473	1.205
1.0399	1.4439	1.3008	1.792	1.2861	1.1676
1.08223	1.4807	1.3267	1.801	1.2862	1.133
1.1022	1.4891	1.3783	1.795	1.2792	1.0983

Figure 10.6: The monthly operating capital K^* in λ_F-units for the USA, the Euro-Zone, DE, GRC, PRT, GBR (1995–2011).

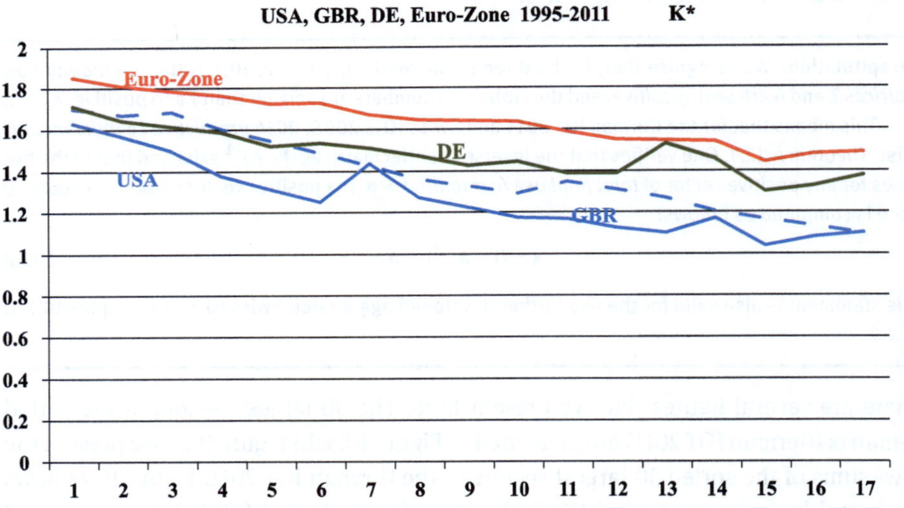

Figure 10.7: The monthly operating capital K^* in λ_F-units for the USA, GBR, DE (Germany) and the Euro-Zone (1995–2011).

10.3.1 The monthly operating capital K^* (λ_F-units): USA, GBR, DE, Euro-Zone 1995 to 2011

In this subsection, we use the data for the input-output matrix **Z** from the "World Input-Output Database" (see World Input-Output Database (WIOD), ([122], 2015)). More than 40 national and regional Input-Output Tables together with the import-export are available. The analysis is applied on the Input-Output Tables for the USA, GBR, DE (Germany), GRC, PRT and the world regional Input-Output Table for the EZ (Euro-Zone). This database has the advantage that one can extend the analysis to other nations or world regions.

Our exploration of the data (Figure 10.6) has yielded an interesting result: The λ_F-normalized circulating capital unveils the disorder in GRC, already in the year about 2004, where the Greek (GRC) peak appears. The beginning of the monetary problems of Greece is obviously in the years 1998 and 1999 (Figure 10.8 and 10.9). The GRC (Greece) performance differs remarkably from the USA, the Euro–Zone, DE or PRT over the years 1995–2011. Current economy/finance indicators have obviously failed to register. The calculations rely on the official available numbers.

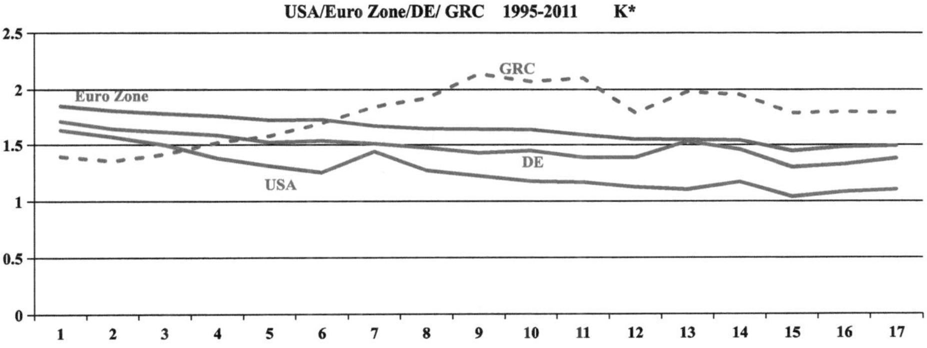

Figure 10.8: The monthly operating capital K^* in λ_F-units for the USA, DE (Germany), GRC and the Euro-Zone (1995–2011).

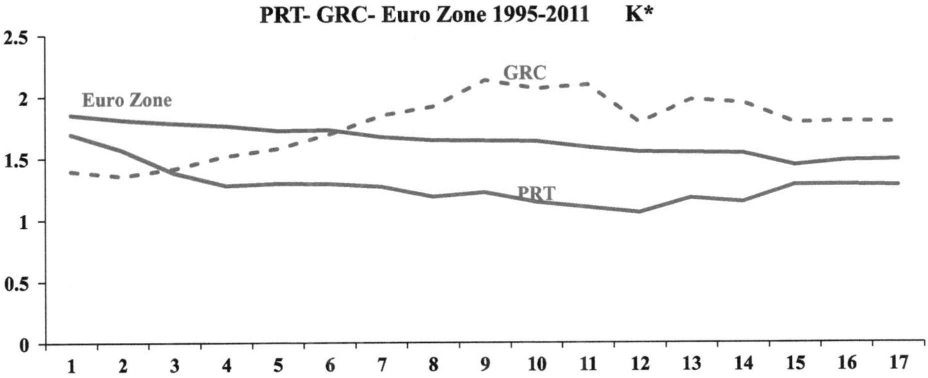

Figure 10.9: We also compare the Euro-Zone, Portugal and Greece! The monthly operating capital K^* in λ_F-units for the PRT, GRC and the Euro-Zone (1995–2011).

Considering the Frobenius number λ_F as an expression of the whole economic power of an economy, the division of the Input-Output *commodity flow matrix* **Z** by $\lambda_F > 0$ incorporates reduction, called the λ_F-normalisation. We expect that this normalisation makes the economies comparable through their normalised IOTs. We continue here this exploration.

Observation. The monthly operating capital K^* for DE and GBR lay intermediate to the Euro-Zone and USA, with a GBR tendency to the USA. It is remarkable that over the years 1995–2011, the monthly operating capital K^* remained between the normalised range 1–2.

10.3.2 The monthly operating capital K^* (λ_F-units) for Switzerland 1995–2011

For Switzerland (CH), only the Input-Output Tables for the years 2001, 2005, 2008, 2011 and 2014, the last one with $n = 49$ sectors, are available, see "Bundesamt für Statistik" ([12], Schweiz, 2016).

Figure 10.10 shows the operating capital K, in monetary terms, and the monthly operating capital of the interindustrial production for Switzerland (2001, 2005, 2008 and 2011). The Frobenius numbers for these years are used to normalise the monthly operating capital and to determine the corresponding monthly operating capital K^* in the λ_F-units.

K* CH 2001-2011

	K	K/12	F-number	K*
2001	363.6	30.02	16.46	1.8412
2005	415.8	34.66	21.95	1.388
2008	510.7	42.57	26.423	1.610
2011	621.9	51.86	24.86	2.084

Figure 10.10: The operating capital K without commodity taxes (billion CHF), the monthly operating capital K_m, the Frobenius number (in thousands) and the monthly operating capital K^* in λ_F-units (CH 2001–2011).

Observation. Figure 10.11 shows graphically the monthly operating capital K, in monetary terms, and the corresponding Frobenius number of the Swiss input-output matrices 2001–2011 (left). The monthly operating capital K^* over the years 2001–2011 is shown (right) in the λ_F-units. This normalised monthly operating capital remains, almost, within the range 1–2. In comparison with Figure 10.7 (USA, GBR, DE, the Euro-Zone), there is a dip ($K^* = 1.388$) in the year 2005 (CHF), see Figure 10.10. Otherwise, the

Swiss K^* remains higher than the value 1.6. This K^* performance of CHF differs from the K^* performances of the USA, GBR, DE (Germany) or the Euro-Zone over the years 2001–2011.

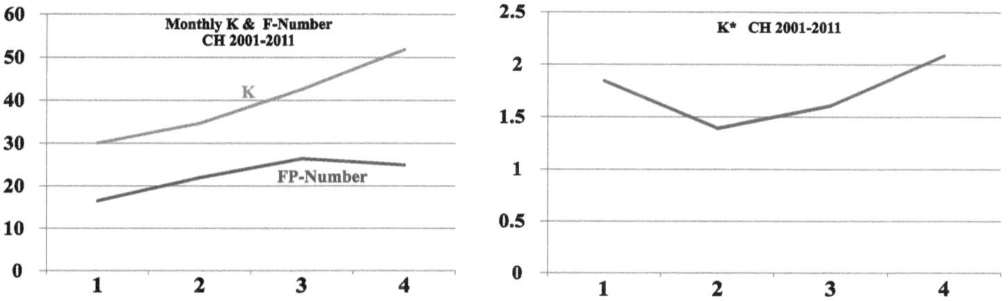

Figure 10.11: The monthly operating capital K without net commodity taxes (billion CHF) and the normalised operating capital K^* for Switzerland 2001–2011.

10.3.3 Switzerland 2008: the six largest sectors of the interindustrial production

Figures 10.12–10.14 show the operating capital K without net commodity taxes, the monthly operating capital, the Frobenius numbers and the monthly operating capital K^* in λ_F-units for the year 2008. At first, the vector of *interindustrial output* $\mathbf{x}$ in value terms is sorted in descending order. Then, the input-output matrix is accordingly sorted by a similarity transformation (row-and-column permutations) so that its Frobenius number remains unaffected.[11]

10.3.3.1 The model-order reduction using the x-sorting process

The operating capital of Switzerland 2008 without net commodity taxes is K_{44} = 510.789 Billion CHF (44 sectors). One by one, the upper-left corner of the sorted matrix $\mathbf{Z}$ is extracted according to the descending dimensions $35, 30, \ldots 8$ and 6.

Figure 10.12 shows the table of the computed operating capital K without net commodity taxes, the monthly operating capital, the corresponding Frobenius number and the corresponding monthly operating capital K^* in λ_F-units for the reduced models with the dimensions $44, 35, 30, \ldots 8$ and 6. Figure 10.13 illustrates graphically the operating capital K and the Frobenius numbers for these reduced-order models. The six largest sectors, corresponding to the 6×6 matrix, have a production value of more than 20 % ((109.857/510.789) × 100 % = 21.5 %) of the interindustrial value of production (Switzerland 2008).

One should notice in Figure 10.13 (right) that the Frobenius number is only reduced from λ_F = 26,424 to λ_F = 23,371 for a model-dimension reduction from n = 44 to n = 6. The Frobenius number has been reduced by about 12 %, while the number of

11 It is known from matrix algebra that similar matrices have the same eigenvalues.

CH 2008		K	Km= (K/12)	FP- number	K*
N= 44sorted	2008	510.789	42.5657	26.4234	1.6109
N=35 sorted	2008	477.0949	39.7579	26.4040	1.5082
N=30sorted	2008	440.4018	36.7001	26.3617	1.3922
N=25 sorted	2008	382.0658	31.8388	26.2769	1.2116
N= 20 sorted	2008	318.7215	26.5600	25.9800	1.1837
N=15 sorted	2008	251.1034	20.9253	25.7448	0.8158
N=8 sorted	2008	143.5614	11.9653	25.4253	0.4706
N=6 sorted	2008	109.8566	9.1547	23.3712	0.3917
6*6 aggregated	2008	510.7824	42.5652	173.4838	0.2454

Figure 10.12: The influence of model reduction (from $n = 44$ to $n = 6$) on the operating capital K without net commodity taxes (Billion CHF), the Frobenius number (in thousands) and the normalised operating capital K^* (CH 2008).

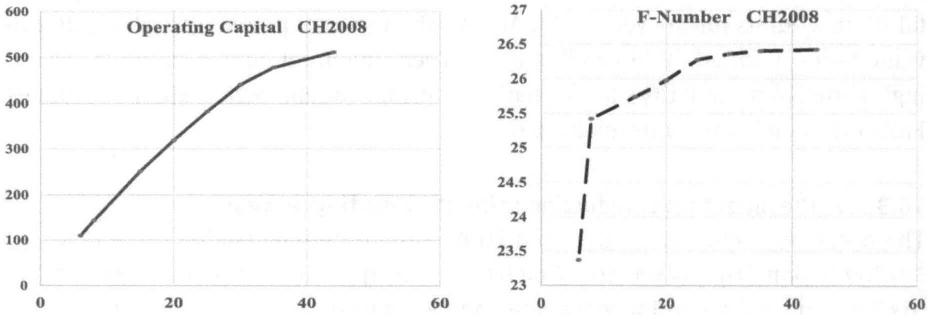

Figure 10.13: The operating capital K without net commodity taxes (billion CHF) and Frobenius number (in thousands) as functions of the largest sectors (from $n = 44$ to $n = 6$, CH 2008).

sectors were reduced from 44 to six sectors. One can conclude therefore that the sorting mechanism used does not affect the Frobenius number seriously whenever the number of sectors is reduced. In comparison, the Frobenius number of the aggregated 6×6 model, matrix (10.2) is $\lambda_F = 173{,}483.8$ (Figure 10.12, last line), while the Frobenius numbers for the sorted Swiss IOT 2008, $n = 44$ and $n = 6$, are respectively $\lambda_F = 26{,}424$ and $\lambda_F = 23{,}371$.[12]

12 The Frobenius number of the $n = 49$ Swiss IOT 2014 is $\lambda_{F_{48}} = 34{,}452$. For the aggregated Swiss IOT 2014 with $n = 6$ aggregated sectors, Example 10.2.1, the Frobenius number is $\lambda_{F_6} = 212{,}342$.

Observation. Figure 10.14 shows the monthly operating capital K^* (in λ_F-units) as a function of the reduced sectors. The number n refers to the largest sectors of the interindustrial market. The monthly operating capital K^* is monotonously reducing as the number of sectors is reduced. It remains between 1 to 2 for the largest 20 sectors. It is reduced linearly from 1.2 to 0.4 whenever the number of sectors is reduced from 20 to six sectors.

Figure 10.13 (right) suggests that the Frobenius numbers of the Swiss IOT 2008 fluctuate around a constant, whereas Figure 10.13 (left) and Figure 10.14 suggest that the operating capitals K or K^* without net commodity taxes follow some linear functions.

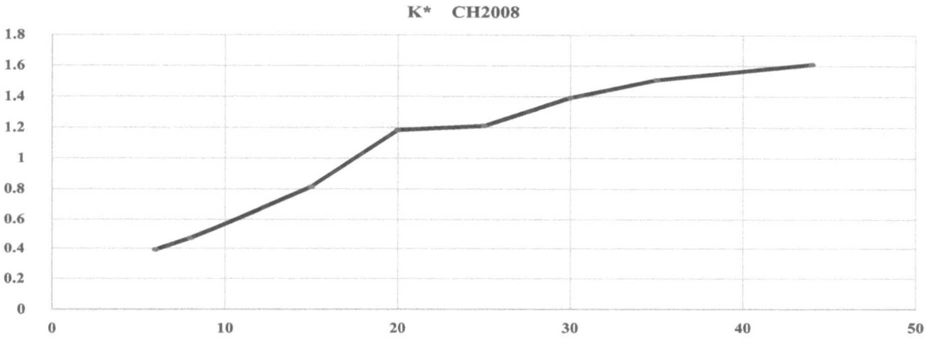

Figure 10.14: The monthly operating capital (λ_F-units) as a function of the largest sectors (from n = 44 to n = 6, CH 2008).

10.3.3.2 The eigenvalue distribution of the matrices E, D and Z (CH 2008)

In an *equal priority* $n \times n$ matrix $\mathbf{E}$, all the entries are $e_{ij} = 1, i, j = 1, \ldots n$. We continue to explore the Swiss IOT 2008. Figure 10.15 shows the eigenvalues of the 44×44 matrix $(1/44)\mathbf{E}$ *(the equal priority* 44×44 matrix $\mathbf{E}$ divided by 44), with the characteristic polynomial $P_{44}(\lambda) = \lambda^{n-1}(\lambda - 1)$ (see also Nour Eldin [73]). There is a unique Perron–Frobenius eigenvalue $\lambda = 1$, and all other eigenvalues are zero eigenvalues. The object production matrix $\mathbf{D}$ changes such equal priority distribution of the eigenvalues to the object production eigenvalue distribution in Figure 10.16. The distribution of these eigenvalues reflects the priority weights of the producers in response to the desired objects for production (market demand).

Example 10.3.1. Compute the eigenvalues of the 3×3 matrix $(1/3)\mathbf{E}$, where $\mathbf{E}$ is the equal priority matrix.

Solution of Example 10.3.1:

Set up the (3×3) equal priority matrix $\mathbf{E}$ and of the matrix $(1/3)\mathbf{E}$,

$$\mathbf{E} = \begin{bmatrix} 1 & 1 & 1 \\ 1 & 1 & 1 \\ 1 & 1 & 1 \end{bmatrix}, \quad \frac{1}{3}\mathbf{E} = \begin{bmatrix} \frac{1}{3} & \frac{1}{3} & \frac{1}{3} \\ \frac{1}{3} & \frac{1}{3} & \frac{1}{3} \\ \frac{1}{3} & \frac{1}{3} & \frac{1}{3} \end{bmatrix}. \tag{10.7}$$

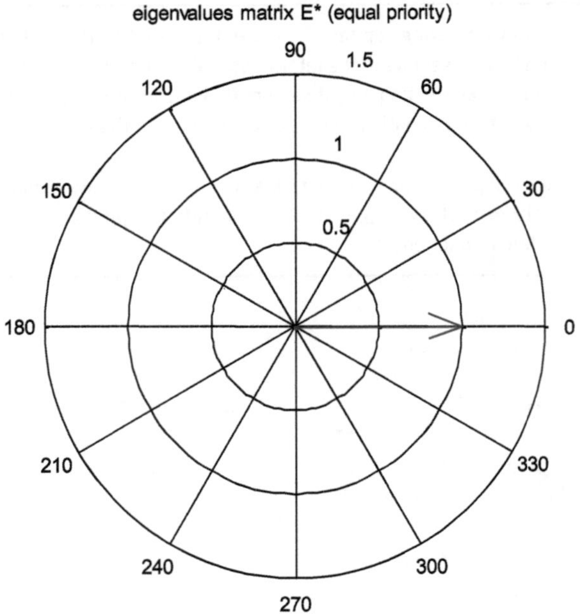

Figure 10.15: The eigenvalues of the equal priority matrix (1/44)**E** in polar form (CH 2008): One observes that with equal priorities the eigenvalues are *totally bundled*.

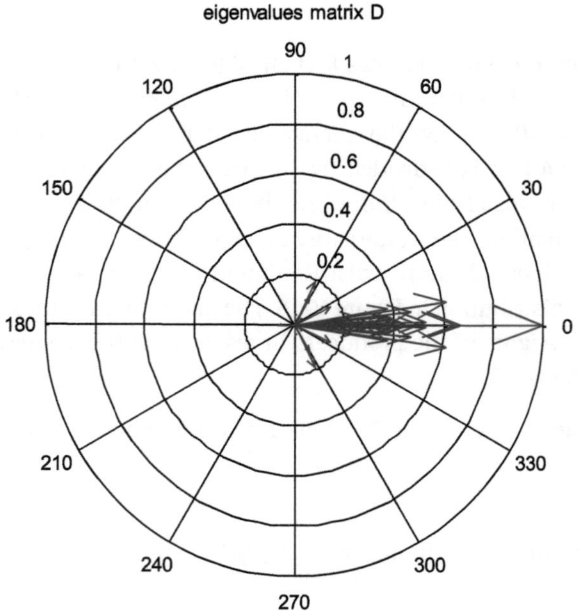

Figure 10.16: The eigenvalues of the production matrix **D** in polar form (CH 2008): One observes that with the distribution matrix **D** the eigenvalues *spread out*.

The characteristic polynomial of matrix (1/3)**E** is:

$$p_3(\lambda) = \lambda^2(\lambda - 1) \tag{10.8}$$

The Frobenius number is $\lambda = 1$ and the two other eigenvalues are $\lambda_2 = \lambda_3 = 0$. ▲

Again, this eigenvalue distribution of the object production matrix **D** is changed to the eigenvalue distribution of the matrix **Z**, see Figure 10.17. Almost all the eigenvalues of the matrix **Z** are now on the real axis in a descending order. This eigenvalue distribution of the matrix **Z** reflects the priority weight assignments by the buyers (buyer demand) to the object prices offered by the producers in the interindustrial market.

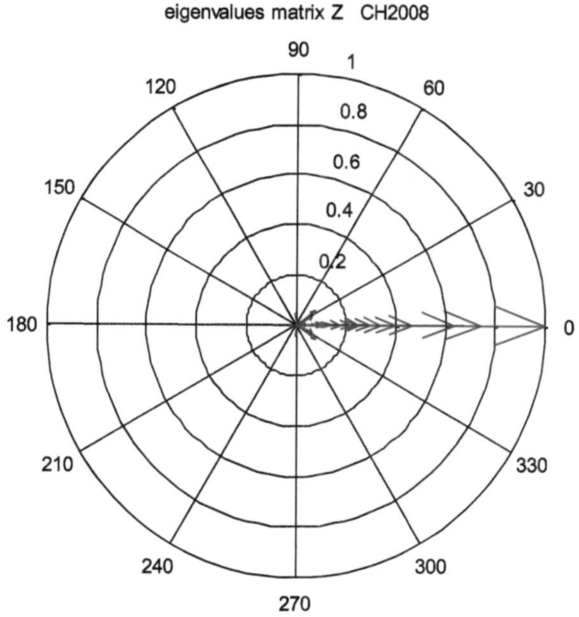

Figure 10.17: The eigenvalues of the input-output matrix **Z** in polar form (CH 2008): One observes that the eigenvalues of the Swiss IOT 2008 matrix **Z** are *nearly all real*.

11 Conclusions and outlook

What have we obtained in theory and practice after some 440 pages of explanations, definitions, formulas and calculations, commenting on a 60-year-old opus of less than 100 pages?

We have analysed in detail Sraffa's initial model centred on single-product industries, which is truly basic. It was presented by Sraffa in PCMC in a quaint, archaic style, mixing examples, formulas and textual explanations with no reference to the mathematical background.

We summarize in six points the essence of Sraffa's price model!

(1) Sraffa's price model determines in a cyclic production process of n sectors and n commodities, measured in physical terms, on the basis of structural economic data, the 'costs of production' of every commodity, termed as 'prices', and brings them with positive wages for workers and with positive profits of entrepreneurs into an *accounting balance*, which is the proposed equilibrium to attain.

(2) Sraffa's price model establishes the *accounting balance* for each of the n industrial sectors of the economy with respect to the *production prices*.

(3) Occurring *accounting balance* equations can be aggregated to the overall accounting equations, i. e., $X = K + (P + W)$ and $Y = P + W$, of the whole economy, which are in analogy to the *national accounting equation*, $Y = C + I + G + (E - X)$, also obtainable from some column-sums of Leontief's Input-Output Tables, e. g., the Swiss IOT 2014.

(4) The Sraffa price model defines a *dynamic system*, under the condition that the *interindustrial market* adopts Sraffa's prices. Then, the production technology, described by the means of production for the actually analysed period, is recreated for the next period, ensuring the same technology, and this process is going on from period to period.

(5) Changes in the technology matrix may be applied to reflect the evolution of the technology and are possible at the start of every period.

(6) In a Sraffa production system which additionally has the properties of a *Standard system*, consisting exclusively of basic commmodities, the national income is determined by the *productiveness* of the represented economy.

The link between Sraffa's price model in *physical terms*, here represented again for *single-product industries*,

$$\boxed{\begin{gathered} \mathbf{S}'\mathbf{p}(1+r) + \frac{\tilde{w}\cdot Y}{L}\cdot \mathbf{L} = \hat{\mathbf{q}}\mathbf{p} = \mathbf{x} \Rightarrow \\ \mathbf{e}'(\mathbf{S}'\mathbf{p}) + r\mathbf{e}'(\mathbf{S}'\mathbf{p}) + \frac{\tilde{w}\cdot Y}{L}\cdot(\mathbf{e}'\mathbf{L}) = \mathbf{e}'(\hat{\mathbf{q}}\mathbf{p}) = \mathbf{e}'\mathbf{x} \Rightarrow \\ K + (P + W) = K + Y = X, \end{gathered}} \tag{11.1}$$

https://doi.org/10.1515/9783110635096-011

and Leontief's Input-Output Tables (IOT) in *monetary terms* becomes thereby clearly visible, especially through the accounting balances, summarized above.

And now we come to more detailed considerations:

In Sraffa's model, by assumption, each of the n industries produce separately one commodity, thus providing a total number of n different commodities. In short, a system of linear equations is constructed in four steps, based on the following approach, enabling one to calculate the n prices of the commodities, the wage rate of the workers and the rate of profits of producers owning the industries:

- the concerned commodities are goods per se, ranging from raw materials to finished goods such as machines;
- industries are constituted of means of production, which are linear combinations of goods completed by labour;
- prices are introduced, but the *numéraire* in which they are measured is not defined; the prices are relative prices and purely numerical quantities. One unit of a given commodity multiplied by its price gives its value;
- the production period considered is short term, typically one year. There are no capacity constraints and no changes in the means of production employed. Workers are paid at the end of the period, and the wage rate is a fixed parameter and is the same for all workers in all industries;
- having established the conditions required for a sustainable production from period to period, the possibility of generating a production surplus is considered; this surplus (tacitly assumed absorbed by demand) is distributed between the producers and the workers;
- the part of the surplus allotted to producers constitutes their profits; it is assumed that this part is distributed to producers in proportion to the value of their means of production (operating capital), so that the rate of profit is uniform for all industries;
- Sraffa introduces the term "national income" equal to the aggregate value of the surplus obtained by summation, which he normalises by convention to one. The quantity of total labour is also normalised to one.
- In fact, in PCMC, Par. 12, Sraffa early introduced the term "national income" in relation with the term *surplus* or *net product*. Indeed, at this moment Sraffa has in mind the *complete production scheme* of an economy, as presented by an *input-output* table (IOT) in physical terms. Then the net product is the "gross national income" and can be associated to the *accounting balance* $X = K + W + P$ (2.145), containing $Y = K + W$, inherent to the Sraffa price model, see Chapter 10.
- The national accounting identity $Y = C + I + G + (E - M)$ is contained in the IOTs.

Presented as such, together with various obscure passages in PCMC due to Sraffa's loathing to follow a more mathematical approach, this model of a production economy appears to be as a theoretical curiosity and has indeed been treated as such, or even ignored, by mainstream economists.

We have now reformulated Sraffa's initial single-product industry model, his standard commodity model and his presentation of joint production in up-to-date notations and concepts based on the systematic use of matrix algebra and graph theory, while working out the complete mathematical properties and clarifying wherever necessary Sraffa's definitions and terms, some of them implicitly implying a closed economy as a general condition of validity, and tidying up loose ends.

The methodology underlying our investigations follows a step-by-step approach that may be summarised as follows, allowing of course for frequent moves backwards and forwards:

- prepare a schematic description of the production processes under consideration;
- elaborate a network representation of those processes;
- formalise the foregoing steps by setting up mathematical models of the corresponding systems of production;
- investigate the properties of the algebraic structures involved and clarification of terms;
- proceed with systematic numerical examples of the models;
- explore practical applications of these models.

In this connection, we summarise some of our modest contributions to a better understanding of PCMC, which have been in the focus of intense research activities, such as:

- analyse the price concepts used in Input-Output Analysis on the basis of a novel systematic analysis, Chapter 2;
- emphasise the central role played by the Perron–Frobenius theorem (presented in Appendix A) in the *economic price models* of Leontief and Sraffa and a Theorem presented by Ashmanov ([2], Theorem 1.5, p. 39) for *productive Leontief models*;
- underline the importance of measurement units and the numéraire, Section 2.8;
- bring out the technology matrix $\mathbf{S}'$, with and without inclusion of "subsistence wages", Section 3.6, and more generally the fundamental importance of the conditions of production and of the surplus ratio;
- calibrate the national income Y as usual by an appropriate currency in *monetary units*, typically in Section 4.9, i. e., no systematic normalization to $Y = 1$, as in PCMC;
- prove formally that, in the general *Sraffa price model* of *single-product industries*, the price vector is positive, $\mathbf{p} > \mathbf{o}$, Section 4.3;
- read the general relationships between profits and wages as $r = \tilde{R}(1 - \tilde{w})$, Section 4.1.2, where $\tilde{R} = Y/K$;
- establish the fundamental relations $R = d_i/(q_i - d_i)$ (5.7), characterising a *Standard system*, Definition 5.1.1 for single-product industries, Section 5.1, and show that, in a *Standard system* the prices, but not labour, are independent of r and w;

- introduce the orthogonal Euler affinity to work on the concept of the *Standard System*, Section 5.2;
- propose the *Sraffa Network* and its adjacency matrix as a basic tool for the analysis of linear production processes, Section 4.6;
- present the algebra required to understand PCMC, Par. 57–61, in a newly detailed form in connection with the determination of non-basic commodities in joint production;
- reformulate Manara's sufficient conditions for positive prices in joint systems of production, with single-product industries as a special case (Section 6.6);
- develop an explicit model of agricultural production based on Sraffa's PCMC, Chap. XI, Section 6.7;
- innovate with the whole Chapter 7 on ecological economy, contributed by H. Knolle. The application of Sraffa's theory of joint production to price models are developed, including a recycling process for waste products, or price models including alternative CO_2-emissions trading, illustrated by numerical examples. Negative prices may appear for waste products and the effect on the prices of the other commodities is studied;
- extend Sraffa's model to an open economy and a system of services, especially developed in Sections 8.1–8.2;
- drop the assumption of a uniform rate of profit r and of a single basic wage unit w, Section 8.3;
- use of the diagonal mark-up matrix $\hat{\mathbf{k}}$ to define an alternative representation of Sraffa's price equation, Section 8.4;
- set up the incompatibility between Sraffa's price model and the marginalist economists, like Cobb-Douglas, Section 8.7;
- innovate with the whole Chapter 9 on the *interindustrial economy*, contributed by H. A. Nour Eldin. Here, the algebraic structure of such an economy is investigated, and Nour Eldin explores the relationships between the *commodity flow matrices* $\mathbf{Z}$ in monetary terms, respectively $\mathbf{S}$ in physical terms. He continues to investigate the relationships between *input-output coefficients* matrices $\mathbf{A}$, respectively $\mathbf{C}$, together with those of the *distribution coefficients matrix* $\mathbf{D}$, as well as the *price partition* matrix $\mathbf{T}$, whose components present the parts of the commodity prices attributed to the production sectors;
- apply finally this algebra to the *Swiss Input-Output Table (IOT) 2008* and to IOTs of contemporaneous *EURO-Zone Input-Output Table*.

As mentioned, we have dropped conventions introduced by Sraffa, such as the systematic normalisation of national income and total labour to one, inadequate for the practical calculation of effective prices. We make extensive use of *matrix analysis*, especially, matrix inversion, aggregation through the use of the summation vector, replacement of vectors by diagonal matrices and the left and right eigenvalue problems.

At the formal algebraic level, we have insisted on the attributes a) *positive*, b) *semi-positive* c) *non-negative* on one hand, and d) *reducible*, e) *irreducible* on the other hand, of matrices in the case of *Leontief quantity models* and single-product Sraffa systems. The central role of the Perron–Frobenius theorem, see Perron [86], Frobenius [31], Gantmacher [34], Varga [118], Young [125], Horn and Johnson [43] is developed, as well as a theorem on *productive Leontief models* presented by Ashmanov ([2], p. 39) for which we present a more general new proof.

The use of graphs, or networks if one aims at applications, greatly facilitates the introduction to the conceptual structures of Sraffa's model, its circular nature and its close connections to Input-Output Analysis. Indeed, networks have become an indispensable interdisciplinary tool of analysis and modeling in the social sciences (see, in particular, Newman: *Networks, An Introduction* [70] and Ormerod: *How Networks Can Revolutionise the World* [76]). They nicely highlight Sraffa's presentation as a refinement of Leontief's approach to a circular economy by the explicit introduction of wages and labour, complementary to the usual presentation of the Input-Output Tables where these items are normally tucked up in the value-added component. Furthermore, his model is consistent with the fundamental national accounting identities once a monetary numéraire has been adequately put to use.

Finally, as clearly emerges from our various comments and footnotes with historical references, Sraffa's conceptual framework sheds light on the types of practical economic problems the early macroeconomists Petty, Law, Cantillon and Quesnay were grappling with during the 17th and 18th centuries.

Sraffa's breakthrough lies in the device of a model of a production economy constructed as follows. Given:
- a technology matrix **S** and the conditions of production for a sustainable production economy;
- a vector of final demand **d**, defining the social surplus in quantity terms;
- a vector of employed labour **L**;
- two parameters: the rate of entrepreneurial profits r and the basic unit wage w, which fix value added;

one defines a system of linear equations that completely determines prices which should be compared with the Walrasian approach central to neo-classical economics. That is the meaning of Sraffa's subtitle for PCMC: *Prelude to a Critique of Economic Theory*. We must however point out that the foregoing is in fact a prelude centered on systems of production involving basic commodities, fulfilling the conditions of production that ensure sustainability of the system. When non-basic commodities also participate in production, further effects appear, and possible market clearing mechanisms enter the picture, but then sustainability becomes problematic. Other economic approaches and Sraffa may then usefully complete each other.

With this behind us, making use of the extended definition of commodities going beyond Sraffa's original approach, we are now in presence of a structured mathematical tool revealing the versatility of the Sraffian economic model. This instrument for the investigation of circular processes, applying the powerful mathematical tool of linear operators (in this context real $n \times n$ matrices), are destined to be discovered by economists as a creative complement to the mainstream marginalist approach for investigations in economics. In particular, we have already indicated how to start incorporating money as such in the model, how to cope with land (only outlined by Sraffa) and how to extend the model to an open economy. And as regards applications, we have mentioned possibilities in ecological economics, with examples, and presented the numerical results of an application of Sraffa's model to the Swiss economy.

Without exaggeration, we can accordingly say that Sraffa's model has now been brought up to the required mathematical standards indispensable for further research in Sraffa studies, which are no idle pastime but have much to contribute in new approaches to political economy. Although still incomplete, as it does not integrate monetary macroeconomics at present, the model is sufficiently flexible for opening up new avenues of research. In particular, a step towards the full potential of the model is facilitated by assuming variable rates of profit and variable wage units throughout the concerned industries. In this connection, we must once again insist on the fact that Sraffa's model is a short-term approach simulating the "market", and must not be confused with an equilibrium model. His prices must not be confused with A. Smith's natural prices, which are a long-term equilibrium concept.

This being said and bearing in mind that Sraffa's model is a particular version of Input-Output Analysis, the following are some such avenues for theory and applications of Sraffa's *price model*, without regard to the complexities involved. More elaborate techniques and concepts such as those presented in Takayama [116], Chap. 1; 4–7 and Welsh [121], could also be efficiently put to use in this context following a resolutely heuristic approach (see Taleb [117], *Antifragile. Things that Gain from Disorder*, pp. 217–223):

- First of all, Chapters VI, IX, X, XII, and Appendices A–C of PCMC, which we have not addressed, should be analysed using the Sraffa Network approach and the algebraic and computational apparatus now available.
  Then in a larger perspective:
- Pursue avenues described in Miller and Blair ([65] Chapter 13 and 14), and EUROSTAT Manual ([72], Chapter 15), and, in particular:
- Elaborate detailed models of open economies with varying rates of profits for entrepreneurs and varying wage rates in accordance with the industry and, in the case of joint production, prices of identical commodities depending on the industry producing them;
- incorporate explicitly investment, savings, taxes and government spending and the balance of payments into the single-product Sraffa system for an open production economy;
- embed the model into a fully fledged monetary economy of production and exchange, meaning *accounting for credit and debt and other monetary effects* gener-

ated by a sophisticated banking system (referring to PCMC Par. 56, this is a necessary requirement if one intends to examine the detailed construction of the *Standard system* and the *Standard Commodity* in joint systems of production);
- investigate further basic and *non-basic* commodities. The central importance of basic commodities lies in the fact that they satisfy Sraffa's conditions of production and thus ensure sustainability, favoring medium- to long-term growth. They are essential for the viability of circular economic systems.
  Non-basic commodities, beyond their possible role in subsystems where they may act more or less as basics, disrupt circularity and are usually detrimental to sustainability. They characterise linear economic systems and, for example, provide a fertile ground for marketing and short-term consumerism;
- introduce capacity constraints and institutional constraints into the Sraffa production processes, with a new look at optimisation analysis;
- explore of other links with neoclassical models, in particular comparison of the equilibrium concepts in Sraffian and Walrasian models and their practical utility;
- investigate further the *practical significance* of the Standard system and the Standard commodity;
- study further the mathematical properties of joint production processes, including the effects of rapid depreciation and amortisation of technologies in the present age of digitalisation;
- introduce variable prices for identical commodities produced in different joint-production processes simultaneously producing the same commodity in their output;
- include information, financial capital and patents as further species of commodities. In fact, consider everything which has a price, distinguishing between commodities generating;
  - value added during the period under consideration;
  - possible value added in future periods;
  - no value added.
- As regards the latter, investigations into non-productive economic processes, using the Sraffian approach without surplus. For example, in the context of financial markets: operations based on information and virtual commodities alone and deficient conditions of production of financial techniques leading to Ponzi finance and non-sustainable financial systems.
- Apply Sraffa's model to analyse the globalised world economy considered as a closed system;
- use this model, a factual representation of the economic production process devoid of ideological preconceptions, as a tool in devising reindustrialisation policies;
- adapt Sraffa' price model to possible changes in the notion of social surplus, replacing the present use of surplus equal to GDP.

But entirely new issues are rapidly entering the scene. The world of digitalisation, algorithms and Big Data is in the course of fundamentally modifying the technology matrix **S** at the heart of Sraffa's model of a production economy. The social consequences are disturbing, to say the least. In *Weapons of Math Destruction*, Cathy O'Neil [75], a former Wall Street quant, gives a vivid description of how all this is increasing inequality and threatening the basis of our democratic societies. This is just an antipasta; the looming consequences of successive surges in artificial intelligence (deep learning), robotics and neurotechnologies will be even more dramatic, as systematically described by Laurent Alexandre [1], surgeon and neurobiologist, in *La Guerre des Intelligences*. The changes for labour and employment in Sraffa's model will be drastic and far reaching. For the moment, nobody knows where we are heading: towards the Garden of Eden, as optimistic technology pundits would have it, or as anticipated decades ago following the darker sides of human nature, towards the world of Fahrenheit 451 or that of Zardoz,[1] the world of underdogs dominated by a technological elite controlling basic commodities. Sraffa's approach cannot provide ready-made solutions, but it may help decision makers in analysing the economic consequences of future developments.

In a certain sense, we have deconstructed and reconstructed, with emphasis on numerical results, Sraffa's *Production of Commodities by Means of Commodities*, starting out from Leontief's Input Output Analysis. We now have a sound mathematical basis for a theory of value in economics and results can be expected from investigations along the avenues indicated here should thus open the way to a *Constructive Critique of Neoclassical Economic Theory*. The central issue focuses on the creation of value added and its equitable distribution as income between entrepreneurs, capital owners and wage earners, with the accompanying aim of overcoming present-day trends in poverty and unemployment.

We conclude that *Input-Output Analysis* and Sraffian models have further theoretical and practical potential and can effectively participate in bringing new blood and substance into the economic curricula of all institutions of higher academic and professional learning. That is our final message.

When the facts change, I change my mind. What do you do, sir?
(quotation attributed to J. M. Keynes)

1 Film by John Boorman, 1974, starring Sean Connery and Charlotte Rampling.

A Mathematical tools

A.1 A short guide to algebraic concepts and techniques

The reader is encouraged to get comfortably acquainted with the following concepts and techniques used again and again throughout our exposition (all the more elaborate mathematical details are given below).

We draw the reader's attention to the fact that the presentation of these concepts may vary from one author to another, e. g., in Bortis, Kurz and Salvadori, Miller and Blair, Pasinetti, Schefold, thus complicating understanding and comparisons:

Vectors

- column $\mathbf{\upsilon}$ and row $\mathbf{\upsilon}'$ representations of a vector
- left and right multiplication of a matrix with a vector
- use of the summation vector to obtain sums of vector and matrix components
- normalized vectors, like $\mathbf{\upsilon}$, with component sum equal to 1, $\mathbf{e}'\mathbf{\upsilon} = 1$

Matrices

- a matrix $\mathbf{M}$ and its transpose $\mathbf{M}'$
- the inverse matrix $\mathbf{M}^{-1}$
- diagonal matrices and diagonalization
- the identity matrix
- left and right multiplication of a matrix with another matrix
- formulation of eigenvalue problems; eigenvalues and eigenvectors

Theorems

- the Perron–Frobenius theorem, ensuring positive prices for *non-negative* and *irreducible* matrices
- a theorem published by Ashmanov [2], ensuring that a Leontief model is productive

The reader should distinguish between formal algebraic manipulations of matrices and vectors and numerical calculations with such entities. For the latter, calculations by hand on 2×2 matrices are extremely helpful in checking formal results and to become a trained user of numbers through numerical examples. Most results can be applied directly to $n \times n$ matrices unless explicitly mentioned otherwise in the text. By the way, 2×2 matrices are not as trivial as one may think. In fact they crop up as fundamental tools for example in plane geometry, complex function theory and in specialized areas of physics such as special relativity and quantum mechanics!

https://doi.org/10.1515/9783110635096-012

A.2 Some elementary matrix algebra

It is now necessary to present some elementary *matrix algebra*, see also Miller & Blair [65], Appendix A, and Kurz & Salvadori [52], Mathematical Appendix, as well as Gantmacher [34] and Zurmühl [126]. Basic knowledge is indispensable to correctly manipulate all the entities used in Input-Output Analysis.

(1) A vector $\mathbf{x}$, consisting of m components, generally $x_i \in \mathbb{R}$, $i = 1, \ldots, m$, is a column vector. We note $\mathbf{x}_{m\times 1}$, if it is necessary to show the number of elements m. It is also a $m \times 1$ matrix. Its transpose $\mathbf{x}'$ is a row vector,[1] conversely the transpose of the transpose $(\mathbf{x}')'$ of a column vector is again the same column vector $\mathbf{x}$. This operation is an involution,

$$\mathbf{x} = \begin{bmatrix} x_1 \\ x_2 \\ \ldots \\ x_m \end{bmatrix}, \quad \mathbf{x}' = [x_1, x_2, \ldots, x_m], \quad (\mathbf{x}')' = \mathbf{x}. \tag{A.1}$$

(2) We need subsets of unit vectors, denoted by as $\mathbf{e}_i = [0, 0, \ldots, 1, \ldots, 0]'$, where the *i-th* component is equal to 1 and the others vanish. The set of vectors $\mathcal{E} = \{\mathbf{e}_1, \ldots, \mathbf{e}_n\}$ is a basis of the n-dimensional Euclidean vector space $\mathbb{R}^n$, $i \in \{1, 2, \ldots, n\}$.

(3) An $m\times n$ matrix $\mathbf{A} = (a_{ij})$ is presented as a table and consists of mn matrix elements $a_{ij} \in \mathbb{R}$, $i = 1, \ldots, m, j = 1, \ldots, n$, We note also $\mathbf{A}_{m\times n}$, if it is necessary to emphasize the number of rows m and the number of columns n. The transpose of a matrix $\mathbf{A}$, written $\mathbf{A}' = (a_{ji})$, is obtained by making the rows to columns and the columns to rows. Notice the rule $(\mathbf{A}')' = \mathbf{A}'$. The transposition of a matrix is an involution,

$$\mathbf{A} = \begin{bmatrix} a_{11} & a_{12} & \ldots & \ldots & a_{1n} \\ a_{21} & a_{22} & \ldots & \ldots & a_{2n} \\ \ldots & \ldots & \ldots & \ldots & \ldots \\ a_{m1} & a_{m2} & \ldots & \ldots & a_{mn} \end{bmatrix}, \quad \mathbf{A}' = \begin{bmatrix} a_{11} & a_{21} & \ldots & a_{m1} \\ a_{12} & a_{22} & \ldots & a_{m2} \\ \ldots & \ldots & \ldots & \ldots \\ \ldots & \ldots & \ldots & \ldots \\ a_{1n} & a_{2n} & \ldots & a_{mn} \end{bmatrix}. \tag{A.2}$$

Matrices $\mathbf{A}$ with $n = m$ are called square matrices. A matrix $\mathbf{A}$ with $\mathbf{A}' = \mathbf{A}$ is called symmetric.

(4) There are important special matrices and vectors. An $m \times n$ *zero matrix*, consisting only of zeros. There further is the $n \times n$ *unit* or *identity matrix*:

$$\mathbf{0} = \mathbf{0}_{m\times n} = \begin{bmatrix} 0 & 0 & 0 & \ldots & 0 \\ 0 & 0 & 0 & \ldots & 0 \\ \ldots & \ldots & \ldots & \ldots & \ldots \\ 0 & 0 & 0 & \ldots & 0 \end{bmatrix}, \quad \mathbf{I}_n = \mathbf{I} = \begin{bmatrix} 1 & 0 & 0 & \ldots & 0 \\ 0 & 1 & 0 & \ldots & 0 \\ \ldots & \ldots & \ldots & \ldots & \ldots \\ 0 & 0 & 0 & \ldots & 1 \end{bmatrix}. \tag{A.3}$$

1 The row vector $\mathbf{x}$ can also been written as $\mathbf{x}' = [x_1\ x_2\ \ldots\ x_m]$, without separating commas between its elements, if we want to emphasize the matrix nature of this vector, namely a $1 \times m$ matrix.

The $m \times n$ *zero matrix* is necessary when defining the addition of matrices. The *identity matrix* is necessary in the context of multiplications of square matrices.

The zero vector is written here with a minor bold **o**. An other important vector is the $n \times 1$ summation vector $\mathbf{e} = [1, 1, \dots, 1]'$, consisting of '1':

$$\mathbf{o} = \begin{bmatrix} 0 \\ 0 \\ \dots \\ 0 \end{bmatrix}, \quad \mathbf{e} = \begin{bmatrix} 1 \\ 1 \\ \dots \\ 1 \end{bmatrix}. \tag{A.4}$$

(5) The addition of matrices is obtained by adding element by element.

Definition A.2.1 (Addition of two matrices). Consider the $m \times n$ matrices with m rows and n columns,

$$\mathbf{A} = \begin{bmatrix} a_{11} & a_{12} & \dots & \dots & a_{1n} \\ a_{21} & a_{22} & \dots & \dots & a_{2n} \\ \dots & \dots & \dots & \dots & \dots \\ a_{m1} & a_{m2} & \dots & \dots & a_{mn} \end{bmatrix}, \quad \mathbf{B} = \begin{bmatrix} b_{11} & b_{12} & \dots & \dots & b_{1n} \\ b_{21} & b_{22} & \dots & \dots & b_{2n} \\ \dots & \dots & \dots & \dots & \dots \\ b_{m1} & b_{m2} & \dots & \dots & b_{mn} \end{bmatrix}. \tag{A.5}$$

The sum of matrices **A** and **B** is the $m \times n$ matrix,

$$\mathbf{C} = \mathbf{A} + \mathbf{B} = \begin{bmatrix} a_{11} + b_{11} & a_{12} + b_{12} & \dots & \dots & a_{1n} + b_{1n} \\ a_{21} + b_{21} & a_{22} + b_{22} & \dots & \dots & a_{2n} + b_{2n} \\ \dots & \dots & \dots & \dots & \dots \\ a_{m1} + b_{m1} & a_{m2} + b_{m2} & \dots & \dots & a_{mn} + b_{mn} \end{bmatrix}. \tag{A.6}$$

Example A.2.1. Consider the 3×3 matrices,

$$\mathbf{A} = \begin{bmatrix} 3 & 1.5 & 2 \\ 1 & 4 & 7 \\ 1 & -2 & 5 \end{bmatrix}, \quad \mathbf{B} = \begin{bmatrix} 1 & 0 & -2 \\ 1 & 2 & -3 \\ 2 & 2 & 3 \end{bmatrix}. \tag{A.7}$$

Then the sum is

$$\begin{aligned} \mathbf{C} = \mathbf{A} + \mathbf{B} &= \begin{bmatrix} 3 & 1.5 & 2 \\ 1 & 4 & 7 \\ 1 & -2 & 5 \end{bmatrix} + \begin{bmatrix} 1 & 0 & -2 \\ 1 & 2 & -3 \\ 2 & 2 & 3 \end{bmatrix} \\ &= \begin{bmatrix} 3+1 & 1.5+0 & 2+(-2) \\ 1+1 & 4+2 & 7+(-3) \\ 1+2 & -2+2 & 5+3 \end{bmatrix} = \begin{bmatrix} 4 & 1.5 & 0 \\ 2 & 6 & 4 \\ 3 & 0 & 8 \end{bmatrix}. \quad \blacktriangle \end{aligned} \tag{A.8}$$

The following rules for the addition of $m \times n$ matrices $\mathbf{A} = (a_{ij})$, $\mathbf{B} = (b_{ij})$, $\mathbf{C} = (C_{ij})$, $i = 1, \dots, m, j = 1, \dots, n$ apply.

Lemma A.2.1 (Rules for addition of $m \times n$ matrices).
(1) *Associative law:* $(\mathbf{A} + \mathbf{B}) + \mathbf{C} = \mathbf{A} + (\mathbf{B} + \mathbf{C})$;
(2) *Null element:* $\mathbf{A} + \mathbf{O} = \mathbf{O} + \mathbf{A} = \mathbf{A}$;

(3) *Inverse element* -**A**: $\mathbf{A} + (-\mathbf{A}) = (-\mathbf{A}) + \mathbf{A} = \mathbf{0}$;
(4) *Commutative law:* $\mathbf{A} + \mathbf{B} = \mathbf{B} + \mathbf{A}$;
(5) *For any* $k \in \mathbb{R}$, *one has* $k \cdot (\mathbf{A} + \mathbf{B}) = k \cdot \mathbf{B} + k \cdot \mathbf{A}$.

(6) The scalar product. Consider the $1 \times n$ matrix $\mathbf{A} = [a_{11}, a_{12}, \dots, a_{1n}] := \mathbf{u}' = [u_1, u_2, \dots, u_n]$ and the $n \times 1$ matrix $\mathbf{B} = [b_{11}, b_{21}, \dots, b_{n1}]' := \mathbf{v} = [v_1, v_2, \dots, v_n]'$ that are also written as vectors. The *scalar product* of the vectors **u** and **v** is the product,

$$\mathbf{u}'\mathbf{v} = \mathbf{u}' \cdot \mathbf{v} = [u_1, u_2, \dots, u_n] \begin{bmatrix} v_1 \\ v_2 \\ \dots \\ \dots \\ v_n \end{bmatrix} := \sum_{k=1}^{n} u_k v_k. \tag{A.9}$$

The scalar product of the *summation vector* **e** by itself is

$$\mathbf{e}' \cdot \mathbf{e} = [1, 1, \dots, 1] \begin{bmatrix} 1 \\ 1 \\ \dots \\ \dots \\ 1 \end{bmatrix} = n. \tag{A.10}$$

Consider the semi-positive vector $\mathbf{d} \geq \mathbf{o}$.

By definition let $d_k > 0$ be a positive component of $\mathbf{d} = [d_1, \dots, d_k > 0, \dots, d_n]'$. The following implication holds (the vector $\mathbf{e} > \mathbf{o}$ can be replaced by any other positive vector $\mathbf{p} > \mathbf{o}$),

$$\mathbf{d} \geq \mathbf{o} \Rightarrow \mathbf{e}'\mathbf{d} = [1, 1, \dots, 1] \begin{bmatrix} d_1 \\ \dots \\ d_k \\ \dots \\ d_n \end{bmatrix} \geq 1 \cdot d_k = d_k > 0. \tag{A.11}$$

(7) Then, one defines the multiplication of matrices and the power operation.

Definition A.2.2 (Products of two matrices). Consider[2] the case of an $m \times n$ matrix **A** and an $n \times p$ matrix **B**,

$$\mathbf{A}_{m \times n} = \begin{bmatrix} a_{11} & a_{12} & \dots & \dots & a_{1n} \\ a_{21} & a_{22} & \dots & \dots & a_{2n} \\ \dots & \dots & \dots & \dots & \dots \\ a_{m1} & a_{m2} & \dots & \dots & a_{mn} \end{bmatrix}, \quad \mathbf{B}_{n \times p} = \begin{bmatrix} b_{11} & b_{12} & \dots & b_{1p} \\ b_{21} & b_{22} & \dots & b_{2p} \\ \dots & \dots & \dots & \dots \\ \dots & \dots & \dots & \dots \\ b_{n1} & b_{n2} & \dots & b_{np} \end{bmatrix}. \tag{A.12}$$

2 Note that it is convenient here to indicate the number of rows and column of a matrix as an index symbol, when describing the multiplication of matrices.

The scalar product of the i-th row of matrix **A** with the j-th column of matrix **B** gives the (i,j)-element c_{ij} of a new $m \times m$ matrix **C**:

$$c_{ij} = \sum_{k=1}^{n} a_{ik} b_{kj} \Rightarrow$$

$$\mathbf{C}_{m\times p} = \mathbf{A}_{m\times n} \cdot \mathbf{B}_{n\times p} = \begin{bmatrix} c_{11} & c_{12} & \dots & \dots & c_{1p} \\ c_{21} & c_{22} & \dots & \dots & c_{2p} \\ \dots & \dots & \dots & \dots & \dots \\ \dots & \dots & \dots & \dots & \dots \\ c_{m1} & c_{m2} & \dots & \dots & c_{mp} \end{bmatrix}. \tag{A.13}$$

Note that for a multiplication of two matrices, the number of elements in the rows of the first matrix, here **A**, must be equal to the number of elements in the columns of the second matrix, here **B**.

Example A.2.2. Consider the 3×3 matrices,

$$\mathbf{A}_{3\times 3} = \begin{bmatrix} 3 & 1.5 & 2 \\ 1 & 4 & 7 \\ 1 & -2 & 5 \end{bmatrix}; \quad \mathbf{B}_{3\times 3} = \begin{bmatrix} 1 & 0 & -2 \\ 1 & 2 & -3 \\ 2 & 2 & 3 \end{bmatrix}. \tag{A.14}$$

Then compute the product,

$$\begin{aligned} \mathbf{C}_{3\times 3} = \mathbf{A}_{3\times 3} \cdot \mathbf{B}_{3\times 3} &= \begin{bmatrix} 3 & 1.5 & 2 \\ 1 & 4 & 7 \\ 1 & -2 & 5 \end{bmatrix} \begin{bmatrix} 1 & 0 & -2 \\ 1 & 2 & -3 \\ 2 & 2 & 3 \end{bmatrix} \\ &= \begin{bmatrix} 3\cdot 1 + 1.5\cdot 1 + 2\cdot 2 & 3\cdot 0 + 1.5\cdot 2 + 2\cdot 2 & 3\cdot(-2) + 1.5\cdot(-3) + 2\cdot 3 \\ 1\cdot 1 + 4\cdot 1 + 7\cdot 2 & 1\cdot 0 + 4\cdot 2 + 7\cdot 2 & 1\cdot(-2) + 4\cdot(-3) + 7\cdot 3 \\ 1\cdot 1 + (-2)\cdot 1 + 5\cdot 2 & 1\cdot 0 + (-2)\cdot 2 + 5\cdot 2 & 1\cdot(-2) + (-2)\cdot(-3) + 5\cdot 3 \end{bmatrix} \\ &= \begin{bmatrix} 8.5 & 7 & -4.5 \\ 19 & 22 & 7 \\ 9 & 6 & 19 \end{bmatrix}. \end{aligned} \tag{A.15}$$

We present the rules of multiplication only for square matrices. Here the first important rules with transposed matrices appears. Note again that the transpose of a matrix **A** is another matrix, **A'**, created by writing the columns of $\mathbf{A}'$ as the rows of **A** (A.2).

Lemma A.2.2 (Rules for multiplication of $n \times n$ matrices).

(1) *Multiplication: With two $n \times n$ matrices* **A**, **B**, *the matrix* $\mathbf{A} \cdot \mathbf{B}$ *(also written* **AB***) is again an $n \times n$ matrix;*

(2) *Associative law:* $(\mathbf{A} \cdot \mathbf{B}) \cdot \mathbf{C} = \mathbf{A} \cdot (\mathbf{B} \cdot \mathbf{C})$*;*

(3) *Neutral element:* $\mathbf{A} \cdot \mathbf{I} = \mathbf{I} \cdot \mathbf{A} = \mathbf{A}$*;*

(4) *Distributive law:* $\mathbf{A} \cdot (\mathbf{B} + \mathbf{C}) = \mathbf{A} \cdot \mathbf{B} + \mathbf{A} \cdot \mathbf{C}$*;*
(5) *In general:* $\mathbf{A} \cdot \mathbf{B} \neq \mathbf{B} \cdot \mathbf{A}$*;*
(6) $(\mathbf{A} \cdot \mathbf{B})' = \mathbf{B}' \cdot \mathbf{A}'$*;*
(7) $\mathbf{A} \cdot \mathbf{A}'$ *and* $\mathbf{A}' \cdot \mathbf{A}$ *define symmetric matrices.*

Definition A.2.3 (Powers of a matrix). Consider now a $n \times n$ square matrix $\mathbf{A}$ and $k \in \mathbb{N}$. $\mathbf{A}^k$ is the product of k copies of matrix $\mathbf{A}$,

$$\mathbf{A}^k = \underbrace{\mathbf{A} \cdot \mathbf{A} \cdot \ldots \cdot \mathbf{A}}_{\mathrm{k}}. \tag{A.16}$$

(8) The *diagonal operator* of a vector is an important tool. In the present application, vectors may be represented in two equivalent ways:

Consider the vector $\mathbf{x} = [x_1, x_2, \ldots, x_n]'$. It can be converted into a diagonal matrix. We denote $\operatorname{diag}(\mathbf{x}) = \hat{\mathbf{x}}$, a *diagonal operator* on a vector which delivers a matrix, where the diagonal elements x_{ii} of matrix $\operatorname{diag}(\mathbf{x})$ are equal to the elements x_i of vector $\mathbf{x}$, following the rule $x_{ii} = x_i$, $i = 1, \ldots, n$. Here, due to the notations of Miller & Blair [65], the diagonal matrix is denoted by $\hat{\mathbf{x}}$, i. e.,

$$\hat{\mathbf{x}} = \begin{bmatrix} x_1 & 0 & \ldots & 0 \\ 0 & x_2 & \ldots & 0 \\ \ldots & \ldots & \ldots & \ldots \\ 0 & 0 & \ldots & x_n \end{bmatrix}. \tag{A.17}$$

Note that $\hat{\mathbf{x}}' = \hat{\mathbf{x}}$.

8.1 In the *contravariant* form with $\mathbf{x} = [x_1, x_2, \ldots, x_n]'$ as column vector, the transposition $\mathbf{x}'$ is a row vector. We then have for $\mathcal{X} = x_1 + x_2 + \cdots + x_n$:

$$\mathcal{X} = \mathbf{e}'\mathbf{x}. \tag{A.18}$$

8.2 In the *covariant* form, the vector elements x_i become diagonal elements of the diagonal matrix $\hat{\mathbf{x}}$ (A.17). We then have for summation in compact form:

$$\mathcal{X} = \mathbf{e}'\hat{\mathbf{x}}\mathbf{e}. \tag{A.19}$$

8.3 With $\hat{\mathbf{x}}$ (A.17) there is the important rule

$$\hat{\mathbf{x}}\mathbf{e} = \mathbf{x} \Leftrightarrow \mathbf{x}' = \mathbf{e}'\hat{\mathbf{x}}. \tag{A.20}$$

8.4 Consider a positive vector $\mathbf{p}' = [p_1, \ldots, p_n] > \mathbf{o}'$ and build the diagonal matrices $\hat{\mathbf{p}}$ and $\hat{\mathbf{p}}^{-1}$; then one has the identity

$$\mathbf{I} = \hat{\mathbf{p}}\hat{\mathbf{p}}^{-1} = \hat{\mathbf{p}}^{-1}\hat{\mathbf{p}}. \tag{A.21}$$

A square matrix is regular, respectively, singular, if its determinant is different from zero, respectively, equal to zero, Definition A.4.2.

(9) We define the important notion of the rank of a matrix:

Definition A.2.4 (The rank of a matrix). The $n \times m$ matrix $\mathbf{A} \neq \mathbf{0}$ has rank $\text{rank}(\mathbf{A}) = \rho$, if it has at least a regular $\rho - row$ submatrix and all its submatrices of more than ρ rows are singular.

A.3 Determinant of a matrix

For this section, see Lütkepohl [60], pp. 451–452. The determinant of matrices are only defined for square matrices. For a 1×1 matrix $\mathbf{A} = [a_{11}]$ the determinant equals the single element a_{11}, written as $\det(\mathbf{A}) = |\mathbf{A}| = a_{11}$. For an $m \times m$ matrix the determinant may be defined recursively. Suppose $m = 2$; then the determinant is defined as follows:

$$\mathbf{A} = \begin{bmatrix} a_{11} & a_{12} \\ a_{21} & a_{22} \end{bmatrix} \Rightarrow \det(\mathbf{A}) = |\mathbf{A}| := a_{11}a_{22} - a_{12}a_{21}. \tag{A.22}$$

For instance,

$$\mathbf{A} = \begin{bmatrix} 3 & 1.5 \\ 4 & 7 \end{bmatrix} \Rightarrow \det(\mathbf{A}) = |\mathbf{A}| = 3 \cdot 7 - 1.5 \cdot 4 = 15. \tag{A.23}$$

To specify the determinant of a $m \times m$ matrix $\mathbf{A} = (a_{ij})$, $i, j = 1, \ldots, m$, we define the minor $|\mathbf{A}_{ij}|$ corresponding to the indices pair (i, j) of the matrix element a_{ij} as the determinant of the $(m-1) \times (m-1)$ matrix $\mathbf{A}_{ij} = (a_{kl})$, $k, l = 1, \ldots, m-1$, $k \neq i$, $l \neq j$, that is obtained by deleting the i-th row and the j-th column from $\mathbf{A}$; there are m^2 minors. The cofactor of a_{ij} denoted by A_{ij}, is the minor multiplied by $(-1)^{i+j}$, i. e., $A_{ij} = (-1)^{(i+j)} \cdot |\mathbf{A}_{ij}|$. Now by defintion,

$$\det(\mathbf{A}) = |\mathbf{A}| := \sum_{j=1}^{m} a_{ij} \cdot A_{ij} = \sum_{i=1}^{m} a_{ij} \cdot A_{ij}, \tag{A.24}$$

for any i or j, $i, j \in \{1, \ldots, m\}$. It can be shown that it does not matter which row or column is chosen in (A.24) for the development, because the determinant of a matrix is a unique number.

Example A.3.1. Consider the 3×3 matrix,

$$\mathbf{A} = \begin{bmatrix} 1 & 1 & -1 \\ 2 & 0 & -1 \\ 1 & -2 & 1 \end{bmatrix}. \tag{A.25}$$

We expand the matrix **A**, for example by taking the 2nd row, to obtain its determinant. First, we determine the three cofactors,

$$A_{21} = (-1)^{(2+1)} \left| \begin{bmatrix} 1 & -1 \\ -2 & 1 \end{bmatrix} \right| = (-1)(1 \cdot 1 - (-1) \cdot (-2)) = 1,$$
$$A_{22} = (-1)^{(2+2)} \left| \begin{bmatrix} 1 & -1 \\ 1 & 1 \end{bmatrix} \right| = (1 \cdot 1 - (-1) \cdot 1) = 2,$$
$$A_{23} = (-1)^{(2+3)} \left| \begin{bmatrix} 1 & 1 \\ 1 & -2 \end{bmatrix} \right| = (-1)(1 \cdot (-2) - 1 \cdot 1) = 3. \tag{A.26}$$

Then the determinant is

$$\det(\mathbf{A}) = |\mathbf{A}| = a_{21}A_{21} + a_{22}A_{22} + a_{23}A_{23} = 2 \cdot 1 + 0 \cdot 2 + (-1) \cdot 3 = -1. \quad \blacktriangle \tag{A.27}$$

Then, the following rules apply for the calculation of determinants of matrices $\mathbf{A} = (a_{ij})$, $\mathbf{B} = (b_{ij})$, $i, j = 1, \dots, m$.

Lemma A.3.1 (Rules for the calculation of determinants of matrices).
(1) *The determinant of the identity matrix is 1,* $\det(\mathbf{I}) = \det(\mathbf{I}_m) = 1$.
(2) *If* **A** *is a diagonal matrix, then* $\det(\mathbf{A}) = a_{11}a_{22} \cdot \ldots \cdot a_{mm}$.
(3) $\det(c\mathbf{A}) = c^m \det(\mathbf{A})$ *for* $c \in \mathbb{R}$.
(4) $\det(\mathbf{AB}) = \det(\mathbf{A}) \det(\mathbf{B})$.
(5) *If* **A** *contains a row or a column of zeros, then* $\det(\mathbf{A}) = 0$.

A.4 Inverse matrices and further elementary matrix algebra

(1) The inverse matrix of a square matrix has the role of the reciprocal $1/x$ of a real number $x \neq 0$.

Consider a general $m \times m$ matrix $\mathbf{A} = (a_{kl})$, $k, l = 1, \dots, m$. To define its *inverse matrix* $\mathbf{A}^{-1}$ we need to return to the cofactors $A_{ij} = (-1)^{i+j}|\mathbf{A}_{ij}|$ of a matrix **A**, defined in the above Section A.3. With these cofactors we build the *adjugate matrix* as follows.

Definition A.4.1 (The adjugate or adjunct matrix).

$$\mathbf{A} = \begin{bmatrix} a_{11} & a_{12} & \dots & a_{1m} \\ a_{21} & a_{22} & \dots & a_{2m} \\ \dots & \dots & \dots & \dots \\ a_{m1} & a_{m2} & \dots & a_{mn} \end{bmatrix} \Rightarrow \operatorname{adj}(\mathbf{A}) = \begin{bmatrix} A_{11} & A_{12} & \dots & A_{1m} \\ A_{21} & A_{22} & \dots & A_{2m} \\ \dots & \dots & \dots & \dots \\ A_{m1} & A_{m2} & \dots & A_{mm} \end{bmatrix}'. \tag{A.28}$$

An $m \times m$ matrix $\mathbf{A} = (a_{ij})$, $i, j = 1, \dots, m$ is regular, if its determinant does not vanish, $\det(\mathbf{A}) \neq 0$. For regular matrices the *inverse matrix* can be defined.

Definition A.4.2 (The inverse matrix of a regular matrix). The inverse matrix of a square regular matrix $\mathbf{A}$ is the adj($\mathbf{A}$) divided by the determinant det($\mathbf{A}$) $\neq 0$,

$$\mathbf{A}^{-1} = \frac{1}{\det(\mathbf{A})} \cdot \operatorname{adj}(\mathbf{A}). \tag{A.29}$$

The following rules apply for calculating with inverse matrices, determinants and transpose of matrices. Consider again the $m \times m$ matrices $\mathbf{A} = (a_{ij})$, $\mathbf{B} = (b_{ij})$, $i, j = 1, \ldots, m$.

Lemma A.4.1 (Rules for calculation with inverse matrices).

(1) $(\mathbf{A}')^{-1} = (\mathbf{A}^{-1})'$;

(2) $(c\mathbf{A})^{-1} = \frac{1}{c}\mathbf{A}^{-1}, c \in \mathbb{R} \setminus \{0\}$;

(3) *If* $\mathbf{A}$ *is diagonal, then* $\mathbf{A}^{-1}$ *is also diagonal with diagonal elements* $\frac{1}{a_{ii}}$.

Now we illustrate these notions with a concrete example

Example A.4.1. Consider the 3×3 matrix (A.25).

Compute all the cofactors of $\mathbf{A}$, then construct adj($\mathbf{A}$) and calculate the inverse matrix $\mathbf{A}^{-1}$.

The nine cofactors are:

$$A_{11} = (-1)^{(1+1)} \left| \begin{bmatrix} 0 & -1 \\ -2 & 1 \end{bmatrix} \right| = (0 \cdot 1 - (-1) \cdot (-2)) = -2,$$

$$A_{12} = (-1)^{(1+2)} \left| \begin{bmatrix} 2 & -1 \\ 1 & 1 \end{bmatrix} \right| = (-1)(2 \cdot 1 - (-1) \cdot 1) = -3,$$

$$A_{13} = (-1)^{(1+3)} \left| \begin{bmatrix} 2 & 0 \\ 1 & -2 \end{bmatrix} \right| = (2 \cdot (-2) - 0 \cdot 1) = -4,$$

$$A_{21} = (-1)^{(2+1)} \left| \begin{bmatrix} 1 & -1 \\ -2 & 1 \end{bmatrix} \right| = (-1)(1 \cdot 1 - (-1) \cdot (-2)) = 1,$$

$$A_{22} = (-1)^{(2+2)} \left| \begin{bmatrix} 1 & -1 \\ 1 & 1 \end{bmatrix} \right| = (1 \cdot 1 - (-1) \cdot 1) = 2, \tag{A.30}$$

$$A_{23} = (-1)^{(2+3)} \left| \begin{bmatrix} 1 & 1 \\ 1 & -2 \end{bmatrix} \right| = (-1)(1 \cdot (-2) - 1 \cdot 1) = 3,$$

$$A_{31} = (-1)^{(3+1)} \left| \begin{bmatrix} 1 & -1 \\ 0 & -1 \end{bmatrix} \right| = (1 \cdot (-1) - (-1) \cdot 0) = -1,$$

$$A_{32} = (-1)^{(3+2)} \left| \begin{bmatrix} 1 & -1 \\ 2 & -1 \end{bmatrix} \right| = (-1)(1 \cdot (-1) - (-1) \cdot 2) = -1,$$

$$A_{33} = (-1)^{(3+3)} \left| \begin{bmatrix} 1 & 1 \\ 2 & 0 \end{bmatrix} \right| = (1 \cdot 0 - 1 \cdot 2) = -2. \tag{A.31}$$

As we have calculated the determinant $\det(\mathbf{A}) = -1$ in Example A.3.1, we easily get the adjunct matrix and the inverse matrix, as

$$\operatorname{adj}(\mathbf{A}) = \begin{bmatrix} -2 & 1 & -1 \\ -3 & 2 & -1 \\ -4 & 3 & -2 \end{bmatrix} \Rightarrow$$

$$\mathbf{A}^{-1} = \frac{\operatorname{adj}(\mathbf{A})}{\det(\mathbf{A})} = \frac{1}{-1} \cdot \begin{bmatrix} -2 & 1 & -1 \\ -3 & 2 & -1 \\ -4 & 3 & -2 \end{bmatrix} = \begin{bmatrix} 2 & -1 & 1 \\ 3 & -2 & 1 \\ 4 & -3 & 2 \end{bmatrix}, \tag{A.32}$$

and of course,

$$\mathbf{A} \cdot \mathbf{A}^{-1} = \begin{bmatrix} 1 & 1 & -1 \\ 2 & 0 & -1 \\ 1 & -2 & 1 \end{bmatrix} \begin{bmatrix} 2 & -1 & 1 \\ 3 & -2 & 1 \\ 4 & -3 & 2 \end{bmatrix} = \begin{bmatrix} 1 & 0 & 0 \\ 0 & 1 & 0 \\ 0 & 0 & 1 \end{bmatrix}. \tag{A.33}$$

(2) For the matrix $\hat{\mathbf{x}}$ the diagonal matrix $\hat{\mathbf{x}}^{-1}$ is

$$\hat{\mathbf{x}} = \begin{bmatrix} x_1 & 0 & \dots & 0 \\ 0 & x_2 & \dots & 0 \\ \dots & \dots & \dots & \dots \\ 0 & 0 & \dots & x_n \end{bmatrix}, \quad \hat{\mathbf{x}}^{-1} = \begin{bmatrix} \frac{1}{x_1} & 0 & \dots & 0 \\ 0 & \frac{1}{x_2} & \dots & 0 \\ \dots & \dots & \dots & \dots \\ 0 & 0 & \dots & \frac{1}{x_n} \end{bmatrix}, \tag{A.34}$$

and as usual the matrix times its inverse yields the identity matrix,

$$\hat{\mathbf{x}}\hat{\mathbf{x}}^{-1} = \hat{\mathbf{x}}^{-1}\hat{\mathbf{x}} = \mathbf{I}_n = \begin{bmatrix} 1 & 0 & \dots & 0 \\ 0 & 1 & \dots & \dots \\ \dots & \dots & \dots & \dots \\ 0 & 0 & \dots & 1 \end{bmatrix}. \tag{A.35}$$

When a matrix $\mathbf{S}$ is multiplied by a matrix $\mathbf{T}$, then the resulting matrix $\tilde{\mathbf{S}} = \mathbf{ST}$ is called the "matrix *mapped* by matrix $\mathbf{T}$" or shortly the "*mapped* matrix".

(3) We now establish an anti-symmetry property for the transpose of *commodity flow matrix* $\tilde{\mathbf{S}}$, mapped by the diagonal matrix $\hat{y}$,

$$\tilde{\mathbf{S}}' := \hat{y}\mathbf{S}' = \begin{bmatrix} y_1 & 0 & 0 \\ 0 & y_2 & 0 \\ 0 & 0 & y_3 \end{bmatrix} \begin{bmatrix} s_{11} & s_{21} & s_{31} \\ s_{12} & s_{22} & s_{32} \\ s_{13} & s_{23} & s_{33} \end{bmatrix} = \begin{bmatrix} y_1 s_{11} & y_1 s_{21} & y_1 s_{31} \\ y_2 s_{12} & y_2 s_{22} & y_2 s_{32} \\ y_3 s_{13} & y_3 s_{23} & y_3 s_{33} \end{bmatrix} \Rightarrow$$

$$\tilde{\mathbf{S}} = \mathbf{S}\hat{y} = \begin{bmatrix} s_{11} & s_{12} & s_{13} \\ s_{21} & s_{22} & s_{23} \\ s_{31} & s_{32} & s_{33} \end{bmatrix} \begin{bmatrix} y_1 & 0 & 0 \\ 0 & y_2 & 0 \\ 0 & 0 & y_3 \end{bmatrix} = \begin{bmatrix} s_{11} y_1 & s_{12} y_2 & s_{13} y_3 \\ s_{21} y_1 & s_{22} y_2 & s_{23} y_3 \\ s_{31} y_1 & s_{32} y_2 & s_{33} y_3 \end{bmatrix},$$

$$\tilde{\mathbf{S}}' := \hat{y}\mathbf{S}' \Leftrightarrow \tilde{\mathbf{S}} = \mathbf{S}\hat{y}. \tag{A.36}$$

(4) There is also a rule of symmetry for the mapped vector of *total output*,

$$\tilde{\mathbf{q}} := \hat{\mathbf{y}}\mathbf{q} = \begin{bmatrix} y_1 & 0 & 0 \\ 0 & y_2 & 0 \\ 0 & 0 & y_3 \end{bmatrix} \begin{bmatrix} q_1 \\ q_2 \\ q_3 \end{bmatrix} = \begin{bmatrix} y_1 q_1 \\ y_2 q_2 \\ y_3 q_3 \end{bmatrix}, \quad \text{and}$$

$$\hat{\mathbf{q}}\mathbf{y} = \begin{bmatrix} q_1 & 0 & 0 \\ 0 & q_2 & 0 \\ 0 & 0 & q_3 \end{bmatrix} \begin{bmatrix} y_1 \\ y_2 \\ y_3 \end{bmatrix} = \begin{bmatrix} q_1 y_1 \\ q_2 y_2 \\ q_3 y_3 \end{bmatrix} = \tilde{\mathbf{q}}, \quad \text{therefore}$$

$$\tilde{\mathbf{q}} := \hat{\mathbf{y}}\mathbf{q} = \hat{\mathbf{q}}\mathbf{y}. \tag{A.37}$$

(5) Diagonalizing the mapped vector of *total output* we get the useful identity $\tilde{\mathbf{q}} = \hat{\mathbf{y}}\mathbf{q}$:

$$\hat{\tilde{\mathbf{q}}} = \begin{bmatrix} y_1 q_1 & 0 & \dots & 0 \\ 0 & y_2 q_2 & \dots & 0 \\ \dots & \dots & \dots & \dots \\ 0 & 0 & \dots & y_n q_n \end{bmatrix}$$

$$= \begin{bmatrix} y_1 & 0 & \dots & 0 \\ 0 & y_2 & \dots & 0 \\ \dots & \dots & \dots & \dots \\ 0 & 0 & \dots & y_n \end{bmatrix} \begin{bmatrix} q_1 & 0 & \dots & 0 \\ 0 & q_2 & \dots & 0 \\ \dots & \dots & \dots & \dots \\ 0 & 0 & \dots & q_n \end{bmatrix} = \hat{\mathbf{y}}\hat{\mathbf{q}} \Leftrightarrow \hat{\tilde{\mathbf{q}}} = \widehat{\hat{\mathbf{y}}\mathbf{q}} = \hat{\mathbf{y}}\hat{\mathbf{q}}. \tag{A.38}$$

(6) Another important identity is

$$\hat{\mathbf{x}}^{-1}\mathbf{x} = \begin{bmatrix} \frac{1}{x_1} & 0 & \dots & 0 \\ 0 & \frac{1}{x_2} & \dots & 0 \\ \dots & \dots & \dots & \dots \\ 0 & 0 & \dots & \frac{1}{x_n} \end{bmatrix} \begin{bmatrix} x_1 \\ x_2 \\ \dots \\ x_n \end{bmatrix} = \begin{bmatrix} 1 \\ 1 \\ \dots \\ 1 \end{bmatrix} = \mathbf{e}. \tag{A.39}$$

One also uses the transpose of this identity: $\mathbf{e}' = \mathbf{x}'\hat{\mathbf{x}}^{-1}$.

(7) The *diagonal-operator* applied to a vector $\mathbf{x}$, getting $\hat{\mathbf{x}}$ (A.17), is important for the representation of Sraffa's price model (PCMC) in matrix form.

Consider for instance the $n \times 1$ vector of ones, $\mathbf{e} = [1, \dots, 1]'$, and a $n \times 1$ price vector $\mathbf{p} = [p_1, \dots, p_n]'$, see Miller ([65], p. 12). The following identity is useful:

$$\hat{\mathbf{p}}\mathbf{e} = \begin{bmatrix} p_1 & 0 & \dots & 0 \\ 0 & p_2 & \dots & 0 \\ \dots & \dots & \dots & \dots \\ 0 & 0 & \dots & p_n \end{bmatrix} \begin{bmatrix} 1 \\ 1 \\ \dots \\ 1 \end{bmatrix} = \begin{bmatrix} p_1 \\ p_2 \\ \dots \\ p_n \end{bmatrix} = \mathbf{p}. \tag{A.40}$$

Now let us summarize some results concerning the *diagonal-operator*.

Lemma A.4.2 (Rules for calculation with the *diagonal-operator*).
(1) $\mathbf{I}_n = \hat{\mathbf{x}}\hat{\mathbf{x}}^{-1} = \hat{\mathbf{x}}^{-1}\hat{\mathbf{x}}$, *as equation* (A.35);
(2) $\mathbf{e}' = \mathbf{x}'\hat{\mathbf{x}}^{-1}$, *as equation* (A.39);
(3) $\mathbf{p} = \hat{\mathbf{p}}\mathbf{e}$, *as equation* (A.40).

At this stage it will be useful to illustrate in connection with Leontief and Sraffa, the transposition and left/right multiplications of matrices, which means the non-commutativity of multiplication of matrices (see Lemma A.2.2 (5)).

Example A.4.2. Take a positive commodity flow matrix $\mathbf{S} = \left[\begin{smallmatrix} 1 & 2 \\ 3 & 4 \end{smallmatrix}\right]$, compute the vector of total output $\mathbf{q} = \mathbf{Se}$, build up the diagonal matrix $\hat{\mathbf{q}}$ and calculate,

Solution to Example A.4.2:

$$\mathbf{q} = \mathbf{Se} = \begin{bmatrix} 1 & 2 \\ 3 & 4 \end{bmatrix}\begin{bmatrix} 1 \\ 1 \end{bmatrix} = \begin{bmatrix} 3 \\ 7 \end{bmatrix}, \quad \hat{\mathbf{q}} = \begin{bmatrix} 3 & 0 \\ 0 & 7 \end{bmatrix} \Rightarrow \hat{\mathbf{q}}^{-1} = \begin{bmatrix} \frac{1}{3} & 0 \\ 0 & \frac{1}{7} \end{bmatrix}, \tag{A.41}$$

which gives four different results

$$\begin{aligned}
\mathbf{S}\hat{\mathbf{q}}^{-1} &= \begin{bmatrix} 1 & 2 \\ 3 & 4 \end{bmatrix}\begin{bmatrix} \frac{1}{3} & 0 \\ 0 & \frac{1}{7} \end{bmatrix} = \begin{bmatrix} \frac{1}{3} & \frac{2}{7} \\ 1 & \frac{4}{7} \end{bmatrix}, \\
\hat{\mathbf{q}}^{-1}\mathbf{S} &= \begin{bmatrix} \frac{1}{3} & 0 \\ 0 & \frac{1}{7} \end{bmatrix}\begin{bmatrix} 1 & 2 \\ 3 & 4 \end{bmatrix} = \begin{bmatrix} \frac{1}{3} & \frac{2}{3} \\ \frac{3}{7} & \frac{4}{7} \end{bmatrix}, \\
\mathbf{S}'\hat{\mathbf{q}}^{-1} &= \begin{bmatrix} 1 & 3 \\ 2 & 4 \end{bmatrix}\begin{bmatrix} \frac{1}{3} & 0 \\ 0 & \frac{1}{7} \end{bmatrix} = \begin{bmatrix} \frac{1}{3} & \frac{3}{7} \\ \frac{2}{3} & \frac{4}{7} \end{bmatrix}, \\
\hat{\mathbf{q}}^{-1}\mathbf{S}' &= \begin{bmatrix} \frac{1}{3} & 0 \\ 0 & \frac{1}{7} \end{bmatrix}\begin{bmatrix} 1 & 3 \\ 2 & 4 \end{bmatrix} = \begin{bmatrix} \frac{1}{3} & 1 \\ \frac{2}{7} & \frac{4}{7} \end{bmatrix}. \quad \blacktriangle
\end{aligned} \tag{A.42}$$

(8) Consider the $n \times n$ *commodity flow matrix* $\mathbf{Z}$ of all *interindustry sales* z_{ij} by sector i to all sectors j, the *input-output coefficients* matrix $\mathbf{A}$ from (2.9) and the vector of total production $\mathbf{x} = [x_1, x_2, \ldots, x_n]'$. From equation (2.8) we know that $z_{ij} = a_{ij}x_j$, $i, j = 1, \ldots, n$. We find an equation relating $\mathbf{Z}$ and $\mathbf{A}$ with the diagonal matrix $\hat{\mathbf{x}}$:

$$\begin{aligned}
\mathbf{Z} &= \begin{bmatrix} z_{11} & z_{12} & \ldots & z_{1n} \\ z_{21} & z_{22} & \ldots & z_{2n} \\ \ldots & \ldots & \ldots & \ldots \\ z_{n1} & z_{n2} & \ldots & z_{nn} \end{bmatrix} = \begin{bmatrix} a_{11}x_1 & a_{12}x_2 & \ldots & a_{1n}x_n \\ a_{21}x_1 & a_{22}x_2 & \ldots & a_{2n}x_n \\ \ldots & \ldots & \ldots & \ldots \\ a_{n1}x_1 & a_{n2}x_2 & \ldots & a_{nn}x_n \end{bmatrix} \\
&= \begin{bmatrix} a_{11} & a_{12} & \ldots & a_{1n} \\ a_{21} & a_{22} & \ldots & a_{2n} \\ \ldots & \ldots & \ldots & \ldots \\ a_{n1} & a_{n2} & \ldots & a_{nn} \end{bmatrix}\begin{bmatrix} x_1 & 0 & \ldots & 0 \\ 0 & x_2 & \ldots & 0 \\ \ldots & \ldots & \ldots & \ldots \\ 0 & 0 & \ldots & x_n \end{bmatrix} = \mathbf{A}\hat{\mathbf{x}}.
\end{aligned} \tag{A.43}$$

(9) Let us introduce (A.43) into equation $\mathbf{x} = \mathbf{Ze} + \mathbf{f}$ (2.5), to obtain

$$\mathbf{x} = \mathbf{A}\hat{\mathbf{x}}\mathbf{e} + \mathbf{f} = \mathbf{A}(\hat{\mathbf{x}}\mathbf{e}) + \mathbf{f} = \mathbf{Ax} + \mathbf{f}, \tag{A.44}$$

which implies

$$\mathbf{x} = \mathbf{Ix} = \mathbf{Ax} + \mathbf{f} \Rightarrow \mathbf{f} = \mathbf{Ix} - \mathbf{Ax} = (\mathbf{I} - \mathbf{A})\mathbf{x}. \tag{A.45}$$

The matrix $\mathbf{L} := (\mathbf{I} - \mathbf{A})^{-1}$ is called the *Leontief Inverse* (see Miller & Blair [65], p. 21) and allows to calculate the vector of total supply $\mathbf{x}$ from the vector of total demand $\mathbf{f}$,

$$\mathbf{x} = (\mathbf{I} - \mathbf{A})^{-1}\mathbf{f} = \mathbf{Lf}. \tag{A.46}$$

(10) Miller & Blair ([65], p. 188) discuss equation (A.45) and state that the exogenous demand $\mathbf{f}$ is the driving force of supply $\mathbf{x}$. The following development exists if there is a non-negative matrix $\mathbf{A} \geqq \mathbf{0}$ with Frobenius number $\lambda_A < 1$, see Theorem A.12.1:

$$\begin{aligned} \mathbf{L} &= (\mathbf{I} - \mathbf{A})^{-1} = \mathbf{I} + \mathbf{A} + \mathbf{A}^2 + \mathbf{A}^3 + \cdots \\ \mathbf{Lf} &= (\mathbf{I} - \mathbf{A})^{-1}\mathbf{f} = \mathbf{f} + \mathbf{Af} + \mathbf{A}^2\mathbf{f} + \cdots. \end{aligned} \tag{A.47}$$

One observes that this condition is analogous to the convergence condition of geometric series in elementary analysis. The development (A.47) can be interpreted as follows: $\mathbf{f}$ initiates a residual demand of the same type $\mathbf{Af}$, which itself initiates $\mathbf{A}(\mathbf{Af})$ etc., the producer must account for this in his production planning.

(11) Yang [124] and Schefold ([103], p. 44) describe matrices as column vectors of row vectors, respectively row vectors of column vectors. This representation will be useful in treating some matrix operations. For this reason we set the following definition:

Definition A.4.3 (Decomposition of a square matrix in column or row vectors). Consider the $n \times n$ matrix $\mathbf{X} = (x_{ij})$, $i, j = 1, \ldots, n$ and its column vectors $\mathbf{x}_{\cdot j} = [x_{1j}, x_{2j}, \ldots, x_{nj}]'$, $j = 1, \ldots, n$, respectively its row vectors $\mathbf{x}'_{i\cdot} = [x_{i1}, x_{i2}, \ldots, x_{in}]$, $i = 1, \ldots, n$. Then the matrix $\mathbf{X}$ may be written as a matrix of column vectors or of row vectors as follows:

$$\mathbf{X} = \begin{bmatrix} x_{11} & x_{12} & \cdots & \cdots & x_{1n} \\ x_{21} & x_{22} & \cdots & \cdots & x_{2n} \\ \cdots & \cdots & \cdots & \cdots & \cdots \\ \cdots & \cdots & \cdots & \cdots & \cdots \\ x_{n1} & x_{n2} & \cdots & \cdots & x_{nn} \end{bmatrix} = [\mathbf{x}_{\cdot 1}, \mathbf{x}_{\cdot 2}, \ldots, \ldots, \mathbf{x}_{\cdot n}] = \begin{bmatrix} \mathbf{x}'_{1\cdot} \\ \mathbf{x}'_{2\cdot} \\ \cdots \\ \cdots \\ \mathbf{x}'_{n\cdot} \end{bmatrix}. \tag{A.48}$$

A.5 Eigenvalues and eigenvectors of matrices

This section treats the eigenvalues and eigenvectors of square matrices, see Lütkepohl [60], pp. 455–456. The m solutions of the equation $P_m(\lambda) = \det(\mathbf{A} - \lambda\mathbf{I}) = 0$ are called the *eigenvalues* λ and $P_m(\lambda)$ is called the *characteristic polynomial* of the $m \times m$ matrix $\mathbf{A}$. The m roots are usually complex numbers.

A vector $\mathbf{v}_i$ associated with one of the eigenvalues λ_i of matrix **A**, unique up to a *multiplicative constant*, is called an *eigenvector* corresponding to the eigenvalue λ_i, if and only if

$$\mathbf{A}\mathbf{v}_i = \lambda_i \mathbf{v}_i. \tag{A.49}$$

Example A.5.1. Consider the matrix

$$\begin{aligned}\mathbf{A} &= \begin{bmatrix} 3 & 1.25 \\ 4 & 7 \end{bmatrix} \Rightarrow P_2(\lambda) = |\det(\mathbf{A} - \lambda \mathbf{I})| \\ &= \begin{bmatrix} 3-\lambda & 1.5 \\ 4 & 7-\lambda \end{bmatrix} = \lambda^2 - 10\lambda + 16 = (\lambda - 8)(\lambda - 2) = 0.\end{aligned} \tag{A.50}$$

Hence $\lambda_1 = 8$ and $\lambda_2 = 2$ are the eigenvalues of matrix **A**. Corresponding eigenvectors are obtained by solving

$$\begin{bmatrix} 3 & 1.25 \\ 4 & 7 \end{bmatrix}\begin{bmatrix} v_{11} \\ v_{21} \end{bmatrix} = 8 \cdot \begin{bmatrix} v_{11} \\ v_{21} \end{bmatrix}, \quad \begin{bmatrix} 3 & 1.25 \\ 4 & 7 \end{bmatrix}\begin{bmatrix} v_{12} \\ v_{22} \end{bmatrix} = 2 \cdot \begin{bmatrix} v_{12} \\ v_{22} \end{bmatrix}. \tag{A.51}$$

Thus

$$\begin{bmatrix} v_{11} \\ v_{21} \end{bmatrix} = \begin{bmatrix} 1 \\ 4 \end{bmatrix} \quad \text{and} \quad \begin{bmatrix} v_{12} \\ v_{22} \end{bmatrix} = \begin{bmatrix} -5 \\ 4 \end{bmatrix} \tag{A.52}$$

are eigenvectors corresponding to $\lambda_1 = 8$ and $\lambda_2 = 2$, respectively. ▲

It may happen that we encounter triangular $m \times m$ matrices. They have one of the following forms

$$\begin{bmatrix} a_{11} & 0 & \dots & 0 \\ a_{21} & a_{22} & \dots & 0 \\ \dots & \dots & \dots & \dots \\ a_{m1} & a_{m2} & \dots & a_{mm} \end{bmatrix}, \quad \begin{bmatrix} a_{11} & a_{12} & \dots & a_{1m} \\ 0 & a_{22} & \dots & a_{2m} \\ \dots & \dots & \dots & \dots \\ 0 & 0 & \dots & a_{mm} \end{bmatrix}. \tag{A.53}$$

The matrix on the left of equation (A.53) is a *lower triangular* matrix, the matrix on the right of equation (A.53) is a *upper triangular* matrix.

The following rules apply for the calculation of eigenvalues of $m \times m$ matrices $\mathbf{A} = (a_{ij})$, $i, j = 1, \dots, m$.

Lemma A.5.1 (Rules for the calculation of eigenvalues).

(1) *When matrix **A** is symmetric, then all its eigenvalues are real numbers.*

(2) *The eigenvalues of a diagonal matrix are its diagonal elements.*

(3) *The eigenvalues of a triangular matrix are diagonal elements.*

(4) *When $\lambda_1, \lambda_2, \dots, \lambda_m$ are the eigenvalues of **A**, then the determinant is $\det(\mathbf{A}) = \lambda_1 \cdot \lambda_2 \cdot \dots \cdot \lambda_m$.*

(5) *If λ is an eigenvalue of **A**, then λ^k is an eigenvalue of matrix $\mathbf{A}^k$, $\mathbf{A}\mathbf{v} = \lambda\mathbf{v} \Rightarrow \mathbf{A}^k\mathbf{v} = \lambda^k\mathbf{v}$.*

(6) *Similar matrices have the same eigenvalues.*

A.6 Coefficients matrices with identical eigenvalues

Consider a semi-positive *commodity flow matrix* $\mathbf{S} \geq \mathbf{0}$, a non-negative vector of *surplus* $\mathbf{d} \geqq \mathbf{o}$, resulting in a positive vector of *total output* $\mathbf{q} = \mathbf{Se} + \mathbf{d} > \mathbf{o}$ in physical terms (Assumption 2.2.1). Consider the positive price vector $\mathbf{p} > \mathbf{o}$ in monetary terms, arising from the solution of a complete *single product Sraffa system* (4.82). One chooses any currency.

Lemma A.6.1 (Similar matrices). *Given a matrix* $\mathbf{S}$ *and vectors* $\mathbf{q}, \mathbf{p}$, *as defined above, consider*

$$\mathbf{Z} = \hat{\mathbf{p}}\mathbf{S}, \quad \mathbf{C} = \mathbf{S}\hat{\mathbf{q}}^{-1}, \quad \mathbf{D} = \hat{\mathbf{x}}^{-1}\mathbf{Z} = \hat{\mathbf{q}}^{-1}\mathbf{S}, \quad \mathbf{A} = \mathbf{Z}\hat{\mathbf{x}}^{-1}, \quad \hat{\mathbf{x}} = \hat{\mathbf{p}}\hat{\mathbf{q}}. \tag{A.54}$$

The matrices $\mathbf{A}, \mathbf{C}, \mathbf{D}$ *are similar*[3] *and have identical eigenvalues.*

Proof. We set up the vector of *total output* in value terms $\mathbf{x} = \mathbf{Ze}+\mathbf{f} > \mathbf{o}$ (2.7) the positive price vector $\mathbf{p} > \mathbf{o}$, the matrices $\mathbf{Z} = \hat{\mathbf{p}}\mathbf{S}$ and $\mathbf{S} = \hat{\mathbf{p}}^{-1}\mathbf{Z}$ (2.18) the vector of *total output* in physical terms $\mathbf{q} = \mathbf{Se} + \mathbf{d}$, the relation between vectors of output, $\hat{\mathbf{x}} = \hat{\mathbf{p}}\hat{\mathbf{q}}$ (2.18).

Then we set up the equations $\mathbf{Z} = \hat{\mathbf{x}}\mathbf{D} = \mathbf{A}\hat{\mathbf{x}}$ (2.12) and (2.8) and the equations $\mathbf{S} = \hat{\mathbf{q}}\mathbf{D} = \mathbf{C}\hat{\mathbf{q}}$ (2.19) and (2.17).

Finally we get $\mathbf{A} = \hat{\mathbf{x}}\mathbf{D}\hat{\mathbf{x}}^{-1} = \hat{\mathbf{p}}(\hat{\mathbf{q}}\mathbf{D}\hat{\mathbf{q}}^{-1})\hat{\mathbf{p}}^{-1} = \hat{\mathbf{p}}\mathbf{C}\hat{\mathbf{p}}^{-1}$.

This means that $\mathbf{A} = \hat{\mathbf{x}}\mathbf{D}\hat{\mathbf{x}}^{-1}$ and $\mathbf{C} = \hat{\mathbf{q}}\mathbf{D}\hat{\mathbf{q}}^{-1}$. Matrices $\mathbf{A}, \mathbf{C}, \mathbf{D}$ are similar and therefore all have the same eigenvalues. Indeed, if two matrices are similar to a third one all three are similar. □

Lemma A.6.1 is also applied in the *interindustrial economy* (9.22).

We investigate now a further eigenvector property of similar matrices $\mathbf{C}$ and $\mathbf{D}$, fulfilling Lemma A.6.1. We also refer to vectors $\mathbf{q}, \mathbf{e}, \mathbf{d}$ and matrix $\mathbf{S}$, mentioned in Lemma A.6.1.

Lemma A.6.2. *Let* λ *be an eigenvalue common to the matrices* $\mathbf{C}$ *and* $\mathbf{D}$. *Consider a right eigenvector* γ *of* $\mathbf{D}$, $\mathbf{D}\gamma = \lambda\gamma$. *Set up the diagonal matrix* $\hat{\gamma}$. *Then, with the output vector* $\mathbf{q} = \mathbf{Se} + \mathbf{d}$, *the vector* $\hat{\mathbf{q}}\gamma$ *is an eigenvector of the matrix* $\mathbf{C}$, $\mathbf{C}(\hat{\mathbf{q}}\gamma) = \lambda(\hat{\mathbf{q}}\gamma)$.

Proof. We have $\mathbf{D} = \hat{\mathbf{q}}^{-1}\mathbf{S} = \hat{\mathbf{q}}^{-1}\mathbf{C}\hat{\mathbf{q}}$ and $\mathbf{D}\gamma = \lambda\gamma$. Then we find $\hat{\mathbf{q}}^{-1}\mathbf{C}\hat{\mathbf{q}}\gamma = \lambda\gamma$, leading to $\mathbf{C}(\hat{\mathbf{q}}\gamma) = \lambda(\hat{\mathbf{q}}\gamma)$. □

We now illustrate Lemma A.6.1 by an example.

Example A.6.1. Consider Example 5.1.2 with the positive *commodity flow* matrix $\mathbf{S}$, the semi-positive vector of *final demand* $\mathbf{d}$ and the positive price vector $\mathbf{p} = [3, 4, 2]'$.

Calculate the vector of *total output* $\mathbf{q}$ and the diagonal matrix $\hat{\mathbf{q}}^{-1}$. Compute the vector $\mathbf{x} = \hat{\mathbf{p}}\mathbf{q}$, the commodity flow matrix $\mathbf{Z}$, the input-output coefficients matrices $\mathbf{C}$ and $\mathbf{A}$ and the distribution coefficients matrix $\mathbf{D}$.

3 Two square matrices $\mathbf{A}$ and $\mathbf{B}$ are similar, if there exists a regular $n \times n$ matrix $\mathbf{P}$, so that equation $\mathbf{A} = \mathbf{P}^{-1}\mathbf{BP}$ holds.

Compute the common characteristic polynomial $P_3(\lambda)$ of the matrices **A**, **C**, **D**, their eigenvalues and the Frobenius number. (For the definition of the Frobenius number, see Section A.9.)

Solution to Example A.6.1:
Identify the matrix **S**, the vectors **p**, **d** and calculate vector **q**:

$$\mathbf{S} = \begin{bmatrix} 90 & 50 & 40 \\ 120 & 125 & 40 \\ 60 & 150 & 200 \end{bmatrix}, \quad \mathbf{p} = \begin{bmatrix} 3 \\ 4 \\ 2 \end{bmatrix}, \quad \mathbf{d} = \begin{bmatrix} 0 \\ 165 \\ 70 \end{bmatrix}, \quad \mathbf{q} = \mathbf{Se} + \mathbf{d} = \begin{bmatrix} 180 \\ 450 \\ 480 \end{bmatrix}. \tag{A.55}$$

Then calculate the matrices $\hat{\mathbf{q}}^{-1}$, $\hat{\mathbf{x}}^{-1}$ and **Z**,

$$\hat{\mathbf{q}}^{-1} = \begin{bmatrix} \frac{1}{180} & 0 & 0 \\ 0 & \frac{1}{450} & 0 \\ 0 & 0 & \frac{1}{480} \end{bmatrix}, \quad \hat{\mathbf{x}}^{-1} = \hat{\mathbf{p}}^{-1}\hat{\mathbf{q}}^{-1} = \begin{bmatrix} \frac{1}{540} & 0 & 0 \\ 0 & \frac{1}{1{,}800} & 0 \\ 0 & 0 & \frac{1}{960} \end{bmatrix},$$

$$\mathbf{Z} = \hat{\mathbf{p}}\mathbf{S} = \begin{bmatrix} 3 & 0 & 0 \\ 0 & 4 & 0 \\ 0 & 0 & 2 \end{bmatrix} \begin{bmatrix} 90 & 50 & 40 \\ 120 & 125 & 40 \\ 60 & 150 & 200 \end{bmatrix} = \begin{bmatrix} 270 & 150 & 120 \\ 480 & 500 & 160 \\ 120 & 300 & 400 \end{bmatrix}. \tag{A.56}$$

Continue with calculating the matrices **C** and **D**,

$$\mathbf{C} = \mathbf{S}\hat{\mathbf{q}}^{-1} = \begin{bmatrix} 90 & 50 & 40 \\ 120 & 125 & 40 \\ 60 & 150 & 200 \end{bmatrix} \begin{bmatrix} \frac{1}{180} & 0 & 0 \\ 0 & \frac{1}{450} & 0 \\ 0 & 0 & \frac{1}{480} \end{bmatrix} = \begin{bmatrix} \frac{1}{2} & \frac{1}{9} & \frac{1}{12} \\ \frac{2}{3} & \frac{5}{18} & \frac{1}{12} \\ \frac{1}{3} & \frac{1}{3} & \frac{5}{12} \end{bmatrix},$$

$$\mathbf{D} = \hat{\mathbf{q}}^{-1}\mathbf{S} = \begin{bmatrix} \frac{1}{180} & 0 & 0 \\ 0 & \frac{1}{450} & 0 \\ 0 & 0 & \frac{1}{480} \end{bmatrix} \begin{bmatrix} 90 & 50 & 40 \\ 120 & 125 & 40 \\ 60 & 150 & 200 \end{bmatrix} = \begin{bmatrix} \frac{1}{2} & \frac{5}{18} & \frac{2}{9} \\ \frac{4}{15} & \frac{5}{18} & \frac{4}{45} \\ \frac{1}{8} & \frac{5}{16} & \frac{5}{12} \end{bmatrix}. \tag{A.57}$$

Then compute the matrix **A**:

$$\mathbf{A} = \mathbf{Z}\hat{\mathbf{x}}^{-1} = \begin{bmatrix} 270 & 150 & 120 \\ 480 & 500 & 160 \\ 120 & 300 & 400 \end{bmatrix} \begin{bmatrix} \frac{1}{540} & 0 & 0 \\ 0 & \frac{1}{1{,}800} & 0 \\ 0 & 0 & \frac{1}{960} \end{bmatrix} = \begin{bmatrix} \frac{1}{2} & \frac{1}{12} & \frac{1}{8} \\ \frac{8}{9} & \frac{5}{18} & \frac{1}{6} \\ \frac{2}{9} & \frac{1}{6} & \frac{5}{12} \end{bmatrix}. \tag{A.58}$$

The characteristic polynomial of **A**, **C** and **D** is

$$\begin{aligned} P_3(\lambda) &= \det(\mathbf{A} - \lambda\mathbf{I}_3) = \det(\mathbf{C} - \lambda\mathbf{I}_3) = \det(\mathbf{D} - \lambda\mathbf{I}_3) \\ &= \frac{35}{1{,}296} - \frac{1}{3}\lambda + \frac{43}{36}\lambda^2 - \lambda^3 = -\left(\lambda - \frac{5}{6}\right)\left(\lambda - \frac{1}{6}\right)\left(\lambda - \frac{7}{36}\right) = 0. \end{aligned} \tag{A.59}$$

The eigenvalues of **A**, **C** and **D** are: $\lambda_1 = \frac{5}{6}$, $\lambda_2 = \frac{7}{36}$, $\lambda_3 = \frac{1}{6}$. The Frobenius number is $\lambda_A = \lambda_C = \lambda_D = \frac{5}{6}$. ▲

A.7 The orthogonal Euler affinity and dilatation

Let there be $k \in \mathbb{N}$ numbers y_i, $i = 1, \ldots, k$, setting up the vector $\mathbf{y} = [y_1, y_2, \ldots, y_k]'$. Consider the $k \times k$ diagonal matrix,

$$\hat{\mathbf{y}} = \begin{bmatrix} y_1 & 0 & 0 & 0 \\ 0 & y_2 & 0 & 0 \\ \cdots & \cdots & \cdots & \cdots \\ 0 & 0 & 0 & y_k \end{bmatrix}. \tag{A.60}$$

Definition A.7.1 (Orthogonal Euler affinity). The mapping which associates to a vector $\mathbf{x} \in \mathbb{R}^k$ the vector $\tilde{\mathbf{x}} = \hat{\mathbf{y}}\mathbf{x}$ is called an *orthogonal Euler affinity* or *orthogonal Euler map.*[4]

Let us consider the following example to illustrate the orthogonal Euler affinity.

Example A.7.1. Take dimension $k = 2$. Consider the square $A(0.5, 0.5)$, $B(2, 0.5)$, $C(2, 2)$, $D(0.5, 2)$ in the plane $\mathbb{R}^2$ and the diagonal matrix

$$\hat{\mathbf{y}} = \begin{bmatrix} 3 & 0 \\ 0 & 2 \end{bmatrix}. \tag{A.61}$$

Apply the corresponding orthogonal Euler affinity to the vertices of the square $ABCD$, leading to the image rectangle $A'(1.5, 1)$, $B'(6, 1)$, $C'(6, 4)$, $D'(1.5, 4)$, see Figure A.1 (left).

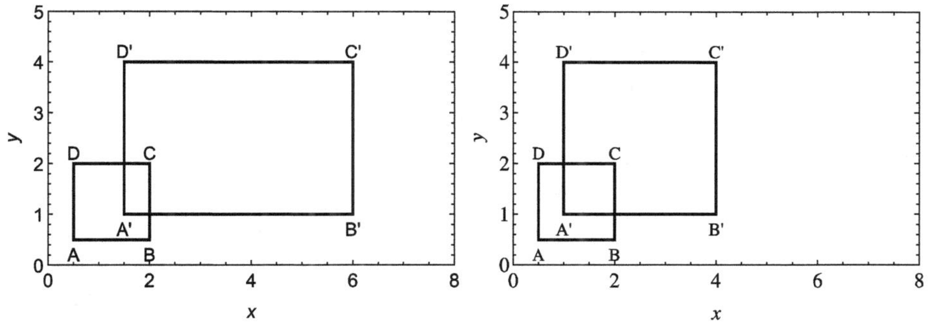

Figure A.1: The orthogonal Euler affinity (left) and dilatation (right).

Definition A.7.2 (Dilatation of a vector space). If all the coefficients of the diagonal matrix are equal, $y_i = y$, $y \in \mathbb{R}$, $i = 1, \ldots, k$, then the orthogonal Euler affinity is a *dilatation*. We denote such a vector by $\mathbf{y}_d$ and the associated diagonal matrix by $\hat{\mathbf{y}}_d$. The defining equation becomes $\tilde{\mathbf{x}} = \hat{\mathbf{y}}_d\mathbf{x} = \mathbf{x}\hat{\mathbf{y}}_d$.

4 Consider a $(k \times k)$ matrix $\mathbf{X} = \{x_{ij}\}, i, j = 1, \ldots, k$, made of the k column-vectors $\mathbf{x}_{.j} = [x_{1j}, x_{2j}, \ldots, x_{kj}]'$, $\mathbf{X} = \{\mathbf{x}_{.1}, \mathbf{x}_{.2}, \ldots, \mathbf{x}_{.k}\}$, then the images resulting from the *orthogonal Euler map* $\tilde{\mathbf{X}} = \hat{\mathbf{y}}\mathbf{X} = \{\tilde{\mathbf{x}}_{.1}, \tilde{\mathbf{x}}_{.2}, \ldots, \tilde{\mathbf{x}}_{.k}\}$ are the k column-vectors $\tilde{\mathbf{x}}_{.j} = \hat{\mathbf{y}}\mathbf{x}_{.j} = [\tilde{x}_{1j}, \tilde{x}_{2j}, \ldots, \tilde{x}_{kj}]' = [y_1x_{1j}, y_2x_{2j}, \ldots, y_kx_{kj}]'$.

Thus the *dilatation map* is described by the diagonal matrix

$$\hat{y}_d = \begin{bmatrix} y & 0 & 0 & 0 \\ 0 & y & 0 & 0 \\ \dots & \dots & \dots & \dots \\ 0 & 0 & 0 & y \end{bmatrix}. \tag{A.62}$$

Example A.7.2. Take the dimension $k = 2$. Consider the square $A(0.5, 0.5)$, $B(2, 0.5)$, $C(2, 2)$, $D(0.5, 2)$ in the plane $\mathbb{R}^2$ and the diagonal matrix

$$\hat{y}_d = \begin{bmatrix} 2 & 0 \\ 0 & 2 \end{bmatrix}. \tag{A.63}$$

Compute the image of the given square under the dilatation $\hat{y}_d$.

Solution of Example A.7.2:
Note that the multiplication with the diagonal matrix $\hat{y}$ can be replaced by a multiplication with the scalar y. One gets the rectangle $A'(1, 1)$, $B'(4, 1)$, $C'(4, 4)$, $D'(1, 4)$ in the plane $\mathbb{R}^2$, see Figure A.1 (right).

A.8 The theory of non-negative matrices

In this section, we focus on the theory of *non-negative* matrices[5]. We will also have to pay attention to the aspect that these matrices are either *reducible* or *irreducible.* Then a version of the theorem of Georg Frobenius ([31], 1912) is presented. The theorem of Frobenius is a generalization of the theorem of Oskar Perron ([86], 1907). For this reason, today it is mostly called *Perron–Frobenius theorem.*

At present it is known that the work of Perron and Frobenius was anticipated by other mathematicians and economists. Kenji Mori writes ([66], p. 1) "Georg von Charasoff (1877–1931) was one of the first economic theorists to recognize that the price of production is an eigenvector of the input matrix, and to determine the rate of profit using its eigenvalues." Mori continues: "Since Georg von Charasoff, a Russian mathematician and economist from Tblissi, was rediscovered more than 70 years after his main work, *Das System des Marxismus, Darstellung und Kritik (1910),* he has been acknowledged in many articles on the history of economic thought as a pioneer in *linear economic theory.*[6]" Parys ([79], p. 4) says that Charasoff ([15]: preface, xii) claims

5 Kurz & Salvadori ([52], p. 104) give the definition of *semi-positive* matrices. Pasinetti ([80], p. 26) makes the remark in parentheses: ...a non-negative matrix in which at least one element is strictly positive (and is also called a *semi-positive matrix*).

6 *Linear production economics* is a technical term, used in L. J. Bortkiewicz's (1868–1931) and Dimitriev's (1868–1913) concepts of production economics and described by linear equations or processes, as it is the case in Input-Output Analysis, where typically matrix algebra is in use.

to have presented a "definitive solution" to the main problems of the *classical theory of value*, thanks to his theory of *Urkapital* ('original capital'), a composite commodity similar to Sraffa's standard composite commodity in PCMC. In the language of modern mathematical economics, *Urkapital* corresponds to an eigenvector of the augmented (or closed) input matrix.

The **Perron–Frobenius theorem** is today recognized as governing a part of *linear production economics*, as Mori ([66], p. 1) has stated. This is the case for PCMC's *single product industries*.

We will start with a short review of the elementary notions on *non-negative* matrices (see Gantmacher [34], Chap. 13, and Takayama [116], pp. 367–380). Usually such an overview begins by setting a real or complex square matrix $\mathbf{A} = (a_{ij})$, $i, j = 1, \dots, n$, with n rows and n columns.

If necessary we shall also refer to positive matrices **A**, as those with only positive elements, that is $a_{ij} > 0$ for all $i, j = 1, \dots, n$, compactly written, $\mathbf{A} > \mathbf{0}$.

But we will specifically consider in this section *square, non-negative real* $n \times n$ *matrices* **A**. One defines such a matrix as *non-negative*, if all its elements are not negative, i. e., if for its elements one has $a_{ij} \geq 0$ for all $i, j = 1, \dots, n$, compactly written $\mathbf{A} \geq \mathbf{0}$. If there is at least one positive element $a_{ij} > 0$, the matrix **A** is called *semi-positive*.

(1) The notion of permutation. This paragraph covers the notion of *permutation* and *permutation matrix*.

Definition A.8.1 (Takayama [116], p. 368). A *permutation*, is a one-to-one function from the set $\{1, 2, 3, \dots, n\}$ onto itself, denoted by

$$\sigma = \begin{pmatrix} 1 & 2 & 3 & \dots & n \\ j_1 & j_2 & j_3 & \dots & j_n \end{pmatrix}, \quad \sigma(i) = j_i,\ i = 1, \dots, n. \tag{A.64}$$

We give an illustration.

Example A.8.1. The permutation $1 \to 1$, $2 \to 3$, $3 \to 2$ is usually described by

$$\sigma = \begin{pmatrix} 1 & 2 & 3 \\ 1 & 3 & 2 \end{pmatrix}. \tag{A.65}$$

A *permutation matrix*, usually denoted by **P**, is one which is obtained by permuting the columns (or rows) of the identity matrix **I**. This means that a permutation matrix has exactly one '1' in each column and in each row. The other elements are '0'. ▲

Definition A.8.2 (Takayama [116], p. 368, or Horn & Johnson [43], p. 32). An $n \times n$ matrix $\mathbf{P} = (p_{ij})$ is called a **permutation matrix** if $p_{\sigma(j)j} = 1$, $j = 1, \dots, n$, respectively $p_{i\sigma(i)} = 1$, $i = 1, \dots, n$, and if $p_{ij} = 0$ for all $i \neq \sigma(j)$, respectively $p_{ij} = 0$, for all $j \neq \sigma(i)$.

Example A.8.2. The matrix $\mathbf{P}_\sigma = \begin{bmatrix} 1 & 0 & 0 \\ 0 & 0 & 1 \\ 0 & 1 & 0 \end{bmatrix}$ is the representation of the permutation σ in (A.65).

If σ is some permutation (A.64), we denote by $\mathbf{P}_\sigma$ the permutation matrix obtained by permuting the *columns* of the identity matrix $\mathbf{I}$ by σ. Similarly, the transpose $\mathbf{P}'_\sigma$ of $\mathbf{P}_\sigma$ is obtained by permuting the *rows* of the identity matrix by σ, see Takayama [116], p. 369. ▲

Lemma A.8.1 (Takayama [116], p. 369). *Every permutation matrix $\mathbf{P}_\sigma$ is orthogonal, which means that the transposed matrix $\mathbf{P}'_\sigma$ is equal to the inverse matrix $\mathbf{P}_\sigma^{-1}$, $\mathbf{P}'_\sigma = \mathbf{P}_\sigma^{-1}$. Note that the similar matrix*

$$\widetilde{\mathbf{A}} = \mathbf{P}_\sigma^{-1}\mathbf{A}\mathbf{P}_\sigma = \mathbf{P}'_\sigma\mathbf{A}\mathbf{P}_\sigma \tag{A.66}$$

is the matrix obtained by permuting the rows and the columns of matrix $\mathbf{A}$ by the permutation σ.

We now give an illustration of an application of a permutation matrix and an exercise to apply Lemma A.8.1.

Exercise A.8.1. Consider the matrix $\mathbf{Z}_1 = \begin{bmatrix} 1 & 0 & 2 \\ 0 & 5 & 0 \\ 3 & 0 & 4 \end{bmatrix}$ and the permutation described by $\sigma = \begin{pmatrix} 1 & 2 & 3 \\ 1 & 3 & 2 \end{pmatrix}$. First, permute the *rows* of matrix $\mathbf{Z}_1$, then separately permute the *columns* of matrix $\mathbf{Z}_1$ with the permutation σ. Then apply Lemma A.8.1.

Solution of Exercise A.8.1:
In the first three steps we apply the permutation σ on the matrix $\mathbf{Z}_1$, then we directly use Lemma A.8.1 on $\mathbf{Z}_1$ to obtain the "canonical form" $\widetilde{\mathbf{Z}}_1$.

(1) Applying the permutation σ on matrix $\mathbf{Z}_1 = \begin{bmatrix} 1 & 0 & 2 \\ 0 & 5 & 0 \\ 3 & 0 & 4 \end{bmatrix}$ to permute successively the second and third rows, one gets $\begin{bmatrix} 1 & 0 & 2 \\ 3 & 0 & 4 \\ 0 & 5 & 0 \end{bmatrix}$. One obtains the same resulting matrix, by computing

$$\mathbf{P}_\sigma^{-1}\mathbf{Z}_1 = \begin{bmatrix} 1 & 0 & 0 \\ 0 & 0 & 1 \\ 0 & 1 & 0 \end{bmatrix} \begin{bmatrix} 1 & 0 & 2 \\ 0 & 5 & 0 \\ 3 & 0 & 4 \end{bmatrix} = \begin{bmatrix} 1 & 0 & 2 \\ 3 & 0 & 4 \\ 0 & 5 & 0 \end{bmatrix}. \tag{A.67}$$

In words: the product $\mathbf{P}_\sigma^{-1}\mathbf{Z}_1$ gives the same result as the direct application of the permutation σ on the rows and the columns of $\mathbf{Z}_1$.

(2) Applying the permutation σ on matrix $\mathbf{Z}_1 = \begin{bmatrix} 1 & 0 & 2 \\ 0 & 5 & 0 \\ 3 & 0 & 4 \end{bmatrix}$ to permute the 2nd and 3rd columns, one gets $\begin{bmatrix} 1 & 2 & 0 \\ 0 & 0 & 5 \\ 3 & 4 & 0 \end{bmatrix}$. One obtains the same matrix, when one computes

$$\mathbf{Z}_1\mathbf{P}_\sigma = \begin{bmatrix} 1 & 0 & 2 \\ 0 & 5 & 0 \\ 3 & 0 & 4 \end{bmatrix} \begin{bmatrix} 1 & 0 & 0 \\ 0 & 0 & 1 \\ 0 & 1 & 0 \end{bmatrix} = \begin{bmatrix} 1 & 2 & 0 \\ 0 & 0 & 5 \\ 3 & 4 & 0 \end{bmatrix}. \tag{A.68}$$

The multiplication $\mathbf{Z}_1\mathbf{P}_\sigma$ gives the same result as the direct application of the permutation σ on the columns of matrix $\mathbf{Z}_1$.

(3) Applying the permutation σ on matrix $\mathbf{Z}_1 = \left[\begin{smallmatrix} 1 & 0 & 2 \\ 0 & 5 & 0 \\ 3 & 0 & 4 \end{smallmatrix}\right]$ to permute successively the second and third columns and then the second and third rows (or inversely to permute the second and third rows and then the second and third columns), one gets $\tilde{\mathbf{Z}}_1 = \left[\begin{smallmatrix} 1 & 2 & 0 \\ 3 & 4 & 0 \\ 0 & 0 & 5 \end{smallmatrix}\right]$. One obtains the same matrix, when one computes

$$\tilde{\mathbf{Z}}_1 = \mathbf{P}_\sigma^{-1}\mathbf{Z}_1\mathbf{P}_\sigma = \begin{bmatrix} 1 & 0 & 0 \\ 0 & 0 & 1 \\ 0 & 1 & 0 \end{bmatrix}\begin{bmatrix} 1 & 0 & 2 \\ 0 & 5 & 0 \\ 3 & 0 & 4 \end{bmatrix}\begin{bmatrix} 1 & 0 & 0 \\ 0 & 0 & 1 \\ 0 & 1 & 0 \end{bmatrix} = \begin{bmatrix} 1 & 2 & 0 \\ 3 & 4 & 0 \\ 0 & 0 & 5 \end{bmatrix}. \tag{A.69}$$

The multiplication $\tilde{\mathbf{Z}}_1 = \mathbf{P}_\sigma^{-1}\mathbf{Z}_1\mathbf{P}_\sigma$ results in the application of the permutation σ on the rows and the columns of $\mathbf{Z}_1$, resulting in the matrix $\tilde{\mathbf{Z}}_1 = \left[\begin{smallmatrix} 1 & 2 & 0 \\ 3 & 4 & 0 \\ 0 & 0 & 5 \end{smallmatrix}\right]$. On the other hand, the matrix $\mathbf{P}_\sigma = \left[\begin{smallmatrix} 1 & 0 & 0 \\ 0 & 0 & 1 \\ 0 & 1 & 0 \end{smallmatrix}\right]$ is the permutation matrix of Example A.8.2 corresponding to the permutation σ (A.65).

Summarizing, the permutation σ (A.65) to apply to the *rows* and to the *columns* of matrix $\mathbf{Z}_1$ one may compute $\tilde{\mathbf{Z}}_1 = \mathbf{P}_\sigma^{-1}\mathbf{Z}_1\mathbf{P}_\sigma$.

(4) Now we apply Lemma A.8.1. First we need $\mathbf{P}_\sigma^{-1}$ and check the orthogonality, $\mathbf{P}_\sigma^{-1} = \mathbf{P}'_\sigma$. We calculate for this purpose,

$$\mathbf{I} = \mathbf{P}_\sigma^{-1}\mathbf{P}_\sigma = \mathbf{P}_\sigma\mathbf{P}'_\sigma = \begin{bmatrix} 1 & 0 & 0 \\ 0 & 0 & 1 \\ 0 & 1 & 0 \end{bmatrix}\begin{bmatrix} 1 & 0 & 0 \\ 0 & 0 & 1 \\ 0 & 1 & 0 \end{bmatrix} = \begin{bmatrix} 1 & 0 & 0 \\ 0 & 1 & 0 \\ 0 & 0 & 1 \end{bmatrix}. \tag{A.70}$$

Then we may compute

$$\tilde{\mathbf{Z}}_1 = \mathbf{P}_\sigma^{-1}\mathbf{Z}_1\mathbf{P}_\sigma = \begin{bmatrix} 1 & 0 & 0 \\ 0 & 0 & 1 \\ 0 & 1 & 0 \end{bmatrix}\begin{bmatrix} 1 & 0 & 2 \\ 0 & 5 & 0 \\ 3 & 0 & 4 \end{bmatrix}\begin{bmatrix} 1 & 0 & 0 \\ 0 & 0 & 1 \\ 0 & 1 & 0 \end{bmatrix} = \begin{bmatrix} 1 & 2 & 0 \\ 3 & 4 & 0 \\ 0 & 0 & 5 \end{bmatrix}. \tag{A.71}$$

This means that the permutation σ (A.65) applied on the rows and columns (or on the columns and rows) of the matrix $\mathbf{Z}_1$ results in the similar matrix $\tilde{\mathbf{Z}}_1$ (A.66). ▲

(2) Reducible and irreducible matrices. We introduce the notion of *irreducible matrices*. A matrix is *reducible* (or *decomposable*) if it is not *irreducible*. These notions will be used to distinguish between *basic* and *non-basic* commodities in the case of *single product industries*, as an alternative to Sraffa's definition (PCMC, Par. 6) : *"The criterion is whether a commodity enters (no matter whether directly or indirectly) into the production of all commodities. Those that do, we shall call basic, and those that do not, non-basic products."*

Definition A.8.3 (Reducible and irreducible matrices). See Schefold [103], p. 54, Kurz & Salvadori [52], p. 516, and Gantmacher [34], p. 417. A square matrix $\mathbf{A}$ is *reducible* if there exits a permutation matrix **P** with the result that

$$\widetilde{\mathbf{A}} = \mathbf{P}^{-1}\mathbf{A}\mathbf{P} = \begin{bmatrix} \widetilde{\mathbf{A}}_{11} & \mathbf{0} \\ \widetilde{\mathbf{A}}_{21} & \widetilde{\mathbf{A}}_{22} \end{bmatrix}, \tag{A.72}$$

where $\widetilde{\mathbf{A}}_{11}$ and $\widetilde{\mathbf{A}}_{22}$ are square submatrices, *reducible* or *irreducible*. The reduction process may continue and a *reducible* matrix **A** may always be reduced to the form

$$\widetilde{\mathbf{A}} = \mathbf{P}^{-1}\mathbf{A}\mathbf{P} = \begin{bmatrix} \widetilde{\mathbf{A}}_{11} & \mathbf{0} & \dots & \mathbf{0} \\ \widetilde{\mathbf{A}}_{21} & \widetilde{\mathbf{A}}_{22} & \dots & \mathbf{0} \\ \dots & \dots & \dots & \dots \\ \widetilde{\mathbf{A}}_{s1} & \widetilde{\mathbf{A}}_{s2} & \dots & \widetilde{\mathbf{A}}_{ss} \end{bmatrix}, \tag{A.73}$$

where $\widetilde{\mathbf{A}}_{11}, \widetilde{\mathbf{A}}_{22}, \dots, \widetilde{\mathbf{A}}_{ss}$ now are irreducible square matrices, not necessarily of the same order (note that the zero matrix of order 1 is irreducible). Form (A.73) of matrix **A** is called the "canonical form" of **A**.

In the case of **joint production processes,** $(\mathbf{A}, \mathbf{L}) \Rightarrow (\mathbf{B})$, the reducible matrices **A**, **B** are brought into in "canonical form" with two permutation matrices **P**, **Q**, giving $\tilde{\mathbf{A}} = \mathbf{P}^{-1}\mathbf{A}\mathbf{Q}$ and $\tilde{\mathbf{B}} = \mathbf{P}^{-1}\mathbf{B}\mathbf{Q}$ as in (A.72), see Schefold [103], p. 54. The product $\mathbf{P}^{-1}\mathbf{A}$ permutes the rows of **A** by a permutation described by **P**. The product **AQ** permutes the columns of **A** by a permutation described by **Q**.

Observation: The transformation $\mathbf{P}_{\alpha}^{-1}\mathbf{Z}$ of matrix **Z** by the permutation matrix $\mathbf{P}_{\alpha}$ interchanges the rows by permutation α (A.64).

The transformation $\mathbf{Z}\mathbf{Q}_{\beta}$ of matrix **Z** by the permutation matrix $\mathbf{Q}_{\beta}$ interchanges the columns by permutation β (A.64).

The resulting matrices $\widetilde{\mathbf{A}}$ (A.72) and (A.73) are called *triangular block matrices*, issued from a *reducible* matrix **A**.

Remark. Definition A.8.3 is stated in this context for *reducible* and *semi-positive* matrices describing production processes. Obviously, in this case, if the initial matrix **A** is *reducible* and *semi-positive*, then the resulting matrices $\widetilde{\mathbf{A}}_{11}$, $\widetilde{\mathbf{A}}_{12}$ and $\widetilde{\mathbf{A}}_{22}$ are also *semi-positive*, because the transformation (A.72) only operates permutations on matrix **A**.

Definition A.8.4. An square matrix **A** is called *completely reducible* (or *completely decomposable*) if there exits a permutation matrix **P** such that

$$\widetilde{\mathbf{A}} = \mathbf{P}^{-1}\mathbf{A}\mathbf{P} = \begin{bmatrix} \widetilde{\mathbf{A}}_{11} & \mathbf{0} & \dots & \mathbf{0} \\ \mathbf{0} & \widetilde{\mathbf{A}}_{22} & \dots & \mathbf{0} \\ \dots & \dots & \dots & \dots \\ \mathbf{0} & \mathbf{0} & \dots & \widetilde{\mathbf{A}}_{ss} \end{bmatrix}, \tag{A.74}$$

where the matrices $\mathbf{A}_{11}$, $\mathbf{A}_{22}$ and $\mathbf{A}_{ss}$ are square (see Takayama, [116], p. 370). The resulting matrix $\widetilde{\mathbf{A}}$ (A.74) is called a *block diagonal matrix*, generated from a *completely reducible* matrix **A** (see Horn & Johnson [43], p. 30).

We observe that the matrix $\tilde{\mathbf{Z}}_1$ (A.71) is the *completely reducible* matrix, resulting from the permutation σ in (A.65), applied to matrix $\mathbf{Z}_1$, Example A.8.1 and gives rise to a *block diagonal matrix*.

This result is of great importance in the field of *linear production economics*. It however requires an algorithm to calculate a permutation matrix **P** and to transform a *reducible* input-output coefficients matrix **A** in (A.66) or an input-output coefficients matrix **C** into its canonical forms $\tilde{\mathbf{A}}$, respectively $\tilde{\mathbf{C}}$. If existent, such a procedure allows to separate the *basic* commodities from the *non-basic* commodities in Input-Output Tables, such as the Swiss IOT 2008 (see Chapter 10).

Example A.8.3. Explain why the matrix $\mathbf{Z}_1$, from Exercise A.8.1 is *completely reducible.*

Solution to Example A.8.3:
When a permutation matrix **P** is known, one can directly compute the similar matrix $\tilde{\mathbf{Z}}_1$ to get a *canonical form*. In this case the resulting matrix $\tilde{\mathbf{Z}}_1$ (A.71) is a *block diagonal matrix*, arising from the *reducible* matrix $\mathbf{Z}_1$. Thus, in this case, the resulting *triangular block matrix* has become a *completely reduced* form of the initial matrix. ▲

With Example A.8.3 we see that we need a mathematical tool how to decide about the *reducibility* of a matrix. The next *Lemma* gives a direct method for checking the *reducibility* of a matrix.

Lemma A.8.2 (Horn & Johnson [43], p. 533). *Let* **A** *be a non-negative* $n \times n$ *matrix and* **I** *the identity matrix.* **A** *is irreducible if and only if* $[\mathbf{I} + \mathbf{A}]^{n-1} > \mathbf{0}$.

We give an illustration of Lemma A.8.2.

Example A.8.4. Consider the positive matrix $\mathbf{A} = \begin{bmatrix} 2 & 1 & 2 \\ 1 & 6 & 1 \\ 3 & 1 & 5 \end{bmatrix}$, which obviously is *irreducible*, because there are no vanishing elements. According to Lemma A.8.2, we obtain

$$[\mathbf{I} + \mathbf{A}]^{3-1} = \begin{bmatrix} 3 & 1 & 2 \\ 1 & 7 & 1 \\ 3 & 1 & 6 \end{bmatrix}^2 = \begin{bmatrix} 16 & 12 & 19 \\ 13 & 51 & 15 \\ 28 & 16 & 43 \end{bmatrix} > \mathbf{0}. \quad \blacktriangle \tag{A.75}$$

Example A.8.5. Answer the following questions by applying Lemma A.8.2.

(1) Given the *non-negative* matrix $\mathbf{Z}_1 = \begin{bmatrix} 1 & 0 & 2 \\ 0 & 5 & 0 \\ 3 & 0 & 4 \end{bmatrix}$ of Example A.8.1, show that $\mathbf{Z}_1$ is *reducible.*
(2) Justify that the submatrix $\mathbf{Z}_{1(22)} = \begin{bmatrix} 1 & 2 \\ 3 & 4 \end{bmatrix}$, which results from an extraction of the second row and the second column of matrix $\mathbf{Z}_1$ in Example A.8.1 is *irreducible.*
(3) What is about *reducibility* or *irreducibility* of matrix $\mathbf{Z}_2 = \begin{bmatrix} 1 & 0 & 2 \\ 1 & 5 & 0 \\ 3 & 1 & 4 \end{bmatrix}$?

Solution to Example A.8.5:
(1) One computes

$$[\mathbf{I}+\mathbf{Z}_1]^{3-1} = \begin{bmatrix} 2 & 0 & 2 \\ 0 & 6 & 0 \\ 3 & 0 & 5 \end{bmatrix}^2 = \begin{bmatrix} 2 & 0 & 2 \\ 0 & 6 & 0 \\ 3 & 0 & 5 \end{bmatrix}\begin{bmatrix} 2 & 0 & 2 \\ 0 & 6 & 0 \\ 3 & 0 & 5 \end{bmatrix} = \begin{bmatrix} 10 & 0 & 14 \\ 0 & 36 & 0 \\ 21 & 0 & 31 \end{bmatrix}.$$

Obviously, $[\mathbf{I}+\mathbf{Z}_1]^2$ is not *positive* and therefore $\mathbf{Z}_1$ is *reducible*.
(2) Consider the submatrix $\mathbf{Z}_{1(22)} = \left[\begin{smallmatrix} 1 & 2 \\ 3 & 4 \end{smallmatrix}\right]$ of the matrix $\mathbf{Z}_1 = \left[\begin{smallmatrix} 1 & 0 & 2 \\ 0 & 5 & 0 \\ 3 & 0 & 4 \end{smallmatrix}\right]$ in Example A.8.1. One computes $[\mathbf{I}+\mathbf{Z}_{1(22)}]^{2-1} = \left[\begin{smallmatrix} 2 & 2 \\ 3 & 5 \end{smallmatrix}\right] > 0$ which is positive. Therefore, the matrix $\mathbf{Z}_{1(22)}$ is *irreducible*. This was clear on the onset, since matrix $\mathbf{Z}_{1(22)}$ contains no zeros.
(3) Finally compute the matrix,

$$\begin{aligned}[\mathbf{I}+\mathbf{Z}_2]^{3-1} &= \begin{bmatrix} 2 & 0 & 2 \\ 1 & 6 & 0 \\ 3 & 1 & 5 \end{bmatrix}^2 \\ &= \begin{bmatrix} 2 & 0 & 2 \\ 1 & 6 & 0 \\ 3 & 1 & 5 \end{bmatrix}\begin{bmatrix} 2 & 0 & 2 \\ 1 & 6 & 0 \\ 3 & 1 & 5 \end{bmatrix} = \begin{bmatrix} 10 & 2 & 14 \\ 8 & 36 & 2 \\ 22 & 11 & 31 \end{bmatrix} > 0, \end{aligned} \quad (A.76)$$

which is positive. Therefore $\mathbf{Z}_2$ is *irreducible*. This example shows that Lemma A.8.2 is not trivial. Matrix $\mathbf{Z}_2$ has two zero elements, $z_{12} = 0$, $z_{23} = 0$. One would expect that a permutation matrix $\mathbf{P}$ exists, giving $\widetilde{\mathbf{Z}}_2 = \mathbf{P}^{-1}\mathbf{Z}_2\mathbf{P}$, which puts these two zeros at the places $\tilde{z}_{31} = 0$, $\tilde{z}_{32} = 0$, bringing $\mathbf{Z}_2$ into "normal form". As $\mathbf{Z}_2$ is *irreducible*, according to Lemma A.8.2, such a permutation matrix $\mathbf{P}$ does not exist and these moves of elements cannot be obtained by a permutation matrix $\mathbf{P}$! ▲

Example A.8.1 has shown that the matrix $\mathbf{Z}_1$ can be transformed into its "canonical form", if one knows a permutation matrix $\mathbf{P}_\sigma$ able to perform this transformation.

(3) The adjacency matrix. Within the Leontief Input-Output models and the Sraffa models we are confronted to five matrices: the *commodity flow* matrix $\mathbf{Z}$ in (2.4) in *monetary terms*, respectively the *commodity flow* matrix $\mathbf{S}$ in (2.13) in *physical terms*. Then we have derived the dimensionless *input-output coefficients* matrices $\mathbf{A}$ in (2.9) the *distribution coefficients* matrix $\mathbf{D}$ (2.12) defined twice, (2.20), both dimensionless, respectively the *input-output coefficients* matrix $\mathbf{C}$ (2.14) in *physical terms*. Matrices $\mathbf{Z}$ and $\mathbf{S}$ are called *transaction* matrices, whereas $\mathbf{A}$, $\mathbf{C}$, $\mathbf{D}$ are called *coefficients* matrices.

Now, we will concentrate on a useful tool to visualize properties of matrices in relation with the considered production processes, like *commodity flows* and *payment flows*. For this purpose Miller & Blair ([65], p. 675) introduce the notion of *binary* or *Boolean* adjacency matrices.

Definition A.8.5 (The adjacency matrix). Given any $n \times n$ matrix $\mathbf{T} = \{t_{ij}\}$, another $n \times n$ matrix, frequently represented by the letter $\mathbf{W} = \{w_{ij}\}$, called the (Boolean) *adjacency* *matrix*, is generated as $w_{ij} := 1$ in cells for which $t_{ij} \neq 0$, and $w_{ij} = 0$ in cells for which $t_{ij} := 0$, $i, j \in \{1, \dots, n\}$.

We shall in particular work out adjacency matrices for the aforementioned transaction or coefficients matrices **Z**, **S**, **A**, **C**, **D**.

Example A.8.6. We present as an illustration the *adjacency matrices* of the matrices $\mathbf{Z}_1$ and $\mathbf{Z}_2$, Example A.8.5.

$$\mathbf{Z}_1 = \begin{bmatrix} 1 & 0 & 2 \\ 0 & 5 & 0 \\ 3 & 0 & 4 \end{bmatrix} \Rightarrow \mathbf{W}_1 = \begin{bmatrix} 1 & 0 & 1 \\ 0 & 1 & 0 \\ 1 & 0 & 1 \end{bmatrix},$$
$$\mathbf{Z}_2 = \begin{bmatrix} 1 & 0 & 2 \\ 1 & 5 & 0 \\ 3 & 1 & 4 \end{bmatrix} \Rightarrow \mathbf{W}_2 = \begin{bmatrix} 1 & 0 & 1 \\ 1 & 1 & 0 \\ 1 & 1 & 1 \end{bmatrix}. \qquad \text{(A.77)}$$

A.9 The Perron–Frobenius Theorems

We now consider the seminal theorem of Oskar Perron (1880–1975), published in 1907. Ferdinand Georg Frobenius (1849–1917) generalized Perron's results in 1912 (see Horn & Johnson [43], p. 534, and Takayama [116], p. 364). These are results about *positive* matrices and *non-negative matrices*. The theorems (for both versions see also Gantmacher [34], p. 398) are today usually presented in one form as the so-called *Perron–Frobenius theorem*.

Consider the distances of all the real or complex roots of a polynomial to the origin of the complex plane $\mathbb{C}^n$. The maximum of all these distances is called the spectral radius, more precisely:

Definition A.9.1 (The spectral radius). Let **A** be a $n \times n$ matrix. The spectral radius of **A** is $\rho(\mathbf{A}) := \max\{|\lambda| : \lambda \in \sigma(\mathbf{A})$, where $\sigma(\mathbf{A})$ is the set of all eigenvalues (real or complex) of $\mathbf{A}\}$ (see Horn and Johnson [43], p. 45 and p. 52).

(1) The Perron theorem. As noticed earlier, *positive* matrices are less general than *irreducible* and *non-negative* matrices. We start with Perron's theorem, the most elementary form of the three theorems of that type presented here.

Theorem A.9.1 (Perron theorem (Perron [86] and Gantmacher [34], p. 398)). *A positive matrix* $\mathbf{A} = (a_{ij}), i, j = 1, \dots, n$, *always has a unique real and positive characteristic root (= eigenvalue)* λ_A *of the characteristic polynomial whose module exceeds the module of all other eigenvalues of that polynomial, Figure* A.2, *left. This eigenvalue is called the Frobenius number of matrix* **A** *and is associated with an eigen-*

vector $\mathbf{x} = [x_1, x_2, \ldots, x_n]'$ *of matrix* $\mathbf{A}$*, which has only positive components*[7] $x_i > 0$, $i = 1, 2, \ldots, n$.

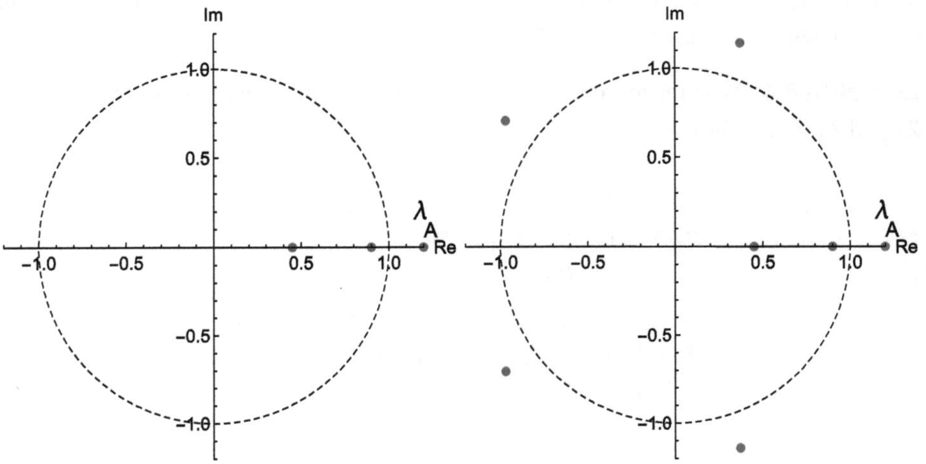

Figure A.2: Perron theorem (left), Frobenius theorem (right).

We note that the Perron theorem only works for *positive matrices*. We illustrate it by the following example

Example A.9.1. Consider the positive matrix $\mathbf{S} = \begin{bmatrix} 1 & 1 & 2 \\ 2 & 5 & 6 \\ 7 & 1 & 4 \end{bmatrix}$. It could be a hypothetical transaction matrix in physical terms related to a just-viable economy, see Definition 2.2.1, consisting of three sectors with no surplus (Garden of Eden economy), therefore $\mathbf{d} = \mathbf{o}$. The matrix $\mathbf{S}$ is evidently irreducible.

Compute the total output vector $\mathbf{q} = \mathbf{Se}$, the corresponding input coefficients matrix $\mathbf{C} = \mathbf{S}\hat{\mathbf{q}}^{-1}$ and the distribution coefficients matrix $\mathbf{D} = \hat{\mathbf{q}}^{-1}\mathbf{S}$.

Then compute the eigenvalues of matrices $\mathbf{S}$, $\mathbf{C}$, $\mathbf{D}$. Analyse the Frobenius numbers λ_S, λ_C and λ_D. Verify: $\lambda_C = \lambda_D = 1$. Compute the left eigenvectors $\mathbf{p}_S$ of matrix $\mathbf{S}$ associated with λ_S, respectively, the left eigenvectors $\mathbf{p}_C$ of matrix $\mathbf{C}$ associated to λ_C.

Solution to Example A.9.1:
We easily calculate $\mathbf{q} = \mathbf{Se} = [4, 13, 12]'$. We then get

$$\mathbf{C} = \mathbf{S}\hat{\mathbf{q}}^{-1} = \begin{bmatrix} \frac{1}{4} & \frac{1}{13} & \frac{1}{6} \\ \frac{1}{2} & \frac{5}{13} & \frac{1}{2} \\ \frac{7}{4} & \frac{1}{13} & \frac{1}{3} \end{bmatrix}, \quad \mathbf{D} = \hat{\mathbf{q}}^{-1}\mathbf{S} = \begin{bmatrix} \frac{1}{4} & \frac{1}{4} & \frac{1}{2} \\ \frac{2}{13} & \frac{5}{13} & \frac{6}{13} \\ \frac{7}{12} & \frac{1}{12} & \frac{1}{3} \end{bmatrix}, \quad \mathbf{De} = \mathbf{e}. \tag{A.78}$$

7 The eigenvector $\mathbf{x}$ is determined up to a scalar factor. This means that, if $\mathbf{x}$ is an eigenvector of $\mathbf{A}$, then, with $k \in \mathbb{R}$, $k \cdot \mathbf{x}$ is also an eigenvector of $\mathbf{A}$.

Then we calculate the characteristic polynomials of the matrices **S** and **C**

$$
\begin{aligned}
Q_3(\lambda) &= \det(\mathbf{S} - \lambda\mathbf{I}) = -18 - 7\lambda + 10\lambda^2 - \lambda^3 = (\lambda - 9)(\lambda - 2)(\lambda + 1), \\
P_3(\lambda) &= \det(\mathbf{C} - \lambda\mathbf{I}) = -\frac{3}{104} + \frac{19}{312}\lambda + \frac{151}{156}\lambda^2 - \lambda^3 \\
&= (\lambda - 1)(\lambda - 0.1546)(\lambda + 0.1866).
\end{aligned} \tag{A.79}
$$

The left eigenvector of the positive transaction matrix $\mathbf{S} > \mathbf{0}$ corresponding to the Frobenius number $\lambda_S = 9$ is positive, $\mathbf{p}_S = [2, 1, 2]' > 0$. The left eigenvector of the positive coefficients matrix **C** corresponding to the Frobenius number $\lambda_C = 1$ is also positive, $\mathbf{p}_C = [\frac{29}{11}, \frac{5}{11}, 1]' > 0$.

Note that the Frobenius number of **S** is greater than 1, $\lambda_S = 9 > 1 > 0$! We realise that $\lambda_D = 1$, since **D** is stochastic, see Lemma A.11.1. The eigenvalues of the matrices **C** and **D** are identical, $\lambda_C = \lambda_D = 1$, according to Lemma A.6.1. There is no surplus and the economy is *just viable.* ▲

(2) The Frobenius theorem generalises Perron's theorem.

Theorem A.9.2. *Frobenius theorem (Frobenius [31] and Gantmacher [34], p. 398) An irreducible and non-negative matrix* $\mathbf{A} = (a_{ij}), i, j = 1, \dots, n$ *always has a real and positive characteristic root (=eigenvalue)* $\lambda_A > 0$, *called the Frobenius number, that is a simple root of the characteristic polynomial.*

The modulus of all other characteristic roots does not exceed this number λ_A, *Figure A.2, right.*

To the Frobenius number, there corresponds an eigenvector $\mathbf{x} = [x_1, x_2, \dots, x_n]'$ *with only positive components* $x_i > 0, i = 1, 2, \dots, n$ *that is unique up to a scalar factor.*

If **A** *has altogether h roots* $\lambda_0 = \lambda_A, \lambda_1, \dots, \lambda_{h-1}$ *of modulus* λ_A, *the complex numbers* $\lambda \in \mathbb{C}$ *are all different from each other and are roots of the circle equation*

$$\lambda^h - \lambda_A^h = 0. \tag{A.80}$$

Consider then all characteristic roots $\lambda_0, \dots, \lambda_i, \dots, \lambda_{n-1}, i = 0, 2, \dots, n-1$ *of matrix* $\mathbf{A} = (a_{ij}), i, j = 1, \dots, n$ *as points of the complex plane. These points extend to themselves through a rotation of the plane with the center at the origin and angle* $2\pi/h$. *For* $h > 1$, *matrix* **A** *can be transformed in "cyclic" form via a permutation, i. e.,*

$$
\mathbf{A} = \begin{bmatrix}
\mathbf{0} & \mathbf{A}_{12} & \mathbf{0} & \dots & \dots & \mathbf{0} \\
\mathbf{0} & \mathbf{0} & \mathbf{A}_{23} & \dots & \dots & \mathbf{0} \\
\dots & \dots & \dots & \dots & \dots & \dots \\
\mathbf{0} & \mathbf{0} & \dots & \dots & \mathbf{A}_{h-2,h-1} & \mathbf{0} \\
\mathbf{0} & \mathbf{0} & \dots & \dots & \dots & \mathbf{A}_{h-1,h} \\
\mathbf{A}_{h1} & \mathbf{0} & \mathbf{0} & \dots & \dots & \mathbf{0}
\end{bmatrix}, \tag{A.81}
$$

where all the diagonal elements are square matrices.

Proof. See Gantmacher ([34], p. 398). □

We illustrate the Frobenius Theorem by the following example, focusing on its numerical consequences.

Example A.9.2. Let the non-negative and irreducible matrix $\mathbf{Z} = \begin{bmatrix} 1 & 0 & 1 \\ 1 & 1 & 0 \\ 1 & 1 & 1 \end{bmatrix}$ be given. It could be a hypothetical transaction matrix in monetary terms related to an economy consisting of three sectors with a final demand $\mathbf{f} = [18, 1, 1]'$.

Compute the total output vector $\mathbf{x} = \mathbf{Ze}+\mathbf{f}$ and the corresponding input coefficients matrix $\mathbf{A} = \mathbf{Z}\hat{\mathbf{x}}^{-1}$.

Then compute the eigenvalues and eigenvectors of both matrices $\mathbf{Z}$ and $\mathbf{A}$ and analyse the Frobenius numbers λ_Z, λ_A and the associated eigenvectors $\mathbf{x}_Z$, $\mathbf{x}_A$.

Solution to Example A.9.2:
Applying Lemma A.8.2, $(\mathbf{I} + \mathbf{Z})^2 = \begin{bmatrix} 2 & 0 & 1 \\ 1 & 2 & 0 \\ 1 & 1 & 2 \end{bmatrix}^2 = \begin{bmatrix} 5 & 1 & 4 \\ 4 & 4 & 1 \\ 5 & 4 & 5 \end{bmatrix} > 0$, we conclude that the matrix $\mathbf{Z}$ is irreducible. We then easily calculate $\mathbf{x} = \mathbf{Ze} + \mathbf{f} = [20, 3, 4]'$ and get

$$\hat{\mathbf{x}}^{-1} = \begin{bmatrix} \frac{1}{20} & 0 & 0 \\ 0 & \frac{1}{3} & 0 \\ 0 & 0 & \frac{1}{4} \end{bmatrix}, \quad \mathbf{A} = \mathbf{Z}\hat{\mathbf{x}}^{-1} = \begin{bmatrix} \frac{1}{20} & 0 & \frac{1}{4} \\ \frac{1}{20} & \frac{1}{3} & 0 \\ \frac{1}{20} & \frac{1}{3} & \frac{1}{4} \end{bmatrix}, \tag{A.82}$$

whose zero elements are at the same place in matrix $\mathbf{Z}$. Then we calculate the characteristic polynomials of matrices $\mathbf{Z}$ and $\mathbf{A}$,

$$\begin{aligned} Q_3(\lambda) &= \det(\mathbf{Z} - \lambda\mathbf{I}) = 1 - 2\lambda + 3\lambda^2 - \lambda^3 \\ &= -(\lambda - 2.325)(\lambda - 0.338 - 0.562i)(\lambda - 0.338 + 0.562i), \\ P_3(\lambda) &= \det(\mathbf{A} - \lambda\mathbf{I}) = \frac{1}{240} - \frac{1}{10}\lambda + \frac{19}{30}\lambda^2 - \lambda^3 \\ &= (\lambda - 0.4179)(\lambda - 0.1482)(\lambda - 0.0673). \end{aligned} \tag{A.83}$$

We notice in this example that a pair of eigenvalues are conjugate complex,[8] being the roots of the real three-degree polynomial $Q_3(\lambda)$.

The eigenvector of the transaction matrix $\mathbf{Z}$ corresponding to the Frobenius number $\lambda_Z = 2.325$ is positive, $\mathbf{x}_Z = [0.755, 0.570, 1]' > 0$, as warranted by the **Frobenius theorem**. The conjugate complex eigenvalues $\lambda_{1/2} = -0.338 \pm 0.562i$ are associated to the eigenvectors, $\mathbf{x}_{Z1/2} = [-0.877 \mp 0.745i, 0.215 \pm 1.307i, 1]'$.

The eigenvector of the coefficients matrix $\mathbf{A}$ corresponding to the Frobenius number λ_A, $0 < \lambda_A = 0.4179 < 1$, is positive, $\mathbf{x}_A = [0.6795, 0.4018, 1]' > 0$. The eigenvalue

8 A complex number is usually written as $z = a + bi \in \mathbb{C}$, with $i = \sqrt{-1}$. Its conjugate complex is $\bar{z} = a - bi \in \mathbb{C}$.

$\lambda_1 = 0.1481$ is associated with the eigenvector $\mathbf{x}_{A1} = [2.5479, -0.6878, 1]'$. The eigenvalue $\lambda_2 = 0.0673$ is accordingly associated with a *non-positive* eigenvector, $\mathbf{x}_{A1} = [14.4392, -2.7139, 1]'$.

Note that for $\mathbf{Z}$ the Frobenius number is larger than 1, $\lambda_Z = 2.325 > 1$. The Frobenius number of $\mathbf{A}$ is smaller than 1, $\lambda_A = 0.4179 < 1$, corresponding to a *productive Leontief model*, as will be defined later (see Definition A.12.1). ▲

We conclude this part with a combined version of the above theorems.

(3) The Perron–Frobenius theorem. We formulate here a combined version of the **Perron theorem** and the **Frobenius theorem** as it is presented in textbooks. This theorem is required in order to handle *transaction* and *coefficients* matrices containing also zero entries and presents in particular the key statement to treat *Sraffa's full price model* for *single-product industries*, expressed for *irreducible* and *non-negative* square matrices.

Theorem A.9.3 (Perron–Frobenius theorem). *Suppose $n \geq 2$ and consider an $n \times n$ irreducible and non-negative matrix* $\mathbf{A}$.
(1) $\rho(\mathbf{A}) > 0$;
(2) *$\rho(\mathbf{A})$ is an algebraically simple eigenvalue of* $\mathbf{A}$;[9]
(3) *There is a unique real vector $\mathbf{x} = [x_1, x_2, \ldots, x_n]'$ such that $\mathbf{Ax} = \rho(\mathbf{A})\mathbf{x}$ and $x_1 + x_2 + \cdots + x_n = 1$. This vector is positive, i. e., all its components are positive, $x_i > 0$, $i = 1, \ldots, n$;*
(4) *There is a unique real vector $\mathbf{y} = [y_1, y_2, \ldots, y_n]'$ such that $\mathbf{y}'\mathbf{A} = \rho(\mathbf{A})\mathbf{y}'$ and $x_1y_1 + x_2y_2 + \cdots + x_ny_n = 1$. This vector is positive, which means all its components are positive, $y_i > 0$, $i = 1, \ldots, n$.* ▲

Proof. See Horn and Johnson ([43], p. 534) or Gantmacher ([34], p. 398). □

Thus, there is also in this third formulation the Frobenius number corresponding to a unique, real, maximal, positive eigenvalue of matrix $\mathbf{A}$, denoted $\lambda_A = \rho(\mathbf{A})$. Thus, the right and left eigenspaces associated with the Frobenius number λ_A are *one*-dimensional. The absolute values of all the other characteristic roots of the polynomial $P_n(\lambda)$ are less than or equal to λ_A. The matrix $\mathbf{A}$ has a left eigenvector $\mathbf{y}$, corresponding to the Frobenius number λ_A, whose components are all positive. In the same way, $\mathbf{A}$ has a right eigenvector $\mathbf{x}$, corresponding to the eigenvalue λ_A, whose components are also all positive.

(4) Primitive matrices. We classify here *irreducible matrices*, separating them into *primitive* and *imprimitive* matrices (see Gantmacher [34], p. 422).

---

9 This means, there is exactly one real, positive, maximal eigenvalue of modulus $\lambda_A = \rho(\mathbf{A})$.
In this case, the spectral radius $\rho(\mathbf{A})$ is called a simple root of the characteristic polynomial $P_n(\lambda)$ of $\mathbf{A}$ and corresponds to the **Frobenius number** (**Frobenius eigenvalue**) λ_A. Notations vary for the Frobenius number, e. g., $\lambda_A = \lambda_{\max}$.

Definition A.9.2 (The primitive matrix). Consider an irreducible and non-negative $n \times n$ matrix $\mathbf{A} \geq 0$ whose characteristic polynomial $P_n(\lambda)$ has $h \leq n$ characteristic roots $\lambda_1, \lambda_2, \ldots, \lambda_h$ of maximal modulus a, ($|\lambda_1| = |\lambda_2| = \cdots = |\lambda_h| = a$), then we call the matrix primitive, if $h = 1$ or imprimitive if $h > 1$.

Lemma A.9.1 (A property of primitive matrices). *A non-negative $n \times n$ matrix $\mathbf{A} \geq \mathbf{0}$ is primitive, if and only if there exists a certain power $m \in \mathbb{N}$, such that $\mathbf{A}^m > 0$ is positive.*

Recall that the *Perron theorem* A.9.1 applies only to *positive* matrices **A** and ensures that there is a unique, real, maximal and positive eigenvalue λ_A, the Frobenius number. On the other hand, the **Frobenius theorem**, formulated as Theorem A.9.2 or Theorem A.9.3 is applied to *non-negative* and *irreducible* matrices and ensures the existence of a *simple*, real, positive eigenvalue λ_A, the Frobenius number. In this case, there usually are $h \leq n$ other complex eigenvalues of matrix **A** of the same modulus λ_A. The uniqueness of the eigenvalue of modulus λ_A is guaranteed if the matrix **A** is primitive.

(5) Uniqueness property: the Frobenius number and the subspace generated by the associated positive eigenvectors. Given a constant vector **x**, every family of vectors $k \cdot \mathbf{x}$, $k \in \mathbb{R}$, generates a unique one-dimensional subspace $V(\mathbf{x})$. Clearly, for the Frobenius number λ_C of matrix **C**, there is one and only one one-dimensional subspace $V(\mathbf{p})$, where **p** is a positive price vector associated to the Frobenius number λ_C, according to the *Perron–Frobenius theorems* A.9.1–A.9.3.

The following examples deal with matrices that do not meet the conditions of the **Perron–Frobenius theorem** to get a better grasp of the essence of this important result.

Example A.9.3. Consider the reducible matrix $\mathbf{Z}_1 = \begin{bmatrix} 1 & 0 & 2 \\ 0 & 5 & 0 \\ 3 & 0 & 4 \end{bmatrix}$, Example A.8.5. Compute $\mathbf{x} = \mathbf{Z}_1\mathbf{e}$ and the reducible coefficients matrix $\mathbf{A}_1 = \mathbf{Z}_1\hat{\mathbf{x}}^{-1}$. Then compute the eigenvalues and eigenvectors of $\mathbf{Z}_1$ and $\mathbf{A}_1$ and describe in what sense the obtained results differ from the statements of the *Perron–Frobenius theorem.*

Solution to Example A.9.3:

We easily calculate $\mathbf{x} = \mathbf{Z}_1\mathbf{e} = [3, 5, 7]'$. We then get

$$\hat{\mathbf{x}}^{-1} = \begin{bmatrix} \frac{1}{3} & 0 & 0 \\ 0 & \frac{1}{5} & 0 \\ 0 & 0 & \frac{1}{7} \end{bmatrix}, \quad \mathbf{A}_1 = \mathbf{Z}_1\hat{\mathbf{x}}^{-1} = \begin{bmatrix} \frac{1}{3} & 0 & \frac{2}{7} \\ 0 & 1 & 0 \\ 1 & 0 & \frac{4}{7} \end{bmatrix}. \tag{A.84}$$

The characteristic polynomials of the matrices $\mathbf{Z}_1$ and $\mathbf{A}_1$ are

$$Q_3(\lambda) = \det(\mathbf{Z}_1 - \lambda\mathbf{I}) = -10 - 23\lambda + 10\lambda^2 - \lambda^3$$
$$= (\lambda - 5)(\lambda - 5.372)(\lambda + 0.372),$$
$$P_3(\lambda) = \det(\mathbf{A}_1 - \lambda\mathbf{I}) = -\frac{2}{21} - \frac{17}{21}\lambda + \frac{40}{21}\lambda^2 - \lambda^3 = (\lambda - 1)^2\left(\lambda + \frac{2}{21}\right). \quad \text{(A.85)}$$

Consider the three eigenvalues $\lambda_1 = 5.372$, $\lambda_2 = 5$, $\lambda_3 = -0.372$ of the transaction matrix $\mathbf{Z}_1$. The positive eigenvalues $\lambda_1 = 5.372$, and $\lambda_2 = 5$, correspond to *semi-positive* eigenvectors $\mathbf{z}_1 = [0.429, 0, 1]'$, respectively $\mathbf{z}_2 = [0, 1, 0]'$, i. e., $\mathbf{Z}_1\mathbf{z}_i = \lambda_i\mathbf{z}_i$, $i = 1, 2$.

Consider the three eigenvalues $\lambda_1 = \lambda_2 = 1$, $\lambda_3 = -0.095$ of the coefficients matrix $\mathbf{A}_1$. The identical positive eigenvalues $\lambda_1 = \lambda_2 = 1$ correspond to the *semi-positive* eigenvectors $\mathbf{y}_1 = [1/3, 0, 2/7]'$, respectively $\mathbf{y}_2 = [0, 1, 0]'$, i. e., $\mathbf{A}_1\mathbf{y}_i = \lambda_i\mathbf{y}_i$, $i = 1, 2$.

We conclude. For the *semi-positive* matrix $\mathbf{Z}_1$, we may speak of a *Frobenius number*, $\lambda_1 = 5.372$. Here it corresponds to a *semi-positive* eigenvector. On the other hand, $\mathbf{A}_1$ has two (non-unique) maximal eigenvalues corresponding to two different *semi-positive* eigenvectors. Here we keep the term *maximal eigenvalues*. There is no Frobenius number for $\mathbf{A}_1$. Theorem A.10.1, further treats this extension of the Frobenius theorem. ▲

The following example illustrates the properties (3) and (4) of the **Perron–Frobenius Theorem A.9.3.**

Example A.9.4. Consider the semi-positive matrix **A**, which is irreducible, because $\mathbf{A}^2$ is positive,

$$\mathbf{A} = \begin{bmatrix} \frac{4}{13} & \frac{2}{9} & \frac{3}{4} \\ \frac{1}{13} & \frac{4}{9} & \frac{1}{2} \\ \frac{2}{13} & 0 & \frac{3}{4} \end{bmatrix}, \quad \mathbf{A}^2 = \begin{bmatrix} \frac{691}{3{,}042} & \frac{176}{1{,}053} & \frac{1693}{1{,}872} \\ \frac{205}{1{,}521} & \frac{226}{1{,}053} & \frac{613}{936} \\ \frac{55}{338} & \frac{4}{117} & \frac{141}{208} \end{bmatrix} > \mathbf{0}. \quad \text{(A.86)}$$

Compute a positive right eigenvector **x** such that $\sum_{i=1}^{3} x_i = 1$. Then compute a right eigenvector **x** and a left eigenvector **y** of **A** such that their scalar product is $\mathbf{x}'\mathbf{y} = 1$.

Solution to Example A.9.4:
Because of (A.86), the matrix **A** is primitive (see Lemma A.9.1), confirming the irreducibility.

There is one real, maximal eigenvalue $\lambda_A = 0.9800$, the Frobenius number, corresponding to right eigenvectors $\mathbf{x} = a[1.4951, 1.1483, 1]'$ and to the left eigenvectors $\mathbf{y} = b[0.2402, 0.0997, 1]'$. The remaining eigenvalues are $\lambda_2 = 0.3681$ and $\lambda_3 = 0.1540$.

Setting for the sum of the components of the right eigenvectors: $a \cdot \sum_{i=1}^{3} x_i = 1$, one gets $a = 0.2745$, and the eigenvector $\mathbf{x} = [0.4104, 0.3152, 0.2745]'$.

Setting then for the scalar product $\mathbf{x}'\mathbf{y} = 1$, we get $b = 2.4724$ and the other specific eigenvector $\mathbf{y} = [0.5995, 0.2465, 2.4724]'$.

Thus we have found an example satisfying the two properties (3) and (4) of *Theorem* A.9.3. ▲

The following example further illustrates the properties of primitive matrices (see Definition A.9.2).

Example A.9.5. Consider the semi-positive matrices,

$$\mathbf{A} = \begin{bmatrix} 1 & 1 & 0 & 0 \\ 1 & 1 & 0 & 0 \\ 1 & 1 & 1 & 1 \\ 1 & 1 & 1 & 1 \end{bmatrix}, \quad \mathbf{B} = \begin{bmatrix} 1 & 1 & 0 & 0 \\ 1 & 1 & 1 & 0 \\ 1 & 1 & 1 & 1 \\ 1 & 1 & 1 & 1 \end{bmatrix}. \tag{A.87}$$

Verify that **A** is imprimitive and that **B** is a primitive matrix.

Solution to Example A.9.5:
We compute,

$$(\mathbf{A}+\mathbf{I})^3 = \begin{bmatrix} 14 & 13 & 0 & 0 \\ 13 & 14 & 0 & 0 \\ 27 & 27 & 14 & 13 \\ 27 & 27 & 13 & 14 \end{bmatrix} \geq \mathbf{0}, \quad \mathbf{B}^3 = \begin{bmatrix} 5 & 5 & 3 & 1 \\ 9 & 9 & 6 & 3 \\ 13 & 13 & 9 & 5 \\ 13 & 13 & 9 & 5 \end{bmatrix} > \mathbf{0}. \tag{A.88}$$

According to Lemma A.8.2, matrix **A** is reducible and therefore *imprimitive*; one computes its eigenvalues as $\lambda_1 = \lambda_2 = 2, \lambda_3 = \lambda_4 = 0$.

Looking for the smallest $m \in \mathbb{N}$ with $\mathbf{B}^m > \mathbf{0}$, one finds $\mathbf{B}^2 \geq \mathbf{0}$ and $\mathbf{B}^3 > \mathbf{0}$, and, according to Lemma A.9.1 the non-negative matrix **B** is *primitive* and consequently *irreducible*. Its eigenvalues are $\lambda_1 = 3, \lambda_2 = 1, \lambda_3 = \lambda_4 = 0$. Thus, there is a unique maximal, real and positive eigenvalue, $\lambda_1 = 3$, the Frobenius number.

The application of Lemma A.9.1 is an easy way to determine if *non-negative* matrices are *irreducible*, especially in *Input-Output analyses*. ▲

Remember that the matrices describing Leontief quantity models and Sraffa economies for *single-product industries* are *semi-positive* and fulfill Assumption 2.2.1, and Assumption 2.2.2. Matrices describing Sraffa economies for *joint production* fulfill Assumption 6.1.1 to Assumption 6.1.3 and are also *semi-positive*.

In the next example, we will analyse borderline cases of *non-positive* matrices, as for example of the *zero matrix*.

Example A.9.6. Verify following statements:
(1) the $n \times n$ zero matrix **O** is the only *non-negative matrix* which is not *semi-positive*;
(2) the $n \times n$ zero matrix **O** is *reducible* for $n > 1$;
(3) the $n \times n$ identity matrix **I** is *reducible*;
(4) The matrices in "canonical form" or in "normal form" do not fulfill the conditions of the **Perron–Frobenius theorem A.9.3**.

Solution to Example A.9.6:
(1) In a non-negative matrix $\mathbf{A} = (a_{ij}), a_{ij} \geqq 0, \forall i,j \in \{1,\ldots,n\}$, all elements are non-negative. But here as all elements are moreover *non-positive*, then, $a_{ij} \leqq 0, \forall i,j \in \{1,\ldots,n\}$, so $a_{ij} = 0, \forall i,j \in \{1,\ldots,n\}$, therefore $\mathbf{A} = \mathbf{0}$.
(2) For $n > 1$ the matrix powers $(\mathbf{I} + \mathbf{O})^{n-1} = \mathbf{I}$ are non-negative; therefore the *non-negative* matrix **O** is reducible, according to Lemma A.8.2.

(3) The matrix powers $(\mathbf{I}+\mathbf{I})^{n-1} = (2^{n-1})\mathbf{I}$ are non-negative; therefore the *non-negative* matrix **I** is reducible, according to Lemma A.8.2.

(4) Consider a $n \times n$ matrix **A** in "canonical form" or in "normal form", then one shows that $(\mathbf{A} + \mathbf{I})^{n-1}$ has the same "canonical form" as **A**, as is illustrated by Example A.9.5, and matrix **A** is reducible, according to Lemma A.8.2. ▲

Recapitulation A.9.1.

- The **Perron theorem** requires a *positive* matrix, whereas the **Frobenius theorem** requires that a matrix be *irreducible* and *non-negative*. In both cases, the conditions for the existence of a real, positive, maximal and unique eigenvalue, called the Frobenius number of that matrix, are fulfilled and ensure the existence of associated positive left and right eigenvectors.
- In the case of a *positive* matrix, the Frobenius number is also called a simple eigenvalue. In the case of an *irreducible* and *non-negative* matrix, there may exist, besides the guaranteed Frobenius number, additional complex eigenvalues of modulus equal to that Frobenius number.

Lemma A.9.2. *Let* **C** *be a square* $n \times n$ *matrix with eigenvalues* λ *and* **I** *the* $n \times n$ *identity matrix. For any real* h*, the eigenvalues of* $h\mathbf{I} + \mathbf{C}$*, respectively* $h\mathbf{I} - \mathbf{C}$*, are the numbers* $h+\lambda$*, respectively* $h-\lambda$*, and every eigenvector of* **C** *is an eigenvector of* $h\mathbf{I}+\mathbf{C}$*, respectively* $h\mathbf{I} - \mathbf{C}$.

Proof. Let λ be an eigenvalue of matrix **C** with eigenvector **x**. Then, for every real h, there holds: (a) $(h\mathbf{I} + \mathbf{C})\mathbf{x} = h\mathbf{x} + \mathbf{C}\mathbf{x} = h\mathbf{x} + \lambda\mathbf{x} = (h + \lambda)\mathbf{x}$. Hence, $h + \lambda$ is an eigenvalue of $h\mathbf{I}+\mathbf{C}$ and **x** is an associated eigenvector; (b) $(h\mathbf{I}-\mathbf{C})\mathbf{x} = h\mathbf{x}-\mathbf{C}\mathbf{x} = h\mathbf{x}-\lambda\mathbf{x} = (h-\lambda)\mathbf{x}$. Hence, $h - \lambda$ is an eigenvalue of $h\mathbf{I} - \mathbf{C}$ and **x** is an associated eigenvector. □

Although the matrix **C** may have one or more negative elements, the Frobenius theory of positive matrices can still be applied in certain cases.

Lemma A.9.3. *The Frobenius theory applies to non-positive matrices* **C***, if*

(a) *negative elements occur only in the main diagonal;*

(b) *all elements outside the main diagonal are negative.*

Proof. With suitable h in case (a) the matrix $\mathbf{C}+h\mathbf{I}$ is positive, and in case (b) the matrix $h\mathbf{I} - \mathbf{C}$ is positive. Thus, Lemma A.9.2 applies. □

A.10 Theorems on non-negative matrices

There exist important theorems on *non-negative matrices* relaxing the conditions of the Perron–Frobenius theorem A.9.3. Some spectral properties of *non-negative* and *irreducible* matrices proven in the preceding section are usually not valid for *reducible* matrices. As any *non-negative* matrix $\mathbf{A} \geqq \mathbf{0}$ can be viewed as a limit of a sequence $(\mathbf{A}_m > 0)$ of *positive* and *irreducible* matrices (see Gantmacher [34], (42), p. 409), one can write

$$\mathbf{A} = \lim_{m\to\infty} \mathbf{A}_m, \quad (\mathbf{A}_m > 0, \quad m = 1, 2 \ldots). \tag{A.89}$$

Some spectral properties of *irreducible* matrices are retained in weakened form for *reducible* matrices.

We present here some important results valuable for any *non-negative* matrices.

Theorem A.10.1. *Every non-negative matrix $\mathbf{A} = (a_{ij}) \geqq \mathbf{0}$, $i, j = 1, \ldots, n$, always has a maximal, real, non-negative characteristic root or eigenvalue $\lambda_A \geq 0$ that dominates the moduli of all other characteristic roots of matrix* **A**. *The maximal characteristic root λ_A is then associated with a semi-positive eigenvector, $\mathbf{x}_A \geq \mathbf{o}$. (Remark: No eigenvector can ever be the null vector.)*

$$\mathbf{A}\mathbf{x}_A = \lambda_A \mathbf{x}_A, \quad (\mathbf{x}_A \geq \mathbf{o},\ \mathbf{x}_A \neq \mathbf{o}). \tag{A.90}$$

Proof. See Gantmacher ([34], p. 409). □

The *non-negative* and *reducible* matrix **A** of Example A.9.5 has maximal eigenvalue $\lambda_A = 2$, according to Theorem A.10.1. A *non-negative* and *reducible* matrix **A** can be brought with permutations of the rows and columns to "canonical form", see Gantmacher ([34], p. 411):

$$\widetilde{\mathbf{A}} = \mathbf{P}^{-1}\mathbf{A}\mathbf{P} = \begin{bmatrix} \widetilde{\mathbf{A}}_{11} & \mathbf{0} \\ \widetilde{\mathbf{A}}_{21} & \widetilde{\mathbf{A}}_{22} \end{bmatrix}. \tag{A.91}$$

Lemma A.10.1.

(a) *The maximal eigenvalue $\lambda_{\tilde{A}}$ of a non-negative reducible matrix $\tilde{\mathbf{A}}$ (A.91) is equal to the maximal eigenvalue of one of the submatrices $\widetilde{\mathbf{A}}_{11}, \widetilde{\mathbf{A}}_{22}$.*

(b) *If the reduction process is terminated to attain the block diagonal form (A.73) then the submatrices $\widetilde{\mathbf{A}}_{11}, \ldots, \widetilde{\mathbf{A}}_{ss}$ are irreducible square matrices. Note that the zero matrix consisting only of one element is irreducible (Kurz & Salvadori [52], p. 104). The maximal eigenvalue $\lambda_{\tilde{A}}$ of matrix $\widetilde{\mathbf{A}}$ (A.73) is then equal to the maximal eigenvalue of one of the submatrices $\widetilde{\mathbf{A}}_{11}, \ldots, \widetilde{\mathbf{A}}_{ss}$.*

Proof. (a) For the equality $\lambda_{\tilde{A}_{ii}} = \lambda_{\tilde{A}}$, $i \in \{1, 2\}$, see Gantmacher ([34], pp. 411–412).

(b) There exists a non-negative *irreducible* submatrix $\tilde{A}_{kk}$, $k \in \{1, \ldots, s\}$ for which $\lambda_{\tilde{A}} = \lambda_{\tilde{A}_{ss}} \geq 0$, see Gantmacher ([34], pp. 411–412). □

We illustrate that a permutation must be found and performed on matrix **A** (A.91) to obtain $\widetilde{\mathbf{A}}_{11}$ and $\widetilde{\mathbf{A}}_{22}$. See the following example.

Example A.10.1. Let a reducible matrix **A** and a permutation matrix **P** be given. Establish the "canonical form", Definition A.8.3, (A.73), the maximal eigenvalue $\lambda_{\tilde{A}}$ of $\tilde{\mathbf{A}}$ and the maximal eigenvalue $\lambda_{\tilde{A}_{11}}$ of $\widetilde{\mathbf{A}}_{11}$,

$$\mathbf{A} = \begin{bmatrix} 30 & 20 & 10 & 40 \\ 0 & 50 & 20 & 40 \\ 0 & 30 & 50 & 0 \\ 10 & 30 & 20 & 40 \end{bmatrix}, \quad \mathbf{P} = \begin{bmatrix} 0 & 0 & 1 & 0 \\ 1 & 0 & 0 & 0 \\ 0 & 1 & 0 & 0 \\ 0 & 0 & 0 & 1 \end{bmatrix}. \tag{A.92}$$

Solution to Example A.10.1:
We compute the inverse matrix $\mathbf{P}^{-1}$ and the product $\mathbf{P}^{-1}\mathbf{AP}$:

$$\mathbf{P}^{-1} = \begin{bmatrix} 0 & 1 & 0 & 0 \\ 0 & 0 & 1 & 0 \\ 1 & 0 & 0 & 0 \\ 0 & 0 & 0 & 1 \end{bmatrix}, \quad \widetilde{\mathbf{A}} = \mathbf{P}^{-1}\mathbf{AP} = \begin{bmatrix} 50 & 20 & 0 & 0 \\ 30 & 50 & 0 & 0 \\ 20 & 10 & 30 & 40 \\ 30 & 20 & 10 & 40 \end{bmatrix}. \tag{A.93}$$

The submatrix $\widetilde{\mathbf{A}}_{11} = \left[\begin{smallmatrix} 50 & 20 \\ 30 & 50 \end{smallmatrix}\right] > \mathrm{O}$ is irreducible. Moreover, the reduction process has attained block diagonal form and submatrix $\widetilde{\mathbf{A}}_{22} = \left[\begin{smallmatrix} 30 & 40 \\ 10 & 40 \end{smallmatrix}\right] > \mathrm{O}$ is also irreducible.

We now determine the characteristic polynomials of $\tilde{\mathbf{A}}$ and $\widetilde{\mathbf{A}}_{11}$,

$$\begin{aligned} P_4(\lambda) &= \det(\tilde{\mathbf{A}} - \lambda \mathbf{I}_4) = 1{,}520{,}000 - 213{,}000\lambda + 9{,}700\lambda^2 - 170\lambda^3 + \lambda^4, \\ P_2(\lambda) &= \det(\tilde{\mathbf{A}}_{11} - \lambda \mathbf{I}_2) = 1{,}900 - 100\lambda + \lambda^2. \end{aligned} \tag{A.94}$$

The maximal eigenvalue $\lambda_{\tilde{A}} = 10(5 + \sqrt{6}) = 74.4949$ of $\tilde{\mathbf{A}}$ is unique, equal to the Frobenius number of $\widetilde{\mathbf{A}}_{11}$, according to Lemma A.10.1, $\lambda_{\tilde{A}} = \lambda_{\tilde{A}_{11}} > 0$. The Frobenius number $\lambda_{\tilde{A}_{22}} = 5(7 + \sqrt{17})$ is less than $\lambda_{\tilde{A}}$. ▲

Consider now a **first limit case** of a *non-negative* matrix, the zero matrix **O**.

Example A.10.2. Determine the maximal eigenvalues and the associated eigenvectors of the zero matrix **O**. Comment on the results.

Solution to Example A.10.2:
We compute the maximal eigenvalues of the zero matrix **O**,

$$P_3(\lambda) = \det(\mathbf{O} - \lambda \mathbf{I}) = \det\left(\begin{bmatrix} 0 & 0 & 0 \\ 0 & 0 & 0 \\ 0 & 0 & 0 \end{bmatrix} - \lambda \begin{bmatrix} 1 & 0 & 0 \\ 0 & 1 & 0 \\ 0 & 0 & 1 \end{bmatrix} \right) = -\lambda^3. \tag{A.95}$$

We set $P_3(\lambda) = 0$ and get the maximal eigenvalue $\lambda_0 = 0$. So we compute the (right) eigenvectors, $\mathbf{O} \cdot \mathbf{x}_O = \lambda_0 \mathbf{x}_O = \mathbf{o}$, where $\mathbf{x}_O = [x_1, x_2, x_3]' \in \mathbb{R}^3 \setminus \{\mathbf{o}\}$, because an eigenvector is never the zero vector.

In particular, the unit vectors $[1, 0, 0]'$, $[0, 1, 0]'$, $[0, 0, 1]'$ are eigenvectors of the zero matrix and usually presented as such in the textbooks. ▲

We recognise that the zero matrix **0** fulfills the assumptions of Theorem A.10.1. But, as it is not *semi-positive*, **0** does not fulfill Assumption 2.2.2. So evidently, **0** cannot describe a Sraffa production system or an Input-Output Table (IOT)!

We contunue with a **second limit case** of a *semi-positive* and *reducible* matrix in "canonical form", which also does not describe a Sraffa production system, because Assumptions 2.2.1 and 2.2.2, are not fulfilled.

Example A.10.3. Let be given the following semi-positive reducible matrices $\widetilde{\mathbf{A}}$ and $\widetilde{\mathbf{B}}$ in "canonical form", $b > 0$, according to Definition A.8.3, (A.73). Compute the maximal eigenvalues $\lambda_{\widetilde{A}}$ of $\widetilde{\mathbf{A}}$ and $\lambda_{\widetilde{B}}$ of $\widetilde{\mathbf{B}}$ and the Frobenius number of the 1×1 positive submatrix $\widetilde{\mathbf{A}}_{11} > \mathbf{0}$ of matrix $\widetilde{\mathbf{A}}$,

$$\tilde{\mathbf{A}} = \begin{bmatrix} 20 & 0 & 0 & 0 \\ 0 & 0 & 0 & 0 \\ 0 & 0 & 0 & 0 \\ 0 & 0 & 0 & 0 \end{bmatrix} \geq \mathbf{0}, \quad \tilde{\mathbf{B}} = \begin{bmatrix} b & b & 0 & 0 \\ b & b & 0 & 0 \\ 0 & 0 & 0 & 0 \\ 0 & 0 & 0 & 0 \end{bmatrix} \geq \mathbf{0}, \quad \tilde{\mathbf{A}}_{11} = [20] > \mathbf{0}. \tag{A.96}$$

Solution to Example A.10.3:
We directly determine the characteristic polynomials of matrices $\widetilde{\mathbf{A}}$, $\widetilde{\mathbf{B}}$ and $\widetilde{\mathbf{A}}_{11}$,

$$\begin{aligned} f_4(\lambda) &= \det(\tilde{\mathbf{A}} - \lambda\mathbf{I}_4) = -20\lambda^3 + \lambda^4, \\ g_4(\lambda) &= \det(\tilde{\mathbf{B}} - \lambda\mathbf{I}_4) = \lambda^2(-2b\lambda + \lambda^2), \\ f_1(\lambda) &= \det(\tilde{\mathbf{A}}_{11} - \lambda\mathbf{I}_1) = 20 - \lambda. \end{aligned} \tag{A.97}$$

There are unique, maximal, real, positive eigenvalues $\lambda_{\tilde{A}} = 20$ and $\lambda_{\tilde{B}} = 2b > 0$. The Lemma A.10.1 applies and $\lambda_{\tilde{A}_{11}} = 20$ is the Frobenius number of the *positive* submatrix $\widetilde{\mathbf{A}}_{11}$, $\lambda_{\widetilde{A}} = \lambda_{\widetilde{A}_{11}} = 20 > 0$.

The following generalisation is found. Let be $n \geq m > 0$. Consider further a matrix **D** set up as matrix **B**, but with an $m \times m$ submatrix exclusively of identical elements b in the left-upper corner, whereas elsewhere all components are 0's. Then the characteristic polynomial of matrix **D** is

$$h_n(\lambda) = \det(\widetilde{\mathbf{D}} - \lambda\mathbf{I}_n) = \lambda^{n-1}(\lambda - mb). \tag{A.98}$$

There is a unique, maximal, real and positive eigenvalue $\lambda_{\widetilde{D}} = mb > 0$. ▲

The next lemma is in Gantmacher [34], Theorem 6, p. 416.

Lemma A.10.2. *If any component of a non-negative matrix* **A** *is increased, then its Frobenius number cannot decrease. If matrix* **A** *is irreducible, then the Frobenius number increases effectively.*

Example A.10.4. Set $b = 1$. Ceteris paribus, replace the element $b_{11} = b = 1 > 0$ of the semi-positive matrix $\tilde{\mathbf{B}}$ by $e_{11} = 1.1$ leading to matrix $\tilde{\mathbf{E}}$. Compare the maximal eigenvalues of $\tilde{\mathbf{B}}$ and $\tilde{\mathbf{E}}$, respectively, the Frobenius numbers of the positive 2×2 submatrices $\tilde{\mathbf{B}}_{11}$ and $\tilde{\mathbf{E}}_{11}$ in its upper left corners. Apply Lemma A.10.2.

$$\mathbf{0} \leq \tilde{\mathbf{B}} = \begin{bmatrix} b_{11} = 1 & 1 & 0 & 0 \\ 1 & 1 & 0 & 0 \\ 0 & 0 & 0 & 0 \\ 0 & 0 & 0 & 0 \end{bmatrix} \leq \tilde{\mathbf{E}} = \begin{bmatrix} e_{11} = 1.1 & 1 & 0 & 0 \\ 1 & 1 & 0 & 0 \\ 0 & 0 & 0 & 0 \\ 0 & 0 & 0 & 0 \end{bmatrix}. \tag{A.99}$$

Solution to Example A.10.4:
Matrices $\tilde{\mathbf{B}}$ an $\tilde{\mathbf{E}}$ have the maximal eigenvalues $\lambda_{\tilde{B}} = 2 > 0$ and $\lambda_{\tilde{E}} = 2.05125 > 0$. The Frobenius numbers of matrices $\tilde{\mathbf{B}}_{11} = \left[\begin{smallmatrix} 1 & 1 \\ 1 & 1 \end{smallmatrix}\right] > 0$ and $\tilde{\mathbf{E}}_{11} = \left[\begin{smallmatrix} 1.1 & 1 \\ 1 & 1 \end{smallmatrix}\right] > 0$ are $\lambda_{\tilde{B}} = \lambda_{\tilde{B}_{11}} = 2$ and $\lambda_{\tilde{E}} = \lambda_{\tilde{E}_{11}} = 2.05125$. This is exactly the statement of Lemma A.10.2 applied on the semi-positive matrices $\tilde{\mathbf{B}} \leq \tilde{\mathbf{E}}$ and the irreducible submatrices $\tilde{\mathbf{B}}_{\mathbf{11}} \leq \tilde{\mathbf{E}}_{\mathbf{11}}$, leading to $\lambda_{\tilde{B}} = \lambda_{\tilde{B}_{11}} = 2 \leq \lambda_{\tilde{E}} = \lambda_{\tilde{E}_{11}} = 2.05125$. ▲

Relying now on Theorem A.10.1, we discuss the development of *non-negative reducible* and also *irreducible* matrices into convergent infinite series.[10]

Theorem A.10.2. *Consider a non-negative matrix* $\mathbf{A} \geqq \mathbf{0}$ *with maximal eigenvalue* $\lambda_A = \rho(\mathbf{A}) \geq 0$. *For any* $\lambda > \lambda_A$, *there is*

$$(\lambda\mathbf{I} - \mathbf{A})^{-1} \geqq \mathbf{0} \quad \text{and} \quad \frac{d}{d\lambda}(\lambda\mathbf{I} - \mathbf{A})^{-1} \leqq \mathbf{0}. \tag{A.100}$$

Furthermore, for $\lambda > \lambda_A \geq 0$, *we have the development* ($\mathbf{A}^0 = \mathbf{I}$),

$$(\lambda\mathbf{I} - \mathbf{A})^{-1} = \sum_{j=0}^{\infty} \frac{\mathbf{A}^j}{\lambda^{j+1}} \geqq \mathbf{0}, \tag{A.101}$$

and also

$$\frac{d}{d\lambda}(\lambda\mathbf{I} - \mathbf{A})^{-1} = -\sum_{j=0}^{\infty} \frac{(j+1)\mathbf{A}^j}{\lambda^{j+2}} \leqq \mathbf{0}. \tag{A.102}$$

If $\mathbf{A}$ *is non-negative and irreducible with maximal eigenvalue* λ_A, *then:*

$$(\lambda\mathbf{I} - \mathbf{A})^{-1} > 0 \quad \textit{and} \quad \frac{d}{d\lambda}(\lambda\mathbf{I} - \mathbf{A})^{-1} < \mathbf{0}. \tag{A.103}$$

Proof. See Gantmacher, [34], p. 410, Section: 13.3 Reducible matrices. □

10 Note that *positive* matrices are *non-negative* and *irreducible*.

Finally, Theorem A.10.1 can be sharpened for semi-positive matrices, describing the Sraffa price models. More specifically, there is an important proposition for semi-positive matrices, fulfilling Assumption 2.2.1 and Assumption 2.2.2 which will be used to compute the *productiveness* $R = (1/\lambda_C) - 1$ and to ensure its existence by the statement $\lambda_C > 0$.

Lemma A.10.3. *Consider an n sector Sraffa production scheme* $(\mathbf{S}', \mathbf{L}) \to (\mathbf{q})$ *with a semi-positive and reducible commodity flow matrix* $\mathbf{S} \geq \mathbf{0}$*, fulfilling Assumption 2.2.1 and Assumption 2.2.2. Therefore, there is a positive output vector* $\mathbf{q} = \mathbf{S}\mathbf{e} + \mathbf{d} > \mathbf{o}$*. Assume a semi-positive vector of surplus* $\mathbf{d} \geq \mathbf{o}$*. Then the semi-positive and reducible input-output coefficients matrix* $\mathbf{C} = \mathbf{S}\hat{\mathbf{q}}^{-1} \geq \mathbf{0}$ *(2.17) has a positive, maximal, real eigenvalue* $\lambda_C > 0$*.*

Proof. The semi-positive reducible matrix $\mathbf{C} \geq \mathbf{0}$ can be brought by permutations in completely reduced block diagonal form (A.74) with now irreducible semi-positive matrices $\tilde{\mathbf{A}}_{11}, \ldots, \tilde{\mathbf{A}}_{ss}$. Note that the zero matrix of order one is irreducible. With $m \geq 1$, the semi-positivity implies the existence of at least one positive $m \times m$ submatrix $\tilde{\mathbf{A}}_{kk} > \mathbf{0}$, $k \in \{1, \ldots, s\}$, exhibiting a positive Frobenius number $\lambda_{\tilde{A}_{kk}} > 0$. With Lemma A.10.1 one concludes that one of these Frobenius numbers $\lambda_{\tilde{A}_{kk}} = \lambda_C > 0$ is the requested maximal eigenvalue of the semi-positive matrix $\mathbf{C}$. □

A.11 Stochastic matrices

In this section, we treat a special category of *non-negative* matrices, the stochastic matrices, on which all the properties encountered in Section A.10 apply. They play an important role in the description of economies with no surplus or in the description of *interindustrial economies*.

Definition A.11.1. An $n \times n$ matrix $\mathbf{D} = (d_{ij})$ is stochastic if it is a non-negative matrix and if the sum in each of its rows is equal to 1,

$$d_{ij} \geq 0, \quad \sum_{j=1}^{n} d_{ij} = 1, \quad i \in \{1, \ldots, n\}, \quad \Rightarrow \mathbf{D}\mathbf{e} = \mathbf{e} \tag{A.104}$$

Stochastic matrices have the following properties that are important in our context:

1 If $\mathbf{D}$ is stochastic, then for $m = 1, 2, \ldots$ the matrices $\mathbf{D}^m$ are also stochastic.
2 For an $n \times n$ stochastic matrix $\mathbf{D}$, there is $\mathbf{e}'\mathbf{D}\mathbf{e} = n$ and $\mathbf{e}'\mathbf{D}^m\mathbf{e} = n$.

From equation (A.104), it follows that $\lambda_D = 1$ is an eigenvalue of matrix $\mathbf{D}$ to which the $n \times 1$ eigenvector $\mathbf{e} = [1, \ldots, 1]'$ is associated.

Lemma A.11.1. *A non-negative n×n matrix* $\mathbf{D} \geqq \mathbf{O}$ *is stochastic if and only if its eigenvalue* $\lambda_D = 1$ *is associated with the* $n \times 1$ *eigenvector* $\mathbf{e} = [1, \ldots, 1]'$*. The eigenvalue* $\lambda_D = 1$ *is a maximal eigenvalue of the stochastic matrix* $\mathbf{D}$*.*

Proof. See Gantmacher [34], p. 427. □

In the present context, we observe that the *distribution coefficients* matrices of the systems of production with no surplus are stochastic.

Example A.11.1. Consider Example 5.1.2 and set up the *commodity flow* matrix $\mathbf{S}$. Consider that there is no surplus, thus the vector of *final demand* is therefore $\mathbf{d} = \mathbf{o}$ and the vector of total output is $\mathbf{q}_I = \mathbf{Se}$. Calculate the inverse of the diagonal matrix $\hat{\mathbf{q}}_I$,

$$\mathbf{S} = \begin{bmatrix} 90 & 50 & 40 \\ 120 & 125 & 40 \\ 60 & 150 & 200 \end{bmatrix}, \quad \mathbf{q}_I = \mathbf{Se} = \begin{bmatrix} 180 \\ 285 \\ 410 \end{bmatrix}, \quad \hat{\mathbf{q}}_I^{-1} = \begin{bmatrix} \frac{1}{180} & 0 & 0 \\ 0 & \frac{1}{285} & 0 \\ 0 & 0 & \frac{1}{410} \end{bmatrix}, \tag{A.105}$$

and calculate the matrices $\mathbf{C}$ and $\mathbf{D}$ and their eigenvalues.

Solution to Example A.11.1:

$$\mathbf{C} = \mathbf{S}\hat{\mathbf{q}}_I^{-1} = \begin{bmatrix} 90 & 50 & 40 \\ 120 & 125 & 40 \\ 60 & 150 & 200 \end{bmatrix} \begin{bmatrix} \frac{1}{180} & 0 & 0 \\ 0 & \frac{1}{285} & 0 \\ 0 & 0 & \frac{1}{410} \end{bmatrix} = \begin{bmatrix} \frac{1}{2} & \frac{10}{57} & \frac{4}{41} \\ \frac{2}{3} & \frac{25}{57} & \frac{4}{41} \\ \frac{1}{3} & \frac{10}{19} & \frac{20}{41} \end{bmatrix}, \tag{A.106}$$

$$\mathbf{D} = \hat{\mathbf{q}}_I^{-1}\mathbf{S} = \begin{bmatrix} \frac{1}{180} & 0 & 0 \\ 0 & \frac{1}{285} & 0 \\ 0 & 0 & \frac{1}{410} \end{bmatrix} \begin{bmatrix} 90 & 50 & 40 \\ 120 & 125 & 40 \\ 60 & 150 & 200 \end{bmatrix} = \begin{bmatrix} \frac{9}{18} & \frac{5}{18} & \frac{4}{18} \\ \frac{24}{57} & \frac{25}{57} & \frac{8}{57} \\ \frac{6}{41} & \frac{15}{41} & \frac{20}{41} \end{bmatrix}. \tag{A.107}$$

Establish the characteristic polynomial of matrices $\mathbf{C}$ and $\mathbf{D}$,

$$P_3(\lambda) = \det(\mathbf{C} - \lambda\mathbf{I}_3) = \det(\mathbf{D} - \lambda\mathbf{I}_3) = (1 - \lambda)\left(\frac{350}{7{,}011} - \frac{1{,}993}{4{,}674}\lambda + \lambda^2\right). \tag{A.108}$$

The eigenvalues are $\lambda_1 = \lambda_D = 1$, $\lambda_2 = 0.2132 + 0.0668 \cdot i$, $\lambda_3 = 0.2132 - 0.0668 \cdot i$ and we have the property that $\mathbf{e} = [1, 1, 1]'$ is the eigenvector corresponding to $\lambda_D = 1$,

$$\mathbf{De} = \begin{bmatrix} \frac{9}{18} & \frac{5}{18} & \frac{4}{18} \\ \frac{24}{57} & \frac{25}{57} & \frac{8}{57} \\ \frac{6}{41} & \frac{15}{41} & \frac{20}{41} \end{bmatrix} \begin{bmatrix} 1 \\ 1 \\ 1 \end{bmatrix} = \begin{bmatrix} 1 \\ 1 \\ 1 \end{bmatrix} = \mathbf{e} \Rightarrow \mathbf{De} = \mathbf{e}. \tag{A.109}$$

We see that $\mathbf{D}$ is stochastic, i. e., the sums of the three rows are all equal to one. We conclude with the eigenvalue equation, $\mathbf{De} = \mathbf{e}$, that the eigenvector $\mathbf{e}$ is associated with the eigenvalue $\lambda_D = 1$ of the *distribution coefficients* matrix $\mathbf{D}$. ▲

A.12 The productive Leontief model

In this section, we treat the *Leontief model* and the *productive Leontief model*. We rely therefore directly on the *non-negative input-output coefficients* matrix $\mathbf{A} \geqq \mathbf{0}$, a *non-*

negative vector of *final demand* $\mathbf{f} \geqq \mathbf{o}$ and a *non-negative* vector of *total output* $\mathbf{x} \geqq \mathbf{o}$ (see Ashmanov [2], p. 24). In other words, we return to the most general assumptions of *non-negativeness*. The restrictions of *semi-positivity* will be introduced when needed. We do not refer to the *commodity flow* matrix $\mathbf{Z}$.

It is necessary to comment on the terminology used in this text and used by various authors. We adopt the term *Leontief model* for equation (2.30) expressing the vector of *total output* $\mathbf{x} \geqq \mathbf{o}$, given the matrix $\mathbf{A} \geqq \mathbf{0}$ and the vector $\mathbf{d} \geqq \mathbf{o}$, representing the initial *input-output model* of Leontief [55], [56]. It has also been called the *Leontief quantity model* (see Oosterhaven [77], p. 750). The *Leontief model* (2.30) has to be clearly distinguished from the various Leontief price models, as they are treated in Section 2.5.

Finally, we define the notion of *productive models* (see Ashmanov [2], p. 24, and Gale [33], p. 296). Considering matrix $\mathbf{A} = \mathbf{Z}\hat{\mathbf{x}}^{-1}$, it is known that matrix $\mathbf{Z}$ is obtained from prices and quantities in *physical units* (2.18) thus, what follows can also be applied to Leontief quantity models in physical terms.[11]

It is important to mention that the original version of the notion of the *productive Leontief model*, Definition A.12.1, and the corresponding Theorem A.12.1, the statement concerning the *Frobenius number*, are presented here, requiring a *non-negative* matrix $\mathbf{A} \geqq \mathbf{0}$ and *non-negative* vectors $\mathbf{f} \geqq \mathbf{o}$, $\mathbf{x} \geqq \mathbf{o}$. Then, throughout the applications in this text it is admitted that the economic Assumption 2.2.1 and Assumption 2.2.2 prevail, where the matrix $\mathbf{A} \geq \mathbf{0}$ and the vector $\mathbf{f} \geq \mathbf{o}$ are *semi-positve*, respectively, the vector $\mathbf{x} > \mathbf{o}$ is positive, what is necessary because one generally starts from the *commodity flow* matrix $\mathbf{Z}$.

Definition A.12.1 (Leontief model, productive Leontief model). Let a non-negative input-output coefficients matrix $\mathbf{A} \geqq \mathbf{0}$ (2.9) and any non-negative vector of final demand $\mathbf{f} \geqq \mathbf{o}$ be given. If there exists a unique non-negative output vector $\mathbf{x} \geqq \mathbf{o}$, so that the equation

$$\mathbf{x} = \mathbf{A}\mathbf{x} + \mathbf{f}, \quad \mathbf{x} \geqq \mathbf{o}, \tag{A.110}$$

holds, then one calls (A.110) a **Leontief model.** In the case that the vector of final demand is semi-positive, $\mathbf{f} \geq \mathbf{o}$, the *Leontief model* (A.110) is referred to as a **productive Leontief model.**[12]

Ashmanov ([2], Theorem 1.5, p. 39) formulated the condition for a *productive Leontief model*. This theorem belongs to the group of theorems on *non-negative* matrices. It is the basis for the understanding of the inner structure of productive Leontief models and the related models.

---

11 This application is treated in Section 2.4.2.

12 Analogously, one defines a productive Sraffa model (3.52), for the case of measurement in physical terms. In this case we admit that Assumption 2.2.1 and Assumtion 2.2.2 always prevail.

Theorem A.12.1. *(Condition for a productive Leontief model). A Leontief model* (A.110) *is productive if and only if the Frobenius number λ_A of the non-negative coefficients matrix* $\mathbf{A} \geqq \mathbf{0}$ (2.30) *is less than one,* $\lambda_A < 1$.[13]

Proof. Sufficient condition:
Because all the moduli of eigenvalues λ_a of $\mathbf{A}$ are less than or equal to λ_A, $|\lambda_a| \leq \lambda_A < 1$, the equality $\det(\lambda_a \mathbf{I} - \mathbf{A}) = 0$ holds only for the eigenvalues of $\mathbf{A}$. Consequently, the matrix $\mathbf{A}$ is regular, $\det(\mathbf{I} - \mathbf{A}) \neq 0$. With Theorem A.10.2 one substitutes $\lambda = 1$ into equation (A.101). For *non-negative* matrices $\mathbf{A} \geqq \mathbf{0}$, we obtain the *non-negative Leontief Inverse*, expressed as a convergent geometric series,

$$(\mathbf{I} - \mathbf{A})^{-1} = \sum_{k=0}^{\infty} \mathbf{A}^k \geqq \mathbf{0}, \tag{A.111}$$

with matrix powers converging to $\mathbf{0}$, $\lim_{k\to\infty} \mathbf{A}^k = \mathbf{0}$. Then, for any specific vector of *final demand* $\mathbf{f} \geqq 0$, one considers the product with the inverse matrix (A.111), obtaining the unique *non-negative* solution of the system (A.110),

$$\mathbf{x} = (\mathbf{I} - \mathbf{A})^{-1}\mathbf{f} \geqq \mathbf{o}. \tag{A.112}$$

Necessary condition:
There exists a unique *non-negative* solution $\mathbf{x} \geqq \mathbf{o}$ to the productive Leontief model (A.110). This means in terms of linear algebra[14]:

$$\begin{aligned} &\mathbf{x} = \mathbf{A}\mathbf{x} + \mathbf{f} \Leftrightarrow \mathbf{x}(\mathbf{I} - \mathbf{A}) - \mathbf{f} = \mathbf{o} \Leftrightarrow \\ &\operatorname{rank}(\mathbf{I} - \mathbf{A}, \mathbf{f}) = \operatorname{rank}(\mathbf{I} - \mathbf{A}) = n \Leftrightarrow \det(\mathbf{I} - \mathbf{A}) \neq 0. \end{aligned} \tag{A.113}$$

Consequently, the *Leontief Inverse* $(\mathbf{I} - \mathbf{A})^{-1}$ exists. With Theorem A.10.2, equation (A.101), one performs the development of the *Leontief Inverse* into the series (A.111) having to set $\lambda = 1$. This excludes Frobenius numbers λ_A greater or equal than one, see condition $\lambda > \lambda_A$ of Theorem A.10.2, so the Frobenius number must be less than $\lambda = 1$, $\lambda_A < \lambda = 1$. □

We treat now a **third limit case:** semi-positive *reducible* matrices describing jus-viable economies (no surplus); Assumption 2.2.1 and Assumption 2.2.2 hold.

Example A.12.1. Given a semi-positive reducible commodity flow matrix $\mathbf{S}$, compute the output vector $\mathbf{q}_I = \mathbf{S}\mathbf{e}$, the *input-output coefficients* matrix $\mathbf{C} = \mathbf{S}\hat{\mathbf{q}}_I^{-1}$ and the *distribution coefficients* matrix $\mathbf{D} = \hat{\mathbf{q}}_I^{-1}\mathbf{S}$. Compute the eigenvalues of $\mathbf{C}$ and $\mathbf{D}$, choose the maximal eigenvalues and compute the corresponding eigenvectors.

13 As there is here a *non-negative* matrix, the Theorem A.10.1 states the existence of a maximal eigenvalue. In the present case of Theorem A.12.1, also with a non-negative matrix, this maximal eigenvalue is usually called the Frobenius number, as is the case for the Perron–Frobenius theorem A.9.3.

14 See Nef [69], pp. 122–123, Theorem 6, which states following equivalent propositions in al. 3: The linear system (A.110) has exactly one solution: al. 4: $\det(\mathbf{A} - \mathbf{I}) \neq 0 \Leftrightarrow \operatorname{rank}(\mathbf{I} - \mathbf{A}, \mathbf{f}) = \operatorname{rank}(\mathbf{I} - \mathbf{A}) = n$ (Theorem 3).

$$\mathbf{S} = \begin{bmatrix} 20 & 10 & 0 \\ 0 & 0 & 10 \\ 0 & 30 & 0 \end{bmatrix}, \quad (\mathbf{I}+\mathbf{S})^2 = \begin{bmatrix} 441 & 220 & 100 \\ 0 & 301 & 20 \\ 0 & 60 & 301 \end{bmatrix} \geq \mathbf{0}. \tag{A.114}$$

Solution to Example A.12.1:
We compute the output vector and the matrices:

$$\mathbf{q}_I = \begin{bmatrix} 30 \\ 10 \\ 30 \end{bmatrix}, \quad \mathbf{C} = \mathbf{S}\hat{\mathbf{q}}_I^{-1} = \begin{bmatrix} \frac{2}{3} & 1 & 0 \\ 0 & 0 & \frac{1}{3} \\ 0 & 3 & 0 \end{bmatrix}, \quad \mathbf{D} = \hat{\mathbf{q}}_I^{-1}\mathbf{S} = \begin{bmatrix} \frac{2}{3} & \frac{1}{3} & 0 \\ 0 & 0 & 1 \\ 0 & 1 & 0 \end{bmatrix}. \tag{A.115}$$

We now establish the characteristic polynomial of both matrices **C** and **D**,

$$f_3(\lambda) = \det(\mathbf{C} - \lambda\mathbf{I}_3) = \det(\mathbf{D} - \lambda\mathbf{I}_3) = \frac{1}{3}(-2 + 3\lambda + 2\lambda^2 - 3\lambda^3). \tag{A.116}$$

The eigenvalues of matrix **C** and the stochastic matrix **D**, $\mathbf{De} = \mathbf{e}$, are $\{1, 2/3, -1\}$. The maximal eigenvalue of both matrices is $\lambda_C = \lambda_D = 1$, associated to the positive eigenvector $\mathbf{c}_1 = [1, 1/3, 1] > \mathbf{o}$ of **C**, respectively $\mathbf{d}_1 = [1, 1, 1] > \mathbf{o}$ of **D**. The eigenvalue $\lambda_C = \lambda_D = 1$ is maximal (see Lemma A.10.1). ▲

It is necessary to establish the terminology. We say that an *economy is productive if this economy produces a surplus*. In this case, we are in presence of *"productive economies or surplus economies"* (see Hall [39], p. 78).

Further, an economy is *just viable* if it produces no surplus and *viable* if it produces either a surplus or has no surplus; in other words, a *viable economy* is either *productive* or *just viable* (see Kurz and Salvadori [52], pp. 96–97).

Consequently, a *productive Leontief model* or *productive Sraffa model* with Frobenius number $\lambda = 1/(1+R) < 1$ results in a *productive economy* and vice versa.

Helmut Knolle [49] proposed to term $R = (1/\lambda_C) - 1 > 0$ as a measure of the *productiveness* of an economy. The number $R > 0$, also called the *maximal rate of profits*, captures the *productivity of an economy*, a notion dating back to Krugman [47], expressing the state of being in the presence of a *productive model*.

There is accordingly another approach to get this result: the *Hawkins–Simon condition* (see Theorem A.12.2 hereafter).

Notation A.12.1 (Principal minors of a matrix). We define the notion of the n principal minors of matrix $(\mathbf{I}-\mathbf{A})$. They are the determinants of the $m \times m$ sub-matrices of matrix $(\mathbf{I}-\mathbf{A})$, keeping the first m rows and columns, $1 \leq m \leq n$, after deleting the last $n-m$ rows and columns.

Takayama [116], p. 360, then formulates the following questions:

(1) **(The existence problem)** For any given $\mathbf{f} \geqq \mathbf{o}$, can we guarantee that there exists an $\mathbf{x} \geqq \mathbf{o}$, such that $(\mathbf{I}-\mathbf{A})\mathbf{x} = \mathbf{f}$? If so, is such an $\mathbf{x}$ unique?
(2) **(The non-singularity problem)** Is the matrix $(\mathbf{I}-\mathbf{A})$ non-singular? If so, is $(\mathbf{I}-\mathbf{A})^{-1} \geq \mathbf{0}$?

Resorting to the notion of principal minors, these two questions can be answered affirmatively as follows:

Theorem A.12.2 (The Hawkins–Simon condition for the viability of an economy). *The questions (1) and (2) receive an affirmative answer if the necessary and sufficient condition is fulfilled that all the n principal minors of matrix* $(\mathbf{I}-\mathbf{A})$ *are positive.*

We illustrate the *Hawkins–Simon condition* with the following example.

Example A.12.2. Given the semi-positive and reducible matrix in canonical form,

$$\tilde{\mathbf{A}} = \begin{bmatrix} \frac{5}{12} & \frac{1}{4} & 0 & 0 \\ \frac{1}{8} & \frac{5}{16} & 0 & 0 \\ \frac{2}{11} & \frac{3}{11} & \frac{5}{11} & \frac{1}{11} \\ \frac{1}{4} & \frac{1}{2} & \frac{1}{2} & \frac{3}{4} \end{bmatrix} \geq \mathbf{0}, \quad \mathbf{I}-\tilde{\mathbf{A}} = \begin{bmatrix} \frac{7}{12} & -\frac{1}{4} & 0 & 0 \\ -\frac{1}{8} & \frac{11}{16} & 0 & 0 \\ -\frac{2}{11} & -\frac{3}{11} & \frac{6}{11} & -\frac{1}{11} \\ -\frac{1}{4} & -\frac{1}{2} & -\frac{1}{2} & \frac{1}{4} \end{bmatrix}. \tag{A.117}$$

Compute the $n=4$ principal minors of $(\mathbf{I}-\tilde{\mathbf{A}})^{-1}$, the Leontief Inverse and the canonical form $\tilde{\mathbf{A}}$. Take the vector of final demand $\mathbf{f}=[20,0,0,0]'$ and compute the corresponding vector of total output $\mathbf{x}$. Comment on the obtained results.

Solution to Example A.12.2:
The principal minors are:

$$\det\left(\begin{vmatrix} \frac{7}{12} & -\frac{1}{4} & 0 \\ -\frac{1}{8} & \frac{11}{16} & 0 \\ -\frac{2}{11} & -\frac{3}{11} & \frac{6}{11} \end{vmatrix}\right) = \frac{71}{352} > 0, \quad \det\left(\begin{vmatrix} \frac{7}{12} & -\frac{1}{4} \\ -\frac{1}{8} & \frac{11}{16} \end{vmatrix}\right) = \frac{71}{192} > 0,$$

$$\det\left(\left|\frac{7}{12}\right|\right) = \frac{7}{12} > 0, \quad \det\left(\begin{vmatrix} \frac{7}{12} & -\frac{1}{4} & 0 & 0 \\ -\frac{1}{8} & \frac{11}{16} & 0 & 0 \\ -\frac{2}{11} & -\frac{3}{11} & \frac{6}{11} & -\frac{1}{11} \\ -\frac{1}{4} & -\frac{1}{2} & -\frac{1}{2} & \frac{1}{4} \end{vmatrix}\right) = \frac{71}{2{,}112} > 0. \tag{A.118}$$

Then we compute the Leontief Inverse

$$(\mathbf{I}-\tilde{\mathbf{A}})^{-1} = \begin{bmatrix} \frac{132}{71} & \frac{48}{71} & 0 & 0 \\ \frac{24}{71} & \frac{112}{71} & 0 & 0 \\ \frac{129}{71} & \frac{176}{71} & \frac{11}{4} & 1 \\ \frac{438}{71} & \frac{624}{71} & \frac{11}{2} & 6 \end{bmatrix} \geq \mathbf{0}, \quad \mathbf{x} = (\mathbf{I}-\tilde{\mathbf{A}})^{-1}\begin{bmatrix} 20 \\ 0 \\ 0 \\ 0 \end{bmatrix} = \begin{bmatrix} \frac{2{,}640}{71} \\ \frac{480}{71} \\ \frac{2{,}580}{71} \\ \frac{8{,}760}{71} \end{bmatrix} > \mathbf{o}. \tag{A.119}$$

The minors (A.118) are positive, the *Hawkins–Simon condition* is fulfilled. The Leontief Inverse is semi-positive $(\mathbf{I}-\tilde{\mathbf{A}})^{-1} \geq \mathbf{0}$, and for any *non-negative* vector of final demand, $\mathbf{f} \geqq \mathbf{o}$, there is a *non-negative* vector of total output, $\mathbf{x} \geqq \mathbf{o}$. We have taken a *semi-positive* vector $\mathbf{f} \geq \mathbf{o}$ of final demand, and obtained a *positive* vector $\mathbf{x} > \mathbf{o}$ of total output. This economy is viable. ▲

A.13 Power iteration algorithm of Mises–Geiringer

In 1929, Richard von Mises (1883–1953) and Hilda Geiringer (1893–1973) proposed an eigenvalue power-iteration algorithm, called the *Mises–Geiringer Iteration*. Given a *diagonalisable* matrix $\mathbf{A}$, the algorithm will produce the greatest eigenvalue λ (in absolute value) of matrix $\mathbf{A}$ and the corresponding eigenvector $\mathbf{y}$, thus, $\mathbf{Ay} = \lambda\mathbf{y}$ (see Rutishauser [98], p. 168).

The power-iteration algorithm starts with a random vector $\mathbf{y}^{(0)} \neq \mathbf{o}$. The algorithm is presented as a recurrent algorithm.

$$\begin{aligned}
&\text{Initialisation: matrix } \mathbf{A}\text{, vector } \mathbf{y}^{(0)},\ k = 0, \varepsilon > 0,\\
&\textbf{repeat}\\
&\qquad \mathbf{y}^{(k+1)} = \frac{\mathbf{A}\mathbf{y}^{(k)}}{\|\mathbf{A}\mathbf{y}^{(k)}\|}, \quad \lambda^{(k+1)} = \|\mathbf{A}\mathbf{y}^{(k)}\|,\\
&\qquad k := k+1,\\
&\textbf{until}\quad (\text{stop condition: } \|\mathbf{y}^{(k+1)} - \mathbf{y}^{(k)}\| < \varepsilon).
\end{aligned} \tag{A.120}$$

This *power-iteration algorithm* is appropriated to compute the Frobenius number $\lambda_A \approx \|\mathbf{y}^{(k)}\|$ and the corresponding eigenvector $\mathbf{y} \approx \mathbf{y}^{(k+1)}$ when the number n of the sectors is relatively high. It converges rapidly. The experience is that the numerical evaluation of the eigenvalues, using the characteristic equation, is not very accurate when $n \gg 1$. However with the modern software packages, like MATHEMATICA and MATLAB, this statement is softened. In the examples of this book, we have not encountered the aforementioned limits.

A.14 Directed graphs or digraphs

(1) Directed graphs. A directed graph (or a digraph) is a structure amounting to a set of objects in which some pairs of objects are *related* in some way. The objects correspond to mathematical abstractions called *vertices* (or *nodes* or *points*). Each of the related pairs of vertices is called an *edge* (or *arc*). Typically, a graph is depicted by dots for vertices and curves for edges or arrows for directed edges (see also Wagner [119], p. 10).

Definition A.14.1 (Directed graphs or digraphs). A directed graph is formally defined as $G = (\mathcal{N}, \mathcal{A})$, consisting of the set $\mathcal{N}$ of nodes and the set $\mathcal{A}$ of arrows also called directed arcs which are ordered pairs of elements of set $\mathcal{N}$, $i, j \in \mathcal{N}$, designed as $(i \to j) \in \mathcal{A}$.

We will also need the notion of a *complete digraph* (see Illik [44]), an extension of the notion of a *complete graph* (see K. Wagner [119], p. 24). In a complete digraph, two different nodes are connected with exactly two opposed arrows. In a *complete digraph*, there are normally no arrows $(i \to i)$, $i \in \mathcal{N}$. But by also adding to the *complete digraph* arrows $(i \to i)$ for every node $i \in \mathcal{N}$, we obtain a new type of digraph: a *node-complete digraph.*

Definition A.14.2 (Complete digraph). A directed graph $G = (\mathcal{N}, \mathcal{A})$ is called a complete digraph, if there is an arrow $(i \to j)$, from every node $i \in \mathcal{N}$ to any other node $j \in \mathcal{N}$, $i \neq j$, for all $(i,j) \in \mathcal{A} \subset \mathcal{N} \times \mathcal{N}$.

Definition A.14.3 (Node-complete digraph). A directed graph $G = (\mathcal{N}, \mathcal{A})$ is called a node-complete digraph, if there is an arrow $(i \to j)$ from every node $i \in \mathcal{N}$ to every node $j \in \mathcal{N}$, for all $(i,j) \in \mathcal{A} = \mathcal{N} \times \mathcal{N}$.

Complete digraphs have a number of $|\mathcal{N}|(|\mathcal{N}| - 1)$ arrows. *Node-complete digraphs* have a total amount of $|\mathcal{N}|^2$ arrows, whereas *complete graphs* have $\frac{(|\mathcal{N}|-1)|\mathcal{N}|}{2}$ undirected arcs.

Digraphs are used to represent the processes of production of an economy. One also needs to refer to parts of the whole given economy and therefore to parts of the digraph representing it. We thus resort to the notion of *subdigraph*, i. e., a digraph consisting of an arbitrary number of nodes of a given digraph and all the arrows between them defined on that given digraph. Formally, subdigraphs are defined as follows (see Wagner [119], p. 22):

Definition A.14.4 (Subdigraph). Let $G = (\mathcal{N}, \mathcal{A})$ be a digraph. Then a digraph $G' = (\mathcal{N}', \mathcal{A}')$ is a subdigraph of G, if $\mathcal{N}' \subseteq \mathcal{N}$, and $\mathcal{A}' = \mathcal{A} \cap (\mathcal{N}' \times \mathcal{N}')$, symbolically $G' \subseteq G$.

We also resort to the notion of *bidigraphs.*

Definition A.14.5 (Bipartite digraph or bidigraph). A digraph $G = (\mathcal{N}, \mathcal{A})$ is a bipartite digraph (bidigraph) if its node set $\mathcal{N}$ can be partitioned into two disjoint subsets $\mathcal{N}_1$ and $\mathcal{N}_2$, $\mathcal{N}_1 \cup \mathcal{N}_2 = \mathcal{N}$, $\mathcal{N}_1 \cap \mathcal{N}_2 = \emptyset$, with the result that each arrow of $G = (\mathcal{N}, \mathcal{A})$ is strictly directed either from a node of $\mathcal{N}_1$ to a node of $\mathcal{N}_2$ or vice versa. This means that the arrows of the bipartite digraph G are in a set $\mathcal{A} \subseteq \mathcal{N}_1 \times \mathcal{N}_2$.[15] In the present text, one always sets $|\mathcal{N}_1| = |\mathcal{N}_2|$, a property generally not required for a bidigraph.

Now we will define the notions of indegree, outdegree and the degree of a digraph (see Wagner [119], p. 68).

Definition A.14.6 (Indegree, outdegree and degree of a node). For a node of digraph $G = (\mathcal{N}, \mathcal{A})$, the number of head endpoints of arcs pointing on that node is called the *indegree* of the node. The number of tail endpoints of arcs pointing out of that node is its *outdegree.*

For a node $v \in \mathcal{N}$, the *indegree* is denoted $\gamma^-(v)$ and the *outdegree* as $\gamma^+(v)$. A node with $\gamma^-(v) = 0$ is called a *source* because it is the origin of each of its incident arcs.

15 Not to be confused with the definition of a *bipartite graph* (see Skiena [105] 1990, p. 213) and also http://primes.utm.edu/graph/glossary.html

Similarly, a node with $\gamma^+(v) = 0$ is called a *sink*. $\gamma^+(v) + \gamma^-(v) = \gamma(v)$ is called the *degree* of the node.

(2) Digraphs issued from adjacency matrices. In this subsection, the link between *adjacency matrices*, Definition A.8.5, and the associated digraphs is established.

Definition A.14.7 (Adjacency matrix **W** and associated digraph $G(\mathbf{W})$). An *adjacency matrix* $\mathbf{W} = (w_{ij})$, $i, j = 1, \ldots, n$, generates a digraph $G(\mathbf{W})$[16] as follows: To every single $i \in \{1, \ldots, n\}$ corresponds one and exactly one node of the digraph $G(\mathbf{W})$, in this general case noted as i. Then, every entry $w_{ij} = 1$ generates an *arrow* (arc) $(i \to j)$ in $G(\mathbf{W})$, pointing from the node i to the node j, see Figure 4.12.

If the adjacency matrix **W** generates *digraph* $G(\mathbf{W})$, then **W** is associated with $G(\mathbf{W})$. When matrix **T** has generated the (Boolean) adjacency matrix **W**, Definition A.8.5, then we also note $G(\mathbf{T})$ as its *associated digraph*.

(3) The commodity flow represented by a commodity digraph. In the context under discussion, we want to represent the *commodity flow* of a Sraffa production economy by a digraph, called the *commodity digraph*.

Definition A.14.8 (Commodity flow and commodity digraph). Consider a commodity flow matrix $\mathbf{S} = (s_{ij})$, $i, j \in \{1, \ldots, n\}$, of a production economy. Every commodity i is now referred to as exactly one node Ci. The *adjacency matrix* $\mathbf{V} = (v_{ij})$, $i, j = 1, \ldots, n$, of matrix **S** generates a digraph $G(\mathbf{V})$, called the commodity diagraph. The nodes Ci are visualised by a *square node*, which mnemotechnically we think of a storage depot for commodities.

The arrow $(j \to i)$ of the commodity digraph $G(\mathbf{V})$ indicates that *commodity* j is used in the production of commodity i. The commodity j enters the production of i directly if an arrow $(j \to i)$ links node Cj to node Ci. The commodity j enters the production of i indirectly if a succession of equally oriented arrows, linking nodes, thus forming a directed path, leads from node Cj to node Ci. In a commodity digraph, the sectors of production do not appear. The *commodity flow* goes in direction of the arrows $(j \to i)$, opposite to the direction of the *payment flow* or *demand flow*.

Example A.14.1. The commodity digraph of Figure A.3 represents the commodity flow of four sectors: C1 and C2 directly purchase from C3 and indirectly from C4.

(4) Connectivity of digraphs. We return now to the more general concept of connectivity. Connectivity is one of the basic concepts of graph theory. The purpose is to determine the minimum number of elements (nodes or arcs) needed to be removed to disconnect a digraph in one or more separate parts, called *components*.

16 The notation $G(\mathbf{W})$ is proposed by Varga ([118], p. 49), see also Miller and Blair ([65], p. 675).

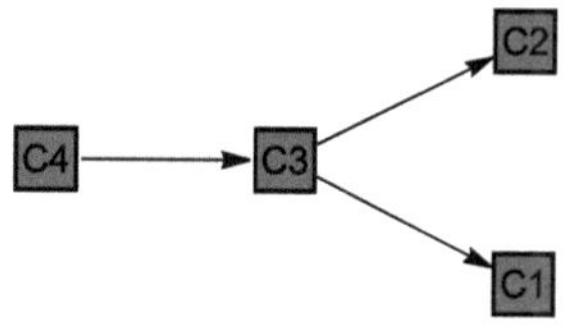

Figure A.3: Commodity digraph representing commodity flow, Example A.14.1.

The notions of graph theory are presented and defined in standard textbooks (consult, e. g., Skiena [105]). There are a lot of intuitive notions in graph theory, starting with the notions of *a graph* which is a digraph, where all the (directed) arcs are replaced by *undirected edges*. We now present the notion of the paths.

An *undirected path* is a sequence of edges that begins at a node of a graph and goes from node to node along edges of the graph. An *undirected graph* is called *connected* if there is a path between every pair of nodes. A component of an *undirected graph* is a subgraph in which any two nodes are connected to each other by paths. A graph that is not connected is said to be disconnected. Then, there exist two nodes in the graph with the result that no path in the graph has those nodes as endpoints. Note that a *complete graph* has an *edge* between each pair of distinct nodes, and a *connected graph* has a *path* between each pair of distinct nodes.

In a *digraph*, a *directed path* consists of arcs of the same direction (see Definition A.14.9). A *digraph* is called *connected* (or *weakly connected*) if replacing all of its arcs with undirected edges produces a connected (undirected) graph. A *digraph* is called *strongly connected* if there is a directed path between every pair of nodes.

We refer here to Varga ([118], Definition 1.6, p. 20) concerning the notions of *directed paths* and of *strongly connected digraphs*, which we will reproduce here.

Definition A.14.9 (Strongly connected digraphs). A directed path from node e_i to node e_j is a sequence of nodes $e_1, e_{l_1}, \ldots, e_{l_{r-1}}, e_{l_r=j}$, with arcs of the same orientation from e_i to e_j written as $P(\overrightarrow{e_i, e_j}) = \{\overrightarrow{e_i e_{l_1}}, \overrightarrow{e_{l_1} e_{l_2}}, \overrightarrow{e_{l_2} e_{l_3}}, \ldots, \overrightarrow{e_{l_{r-1}} e_{l_r=j}}\}$, connecting e_i to e_j. A directed graph $G = (\mathcal{N}, \mathcal{A})$ is strongly connected if for any ordered pair of nodes $e_i, e_j \in \mathcal{N}$ there exists a directed path $P(\overrightarrow{e_i, e_j}) \subset \mathcal{A}$ from node e_i to node e_j.

Definition A.14.10 (Connected or weakly connected digraphs). A digraph is called weakly connected (or just connected), if the undirected underlying graph, obtained by replacing all arrows with undirected arcs, is a connected graph.

We now give some illustrations of digraphs $G(\mathbf{W})$ of adjacency matrices $\mathbf{W}$.

Example A.14.2. The digraph associated to the adjacency matrix $\mathbf{W}_1$ corresponding to matrix $\mathbf{Z}_1$ of Example A.8.6 is presented, see Figure A.4.

Example A.14.3. The associated digraphs $G(\mathbf{W}_1)$ and $G(\mathbf{W}_2)$ of the adjacency matrix $\mathbf{W}_1$ related to the matrix $\mathbf{Z}_1$, respectively of $\mathbf{W}_2$ related to $\mathbf{Z}_2$, Example A.8.5, are presented.

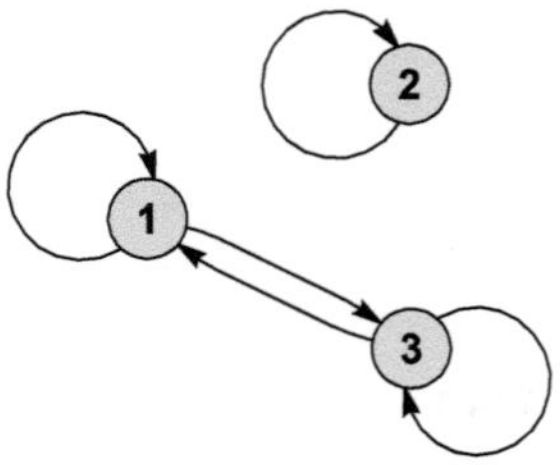

Figure A.4: Associated digraph $G(\mathbf{W}_1)$ of matrix $\mathbf{W}_1$, Example A.14.2.

In Figure A.5 the digraph $G(\mathbf{W}_1)$ is disconnected because it contains two separate components, whereas the digraph $G(\mathbf{W}_2)$ is strongly connected. This one can verify because $G(\mathbf{W}_2)$ contains a directed circuit $1 \to 3 \to 2 \to 1$. ▲

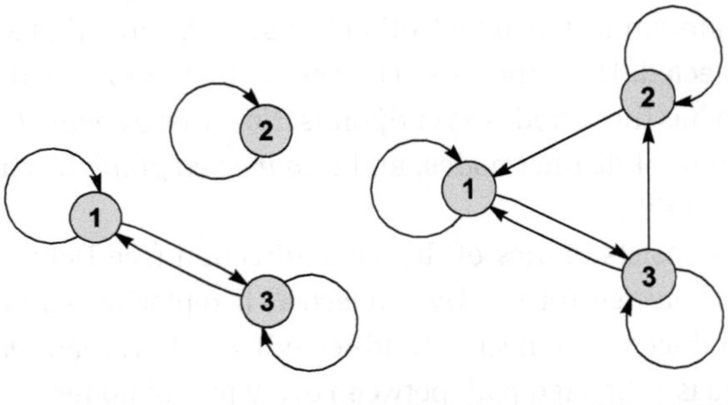

Figure A.5: Associated digraphs $G(\mathbf{W}_1)$ (left) and $G(\mathbf{W}_2)$ (right), Example A.14.3.

(5) Basic and non-basic commodities, entering directly or indirectly the process of production. In single-commodity processes, a commodity that enters directly or indirectly into the production of all commodities is a *basic* commodity, otherwise it is a *non-basic* commodity. We will study how the presence of *basics* and *non-basics* can be visualised in the associated digraph or in the adjacency matrix of such a production process.

We introduce the subject with two production economies, the first contains only *basic* commodities, the second contains *non-basic* commodities.

Example A.14.4. An economy that has only *basic* commodities:

C1 enters C2 directly and enters C3, C4 and itself indirectly,
C2 enters C1 and C3 directly and C4 and itself indirectly,
C3 enters itself and C4 directly and enters C2 and C1 indirectly,
C4 enters C2 directly and enters C1, C3 and itself indirectly.

All commodities C1, C2, C3, C4 enter directly or indirectly into all others. They are all basic (see Figure A.6, left).

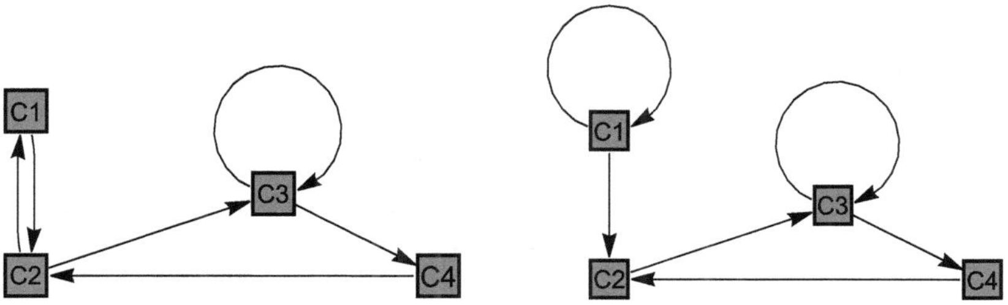

Figure A.6: Example A.7.2 (left) $G(\mathbf{V}_1)$, Example A.14.5 (right) $G(\mathbf{V}_2)$.

We observe that the commodity digraph $G(\mathbf{V}_1)$ is *strongly connected* (Definition A.14.9, above). Its associated *adjacency matrix* is:

$$\mathbf{V}_1 = (v_{ij})_1 = \begin{bmatrix} 0 & 1 & 0 & 0 \\ 1 & 0 & 1 & 0 \\ 0 & 0 & 1 & 1 \\ 0 & 1 & 0 & 0 \end{bmatrix}, \quad i,j = 1,\dots,4. \quad \blacktriangle \tag{A.121}$$

Example A.14.5. An economy that has one *basic* commodity *C*1 and three non-basic commodities C2, C3, C4:

C1 enters into itself and C2 directly and enters into C3 and C4 indirectly,
C2 enters into C3 directly and enters into C4 and itself indirectly,
C3 enters into itself and C4 directly and enters into C2 indirectly,
C4 enters into C2 directly and enters into C3 and itself indirectly.

We observe that the commodity digraph $G(\mathbf{V}_2)$ (Figure A.6, right), is *weakly connected* (Definition A.14.10).

Its *adjacency matrix* is:

$$\mathbf{V}_2 = (v_{ij})_2 = \begin{bmatrix} 1 & 1 & 0 & 0 \\ 0 & 0 & 1 & 0 \\ 0 & 0 & 1 & 1 \\ 0 & 1 & 0 & 0 \end{bmatrix}, \quad i,j = 1,\dots,4. \quad \blacktriangle \tag{A.122}$$

Note that the number of positive entries (ones) in the adjacency matrices (A.121), (A.122) is equal to the number of directed arrows, constituting the commodity digraphs.

(6) Reducibility of matrices and connectivity of digraphs. There exists a theorem relating *irreducibility* and *connectedness of digraphs: If a digraph is strongly connected, its adjacency matrix* **W** *is irreducible and vice versa*, according to the following Theorem (see also Varga ([118], p. 20)).

Theorem A.14.1. *A real $n \times n$ matrix* **A** *is irreducible if and only if its associated digraph $G(\mathbf{A})$ is strongly connected.*

We illustrate the property of *connectedness* of digraphs, stated by Theorem A.14.1, resorting to Lemma A.8.2 concerning *irreducibility* of matrices (see Horn and Johnson [43], p. 533), taking Example A.14.4 and Example A.14.5.

Setting $n = 4$, the connection is described by the (4×4) adjacency matrices $\mathbf{V}_i$ (A.121) and (A.122). We have to form the matrix $\mathbf{I} + \mathbf{V}_i$ and check whether $(\mathbf{I} + \mathbf{V}_i)^3$ is strictly positive. For Example A.14.4, one gets indeed,

$$\mathbf{I} + \mathbf{V}_1 = \begin{bmatrix} 1 & 1 & 0 & 0 \\ 1 & 1 & 1 & 0 \\ 0 & 0 & 2 & 1 \\ 0 & 1 & 0 & 1 \end{bmatrix} \Rightarrow (\mathbf{I} + \mathbf{V}_1)^3 = \begin{bmatrix} 4 & 4 & 4 & 1 \\ 4 & 5 & 8 & 4 \\ 1 & 4 & 9 & 7 \\ 3 & 4 & 4 & 2 \end{bmatrix} > 0, \tag{A.123}$$

being strictly positive and $\mathbf{V}_1$ is *irreducible*. The digraph $G(\mathbf{V}_1)$, Figure A.6 (left), is indeed strongly connected, there are directed paths from each node to another node. For Example A.14.5, one gets the matrices,

$$\mathbf{I} + \mathbf{V}_2 = \begin{bmatrix} 2 & 1 & 0 & 0 \\ 0 & 1 & 1 & 0 \\ 0 & 0 & 2 & 1 \\ 0 & 1 & 0 & 1 \end{bmatrix} \Rightarrow (\mathbf{I} + \mathbf{V}_2)^3 = \begin{bmatrix} 8 & 7 & 5 & 1 \\ 0 & 2 & 7 & 4 \\ 0 & 4 & 9 & 7 \\ 0 & 3 & 4 & 2 \end{bmatrix} \geq 0, \tag{A.124}$$

which is not strictly positive and $\mathbf{V}_2$ is *reducible*. The digraph $G(\mathbf{V}_2)$, Figure A.6 (right), is indeed only weakly connected because there are no directed paths from $C2$, $C3$, $C4$ to $C1$.

Matrix $\mathbf{V}_1$ represents an economy with only *basic* commodities and $\mathbf{V}_2$ represents an economy with one *basic* commodity and three *non-basic* commodities.

(7) Sraffa Network. In Section 4.6, the *Sraffa Network* was used for the representation of *single-product industries*. In Section 6.4 the *Sraffa Network* was used for the representation of *joint production processes*. We will give now a more formal definition of this *bipartite digraph* which we have called *a Sraffa Network* in the context of Sraffa production economies. As an illustration, one may look at the *Sraffa Network* of Example 6.5.1 (see Figure A.7).

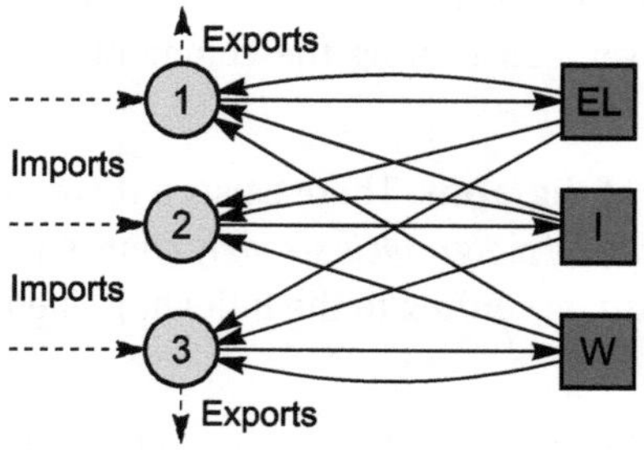

Figure A.7: An example of a *Sraffa Network*.

Definition A.14.11 (Sraffa Network). A Sraffa Network is a bipartite digraph $G = (\mathcal{N}, \mathcal{A})$: its node set $\mathcal{N}$ is partitioned into two disjoint subsets, $\mathcal{F}$ for the n industries and $\mathcal{C}$ for the n commodities, $\mathcal{F} \cup \mathcal{C} = \mathcal{N}$, $\mathcal{F} \cap \mathcal{C} = \emptyset$. The digraph G may be a sub-digraph, representing a part of the economy, whereas the digraph G' represents the whole economy, $G \subseteq G'$. The construction rules are as follows:

- Each **commodity** i is designed by one **square green node.** (Mnemotechnically, think of a storage depot for commodities.) The nodes may be numbered from $n+1$ to $2n$ or designed by letters to designate the commodities. All of them are elements of the node set $\mathcal{C}$.
- Each **productive entity** or industry S_j is designed by one **round yellow node.** (Mnemotechnically, think of a wheel of a production machine.) These nodes may be numbered from one to n, all of them are elements of the node set $\mathcal{F}$.
- An **arrow** $(i \to j)$ of digraph G, pointing from a commodity $i \in \mathcal{C}$ to an industry S_j, $j \in \mathcal{F}$, shows that industry S_j requires (purchases) commodity i. An **arrow** $(j \to i)$ of digraph G, pointing from an industry S_j to a commodity i, shows that industry S_j produces (sells) commodity i.
- **Dashed arrows** $(i \to j)$, pointing from a node $i \in G'$ to a node $j \in G$ of the sub-digraph G, show that only a part of the economy is considered and the present model is imbedded in a greater economy.

Lemma A.14.1. *A Sraffa Network represents a single-product economy if and only if for all the commodity nodes $v \in \mathcal{C}$, the indegrees are $\gamma^-(v) = 1$. A Sraffa Network represents a joint production economy if and only if, for at least one commodity node $v \in \mathcal{C}$, the indegree is $\gamma^-(v) > 1$.*

(8) A Sraffa Network and its associated commodity digraph. The idea is to compress the *Sraffa Network*, a bipartite digraph, into an associated digraph, indicating for one cycle of the production process exclusively the interconnections of any commodity $i \in \{1, \ldots, n\}$ with other commodities $j \in \{1, \ldots, n\}$. Any interconnection from i to j is represented by one and only one directed arc $i \to j$. The resulting digraph is then called *commodity digraph* of that *Sraffa Network*. Considering a production scheme $(\mathbf{S}', \mathbf{L}) \Rightarrow (\mathbf{F}')$, we set up the adjacent matrix (6.116):

$$\mathbf{\Sigma} = \begin{bmatrix} \mathbf{0} & \mathbf{F}' \\ \mathbf{S} & \mathbf{0} \end{bmatrix} \Rightarrow \mathbf{W} = \begin{bmatrix} \mathbf{0} & \mathbf{Q}' \\ \mathbf{V} & \mathbf{0} \end{bmatrix}. \tag{A.125}$$

In a *single-product industry*, a one-to-one relationship exists between the industry and the single commodity produced by this industry, and the adjacency matrix $\mathbf{V}$ directly leads to the requested associated *commodity digraph*, Definition A.14.8. In a *joint production processes*, any industry may produce more than one commodity. For this case, consider commodities $i, j \in \{1, \ldots, n\}$. An entry $v_{il} = 1$ of matrix $\mathbf{V}$ indicates that commodity i is used for production in sector l, $l \in \{1, \ldots, n\}$. Then, an entry $q_{lj} = 1$ of matrix $\mathbf{Q}'$ indicates that the sector S_l produces commodity j. For this reason, the product $v_{il}q_{lj}$ indicates that in the present *joint production process* the commodity i is used in the

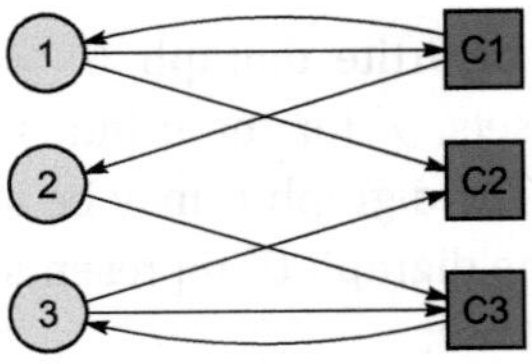

$$\mathbf{W} = \begin{array}{c} \\ 1 \\ 2 \\ 3 \\ C1 \\ C2 \\ C3 \end{array} \begin{array}{c} \begin{array}{ccc|ccc} 1 & 2 & 3 & C1 & C2 & C3 \end{array} \\ \left[\begin{array}{ccc|ccc} 0 & 0 & 0 & 1 & 1 & 0 \\ 0 & 0 & 0 & 0 & 0 & 1 \\ 0 & 0 & 0 & 0 & 1 & 1 \\ \hline 1 & 1 & 0 & 0 & 0 & 0 \\ 0 & 0 & 0 & 0 & 0 & 0 \\ 0 & 0 & 1 & 0 & 0 & 0 \end{array}\right] \end{array} \tag{A.126}$$

Figure A.8: Sraffa Network and adjacency matrix **W**.

production of commodity j. Consequently, when the sum $\sum_{l=1}^{n} v_{il}q_{lj} > 0$ is positive, we know that at least one of the intermediate sectors S_l, $l \in \{1, \dots, n\}$ is joined to produce commodity j, using commodity i during this cycle of the production process.

Describing this, we have now got the definition of the arrow $i \to j$, an element of the *commodity digraph* of this *joint production* process described by a *Sraffa Network*. Continuing, we bring matrix $\mathbf{VQ}'$ in Boolean form $\mathbf{W}_C$ and set up the associated digraph, which is the requested *commodity digraph* $G(\mathbf{W}_C)$.

Example A.14.6. Given the Sraffa Network Figure A.8 and its adjacency matrix (A.126), determine the corresponding associated commodity digraph and provides its economic interpretation.

Solution to Example A.14.6:
We identify matrices **V** and $\mathbf{Q}'$ and calculate the products $\mathbf{VQ}'$:

$$\mathbf{V} = \begin{bmatrix} 1 & 1 & 0 \\ 0 & 0 & 0 \\ 0 & 0 & 1 \end{bmatrix}, \quad \mathbf{Q}' = \begin{bmatrix} 1 & 1 & 0 \\ 0 & 0 & 1 \\ 0 & 1 & 1 \end{bmatrix}, \quad \mathbf{VQ}' = \begin{bmatrix} 1 & 1 & 1 \\ 0 & 0 & 0 \\ 0 & 1 & 1 \end{bmatrix}. \tag{A.127}$$

Then, we establish the digraph $G(\mathbf{W}_C)$ Figure A.9 corresponding to the adjacency matrix $\mathbf{W}_C$:

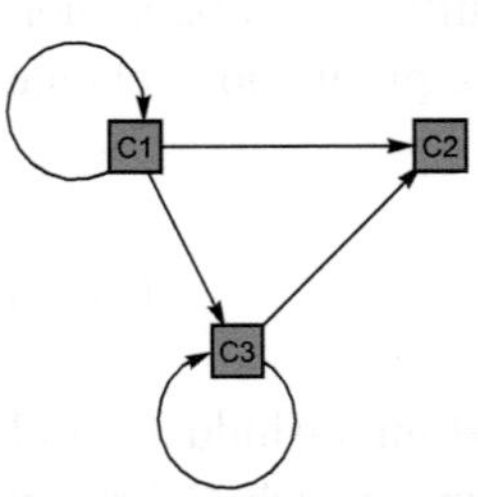

$$\mathbf{W}_C = \begin{array}{c} \\ C1 \\ C2 \\ C3 \end{array} \begin{array}{c} \begin{array}{ccc} C1 & C2 & C3 \end{array} \\ \begin{bmatrix} 1 & 1 & 1 \\ 0 & 0 & 0 \\ 0 & 1 & 1 \end{bmatrix} \end{array}. \tag{A.128}$$

Figure A.9: Adjacency matrix $\mathbf{W}_C$ and commodity digraph $\mathbf{W}_C$.

The Boolean matrix $\mathbf{W}_C$ reveals that the commodity C1 is used for the production of itself and both other commodities C2 and C3. The commodity C2 is not necessary for the production of any of the three commodities, while the commodity C3 is used for the production of itself and commodity C2. ▲

A.15 Algebraic criteria determining the presence of basic products

(1) Motivation.

Sraffa supplies a precise economic meaning to the notions of *basic products* and *non-basic products* (PCMC, Par. 6, 7) in the case of *single-product industries*:

> *"The criterion is whether a commodity enters (no matter whether directly or indirectly) into the production of all commodities. That do so we shall call basic, and those that do not, non-basic commodity. We shall assume throughout that any system contains at least one basic commodity."*

Fortunately, since the publication of PCMC, precise algebraic criteria have been developed to determine if a product is *basic* or *non-basic* in *single-product industries*, entering *directly* or *indirectly* into the production of the commodities. These criteria use the *commodity flow* matrix **S**, its *adjacency matrix* **V** and its *commodity digraph* $G(\mathbf{V})$, Definition A.14.8.

(2) On basic and non-basic products.

We present here some *Lemmas* that identify efficient algebraic criteria concerning the nature of the production processes (see Kurz and Salvadori [52], pp. 94–96, 104).

We use the unit vectors $\mathbf{e}_i = [0, 0, \dots, 1, \dots, 0]'$, where the i-th component is equal to 1. The vector set $\mathcal{E} = \{\mathbf{e}_1, \dots, \mathbf{e}_n\}$ is a basis of the n-dim Euclidean vector space $\mathbf{R}^n$, $i \in \{1, 2, \dots, n\}$.

Lemma A.15.1 (Basic and non-basic commodities). *The next six statements (1)–(6) can also be expressed using the adjacency matrix* **V** *instead of the commodity flow matrix* **S** *because both non-negative matrices have the 0 entries at the same place. Then, the calculated results are then generally shorter and easier to obtain. All the statements are logical equivalences and therefore convenient to apply.*

(1) *Commodity i enters directly into the production of commodity j if and only if the corresponding matrix element is positive*

$$s_{ij} > 0. \tag{A.129}$$

(2) *We can assert that commodity j enters indirectly into the production of commodity i if and only if the following relations apply*

$$\mathbf{e}_i'\mathbf{S}\mathbf{e}_j = 0; \quad \mathbf{e}_i'(\mathbf{S}^2 + \dots + \mathbf{S}^n)\mathbf{e}_j > 0. \tag{A.130}$$

(3) *Thus, commodity j enters directly or indirectly into the production of commodity i if and only if the following matrix is positive,*

$$\mathbf{e}_i'(\mathbf{S} + \mathbf{S}^2 + \dots + \mathbf{S}^n)\mathbf{e}_j > 0. \tag{A.131}$$

(4) *A basic commodity (or, for short, a basic) is a commodity that enters directly or indirectly into the production of all commodities, that is, commodity j is basic if and only if the following column vector is positive,*

$$(\mathbf{S}+\mathbf{S}^2+\cdots+\mathbf{S}^n)\mathbf{e}_j > \mathbf{0}. \tag{A.132}$$

(5) *All commodities are basic if and only if the following matrix is positive,*

$$(\mathbf{S}+\mathbf{S}^2+\cdots+\mathbf{S}^n) > \mathbf{0}. \tag{A.133}$$

(6) *When a system exhibits both basics and some non-basics, the commodity flow matrix* **S** *is reducible and can be transformed by a permutation matrix into the "canonical form" (Definition* A.8.3*).* ▲

(3) Separation of basic from non-basic commodities.

Now we present a useful Lemma of equivalent statements about *reducible* matrices (see Kurz and Salvadori [52], pp. 104–105; Horn and Johnson [43], p. 402; Takayama [116], p. 370) and Definition A.8.3.

We are in presence of a complex situation. We treat methods to distinguish *basics* from *non-basic* commodities. This means that we try to identify the *economic* notion of *basic* commodity, respectively *non-basic* commodity, via mathematical notions. This means that we set up a unique mathematical correspondence between *basic* commodities, respectively *non-basic* commodities, that are *economic* notions and abstract mathematical representations of these notions.[17]

Lemma A.15.2 (Equivalent statements on non-basic commodities). *Consider the matrix order $n \geq k > 1$ and the $n \times n$ square non-negative matrix* **S**. *Then the following statements are equivalent.*

(i) *The matrix* **S** *is reducible.*

(ii) *There exists a permutation matrix* **P** *such that*

$$\widetilde{\mathbf{S}}' = \mathbf{P}^{-1}\mathbf{S}'\mathbf{P} = \begin{bmatrix} \widetilde{\mathbf{S}}'_{11} & \mathbf{0} \\ \widetilde{\mathbf{S}}'_{12} & \widetilde{\mathbf{S}}'_{22} \end{bmatrix}, \tag{A.134}$$

where the $(k \times k)$ submatrix $\widetilde{\mathbf{S}}_{11}$ and the $(n-k \times n-k)$ submatrix $\widetilde{\mathbf{S}}_{22}$ are square matrices. Matrix $\widetilde{\mathbf{S}}_{11}$ (A.134) *is irreducible and represents the basics.*

(iii) *The remaining $(n-k)$ commodities are non-basic and present in matrix $\widetilde{\mathbf{S}}_{22}$.*

Proof. For statements (*i*): Varga ([118], Definition 1.5, p. 18) states for a reducible matrix **S** the existence of a permutation matrix, performing the transformation of **S** into the "canonical form" (A.134),

For statement (*ii*): Steedman ([114], p. 324) says that matrix "$\widetilde{\mathbf{S}}_{11}$ *refers to basics*". Pasinetti ([80], p. 104) says that the $(k\times k)$ matrix "$\widetilde{\mathbf{S}}_{11}$ *is an irreducible square submatrix*

17 This phenomenon occurs frequently in science. We now observe it in *physics*. In *thermodynamics* we have the notion of *ideal gas* which is a mathematical model for the *real gas*.

of order k", and then he says that *"the first k commodities on the rows of* $\tilde{\mathbf{S}}$ *are basic commodities"*,

For statement *(iii)*: Pasinetti ([80], p. 104) says *"the remaining* $(n-k)$ *are non-basic commodities."* □

It can be easily illustrated that the *associated digraph* $G(\tilde{\mathbf{S}})$ is disconnected if $\tilde{\mathbf{S}}_{12} = \mathbf{0}$. In this case, $\tilde{\mathbf{S}}$ is block diagonal, and there are no connections from $G(\tilde{\mathbf{S}}_{11})$ to $G(\tilde{\mathbf{S}}_{22})$. If $\tilde{\mathbf{S}}_{12} > 0$ and $G(\tilde{\mathbf{S}}_{22})$ is connected, then $G(\tilde{\mathbf{S}})$ is also connected because $G(\tilde{\mathbf{S}}_{11})$ and $G(\tilde{\mathbf{S}}_{22})$ are connected by arrows, represented by non-zero elements of $\tilde{\mathbf{S}}_{12}$. Matrix $\tilde{\mathbf{S}}_{12}$ reflects the fact that *basics* enter directly or indirectly into the production of *non-basics*.

(4) Equivalent statements on basic commodities.

We present now a useful Lemma of equivalent statements on *irreducible* matrices in relation to the notion of *basic* commodities (see Kurz and Salvadori [52], p. 122; Varga [118], Theorem 1.6, p. 20).

This is the simple case because there are only *basic* products.

Lemma A.15.3 (Equivalent statements on basic products). *Consider a matrix of order* $n > 1$, *e. g., the* $n \times n$ *square non-negative matrix* $\mathbf{S}$. *Then the following statements are equivalent.*

(i) *The matrix* $\mathbf{S}$ *is irreducible.*
(ii) $\mathbf{I} + \mathbf{S} + \cdots + \mathbf{S}^{n-1} > \mathbf{0}$.
(iii) $\mathbf{S} + \mathbf{S}^2 + \cdots + \mathbf{S}^n > \mathbf{0}$.
(iv) *The associated digraph* $G(\mathbf{S})$ *is strongly connected.*
(v) *All the products are basic.*

Proof. For statements

(i), *(ii)*, *(iii)* see Kurz and Salvadori [52], p. 122;

(iv) see Varga [118], Theorem 1.6, p. 20;

(v) ⇒ If matrix $\mathbf{S}$ is irreducible, *"then all the commodities in the economic system are basic commodities"*;[18]

⇐ Then, when all the products are basic, the matrix $\mathbf{S}$ is irreducible. Otherwise, if some of the commodities are *non-basics*, then they are by definition in submatrix $\mathbf{S}_{22}$ (A.134) and matrix $\mathbf{S}$ is not irreducible. □

If matrix $\mathbf{S}$ is *irreducible*, then no zero matrix $\mathbf{0}$ can be extracted in the left-lower corner of $\mathbf{S}$ and no *non-basic* commodities can be identified. For this reason, all the commodities are *basic* in this case.

---

18 Pasinetti [80], p. 104, says explicitly: *"If the matrix of technical interindustry coefficients is an irreducible matrix, then the commodities in the economic systems are basic commodities."*

B Bertram Schefold's mathematical explanations to PCMC*

In the German translation of PCMC in [109], pp. 216–225, Appendix 6, 'Einige Grundthesen, mathematisch formuliert', Bertram Schefold formulates some mathematical concepts that are at the basis of Sraffa's book [108].[1] His presentation is relatively short at ten pages. He starts with the remark that "Sraffa considers closed systems, where n commodities are produced by n production processes with the help of these commodities and labour."

To our knowledge, Schefold revealed for the first time in this Appendix the connection of Sraffa's PCMC with the famous **Perron–Frobenius theorem A.9.3** which governs the theory of production of Leontief and Sraffa. This is the reason why we present here Schefold's work.

B.1 Schefold's presentation

(1) Schefold defines the *input coefficients matrix* in physical terms, noting it with a non-bold letter $A = (a_i^j)$. Instead of A, we continue to use Miller and Blair's notations ([65], p. 47), $\mathbf{C} = \mathbf{S}\hat{\mathbf{q}}^{-1}$, (2.16) for the *input coefficients matrix*, $\mathbf{C} = (c_{ij})$, $i, j = 1, \ldots, n$.[2]

Then Schefold assumes three important properties of matrix $\mathbf{C}$:

(1) $\mathbf{C}$ is non-negative. (We know that it is no problem to relax this condition to *semi-positivity*, which is what we will continue to do.)
(2) When there is no surplus production, the sums of the columns of $\mathbf{C}$ are equal to *one*.
(3) $\mathbf{C}$ is not completely reducible (Definition A.8.4), meaning that there is at least one *basic commodity* in the production process.

Schefold discusses *reducible* and *irreducible* matrices (see Definition A.8.3) and clarifies that Sraffa uses the notion *basic commodity* instead of *irreducibility* of matrices. Schefold states that *"it is intuitively plausible and it can be proven that a non-negative matrix is irreducible if and only if the matrix corresponds to a system of production where every commodity is a basic commodity".*

It has to be underlined again that the Perron–Frobenius theorem A.9.3 for non-negative and irreducible matrices $\mathbf{C}$, is fundamental to the Sraffa theory. Schefold discusses the property required to develop the matrices $(\mathbf{I}-\mathbf{C}')^{-1}$ and then $(\mathbf{I}-(1+r)\mathbf{C}')^{-1}$,

1 Piero Sraffa, *Warenproduktion mittels Waren*, Edition Suhrkamp 780, Erste Auflage, 1976.

2 Then, Schefold presents the equations for the right and left eigenvectors, but he does not use the *transposition notation* for the matrices and vectors. We draw attention to the fact that in this résumé we use the accent ($'$) to design *transposition* of matrices and vectors. We therefore note the left and right eigenvalue equations as follows ($\mathbf{y}$ is a left eigenvector, and $\mathbf{x}$ is a right eigenvector of matrix $\mathbf{C}$):

$$\mathbf{C}'\mathbf{y} = \lambda\mathbf{y}; \quad \mathbf{C}\mathbf{x} = \lambda\mathbf{x}. \tag{B.1}$$

https://doi.org/10.1515/9783110635096-013

$0 < r < R$, where $\lambda_C = 1/(1+R)$ is the Frobenius number of $\mathbf{C}$, into series (see Theorem A.10.2).

(2) Then Schefold treats the first Sraffa model without a surplus (PCMC, Par. 1–3) and the second Sraffa model (PCMC, Par. 6) with a surplus as an eigenvector problem. In these cases, matrix $\mathbf{C}$ is *positive* and therefore *irreducible*, see Definition A.8.3.

Schefold in PCMC [109], p. 218–220, enlarges the view and treats the case where matrix $\mathbf{C}$ is *reducible* (see again Definition A.8.3). He assumes that generally matrix $\mathbf{C}$ is not *completely reducible* (see Definition A.8.4). This means, for the transformation indicated below (B.2), that $\tilde{\mathbf{C}}_{12} \neq \mathbf{O}$.

In this case where $\mathbf{C}$ is *reducible* (Definition A.8.3), using the notations of Horn and Johnson [43], p. 402, and Takayama, [116], p. 370, we obtain,

$$\tilde{\mathbf{C}}' = \mathbf{P}^{-1}\mathbf{C}'\mathbf{P} = \begin{bmatrix} \tilde{\mathbf{C}}'_{11} & \mathbf{0} \\ \tilde{\mathbf{C}}'_{12} & \tilde{\mathbf{C}}_{22} \end{bmatrix}, \tag{B.2}$$

where $\mathbf{P}$ is a permutation matrix. The following equations hold after permutations of rows and columns of matrix $\mathbf{C}'$,

$$\begin{aligned} (1+R)\tilde{\mathbf{C}}'_{11}\mathbf{p}_1 &= \mathbf{p}_1 \\ (1+R)(\tilde{\mathbf{C}}'_{12}\mathbf{p}_1 + \tilde{\mathbf{C}}'_{22}\mathbf{p}_2) &= \mathbf{p}_2, \end{aligned} \tag{B.3}$$

where $\lambda_{\tilde{C}} = 1/(1+R) > 0$ is the Frobenius number of matrix $\mathbf{C}$ and is also the Frobenius number of one of the *irreducible* matrices $\tilde{\mathbf{C}}'_{kk}$, $k = 1, 2$. Analyse with Lemma A.10.1 and A.10.3 the possibility of complete reduction. One finds with the second equation of (B.3),

$$\mathbf{p}_2 = \left(\mathbf{I} - (1+R)\tilde{\mathbf{C}}'_{22}\right)^{-1}(1+R)\tilde{\mathbf{C}}'_{12}\mathbf{p}_1. \tag{B.4}$$

Then, Schefold argues. If and only if the Frobenius number $\lambda_{\tilde{C}_{22}} > 0$ of matrix $\tilde{\mathbf{C}}_{22}$ is smaller than $\lambda_{\tilde{C}_{11}}$, i. e., $\lambda_{\tilde{C}_{11}} > \lambda_{\tilde{C}_{22}} > 0$, then the inverse matrix in (B.4) and the second price vector exist and are positive, i. e., $(\mathbf{I} - (1+R)\tilde{\mathbf{C}}'_{22})^{-1} > \mathbf{O}$, $\mathbf{p}_2 > \mathbf{o}$. See also Theorem A.10.2 for the existence of non-negative inverses of this type.

Schefold then goes on to describe the conditions for the production process to generate a surplus (see also Bharadwaj [4], and Sraffa ([109], pp. 219–220)).

(3) Then Schefold gives the solution for the Sraffa price model, see PCMC [109], pp. 220–221. Labour is explicitly introduced, and the question of distribution of the *net product* as the profit P of entrepreneurs and wages W of workers is treated.

Without restriction of generality, he assumes that matrix $\mathbf{C}$ is *irreducible*. Schefold directly gives the *normalised Sraffa price model* (5.83) or (4.59) we have developed in our text, using the labour vector per units of commodities $\boldsymbol{\pi} = \hat{\mathbf{q}}^{-1}\mathbf{L} = \hat{\check{\mathbf{q}}}^{-1}\check{\mathbf{L}}$, (4.58), (4.86), (5.81), (5.82) with respect to total output ($\hat{\check{\mathbf{q}}}$ or $\hat{\mathbf{q}}$). As Sraffa also normalises *national income* (5.83), $Y_0 = 1$, this gives,

$$(1+r)\mathbf{C}'\mathbf{p} + w\boldsymbol{\pi} = \mathbf{p}, \tag{B.5}$$

with the vector of *labour* $\boldsymbol{\pi}$ and the wage rate w. Equation (B.5) has two degrees of freedom: one is the determination of the *distribution* between r and w, and the other a *dilatation factor* determining the units of calculation.[3] Equation (B.5) can be solved for the price vector,

$$\mathbf{p} = (\mathbf{I} - (1+r)\mathbf{C}')^{-1} w\boldsymbol{\pi}, \tag{B.6}$$

or,

$$\frac{\mathbf{p}}{w} := \mathbf{p}_c = (\mathbf{I} - (1+r)\mathbf{C}')^{-1} \boldsymbol{\pi}, \tag{B.7}$$

where $\mathbf{p}_c = \mathbf{p}/w$ is called 'commanded labour' because it is for every commodity i the quantity of the *numéraire* per unit of commodity i produced by one unit of labour. (This notion stems from Adam Smith and is also the "wage unit" of Keynes).

(4) For the non-negative matrix $\mathbf{C}$, the following development holds (see Theorem A.10.2), the commanded labour $\mathbf{p}_c$ is understood as a function of the actual profit rate r, $0 \le r \le R$ and $\lambda = 1/(1+R)$ is the Frobenius eigenvalue of the irreducible matrix $\mathbf{C}$,

$$\mathbf{p}_c(r) = (\mathbf{I} - (1+r)\mathbf{C}')^{-1} \boldsymbol{\pi} = \left(\sum_{j=0}^{\infty} (1+r)^j \mathbf{C}'^j \right) \boldsymbol{\pi}. \tag{B.8}$$

It is immediately clear from (B.8) that the prices in the *price vector* $\mathbf{p}$ monotonously increase with increasing profit rate r. For $r = 0$, the prices for 'commanded labour' are the *labour values* $\mathbf{u}$ (7.5),

$$\mathbf{p}_c(r=0) := \mathbf{u} = (\mathbf{I} - \mathbf{C}')^{-1} \boldsymbol{\pi} = \left(\sum_{j=0}^{\infty} \mathbf{C}'^j \right) \boldsymbol{\pi}. \tag{B.9}$$

Schefold discusses the following properties: One can easily see that the equation below holds:

$$\mathbf{p}_c(0) = \mathbf{C}'\mathbf{p}_c(0) + \boldsymbol{\pi}. \tag{B.10}$$

Then, there is the limit

$$\lim_{r\to R} \mathbf{p}_c(r) = \lim_{r\to R} (\mathbf{I} - (1+r)\mathbf{C}')^{-1} \boldsymbol{\pi} = \lim_{r\to R} \left(\sum_{j=0}^{\infty} (1+r)^j \mathbf{C}'^j \right) \boldsymbol{\pi} \to \infty. \tag{B.11}$$

Schefold points out that Sraffa prefers to represent prices in terms of *standard commodity* instead of 'commanded labour'. Consider a non-negative vector $\mathbf{d}$ representing

3 [109], p. 220, *dilatation factor* determining the units of calculation = Recheneinheit.

a basket of commodities of total value 1, corresponding to a price vector expressed in a given *numéraire*, and then one sets the equation,

$$Y_0 = 1 = \mathbf{d}' \cdot \mathbf{p} = \mathbf{d}' \cdot \mathbf{p}_C \cdot w. \tag{B.12}$$

The wage rate is then presented as,

$$w = w(r) = \frac{1}{\mathbf{d}' \cdot \mathbf{p}} = \frac{1}{\mathbf{d}' \cdot (\mathbf{I} - \mathbf{C}')^{-1}\boldsymbol{\pi}}, \tag{B.13}$$

w falling monotonously with increasing profit rate r and disappearing with $r = R$, independently of the chosen *numéraire*,

$$\lim_{r \to R} w(r) = \lim_{r \to R} \frac{1}{\mathbf{d}' \cdot (\mathbf{I} - \mathbf{C}')^{-1}\boldsymbol{\pi}} = 0. \tag{B.14}$$

(5) From here on, Schefold treats Sraffa's Chapter III (PCMC [109], p. 222).

In PCMC, Par. 13, Sraffa says that the wage rate w varies, $0 \le w \le 1$, when the price equations are normalised, for this reason the (normalized) net product (= national income) is $Y_0 = 1$. Schefold starts considering the expression $\mathbf{q}'(\mathbf{I} - \mathbf{C}')\mathbf{p}$, which he calls "net product". We will show this. We have to set the following sequence of well known equations: $\mathbf{C} = \mathbf{S}\hat{\mathbf{q}}^{-1}$, $\mathbf{S} = \mathbf{C}\hat{\mathbf{q}}$, then $\mathbf{q} = \mathbf{Se} + \mathbf{d} = \mathbf{C}\hat{\mathbf{q}}\mathbf{e} + \mathbf{d} = \mathbf{Cq} + \mathbf{d}$, one obtains $\mathbf{d} = \mathbf{q} - \mathbf{Cq} = (\mathbf{I} - \mathbf{C})\mathbf{q}$ and $\mathbf{d}' = \mathbf{q}'(\mathbf{I} - \mathbf{C}')$, we set the intermediate result,

$$\mathbf{d}' = \mathbf{q}'(\mathbf{I} - \mathbf{C}'), \tag{B.15}$$

considering Schefold's next equation in PCMC [109], p. 22, for the normalised net product (= national income),

$$Y_0 = \mathbf{d}'\mathbf{p} = \mathbf{q}'(\mathbf{I} - \mathbf{C}')\mathbf{p} = 1 = w\mathbf{e}'\boldsymbol{\pi} + r\mathbf{e}'\mathbf{C}'\mathbf{p}. \tag{B.16}$$

This last equality corresponds exactly to the *national income* part contained in equation (4.178),

$$\mathbf{C}'\mathbf{p}(1 + r) + \boldsymbol{\pi}w = \mathbf{p} \Rightarrow r\mathbf{e}'\mathbf{C}'\mathbf{p} + w\mathbf{e}'\boldsymbol{\pi} = Y_0 = 1. \tag{B.17}$$

Schefold then considers here the variation of the prices, resulting from the variation of the profit rate r. He differentiates (B.5) with respect to r and gets,

$$\frac{d\mathbf{p}}{dr} = \frac{dw}{dr}\boldsymbol{\pi} + \mathbf{C}'\mathbf{p} + (1 + r)\mathbf{C}'\frac{d\mathbf{p}}{dr}, \tag{B.18}$$

or with the constant $R = (1/\lambda_C) - 1$, obtained from the Frobenius number λ_C of matrix $\mathbf{C}'$, having normed $L = 1$, $Y = 1$, one gets therefore $\tilde{w} = w$:

$$\frac{d\mathbf{p}}{dr} = (\mathbf{I} - (1 + r)\mathbf{C}')^{-1}\left(\frac{dw}{dr}\boldsymbol{\pi} + \mathbf{C}'\mathbf{p}\right), \tag{B.19}$$

i. e., a system of *nonlinear differential equations*, written in vector form, for the prices p_i, $i = 1, \dots, n$, components of the price vector $\mathbf{p}$, and the *wage rate* w.

The differential equation (B.19) can be developed in a series, appliying Theorem A.10.2:

$$\frac{d\mathbf{p}}{dr} = \left(\frac{dw}{dr}\boldsymbol{\pi} + \mathbf{C}'\mathbf{p}\right) + (1+r)\mathbf{C}'\left(\frac{dw}{dr}\boldsymbol{\pi} + \mathbf{C}'\mathbf{p}\right) + (1+r)^2(\mathbf{C}')^2\left(\frac{dw}{dr}\boldsymbol{\pi} + \mathbf{C}'\mathbf{p}\right) + \cdots \quad \text{(B.20)}$$

(6) The series development is not intended to solve equation (B.20) but is used to examine the conditions of price invariance to identify some new equalities. There is price invariance when the condition

$$\frac{d\mathbf{p}}{dr} = \mathbf{o}_{n1} \quad \text{(B.21)}$$

holds. Schefold discusses two cases. First, (B.21) holds, when with (B.7) the factors

$$\mathbf{o}_{n1} = \frac{dw}{dr}\boldsymbol{\pi} + \mathbf{C}'\mathbf{p} = \frac{dw}{dr}\boldsymbol{\pi} + w\mathbf{C}'(\mathbf{I} - (1+r)\mathbf{C}')^{-1}\boldsymbol{\pi} \quad \text{(B.22)}$$

vanish in equation (B.20).

Schefold then argues that $\boldsymbol{\pi}$ is an eigenvector of matrix $\mathbf{C}'$. Further, when the vector $\boldsymbol{\pi}$ is positive, the equation

$$(1+R)\mathbf{C}'\boldsymbol{\pi} = \boldsymbol{\pi} \quad \text{(B.23)}$$

holds (because of the Perron–Frobenius theorem A.9.3).[4] Evidently, with a (given) profit rate r, the *prices* are stationary if and only if they remain constant for all profit rates and equal to the *labour values* (prices equal labour value!). This only happens when the *labour vector* is an eigenvector of matrix $\mathbf{C}'$. This is a special case of condition (B.23), therefore Schefold concludes that this case "is quite unlikely" (PCMC [109], p. 222).

Then Schefold goes on to discuss the case where the equality does not hold, so:

$$\frac{dw}{dr}\boldsymbol{\pi} + \mathbf{C}'\mathbf{p} \neq \mathbf{o}_{n1}. \quad \text{(B.26)}$$

4 Schefold continues (PCMC [109], footnote 121, p. 222): *"There is a theorem in matrix algebra which says that in a matrix ring generated by a matrix, there is an exchangeability of elements". It is indeed a property of commutativity used in (B.24), giving*

$$\frac{dw}{dr}\boldsymbol{\pi} + w\mathbf{C}'\left(\mathbf{I} - (1+r)\mathbf{C}'\right)^{-1}\boldsymbol{\pi} = \frac{dw}{dr}\boldsymbol{\pi} + w\left(\mathbf{I} - (1+r)\mathbf{C}'\right)^{-1}\mathbf{C}'\boldsymbol{\pi} = \mathbf{o}' \quad \text{(B.24)}$$

also

$$\frac{\frac{dw}{dr}}{w}\boldsymbol{\pi} = \left\{\frac{\frac{dw}{dr}}{w}(1+r) - 1\right\}\mathbf{C}'\boldsymbol{\pi} \quad \text{(B.25)}$$

and then $w = 1 - (r/R)$, an equality which is valid for Standard systems.

He discusses the possibility that some of the equations in the system (B.20) may be equal to zero for a given profit rate r. If $\mathbf{C}_i'$ is the i-th row of matrix $\mathbf{C}'$, this gives

$$\frac{dw}{dr}w_i + \mathbf{C}_i'\mathbf{p} = 0. \tag{B.27}$$

If for all $i = 1, \ldots, n$ the price $p_i = const$, then $\frac{d\mathbf{p}}{dr} = 0$ and therefore $(\frac{dw}{dr}\boldsymbol{\pi} + \mathbf{C}'\mathbf{p}) = 0$ and we get the equality

$$-\frac{dw}{dr} = \frac{\mathbf{C}_i'\mathbf{p}}{w_i}. \tag{B.28}$$

Schefold then discusses the terms of first, second, ...order of the development (B.20) with either $\frac{dp_i}{dr} > 0$ or $\frac{dp_i}{dr} < 0$ for any commodity $i = 1, \ldots, n$, in PCMC [109], pp. 223–224.

B.2 Determination of the standard net product according to Schefold

(1) We will summarise here Schefold's ([109], pp. 224–225) construction of a *Standard system* with normalised labour size, $L_1 = 1$, and national income, $Y_0 = 1$, leading to a *Standard net product* with *Standard national income.*

Remember that our procedure, Chapter 5 started from Sraffa's *actual economic system*, expressed as a *production scheme* $(\mathbf{S}', \mathbf{q}, \mathbf{L})$, and we intended by application of a *orthogonal Euler map* to get a *Standard system* $(\check{\mathbf{S}}', \check{\mathbf{q}}, \check{\mathbf{L}})$, which accordingly also enables one to present the elements of *standard commodities*. Then additionally, due to the normalisation of the size of *labour*, $L = L_1 = 1$, we get a further *Standard system* $(\check{\mathbf{S}}_1', \check{\mathbf{q}}_1, \check{\mathbf{L}}_1)$, see equation (5.69) with a *Standard net product* contained in $\check{\mathbf{d}}_1 = \check{\mathbf{S}}_1\mathbf{e} - \check{\mathbf{q}}_1$ (see Recapitulation 5.3.1).

Schefold [109], pp. 224–225, starts from the construction of the *Standard system* with the semi-positive and irreducible or positive matrix $\mathbf{C} = \check{\mathbf{S}}_1\hat{\check{\mathbf{q}}}_1^{-1} = \mathbf{S}\hat{\mathbf{q}}^{-1}$ and[5] the Frobenius number λ_C. He circumvents the calculation of multipliers, Subsection 5.2.2. He sets up directly the eigenvalue equation (5.68) determining the quantity eigenvectors $\check{\mathbf{q}}_1$ of *total output* of the present *Standard system* $(\check{\mathbf{S}}_1', \check{\mathbf{q}}_1, \check{\mathbf{L}}_1)$ with the *Standard net product* contained in $\check{\mathbf{d}}_1$,

$$(1+R)\mathbf{C}\check{\mathbf{q}}_1 = \check{\mathbf{q}}_1 \Leftrightarrow (\mathbf{I} - \mathbf{C})\check{\mathbf{q}}_1 = R \cdot \mathbf{C}\check{\mathbf{q}}_1 \Leftrightarrow \mathbf{C}\check{\mathbf{q}}_1 = \frac{1}{1+R}\check{\mathbf{q}}_1 = \lambda_C\check{\mathbf{q}}_1. \tag{B.29}$$

One sets the total means of production,

$$\check{\mathbf{S}}_1\mathbf{e} = (\check{\mathbf{S}}_1\hat{\check{\mathbf{q}}}_1^{-1})\check{\mathbf{q}}_1 = \mathbf{C}\check{\mathbf{q}}_1. \tag{B.30}$$

5 We note that matrix $\mathbf{C} = \check{\mathbf{S}}_1\hat{\check{\mathbf{q}}}_1^{-1} = \mathbf{S}\hat{\mathbf{q}}^{-1}$ is unique in the Euler class $\mathcal{E} = \{(\mathbf{S}', \mathbf{q}, \mathbf{L})|\hat{\mathbf{y}}\}$ (see Proposition 5.2.1).

Then Schefold ([109], p. 224) determines the vector of surplus (2.55) getting with (B.29), (B.30) and (5.25):

$$\breve{\mathbf{q}}_1 = \breve{\mathbf{S}}_1\mathbf{e} + \breve{\mathbf{d}}_1 = \mathbf{C}\breve{\mathbf{q}}_1 + \breve{\mathbf{d}}_1 \Rightarrow \breve{\mathbf{d}}_1 = (\mathbf{I} - \mathbf{C})\breve{\mathbf{q}}_1 = R \cdot \mathbf{C}\breve{\mathbf{q}}_1. \tag{B.31}$$

We also find:

$$\mathbf{C}\breve{\mathbf{q}}_1 = \frac{1}{R}(\mathbf{I} - \mathbf{C})\breve{\mathbf{q}}_1 \Rightarrow r\breve{\mathbf{q}}_1'\mathbf{C}' = \frac{r}{R}\breve{\mathbf{q}}_1'(\mathbf{I} - \mathbf{C})'. \tag{B.32}$$

Then Schefold applies the Sraffa normalisation of the *national income*, leading to the *Standard national income*,

$$Y_0 = \breve{\mathbf{d}}_1'\mathbf{p} = \breve{\mathbf{q}}_1'(\mathbf{I} - \mathbf{C}')\mathbf{p} = 1. \tag{B.33}$$

Then he normalised the size of *labour*, $L_1 = 1$, calibrating an eigenvector $\breve{\mathbf{q}}_1$, using the vector of *labour per unit of commodities* $\boldsymbol{\pi} = \hat{\mathbf{q}}^{-1}\mathbf{L}$, equation (4.58),

$$L_1 = 1 = \breve{\mathbf{q}}_1'\boldsymbol{\pi}. \tag{B.34}$$

Consider now with this eigenvector $\breve{\mathbf{q}}_1$ (B.34) the *Standard system* $(\breve{\mathbf{S}}_1, \breve{\mathbf{q}}_1, \breve{\mathbf{L}}_1)$ and write down the *single-commodity Sraffa system* (4.174) expressed by the commodity flow matrix $\breve{\mathbf{S}}_1$, the vectors $\breve{\mathbf{q}}_1$ and $\breve{\mathbf{L}}_1$,

$$\breve{\mathbf{S}}_1'\mathbf{p}(1 + r) + w \cdot \breve{\mathbf{L}}_1 = \hat{\breve{\mathbf{q}}}_1\mathbf{p}. \tag{B.35}$$

When equation (B.35) is multiplied from the left by the diagonal matrix $(\hat{\breve{\mathbf{q}}}_1)^{-1}$, we know that we get the *single-commodity Sraffa system* with normalised *labour per unit of commodity* (B.5) described by the *input-output coefficients* matrix $\mathbf{C}$, which we transcribe in an appropriate form for further calculations,

$$\mathbf{C}'\mathbf{p}(1 + r) + w \cdot \boldsymbol{\pi} = \mathbf{p} \quad \Leftrightarrow \quad r \cdot \mathbf{C}'\mathbf{p} + w \cdot \boldsymbol{\pi} = (\mathbf{I} - \mathbf{C}')\mathbf{p}. \tag{B.36}$$

Starting again from the definition of the *Standard national income* (B.33), taking into account the normalised labour size $L_1 = \breve{\mathbf{q}}_1'\boldsymbol{\pi} = 1$ (B.34), the conclusions in (B.32) and (B.36), the known rule of proportionality (5.142) reappears:

$$\begin{aligned} Y_0 = 1 = \breve{\mathbf{d}}_1'\mathbf{p} &= \breve{\mathbf{q}}_1'(\mathbf{I} - \mathbf{C}')\mathbf{p} = \breve{\mathbf{q}}_1'(r \cdot \mathbf{C}'\mathbf{p} + w \cdot \boldsymbol{\pi}) = (r \cdot \mathbf{q}_1'\mathbf{C}')\mathbf{p} + w(\breve{\mathbf{q}}_1'\boldsymbol{\pi}) \\ &= \frac{r}{R}\breve{\mathbf{q}}_1'(\mathbf{I} - \mathbf{C}')\mathbf{p} + w\breve{\mathbf{q}}_1'\boldsymbol{\pi} = \frac{r}{R} + w. \end{aligned} \tag{B.37}$$

Thus, the parallelism of the vectors of *total output*, *surplus* and *means of production* is confirmed: $\breve{\mathbf{q}}_1 \parallel \breve{\mathbf{d}}_1 \parallel \breve{\mathbf{S}}_1\mathbf{e}$, for the obtained Standard system, because $r = R(1 - w)$.[6]

(2) We now illustrate Schefold's method.

6 Schefold (see Sraffa [109], p. 225) in fact works with the vector of labour per units of commodities, $L_1 = \breve{\mathbf{q}}_1'\boldsymbol{\pi} = 1$, and the normalisation of national income, $Y_0 = \breve{\mathbf{d}}_1'\mathbf{p} = 1$.

Example B.2.1. Consider *Example* 5.4.1 (with $\mathbf{L} = [152, 152]'$). Choose now the labour vector $\mathbf{L} = [22/34, 12/34]'$. Apply Schefold's approach to determine the Standard net product, defined by $L_1 = 1$, with standard national income $Y_0 = 1$ GDP. Start computing the input-output coefficients matrix $\mathbf{C}$, its Frobenius eigenvalue λ_C and associated right eigenvectors $\breve{\mathbf{q}}_1$. Set up and solve the single commodity Sraffa system (B.36), compute the total wages W, the total profits P circulating capital K and the total output X.

Solution to Example B.2.1:

We start by identifying the essential elements of the actual economic system.

- The *commodity flow* matrix $\mathbf{S} = \left[\begin{smallmatrix} 280 & 120 \\ 12 & 8 \end{smallmatrix}\right]$ in *physical terms* is measured in qr. wheat; the second row is measured in t. iron.
- The vector of *normalised labour* $\mathbf{L} = [L_1, L_2]' = [22/34, 12/34]'$ is measured in TAL (= total amount of labour).
- The vector of *surplus* $\mathbf{d} = [d_1, d_2]' = [175, 0]'$ in *physical terms* has the mixed units: $[d_1]$ = qr. wheat, $[d_2]$ = t. iron.
- The vector of *total output* $\mathbf{q}$ (2.15) in *physical terms*, where the units are mixed: $[q_1]$ = qr. wheat, $[q_2]$ = t. iron.

One gets the vector of *total output* in physical terms,

$$\mathbf{q} = \mathbf{Se} + \mathbf{d} = \begin{bmatrix} q_1 \\ q_2 \end{bmatrix} = \begin{bmatrix} 280 & 120 \\ 12 & 8 \end{bmatrix} \begin{bmatrix} 1 \\ 1 \end{bmatrix} + \begin{bmatrix} 175 \\ 0 \end{bmatrix} = \begin{bmatrix} 575 \\ 20 \end{bmatrix}. \tag{B.38}$$

The *normalised labour* vector $\mathbf{L}$ in TAL gives the total amount of labour,

$$\mathbf{L}'\mathbf{e} = [L_1, L_2] \begin{bmatrix} 1 \\ 1 \end{bmatrix} = \begin{bmatrix} \frac{22}{34}, \frac{12}{34} \end{bmatrix} \begin{bmatrix} 1 \\ 1 \end{bmatrix} = 1 \ \text{ TAL}. \tag{B.39}$$

These calculations are summarised in Table B.1: After the preliminary identification of the matrices, we perform Schefold's calculations to determine a *Standard net product*, computing the matrices $\mathbf{C}$ and $\mathbf{D}$,

$$\begin{aligned} \mathbf{C} = \mathbf{S}\hat{\mathbf{q}}^{-1} &= \begin{bmatrix} 280 & 120 \\ 12 & 8 \end{bmatrix} \begin{bmatrix} \frac{1}{575} & 0 \\ 0 & \frac{1}{20} \end{bmatrix} = \begin{bmatrix} \frac{56}{115} & 6 \\ \frac{12}{575} & \frac{2}{5} \end{bmatrix}, \\ \mathbf{D} = \hat{\mathbf{q}}^{-1}\mathbf{S} &= \begin{bmatrix} \frac{1}{575} & 0 \\ 0 & \frac{1}{20} \end{bmatrix} \begin{bmatrix} 280 & 120 \\ 12 & 8 \end{bmatrix} = \begin{bmatrix} \frac{56}{115} & \frac{24}{115} \\ \frac{3}{5} & \frac{2}{5} \end{bmatrix}, \end{aligned} \tag{B.40}$$

then set up the eigenvalue equations, $\mathbf{C}\breve{\mathbf{q}}_1 = \lambda\breve{\mathbf{q}}_1$ (B.29) and $\mathbf{D}\boldsymbol{y} = \lambda\boldsymbol{y}$, and determine the characteristic polynomial,

$$P_2(\lambda) = \det(\mathbf{C} - \lambda\mathbf{I}_2) = \lambda^2 - \frac{102}{115}\lambda + \frac{8}{115} = \left(\lambda - \frac{4}{5}\right)\left(\lambda - \frac{2}{23}\right) = 0. \tag{B.41}$$

The Frobenius number is $\lambda_C = 4/5$, associated with the *right eigenvectors* of matrix $\mathbf{C}$, which are $\breve{\mathbf{q}}_1 = k[115/6, 1]'$, $k \in \mathbb{R}^+$. The *productiveness* is $R = (1/\lambda_C) - 1 = (5/4) - 1 = 0.25$. The right eigenvectors of the distribution coefficients matrix $\mathbf{D}$ are parallel to vector $\boldsymbol{y} = [\frac{2}{3}, 1]'$.

Table B.1: Input-Output Table of Sraffa's model [108], Par. 5 (with labour).

	Buying sectors		Final	Total
	wheat	iron	demand	output
wheat (*qr.* wheat)	$s_{11} = 280$	$s_{12} = 120$	$d_1 = 175$	$q_1 = 575$
iron (*t.* iron)	$s_{21} = 12$	$s_{21} = 8$	$d_2 = 0$	$q_2 = 20$
labour (TAL)	$L_1 = \frac{22}{34}$	$L_2 = \frac{12}{34}$		$L = 1$
	↓	↓		
production (*unit*)	$q_1 = 575$	$q_2 = 20$		

We then calculate the normalised vector of *labour per unit of commodity* (4.58), and the Euler transformation matrix $\mathbf{G} = \hat{\mathbf{y}}$,

$$\mathbf{G} = \begin{bmatrix} \frac{2}{3} & 0 \\ 0 & 1 \end{bmatrix}, \quad \boldsymbol{\pi} = \hat{\mathbf{q}}^{-1}\mathbf{L} = \begin{bmatrix} \frac{1}{575} & 0 \\ 0 & \frac{1}{20} \end{bmatrix} \begin{bmatrix} \frac{22}{34} \\ \frac{12}{34} \end{bmatrix} = \begin{bmatrix} \frac{11}{9{,}775} \\ \frac{3}{170} \end{bmatrix}. \tag{B.42}$$

Following Schefold, we consider the normalised size of *labour*, $L_1 = 1$, calibrating the eigenvectors $\breve{\mathbf{q}}_1$ of matrix $\mathbf{C}$, using the previously obtained vector of *labour per unit of commodities* $\boldsymbol{\pi}$ (B.42), applying the norming rule (B.34)

$$\begin{aligned} L_1 &= \breve{\mathbf{q}}_1'\boldsymbol{\pi} = k[(115/6), 1] \begin{bmatrix} \frac{11}{9{,}775} \\ \frac{3}{170} \end{bmatrix} = k\left(\frac{115}{6} \cdot \frac{11}{9{,}775} + 1 \cdot \frac{3}{170}\right) \\ &= k \cdot \frac{2}{51} = 1 \Rightarrow k = \frac{51}{2} \Rightarrow \breve{\mathbf{q}}_1 = \frac{51}{2} \begin{bmatrix} \frac{115}{6} \\ 1 \end{bmatrix} = \begin{bmatrix} \frac{1{,}955}{4} \\ \frac{51}{2} \end{bmatrix}. \end{aligned} \tag{B.43}$$

Then, we can immediately calculate the other vectors of this *standard net product*, starting with the surplus vectors $\breve{\mathbf{d}}_1$. Following (B.31), we obtain (expressed without physical units),

$$\breve{\mathbf{d}}_1 = (\mathbf{I} - \mathbf{C})\breve{\mathbf{q}}_1 = \left(\begin{bmatrix} 1 & 0 \\ 0 & 1 \end{bmatrix} - \begin{bmatrix} \frac{56}{115} & 6 \\ \frac{12}{575} & \frac{2}{5} \end{bmatrix}\right) \begin{bmatrix} \frac{1{,}955}{4} \\ \frac{51}{2} \end{bmatrix} = \begin{bmatrix} \frac{391}{4} \\ \frac{51}{10} \end{bmatrix}. \tag{B.44}$$

Finally, we calculate directly the vector of *means of production* (again expressed without physical units) with (B.30),

$$\breve{\mathbf{S}}_1\mathbf{e} = \mathbf{C}\breve{\mathbf{q}}_1 = \begin{bmatrix} \frac{56}{115} & 6 \\ \frac{12}{575} & \frac{2}{5} \end{bmatrix} \begin{bmatrix} \frac{1{,}955}{4} \\ \frac{51}{2} \end{bmatrix} = \begin{bmatrix} 391 \\ \frac{102}{5} \end{bmatrix}. \tag{B.45}$$

The three calculated vectors $\breve{\mathbf{S}}_1\mathbf{e}$, $\breve{\mathbf{q}}_1$, $\breve{\mathbf{d}}_1$ are all parallel. They are eigenvectors of the *input-output coefficients* matrix $\mathbf{C}$ and constitute proxies for a *standard commodity*, that is a *Standard net product* of the *Standard system* $(\breve{\mathbf{S}}_1, \breve{\mathbf{q}}_1, \breve{\mathbf{L}}_1)$.

Without detailed calculation, we present the steps to attain the transformed vector of labour: $\breve{\mathbf{L}} = \mathbf{G} \cdot \mathbf{L} = [\frac{22}{51}, \frac{6}{17}]'$.

Its norm is $\mathbf{e}\check{\mathbf{L}} = 40/51$, leading to $\check{\mathbf{L}}_1 = (51/40)\check{\mathbf{L}} = [\frac{11}{20}, \frac{9}{20}]'$.

We also compute the matrix

$$\check{\mathbf{S}}_1 = \frac{51}{40}(\mathbf{S}\cdot\mathbf{G}) = \frac{51}{40}\begin{bmatrix} 280 & 120 \\ 12 & 8 \end{bmatrix}\begin{bmatrix} \frac{2}{3} & 0 \\ 0 & 1 \end{bmatrix} = \begin{bmatrix} 238 & 153 \\ \frac{51}{5} & \frac{51}{5} \end{bmatrix}. \tag{B.46}$$

Then we take the *complete single-product Sraffa price model* (B.36), that we reproduce here with the specified values for the profit rate, $r = 0.15$, and the wage rate $w = 1 - (r/R) = 1 - (0.15/0.25) = 0.4$, presented in matrix form,

$$\mathbf{C}'\mathbf{p}\cdot 1.15 + 0.4\boldsymbol{\pi} = \mathbf{p} \Leftrightarrow$$

$$1.15\cdot\begin{bmatrix} \frac{56}{115} & \frac{12}{575} \\ 6 & \frac{2}{5} \end{bmatrix}\begin{bmatrix} p_1 \\ p_2 \end{bmatrix} + 0.4\begin{bmatrix} \frac{11}{9{,}775} \\ \frac{3}{170} \end{bmatrix} = \begin{bmatrix} p_1 \\ p_2 \end{bmatrix}, \tag{B.47}$$

and calculate the prices p_1 and p_2, components of the vector $\mathbf{p} = [p_1, p_2]'$. We get the price vector $\mathbf{p} = [p_1 = (56/9{,}775)\frac{\text{GDP}}{\text{qr. wheat}}, p_2 = (22/255)\frac{\text{GDP}}{\text{t. iron}}]'$.

Let us conclude with the aggregate economic variables of the *Standard system*.

$$Y_0 = \check{\mathbf{d}}_1'\mathbf{p} = \left[\frac{391}{4}, \frac{51}{10}\right]\cdot\begin{bmatrix} \frac{56}{9{,}775} \\ \frac{22}{255} \end{bmatrix} = \frac{14}{25} + \frac{11}{25} = 1 \quad \text{GDP}. \tag{B.48}$$

Consequently, the *total wages* and *total profits* are accordingly:

$$W = w\cdot Y_0 = 0.4\cdot 1 = 0.4\ \text{GDP},$$
$$P = Y_0 - W = 0.6\ \text{GDP}, \tag{B.49}$$

the *total output*

$$X = \check{\mathbf{q}}_1'\mathbf{p} = \left[\frac{1955}{4}, \frac{51}{2}\right]\cdot\begin{bmatrix} \frac{56}{9{,}775} \\ \frac{22}{255} \end{bmatrix} = 5\ \text{GDP} \tag{B.50}$$

and the *total circulating capital*

$$K = X - Y_0 = 4\ \text{GDP}. \tag{B.51}$$

Thus, Schefold's approach leads to the *Standard system*, to the *Standard net product* with *Standard national income*, as does the methododology based on the *orthogonal Euler map*. ▲

Let us summarize: Consider a *semi-positive* and *irreducible* or *positive* matrix $\mathbf{S}$, forming an *actual economic system* $(\mathbf{S}', \mathbf{q}, \mathbf{L})$ of basic commodities. Compute the *input-output coefficients* matrix $\mathbf{C} = \mathbf{S}\hat{\mathbf{q}}^{-1}$ and the *distribution coefficients* matrix $\mathbf{D} = \hat{\mathbf{q}}^{-1}\mathbf{S}$. We know that both matrices $\mathbf{C}$ and $\mathbf{D}$ have the same eigenvalues λ, see Lemma A.6.1. We then recognise that there are two methods to obtain a *Standard system*. There is the method of the *orthogonal Euler map*, Subsection 5.2.2 and the Schefold approach, Section B.2. Both methods rely indeed on a mathematical basis, Lemma A.6.2.

C Glossary of terms as they are used in this book

This glossary defines and comments the essential terms and concepts used in this textbook. A clear understanding of these terms is important for the understanding of the applicability and limitations imposed by the models of production presented.

Aggregation

Designates the process of regrouping or summing up entities to form a whole.

Comment. In the context of Leontief and Sraffa, where one operates in accordance with the rules of linear algebra, aggregation means adding up linear terms and equations. Given also that constant returns to scale, which amount here to the multiplication of all terms of an equation by a constant, are thus, whenever implicitly assumed, compatible with linear structures, this summation process does not lead to the fallacy of composition that in most cases seriously impedes aggregation of microeconomic ensembles to form a coherent global macroeconomic whole. While this is valid at the theoretical level, in practice however measurement problems and statistical uncertainties may nevertheless lead to aggregation biases ([65], Par. 4.9: The Aggregation Problem), a special case of the fallacy of composition. Where no confusion will arise, we shall use indifferently the terms aggregate and total.

– **aggregate profits/aggregate wages**

Total profits P, respectively total wages W, in the economy obtained by summing up the profits, respectively wages, realised industry by industry.

– **aggregate surplus**

Or simply: surplus, is the total value of commodities produced in addition to interindustry transactions required to maintain the conditions of production. In other words: National Income or Gross Domestic Product (GDP) Y, which by definition is equal to total value added generated by the economy during the reference period.

– **autonomous (exogenous) variables**

Contrary to induced (endogenous) variables such as consumption C, investment I and imports M, which figure in $Y = C + I + G + E - M$, and which are generated by the various agents (households, firms, etc) operating in a given economy, government expenditure G and exports E have no causal links with the endogenous variables and are labeled autonomous variable, see Bortis ([8], pp. 144–146).

Base year/Year under review (or current year)

Without indication to the contrary, prices in Leontief and Sraffa are expressed in nominal values, i. e., commodity prices registered at then-moment prices during the *year under review*. Generally speaking, "real" prices in economic terms are obtained by ex-

https://doi.org/10.1515/9783110635096-014

pressing nominal prices in terms of prices at some given *base year* of reference (base year prices) using a price index (Price Index), usually set at one or 100 for that year. A *price index* is a measure that permits a comparison of the prices of the present, actual year, the *year under review* (or the *current year*) to the former prices of the *base year*.

Calibration

The term calibration comes from metrology and means that a variable is measured in a unit of a general or a universal reference, like time in *seconds* or distance in *metres*. In our context, the term is used in a Sraffa price model composed of n equations and $n+k$ variables (normally $k = 2$), giving precise numerical values to the k excess variables (such as the rate of profits, the wage rate, one of the prices, national income, etc.) based on the economic production process under consideration, and further defining the *appropriate numéraire* or choosing a *currency* as reference for the measurement of value.

Capitalists

This technical term designates here, without any political or ideological undertone, the owners of the technical means of production, i. e., entrepreneurs and shareholders.

Capital

Designates in general the value of the means of production in terms of land, raw material, intermediate products, instruments of production (i. e., capital goods), financial goods and services etc., excluding wages and salaries, that entrepreneurs respectively businesses (also referred to as capitalists, for example by Sraffa) require to produce the goods and services they intend to sell.

Circulating capital

Will be used here to designate the aggregate value K of what Sraffa, PCMC Par. 4, has termed the *means of production*. The term *means of production*, here excluding fixed capital, includes short-lived commodities that are used up in a single period of production (raw materials, intermediate materials, combustibles, energy, services, etc.). In accounting, *circulating capital* comes under the heading current assets and is not to be confused with *working capital* (operating capital) calculated as current assets minus current liabilities.

Closure

The term closure has various meanings:

(a) Macroeconomically this expression refers to closed economies, i. e., economies without external exchanges of goods and services, meaning no exports E ($E = 0$) and no imports M ($M = 0$).

(b) The system of national accounts is closed in the sense that on its basis exists an equation called the *national income accounts identity* (see Mankiw [63], p. 28), $Y = C + I + G + (E - M) = GDP$. This means that the national income Y or gross domestic product GDP is equal to the sum of consumption C plus investment I plus government purchase G plus net exports $(E - M)$.

(c) Consider input-output tables, as presented by Miller and Blair [65], Chapter 2, pp. 10–68. There, the *commodity flow* matrix $\mathbf{Z}$ in monetary terms, governing inter-industry requirements, is extended by inclusion of additional rows containing *value added* elements, in particular labour, and additional columns containing *final demand* elements, such as household consumption expenditures and gross private domestic investment. This leads to an accounting identity where the *total outlays* are equal to the *total inputs* (see equation (2.7)).

Commodities

This term designates both goods, including composite products, services, and, at least by a purely formal extension, also money.

Comment. The following are specific to Sraffa [108]: Commodities (excluding "money") which enter, either directly or indirectly, into the production of all commodities are **basic** commodities; those that do not are called **non-basic**. Human labour often plays an important part in the latter: take for example jewelery, which requires some commodities but more importantly hours of specialized labour, but enters into no other industrial products.

Conditions of production

Generally speaking, in a closed economic system involving several industries, the conditions of production mean that, to ensure sustainability of the system from period to period without production of a surplus (in other words maintaining the system in a self reproducing state), the total quantity of a specific commodity produced by one or several industries must at least equal the total quantity of that commodity required for production by all the industries, including the industries producing the commodity. The totals are explicit sums, like quantities.

This definition may be extended from quantities to the values of the quantities involved by multiplication with a commodity price, leading to a system that conditions prices (see also the item Equilibrium).

Constant returns to scale

This concept may be applied to individual industries or to the system of production taken as a whole.

A system of production presents constant returns to scale if an increase of an equal percentage in all factors of production causes an increase in output of the same percentage. Mathematically speaking, a production function F with output Q has constant returns to scale if $zQ = F(zK, zL)$ where K is the amount of capital and L the amount of labour.

Comment. This assumption is routinely used by firms in their standard cost calculations and marginal cost optimisation using linear programming and medium-term business planning.

Demand

We will from time to time refer to: final exogenous demand, the technical term used in Input-Output analysis. Depending on circumstances, it is equal to: effective demand, in the Keynesian terminology, which is the driving force inciting the owners of the system of production to generate a surplus which may be sold to pay wages and to provide profits.

Comment. A production surplus without demand generates no profit; it just increases the producer's stocks. We draw the reader's attention to the fact that Sraffa in PCMC makes no reference whatsoever to demand, but does mention markets.

Economic systems: linear vs. circular

A linear economic system corresponds in this text to an individual row in Sraffa's price equations. Productive entities in agriculture, manufacturing and services operate linearly at the microeconomic level following this type of scheme, from purchasing, through production and distribution unto sales. Such systems may be represented by directed networks without circuits (except for feedback loops specific to each productive entity).

Comment. Linear economic systems are driven by the "bigger-better-faster" competitive syndrome and are typical of *non-basic* commodities such as fast-moving consumer goods (FMCG). The Walrasian model follows this logic bottom up, aiming at a macroeconomic equilibrium. The emphasis is on the final exchange, i. e., the "market", at the end of a given period.

Generally speaking, circular economic systems correspond here to macroeconomic ensembles of interrelated processes of production and exchange defined by the systems of equations encountered in the Leontief and Sraffa models. Such systems may be represented by directed networks containing many circuits.

Comment. The foregoing highlights the fallacy of composition that lurks in Walrasian aggregation. Circular systems that fulfill Sraffa's conditions of production are self-replacing and are typical of basic commodities. The emphasis of the Sraffian model is on continuous production and exchange ensuring sustainability from period to period. Note that the term "circular economies" has in the recent literature been taken, for better or for worse, as synonymous to ecological economic systems reprocessing commodities, saving energy while reducing resource consumption and waste.

Economics

The Webster New Collegiate Dictionary, p. 260, defines the term "Economics" as follows: *"The science that investigates the conditions and laws affecting the production, distribution and consumption of wealth, or material means of satisfying human desires; political economy."*

Equilibrium

Consider a closed economy of production represented by a system of Leontief or of Sraffa price equations. If the conditions of production in terms of physical units are fulfilled, meaning that the production process is self-replacing (i. e., sustainable) in the sense that the total input of each commodity required for production is equal to the total output of that commodity, the process is said to be in a quantitative equilibrium. If one considers the conditions of production expressed in value terms, for example in monetary units, the process is said to be in an accounting equilibrium.

This is an equilibrium defined in a system of production governed by input and output; it must not be confused with the notion of an equilibrium in a market system governed by supply and demand.

Growth

This term is used in macroeconomics as an indication of the increase in inflation-adjusted GDP (respectively National Income, or surplus as in PCMC) over time, based on national accounting data. It is related to the expansion of production capacity. This measure must not be confused with the ratio of the surplus generated by the means of production to operational capital, often wrongly referred to as "growth".

Comment: Say, there is a times series $Q_t, t = 0, 1, \ldots, n$, measured at equidistant time intervals of the given period of $\Delta t = 1\,year$, then the discrete *annual growth rate* is defined as $r_t = \frac{Q_t - Q_{t-1}}{Q_{t-1}}$, being itself a time series and clearly the dimension of this discrete growth rate is $[r] = 1$. The time does not appear explicitly in the discrete growth rate, which must always be indicated with the chosen time interval. Indeed, the indication of a discrete growth rate is given together with the time interval. We say: the *annual discrete growth rate*, the *monthly discrete growth rate.*

On the other hand, the *continuous growth rate* is defined as follows for a continuous function $Q(t), t \in \mathbb{R}$ for every continuous time point t: $c(t) = \frac{dQ(t)}{dt}\frac{1}{Q(t)}$, $c(t)$ is in itself a continuous function. The dimension of the *continuous growth rate* is $[c(t)] = \frac{1}{time}$.

The relationship between the *discrete growth rate* r_t and the *continuous growth rate* $c(t)$ can be established as follows: $c(t) = \frac{dQ(t)}{dt}\frac{1}{Q(t)} \sim \frac{Q(t)-Q(t-1)}{\Delta t}\frac{1}{Q(t-1)} \sim \frac{Q(t)-Q(t-1)}{Q(t-1)}\frac{1}{\Delta t} =: \frac{r(t)}{\Delta t}$.

Clearly, the relationship of the dimension is again: $[c(t)] = [\frac{r(t)}{\Delta t}] = \frac{[r(t)]}{\Delta t} = \frac{1}{time}$.

Graphs and networks

In the mathematical literature, a graph is a collection of vertices (nodes) joined by edges (links or connections). When a direction is given on the links (arrows), the graph is directed. When applied in modeling practical problems, such as the systems of production examined here, the directed graphs are termed networks, see Newman ([70], Chap. 6: *Mathematics of Networks*, pp. 109–157.)

Inflation (Deflation)

Simply stated, following Mankiw [63], the generic term inflation (deflation) is an increase (decrease) in the overall level of prices measured using a price index (see Price Index).

Such indices give no clue as to the causes of changes in prices. In the context of the extended Leontief and Sraffa models discussed in this text, "price inflation/deflation" may thus have various, even simultaneous, causes:

- in increase in the money supply;
- changes in the level of wages;
- changes in corporate policies regarding EBITDA;
- changes in final demand of commodities;
- changes in interindustry transactions due to technology;
- modifications in the prices of exports/imports.

Labour (Labor)

Labour is expressed in this text generally in *man-hours* during a given reporting period, e. g., annually. It is accordingly a measure in units of time of human hours of labour by men and women alike, required to execute a production task. It is the *required time of work*.

This measure must not be confused with the notion of "labour force". But *labour* may also mean the *number of workers* required for the execution of such a task.

Sraffa normalises the total quantity of labour L, in this case giving $L = 1$, with the artificial physical unit *Total Amount of Labour* (TAL), the reference to the initial natural physical unit *man-years* vanishes. In the context of Leontief and Sraffa, *labour* L_j can have a second signification, namely the required number of workers of sector S_j, *quantity of labour* being the main and *number of workers* the second signification.

Land

It is an ambivalent item. Every productive entity has a labour force and land attached to it. As such, land does not have the status of a commodity but is an item which will, according to the context investigated, generate rents or income from concessions which are then included in national income, in addition to profits and wages.

But again, depending on the system of production being modelled, land can also appear under the label of commodities, for example as surveyed or equipped land being sold for building or exploitation purposes.

Micro/Macro-Economics

Microeconomics is understood here as the study of ensembles of economic agents (persons, households, entrepreneurs, firms, etc.) especially regarding their activity in determining prices based on the problem of allocating scarce resources to comply with market demands. These economic agents, together with institutions that govern their activities, are the basic constituents of vast systems of monetary economies of production and exchange (national economies).

Macroeconomics is then the study of such economies considered as a whole. At such a global level, a common measure of value must be applied in theory and practice to quantify problems and decisions: money, the basis of fixing prices for production, for exchange, for future payment promises etc.

Comment. The Leontief and Sraffa models of production examined here represent some of the constituents of such macroeconomic processes that in the end are expressed in monetary terms obeying a number of fundamental national accounting identities.

Mark-up factor k

This parameter has been used in economic theory by Kalecki ([46], Chapter 2) and by Weintraub ([120], pp. 44–47) who replaces at the macroeconomic level the identity $P + W = Y$, by $kW = kwL = Y$ which defines k. The idea is that entrepreneurs in setting their profit targets will find it straightforward to aim at obtaining adequate earnings to cover various operating costs other than wages by fixing an income target exceeding by an appropriate amount their costs of labour.

Matrix

In the context under discussion, one distinguishes between two conceptual types of matrices that are each representative of the level of technology attained:

(a) matrices whose components represent quantities;
   - *the commodity flow matrix (also termed the commodity requirement matrix)*;
   - *the input matrix*;

(b) matrices whose components are pure numbers (ratios);
   - *the input coefficient matrix*;
   - *the technical coefficient matrix.*

Means of production

In input-output analysis and in PCMC the means of production are incorporated in the commodity flow matrix.

Comment. In Sraffa the components of the matrix are constants over the time period concerned, this means we are dealing with a short-term time horizon. Wages are either included as *sustainable wages* PCMC, Par. 1, or excluded from this definition of the means of production PCMC, Par. 8, in the case of explicitly indicated *quantities of labour*. In Input-Output analysis, labour is incorporated under *value added*, see Miller and Blair ([65], Par. 4.4).

Measurement and dimensional analysis

If various commodities are to be added together or their values compared, or if wages enter expressions also involving means of production composed of commodities, all these items must be expressed in some common measure, a *numéraire* or *money*, to give meaningful relations.

Comment. Indeed, quantities expressed, e. g., in *kg*, *lt*, *kWh*, barrels of petrol, heads of animals etc. cannot be meaningfully added together as such. Furthermore, if economic equations are involved, the left and right hand sides of the equations must be expressed using the same measurement units, in technical terms: they must reduce to the same dimensions (see Chable [13] and de Jong [24] regarding dimensional analysis.)

As an example, the reader is invited to interpret in measurement terms the quantity equation of money: $M \cdot v = P \cdot Q = Y$ to ascertain its dimensional consistency.

Finally to be consistent, measurement units must be calibrated, e. g., one cannot measure in the same model a numéraire in USD and EURO, or if quantities are involved, for example in barrels of petrol and tons of petrol. The instrument of measurement that must be calibrated in our monetary economies of production and exchange is national accounting.

Money

Money is not a commodity and accordingly cannot be considered as a numéraire in the usual sense. Briefly stated, in modern monetary economies of production and exchange, the banking system generates a twofold composite monetary entity formed of money as such (the form) and bank deposits (the substance); the two cannot be dissociated.

Comment. Money as such is just a numerical counter, and is in this sense a formal "dimensionless numéraire". The corresponding measurement unit is then labelled in EURO, USD, CHF, etc., in which for example prices and wages are expressed, depending on the banking system involved. Bank deposits are measured in money units and represent physical output in its monetary form as registered in bank accounts, following payments on the factor market for the production of commodities. On further analysis, notes and coins in circulation issued by the banking system follow the same logic: they just constitute a particular monetary service, officialised by a metal or paper commodity, offered to facilitate certain economic transactions.

We invite the reader to consult Rossi and Rochon et al.: *Money and Banking in a Monetary Theory of Production* ([93], Chap. 16) to obtain more details on further fundamental aspects of money in theory and practice (see also Rossi [95]).

National accounting and national income

On the macroeconomic level, monetary economies conform to a number of well-known national accounting identities linking national income Y, respectively Gross Domestic Product (GDP), to such variables as consumption, investments, profits, wages, savings, taxes, government purchases, imports and exports. In this textbook, we systematically refer to a simplified, but fundamental, accounting identity given by $Y = P + W$, where P are aggregate entrepreneurial profits and W aggregate wages registered in the economy during the reference period. In PCMC, national income is the surplus, if any, generated by the production process.

Normalisation

Will mean here setting a parameter $c = 1$ or a sum of vector components $\mathbf{e}'\boldsymbol{\upsilon} = 1$ (when dealing with components $\upsilon_i \geq 0$).

Comment. This can be illustrated on the equation $ax + by = c$, given a, b, c. We normalise c in two ways. Either

(1) by dividing the left and right-hand sides by c: $\rightarrow (a/c)x + (b/c)y = 1$ (see Sraffa ([108], Par. 10)), or
(2) by positing *ab initio* $c = 1$, leading to the linear equation $ax + by = 1$ (see Sraffa ([108], Par. 11–12)).

Numéraire

Quite generally it is defined as a measurement unit chosen for pricing. Numéraire must not be confused with money.

Comment. Thus in the context quantified by PCMC, the numéraire is expressed in terms of a given commodity, for example, as a *quarter of* wheat or a *ton of* iron (not wheat or iron as such!); such a *real numéraire* accordingly has a physical dimension. The numéraire may then be multiplied by a number to give a value as a measure of exchange. For example, certain coins such as Kruger Rands, Double Eagles, etc. are a numéraire expressed in units of the commodity gold; bitcoins on the other hand are a purely numerical accounting numéraire. The numéraire can also be a composite entity, as in Sraffa's standard commodity. However, for consistency with input-output analysis and national accounting in a monetary economy of production and exchange, the numéraire must be replaced by money.

We invite the reader to consult Schmitt in Gnos and Rossi ([95], pp. 22–37) Pasinetti ([80], pp. 63–67) for further comments.

Outlays
This is a monetary concept. Outlays simply equal the expenditures in monetary terms generated by the production process of an industry (acquisition of commodities, labour, imports, other value-added items). The simplest examples of outlays are expenses linked to purchased commodities and labour. Total outlays are equal to total output in monetary terms, see Miller and Blair ([65], Subsection 2.2.1, p. 14).

Price
Is the number of units of numéraire or of money required to obtain one unit of a given commodity.

Comment: Many factors may enter the determination of prices, such as scarcity of certain commodities, the characteristics of the production process, the cost of labour, the level of demand, government policies etc.

Price index
A price index is a dimensionless measure that enables the comparison of a given set of prices as they change over successive reporting periods starting from a base reference period (see Base year/Year under review). Accordingly, such an index usually indicates how the prices of a given bundle of commodities (goods and services) have evolved; examples are the Consumer Price Index, the Producer Price Index, the GDP deflator (see the item Inflation) which then give measures of the increase in the cost of living. Price indices are based on various mathematical formulas which incorporate the quantities and the prices of the commodities concerned. The most commonly used are the Paasche Index and the Laspeyres Index.

Prices indices may also be defined for each relevant commodity taken individually; in that case, the price of the commodity is a dimensionless entity called an *index price*.

Productive entity
This generic term is understood here in a large sense as encompassing entities providing goods and services, such as industries (single-product industries and joint-production industries), industrial sectors, human resources (providing labour) or landowners (providing land for basic requirements like agriculture, extraction of raw materials and housing).

Production system (system of production, production scheme)
Having ascertained the quantities of various commodities required by the productive entities for production (with labour included directly or indirectly), and the quantities

of commodities these entities produce, and having defined an appropriate *numéraire* as a uniform unit of measurement, the production process can then be represented by a system of production equations constituting a system of production, represented by a *production scheme*.

Production economy (Economy of production)

The production process is termed a production economy if it is understood that there exists a price for each commodity (wages considered as the "price" of labour) and a market for the sale or exchange of the commodities produced, and if furthermore the corresponding system of production is consistent with the national accounting identities. If prices are expressed in money as *numéraire*, the economy is then a monetary economy of production and exchange.

Production process

A production process regroups one or several productive entities. Every production process has an associated Sraffa Network and its adjacency matrix **W** characterising the underlying logical connections between commodities and productive entities entering that process.

Production of commodities

- *conditions of production* simply reflect the fact that, from one production period to another, the process of production must reconstitute the commodities required to sustain the apparatus of industrial production. This is represented in the input-output model by *interindustry* purchases. PCMC, Par. 3, specifies these conditions by a system of linear equations.
- *joint production* Sraffa's initial model, PCMC, Part I, considers industries producing one commodity only, i. e., single-product industries. In PCMC, Part II, he expands his model to consider industries producing more than one commodity, i. e., *multiple-product industries*, analysing in particular the mathematical properties of the systems describing such *joint production*.

Productivity, productiveness

Productivity is commonly defined as a ratio between the output volume and the volume of inputs. In other words, it measures how efficiently production inputs, such as labour and capital, are being used in an economy to produce a given level of output. Productivity is considered a key source of economic growth and competitiveness and, as such, is basic statistical information for many international comparisons and national performance assessments, see Krugman [47].

We term in this text the variable $R = (1/\lambda_C)-1$ as *"productiveness"*. It is a characterisation of the *productivity* of a Sraffa production system, where $\lambda_C > 0$ is the Frobenius number of the corresponding *semi-positive input-output coefficients* matrix **C**.

Profits

The earnings entrepreneurs, or capitalists to use Sraffa's vocabulary, obtain from sales of their production surplus after payment of wages to labour.

Comment. When referring to the surplus produced in Sraffa's model, this implicitly assumes that there exists a demand that will absorb the surplus via sales, thus generating a gross profit on production after payment of wages, i. e., in business terminology, EBITDA: earnings before interest, taxes, depreciations and amortization (see comments in Godeley and Lavoie [37], Chap. 8.2). Realised profits thus depend crucially *ex post* on final demand and not on the surplus level of production.

- *share of total profits* $\tilde{r}$

This is the ratio of total aggregate profits P to national income Y: $\tilde{r} = P/Y$.

- *uniform rate of profits* r

The uniform rate of profits is defined as $r = P/K$, where P is total aggregate profits registered by the national accounts and K the total value of circulating capital (the value of the means of production in Sraffa's terms). Sraffa (PCMC, Par. 4) posits that the surplus is distributed in proportion to circulating capital in all industries, thus necessarily implying a uniform rate of profits for all the industries.

- *productiveness or maximum rate of profits* R

The productiveness (maximum rate of profits) is obtained from the Sraffa price equations with *semi-positive input-output coefficients* matrix $\mathbf{C}$ by setting the *wage rate* w or the *total wages to national income* $\tilde{w}$ to zero, $w = \tilde{w} = 0$. For example, in single-commodity processes one has the eigenvalue equation $\mathbf{C}'(1 + R)\mathbf{p} = \mathbf{p}$, the Frobenius number being $\lambda_C = 1/(1+R) > 0$, and the uniform rate of profits r is limited to $0 \leq r \leq R$. Mathematically speaking, R is just an upper bound on the aggregate uniform rate of profits r. The *productiveness* R must clearly be distinguished from the *surplus ratio* $\tilde{R} = Y/K$. In general $R \neq \tilde{R}$.

Comment. For a *Standard system* Sraffa terms: "*R* is the Standard *Standard ratio* or Maximum rate of profits" (PCMC, Par. 30). In *Standard systems* there is the equality $R = \tilde{R}$. Therefore there is in *Standard systems* the rule $r = R(1 - \tilde{w})$.

The eigenvalues of its distribution coefficients matrix $\mathbf{D}$ and the eigenvalues of the input-output coefficients matrix $\mathbf{C}$ are identical. For an economy, where a surplus exists, $\mathbf{d} \geq \mathbf{o}$, the corresponding Frobenius number is smaller than 1, $0 < \lambda_D = \lambda_C < 1$. The Frobenius number λ_C gives rise to the determination of the *productiveness* R. There is $R = (1/\lambda_C) - 1 > 0$ because we also have $0 < \lambda_C = 1/(1 + R) < 1$. The *rate of profits* r is then contained in an interval, $0 \leq r \leq R$. For a economy with no surplus, $\mathbf{d} = \mathbf{o}$, there is $\lambda_C = 1$ and the *productiveness* and *rate of profits* are zero, $r = R = 0$.

Rate of profits (continuous and discrete measurements)

The term "rate of profits" implicitly assumes the ensuring surplus is absorbed by exogenous demand; if this is not the case, one should use the term "surplus ratio".

The *rate of profits* used in this context is a *discrete rate*, defined for a given constant discrete time interval of a given period, which in this case can be weekly, monthly or annuall.

Rates and ratios

There is some confusion in the use of these terms in the literature, in particular in Sraffa [108]. We shall use the term ratio ("Quote" in German; "quote-part" in French) to designate the percentage part of a whole: the ratio P/Y for example designates the part of total profits in national income according to the national accounts. Other relationships of this type between different variables will be termed rates: Y/K, for example, is the percentage relation between national income (the surplus) and operational capital and is a surplus rate. Similarly, P/K is a rate of profit.

We shall also refer to rates when considering an increase in macroeconomic variables: $\Delta Y/Y$, for example, would be the rate of change in GDP (growth or decrease) from one accounting period to another.

Comment. In international financial transactions, the terms even vary from one language to another: a currency "exchange rate" in English, is "cours de change" or "taux de change" in French , "Wechselkurs" in German; but the term remains literally the same for interest rate.

Short term/Long term

In planning their production, firms must make many decisions. In order to reduce these to theoretically manageable decisions, one can distinguish three time periods and group production decisions into one of these (see Lipsey ([58], *An Introduction to Positive Economics*, pp. 218–219)). These time periods vary from one industry to another. We apply this approach in this text.

The short term: Defined as the period of time over which the inputs of some factors cannot be varied. The firm cannot get the use of more of the fixed factors of production (such as plant and equipment or land) than it has on hand; the factors that can be varied in the short term (such as cheap labour and raw materials) are called variable factors.

The long term (medium term) is defined as the period long enough for inputs of fixed factors of production to vary without changes in the production technology. This is where the notion of constant returns to scale enters the picture.

The very long term is concerned with situations in which the technological possibilities open to firms are subject to change, leading to new and improved products and new methods of production.

Governments in defining their economic policies are also confronted with these differing time horizons, but at the macroeconomic level and over a very long term, Keynes quipped: "In the long run we are all dead".

Standard system
In Sraffa's initial model (PCMC, Part I), each industry produces as output just one commodity. In this connection, Sraffa's Standard system (PCMC, Chap. IV and V) is a theoretical construct describing a system of production in which the various commodities entering the means of production are in the same proportion as they figure in the total output. The national income Y in such a system is a composition of all the commodities, i. e., a composite commodity, called the standard commodity taken as the basis to define the *numéraire*.

Comment. In order to avoid logical inconsistencies in the use of this numéraire, Sraffa (PCMC, Par. 35), then posits that only basic commodities be included in the corresponding Standard system. Note that Sraffa's conditions of production (PCMC, Par. 3) represent a limit case of a Standard system.

Surplus and total output
The global economic process of production generates a total output of commodities which is split between,
(a) interindustry purchases required to maintain the means of production at their attained level (raw materials, replacements, etc.);
(b) a production surplus, fueled by exogenous demand, which will be distributed between wage earners (labour) as income and entrepreneurs in the form of gross profits.

Surplus ratio – Standard ratio
In the context of Sraffa's model, the Standard ratio is the ratio in value terms between the surplus generated by a given industry and the means of production (circulating capital). This ratio is in general specific to each industry. For aggregate Sraffa economies, the *surplus ratio* or *national income to circulating capital ratio* is written in this text $\tilde{R} = Y/K$.

In Sraffa's Standard system, however, the surplus ratio is the same for all industries and is called the *Standard Ratio R*, with $R = \tilde{R}$. It is a fundamental characteristic of a Standard system.

Value
Goods and services are measured in terms of their value expressed in numerical units of some adopted *numéraire* or in terms of money. In monetary terms, value is thus defined as multiplying a physical quantity by a price and thus obtaining a number labelled in a given currency.

Value added
Is the value of a firm's output minus the value of the intermediate commodities that the firm has purchased for production, i. e., to replace used *means of production*. Value

added comprises such items as wages, profits, interest, taxes paid, adjustments to inventories, imports, etc. Within the models examined here, we will concentrate essentially on wages and profits. In a closed economy, national income, or equivalently GDP, equals the total value added, summed over the whole economy.

Vertical integration versus horizontal integration

In *microeconomics* and management, *vertical integration* describes companies/ industries in which the whole supply (production) chain is controlled by a common owner. Each constituent of the production chain produces a different commodity, and these commodities combine to produce a final commodity to satisfy demand.

By analogy, this expression was taken up in *macroeconomics* by Pasinetti [80] in connection with Sraffa's model where the rows of the transpose of the supply matrix $\mathbf{S}'$ are the columns of the matrix $\mathbf{S}$ in Leontief's input-output representation, i. e., the *vertical components* corresponding to commodities entering production. In this connection, vertical integration is just a mathematical artefact.

In microeconomics, *horizontal integration* is a procedure where businesses/industries create or acquire production units of commodity outputs that are alike. Where competitors are acquired following this goal, this leads to market monopolies.

In Sraffa's model at the macroeconomic level, horizontal integration then logically refers to the regrouping of the column components of $\mathbf{S}'$, i. e., the *row elements* of $\mathbf{S}$ in Leontief's representation.

Wages

- *wage ratio* $\tilde{w}$

This is the ratio of aggregate wages W to national income Y: $\tilde{w} = W/Y$.

- *unit wage* w

In Sraffa's model, this is the "wage per unit of labour" and has in PCMC the dimension *numéraire* per *unit of time*, $[w] = (numeraire/unit\ of\ time)$.

Comment: This definition is however ambiguous: it can be interpreted as either (a) the weighted average wage of the labour force or, as is often implied, (b) the wage of the lowest paid unit of labour in the economy. In the latter case, however (a managing director would receive for example per annum a salary 50 times higher than that of an unqualified manual worker and would count as 50 units of labour) this biases the count of persons employed and would invalidate unemployment statistics. For our mathematical purposes, we shall consider w as a given parameter and take the Sraffa definition at face value.

- *subsistence wage*

There exists some confusion about this antiquated concept which goes back to 19th-century writings and earlier. Sraffa (PCMC, Par. 1) uses this notion of a minimal means workers require for survival and implicitly includes such wages, payable as a bundle of commodities, as components in the commodity flow matrix defining the conditions of production. From Sraffa's PCMC, Par. 8, onwards however, subsistence wages are no longer included in the commodity flow matrix and are incorporated in the basic wages entering his model, Sraffa being no doubt aware of the disputed nature of the concept of subsistence wages in a modern economy.

The reader is warned that variations in definitions and notations from one author to another can generate much confusion in economic discussions on matters such as those treated in this book. In addition, we encourage the reader to look up translations of these terms in other languages, notably in French, German, Italian, Russian and Spanish, which will reveal conceptual differences due to differing cultural approaches and of course to different economic schools of thought.

Bibliography

[1] Alexandre, L., *La Guerre des Intelligences*, JCLattès, Paris (2017).

[2] Ashmanov, S. A., *Introduction in Mathematical Economics* (in Russian language), Nauka, Moscow (1984), 296 pages.

[3] Bailly, J.-L., *On the Macroeconomic Foundations of the Wage-Price Relationship*, in: Rochon L.Ph. and S. Rossi (eds.), Modern Theories of Money: The Nature and Role of Money in Capitalist Economies, Edward Elgar Publishing, Cheltenham, UK (2003).

[4] Bharadwaj, K., *On the Maximum Number of Switches Between Two Production Systems*, Schweizerische Zeitschrift für Volkwirthschaft und Statistik, Vol. 106, No. 4, pp. 409–29 (December 1970).

[5] Birner, J., *The Cambridge Controversies in Capital Theory*, Routledge, London (2002).

[6] Bortis H., *Keynes and the Classics - Notes on the Monetary Theory of Production*, in: L.-Ph. Rochon and S. Rossi (eds.): Modern Theories of Money – The Nature and Role of Money in Capitalist Economies, Edward Elgar, Cheltenham, UK and Northampton, MA, USA (2003).

[7] Bortis, H., *Piero Sraffa and Shackle's 'Years of High Theory' - Sraffa's Significance in the History of Economic Theories, 3. Chapter*, in: Enrico Sergio Levrero, Antonella Palumbo and Antonella Stirati (eds.): Sraffa and the Reconstruction of Economic Theory, volume 3, Antonella Stirati (ed.): Sraffa's Legacy: Interpretations and Historical Perspectives, Palgrave-Macmillan, London, pp. 55–83 (2013).

[8] Bortis, H., *Institutions, Behaviour and Economic Theory*, Cambridge University Press, 435 pages (1997).

[9] Bortkiewicz, L., *Zur Berichtigung der grundlegenden theoretischen Konstuktion von Marx im 3. Band des 'Kapitals'*, Jahrbücher für Nationalökonomie und Statistik, pp. 319–35 (1907).

[10] Bottazzi, Patrick et al., *Assessing sustainable forest management under REDD+ : A community-based labour perspective*, Ecological Economics, Vol. 93, pp. 94–103 (2012).

[11] Branson, W. H., *Macroeconomic theory and policy*, Grand Rapids [etc.]: Harper & Row, cop., New York (1989).

[12] Bundesamt für Statistik, *Input-Output Tables*, retrieved from Input-Output Tables: http://www.bfs.admin.ch, Schweiz (2016, 03).

[13] Chable, D., *Economics, Thermodynamics, and Quantum Theory: Rekindling Political Economy*, paper presented at the Dijon Conference on Post-Keynesian Economic Policies, University of Burgundy, December (2007).

[14] Champernowne, D. G. *Unemployment, Basic and Monetary: The Classical Analysis and the Keynesian*. Review of Economic Studies, Vol. 3, pp. 201–16 (1936).

[15] Charasoff, G., *Das System des Marxismus. Darstellung und Kritik*, Hans Bondy, Berlin (1910).

[16] Carré H., *NACE Rev. 2, Statistical classification of economic activities in the European Community*, General and regional statistics, Methodologies and working papers, European Community (2008).

[17] Carsten N., Schmid Chr., van Nieuwkopp R., *Schätzung einer Input-Output-Tabelle der Schweiz 2008*, Schlussbericht an das Bundesamt für Statistik, Rütter + Partner, sozioökonomisdche Forschung und Beratung, Rüschlikon/Bern (Mai 2011).

[18] Chesney, M., *De la Grande Guerre à la Crise Permanente*, Presses polytechniques et universitaires romandes, Lausanne (2015).

[19] Chiang, A. C. and Wainwright K., *Fundamental Methods of Mathematical Economics*, 4th edition, McGrawHill Int. Editions (2005).

https://doi.org/10.1515/9783110635096-015

[20] Cobb Ch.-W., Douglas P. H., *A Theory of Production*, The American Economic Review, Vol. 18, No. 1, Supplement, Papers and Proceedings of the Fortieth Annual Meeting of the American Economic Association, pp. 139–65 (1928).

[21] Daly, H. E., Farlay, J., *Ecological Economics, Principles and Applications*, Pan-American Copyright Conventions, Island Press, Suite 300, 1718 Connecticut, Ave., NW, Washington, DC 20009 (2004).

[22] Dietzenbacher, E., *In Vindication of the Ghosh Model: A Reinterpretation as a Price Model*, Journal of Regional Science, Vol. 41, pp. 185–96 (1997).

[23] Duchin, F., Steenge, A. E., *Mathematical Models in Input-Output Economics*, in Encyclopedia of Life Support Systems (EOLSS) and, Rensselaer Polytechnic Institute, Working Papers in Economics, Nr. 0703, New-York (2007).

[24] de Jong, F. J., *Dimensional Analysis for Economists*, North-Holland Publishing Co. (1967).

[25] de Mesnard, Louis, *About the reinterpretation of the Gosh model as a price model*, LATEC (UMR CNRS 5118), Faculty of Economics, University of Burgundy, Working paper (2001).

[26] Emmenegger, J.-F., *Rehabilitierung der Input-Output Analyse in den universitären Curricula*, Hrsg. IWH (Ludwig Udo, Brautzsch Hans-Ulrich), Neuere Anwendungsfelder der Input-Output-Analyse – Tagungsband – Beiträge zum Halleschen Input-Output-Workshop 2012, IWH-Sonderheft 1/2013, Halle (Saale), pp. 65–79 (2013).

[27] Emmenegger, J.-F., Chable, D., Knolle, H., *Algorithmen und Berechnungen zu Piero Sraffa's Standardware*, Hrsg. GWS (Gerd Ahlert), Neuere Anwendungsfelder der Input-Output-Analyse, Tagungsband, Beiträge zum Input-Output-Workshop 2014 in Osnabrück, GWS Research Report (2014/2).

[28] Emmenegger, J.-F., Chable, D., *Sraffa Preismodelle mit Profit- und Lohnverteilungen*, pp. 15–24, Hrsg. GWS (Mönnig Anke), Tagungsband zum 8. Input-Output-Workshop 2016, Osnabrück, GWS Specialists in Empirical Economic Research, Osnabrück (Januar 2017).

[29] Emmenegger, J.-F., Chable, D., *Die Rolle des Satzes von Perron-Frobenius in Leontief und Sraffa Modellen*, Hrsg. GWS (Mönnig Anke), pp. 14–25, Tagungsband zum 9. Input-Output-Workshop 2017, Osnabrück, GWS Specialists in Empirical Economic Research, Osnabrück (Januar 2018).

[30] Emmenegger, J.-F., Chable, D., *Konvexe Kegel in Sraffas Ökonomien der Kuppelproduktion*, Hrsg. GWS (Mönnig Anke), pp. 10–20, Tagungsband zum 10. Input-Output-Workshop 2018, Bremen, GWS Specialists in Empirical Economic Research, Osnabrück (Januar 2019).

[31] Frobenius, G., *Über Matrizen aus nicht negativen Elementen*, Sitzungsberichte der Königlich Preussischen Akademie der Wissenschaften 26, Berlin, pp. 456–77 (1912).

[32] Forrester, J. W., *A New View of Business Cycle Dynamics*, The Journal of Portfolio Management (Fall 1976).

[33] Gale, D., *The Theory of Linear Economic Models*, McGraw-Hill Book Company (1960).

[34] Gantmacher, Felix R. (ed.): *Matrizentheorie*, Springer-Verlag, Berlin, Heidelberg, New-York, Tokyo (1986).

[35] Georgescu-Roegen, N., *The Entropy Law and the Economic Process*, Harvard University Press (1999).

[36] Gilibert G., *Circular Flow*, in The New Palgrave Dictionary of Economics, Vol I, pp. 742–5 (2013).

[37] Godeley, W., Lavoie, M., *Monetary Economics*, Palgrave MacMillan, London (2006).

[38] Gnos, Cl., Rossi, S., *Modern Monetary Macroeconomics: A New Paradigm for Economic Policy*, Edward Elgar, Cheltenham, UK et Northampton, MA, USA, xiii, 315 pages (co-editor, avec Claude Gnos, Université de la Bourgogne, France) (2012).

[39] Hall, Luis J., *A Mathematical Study of the Sraffa Model*, Ciencias Econónicas, Vol. 30, No. 2:2012, pp. 77–88 (2012). ISSN: 0252-95211.

[40] Hand, D. J., *Measurement, Theory and Practice. The World Through Quantification*, Arnold

Publishers, London (2004).
[41] Hobson J. A., *The Industrial System* (1909); Reprint: Augustus M. Kelley Publishers, USA (1969).
[42] Holub, H.-W., Schnabl, H., *Input-Output-Rechnung: Input-Output-Analyse, Einführung*, R. Oldenbourg Verlag, München, Wien (1994).
[43] Horn, R. A., Johnson, Ch. R., *Matrix Analysis*, second ed., Cambridge University Press (2013).
[44] Illik, J. Anton, *Formale Methoden der Informatik*, Expert Verlag, Renningen, Deutschland (2009).
[45] Kaldor, N., *Alternative Theories of Distribution*, The Review of Economic Studies Ltd, Vol. 23, No. 2, pp. 83–100 (1955–1956).
[46] Kalecki, Michal, *Theory of Economic Dynamics*, Unwin University Books, London (1954).
[47] Krugman, P., *The Age of Diminishing Expectations, U.S. Economic Policy in the 1990s*, Orell Füssli, Zürich (1994).
[48] Kowalsky, Hans-Joachim, *Lineare Albegra*, de Gruyter Lehrbuch, Berlin, New York (1972).
[49] Knolle, Helmut, *Vollbeschäftigung bei Nullwachstum: Für eine soziale und ökologische Wirtschaft*, Im Jahrbuch Denknetz/Réseau de Réflexion, hrsg. Hans Baumann & all, Verlag: Edition 8, Postfach 3522, 8021 Zürich (2010).
[50] Knolle, Helmut, *Und erlöse uns von dem Wachstum*, Eine historische und ökonomische Kritik der Wachstumsideologie, Pahl-Rugenstein, 2. erw. Auflage (2011).
[51] Knolle, Helmut, *Die Wachstumsgesellschaft, Aufstieg, Niedergang und Veränderung*, PapyRossa Verlags GmbH & Co. KG, Köln (2016).
[52] Kurz, H. D., Salvadori, N., *Theory of Production, A Long-Period Analysis*, Cambridge University Press, paperback (1995 and 2007).
[53] Kurz, H. D., Salvadori, N., *Interpreting Classical Economics: Studies in Long-Period Analysis*, Routledge Studies in the History of Economics, 269 pages (2007).
[54] Kurz, H. D., Salvadori, N., *Classical Economics and Modern Theory*, Routledge Studies in the History of Economics, 333 pages (2003).
[55] Leontief, W. W., *Die Wirtschaft als Kreislauf*, Archiv für Sozialwissenschaft und Sozialpolitik, Vol. 60, pp. 577–623 (1928).
[56] Leontief, W. W., *The Structure of American Economy 1919–1929*, Oxford University Press, Fair Lawn, New Jersey (1941).
[57] Leontief, W. W., *Input-Output Economics*, second edition, Oxford University Press, New-York (1986).
[58] Lipsey, R. G., *An Introduction to Positive Economics*, Weidenfeld & Nicolson, London (1979).
[59] Ludwig, U., Brautzsch, H.-U., *Neuere Anwendungsfelder der Input-Output-Analyse. Beiträge zum Input-Output-Workshop 2012 in Halle, Tagungsband*, Institut für Wirtschaftsforschung Halle IWH, Halle (2012).
[60] Lütkepohl, H, *Introduction to Multiple Time Series Analysis*, Springer Verlag, Berlin, Heidelberg, New York (1991).
[61] Manara, C. F., *Sraffa's Model for the Joint Production of Commodities by Means of Commodities*, in Pasinetti (Ed.) [83], hereunder, Chap. 1, pp. 1–15 (1980).
[62] Manara, C. F., *Il modello di Sraffa per la produzione conguinta di merci a mezzo di merci*, L'industria, No. 1 (1968).
[63] Mankiw, N. G., *Macroeconomics*, 5th edition, Worth Publishers (2003).
[64] Michaelova, A. and Dutschke, M., *Will credits from avoided deforestation jeopardize the balance of the carbon market?* In: Palmer, Ch. and Engel, St. (eds.) Avoided deforestation, Routledge, New York (2009).
[65] Miller, R. E., Blair, P. D., *Input-Output Analysis, Foundations and Extensions*, 2nd edition, Cambridge University Press (2009).

[66] Mori, Kenji, *Charasoff and Dimitriev: An Analytical Characterisation of Origins of Linear Economics*, Discussion Paper No 249, Tohoku Economics Research Group, Graduate School of Economics and Management Tohoku University, Kawauchi, Aoba-Ku, Sendai, Japan, January (2010).

[67] Murphy, A. E., *The Genesis of Macroeconomics*, New ideas from William Petty to Henry Norton, Oxford University Press, New York (2009).

[68] Nathani, C., Schmid, Ch., van Nieuwkoop, R., *Schätzung einer Input-Output-Tabelle der Schweiz 2008*, Schlussbericht an das Bundesamt für Statistik, with the Swiss Input-Output-Table, Schweizerisches Bundesamt für Statistik, Rüschlikon/Bern (2011).

[69] Nef, W., *Lehrbuch der Linearen Algebra*, Birkhäuser Verlag, Basel und Stutgart (1966).

[70] Newman, M. E. J., *Networks, An Introduction*, Oxford Univ. Press (2012).

[71] Newman, P., *Production of Commodities by means of Commodities*, Schweizerische Zeitschrift für Volkswirtschaftslehre und Statistik, 98. Jg., pp. 58–75 (1962).

[72] Norlund, Laurs, *Eurostat Manual of Supply, Use and Input-Output Tables*, Office for Official Publications of the European Community, Luxembourg (2008).

[73] Nour Eldin, H. A., Emmenegger, J.-F., Nabout, A. A., *Currency and Market Decisions in a Decision-Based Economy* (paper to be published) (2020).

[74] Nour Eldin, H. A., Heister, M., *Two state representation forms with explizit Kronecker-Indexes, Decoupling-Indexes and prime-matrix product representation*, in German, Regelungstechnik 28, Heft 12 (1980).

[75] O'Neil, Cathy, *Weapons of Math Destruction*, Crown Publishers, New York (2016).

[76] Ormerod, P., *Positive Linking: How Networks Can Revolutionise the World*, Paperback, Kindle Edition (2012).

[77] Oosterhaven, J., *Leontief versus Ghoshian Price and Quantity Models*, Southern Economic Journal (1996).

[78] Oetsch, W., *Die Kapitalkontroverse heute*, in Zeitschrift für Wirtschafts- und Sozialwissenschaften (ZWS) 112, Duncker & Humbolt, Berlin, pp. 419–32 (1992).

[79] Parys, Wilfried, *All but one: How pioneers of linear economics overlooked Perron-Frobenius mathematics*, in Conference at the University of Paris Ouest, Nanterre, 17–18. January 2013, "The Pioneers of Linear Models of Production" (2013).

[80] Pasinetti, L. L., *Lectures on the Theory of Production*, The Macmillan Press Ltd, London and Basingstoke (1977).

[81] Pasinetti, L. L., *A Mathematical Formulation of the Ricardian System*, in The Review of Economic Studies, 1959–1960, vol. 27, pp. 78–98 (1960).

[82] Pasinetti, L. L., *Structural Economic Dynamics, A Theory of the Economic Consequences of Human Learning*, Cambridge University Press (1993).

[83] Pasinetti, L. L. (ed.): *Essays on the Theory of Joint Production*, Columbia Univ. Press, New York (1980).

[84] Pasinetti, L. L.: *Theory of Value - a Source of alternative Paradigms in Economic Analysis*, in: Baranzini Mauro and Scazzierri Roberto (eds.) Foundations of Economics: Structure of Inquiry and Economic Theory, Basil Blackwell, pp. 409–31 (1986).

[85] Pasinetti, L. L.: *The notion of Vertical Integration in Economic Analysis*, Metroeconomica, pp. 1–29 (1973).

[86] Perron, O., *Zur Theorie der Matrices*, Mathematische Annalen, Vol. 64, No. 2, pp. 248–63 (1907).

[87] Plotnitsky, A., *The Knowable and the Unknowable*, University of Michigan Press, Ann Arbor (2005).

[88] Quesnay, François, *Oeuvres économiques complètes et autres textes*, éd. par Christine Théré, Loïc Charles et Jean-Claude Perrot, Paris, Ined, 2 vol., p. 618 (2005).

[89] Reich, U.-P., *Cambridge, England gegen Cambridge, Massachussets; Input-Output Analyse the Kapitalkontroverse*, 9. Input-Output Workshop, Bremen (2018).

[90] Reich, U.-P., *Accounting for international trade in value added: a comment on the OECD-WTO prject*, Economic Systems Research, Journal of the International Input-Output Association, Vol. 30, No. 4, December (2018).

[91] Ricardo, D., *Über die Grundsätze der Politischen Ökonomie und der Besteuerung* (mit einer Einleitung von Piero Sraffa, hrsg. von Heinz D. Kurz), Marburg (1994).

[92] Robinson, J., *The Production Function and the Theory of Capital*, Review of Economic Studies, Vol. 21, No. 2, pp. 81–106 (1953–1954).

[93] Rochon, L.-Ph., and Rossi, S. (eds.), *Modern Theories of Money*, Edward Elgar Publishing, 588 pages (2003).

[94] Rochon, L.-Ph., and Rossi, S. (Eds.), *An Introduction to Macroeconomics. A heterodox approach to economic analysis*. Ed. Elgar Publishing, Cheltenham (2016).

[95] Rossi, S., *Money and Inflation*, Ed. Elgar. Publishing, Cheltenham (2001).

[96] Roncaglia, A., *Piero Sraffa. His Life, Thought and Cultural Heritage*, New York, Routledge Studies in the History of Economics (2000).

[97] Roncaglia, A., *Piero Sraffa*, Series Editor A.P. Thirlwall: *Great Thinkers in Economics*, Palgrave Macmillan (2009).

[98] Rutishauser, H., *Vorlesungen über numerische Mathematik*, Birkhäuser Verlag, Basel, Stuttgart (1976).

[99] Saaty, T. L., *The Analytic Hierarchy Process*, MacGraw Hill, New York (1980).

[100] Salvadori, N., and Steedman, I., *Joint Production Analysis in a Sraffian Framework*, Bulletin of Economic Research, Vol. 40, No. 3, pp. 165–95, 0307-3378 $2.0 (1988).

[101] Samuelson, P. A., Nordhaus, W. D., *Economics*, 5th edition, McGraw-Hill, Inc., New-York (1995).

[102] Schefold, B., *Sraffas Theorie der Kuppelproduktion. Ein Überblick.*, Zeitschrift für Wirtschafts- und Sozialwissenschaften, pp. 315–40 (1983/4).

[103] Schefold, B., *Mr Sraffa on Joint Production and Other Essays*, Unwin Hyman, London (1989).

[104] Schumann, J., *Input-Output Analyse*, Springer Verlag, Berlin (1968).

[105] Skiena, S. *Strong and Weak Connectivity*, §5.1.2 in Implementing Discrete Mathematics: Combinatorics and Graph Theory with Mathematica, Addison-Wesley, Reading, MA, pp. 172–4 (1990).

[106] Smith, A. *The Wealth of Nations*, Pelican Books, London, p. 131 (1970).

[107] Sokow, R. M., *Technical Change and the Aggregate Production Function*, The Reviw of Economics and Statistics, Vol. XXXIX, No. 3, pp. 312–20 (August 1957).

[108] Sraffa, P., *Production of Commodities by means of Commodities*, Cambridge University Press, Cambridge (1960).

[109] Sraffa, P., *Warenproduktion mittels Waren* (aus dem Englischen übersetzt mit einem Anhang von B. Schefold), Edition Suhrkamp 780, Erste Auflage (1976).

[110] Steenge, A. E., *Consistency and Composite Numeraires in Joint Production Input-Output Analysis; an Application of Ideas of T. L. Saaty*, Math. Modelling, Vol. 9, No. 3–5, pp. 233–41, Pergamon Journals Ltd. (1987).

[111] Stetsyuk, P. I., Koshlai, L. B., *Optimal Normalized Structure of Demand and Value Added in a Productive Leontief Model*, Cybernetics and Systems Analysis, Vol. 46, No. 5 (2010).

[112] Stetsyuk, P. I., Bondarenko, A. W., *Spectral Leontief Models* (in Russian), Theory of Optimal Solutions ("Theorija Optimalnich Rishen"), No. 10, pp. 84–90 (2011).

[113] Stetsyuk, P. I., Emmenegger, J.-F., *The Maximum Singular Value of a Matrix and its Economic Interpretation* (in Russian), Cybern. Syst. Anal., Vol. 50, No. 3, pp. 51–7 (2014).

[114] Steedman I., *Basics, Non-Basics and Joint Production*, The Economic Journal, Vol. 87, pp. 324–8 (June 1977).

[115] Stern, N., *The Economics of Climate Change*. The Stern Review, Cambridge University Press (2007).

[116] Takayama, A., *Mathematical Economics*, 2nd edition, Cambridge University Press, London, New-York, Cambridge (1985).

[117] Taleb, N. N., *Antifragile. Things That Gain from Disorder*, Penguin Books, London (2014).

[118] Varga, R. S., *Matrix Iterative Analysis*, Prentice-Hall, Inc. Englewood Cliffs, New Jersey (1962).

[119] Wagner, K., *Graphentheorie*, Hochschultaschenbücher, Bibliographisches Institut, Vol. 248/248a*, Manheim-Wien-Zürich (1970).

[120] Weintraub, Sydney, *Capitalism's Inflation and Unemployment Crisis*, Addison Wesley, Massachusetts (1978).

[121] Welsh, D. J. A., *Matroid Theory*, Dover (2010).

[122] World Input-Output Database, *WIOD*, retrieved from World Input-Output Tables: http://www.wiod.org/new_site/database/niots.htm (April, 2015).

[123] Wolfram, St., *The Mathematica Handbook*, Organisation: University of California, Irvine, Department of Physics (2006) and Software Package *Wolfram Mathematica 8*, Wolfram Research, Inc. 100 Trade Center Drive, Champaign, IL 61820, USA.

[124] Yang, M., *Matrix Decomposition*, Electrical and Computer Engineering, Northwestern University Evanston IL 60208, Working paper, USA (2000).

[125] Young, D. M., *Iterative Solution of Large Linear Systems*, Academic Press, Inc. New York (1971).

[126] Zurmühl, R., *Matrizen und ihre technischen Anwendungen*, Springer Verlag, Berlin (1964).

Index